Applied Myrmecology

Westview Studies in Insect Biology
Michael D. Breed, Series Editor

Applied Myrmecology: A World Perspective, edited by Robert K. Vander Meer, Klaus Jaffe, and Aragua Cedeno

The Genetics of Social Evolution, edited by Michael D. Breed and Robert E. Page, Jr.

The Entomology of Indigenous and Naturalized Systems in Agriculture, edited by Marvin K. Harris and Charlie E. Rogers

Interindividual Behavioral Variability in Social Insects, edited by Robert L. Jeanne

Integrated Pest Management on Rangeland: A Shortgrass Prairie Perspective, edited by John L. Capinera

Management of Pests and Pesticides: Farmers' Perceptions and Practices, edited by Joyce Tait and Banpot Napompeth

Fire Ants and Leaf-Cutting Ants: Biology and Management, edited by Clifford S. Lofgren and Robert K. Vander Meer

Applied Myrmecology

A World Perspective

EDITED BY

Robert K. Vander Meer,
Klaus Jaffe, and Aragua Cedeno

Routledge
Taylor & Francis Group
LONDON AND NEW YORK

First published 1990 by Westview Press, Inc.

Published 2018 by Routledge
52 Vanderbilt Avenue, New York, NY 10017
2 Park Square, Milton Park, Abingdon, Oxon OX14 4RN

Routledge is an imprint of the Taylor & Francis Group, an informa business

Library of Congress Cataloging-in-Publication Data
Applied myrmecology : a world perspective / edited by Robert K. Vander Meer, Klaus Jaffe, and Aragua Cedeno.
p. cm.—(Westview studies in insect biology)
Includes index.
ISBN 0-8133-7785-4
1. Ants. I. Vander Meer, Robert K. II. Jaffe, Klaus. III. Cedeno, Aragua. IV. Series.
QL568.F7A58 1990
595.79′6—dc19 89-5666
CIP

ISBN 13: 978-0-367-01324-0 (hbk)

Contents

CONTROL

Foreword

"Man tends to look up and tries to figure out what is happening up above and at the same time he cannot even figure out the ants. I wish I could figure out ants. I suggest let's deal with ants. If we get the chance to understand them, this world would run on a different basis. Without understanding ants, I don't think we can understand other things."

--Lech Walesa

About thirty-one years ago I was assigned to organize a small laboratory in Gulfport, Mississippi, with the sole purpose of developing safe and effective methods to control the imported fire ant, then known as *Solenopsis saevissima* var. *richteri*. The U.S. Congress had just initiated a new control program and the only available control technology required large amounts of the insecticides heptachlor and dieldrin. Little did I know how controversial this new invader from South America would become. I doubt many other insects have presented such a challenge to our ingenuity and resourcefulness. While we have had considerable success in developing methods for their control, heightened environmental awareness and increasingly stringent pesticide regulations have forced us to continually research and develop new control strategies. For example, Mirex bait was the crowning achievement of our research in the early 1960s, and millions of pounds were applied in the southern United States. Unfortunately, application rates as low as 500 milligrams per acre left residues in nontarget animals. Despite scant evidence for its mammalian toxicity, the detection of residues, especially in humans, dictated its withdrawal by the U.S. Environmental Protection Agency. Thus, we had to look elsewhere to provide new, environmentally safe control strategies.

This search made us aware of the parochial nature of our overall research program. We had always tried to satisfy the needs of the federally-supported Imported Fire Ant Control Programs for immediate control techniques. As a result, our research on the fire ant's biology, ecology, physiology and biochemistry lagged far behind. We knew little about pheromones and even less about the ants' natural enemies in South America. In fact, we are only now learning the exact location of their homeland.

For the first time, we realized that maximum efficiency on our part meant a free trade of information with scientists working on other pest ant species. We felt that the resulting data base would give all of us added insight into pest control. A new field was born: "applied myrmecology."

However, there were some initial problems. In 1981, I received an invitation to present a paper on the fire ants' economic impact for a symposium on "Economically Important Social Insects" at the 9th Congress of the International Union for the Study of Social Insects (IUSSI) in Boulder, Colorado. To my dismay, interest and attendance at the symposium were poor. The vast majority of myrmecologists were not studying pest ants; instead, they emphasized the basic aspects of behavior, caste determination and origin of sociality. While these challenging long-range research areas could lead to new control techniques, they did not help solve our immediate problem of how to control the fire ant. Fortunately--as a result of prior contacts with Dr. Alain Kermarrec (INRA, Guadeloupe)--we were able to gather a small but determined group of scientists working on fire ants and leaf-cutting ants. Our purpose? To organize a symposium specifically aimed at an interchange of research objectives and ideas on control and management of these two pest ants. Our efforts eventually culminated in the March 1985 symposium "Fire Ants and Leaf-Cutting Ants," which we hosted in Gainesville, Florida. This highly successful meeting demonstrated how interactions between researchers of the two different ant groups could be greatly beneficial to both. The symposium resulted in the publication of a book entitled *Fire Ants and Leaf-Cutting Ants: Biology and Management*.

Another result of the symposium was the decision to broaden "Attini," the newsletter started by Dr. Cherrett as a vehicle for leaf-cutting ant news. We decided that the newsletter would now publish information on all pest ants. And, although no specific plans were made for a second conference, Klaus Jaffe arranged a "Pest Ants" colloquium in Caracas, Venezuela (October 1988). The papers presented at this meeting--in addition to other solicited manuscripts--are included in this book.

While both conferences successfully provided a medium for exchange of applied myrmecology research data, we must now re-evaluate our accomplishments and decide the direction of future research. Would we be better off working within the framework of already-existing international organizations? The answer is a resounding "No," if the papers presented at the past two IUSSI meetings and the International Congress of Entomology are any indication. IUSSI's 9th Congress in Boulder, Colorado (1982), presented only fourteen papers or posters related to pest ants: five on fire ants, four on carpenter ants, two on leaf-cutting ants and one each on pharaoh and harvester ants. The numbers for the 10th Congress in Munich, Germany (1986), were somewhat better (seventeen total), but they still represented only about 5% of the 350 papers presented. The numbers according to pest ant species were: nine on fire ants, two on leaf-cutting ants, three on carpenter ants, three on harvester ants and one on the Argentine ant.

The July 1988 International Congress of Entomology in Vancouver, Canada, suffered from a shortage of papers specifically on pest ants: There were four papers on fire ants (all presented by USDA/ARS scientists) and one each on leaf-cutting, Argentine and carpenter ants. Either there is a lack of interest in pest ant research or pest ant researchers are uninterested in presenting their research at meetings.

Whatever the cause, this lack of representation does not reflect the large volume of pest ant literature. Cherrett (1986) stated that some 1300 papers and four books have been published on leaf-cutting ants. Wojcik (in a personal communication) recorded the following publications on fire ants: 973 research papers, 88 theses and dissertations, 220 review articles or bibliographies and 214 popular articles or extension and regulatory bulletins. Numerous papers have been published on other pest ants. It appears to me that, at some level, researchers in applied myrmecology are not effectively organized and that improved communication needs to be a primary goal. A structured approach to pest ant meetings and an emphasis on "Attini" as an instrument for dialogue and discussion may be the answer.

Fortunately, Harold Fowler organized a third pest ant conference (Brazil, 1989) and Robert Vander Meer is organizing a "Pest Ant" symposium for the 11th International IUSSI meeting (Bangalore, India, 1990). Perhaps we have awakened a broad interest in pest ants.

We should also consider broadening our scope of interest to include researchers who study ants as biocontrol agents. With the current trend toward reduced pesticide usage and management of insect populations, we need to emphasize ants as predators of pest insects. This is a real factor in fire ant control strategies since *S. invicta* is being promoted for control of several pest species. (Majer, 1982, has published an interesting review of the manipulation of ant species in agricultural and forest ecosystems.)

Applied Myrmecology: A World Perspective represents the first effort to bring together a broad range of pest ant articles in a single volume. For this, the editors need to be commended. I urge the contributors and others interested in applied myrmecology to continue these essential interactions.

C. S. Lofgren

REFERENCES

CHERRETT, J.M. 1986. The biology, pest status and control of leaf-cutting ants. Agric. Zool. Rev. 1: 1-37.

MAJER, J.D. 1982. Ant manipulation in agro- and forest ecosystems. pp. 90-97. In The biology of social insects. M.D. Breed, C.K. Michener and H.E. Evans (eds.). Westview Press, Boulder, Colo. 420p.

Preface

Ants have always fascinated the nature observer. Reports from ancient Egypt and Mesopotamia indicate that ants interested humans long ago. Myrmecology as a science had its beginning in the last century with great naturalists like Andre, Darwin, Emery, Escherich, Fabre, Fields, Forel, Janet, Karawaiew, McCook, Mayr, Smith, Wasmann and Wheeler. They studied ants as an interesting biological phenomenon, with little thought of the possible beneficial or detrimental effects ants could have on human activities (see Wheeler 1910 as an example).

When Europeans began colonizing the New World, serious ant problems occurred. The first reports of pest ants came from Spanish and Portuguese officials of the fifteenth and sixteenth centuries in Trinidad, The West Indies, Central America and South America. Leaf-cutting ants were blamed for making agricultural development almost impossible in many areas. These ants, *Atta* and *Acromyrmex* species, are undoubtedly the first ants identified as pests and may be considered to have initiated interest and research in applied myrmecology (Mariconi 1970).

Ant problems other than those associated with leaf-cutting ants have been reported mainly in this century. In fact, applied myrmecology owes its rapid growth in the past few decades mainly to the fire ant, *Solenopsis invicta*. This ant was accidentally introduced to the United States from Brazil or Argentina and has caused medical and agricultural problems to such an extent that the affected states and the U.S. Federal government have invested heavily in fire ant research programs.

No efficient measures against pest ants were available until recently. In the 1960s and 1970s large quantities of insecticide were used to control *Solenopsis invicta* in the United States. Environmental contamination ultimately led to the prohibition of Mirex and other chlorinated hydrocarbon pesticides in the United States and other countries. These prohibitions sparked research on alternative pest management strategies and opened new areas of applied myrmecology. Much of this early work was brought together in the book *Fire Ants and Leaf-Cutting Ants: Biology and Management* (Lofgren and Vander Meer 1986). Articles in this volume discussed various aspects of the biology, pest problems and control of leaf-cutting ants and fire ants.

On many occasions, the technological advances of European-originated society has created an arrogant attitude that blinds us to opportunities presented to us by nature. Several societies have been more attentive to ants than has Western science. A comparison of two sympatric Amerindian societies inhabiting the Orinoco-Amazon basin illustrates how important the society is in understanding and using nature. The Yanomami--an isolated society of hunter-gatherers possessing only the rudiments of agricultural practice--recognize at least thirty-nine distinct ant species, but apparently have no use for any of them (Herzog 1979). The Yecuana, a culture also dedicated to hunting and gathering, but with more evolved agricultural practices as well as knowledge of river navigation, recognize thirty-six ant species and have at least eighteen different medicinal and mythical uses for them (Jaffe and Muller 1979). These examples demonstrate how cultural attitudes affect our perception of nature.

In this respect, the subject area of applied myrmecology is increasing its scope, recognizing that many ant species have an important role in their ecosystems. These roles may eventually make them harmful to humans, or alternatively make them attractive to pharmacology or even integrated pest management programs. Thus, they may be exploited by humans. The present book, *Applied Myrmecology: A World Perspective*, reflects this developing consciousness. We hope that our efforts and those of all the contributors will stimulate further research and demonstrate that applied myrmecology has matured into a distinct scientific discipline.

Robert K. Vander Meer
Klaus Jaffe
Aragua Cedeno

REFERENCES

HERZOG, H., M.C. MULLER, M. GRANIER, J. LATKE and K. JAFFE. 1979. Notas etnomirmecologicas Yanomani. Acta Terramaris 1:51-57.

JAFFE, K. and M.C. MULLER. 1979. Notas etnomirmecologicas Yecuana. Acta Terramaris 1:43-50.

MARICONI, F.A.M. 1970. As Saúvas, Editorial Agronómica Ceres, Sao Paulo, 167 pp.

LOFGREN, C.S. and R.K. VANDER MEER. 1986. Fire Ants and Leaf-Cutting Ants: Biology and Management. Westview Press, Boulder, Colo. 435 pp.

Acknowledgments

The editors wish to thank all of the contributors, whose patience and timely responses to editorial queries made this mammoth undertaking possible. The overview writers all did an excellent job of synthesizing their sections and presenting questions for the future. John Latke, Luc Passera, Sanford Porter, David Morgan, Dave Williams and Malcolm Cherrett deserve a special thanks. The concept for the book came from a "Pest Ants" colloquium held in Caracas, Venezuela, October 1988. We thank the following institutions, which financially supported that meeting and the preparation of this book: the Universidad Simón Bolivar (Venezuela), the Universidad Central de Venezuela, CVG-Proforca (Venezuela), Fundación Polar (Venezuela), Griffin Corporation (USA), Industrias Pampero (Venezuela) and Shulton de Venezuela. We thank the U.S. Department of Agriculture, Agricultural Research Service, for approval for Robert K. Vander Meer to work on this project and for the use of computer facilities. Jeffrey S. Vander Meer assisted in the editing and helped keep the language as concise and correct as possible, although a couple of split infinitives almost snuck through.

Special thanks go to a special artist, Eduardo Perez, from the Universidad Simón Bolivar, Caracas, who allowed us to enhance the book with several of his beautiful ant drawings.

Thanks to the Westview Press staff for their encouragement throughout the past year and for the confidence they and Scientific Advisor Mike Breed had in us. Finally, we thank all others who had a role in making this book possible.

R.K.V.M.
K.J.
A.C.

Applied Myrmecology

Ant Pests of the World

1

Major Ant Problems of South America

H.G. Fowler, J.V.E. Bernardi, J.C. Delabie, L.C. Forti and V. Pereira-da-Silva

INTRODUCTION

South America has one of the world's richest ant faunas (Kempf 1972), and consequently should also have its share of ant problems. However, except for the notoriety of the leaf-cutting ants, little information is available on other economically-important ants. After reviewing the available literature, a list of the principal ant problems in South America has been compiled for this study. This review is not definitive but, rather, representative. The majority of the South American ant literature is not indexed, nor in what is considered periodical scientific publications, and much of this information is anecdotal--weak in scientific merit. This has caused additional problems when misleading information is taken seriously (Zenner de Polania and Ruiz Bolanos 1983). For brevity's sake, only non leaf-cutting ant pests (see Cherrett 1986, Fowler et al. 1986) have been reviewed.

Control of *Solenopsis invicta* in the United States has a long and well-documented history (Banks et al. 1985); however, South American ant control has only a fragmentary history. The evolution of control methodologies pieced together (Table 1) suggest that ant control in South America is changing slowly due to insufficient research funds. Even when baits are used for ant control, as for leaf-cutting ants, little attention has been directed towards environmental concerns, such as bait specificity (Williams 1986). For ants other than leaf-cutting ants, control measures seem to be almost casual extensions of leaf-cutter control, and utilize unsuitable leaf-cutting ant toxicants (Costa 1937, Monte 1931, Fonseca 1936, 1953, Fowler 1981, Anonymous 1953).

Based upon available literature and personal experience, a list of South America's problem ants (Table 2) has been drawn up. Each of these shall be discussed briefly.

ACROPYGA SPP.

This genus has global representation as do all species associated with root scales (Pseudococcidae). In South America, several *Acropyga* species affect coffee (Bunzli 1935, Weber 1944, Costa Lima 1931, 1932) and cocoa (Strickland 1945, Weber 1944, Delabie and Mantovani 1988). The scales

TABLE 1. The evolution of ant control, especially leaf-cutting ant control, in South America, fire ant control in the United States and the development of control products in the world.

Year:	1900		1920		1940		1960		1980
Leaf-cutting ant control	A	A	A	B	B	C	C	C	C
Fire ant control (United States)				B	C	C	C	CD	DEF
Product development	A	A	A	B	B	C	D	E	F

Control groups:
A = arsenicals; lime-sulfur; petroleum oils; nicotine; cyanides
B = flourine CMPS; certain synthetic organics (e.g., dinitro CMPDS; thiocyanates); rotenone; pyrethrum
C = cholorinated hydrocarbons (e.g., DDT, BHC, toxaphene, chlordane, aldrin, dieldrin); organo-phosphates (e.g., parathion)
D = synthetics (e.g. allethrin) pyrethrum + synergists; carbamates
E = chemosterilization; juvenile hormones
F = insect growth regulators; avermectins

protected by *Acropyga* are known vectors of several root diseases and are true trophobionts of these ants, forming symbiotic associations in which both partners are obliglately linked throughout their entire life cycles. The ants rear their brood with the scales. Galleries are constructed around roots to house the scales; in return, the scales supply the ants with honeydew, apparently their only food source. However, the ants may crop excess scales for food and thus regulate scale populations (Delabie and Mantovani 1988). The symbiosis between *Acropyga* and scales is so complete that during swarming, female Acropyga carry reproductively active adult female scales in their mandibles (Eberhard 1978, Campos and Morais 1986, Delabie, this volume). Both ants and scales are hypogaeic. Ant nests are formed by anamostising galleries filled with "rearing cells" for scales, which are only found there as nymphs (Delabie and Mantovani 1988, Bunzli 1935). *Acropyga* species are polygenic, a trait which may explain their exceedingly high densities in certain regions (Delabie and Mantovani 1988, Strickland 1945). This trait also creates additional problems for their control. Currently no products or recommendations are available for control of this underestimated pest.

AZTECA SPP.

Of the rich *Azteca* fauna in South America, *A. paraensis* and *A. chartifex* have been the most studied. These Dolichoderinae are exclusively arboreal and restricted to forested areas from the Amazon basin throughout Brazil's north-eastern region. Economically important crops, such as cocoa and cashew nuts, are cultivated in this area, generally under a canopy of emergent vegetation. *A. chartifex*'s nest is of carton and forms an extended cone, similar to the nest of the arboreal termite, *Nasutitermes*. *A. paraensis* does not construct carton nests, but instead associates with certain epiphytic plants of the families Araceae, Bromeliaceae, Orthidaceae and Gesneriaceae, to produce ant gardens. These ant gardens are tended and planted by ants, which in turn receive a nesting site in the rounded root mass created by the ant gardens. *A. paraensis* apparently favors the growth of their ant gardens by removing leaves, which opens the tree canopy. Leaf sap is used to physically support the ant garden.

Both *Azteca* species depend heavily upon Homopteran honeydew, especially the dew produced by fruit-attacking Membracids. A swarming mass of ant workers protects these honeydew producers, and they are often further protected by an ant-built carton shelter. In cocoa plantations, the heavy nests of *A. chartifex* break branches. By building carton galleries on branches, the ants interfere with flowering and reduce potential fruit yield. On some cocoa plantations, A. paraensis tends ant gardens located on the largest branches and also interferes with flowering, although it does not produce carton galleries (Bondar 1923, 1934a, 1934b).

Both species are aggressive, and rural workers are wary about working on plantations where these ants are present, thus making fruit harvest and plantation maintenance more costly and irregular. Often these ants are associated with nesting social wasps, aggressive and active stingers of the genera *Polybia* and *Apoica*. This association further hinders rural workers (Bondar 1925, 1939, 1920, 1930, Silva 1945, 1955, Silva and Barbosa 1948).

From 1950 to 1960, countless *Azteca* nests were destroyed in insecticidal control programs (Silva and Barbosa 1948, Silva 1955, Gallo et al. 1978). These control programs apparently favored the expansion of other ants in Brazil's cocoa-growing region, especially *Wasmannia auropunctata* (Delabie 1988). However, by comparing production records from trees with Azteca nests to trees without nests, Vello and Magalhaés (1971) proposed that the odors of these ants attracted pollinators, increasing flower visitation and fruit set. However, as this ant dominates the canopy, increased fruit production may be due to its role as a predator of phytophagous insects. As a result of the new evidence, several growers have recently established *A. chartifex* nests in their cocoa plantations.

More evidence is needed to determine if Azteca species are pestiferous or beneficial. Nevertheless, the history of their control in cocoa demonstrates that ill-informed massive control programs based upon easily-applied insecticides may be unnecessary and ecologically harmful.

WASMANNIA AUROPUNCTATA

This Myrmicinae is one of the most prevalent Neotropic ants. The small ant is about 1.5 mm in length and yellowish. However, *W. auropunctata* is polygynous, a good colonizer, quite aggressive and is an active protector of Homoptera.

Currently, the species has a pan-tropical distribution, and is found in almost all South American countries (Kempf 1972). Because of its good colonizing ability and rapid colony growth due to the presence of multiple queens, colonies expand rapidly - they have recently invaded and decimated the fauna of the Galapagos Islands (Clark et al. 1982, Lubin 1984). This species prefers disturbed regions. In the cocoa-growing region of Bahia, Brazil, *W. auropunctata* lives in the cocoa canopy, and has apparently expanded following massive control programs against *Azteca* spp. in the 1950s and 1960s (Delabie 1988).

Because of its polygyne nature, *W. auropunctata* colonies frequently and easily divide. Where it occurs, this ant dominates and is an active predator of herbivorous insects and other ants, with the exception of *Camponotus* species (Delabie 1988). However, *W. auropunctata* also depends upon honeydew of scales, especially *Plannococcus citri*, which breed and multiply on the fruit of many tree species, especially cocoa and Citrus. These scales can reach populations which completely cover fruits and promote their early abortion. Because of their aggressivity and strong stings, which can result in allergic reactions, fruit-pickers are reluctant to work in areas in which these ants occur, and this further complicates management practices (see Ulloa-Chacon and Cherix this volume).

TAPINOMA MELANOCEPHALUM

This introduced Dolichoderinae occurs throughout the Americas (Kempf 1972), in locally abundant populations. Because of its small size, this ant, like other small species such as *W. auropunctata* and *Monomorium* spp., is also a domestic pest, and may transmit hospital infections. *T. melanocephalum* also associates with root scales (Smith 1936), and--in Sao Paulo's banana-growing region--fruit scales. It is the most abundant ant present in these heavily-sprayed crops. Because *T. melanocephalum* is also polygenic, with easily-dividing colonies, control is difficult and, apparently, like *W. auropunctata*, *T. melanocephalum* is favored by insecticidal control of other pests (see Harada this volume).

CAMPONOTUS SPP.

This is one the largest and most heterogeneous ant genera in the Neotropics (Kempf 1972) and, consequently, it is difficult to determine which species may be of economic importance, especially since identification of species is difficult. In South America, the subgenus *Camponotus*, notorious for Holarctic wood destruction is not present, and the majority of pest citations deal with honeydew collections by representative species of other subgenera (Trujillo Pelluffo 1942, Anonymous 1950, Moreira 1918, 1919). These ants can be household pests and some species have also been cited as apiary pests (Emelen 1924).

Of the larger species, such as *Camponotus rufipes* and *C. brasilensis*, ants protecting associated Homoptera may create problems in fruit producing areas due to their aggressiveness (Scarpellini et al. 1986). See related chapters in this volume.

IRIDOMYRMEX HUMILIS

The Argentine ant, known as the Brazilian ant in Argentina, has spread from South America throughout the world. Even though exotic species are generally more problematic than native species, *I. humilis* can be pestiferous in South America. This polygynous Dolichoderinae is a recorded tender of Homoptera (Trujillo Peluffo 1942, Nasca et al. 1982, Gallo et al. 1978). In Caribbean areas where it was accidentally introduced, *I. humilis* has displaced other ant species (Erickson 1971, Crowell 1968, Haskins and Haskins 1965), due partly to its good colonizing ability and polygynous nest structure.

Because of its small size and the capacity of colonies to divide easily, *I. humilis* is a common household pest (Dade 1974), and has been shown to be an important vector of hospital infections in Chile (Ipinza et al. 1981). See related chapters in this volume.

PHEIDOLE SPP.

This genus is also represented by many South American species (Kempf 1972). Some of the smaller *Pheidole* species can be of occasional importance in households as domestic pests, but the introduced *Pheidole megacephala* is undoubtedly the genus' most important representative pest, as it is in other tropical regions (Chang et al. 1980). This ant is known as an active displacer of native fauna in regions where it was introduced (Haskins and Haskins 1965, Fowler 1988b). Densities may reach such levels in these modified faunas that *P. megacephala* becomes a domestic and household pest, often damaging structures such as electrical cables (Fowler 1988b).

MONOMORIUM SPP.

The genus' three exotic species (Table 2) are generally associated only with human structures. Like *Wasmannia auropunctata* and *Iridomyrmex*

TABLE 2. Ant pests of South America, given in alphabetical order, with their origin and principle activity of economic importance.

	Origin	Principle activity of the pest status						
Ant Taxa	Native (N) or Introduced (I)	cut leaves or flowers	other plant damage	tending of Homoptera	aggres-sive?	public health importance	domestic pest	material damage
Acropyga spp.	N			x				
Acromyrmex spp.	N	x	x					
Atta spp.	N	x						x
Azteca spp.	N		x	x	x			
Camponotus spp.	N		x	x	x		x	x
Crematogaster spp.	N		x	x		?	x	
Iridomyrmex humilis	N/I			x		x	x	
Monomorium pharaonis	I					x	x	
Momomorium destructor	I					x	x	
Monomorium floricolis	I					?	x	
Paratrechina fulva	I/N			x		?	x	
Pheidole spp.	N					?	x	
Pheidole megacephala	I			x			x	?
Solenopsis spp.	N/I		x	x	x	?	x	x
Tapinoma melanocepalum	I			x		?	x	
Wasmannia auropunctata	N/I			x	x	?	x	?

humilis, these ants form polygynous colonies and fragment easily. Recent studies (Fowler unpublished) reveal that *M. pharaonis* is common in hospitals and health care centers, and is undoubtedly associated with hospital infections, as has been found in the rest of the world. Being restricted to structures, these *Monomorium* species are notorious household pests, difficult to control due to their colony organization. Colonies are often transported unknowingly from one local to another in boxes for the supermarket, etc., as occurs with *W. auropunctata* and *I. humilis* (see Eichler, this volume; Williams this volume).

PARATRECHINA FULVA

This ant, sometimes called *Nylanderia fulva*, has an interesting history. Even though it has long been considered a pest (Ihering 1933, Mello 1933, Monte 1933) of both households and agriculture, due to its propensity for protecting Homoptera, some of Brazil's popular literature during the 1930s and 1940s suggests that this ant was negatively associated with leaf-cutting ants of the genus *Atta*, and that it effectively controlled colonies of *Atta*. *Paratrechina fulva* was apparently taken into Colombia for *Atta* control, and while it did not produce any noticeable effects on *Atta* populations, it became a major coffee pest through its protection of mealy-bugs (Zenner de Polania and Ruiz Bolaños 1983). This species has probably been an under-estimated problem in many areas. See Zenner-Polania for details, this volume.

SOLENOPSIS spp.

Fire ants cause problems not only in North America, but occasionally in their homeland of South America. Some species are known to be detrimental because of their association with Homoptera (Nasca et al. 1982, Hays 1958), as well as their aggressiveness towards people (Monte 1941, Santos 1931). *Solenopsis* species can also damage vegetable gardens directly, crippling plants such as potato (Monte 1937, Mendes 1935, Ribeiro 1954, 1955a, 1955b, Anonymous 1940).

As in the United States, some *Solenopsis* species can damage structures - in particular, electrical installations (Galli and Fernandes 1988, MacKay and Vinson, this volume). Often, *Solenopsis* and *Wasmannia* are confused by even economic entomologists (Delabie 1988). *Solenopsis*, nevertheless, is at best a local and occasional problem, and never a problem on the scale encountered in the United States.

LEAF-CUTTING ANTS

The most problematic South American ants are undoubtedly leaf-cutting ants. Damages and control procedures will not be reviewed here, as these will be discussed elsewhere. However, some comment on their pest status and current management practices is necessary. Of the known leaf-cutting ant taxa, there is little hard evidence that the majority of the taxa are of significant economic importance. Furthermore, through the indiscriminate application of insecticides against all leaf-cutting colonies, human beings may have accelerated the extinction of some unique endemic species, such as *Atta robusta* (Table 3).

TABLE 3. The species of leaf-cutting ants, their distribution and their relative pest status, as assessed by citations

Leaf-cutting species	Vegetation cut (monocots m/ dicots-d) (amplitude)$	Distribution	Relative pest status§
Acromyrmex (Acromyrmex)			
ambiguus	d	pandemic	weak
aspersus	d	pandemic	dubious
coronatus	d	pandemic	weak
crassispinus	d	pandemic	moderate
diasi	d	endemic	dubious
disciger	d	pandemic	weak
*gallardoi**	d?	endemic	dubious
hispidus	d	pandemic	weak
hystrix	d	pandemic	weak
laticeps	d	pandemic	weak
lobicornis	d/m	pandemic	moderate
lundi	d	pandemic	weak
niger	d	pandemic?	weak
nobilis	d	endemic?	dubious
octospinosus	d	pandemic	moderate
rugosus	d	pandemic	dubious
subterraneus	d	pandemic	moderate
Acromyrmex (Moellerius)			
heyeri	m	pandemic	weak
*landolti***	m	pandemic	moderate
*mesopotamicus***	m?	endemic	dubious
pulvereus	m?	endemic	dubious
silvestrii	m	pandemic?	weak
striatus	m	pandemic	weak
versicolor	d/m?	endemic	dubious
Atta			
bispaerica	m	pandemic	moderate
capiguara	m	pandemic	strong
cephalotes	d	pandemic	moderate
colombica	d	pandemic	weak
*goiana****	m	endemic	dubious
*insularis****	d	endemic	weak
laevigata	d/m	pandemic	moderate
mexicana	d/m	pandemic	weak
*opaciceps****	d	endemic	weak
*robusta****	d	endemic	dubious
saltensis	d	pandemic	weak
*silvai****	d	endemic	dubious

TABLE 3. Continued.

Leaf-cutting species	Vegetation cut (monocots m/ dicots-d)	Distribution (amplitude)$	Relative Pest status§
sp. nov.***	m	endemic	dubious
texana	d	pandemic	moderate
vollenweideri	m	pandemic	weak

$Based upon km of geographic distribution, or restriction to certain vegetation types. §Data derived from Cherrett (1986a); Fowler, et al. (1986) and Fowler & Robinson (1979); *Species of unknown taxonomic status. **Species under taxonomic changes in the near future (Fowler, 1988a). ***Species in danger of extinction.

Although these ants are generally considered damaging (Nasca et al. 1982, Rizo 1977, Bruner et al. 1945, Gallo et al. 1978), economic threshold data do not exist, especially for large, extensive pulp-wood plantations in Brazil. In these areas, given the ants' size, bait is generally spread at regular intervals throughout the area, often in polyethylene bags, which the ants cut open to get the bait. No studies relating to leaf-cutting ant damage to crop growth stage or to ant density were located and no evidence exists that extensive control program are warranted on a cost/benefit basis. Even those studies which have been cited to demonstrate the amount of damage caused by these ants are apparently flawed (see other chapters in this volume). Currently, research is directed toward identifying novel toxicants or more attractive baits, and no research is directed to the basic question of economic returns or of when, how much, or where to apply control strategies. These studies are of utmost importance to establish more rational control strategies for leaf-cutting ants (Table 1), that will also prove beneficial to solving other South American ant problems (Table 2).

REFERENCES CITED

ANONYMOUS, 1940. A debelacõa da formiga "ruiva" (*Solenopsis saevissima* Sm.). Campo, Rio de Janeiro, 11(124): 40.

ANONYMOUS, 1950. Combate às formigas doceiras, sara-sara (*Camponotus*). Chacaras e Quintais, São Paulo, 81(4): 467-477.

ANONYMOUS, 1953. Meios de Combate às formigas caseiras. Sel. Agric., Rio de Janeiro, 8(89): 72-74.

BANKS, W.A., C.S. LOFGREN and D.F. WILLIAMS. 1985. Development of toxic baits control of imported fire ants. In, Kaneko, T.M. and Spicer, L.D. (eds.), Pesticide Formulations and Application Systems. American Society for Testing and Materials, Philadelphia, pp. 133-143.

BONDAR, G. 1922. As lendas e a verdade sobre a formiga cacarema da Bahia e seu papel na lavoura. Chacaras e Quintais, São Paulo, 26(5): 369-371.

BONDAR, G. 1923. A formiga cacarema e o seu combate na lavoura. Correio Agric. Bahia, 1(1): 4.

BONDAR, G. 1930. A formiga de enxerto, praga dos cacaueiros na Bahia. Campo, Rio de Janeiro, 1(3): 50-52.
BONDAR, G. 1934a. A formiga cacarema nos cacauais e das matas da Bahia. Campo, Rio de Janeiro, 5(9): 36-37.
BONDAR, G. 1934b. A formiga do enxerto, a maior praga dos cacauais e das matas da Bahia. Campo, Rio de Janeiro, 5(5): 33-36.
BRUNER, S.C., L.C. SCARAMUZZA and A.R. OTERO. 1945. Catalogo de los insectos que atacam a las plantas economicas de Cuba. Bol. Est. Exp. Agron. Santiago de La Vega, 65: 1-246.
BUNZLI, G.H. 1935. Untersuchungen uber coccidophile Ameisen aus den Kaffee feldern von Surinam. Mitt. Schweiz. Entomol. Ges., 16: 453-593.
CAMPOS, L.A. de O. and H.C. de MORAIS. 1986. Transporte de homoptero por fêmeas de formigas do gênero *Acropyga* (Formicinae) em Vicosa, MG. Anais do VII Encontro de Mirmecologia do Estado de São Paulo. F. Caetano ed. UNESP - FAPESP - CNPq. 52-53.
CHANG, V.C.S., A.K. OTA, and D. SANDERS. 1980. Parallel ridge barrier to control ant damage to orifices of drip irrigation tubes. J. Econ. Entomol., 73: 403-406.
CHERRETT, J.C.M. 1986. The economic importance and control of leaf-cutting ants. In, Vinson, S.B. (ed.), Economic Impact and Control of Social Insects. Praeger, New York, pp. 165-192.
CLARK, D.P., C. GUYASAMIN, O. PAZMINO, C. DONOSO and Y. PAEZ de VILLACIS. 1982. The tramp ant, *Wasmannia auropunctata*: autoecology and effects on ant diversity and distribution on Santa Cruz Island, Galapagos. Biotropica, 14: 196-207.
COSTA LIMA, A.M. da. 1931. A propósito de *Acropyga pickeli* Borgm., 1927. Bol. Biol., 21: 65-69.
COSTA, R.G. 1937. Ainda as formigas acucareiras. Chacaras e Quintais São Paulo, 55(3): 296.
CROWELL, K.L. 1968. Rates of competitive exclusion by the Argentine ant in Bermuda. Ecology, 49: 551-555.
DADE, W.E. 1974. Hormigas en viviendas y jardines de Lima metropolitana. *Iridomyrmex humilis* (Mayr) *Monomorium minimun* (L.) Rev. Peru. Entomol., 17: 126-127.
DELABIE, J.H.C. 1988. Ocorrência de *Wasmannia auropunctata* (Hymenoptera, Formicidae, Myrmicinae) em cacauais na Bahia. Rev. Theobroma, (in press).
DELABIE, J.H.C. and J.E. MANTOVANI. 1988. Observacões sobre a biologia de *Acropyga* spp. (Formicidae, Plagiolepidini) associdas a rizosfera do cacaueiro. Res. IX Encontro de Mirmecologia, Vicosa, MG (in press).
EBERHARD, W.G. 1978. Mating swams of a South American *Acropyga* (Hymenoptera: Formicidae) Ent. News 89: 14-16.
EMELEN, A.V. 1924. Formigas nocivas as abelhas. Chacaras Quintais, São Paulo, 30(2): 153-155.
ERICKSON, J.M. 1971. The displacement of native ant species by the introduced Argentine ant, *Iridomyrmex humilis* Mayr. Psyche, 78: 257-266.
FONSECA, J.P. da. 1936. Combate às formigas caseiras. O Biol., 2: 68-69.
FONSECA, J.P. da. 1953. Meios de combate às formigas caseiras. Sitios Fazendas, São Paulo, 19(7): 54.

FOWLER, H.G. 1981. Control of noxious ants with locally produced baits. Turrialba, 31: 161-162.

FOWLER, H.G. 1988a. Taxa of the Neotropical grass-cutting ants, (*Acromyrmex*) *Moellerius* (Hymenoptera: Formicidae: Attini). Cientifica (São Paulo) (in press).

FOWLER, H.G. 1988b. Eradication of the native ant fauna by the introduction of an exotic ant in Itapirica, Bahia, Brazil, during hydroelectric dam construction. Environ. Conserv. (in press).

FOWLER, H.G. and S.W. ROBINSON. 1979. Field identification and relative pest status of Paraguayan leaf-cutting ants. Turrialba, 29: 11-16.

FOWLER, H.G., L.C. FORTI, V. PEREIRA-da-SILVA and N.B. SAES. 1986. Economics of grass-cutting ants. In, Lofgren, C.S. and Vander Meer, R.K. (eds.), Fire Ants and Leaf-Cutting Ants: Biology and Management. Westview Press, Boulder, pp. 18-35.

GALLI, J.C. and O.A. FERNANDES. 1988. Dano causado por formigas do gênero *Solenopsis* (Hymenoptera, Formicidae) à fios electricos em Jaboticabal-SP. An. Soc. Entomol. Bras., 17: 225-226.

GALLO, D., O. NAKANO, S. SILVEIRA-NETO, R.P.L. CARVALHO, G.C. de BATISTA, E. BERTI-FILHO, J.R.P. PARRA, R.A. ZUCCHI and S.B. ALVES. 1978. Manual de Entomologia Agricola. Editora Ceres, São Paulo, 531 p.

HASKINS, C.P., and E.F. HASKINS. 1965. *Pheidole megacephala* and *Iridomyrmex humilis* in Bermuda - equilibrium or slow displacement? Ecology, 46: 736-740.

HAYS, K.L. 1958. The present status of the imported fire ant in Argentina. J. Econ. Entomol., 51: 111-112.

IHERING, R. von. A formiga cuibana, um flagello. Campo, Rio de Janeiro, 4(1): 31-32.

IPINZA, I. REGLA, G. FIGUEROA and J. OSORIO. 1981. *Iridomyrmex humilis*, "hormiga argentina," como vector de infecciones intra-hospitilares. I. Estudios bacteriologicos. Folia Entomol. Mex., 50: 81-96.

KEMPF, W.W. 1972. Catalogo abreviado das formigas da Região Neotropical (Hymenoptera: Formicidae). Studia Entomol., 15: 3-344.

LUBIN, Y.D. 1984. Changes in the native fauna of the Galapagos Islands following invasion by the little red fire ant, *Wasmannia auropunctata*. Bol. J. Linn. Soc., 21: 229-242.

MELLO, P. de. 1933. A formiga cuyabana em Pernambuco. Rel. Soc. Ind. Viacao, Recife, 2: 67-69.

MENDES, D. 1935. A debeliacão da formiga ruiva, *Solenopsis saevissima* Sm. Chacaras Quintais, São Paulo, 51(5): 630.

MONTE, O. 1931. Formigas doceiras. Bol. Agric. Zootech. Vet., Minas Gerais, 4: 95.

MONTE, O. 1937 Formiguinhas no batatal. Chacaras Quintais, São Paulo, 56(1): 103.

MONTE, O. 1941. Combate à formiga lavapes. Chacaras Quintais, São Paulo, 63(5): 606.

MONTE, O. 1946. Combate à cuiabana. Chacaras Quintais, São Paulo, 73(3): 334.

MOREIRA, C. 1918. Vida da sara sará e como combate-a. Chacaras Quintais, São Paulo, 17(6): 62-463.

MOREIRA, C. 1919. A sara-sará de pernas ruivas. Lavoura, Rio de Janeiro, 22: 45-51.

NASCA, A.J., A.L. TERÁN, R.V. FERNANDEZ and A.J. PASQUALINI. 1982. Animales perjudiciales e beneficos a los citros en el noroeste argentina. CIPRON, Tucuman, 361 p.

RIBEIRO, J.H.C. 1954. Sobre a formiga ruiva e seu combate. Agron. E. Rio, 13: 227-231.

RIBEIRO, J.H.C. 1955a. Sobre a formiga ruiva e seu combate. Chacaras Quintais, São Paulo, 91(1): 51-53.

RIBEIRO, J.H.C. 1955b. Sobre a formiga ruiva e seu combate. Agric. Pecuar., Abr (398), 44, 46, 48.

RIZO, H.F. 1977. Catalogo de insectos perjudicales en cultivos de la republica argentina. Ed. Hemisfero Sul, Buenos Aires, 65 p.

SANTOS, E. 1931. A praga das pequenas formigas. Campo, Rio de Janeiro, 2(7): 42.

SCARPELLINI, J.R., O. NAKANO and C.S. MOREIRA. 1986. Ocorrêcia e inconveniência de *Camponotus* (*Myrmobrachys*) *brasilensis* Mayr, 1862 (Hymenoptera-Formicidae) em citrus. Rev. Agric., 61: 175-176.

SILVA, P. 1945. A formiga de enxerto. Bol. Divulg. Inst. Cacau, 1: 1-21.

SILVA, P. 1955. A formiga cacarema e o cacaueiro. Junta Exec. Com. Pragas Cacau. Bahia, 2: 1-13.

SILVA, P. and M.da R. BARBOSA. 1948a. Formiga cacarema x formiga de enxerto. Bahia Rural, 16(9): 16-17.

SILVA, P. and M.da R. BARBOSA. 1948b. O Instituto do Cacau e o combate à formiga de enxerto. Bahia Rural, 16(7): 23-26.

SMITH, M.R. 1936. The ants of Puerto Rico. Puerto Rico J. Agric., 20: 819-875.

STRICKLAND, A.H. 1945. A survey of the arthropod soil and litter fauna of some forest reserves and cacao estates in Trinidad, British West Indies. J. Anim. Ecol. 14(1): 1-11.

TRUJILLO PELUFFO, A. 1942. Insectos y otras parasitas de la agricultura y sus produto en la Uruguay. Fac. Agron. Montevideo. 323 p.

WEBER, N.A. 1944. The Neotropical coccid-tending ants of the genus *Acropyga* Roger. Annals Entomol. Soc. Am. 37: 89-122.

WILLIAMS, D.F. 1986. Chemical baits: specificity and effects on other ant species. In, Lofgren, C.S. and Vander Meer, R.K. (eds.), Fire Ants and Leaf-cutting Ants: Biology and Management. Westview Press, Boulder, pp. 378-386.

VELLO, F. and W.S. MAGALHÃES. 1971. Estudos sobre a participacão da formiga cacarema *Azteca chartifex spiritii* Forel na polinizacão do cacaueiro na Bahia. Rev. Theobroma 1(4): 29-44.

ZENNER DE POLANIA, I. and N. RUIZ BOLAÑOS. 1983. Control quimico de la hormiga loca, *Nylanderia fulva* (Mayr). Rev. ICI, 18: 241-250.

2

Pest Ants of India

G.K. Veeresh

INTRODUCTION

The earliest reference to Indian ants is traceable to Susrutha the renowned ancient physician (Ayyar 1939), who classified ants (pipilica) into six groups, based primarily upon their economic importance. However, systematic study and documentation of Indian Ants only began during the last century (Sykes 1936; Kirby, 1837; Emery 1981).

Butler (1985) made the interesting observation that the jungle people in Kanara district (South India) ate red ants (*Oecophylla smaragdina* Fabr.) and Tamil coolies used them as smelling salt to relieve themselves of severe cold symptoms: "The modus operandi is to go up to the nest in the bush, seize it in both the hands, rub ants and the nest together virulently between the palms and then take a few sniffs of the strong ammonia-like fumes" (Aitken 1889).

South Central tribals of Bastar "Pujas" sold crushed *O. smaragdina* workers for use in curry and rice preparations which were believed to give resistance against fatigue and the sun's heat (Long 1898).

From the middle of this century, scientists began to realize the important role ants play in the biological control of insect pests like termites (Veeresh and Shivashankar 1987), bed bugs (Negi et al. 1930), natural enemies of lac insect (Negi 1933), caterpillars (Jayaratnam 1977) and cutworms (Veeresh and Rajanna 1981).

While foraging, several ant species may cause damage to crops, their products, and become harmful to humankind's interests. Based upon their food habits, harmful ants are divided into three groups. These are ants which (1) cause direct damage to crops, (2) indirect damage to crops and (3) non-crop damage.

Ants directly damage crops by attacking plant parts above or below ground. Indirectly, they maintain mutualistic associations with a variety of plant-sucking insect pests. Ants come into conflict with human activities in many other ways, such as stealing seeds sown in the field, attacking beehives, loosening root systems by their excavation, invading dwellings for sugary and oily substances, as well as by their bites and stings. These aspects are dealt with herein under their respective headings.

ANTS AS DIRECT PESTS TO CROPS

Solenopsis geminata F

The fire ant *Solenopsis geminata*, found all over India, must have spread by commerce as it is pantropical in distribution (Bingham. 1903). It is an opportunistic omnivore and, therefore, is considered both pest and predator.

The morphology of soldier, worker, queen and male has been described by Bingham (1903) and Musthak Ali (1981). Based upon the size, *S. geminata* workers are differentiated as minor, medium and major. Minor workers often remain in the nest caring for the brood, while the medium and major are active foragers (Rani and Narendran 1987). The foraging workers move briskly and quickly disappear at the slightest disturbance.

The red ant nests in the soil of open fields and swarms several times, from March to October (Mukerji 1931). Swarming normally occurs from May to November (Musthak Ali 1981; Reddy 1981; Kamatar 1983). The swarming behavior, colony founding and the biology of *S. geminata* are similar to *S. invicta* (Markin et al. 1972). The incubation, larval and pupal periods are around 14, 42 and 12 days respectively (Kamatar 1983), figures which agree with Travis (1939) for *S. geminata* in the U.S.A.

A colony's worker population varies from 4130 to 11,137 (Kamatar 1983) and may contain several queens. Adams et al. (1976) recorded up to 31 queens in a colony.

No deeper than five to six feet, the shallow nests form clusters of 5-10, with 10-15 subsidiary entrances surrounding a main entrance (Musthak Ali 1981; Bolton 1977). The entrances are disc-like, with a raised rim composed of soil particles. Nest building continues around the clock, but peak activity occurs during cool hours in the rainy season. *Solenopsis geminata*, a group forager, follows a trail from nest to food. Upon finding a food source, the scout lays a pheromone trail back to the nest by touching its sting to the ground. Back at the nest, it recruits workers to follow the trail.

The typical red ant nest forms a system of subterranean inter-communicating cavities with one or more openings, surrounded by rings of excavated soil. Wilson (1971) included the genus *Solenopsis* under mound building and Wheeler (1910) classified the nest of *S. geminata* under craters, but careful study reveals them to be between mound building and craters.

Humidity seems to be the chief regulatory factor as the ants prefer moist places with a sufficient supply of water. The optimum temperature zone for foraging activity ranges from 25.5 to 33°C, the critical maximum and minimum temperatures being 49.8° and 2.2°C respectively (Rani and Narendran 1987).

Solenopsis geminata, a "troublesome pest" (Mallis 1938), attacks man and animal, feeds on fruit and vegetable, various types of plants and plant products and encourages sucking pests (Dutta et al. 1982, 1983). However, it has been used in biological control of lac insect enemies (Negi 1933) and cut worms in lawns (Veeresh and Rajanna 1981).

Among the field crops subjected to red ant attack are potato, brinjal (*Solanum melangena*) (Stebbing 1908), Lablab, *Cajanus cajan*, ailanthus, cucumber, tomato (leaves and seedlings; Fletcher 1917), mango, papaya, sapota (fruit), brinjal, bhendi, citrus and banana (flowers and flower buds; Kamatar 1983).

Most *S. geminata* colonies prefer more carbohydrates than protein or lipid foods (Rani and Narendran 1987), but having once tasted flesh, they attack even live animals. A farmer witnessed such behavior. After tasting meat left by the watch dogs, the ants attacked his poultry, ultimately forcing him to abandon the poultry house (Verghese and Veeresh 1976). Fire ants damage stifled and stored silk cocoons resulting in non-reelability of silk that causes 50 to 60% loss (Reddy et al. 1988).

Dorylus orientalis West Wood

Barlow (1899) first reported the large root-cutting ant *Dorylus orientalis* as a plant pest (Potato), a finding immediately disputed by Forel (1899), who maintained that the species was exclusively insectivorous. Subsequently, several entomologists recorded *D. orientalis* as a plant pest in India and Sri Lanka. Roonwal (1975), on the strength of personal observations, supported the view that *D. orientalis* was a serious potato pest. Green (1903) reported that its workers live entirely underground, and attack potatoes, Dhalias tubers and sunflower roots (*Helianthus* sp.). Similarly, Stebbing (1905) found the ant attacked "corn flower plant," in addition to potato. Lefroy (1906, 1909) reported that the ant destroyed healthy cauliflower, cabbage and artichoke plants just below the soil. A number of plant species were added to the host list by Fletcher (1914, 1919), including coconut palms, groundnut pods, seedlings and roots; sugarcane, etc. Similar reports are available from Sri Lanka (Hutson 1933) where *D. orientalis* is considered a pest of potato, carrot, onion, ginger rhizomes, coconut seedlings, roots of tree tomato (*Cyphomandra betacea*) and some young trees like citrus. Beeson (1941) considered it to be "entirely herbivorous" and a pest in gardens on vegetables, bulbs and tubers. The damage, occurs entirely underground, and is perpetrated by the workers, who also eat insects and earthworms (Roonwal 1975).

Dorylus ants - presumably winged males - also attacks beehives, feeding on larvae and pupae (Ghosh 1936; Charian and Ramachandran 1943).

Dorylus orientalis is found everywhere except the extreme northern and northwest sections of India. It also occurs in Southern Nepal, Sri Lanka, Burma and South China to an altitude of 1830 m.

In recent years the *D. orientalis* has become a serious groundnut pest in the coastal belt of Karnataka (South India). Damage is noticed only after the harvest. Supposedly marketable pods contain nothing but black powder when opened; the only indication on the exterior is a small black hole. Up to 40 percent of pods have reportedly been damaged by the Dorylus ant.

D. orientalis' habits and biology are not fully understood. No data on females are available (Roonwal 1975). At night males swarm towards any source of light. The workers, meanwhile, live entirely underground, constructing tunnels and galleries to reach the plant parts.

Myrmecine Ants as Pests

The myrmecine ants, *Myrmicaria brunnea* Saunders, *Phidolgiton diversus* (Jerdon) and *Tetramorium smithi* (Mayr) are serious pests of the Indian bhendi crop (*Abolmoschus esculentus* Moench) (Abraham and

Remamonoy 1973). The feeding habits of all the forms are similar. They cut and feed on the petals, the ovarian tissues, pollen grains, buds, flowers and fruits, causing losses up to 90 per cent (Musthak Ali 1981).

The three species are found throughout India (Bingham. 1903). In addition, *M. brunnea* feeds on *Arctotis grandis* (Fletcher 1919) and bean seedlings (Musthak Ali 1981). It is a frequent intruder into South Indian households (Ayyer 1963). Swarming occurs during the months of July and August between 1700-1900 hrs.

ANTS AS INDIRECT PESTS TO CROPS

As indirect pests, ants cause more harm to Indian crops than as direct pests. The ants' agricultural importance lies in the injury they inflict indirectly on a variety of cultivated crops, ornamental plants, trees and shrubs by their symbiotic associations with a wide range of mealy bugs, scale insects, aphids, fulgorids, psyllids, membracids and lycaenid caterpillars (Ayyer, 1936). Using their antennae, the ants coax the pest insects into yielding drops of "honeydew," which encourages further excretion and injury to the plants. The ants also foster and protect the pests from their natural enemies and facilitate their spread by transplanting them from infected to uninfected areas. Hence ant colonization and movement has a direct bearing on the amount of injury to the plants.

Ant species associated with sucking insect pests come mainly from the Formicinae, Dolichodorinae and Myrmicinae subfamilies. The crops most affected by the sucking insects associated with ants include groundnut, cotton, lablab, brinjal, maize, sorghum, sugarcane, finger millets, coffee, citrus, mango, guava and tamarind.

Among the formicine ants associated with sucking pests, the black ant *Camponotus compressus* Fabrs. is most notorious. Prominent all over India, it is associated with many aphid species; for example *Aphis craccivora* on groundnut, *A. gossipii* on cotton, *A. traversi* on citrus, *A. medicaginis* on lablab; *Tetranerura ulmi* and *T. nigriabdomenalis* on roots of finger millets (Ayyar 1935) and *Forda orientalis* on roots of finger millet (George 1920; Gadiyappanavar 1973; Musthak Ali and Sharat Chandra 1986).

It is also associated with coccids on Citrus, *Icerya aegyptica* and *Coccus viridis*; on guava, *Chloroperlvineria psidii*; on mango, *Rostrococcus mangiferae* (Fletcher 1914; Ayyar 1938; Das 1959; Tandon 1985); on sugarcane, *Pseudococcus saccaricola* (Rao 1942).

With fulgorids, *C. Compressus* is seen on sorghum, *Pundaluova simplicia* (Ayyar 1935); *Perigrinus maidis* (Khan and Rao 1956; Challaiah and Basheer 1965) and on sugarcane. *Pyrilla ourpusilla* (Mishra 1917). It also associates with the membracids *Cajanus indicus* (Ananthasubramanian and Ananthakrishnan 1975).

Other *Camponotus* species associated with sucking insects are *C. sericeaus*; *C. pollutula* and *C. subnudia* (Basalingappa et al. 1986; Kareddy and Basalingappa 1981).

The Indian Red Ant *Oecophylla smaragdina* is commonly in league with coffee and citrus scale *Coccus viridis* (Coleman and Khanna 1918). It prefers moist, well-wooded areas and is, therefore, more abundant in South India (Hingston 1924).

The other formicine ants associated with sucking insects are *Polyrhachis laevissima* with mango aphid *Lesius chrysomelles* (Devi et al. 1986); the crazy ant *Anoplolepis longipes* with green scale *Coccus viridis* on citrus (Das 1959); Coffee (Rao 1986); on Hibiscus and Guava (Soans and Soans 1971); *Paratrechina longicornis* with brinjal pseudococcid (Rawat and Modi 1969) and *Perigrinus maidis* (Verma 1955).

Among the Dolichoderinae, *Dolichoderus affinis*, *D. bituberlatus*, *D. tusms* and *Tapinoma melanocephalum* associate with aphids and coccids of cultivated crops (Devi et al. 1986; Das 1959; Dutt 1912; Mukerji. 1931).

Frequently encountered Myrmicine ants associated with sucking insects include: *Solenopsis geminata* (George 1927). *Crematogaster hodgsoni* (Colemn and Kanna 1918), *C. subnuda* (Mukerji 1931), *C. dohrni* (Fletcher 1914), *Laphomyrmex quadrispinosu* (Devi, et al. 1986); *Meranoplus bicolor* (Ananthasubramanian and Ananthakrishnan 1975); *Monomorium scabriceps* (Mishra 1917); *Myrmicaria brunnae* and *Pheidole* sp. (Devi et al. 1986).

ANTS HARMFUL TO HUMAN INTERESTS

There are several ways in which ants cause harm, other than as crop pests.

Ants carry away seeds sown in fields (Srivastava et al. 1969), nursery beds (Reddy and Puttaswamy 1984) and from "godowns" and dwellings (Fletcher 1919; Lal 1939). They attack woodwork and masonry, nesting near foundation walls and buildings (Pruthi and Singh 1950). They also disrupt irrigation mechanisms by loosening the root system (Veeresh and Gubbaiah 1984; Ayyar 1936). Sometimes the ants invade houses for flour, fats and sugars (Naik 1984) or attack beehives, causing desertion (Eates 1943).

While several species of *Camponotus*, *Monomorium*, *Pheidologeton* and *Pheidole* carry away seeds from storage, nursery or field, some are common enemies of bees (Breithaupt 1919; Ghosh 1936; Charian and Ramachandran 1943).

The ant most involved with human activities is the crazy ant *Anoplolepis longipes*. The crazy ant is a native of China (Wheeler 1910) or Tropical Africa (Wilson and Taylor 1967); it has spread to many tropical countries including Sri Lanka, India and Burma.

The crazy ant population increases enormously if moistened debris and sugary substances are available. During population explosion, deep craters are excavated around trees and plants, particularly those which produce sugary secretion. In recent years, the ants have become endemic over a large area, covering about 2000 hectares, 70 miles from Bangalore (Veeresh and Gubbaiah 1984). In this area, predominated by sugarcane and mulberry, *Anoplolepsis longipes* has made sugarcane cultivation impossible by loosening the root system and thus hindering irrigation. In mulberry gardens, they make a nest around the plant and because of heavy powdery mildew incidence the leaves become unfit for silkworm feeding.

Mealy bugs and scale insect populations flourish in places where the crazy ant is present - for example, crotons, mangoes, guava, citrus.

Predominantly ground-nesting, it lives at the base of trees, under farm waste, beneath stones and foundations. The nests can be shallow or deep (20-45 cm), with a large main entrance 15 to 18 cm in diameter with several small subsidiary entrances.

The colony multiplies by budding, starting 6 to 8 weeks after the first rains. Males and queens develop during a short period before the wet season from March to May, but do not fly.

Under laboratory conditions, the queen lays 7 to 22 eggs per day. The developmental time for egg, larva and pupa is about 19, 20 and 22 days respectively, with a total life cycle of around 60 days. The longevity of the worker, male and queen is about 85, 11 and 138 days respectively.

Foraging takes place throughout day and night in all seasons, except during heavy rains, strong wind and high temperature. Food consists mainly of sugary substances and protenaceous matter, which results in the above-mentioned negative interactions with humans.

MANAGEMENT OF ANT PESTS

No recommended control practices exist for all ant pests. However, several researchers have attempted ant control by banding trees with repellent or sticky substances (Coleman and Kannan 1918; Tandon 1985; Rao 1986), or treating breeding places with insecticides at the time of swarming or budding (Rao. 1986). They may also set out poison baits containing 0.25 per cent aldrin in 20g of jaggery at the nest entrance (Naik 1984), apply insecticide at the time of seed planting (Srivastava et al. 1969) and keep ants away from attractive materials using an appropriate mix of water and repellents. The Hindu custom of sprinkling water on the floor around the dining plate, while eating, might be a ritual developed to prevent ants from reaching the plate.

REFERENCES

ABRAHAM, C.C. and K.S. REMAMONY. 1978. New records of Myrmicine ants as pests of bhendi, *Abelmoschus esculentus* Moench. J. Bombay Nat. Hist. Soc. 75: 242-243.

ADAMS, C.T., W.A. BANKS and J.K. PLUMLEY 1976. Polygyny in the tropical fire ant *Solenopsis geminata* with notes on the imported fire ant *Solenopsis invicta* Fla. Ent. 59: 411-415.

ANANTHASUBRAMANIAN, K.S. and T.N. ANANTHAKRISHNAN. 1975. Taxonomic, biological and ecological studies of some Indian membracids (Insecta: Homoptera), Part I and II. Rec. Zool. Surv. India, 68:161-272.

AITKEN, E.H.. 1889. The red ant. J. Bombay nat. Hist. Soc. 4: 151.

AYYAR, P.N.K. 1935. The biology and economic status of the common black ant of South India, *Camponotus compressus*. Fab. Bull. ent. Res. 26: 575-585.

AYYAR, P.N.K. 1937. A new carton-building species of ant in South India *Crematogaster dohrni artifex* Mayr. J. Bombay nat. Hist. Soc. 39: 291-308.

AYYAR, T.V. 1939. Entomology in India--A retrospect. Indian J. Ent., 1: 9-16.

AYYAR, T.V.R. 1963. Handbook of Economic Entomology for South India. Govt. of Madras 316 pp.

BARLOW, F. 1899. Notes on Insect Pests from the Entomological Section, Indian Museum. Indian Mus. Notes, Calcutta 4: 180-221.

BASALINGAPPA, S., V.B. KAREDDY, S. MATAPATHI and R. MEENA. Gandhi, 1986, Population studies on the Ant *Componotus serieus* Fabricius Comp. Physiol. Ecol. 11:32-37.

BEESON, C.F.C. 1941. The Ecology and control of the forest insects of India and neighbouring countries. 1,007 p.

BINGHAM, C.T. 1903. Hymenoptera, vol.II, Ants cuckoowasps. *In* W.T. Blanford, ed., The fauna of British India, including Ceylon and Burma. Taylor and Francis, London, 506 pp.

BOLTON, 1977. The ant tribe Teramoriini (Hymenoptera: Formicidae). The genus *Tetramorium* Mayr in the Oriental and Indo-Australian regions and in Australia. Bull. Br. Mus. Nat. Hist. (ent.). 36: 67-151.

BREITHAUPT, A.G.H. 1919. Ants attacking Bees. J. Bombay nat. Hist. Soc. 26: 878.

BUTLER, A.L. 1895. Red ant as smelling salts. J. Bombay nat. Hist. Soc. 10: 330.

CHARIAN, M.C. and S. RAMACHANDRAN, 1943. Bee enemies. Indian Fmg. 252-253.

CHELLIAH, S. and M. BASHEER. 1965. Biological studies of *Perigrinus maidis* (Ashmad) (Araeopidae:Homoptera) on Sorghum. Indian J. Ent., 27: 466-471.

COLEMAN, L.C. and K.K. KANNAN. 1918. Some scale insect pests of Coffee in South India, Mysore Dept. Agri. Entom. Serv., Bull No.4, 66 pp.

DAS, G.M. 1959. Observations on the association of ants with the coccids of tea. Bull. Ent. Res. 50: 437-448.

DEVI, C.M., S.S. DEVI, T.D. DEVI and T.K. SINGH. 1986. Aphidocolous ant (Hymenoptera:Formicidae) fauna of Mizoram. Proceedings of the National Symposium on Aphidology in India II held at Agartala 2-4 Nov. 1985. 85-88.

DUTTA, S.K., D. RAYACHAUDHURI and B.K. AGARWALA. 1982. Study on aphid tending ants in India. Entomon. 7: 327-328.

DUTTA, S.K., D. RAYCHAUDHURI and B.K. AGARWALA. 1983. Study on aphid tending ants in India. Entomon. 8: 23-25.

DUTTA, G.R. 1912. Life histories of Indian insects (Hymenoptera). Mem. Dep. Agric. India ent. Ser., 4: 247-253.

EATES, K.R. 1943. Black ant ride honeybees. J. Bombay nat. Hist. Soc. 44: 488.

EMERY, C. 1881. Genera Insectorum. Hymenoptera, Fam. Formicidae subfamily, Dorylinae, Brassels pp. 14-15.

FLETCHER, T.B. 1914. Some South Indian Insects and other Animals of importance, Madras, 274 pp.

FLETCHER, T.B. 1919. Anotated list of Indian Crop pests report. Procd. Third. ent. Meet. Pusa, pp. 33-35.

FOREL, A. 1899. Less Formicides de Lempire des Indes et de Ceylon. J. Bombay nat. Hist. Soc., 14, 679-715.

GADIYAPPANAVAR, R.D. 1970. Bionomics of Ragi (Elusine coracana) Root aphid. *Tetraneura nigriabdomanalis* (Sasaki) M.Sc. thesis, University of Agril. Sciences, Bangalore, 98 pp.

GEORGE, C.J. 1927. South Indian Aphididae. Proc. Asiat. Soc. Bang. Calcutta. 23:1-12.

GHOSH, C.C. 1936. Bee-keeping. Bull. Imp. Counc. Res., Delhi, No.6, 99 pp.

GREEN, E.E. 1903. Notes on *Dorylus orientalis*. West. Indian Mus. Notes, Calcutta, 5: 39.
HINGSTON, R.W.G. 1924. The red ant. J. Bombay nat. History. Soc., 29: 362-372 and 681-693.
HUTSON, J.C. 1933. Pests of garden plants: The root eating ant. Trop Agr. Peradenia, 80: 276-279
JAYARATHNAM, K. 1977. Studies on the population dinamics of diamond back moth *Plutella xylostella* (Linn.) (Lepidoptera: Yponomeutidae) and Crop loss due to pest in cabbage. Ph.D. thesis, Univ. Agril. Sci., Bangalore, 220 pp.
KAMATAR, V. K. 1983. Ants of Raichur district (Karnataka State) with observations on the biology and behaviour of fire ant *Solenopsis geminata* Fabricius. M.Sc. Thesis. Univ. Agric. Sci., Bangalore, 95 pp.
KAREDDY, V.V. and BASALINGAPPA. 1981. Nesting pattern of the ant *Camponotus sericeus*. Progress *in* Soil Biology and Ecology in India, UAS Tech. Series No. 37: 218.
KHAN, M. and A.S. RAO. 1956. The influence of the black ant (*Camponotus compressus* F.) on the incidence of two Homopterous crop pests. Indian J. Ent.. 18: 199-200.
KIRBY, W. 1837. On the History. Habits and Instincts of Animals. (7th Bridge Water Treatise. 2nd Amer. ed) - Philadelphia.
LAL, K.B. 1939. Ants of Delhi. Indian J. Ent. 1:107.
LEFROY, H.M. 1906. Indian Insect Pests. Calcutta. pp. 231-232.
LEFROY, H.M. 1909. Indian Insect Life, Calcutta. pp.227-228.
LONG, A.M. 1898. Red ants as an article of food. J. Bombay nat. Hist. Soc. 13:536.
MALLIS, A. 1938. The California fire ant and its control. Pan. Pacific. Ent. 14: 87-91.
MARKIN, G.P., J.H. DILLIEN and H.L. COLLINS. 1973. Growth and development of colonies of the red imported fire ant *Solenopsis invicta*, Ann. ent. Soc. Am., 66: 803-808.
MISRA, C.S. 1917. The Indian sugarcane leafhopper, *Pyrilla aberrans* Kirby. Ent. Ser., 73: 133.
MUKERJI, D. 1931. Reports on a collection of ants in the Indian Museum, Calcutta. J. Bombay nat. Hist. Soc. 34: 149-163.
MUKERJI, D. 1934. On the anatomy of the worker of the ant *Dorylus* (Alaopone) *orientalis*. W. Zool. An. Leipzig. 105: 97-105.
MUSTHAK ALI, T.M. 1981. Ant fauna (Hymenoptera: Formicidae) of Bangalore with observation on their nesting and foraging habit. M.Sc. Thesis, Univ. Agric. Sci., Bangalore. 208 pp.
NAIK, S. 1984. Investigation on *Tapinoma melanocephalum* Fabricius (Dolichoderinae: Formicidae) with special reference to some aspects of its Biology and Control. M.Sc. Thesis. Univ. Agril. Sci., Bangalore, 91 pp.
NEGI, P.S., M.P. MISHRA and S.N. GUPTA. 1930. Ants and lac insects. J. Bombay nat. Hist. Soc., 34: 182-188.
NEGI, P.S. 1933. The small red ant. *Solenopsis geminata* sub sp. *rufa* Jerdon and its usefulness to man. J. Bombay nat. Hist. Soc., 36, 1081-1025.
PRUTHI, H.S. and M. SINGH. 1950. Pests of stored grain and their control. I.C.A.R., Delhi, 25-27.

RANI, P.J. and T.C. NARENDRAN. 1987. Ecology of *Solenopsis geminata* Fabr. A dominant species of ants in Malabar. Geobios. 14: 200-204.

RAO, N.S. 1986. Bioecology and management of the crazy ant *Anoplolepis longipes* (Jerdon). (Hymenoptera:Formicidae) Ph.D. thesis University of Agril. Sciences, Bangalore 120 pp.

RAO, R.M.S. 1942. Life-history and bionomics of a new pest of sugarcane in India. Proc. Indian Acad. Sci., 16: 79-85.

RAWAT, R.R. and B.R. MODI. 1969. Biology and natural enemies of the brinjal mealy bug, *Centrococcus isolitus* (Green) In Madhya Pradesh. Indian J. Agric. Sci. 39: 112-116.

REDDY, G. 1981. Ant fauna of Dharwad (Karnataka State) with special reference to foraging behaviour of *Myrmicaria brunnea* Saunders, (Hymenoptera: Formicidae Univ. Agric. Sci., Bangalore, 81 pp.

REDDY, D.N.R. and PUTTASWAMY. 1984. Pests infesting chilli (*Capsicum annum* L.) In the nursery. Mysore J. Agric. Sci., 18: 122-125.

REDDY, D.N.R., M. VIJAYENDRA and Y.K. KOTIKAL. 1988. The fire ant *Solenopsis geminata* (Hymenoptera: Formicidae) damage to stored silkworm cocoons. J. Soil Biol. and Ecol., 8: (2 in press).

ROONWAL, M.L. 1975. Plant-pest status of root eating ant, *Dorylus orientalis* with notes on anatomy, distribution and habits (Insecta:Hymenoptera). J. Bombay nat. Hist. Soc. 72: 305-313.

SOANS, A.B. and J.S. SOANS. 1971. Proximity of the colonies of the tending ant species *Anoplolepis longipes* Jerdon, as a factor determining the occurrence of aphids. J. Bombay nat. Hist. Soc., 68: 289.

SRIVASTAVA, B.K., G.G. KUNDU, D. KISHORE and M.G. JOTWANI. 1969. A new record of insect damage to sorghum seeds in the field. Indian J. Agric. Sci., 39: 1135-1136.

STEBBING, E.P. 1905. Insect life in India and how to study it. J. Bombay nat. Hist. Soc., 16: 683.

SYKES, W.H. 1836. Description of new species of Indian ants. Trans. ent. Soc. Lond.. 1: 99-107.

TRAVIS, B.V. 1939. Notes on the biology of the fire ant *Solenopsis geminata* Florida, Georgia, Florida Ent., 24: 15-22.

VERMA. 1955. Biological studies to explain the failures of *Cyrtorhinus mundulus* (Breddin) as egg predator of *Perigrinus maidis* (Ashmead) Proc. Hawaii ent. Soc. 15: 623-34.

VEERESH, G.K. and C. RAJANNA. 1981. Ants in the integrated control of cutworms in lawns. Progress *in* Soil Biology and Ecology in India. UAS Tech. Series No. 37: 218-222.

VEERESH, G.K. and GUBBAIAH. 1984. A report on the crazy ant (*Anoplolepis longipes* Jerd.) menace in Karnataka. J. Soil Biol. Ecol., 4: 65-73.

VEERESH, G.K. 1987. Pest status of crazy ant *Anoplolepis longipes* (Jerdon) in Karnataka, India and cause for its outbreak. *In:* Chemistry and Biology of Social Insects. (Ed) J. Ederd, and H. Rembold Verlag J. Peperny Munchen, p. 667-668.

VEERESH, G.K. and T. SHIVASHANKAR. 1987. Nesting and foraging behaviour of the termite Predator *Leptogenys processionalis* (Jorden) (Formicidae, Ponerinae). *In:* Chemistry and Biology of Social Insects (Ed) J. Ederd H.Rembold. Verlag, J. Pepery. p. 667-668

VERGHESE, A. and G.K. VEERESH. 1976. Ants of India - a brief review. Proceedings of First All India Symposium on Soil Biology and Ecology, UAS, Bangalore, p. 172-181.

WHEELER, W.M. 1910. Ants, their structure, development and behaviour. Columbia University Press, New York, 663 pp.

WILSON, E.O. 1964. The true army ants of the Indo-Australian area (Hymenoptera:Formicidae:Dorylinae). Pacific Insect, 6: 427-483.

3

Pest Ants in Urban and Agricultural Areas of Southern Africa

A.J. Prins, H.G. Robertson and A. Prins

INTRODUCTION

Ants are one of the most ubiquitous animal groups in both natural and man-made terrestrial systems and as such are bound to influence other components of these systems. In southern Africa, the advantageous effects of ants almost certainly outweigh their disadvantages. Their greatest benefit is as predators of insect pests, but they also process dead organic matter. Although this paper is mainly concerned with the detrimental effects of ants in agricultural and urban settings, examples of their beneficial effects will also be mentioned.

More than 600 described species of ant live in southern Africa, with many more awaiting description. However, only 16 figure prominently as pests. The applied myrmecology literature in southern Africa is fragmented and many pest species observations and studies remain unpublished. Much of the information presented here was obtained by the senior author from applied entomologists who sent him ants for identification.

This paper introduces the most common pest species and then considers the importance of ants in biocontrol, as well as in houses, pastures, wood and their direct attacks on humans and other animals. The paper then outlines research priorities in the field of southern African applied myrmecology. The action of ants in citrus orchards is covered by Samways (this volume).

THE MAJOR PEST SPECIES

The 16 most commonly encountered pest-ant species in southern Africa are listed in Table 1, together with a subjective assessment of their pest status. The brown house ant *Pheidole megacephala*, the Argentine ant *Iridomyrmex humilis*, the common pugnaceous ant *Anoplolepis custodiens*, and the black pugnaceous ant *A. steingroeveri* are probably the four most important species.

Pheidole megacephala, like other *Pheidole* species, has major and minor workers. The more rarely encountered major workers are used to taxonomically distinguish *Pheidole* species from one another. Broekhuysen (1948) studied the biology and control of *P. megacephala* in the Pretoria area.

Iridomyrmex humilis is the only dolichoderine ant in southern Africa

TABLE 1. Common pest ant species in southern Africa. A subjective rating is given of their pest status in biocontrol, houses, pastures, wood, and in their aggressiveness towards humans and animals: 0 = of no importance; 1 = of minor or occasional importance; 2 = important.

Species	Pest status				
	Bio-control	Houses	Pasture	Wood	Humans/ animals
MYRMICINAE					
Crematogaster castanea	2	0	0	1	1
Crematogaster liengmei	2	0	0	1	1
Crematogaster perinqueyi	2	0	0	1	1
Messor capensis	0	0	2	0	0
Pheidole megacephala	2	2	1	0	0
Pheidole sculpturata	2	0	0	0	0
Myrmicaria natalensis	1	0	0	0	0
DOLYCHODERINAE					
Iridomyrmex humilis	2	2	0	0	0
Technomyrmex albipes	2	2	0	0	0
DORYLINAE					
Dorylus helvolus	1	1	1	0	1
FORMICINAE					
Acantholepis capensis	1	1	0	0	0
Anoplolepis custodiens	2	0	0	0	2
Anoplolepis stiengroeveri	2	0	0	0	2
Camponotus maculatus	1	1	1	0	0
Camponotus rufoglaucus	1	0	0	0	0
Polyrhachis schistacea	1	0	0	0	0

with a distinct petiolar node (dolichoderines are characterized by having only one petiolar segment and a slit-shaped, rather than a circular, orifice at the end of the gaster).

A. custodiens and *A. steingroeveri* workers vary greatly in size and differ from *Camponotus* species by having 11-segmented rather than 12-segmented, antennae. The refringent hairs on gasters of *A. custodiens* workers have three lines of convergence, which form a checkered pattern, whereas *A. steingroeveri* has only one central line of convergence. Steyn (1954) and Louw (1968a, b) studied aspects of *A. custodiens*' biology.

Some southern African ants are very common but do not qualify as pests. These include *Monomorium albopilosum* Emery, *Tetramorium quadrispinosum* Emery, *T. sericeiventre* Emery and *T. setuliferum* Emery.

These ground nesting species rarely venture into shrubs or trees to harvest honeydew from homopterans.

Of the species listed in Table 1, only *I. humilis* and *T. albipes* are exotic. Table 2 lists all known ant species that have been introduced to southern Africa. *Iridomyrmex humilis*' introduction undoubtedly has resulted in the most serious consequences.

Specimens of *I. humilis* were collected in Stellenbosch in 1901 and in 1908 C.P. Lounsbury reported its presence in the Cape Peninsula. Perhaps it arrived in Cape Town with fodder imported from South America for British cavalry during the Anglo-Boer War (Skaife 1955, 1961, 1962). There has been no thorough survey of its present distribution in southern Africa, although Skaife (1962) recorded it from Cape Town to East London and in inland towns such as Clanwilliam and Graaff Reinet. He also mentioned a few records from Johannesburg and Pretoria but remarked that *I. humilis* did not seem to thrive at high altitudes. However, it has been recorded from Maseru, Lesotho, 1500m above sea level (Prins 1978) and Jackson (1980) recorded it commonly on the Witwatersrand where it is now the most common household ant (H.G. Robertson, pers. obs.).

In addition to being an economic pest, *I. humilis* poses a serious ecological problem in the unique Cape fynbos floral kingdom because it removes elaiosomes from seeds but, unlike the indigenous ants, does not protect the seeds from predators and fire by burying them. It also does not contribute to seed dispersal. However, these studies (Bond and Slingsby 1983, 1984; Bond and Breytenbach 1985) included only a small region of the fynbos. More research is needed to determine habitat preferences of *I. humilis* in the fynbos. De Kock and Giliomee (1989) found its distribution in the fynbos closely correlated to man-made disturbance and attributed this distribution not so much to habitat preference but to *I. humilis*' dependence on humans for long range dispersal (colony foundation is by fission and therefore unaided dispersal is limited). While dispersal by humans has certainly influenced its current distribution in the fynbos, much of the mountain fynbos region of the Cape Peninsula is free of *I. humilis*, even though the human population in this region is high and the ant has had approximately 89 years in which to disperse within the region. The correlation of *I. humilis* with human disturbance may not be due only to dispersal by humans, but also because *I. humilis* prefers disturbed habitats.

A remarkably high proportion of ants introduced into South Africa originated from the East and entered through the seaport of Durban (Table 2). In the earlier part of the century, imports to Durban included jute from India and sugarcane from Mauritius (M.J. Robertson pers. comm.). Ant nests could easily have been imported with these shipments. The Durban area also has a subtropical climate conducive to the subtropical 'tramp' species.

ANTS AND BIOCONTROL

Ants world-wide detrimentally affect the biological control of homopteran pests, and southern Africa is no exception. Ants patrolling homopterans for honeydew prevent wasp parasitoids and coccinellids from ovipositing and hence lessen the effectiveness of control. This problem has been encountered especially in southern Africa's citrus orchards (see Samways, this volume) but is also a problem in the control of mealybugs on

TABLE 2. Species of ant introduced to southern Africa

Species	Origin	Port of Entry	Reference
MYRMICINAE			
Cardiocondyla emeryi	W. Africa	?	Bolton (1982)
Monomorium destructor	India	Durban	Bolton (1982)
Monomorium pharaonis	India	?	Bolton (1982)
Monomorium subopacum	Mediterranean	Durban	Bolton (1982)
Solenopsis geminata	N. America	Durban	
Tetramorium bicarinatum	S.E. Asia	Durban	Bolton (1982)
DOLYCHODERINAE			
Iridomyrmex humilis	S. America	Cape Town	Skaife (1962)
Technomyrmex albipes	?	?	
FORMICINAE			
Anoplolepis longipes	India	Durban	
*Camponotus herculeanus**	Holarctic	Cape Town	
Paratrechina longicornis	?	Durban	
Plagiolepis pygmaea	?	?	

* Eradicated

deciduous fruit (Myburgh et al. 1973; Kriegler and Whitehead 1962) and on granadilla, pineapple and guava (Annecke and Moran 1982); scale insects on olive, guava, figs and garden ornamentals (Annecke and Moran 1982); *Hilda patruelis* (Tettigometridae) on groundnuts and garden ornamentals (Weaving 1980; Annecke and Moran 1982); and black pine aphid *Cinara cronartii* Tissot and Pepper on young pines (van Rensburg 1979).

In vineyards, the Argentine ant *I. humilis* has a particularly detrimental effect on biological control of mealybug by coccinellids (Myburgh et al. 1973). Dieldrin, the insecticide most effective in controlling the ants (Durr 1955; Durr et al. 1955), has been banned along with related insecticides. Farmers have subsequently discarded biological control and resorted to the exclusive use of insecticides to directly control mealybug and other pests (V.B. Whitehead pers. comm.).

Besides affecting the biological control of homopterans, ants have also been a problem in the biological control of southern African weeds. Fifty-four to 74% of *Cactoblastis cactorum* eggs on cactus weeds were preyed upon by ants, and predation of *C. cactorum* pupae by driver ants *Dorylus helvolus* reached 45% (Robertson 1985, 1988; Robertson and Hoffmann 1989). Egg predation by ants also contributed to the failure of *Tucumania tapiacola* Dyar and *Nanaia* sp. moths to become established on jointed cactus *Opuntia aurantiaca* in South Africa (Hoffmann and Zimmermann in press).

ANTS IN PASTURES AND CROPS

Predatory ants such as *A. custodiens*, *A. steingroeveri*, *P. megacephala* and *D. helvolus* can have a beneficial effect in pastures and crops by reducing herbivore populations. *P. megacephala* aids control of the sorghum stemborer *Chilo partellus* (Swinhoe) by preying on the diapausing larvae. In Transvaal lowveld cotton fields *P. megacephala* evidently destroys pupae of the red bollworm *Diparopsis castanea* Hampson (Broekhuysen 1948). Ants also destroy eggs of the sugarcane borer *Eldana saccharina* Walker (Leslie 1982). Despite the beneficial effects of ants in pastures and crops, certain ant species are regarded as pests because they remove seeds or destroy plants.

The seed harvesting ants, in particular *Messor capensis*, might reduce seed densities and change the structure of pasture vegetation. Research on this topic is currently being undertaken at Bloemfontein by the University of the Orange Free State and at Prince Albert by the Percy FitzPatrick Institute of African Ornithology, University of Cape Town. At Prince Albert, *M. capensis* harvests seeds of both palatable and unpalatable plant species in a density-dependent fashion. Consequently, *M. capensis* might have a beneficial, stabilizing effect on the balance of species within the vegetation (S. Dean pers. comm.).

Messor capensis nests can be very large; nest entrances are surrounded by mounds of excavated soil up to 30 cm high. Accumulation of seeds and nutrient-rich soils around the nest entrances promotes germination of shrubs. Grass seedlings, however, are often removed by the ants. In overgrazed semi-arid regions, large bare patches occur between the shrub-covered *M. capensis* mounds, making the soil susceptible to erosion. The extent to which *M. capensis* prevents establishment of shrubs between the nests is not known. In the grasslands of Kenya, *Messor cephalotes* (Emery) harvests grass stalks near nest entrances, creating large, round patches of exposed soil, up to eight meters in diameter (Bequaert 1922).

Messor capensis can be a problem in wheat fields because workers from nests surrounding the fields carry off wheat kernels after sowing. Harvesting can be so extensive that it creates bare patches along the borders of the field (F. Walters pers. comm.).

Besides *M. capensis*' grass-removing habits, there are other records of ants destroying plants. Farmers sometimes implicate *Dorylus helvolus* in damage to potato and dahlia tubers. *P. megacephala*, apparently attracted by the high sugar content in the roots of one to three month old chicory plants, damage them (Annecke and Moran 1982). Similarly, *P. megacephala* sometimes damages the ripened strawberries. In the Northern Transvaal and Zimbabwe, *P. megacephala* collects tobacco seeds in seed beds (Smee 1929 in Broekhuysen 1948). *Camponotus maculatus* and *C. fulvopilosus* (De Geer) damage the leaves and growing tips of young trees in plantations in the more arid areas of the northwestern Cape and Namibia where areas have been cleared of shrubs and grass cover for tree planting. *Camponotus maculatus* has also been blamed for damaging young vegetable leaves.

ANTS IN WOOD

Melissotarsus beccarii Emery and *M. emeryi* Forel are the only southern African species known to bore into live wood. Scale insects live inside the galleries of both these species (Delage-Darchen 1972; Prins et al. 1975). *Melissotarsus* workers are remarkable in that they are able to produce silk which they use to line their tunnels and to block entrances to the outside.

Almost 90% of trees of the pincushion protea, *Leucospermum praemorsum* (Meissner) were infested with *M. emeryi* at a site near Clanwilliam, Cape Province (Prins et al. 1975). Each nest consisted of a complex of tunnels and passages, mostly under the bark but also penetrating deeper into the heartwood, some reaching even the small growing tips. Although these trees occurred naturally, *M. emeryi* could become a problem in areas where proteas are cultivated for their flowers.

In the eastern Transvaal, *M. beccarii* inhabits *Burkea africana* Hook. and marula *Sclerocarya birrea* (Rich.), particularly in the Komatipoort area. In East London, the same species nested in *Ficus sur* Forsk. and boer-bean, *Schotia* sp, along roadsides. Tunnelling by these ants induces parts of the tree to die and this causes branches to fall, endangering pedestrians and property. There is no known effective way of controlling these ants.

A number of ant species, especially in the genera *Crematogaster* and *Camponotus*, nest in pre-existing cavities made by wood-boring beetles, termites and carpenter bees. Existing tunnels are enlarged and, as a result, timbers and fence poles can become weakened and honeycombed.

On three occasions (1943 and twice in 1965), the carpenter ant *Camponotus herculeanus* (Linn.)--a wood-boring species native to North America and Europe--was discovered in poplar stumps on the Cape Town docks. The logs were riddled with passages made by the ants and, as both males and females were collected, the logs, imported from Canada for the manufacture of safety matches, were fumigated with methyl bromide. The ants were thus eradicated and prevented from becoming potential timber and plantation pests in southern Africa.

ANTS IN URBAN ENVIRONMENTS

Table 1 lists the ant species most commonly encountered in urban environments. The Argentine ant *I. humilis* is the most common urban species in the Cape and the Witwatersrand. *I. humilis* occurs in Natal and the eastern Cape, but *P. megacephala*, *T. albipes* and A. capensis are more commonly encountered. It is puzzling that *I. humilis* has not become better established in Natal especially as this region seems particularly vulnerable to tramp species (Table 2).

In the Cape Peninsula, *Tetramorium quadrispinosum* is widespread in urban areas but is not a pest species because it is a ground forager that rarely if ever comes into houses. In many residential properties *T. quadrispinosum* nests below the drive-way paving while *I. humilis* nests around the house's borders. It is not known how *T. quadrispinosum* manages to survive and thrive in the presence of *I. humilis*.

The driver ant *D. helvolus* is a common garden dweller in most southern African regions, although rarely seen because of its subterranean habits. In old houses, workers occasionally come up into rooms through holes in the walls. One author (HGR) observed a large group of *D. helvolus* that surfaced in a sitting-room to release male alates.

ANTS THAT ATTACK HUMANS AND ANIMALS

Although driver ants in Africa probably attack humans and animals more often than other species, the most common driver ant in southern Africa, *D. helvolus*, rarely causes problems due to its subterranean habits. Instead, *Anoplolepis custodiens* and *A. steingroeveri* are probably the region's most troublesome ant species. For instance, during late summer and autumn in western Cape orchards and vineyards, *A. steingroeveri* workers quite frequently attack farm laborers and hinder their work. In the Buthulie area of the Orange Free State, *A. custodiens* kills chickens and therefore chicken runs have to be protected.

Species of *Crematogaster* can inflict a painful bite when disturbed from their nests. For example, *C. peringueyi* nesting in the cavities of vines hinder laborers as they pick grapes or tend the vines (Myburgh et al. 1973).

However, ants can help in preventing other animals from biting humans and livestock. Ants such as *C. castanea* (J. Walker pers. comm.) attack nymphal stages of the sand tampan *Ornithodorus savignyi* Audouin and *Mymicaria natalensis* has been found to attack tampan eggs.

FUTURE DIRECTIONS IN SOUTHERN AFRICAN APPLIED ANT RESEARCH

Southern African applied ant research has tended to be problem-orientated and parochial with little attempt to examine issues from an international or even regional perspective. The research proposals below are aimed at establishing a more holistic approach to the subject.

A long-term study on the urban ant distribution in southern Africa needs to be conducted. The distribution of *I. humilis* is changing over time, but in what ways? We don't know. Neither do we know how its distribution has been influenced by the distribution of aggressive indigenous species such as *P. megacephala*. Various comments have been made about the influence of climate on *I. humilis* distribution, but there are no studies to substantiate these claims.

A key to ant control is an understanding of their habitat requirements. A habitat change can alter an ant's population status. Recently, this was discovered by accident at Zimbabwe's Mazoe Citrus Estates, where herbicidal clearing of the vegetation below the trees resulted in an increase in *Anoplolepis custodiens* numbers (A.J. Gardiner pers. comm.). Although habitat selection by ants has been studied at a community level (eg. Donnelly and Giliomee 1985a,b), no studies on southern African ants have comprehensively examined habitat requirements in a particular ant species.

The present paper emphasized that ants have beneficial effects in urban and agricultural settings, but few studies have examined ants from this point of view. It would be interesting to know the overall effect that ants have in these environments.

Ecological research on ants in Africa has been severely hindered by the poor state of ant taxonomy in the region. Barry Bolton has published a number of major revisions on the region's myrmicine ants (Bolton 1980, 1982, 1987 and references therein) and these are vast improvements. However, important ant genera such as *Crematogaster*, *Pheidole* and *Camponotus* have yet to be given modern taxonomic treatment and are in a state of chaos.

ACKNOWLEDGMENTS

We are grateful to Dr V.B. Whitehead, Dr J.H. Hoffmann and Ms A. de Kock for their comments on the manuscript.

REFERENCES

ANNECKE, D.P. and V.C. MORAN. 1982. Insects and mites of cultivated plants in South Africa. Butterworths, Durban.

BEQUAERT, J. 1922. Ants in their diverse relations to the plant world. Bull. Am. Mus. Nat. Hist. 45: 333-583.

BOLTON, B. 1980. The ant tribe Tetramoriini (Hymenoptera: Formicidae). The genus *Tetramorium* Mayr in the Ethiopian zoogeographical region. Bull. Br. Mus. nat. Hist. (Ent.) 40: 193-384.

BOLTON, B. 1982. Afrotropical species of the myrmicine ant genera *Cardiocondyla, Leptothorax, Melissotarsus*, *Messor*, and *Cataulacus* (Formicidae). Bull. Br. Mus. nat. Hist. (Ent.) 45: 307-370.

BOLTON, B. 1987. A review of the Solenopsis genus-group and revision of Afrotropical *Monomorium* Mayr (Hymenoptera: Formicidae). Bull. Br. Mus. nat. (Nat. Hist.) Ent. ser. 54: 263-452.

BOND, W.J. and P. SLINGSBY. 1983. Seed dispersal by ants in shrublands of the Cape Province and its evolutionary implications. S. Afr. J. Sci. 79: 231-233.

BOND, W.J. and P. SLINGSBY. 1984. Collapse of an ant-plant mutualism: the Argentine ant (*Iridomyrmex humilis*) and myrmecochorous Proteaceae. Ecology 65: 1031-1037.

BOND, W.J. and G.J. BREYTENBACH. 1985. Ants, rodents and seed predation in Proteaceae. S. Afr. J. Zool. 20: 150-154.

BROEKHUYSEN, G.J. 1948. The brown house ant (*Pheidole megacephala* F.). Union of South Africa, Dept. of Agric., Bull. 266: 1-40.

DE KOCK, A.E. and J.H. GILIOMEE. 1989. A survey of the Argentine ant, *Iridomyrmex humilis* (Mayr), (Hymenoptera: Formicidae) in South African fynbos. J. ent. Soc. sth. Afr. 52: 157-164.

DELAGE-DARCHEN, B. 1972. Une fourmi de Cote d'Ivoire: *Melissotarsus tibulans* Del., n. sp. Insectes Soc. 19: 213-226.

DONNELLY, D. and J.H. GILIOMEE. 1985. Community structure of epigaeic ants (Hymenoptera: Formicidae) in fynbos vegetation in the Jonkershoek Valley. J. ent. Soc. sth. Afr. 48: 247-257.

DONNELLY, D. and J.H. GILIOMEE. 1985. Community structure of epigaeic ants in a pine plantation and in newly burnt fynbos. J. ent. Soc. sth. Afr. 48: 259-265.

DURR, H.J.R. 1955. The control of the Argentine ant (*Iridomyrmex humilis*) by means of DDT, Chlordane and Dieldrin. J. ent. Soc. sth. Africa 18: 7-12.

DURR, H.J.R., C.J. JOUBERT and S. WALTERS. 1955. A biological evaluation of the effects, two years after application to the soil, of Aldrin, Chlordane and Dieldrin, on workers of the Argentine ant *Iridomyrmex humilis* (Mayr). J. ent. Soc. sth. Afr. 18: 235-237.

HOFFMANN, J.H. and H.G. ZIMMERMANN. In press. Ovipositional and feeding strategies in cactophagous pyralids: predictions for biological control of cactus weeds in southern Africa. *In* E.S. Delfosse (ed.), Proc. VII Int. Symp. Biol. Contr. Weeds, 6-11 March 1988, Rome, Italy.

JACKSON, M.E. 1980. Occurrence of the Argentine ant on the Witwatersrand. S. Afr. J. Sci. 76: 31-32.

KRIEGLER, P.J. and WHITEHEAD, V.B. 1962. Notes on the biology and control of *Crematogaster peringueyi* var. *angustior* Arnold on grape vines (Hymenoptera: Formicidae). J. ent. Soc. sth. Afr. 25: 287-290.

LESLIE, G.W. 1982. A study of egg predators of *Eldana saccharina* Walker (Lepidoptera: Pyralidae). Proc. S. Afr. Sugar Technol. Assn. 56: 1-3.

LOUW, J.F. 1968a. Die invloed van temperatuur en relatiewe humiditeit op die gewigsverlies van die malmier, *Anoplolepis custodiens* (Smith). S. Afr. Tydskr. Landbouwet. 11: 575-582.

LOUW, J.F. 1968b. Die ekologie van die malmier, *Anoplolepis custodiens* (Smith) (Hymenoptera: Formicidae), in die sentrale Oranje-Vrystaat - 1. Die bogrondse aktiwiteitspeil van die werkers. J. ent. Soc. sth. Afr. 31: 241-248.

MYBURGH, A.C., V.B. WHITEHEAD and C.C. DAIBER. 1973. Pests of deciduous fruit, grapes and miscellaneous other horticultural crops in South Africa. Rep. S. Afr. Dept. Agric. Tech. Serv. Ent. Mem. 27: 1-38.

PRINS, A.J. 1978. Hymenoptera. *In* M.J.A. Werger (ed.), Biogeography and ecology of southern Africa. Junk, Den Haag, pp. 823-875.

PRINS, A.J., Y. BEN-DOV and D.J. RUST. 1975. A new observation on the association between ants (Hymenoptera: Formicidae) and armoured scale insects (Homoptera: Diaspididae). J. ent. Soc. sth. Afr. 38: 211-216.

ROBERTSON, H.G. 1985. Egg predation by ants as a partial explanation of the difference in performance of *Cactoblastis cactorum* on cactus weeds in South Africa and Australia. *In* E.S. Delfosse (ed.), Proc. VI Int. Symp. Biol. Contr. Weeds, 19-25 August 1984, Vancouver, Canada. Agric. Can., pp. 83-88.

ROBERTSON, H.G. 1988. Spatial and temporal patterns of predation by ants on eggs of *Cactoblastis cactorum*. Ecol. Ent. 13: 207-214.

ROBERTSON, H.G. and J.H. HOFFMANN. 1989. Mortality and life-tables of *Cactoblastis cactorum* (Berg) (Lepidoptera: Pyralidae) compared on two host-plant species. Bull. ent. Res. 79 in press.

SKAIFE, S.H. 1955. The Argentine ant, *Iridomyrmex humilis* (Mayr). Trans. R. Soc. S. Afr. 34: 355-377.

SKAIFE, S.H. 1961. The study of ants. Longmans, London.

SKAIFE, S.H. 1962. The distribution of the Argentine ant *Iridomyrmex humilis* Mayr. Ann. Cape Prov. Mus. 2: 297-298.

SMEE, C. 1929. Insects in tobacco seedbeds. Bull. Dept. Agric. Nyasaland, Ent. 5.

STEYN, J.J. 1954. The pugnacious ant (*Anoplolepis custodiens* Smith) and its relation to the control of citrus scales at Letaba. Mem. ent. Soc. sth. Afr. 3: 1-96.

VAN RENSBURG, N.J. 1979. *Cinara cronartii* on the roots of pine trees (Homoptera: Aphididae). J. ent. Soc. sth. Afr. 42: 151-152.

WEAVING, A.J.S. 1980. Observations on *Hilda patruelis* Stal (Homoptera: Tettigometridae) and its infestation of the ground nut crop in Rhodesia. J. ent. Soc. sth. Afr. 43: 151-167.

4

Seed Harvesting Ant Pests in Australia

A.N. Andersen

INTRODUCTION

Ants interact with seeds in both beneficial and harmful ways. Many plant species entice ants to disperse their seeds by offering them a food reward: ants transport the seeds to their nests, where they eat the food body (elaiosome) but leave the seed itself intact. Seed dispersal by ants (myrmecochory) offers several potential advantages to plants, including the targeting of seeds to nutrient-enriched microsites, avoidance of predators, protection from fire and distance dispersal (Berg 1975; Culver and Beattie 1978; O'Dowd and Hay 1980; Andersen 1988). Seed-harvesting, on the other hand, results in seed destruction, and is a form of plant predation (Janzen 1971). It is especially prominent in desert regions, where specialist harvester ants can consume vast quantities of seeds (Brown et al. 1979) and have a marked impact on the structure of desert plant communities (Reichman 1979; Inouye et al. 1980).

Seed-harvesting by ants is unusually widespread in Australia. In addition to the arid zone (Andersen 1982; Briese 1982; Davison 1982; Morton and Davidson 1988), ants harvest seeds in temperate eastern and southeastern Australia (Ashton 1979; Drake 1981; Andersen and Ashton 1985; Andersen 1987), monsoonal northern Australia (Greenslade and Mott 1979) and mediterranean southwestern Australia (Majer 1982). Harvester ants are therefore an important part of Australia's ecological landscape. Their prevalence extends to agriculture (Campbell 1966; McGowan 1969; Campbell and Swain 1973: Johns and Greenup 1976) and forestry (Cremer 1966; Abbott and van Heurck 1985), where they seriously interfere with seeding operations.

Harvester ant pests particularly affect introduced pasture grasses (Campbell 1966; McGowan 1969; Johns and Greenup 1976; Campbell and Gilmour 1979), but can also influence the success of native pasture grasses (Mott and McKeon 1977), pasture forbs such as *Stylosanthes* (Mott and McKeon 1977) and *Medicago*, and crops such as tobacco (Smith and Atherton 1944) and *Sorghum* (Russell et al. 1967; Radford and Allsopp 1987) and sunflower (Radford and Allsopp 1987). The impact of harvester ants is primarily felt during initial sowing, but they can also affect the persistence of desired pasture species (McGowan 1969). Harvester ants may also play an important role in the revegetation of degraded land (see Majer this volume).

HARVESTER ANTS

Elsewhere in the world, harvesting by ants is more-or-less confined to a few myrmicine genera. In Australia, however, formicine and ponerine ants are also important harvesters. Species of *Monomorium* (ex *Chelaner*), *Pheidole, Meranoplus, Melophorus, Tetramorium, Rhytidoponera* and *Prolasius* are the most important and can be grouped according to their distribution. Harvesting species of *Melophorus, Meranoplus* and *Tetramorium* occur almost exclusively in the arid zone (Briese and Macauley 1981; Greenslade 1982; Morton and Davidson 1988), while *Prolasius* and *Rhytidoponera* occur predominantly in cool-temperate regions (Ashton 1979; Andersen and Ashton 1985), and seed-eating *Monomorium* and *Pheidole* are found throughout the continent (Briese and Macauley 1981; Greenslade 1982; Andersen and Ashton 1985).

The species-level taxonomy of Australian ants is poorly known, with the number of valid names representing just a fraction of even the known fauna. This means it is impossible to provide a reliable estimate of the total number of harvester species, although it must be at least several hundred. Most of these belong to *Monomorium* and *Pheidole*, with well over 100 species each. The proportion of harvesting species within a genus varies enormously. For example, the great majority of *Rhytidoponera* species do not eat seeds, whereas seed-harvesting is virtually universal in *Pheidole*.

Australian harvester ants span the full spectrum from generalist omnivore to specialist granivore. The latter inhabit mostly arid environments, where they exploit pulsed seed production (particularly of ephemerals) by storing seeds in large underground granaries. They typically eat a wide range of seeds depending on availability (Briese 1982; Davison 1982), but some specialize on one or a few seed species (Andrew 1986). Most Australian harvester ants, however, including almost all those outside the arid zone, have broadly omnivorous diets. They scavenge for a wide range of food items, and feed on seeds opportunistically; their seed preferences are determined primarily by size, morphology, and availability (Smith and Atherton 1944; Briese and Macauley 1981; Andersen 1982; Andersen and Ashton 1985).

Species of *Pheidole* are easily the most important seed-harvesting pests. This is because they are particularly adept at colonizing disturbed land, where they are free to proliferate without competition from more specialized species. Large populations of harvesting species of *Pheidole* have been reported in agro-ecosystems throughout Australia, including Queensland (Smith and Atherton 1944; Russell et al. 1967), New South Wales (Campbell 1966; Campbell and Swain 1973; Johns and Greenup 1976), Victoria (McGowan 1969), Western Australia (Twigg et al. 1982; Panetta 1988) and the Northern Territory (Mott and McKeon 1977). Densities of *Pheidole* nests can be as high as $8m^{-2}$ (Campbell 1966).

HARVESTING RATES

Rates of seed-harvesting by pest ants vary enormously in space and time, and according to seed species. A study of harvesting by seven pasture species in northeastern Victoria (McGowan 1969) revealed that seed removal was virtually complete within a few days for *Phalaris tuberosa* and *Lolium rigidum*, but less than 20% for *Hordeum leporinum*. Seeds of *L. rigidum* are also removed extremely rapidly from pastures in southwestern Western Australia (62% after 24 hours; Panetta 1988). Seeds of native pasture species were preferentially removed in the Northern Territory (Mott and McKeon 1977), and considerable seed selectivity has been observed in northern New South Wales (Johns and Greenup 1976). Many factors can influence seed selection by harvester ants (Brown et al. 1979), but for most Australian pest species these factors primarily relate to the ease of seed transport and consumption: seed abundance, seed size, the thickness and texture of the seed coat, and the presence of any protective covering.

The foraging activity of most ant species varies markedly in time on both daily and seasonal scales (Briese and Macauley 1981; Andersen 1983), so it is not surprising that seed-harvesting rates show considerable temporal variation. Most species of *Pheidole*, for example, restrict their foraging to mild weather. This means that rates of removal during summer are almost always greater than during winter (Johns and Greenup 1976; Andersen and Ashton 1985), and more seeds are removed during the day than at night. During cooler months, foraging activity - and therefore seed removal - often occurs only in the middle of the day (10:00 a.m. to 3:00 p.m.; Campbell 1968; Andersen 1983), whereas during summer the midday heat is avoided and foraging is typically restricted to early morning and late afternoon (Campbell 1968; Mott and McKeon 1977; Andersen 1983; Panetta 1988). Temperature accounted for up to 87% of the temporal variance in harvesting rates by *Pheidole* in northern New South Wales (Johns and Greenup 1976).

CONTROLLING SEED LOSS

The most common methods used to control seed losses to harvester ants involve insecticides (Campbell 1982), which may be applied by spraying the ground, or by soaking/coating seeds (Champ et al. 1961; Cremer 1966; Victorian Forests Commission 1972; Radford and Allsopp 1987). The marked increase in plant establishment rates that can follow insecticidal treatment (eg. Campbell 1966) emphasizes the importance of harvester ant pests. Poison bait use can also dramatically reduce seed loss (Smith and Atherton 1944; Campbell 1977).

Insecticides control losses to harvester ants but often create their own problems. For example, dieldrin was used to coat seeds during the 1950s and 1960s, however it has since been discontinued because of toxicity to mammals (Campbell 1982). Heptachlor and chlordane are excellent formicides, but as with other organochlorines, serious reservations remain about their general environmental toxicity (Radford and Allsopp 1987). Further, some insecticides, such as chlorophyrifos and terbufus, are apparently phytotoxic, so that reduced harvesting rates can be more than offset by reduced germination rates (Radford and Allsopp 1987). In addition, the indirect effects of insecticides and baits on plant performance are unknown. Harvester ants are seed pests, but they might benefit plants in other ways by accelerating the nutrient cycle, improving soil structure and controlling other insect pests.

Several alternative methods which control seed losses exist, and these can be as effective as insecticides. Seeds can be coated with non-toxic substances to reduce their attractiveness to ants (Russell et al. 1967; Victorian

Forests Commission 1972), or covered with sand to reduce their accessibility to ants (Smith and Atherton 1944). Radford and Allsopp (1987) found that compacting the soil immediately after sowing gave similar or better results than insecticidal seed dressings. Seed losses can also be reduced by sowing during cool or wet weather, periods of low ant activity (Johns and Greenup 1976), or sowing more seeds to satiate the ants (Smith and Atherton 1944).

CONCLUSION

Most studies of harvester ant pests have focused on their impact on seedling establishment in desired plant species. However harvester ants can affect desired plants in other important ways. For example, harvesting species of *Pheidole* have been implicated in the epidemiology of ryegrass toxicity (Twigg et al. 1982). Harvester ants may also affect undesired plant species in crops and pastures (Panetta 1988), as well as in natural ecosystems (Wilson 1989). Selective control of harvester ants might improve the ratio of desirable to weed species. Harvester ants have potentially important indirect effects on plants through their roles as predators and scavengers, and through their nest-building activities. There is a need for more research into the broad role harvester ants play in Australian agro-ecosystems.

REFERENCES

ABBOTT, I. and P. VAN HEURCK. 1985. Comparison of insects and vertebrates as removers of seed and fruit in a Western Australian forest. Aust. J. Ecol. 10: 165-168.

ANDERSEN, A.N. 1982. Seed removal by ants in the mallee of northwestern Victoria. *In* R.C. Buckley (ed.) Ant Plant Interactions in Australia, pp. 31-44. Junk Press, The Hague.

ANDERSEN, A.N. 1983. Species diversity and temporal distribution of ants in the semi-arid mallee region of northwestern Victoria. Aust. J. Ecol. 8: 127-137.

ANDERSEN, A.N. 1987. Effects of seed predation by ants on seedling densities at a woodland site in SE Australia. Oikos 48: 171-174.

ANDERSEN, A.N. 1988. Dispersal distance as a benefit of myrmecochory. Oecologia 75: 507-511.

ANDERSEN, A.N. and D.H. ASHTON. 1985. Rates of seed removal by ants at heath and woodland sites in southeastern Australia. Aust. J. Ecol. 10: 381-390.

ANDREW, M.H. 1986. Granivory of the annual grass *Sorghum intrans* by the harvester ant *Meranoplus* sp. in tropical Australia. Biotropica 18: 344-349.

ASHTON, D.H. 1979. Seed harvesting by ants in forests of *Eucalyptus regnans* F. Muell. in central Victoria. Aust. J. Ecol. 4: 265-277.

BERG, R.Y. 1975. Myrmecochorus plants in Australia and their dispersal by ants. Aust. J. Bot. 23: 475-508.

BRIESE, D.T. 1982. Relationship between the seed-harvesting ants and the plant community in a semi-arid environment. *In* R.C. Buckley, (ed.) Ant-Plant Interactions in Australia, pp. 11-24. W. Junk Press, The Hague.

BRIESE, D.T. and B.J. MACAULEY. 1981. Food collection within an ant community in semi-arid Australia, with special reference to seed harvesters. Aust. J. Ecol. 6: 1-19.

BROWN, J.H., O.J. REICHMAN and D.W. DAVIDSON. 1979. Granivory in desert ecosystems. Ann. Rev. Ecol. Syst. 10: 201-227.

CAMPBELL, M.H. 1966. Theft by harvesting ants of pasture seed broadcast on unploughed land. Aust. J. Exp. Agric. Anim. Husb. 6: 334-338.

CAMPBELL, M.H. 1968. Cops and robbers. Chiasma 6: 93.

CAMPBELL, M.H. 1977. No remedy for seed-stealing ants. Agric. Gaz N.S.W. 88: 44.

CAMPBELL, M.H. 1982. Restricting losses of aerially sown seed due to seed-harvesting ants. *In* R.C. Buckley (ed.) Ant-Plant Interactions in Australia, pp 25-30. Junk Press, The Hague.

CAMPBELL, M.H. and A.R. GILMOUR. 1979. Reducing losses of surface-sown seed due to harvesting ants. Aust. J. Exp. Agric. Anim. Husb. 19: 706-711.

CAMPBELL, M.H. and F.G. SWAIN. 1973. Factors causing losses during the establishment of surface-sown pastures. J. Range. Manage 26: 355-359.

CHAMP, B.R., D.I. SILLAR and H.J. LAVERY. 1961. Seed-harvesting ant control in the Cloncurry District. Qld. J. Agric. Sci. 18: 257-260.

CREMER, K.W. 1966. Treatment of *Eucalyptus regnans* seed to reduce losses to insects after sowing. Aust. For. 30: 162-174.

CULVER, D.C. and A.J. BEATTIE. 1978. Myrmecochory in *Viola*: dynamics of seed-ant interactions in some West Virginia species. J. Ecol. 66: 53-72.

DAVISON, E.A. 1982. Seed utilization by harvester ants. *In* R.C. Buckley (ed.) Ant-plant interactions in Australia, pp 1-6. Junk Press, The Hague.

DRAKE, W.E. 1981. Ant-seed interaction in dry sclerophyll forest on North Stradbroke Island, Queensland. Aust. J. Bot. 29: 293-309.

GREENSLADE, P.J.M. 1982. Diversity and food specificity of seed-harvesting ants in relation to habitat and community structure. *In* K.E. Lee (ed.) Proceedings of the 3rd Australasian Conference on Grassland Invertebrate Ecology, pp 227-233. S.A. Govt. Printer, Adelaide.

GREENSLADE, P.J.M. and J.J. MOTT. 1979. Ants of native and sown pastures in the Katherine area, Northern Territory, Australia (Hymenoptera : Formicidae). *In* T.K. Crosby and R.P. Pottinger (eds.) Proc. 2nd Australasian Conf. Grassl. Invert. Ecol., pp. 153-156. Govt. Printer, Wellington.

INOUYE, R.S., G.S. BYERS, and J.H. BROWN. 1980. Effects of predation and competition on survivorship, fecundity and community structure of desert annuals. Ecology 61: 1344-1351.

JANZEN, D.H. 1971. Seed predation by animals. Ann. Rev. Ecol. Syst. 2: 465-492.

JOHNS, G.G. and L.R. GREENUP. 1976. Pasture seed theft by ants in northern N.S.W. Aust. J. Exp., Ag. Anim. Husb. 16:257-264.

MCGOWAN, A.A. 1969. Effect of seed harvesting ants on the persistence of Wimmera rye grass in pastures of North East Victoria. Aust. J. Exp. Agric. Anim. Husb. 9: 37-40.

MAJER, J.D. 1982. Ant-plant interactions in the Darling Botanical District of Western Australia. *In* R.C. Buckley (ed.) Ant-Plant Interactions in Australia, pp 45-61. Junk Press, The Hague.

MORTON, S.R. and D.W. DAVIDSON. 1988. Comparative structure of harvester ant communities in arid Australia and North America. Ecol. Monogr. 58: 19-38.

MOTT, J.J. and G.M. MCKEON. 1977. A note on the selection of seed types by harvester ants in northern Australia. Aust. J. Ecol. 2: 231-235.

O'DOWD, D.J. and M.E. HAY. 1980. Mutualism between harvester ants and a desert ephemeral: seed escape from rodents. Ecology 61: 531-540.

PANETTA, F.D. 1988. Factors determining seed persistence of *Chondrilla juncea* L. (skeleton weed) in southern Western Australia. Aust. J. Ecol. 13: 211-224.

RADFORD, B.J. and P.G. ALSOPP. 1987. Use of insecticides and a press wheel to control soil insects affecting sorghum and sunflower establishment in southern Queensland. J. Aust. Ent. Soc. 26:161-167.

REICHMAN, O.J. 1979. Desert granivore foraging and its impact on seed densities and distributions. Ecology 60: 1085-1092.

RUSSELL, M.H., J.E. COALDRAKE and A.M. SANDERS. 1967. Comparative effectiveness of some insecticides, repellents and seed-pelleting devices in the prevention of ant removal of pasture seeds. Trop. Grasslands 1: 153-166.

SMITH, J.H. and D.O. ATHERTON. 1944. Seed-harvesting and other ants in the tobacco-growing districts of North Queensland. Qld. J. Agric. Sc. 1: 33-61.

TWIGG, L., J.D. MAJER, and B.A. STYNES. 1982. Influence of seed-harvesting ants in annual ryegrass and their possible effects on the epidemiology of ryegrass toxicity. *In* K.E. Lee (ed.) Proc. 3rd Australasian Conf. Grassl. Invert. Ecol., pp. 235-243. S.A. Govt. Printer, Adelaide.

VICTORIAN FORESTS COMMISSION. 1972. Seed coating eucalypts. Operational Bulletin No. 9, Division of Forest Operations.

WILSON, C.G. 1989. Post-dispersal seed predation of an exotic weed, *Mimosa pigra* L., in the Northern Territory. Aust. J. Ecol. 14: 235-240.

5

Pest Ants in the Hawaiian Islands

N. Reimer, J.W. Beardsley and G. Jahn

INTRODUCTION

Ants are pests throughout Hawaii, in various environments. Endemic terrestrial arthropod populations have been decimated by ant predation at lower elevations throughout the islands. Ants also prey on imported natural enemies which may adversely affect the success of biological control programs in agriculture (Phillips 1934, Carter 1937) and endemic forests (Reimer 1988). The association of some ant species with aphids, scales, and mealybugs has contributed to increased populations of Homoptera and resulted in reduced crop yields. However, ants are sometimes beneficial and have, for example, preyed heavily on fruit flies (Pemberton and Willard 1918, Willard 1927, Wong et al. 1984, Wong and Wong 1988), house flies (Illingworth 1913, Phillips 1934), planthoppers (Muir 1915) and Lepidoptera (Swezey 1909, Illingworth 1915) in Hawaii.

ORIGIN OF HAWAIIAN ANT FAUNA

The Hawaiian Archipelago originated about 70 million years ago (Dalrymple et al. 1973). The endemic terrestrial biota of the islands is believed to have been derived from a relatively small number of successful colonizing organisms which succeeded in the "sweepstakes" of long-range overseas dispersal (Zimmerman 1948).

Ants apparently were not among the successful prehistoric colonizers of Hawaii as the known fauna contains no endemic ant species. The hypogaeic ponerine, *Amblyopone zwaluwenbergi* (Williams) is known only from Hawaii, but Brown (1960) stated that the species was probably introduced from Melanesia or Indonesia.

Frederick Smith first recorded specific ants from the islands in 1879. Of these, one or both small, cryptic ponerines, *Hypoponera opaciceps* (Mayr) and *H. punctatissima* (Roger), could have arrived with the Polynesian voyagers who first populated the islands. At least the former was already widely distributed throughout the islands when the Fauna Hawaiiensis survey was conducted during the 1890s (Forel 1899). According to Smith (1879) - and Blackburn and Kirby (1880) - three aggressive, cosmopolitan, ground-nesting myrmicines, *Pheidole megacephala* (Fabricius), *S. geminata*

(Fabricius) and *Tetramorium quineense* (Fabricius), were well established by the late 1870s or before. Forel listed 20 species, two of which were either listed in error or died out immediately, as they were never recollected, and one which was a misidentification of another listed species. Gulick (1913) listed six additional species, and Wheeler (1934) added nine more. During and shortly after World War II two additional ants of economic importance, *Iridomyrmex humilis* (Mayr) and *Anoplolepis longipes* Jerdon, became established.

Wilson and Taylor (1967) showed that the six species listed by Wheeler as endemic to Hawaii were either synonyms of more widespread species or were known to occur outside of Hawaii. These authors listed 35 species as having been collected in Hawaii.

Huddleston and Fluker (1968) listed seven additional species, including two undetermined *Solenopsis*, in their survey of the ant fauna of Hawaii. These authors found 36 established species, and listed six others which had been previously reported from the islands but were not found by them. Two additional species, believed to be newly arrived immigrants, have been found since their survey was made (Beardsley 1979, 1980). Thus, the present ant fauna of Hawaii (Table 1) consists of a miscellaneous assortment of species derived from scattered parts of the tropics and subtropics, most of which are characteristically vagile forms that are widely distributed outside of the islands.

IMPACT OF ANTS ON ENDEMIC HAWAIIAN INVERTEBRATES

The endemic terrestrial fauna of Hawaii is characterized by 1) adaptive radiation, which has produced several large species complexes, each derived from a single prehistoric immigrant (e.g.: ca 700 species of drosophilid flies) and 2) precinctive distribution, with many species restricted to small areas on single islands (Zimmerman 1948). Although the endemic invertebrate fauna contains a few relatively large, wide ranging species (e.g.: the giant Hawaiian dragonfly, *Anax strenuus* Hagen), most are small, with cryptic habits. Among the endemic insects, flightlessness is common and probably has contributed to precinctiveness. For example, 32 species of flightless carabid beetles in the genus *Mecyclothorax* Sharp only live on Haleakala, east Maui, each usually confined to a relatively small part of that mountain (Britton 1948).

In the absence of ants, Hawaii's endemic invertebrates had no need of specialized adaptations such as ant mimicry and defense chemistry. Such adaptations rarely appear in the endemic Hawaiian invertebrate fauna, although adaptations that apparently arose in response to predation by diurnally active birds are common (e.g.: cryptic coloration, nocturnal feeding habits, etc.).

The impact of ants, particularly the big-headed ant, *Pheidole megacephala*, on the endemic terrestrial invertebrates of Hawaii, was discussed by Perkins (1913). The impact of *P. megacephala* on endemic Coleoptera particularly impressed Perkins. He stated:

> "It may be said that no native Hawaiian Coleoptera insect can resist this predator, and it is practically useless to attempt to collect where it is well established. Just on the limits of its range one may occasionally meet with a few active beetles, e.g.: species of *Plagithmysus*. Often with these

> ants attached to their legs and bodies, but sooner or later they are quite exterminated from these localities."

The attacks of *P. megacephala* are not by any means confined to Coleoptera, and Perkins (loc cit) further stated

> "Below twelve or thirteen hundred feet it (*P. megacephala*) often occupies most of the islands, excepting some extremely arid localities. Though not so utterly destructive to other insects as to the beetles, yet many of them are destroyed by it and, generally speaking, collecting is very poor where it abounds."

As Perkins pointed out, the range of *P. megacephala* in Hawaii is limited to the lowlands, generally below about 600 m elevation. It also appears to be limited by rainfall, and the ant occurs irregularly in dry (<ca 15-20 in. annual rainfall) or wet (>ca 100 in.) areas. Our recent studies have shown that periods of heavy rainfall are often followed by severe reduction of *P. megacephala* populations in pineapple fields (Beardsley et al. 1982). This supports observations by others that high rainfall limits its distribution.

However, since Perkins' time, two additional aggressive polydomous ant species--the Argentine ant *Iridomyrmex humilis* (Mayr) and the long-legged ant, *Anoplolepis longipes* (Jerdon)--have become established in Hawaii. These species, whose ranges are still expanding, occupy areas with somewhat different climatic and physiographic conditions than those most favorable to *P. megacephala. I. humilis* was discovered on Oahu in 1940 (Zimmerman 1941), and the *A. longipes* twelve years later, also on Oahu (Zimmerman 1953). The Argentine ant, although originally found in lowland areas of Oahu, is now largely confined to higher elevations (mostly above 450-600 m elevation) with moderate rainfall. The ant is commonly found at 900 to 1800 m or higher, particularly on Maui and Hawaii. Like *P. megacephala,* this species does not thrive in areas of heavy rainfall (>ca 100 inches annually). *A. longipes,* best adapted to the lowlands, is usually found below about 600 m, in rocky areas of moderate rainfall. The territorial interrelationships between these species on Oahu were discussed by Fluker and Beardsley (1970). They concluded that the Argentine ant was unlikely to completely displace *P. megacephala* in Hawaii as it did in Madeira and the Canary Islands, and that *A. longipes*' distribution would likely be limited by its requirement for rocky nesting sites. Although minor displacements by these more recently arrived ants have occurred, it appears that neither species is likely to permanently displace *P. megacephala* from most of the areas it now occupies.

However, both I. *humilis* and *A. longipes* have the potential, as predators, to profoundly affect the endemic insect fauna in territory already occupied. Data documenting the negative impact of Argentine ants on endemic terrestrial arthropods living in native high elevation shrubland (Haleakala National Park, Maui) have been presented by Medeiros et al. (1986). These authors compared insect populations at otherwise similar ant-infested and uninfested sites, and found extremely low numbers of large endemic noctuid moth larvae (*Agrotis* spp. and *Peridroma* spp.) in ant infested areas. Nests of endemic ground-nesting colletid bees (*Nesoprosopis* spp.) were eliminated from ant infested sites. No comparable data on the effects of *A.*

longipes have been published, but this species is abundant in rocky riparian habitats at low elevations on east Maui. There, Hardy (1979) noted that endemic aquatic and semiaquatic Diptera, among other insects, had disappeared from the lower Kipahulu Stream in areas invaded by *A. longipes*.

Although the majority of endemic Hawaiian insects may have been eliminated from areas permanently occupied by *Pheidole megacephala*, and profoundly depressed by both *I. humilis* and *A. longipes*, a few endemic insect species thrive in the presence of these polydomous ants. On Haleakala, we observed some species of endemic Pseudococcidae (eg. *Pseudococcus nudus* Ferris) being actively tended by Argentine ants for honeydew produced by these mealybugs. In such situations, the ants may benefit the mealybugs, as they do other species of honeydew producers. Interestingly, endemic Hawaiian species of the cosmopolitan lygaeid genus *Nysius* have remained common, sometimes abundant, in lowland areas of Hawaii occupied by *Pheidole* and other ant species. These bugs, like their extra Hawaiian congenerics, possess dermal glands that produce volatile odoriferous secretions which may repel ants and other predators.

ANTS AND SUGARCANE IN HAWAII

Although Hawaii's sugarcane acreage has declined from the levels of 50 years ago, sugar is still the islands' single most important agricultural crop. Until recently, ants were considered minor pests of sugarcane. However, during the past 20 years, a move towards extensive utilization of drip irrigation systems has been accompanied on many sugar plantations by serious ant-inflicted damage to plastic tubing (see Chang and Ota this volume).

Another impact of ants on the Hawaiian sugarcane culture is connected to their care for honeydew-producing Homoptera, particularly mealybugs. Beardsley (1959) discussed the negative impact of the Argentine ant on natural enemies of sugarcane mealybugs. He noted that mealybug populations increased substantially in cane fields invaded by this ant. Fluker et al. (1968) found that the presence of big-headed ants resulted in slightly larger *S. sacchari* populations than those in fields where ants were eliminated. However, the tending of mealybugs by *P. megacephala* did not significantly reduce parasitization of *S. sacchari* by the encyrtid wasp *Anagyrus saccharicola* Timberlake.

Williams (1931) gave a general account of the ant species associated with Hawaiian sugarcane. However, except for casual observations, no attempts have been made to evaluate the impact of ants on other sugarcane insects.

ANTS AND PINEAPPLE IN HAWAII

Carter (1967) listed the ant species common in Hawaiian pineapple fields as *P. megacephala*, *S. geminata*, *Plagiolepis alluaudi* Forel, *Paratrechina bourbonica* (Forel), *Paratrechina longicornis* (Latr.), *Paratrechina vaga* (Forel), and *I. humilis*. The most pestiferous species in pineapple are *P. megacephala*, *S. geminata*, and *I. humilis*. Most commercial fields are situated below 600 m elevation, in areas where *P. megacephala* is the dominant species. None of these ant species cause direct damage to pineapple, but they are considered pests because they tend mealybugs thought to transmit wilt disease. Mealybug

TABLE 1. List of ant species and their distributions in the Hawaiian Islands (1=Hawaii, 2=Kauai, 3=Lanai, 4=Maui, 5=Molokai, and 6=Oahu).[a]

Cerapachyinae	
Syscia silvestrii Wheel	1,2,5,6
Ponerinae	
Amblyopone zwaluwenburgi (Williams)	6
Hypoponera punctatissima (Roger)	1,2,3,4,5,6
Hypoponera opaciceps (Mayr)	1,2,3,4,5,6
Hypoponera zwaluwenburgi (Wheeler)	1,4,6
Ponera swezeyi (Wheeler)	1,6
Leptogenys falcigera Roger	1,2,3,4,5,6
Myrmicinae	
Pheidole fervens Fr. Smith	1,6
Pheidole megacephala (Fabricius)	1,2,3,4,5,6
Solenopsis geminata (Fabricius)	1,3,4,5,6
Solenopsis "a"	3,4,5,6
Solenopsis "b"	6
Monomorium floricola (Jerdon)	1,2,3,4,5,6
Monomorium fossulatum Emery	1,2,3,4,6
Monomorium latinode Mayr	6
Monomorium minutum Mayr	1,2,4,5,6
Monomorium pharaonis (L.)	1,4,6
Monomorium destructor (Jerdon)	1,2,6
Cardiocondyla emeryi Forel	1,2,3,4,5,6
Cardiocondyla nuda (Mayr)	1,2,3,4,5,6
Cardiocondyla wroughtoni (Forel)	1,2,3,4,5,6
Cardiocondyla venustula Wheeler	1,2,3,4,5,6
Tetramorium guineense (Fabricius)	1,2,4,5,6
Tetramorium simillimum (Fr. Smith)	1,2,3,4,6
Tetramorium tonganum Mayr	1,2,6
Strumigenys godeffroyi Mayr	1,2,6
Strumigenys lewisi Cameron	6
Strumigenys rogeri Emery	1,2,4,5,6
Trichoscapa membranifera Emery	1,6
Quadristruma emmae (Emery)	1,6
Dolichoderinae	
Iridomyrmex glaber (Mayr)	6
Iridomyrmex humilis (Mayr)	1,2,3,4,6
Tapinoma melanocephalum (Fabricius)	1,2,3,4,5,6
Technomyrmex albipes (Fr. Smith)	1,2,4,5,6
Formicinae	
Plagiolepis alluaudi Forel	1,2,3,4,5,6
Camponotus variegatus (Fr. Smith)	1,2,3,4,5,6
Anoplolepis longipes (Jerdon)	1,4,6
Paratrechina longicornis (Latreille)	1,2,3,4,5,6
Paratrechina bourbonica (Forel)	1,2,3,4,5,6
Paratrechina vaga (Forel)	1,2,3,4,5,6
Pseudomyrmecinae	
Pseudomyrmex gracilis mexicanus (Roger)	6

a. Distribution data taken from Huddleston & Fluker (1968) and collections by the authors.

wilt reached such devastating proportions in the early 1900s that the survival of Hawaii's pineapple industry was in doubt.

Mealybug wilt disease was controlled in the recent past by eliminating ants from pineapple fields with mirex or heptachlor. These compounds provided effective control until banned by the Environmental Protection Agency in the early 1980s. No effective formicides currently exist for use in pineapple.

A common cultural practice in Hawaiian pineapple allows a field to lie fallow for 6 to 12 months. Shortly before replanting, the field is burned to remove pineapple trash; this also eliminates most infield *P. megacephala* colonies. However, the practice has been prohibited on Maui because of complaints from resort hotels and new residential subdivisions. Other islands on which pineapple is grown will most likely initiate similar laws (see Rohrbach et al. 1988 for details).

Through an elegant series of experiments, Illingworth (1931) demonstrated that *P. megacephala* does not transmit wilt. Pineapple mealybug wilt is induced, and probably transmitted, by the gray pineapple mealybug *Dysmicoccus neobrevipes* Beardsley, the pink pineapple mealybug *Dysmicoccus brevipes* (Cockerell), and, to a lesser extent, the long tailed mealybug *Pseudococcus longispinus* (Targioni-Tozzetti) (Illingworth 1931, Sakimura and Matsumoto 1965, Carter 1967). The ants are attracted to and feed on honeydew excreted by these mealybugs. Mealybugs move into pineapple fields from the perimeter and attain large populations when ants are present; they are never a problem in the absence of ants (Illingworth 1931; Carter 1932, 1933; Ito 1936; Beardsley et al. 1982). Several hypotheses have been proposed to explain the ants' role (Illingworth 1931; Carter 1932, 1967), but the question of why mealybugs thrive under ant attendance has not been answered.

Parasitoids of *P. megacephala* have been recorded in the literature (Reichensperger 1913, Schmitz 1915, Wheeler 1922); however, Hawaii has never attempted biological control of *P. megacephala* or any ant species. Further research in this area is needed.

Control of the big-headed ant in the near future will probably depend on insecticidal baits, since insecticides are most efficient, safe and somewhat selective when used in this form (Lofgren et al. 1963, McEwen et al. 1979). Bait formulations of hydramethylnon, fenoxycarb, and fluoroaliphatic sulfonamide have significantly reduced or eliminated *P. megacephala* populations in experimental field trials (Su et al. 1980; Reimer and Beardsley 1989; Reimer, unpublished data) and Sumitomo's S31183 has shown promise in the laboratory (see Glancey et al. this volume). These are currently the most promising chemicals for *P. megacephala* control and may be suitable for control of other species.

ANTS AND COFFEE IN HAWAII

Coffee is grown commercially on the islands of Hawaii and Kauai. *P. megacephala* and *A. longipes* are the principal pest ants in coffee grown at elevations below 700 m. *I. humilis* can be a problem in coffee grown above 600 m. Ants do not directly damage coffee trees but they do tend green scale (*Coccus viridis* (Green)). All three ant species are attracted to the honeydew excreted by scales. When ants are present, *C. viridis* can build up large populations and produce an abundance of honeydew which coats the leaves and fruits of the trees. Sooty mold flourishes on the honeydew and covers the leaves and fruits. The blighted trees have reduced yields and, under severe conditions, can die.

Bess (1958) hypothesized that removal of honeydew by ants created conditions favorable for scale development. Bess found that, despite similar parasitization rates, green scale populations in Sri Lankan coffee died out when the ant *Technomyrmex albipes* (Fr. Smith) was excluded, but flourished when attended by these ants. However, green scale populations in Hawaii remain high despite the accumulation of honeydew and buildup of sooty mold. The role of ants may differ in Hawaii since different species are involved. In fact, *I. humilis* which is a problem in Hawaii coffee, was found to inhibit parasitization of *Coccus hesperidium* L. in California citrus and had little effect on scale development or survival through honeydew removal (Bartlett 1961).

No pesticides are currently registered for ant control in coffee. The commercially available baits containing hydramethylnon (Amdro) and fenoxycarb (Logic) are being tested for efficacy in coffee. These baits, which contain soy bean oil as an attractant and phagostimulant, are readily accepted by *P. megacephala* but are unattractive to *I. humilis* and *A. longipes*. Different bait attractants need to be developed for these ants.

ANTS AND MISCELLANEOUS CROPS

In addition to sugarcane, pineapple, and coffee, ants have caused problems for other Hawaiian crops. Exportation of Protea and anthurium flowers can be denied if ants, most commonly *P. megacephala* and *I. humilis*, are found in the shipment. The ants are attracted to nectar and aphid honeydew on these plants. In dryland taro, root aphid (*Patchiella reaumuri* Kaltenbach) populations have increased to pestiferous levels when tended by *P. megacephala* or *Pheidole fervens* Fr. Smith. In beans, *S. geminata* has destroyed small fields of seedling plants by chewing the stems. The fire ants presumably chew the plants for the water content because the problem only exists during dry periods, despite ant presence at other times.

URBAN ANT PROBLEMS

Ants in an urban environment are principally nuisance pests. However, at least one species, *Monomorium pharaonis* (L.), can transmit pathogenic organisms (Beatson 1972). Pharaoh's ant has also been found in sterile supplies and intravenous fluid apparatuses (Beatson 1973). Its small size and ability to chew through many materials makes it a serious infection hazard in hospitals and residences (Edwards 1986; Eichler this volume).

Any ant species found in buildings is considered a pest, but the most commonly encountered species in Hawaii are *Camponotus* (*Tanaemyrmex*) *variegatus* (Fr. Smith), *P. megacephala*, *S. geminata*, *I. humilis*, *Iridomyrmex glaber* (Mayr), *Paratrechina bourbonica* (Forel), *Paratrechina longicornis* (Latreille), *A. longipes*, *Monomorium floricola* (Jerdon), *M. pharaonis*, and P. *alluaudi*. *S. geminata* also creates occasional problems in playgrounds and parks because of its stinging behavior. The carpenter ant, *C. variegatus*, a common nocturnal house pest, often develops large colonies in wall voids, hollow doors, and many other sheltered areas. It will not excavate or damage wood but is a large (1 cm), conspicuous species, especially troublesome in the fall with the emergence of large numbers of alates.

Pest control operators and homeowners in Hawaii kill ants with insecticidal sprays and baits. Sprays kill foraging individuals but do not affect the adult or immature ants in the nest. This "short term" solution is appropriate under certain conditions but ant colonies can survive on food reserves in the nest, move to different areas of the building, or forage in areas unsuitable for spray applications (Edwards 1986). These problems can sometimes be eliminated by using baits. Commercial baits are effective for some ant species, but suitable bait attractants have not been found for all urban pest ants in Hawaii.

CONCLUSION

Hawaiian ant fauna, composed entirely of introduced species, has a devastating effect on the endemic arthropod fauna. Many ant species are also pests in agriculture and urban environments. Some ants can be beneficial in certain situations due to their predaceous habits but, in general, are considered to be pests throughout the islands. Chemical control has been successful in agricultural and urban environments but the most effective materials have been proscribed by the EPA. Although biological control is the only feasible method available for management of ants in natural environments, biological control of ants has not been attempted in Hawaii.

REFERENCES

BARTLETT, B. R. 1961. The influence of ants upon parasites, predators, and scale insects. Ann. Entomol. Soc. Am. 54:543-551.

BEARDSLEY, J. W. 1959. Observations on sugar cane mealybugs in Hawaii. Proc. lOth Cong. Internat. Soc. Sugarcane Technol. p. 954-61.

BEARDSLEY, J. W. 1979. Note on *Pseudomyrmex gracilis mexicanus* (Roger). Proc. Hawaii. Entomol. Soc. 23:23.

BEARDSLEY, J. W. 1980. Note on *Iridomyrmex glaber* (Mayr). Proc. Hawaii Entomol. Soc. 23:186.

BEARDSLEY, J. W., T. H. SU, F. L. MCEWEN and D. GERLING. 1982. Field investigations on the interrelationships of the bigheaded ant, the gray pineapple mealybug, and the pineapple mealybug wilt disease in Hawaii. Proc. Hawaii Entomol. Soc. 24:51-67.

BEATSON, S. H. 1972. Pharaoh's ants as pathogen vectors in hospitals. Lancet 1:425-427.

BEATSON, S. H. 1973. Pharaoh's ants enter giving sets. Lancet 1:606-607.

BESS, H. A. 1958. The green scale, *Coccus viridis* (Green) (Homoptera:Coccidae), and ants. Proc. Hawaii Entomol. Soc. 16:349-355.

BLACKBURN, T., and W. F. KIRBY. 1880. Notes on species of aculeate Hymenoptera occurring in the Hawaiian Islands. Entomol. Monthly Mag. 17:85-89.

BRITTON, E. B. 1948. A revision of the Hawaiian species of *Mecyclothorax* (Coleoptera: Carabidae). B. P. Bishop Mus. Occas. Paper 19(4):107-166.

BROWN, W. L. 1960. Contributions toward a reclassification of the Formicidae. III. Tribe Amplyoponini (Hymenoptera). Bul. Mus. Comp. Zool. Harvard 122:145-230.

CARTER, W. 1932. Studies of populations of *Pseudococcus brevipes* (Ckl.) occurring on pineapple plants. Ecology. 13:296-304.

CARTER, W. 1933. The pineapple mealybug, *Pseudococcus brevipes* and wilt of pineapples. Phytopathology 23:207-242.

CARTER, W. 1937. Biological Control. Hawaii Pineapple Prod. Sta. News 11:114-115.

CARTER, W. 1967. Insects and related pests of pineapple in Hawaii. Pineapple Research Institute of Hawaii, Honolulu, Hi. 105pp.

DALRYMPLE, G. B., E. A. SILVER and E. D. JACKSON. 1973. Origin of the Hawaiian Islands. Amer. Sci. 61:294-308.

EDWARDS, J. P. 1986. The biology, economic importance, and control of the Pharaoh's ant, *Monomorium pharaonis* (L.). In Vinson, S. B. [ed.], Economic impact and control of social insects. Praeger Publ., NY, pp. 257-271.

FLUKER, S. S. and J. W. BEARDSLEY. 1970. Sympatric associations of three ants: *Iridomyrmex humilis*, *Pheidole meqacephala*. and *Anoplolepis longipes* in Hawaii. Ann. Entomol. Soc. Amer. 63:1290-1296.

FLUKER, S. S., E. W. HUDDLESTON and J. W. BEARDSLEY. 1968. Some effects of big-headed ants on populations of pink sugarcane mealybug. J. Econ. Entomol. 61:474-477.

FOREL, A. 1899. Formicidae. Fauna Hawaiiensis 1(1):116-122.

GULICK, L. 1913. Synoptic list of ants reported from the Hawaiian Islands. Proc. Hawaii. Entomol. Soc. 2:306-311.

HARDY, D. E. 1979. An ecological survey of Puaaluu Stream. Part III. Report on a preliminary entomological survey of Puaaluu Stream, Maui. Coop. Nat. Park Resources Study Unit. Univ. of Hawaii, Manoa. Tech. Rept. 27:34-39.

HUDDLESTON, E. W. and S. S. FLUKER. 1968. Distribution of ant species of Hawaii. Proc. Hawaii Entomol. Soc. 20:45-69.

ILLINGWORTH, J. F. 1913. Little brown ant doing good work in Hawaii. Hawaii. Forester and Agric. 10:370-371.

ILLINGWORTH, J. F. 1915. Coconut leaf roller (*Omnioides blackburni* (Butl.)) destroyed by ants. Proc. Hawaii Entomol. Soc. 3:142.

ILLINGWORTH, J. F. 1931. Preliminary report on evidence that mealy bugs are an important factor in pineapple wilt. J. Econ. Entomol. 24:877-889.

ITO, K. 1936. Studies on fire ant control in relation to control of internal mealybug population in pineapple fields on Oahu. Pineapple Quarterly 5(3):76-95.

LOFGREN, C. S., F. J. BARTLETT and C. E. STRINGER. 1963. Imported fire ant toxic bait studies: Evaluation of carriers for oil baits. J. Econ. Entomol. 56:62-66.

MCEWEN, F. L., J. W. BEARDSLEY, JR., M. HAPAI and T. H. SU. 1979. Laboratory tests with candidate insecticides for control of the big-headed ant, *Pheidole megacephala* (Fabricius). Proc. Hawaii Entomol. Soc. 8:119-123.

MEDEIROS, A. C., L. L. LOOPE and F. R. COLE. 1986. Distribution of ants and their effects on endemic biota of Haleakala and Hawaii Volcanoes National Parks: a preliminary assessment. Proc. Sixth Conf. in Nat. Sci., Hawaii Volcanoes Nat. Park. Coop. Nat. Park Resources Study Unit, Univ. of Hawaii Manoa, Honolulu. p. 39-52.

MUIR, F. 1915. Review of the autochthonous genera of Hawaiian Delphacidae. Proc. Hawaii Entomol. Soc. 3:180, 203.

PEMBERTON, C. E. and H. F. WILLARD. 1918. A contribution to the biology of fruit fly parasites in Hawaii. J. Agric. Res. 15:419-465.

PERKINS, R. C. L. 1913. Introduction. Fauna Hawaiiensis 1(6):i-ccxxvii (xli-xlii).

PHILLIPS, J. S. 1934. The biology and distribution of ants in Hawaiian pineapple fields. Hawaii Pineapple Prod. Sta. Bull. 15:1-57.

REICHENSPERGER, A. 1913. Zur kenntnis von Myrmecophilen aus Abessinien. Zool. Jahrb. Abt. Syst. 35:185-218.

REIMER, N. J. 1988. Predation on *Liothrips urichi* Karny (Thysanoptera:Phlaeothripidae): A case of biotic interference. Env. Entomol. 17:132-134.

REIMER, N. J. and J. W. BEARDSLEY. 1989. Effectiveness of hydramethylnon and fenoxycarb for control of *Pheidole megacephala* (Hymenoptera: Formicidae), an ant associated with mealybug wilt of pineapple in Hawaii. J. Econ. Entomol. (in press)

ROHRBACH, K. G., J. W. BEARDSLEY, T. L. GERMAN, N. REIMER and W.G. SANFORD. 1988. Mealybug wilt, mealybugs and ants on pineapple. Plant Disease 72:558-565.

SAKIMURA, K. and D. MATSUMOTO. 1965. The long-tailed mealybug returns. Pineapple Research Institute News 13:101-104.

SCHMITZ, H. 1915. Ent. Mitt. Deutsch Ent. Mus. Berlin, IV, p. 281.

SMITH, F. 1879. Descriptions of new species of aculeate Hymenoptera collected by the Rev. Thos. Blackburn in the Sandwich Islands. Jour. Linnean Soc. London 4:674-685.

SU, T. H., J. W. BEARDSLEY and F. L. MCEWEN. 1980. AC-217,300, a promising new insecticide for use in baits for control of the big-headed ant in pineapple. J. Econ. Entomol. 73:755-756.

SWEZEY, 0. H. 1909. Notes on the budmoth of sugarcane, etc. Hawaii Planters Rec. 1:133.

WHEELER, W. M. 1922. Ants of the American Museum Congo expedition. A contribution to the Myrmecology of Africa: Ants of the Belgian Congo. Bulletin of the American Museum of Natural History 45, 1139 pages, p.132.

WHEELER, W. M. 1934. Revised list of Hawaiian ants. B. P. Bishop Mus. Occas. Papers 10:3-21.

WILLARD, H. F. 1927. Some observations in Hawaii on the ecology of the mediteranean fruit fly, *Ceratitis capitata* Wiedeman, and its parasites. Proc. Hawaii Entomol. Soc. 6:505-515.

WILLIAMS, F. X. 1931. Handbook of the insects and other invertebrates of Hawaiian sugarcane fields. Experiment Station of the Hawaiian Sugar Planters Association, Honolulu, Hi. 400pp.

WILSON, E. 0. and R. W. TAYLOR. 1967. The ants of Polynesia (Hymenoptera:Formicidae). Pac. Insects Mon. No. 14, Bishop Museum Press, Honolulu, Hi. 109pp.

WONG, M. A. and T. T. Y. WONG. 1988. Predation of the mediteranean fruit fly and the oriental fruit fly (Diptera: Tephritidae) by the fire ant (Hymenoptera: Formicidae) in Hawaii. Proc. Hawaii Entomol. Soc. 28:169-177.

WONG, T. T. Y., D. 0. MCINNIS, J. I. NISHIMOTO, A. K. OTA, and V.C. S. CHANG. 1984. Predation of the mediteranean fruit fly (Diptera: Tephritidae) by the argentine ant (Hymenoptera: Formicidae) in Hawaii. J. Econ. Entomol. 77:1454-1458.

ZIMMERMAN, E. C. 1941. Argentine ant in Hawaii. Proc. Hawaii. Entomol. Soc. 11:108.

ZIMMERMAN, E. C. 1948. Insects of Hawaii. Vol. 1. Introduction. xv + 206p. Univ. of Hawaii Press, Honolulu.

ZIMMERMAN, E. C. 1953. Note on *Anoplolepis longipes* (Jerdon). Proc. Hawaii. Entomol. Soc. 15:10.

6

Ants That Have Pest Status in the United States

C.R. Thompson

INTRODUCTION

The 600 native species of ants in North America (north of Mexico) are very important parts of the ecosystem. Ants attracted scant attention from the general public until the introduction of the Argentine ant around the turn of the century. But it was the invasion of imported fire ants in the 1920's and their damage to agriculture and public health, which sharply increased the impact of pest ants on the United States. Smith (1965) cataloged eastern United States house ants, listing and discussing 49 species in 18 genera. Though long out of print, this book is still the best single reference on United States pest ants, with excellent drawings of many species.

In this chapter, I have updated the United States pest ant picture since Smith (1965). I have also expanded the focus to include the continental United States. The 17 genera discussed are placed roughly in order of economic importance, but this importance varies from region to region. Beneath each genus heading are common names, a short morphological synopsis and geographical distribution, followed by biological and economic importance. Control methods have not been included.

SOLENOPSIS SPECIES

The most distinct *Solenopsis* morphological characters are the 10-segmented antenna with a two-segmented club and the workers' highly developed polymorphism. Workers are reddish brown to black and 2-6 mm long. Females are similarly colored but larger, about 7-8 mm long.

Three fire ant species are native. The desert fire ant *S. aurea* occurs in southwestern desert areas (Creighton 1950), while the southern fire ant, *S. xyloni*, is found from California through the southern United States into South Carolina. *S. geminata*, the tropical fire ant, ranges from Texas to South Carolina (Smith 1979).

Two foreign fire ant species have invaded the United States. The black imported fire ant, a native of Brazil, Uruguay and Argentina, was introduced into the United States through the port of Mobile around 1918 (Lofgren 1986a). This species, identified as *S. saevissima richteri*, has a limited range, occurring only in northeastern Mississippi and northwestern Alabama (Vinson and Greenberg 1986). But, in the late 1930's a much more successful red fire ant entered the U.S. through Mobile. The two color forms were initially ascribed to intraspecific variation until Buren (1972) showed that the red and black forms were separate species and named the red imported fire ant *S. invicta* (See Vander Meer and Lofgren this volume). The red imported fire ant has nearly displaced the other imported species, *S. richteri*, although the discovery of a viable *S. invicta* x *S. richteri* hybrid has complicated the situation (Vander Meer et al. 1985). *S. invicta* occupies most southern states and infestations are occasionally discovered in Arizona and California. The ant might spread up the west coast as far as Washington (Vinson and Sorensen 1986).

The damage fire ants caused to crops was noticed as early as 1935 (Adams 1986). By 1948, imported fire ants damaged more than $500,000 worth of corn, potatoes, beans and cabbage in Mississippi alone (Adams 1986). Not until the USDA's early 1950's fire ant survey in the South was a link firmly established between the incredibly rapid expansion of the ants and the movement of commercial nursery stock (Lofgren 1986a,b). Fire ants girdling and killing of young citrus trees results in tree replacement costs of $175 to 600/ha (Banks et al. In press). Despite at least $172 million spent on fire ant control from 1957 to 1981 (Lofgren 1986a), fire ants have continued their expansion and damage.

The name "fire ant" alludes to their painful sting. Fire ant venom is composed mostly of alkaloids, but has several allergenic antigens in the protein component (Lofgren 1986b). Those people who are hyperallergic to fire ant venom have the highest risk. There are an estimated 1,460 new fire ant allergy cases each year (Lofgren 1986b).

SOLENOPSIS MOLESTA

This ant is commonly known as the thief ant, grease ant, piss ant or sugar ant. When Thomas Say (1836) found these ants eating meat he had left to thaw on his window sill, he gave them their specific epithet: *molesta*. Although these ants belong to the same genus as the bigger fire ants, their habits and pest status are so different that I have placed them in a separate section. These smallest of American ants are 1.5-2 mm long, yellow to reddish brown, and monomorphic with the same 2-segmented antennal club as their fire ant cousins. Queens are similarly colored but with a length of ca. 4-5 mm.

Members of the thief ant group occur in every state. Their common name derives from some species' habit of living in nests of larger ants and feeding on host larvae.

The thief ant's small size and retiring habits have hampered biological study. Alates occur in Kansas nests in July, while September flights were seen in Massachusetts (McColloch and Hayes 1916). McNamara (1945) reported that queens carried workers on their nuptial flights, possibly to help found new nests. Colonies contain up to a few thousand workers (Smith 1965). In the laboratory, queens laid 27-387 eggs/day, with workers developing from egg to adult in 52 days or more -

much longer during winter (McColloch and Hayes 1916). These ants are highly predacious, and nearly omnivorous (Smith 1965).

S. molesta is often blamed for economic damage caused by other species (Thompson 1988), particularly pharaoh ants, which are mistaken for thief ants. Mallis (1982) called *S. molesta* an important house-infesting ant, difficult to control and persistent, yet McColloch and Hayes (1916) had no house infestation reports for Kansas. Whatever thief ant species McColloch and Hayes (1916) actually studied damaged garden seed, corn, strawberries, vegetables, kafir, milo, cane and feterita. At least some thief ant species tend Homoptera, and act as the intermediate host of poultry tapeworms (Smith 1965).

Thief ants most often invade homes during hot weather, nesting in cracks and crevices or the interior of walls and cabinets where they eat household foods (Smith 1965). There are no reports of bites or stings.

CAMPONOTUS SPECIES

Ants of this genus, carpenter ants or wood ants, are polymorphic with variable coloration even within species. Workers range from six to ten mm long, while the queen of pest species *C. pennsylvanicus*, is one of the largest North America ants (ca. 15 mm long). They are called carpenter ants because many species hollow out galleries in wood.

Carpenter ants occur throughout the United States (Smith 1979) (See Akre and Hansen; Hansen and Akre; and Fowler in this volume for biological and economic information).

MONOMORIUM PHARAONIS

This ant, commonly called the pharaoh ant or little red ant, is 2 mm long and reddish-brown with a mostly darker gaster. Queens have similar coloration, but average 4 mm in length. With the exception of a 3-segmented antennal club, the pharaoh ant resembles a thief ant.

The type specimen described by Linnaeus in 1758 came from Egypt, which may account for its name. The species probably originated in the North African-Middle Eastern region, but has since spread all over the world (Edwards 1986).

We know more about pharaoh ant biology than that of most pest ants. [Unless indicated, the comments below are from Edwards (1986)]. Pharaoh ants build unstructured nests with members of different nests showing no aggression to each other. The polydomous colonies can contain several hundred queens, each queen laying up to 35 eggs/day or about 4,500 eggs in her lifetime. Development from egg to adult worker requires 36 days, while queens and males require 41 days. Workers live 9-10 weeks, males 3-8 weeks and queen longevity is a year or more. Colonies can be large; in six weeks, 350,000 workers and over 800 females and males were collected from one colony.

Since winged pharaoh ant sexual forms can not fly well and mate in the nest; there are no swarms, or mating flights. Females lose their wings shortly after mating. Fertilized single queens produce a few small workers, but never establish colonies. New colonies begin through budding. A group of workers simply carries brood to a new nest site. An accompanying queen is unnecessary since workers can rear new queens and males from brood. Food or water shortages, temperature extremes or overcrowding can cause colony budding or migration.

Pharaoh ants make extensive use of chemical trails to retrieve food sources. Via their sting, workers deposit faranal from the Dufour's gland on the ground. They produce a strong repellent substance from the poison gland and probably have other pheromones and defense secretions. Workers forage 24 hours a day and are not deterred by hot weather.

In non-tropical climates, pharaoh ants cannot live without artificial heating and thus are "invariably associated with human habitation" (Edwards 1986). They are a serious indoor pest in large buildings in the southern United States (Ehmann 1982). Fortunately, pharaoh ants are unable to sting, nor can their mandibles cut human skin (Edwards 1986). This ant can nest almost anywhere and is practically omnivorous even attacking insect collections and chewing holes in silk, rayon or rubber (Smith 1965).

The discovery of pharaoh ants in more than 10% of Britain's hospitals was alarming; at least 12 pathogenic organisms occur on pharaoh ants collected from hospitals, including *Salmonella, Pseudomonas* and *Staphylococcus*. Because of their ability to squeeze through tiny cracks and spaces the ants have infiltrated sterile hospital supplies, intravenous fluid sets and patient wound dressings.

In 1965 M.R. Smith wrote, the pharaoh ant is "without a doubt the most persistent and difficult of all our house-infesting ants to control or eradicate". Almost 25 years later this statement is still true (See Williams; and Eichler this volume, for economic and control information).

MONOMORIUM MINIMUM

Commonly called the little black ant, workers of this species are monomorphic, usually black, sometimes dark brown, 1.5 mm long, and have a 12-segmented antenna ending in a 3-segmented club. The queen is also dark brown to black and about 4 mm long.

Native to the U.S., the little black ant ranges from Ontario south to Florida and west through Montana and Colorado to California. It occurs rarely, if at all, in the Pacific northwest (Smith 1979).

Little black ants construct small crater nests of fine soil (Mallis 1982) in exposed or covered areas and in rotting wood (Smith 1965). The large nests contain many queens, with probable mating flights from June to August (Smith 1965). This ant species successfully competes with fire ants for food and territory (Urbani and Kannowski 1974). This ant is predacious, tends Homoptera for honeydew, feeds on plant exudates (possibly pollen), and can kill bird hatchlings (Smith 1965).

Once a common house-infesting species (Smith 1965), the little black ant's pest status may now be much lower (Truman et al. 1976). Its economic impact derives from its habit of nesting in lawns, woodwork and building foundations, as well as from its feeding on many household foods (Smith 1965). There are no reports of bites or stings.

IRIDOMYRMEX HUMILIS

Workers of this ant, commonly called the Argentine ant, are 2.2-2.8 mm long and light to medium brown with a darker gaster. Queens are similarly colored, but much larger (4.5-5.0 mm long). The bicoloration distinguishes them from the monocolored odorous house ant. They have a small, but not vestigial, petiole which is not hidden by an overhanging gaster.

This ant is native to Brazil or Argentina. It probably hitched a ride in a Brazilian coffee shipment, and was already well established in New Orleans when discovered in 1891 (Newell and Barber 1913). These ants now occur throughout the south, as well as in Arizona, Illinois, Maryland, Missouri, Oregon and Washington (Mallis 1982). In California, pest control personnel encounter Argentine ants more often than any other ant (Ehmann 1982). Commerce, heavy rains and floods have facilitated their spread (Smith 1965).

Knowledge of the Argentine ant life cycle is extensive due to Newell and Barber's (1913) work (unless noted, further comments are from their bulletin). Argentine ants nest anywhere near food and water. These ants will dig to a depth of 32 cm during dry or cold weather. Colonies migrate frequently and can move to a nearby location in a matter of minutes.

Argentine ants have a possibly unique alate production/inhibition system, at least partly based on diet, which results in male production year round (Passera et al. 1988, Bartels 1988). Alates mate in the nest, thus winged forms are seldom seen, although males will fly to lights (Newell and Barber 1913).

Opinions differ as to whether Argentine ant queens can found nests alone. Smith (1965) stated that they could, but Newell and Barber (1913) never observed claustral founding or even brood tending by queens. Keller and Passera (1988) found that Argentine ant queens were light, small, and cheap, energy-wise, to produce while species with claustral founding had heavy, high energy, expensive females. Their observations support the anti-claustral side of the argument.

Large Argentine ant nests have hundreds of queens and most new colonies are founded by migration or budding. Workers have never been observed rearing a new queen from brood. A queen can lay 60 eggs/day, but oviposition stops if temperatures drop below 20°C. The worker develops from egg to adult in about 74 days and lives at least 10-12 months. Although predacious, carnivorous, and granivorous, these ants prefer nectar and honeydew, which comprises 99% of their diet (Markin 1970a). They tend aphids, carrying them from one plant to another. They forage along regular paths, visible even on asphalt (Ribble 1978). During the winter, several colonies may combine and during extreme cold weather become temporarily inactive or dormant (Smith 1965). Workers forage between 5-35°C (Markin 1970b), but only in daylight during cold weather. These ants can out-forage, out-reproduce and out-fight native species in many habitats (Newell and Barber 1913, Erickson 1971), particularly in disturbed areas with permanent water supplies (Ward 1987).

In urban areas during winter, Argentine ants move further down in the soil or closer to heated buildings (Ehmann 1982). Ehmann noted that in hot regions the ants take advantage of regularly watered yards, but excessively dry or wet conditions cause them to invade homes, particularly in the summer (Ehmann 1982). In urban areas with heavy infestations, the ants may infest almost every residence (Truman et al. 1976). Once inside, these ants get into everything, even refrigerators (Newell and Barber 1913). Fortunately, they cannot sting, but they will bite (Newell and Barber 1913).

Argentine ants also attack agricultural products. They decrease the numbers of natural enemies of aphids and other plant pests (Smith 1965; Moreno et al. 1987; Haney et al. 1987). Smith (1965) noted that they attack the buds or fruit of citrus and fig trees. They harvest seeds and adversely affect cabbage, sugar cane, cotton, and corn--mostly because they tend homopteran pests. The ants build protective coverings over homoptera, especially scales. They have been accused of destroying ant-plant mutualisms (Bond and Slingsby 1984), and of nectar theft from flowers normally visited by honeybees (Buys 1987).

TAPINOMA SESSILE

Known as the odorous house ant, workers of this native species have a 12-segmented clubless antenna and are 2.4-3.2 mm long. They are entirely brown, brownish-black or black. Queens are similar in color but larger (3.7-4.3 mm long). This ant has been confused with the Argentine ant, but odorous house ant workers, although nearly the same size, are darker brown with no bicoloration (see key in Harada this volume).

Odorous house ants occur in the lower 48 states (but rarely in southwestern deserts), from sea level to 3,200 m (Wheeler and Wheeler 1973). The ants' common name is derived from a "putrid coconut-like odor" (Ehmann 1982) caused by butyric acid, released when the ant is crushed (Creighton 1950).

Odorous house ants build shallow, impermanent nests beneath objects in virtually all habitats (Smith 1928; Smith 1965). This extremely restless species relocates about every 25 days (Smallwood 1982). Mating generally occurs in the nest, but mating flights have been sighted in May (Mississippi) and June-July (Michigan) (Smith 1965). New colonies are established by a single queen, or by budding (Smith 1928) (See Harada in this volume for further biological and economic information).

TAPINOMA MELANOCEPHALUM

Blackheaded ant workers--whose movements are rapid and erratic--range from 1.3 to 1.5 mm in length, with dark brown head and thorax and with yellow-tan legs and gaster. Queens have the same coloration, but are slightly larger--2.5-3.0 mm long.

Blackheaded ants probably originated in Africa or the Orient, but have been so widely spread by commerce that the exact home is unknown (Smith 1965). In the United States, these ants are found outside only in south Florida, with populations in greenhouses and other structures in northern states (Smith 1979) (See Harada for more information on the biology and economics of this species).

ATTA TEXANA AND *A. MEXICANA*

These ants are leaf-cutting ants, also called the town, night, fungus, red town, cut, parasol, and pack ants. Workers are orange-brown to dark brown, extremely polymorphic and 4-14 mm long. A distinct character is the three or more pairs of spines on the thoracic dorsum. The queens have similar coloration but are extremely large (ca. 18 mm long).

Atta texana and *A. mexicana* are native to the U.S.. They occur in southern Texas, Louisiana and Arizona (Smith 1979).

Leaf-cutting ant nests are usually found only in well-drained sand or loam on slopes facing south or west (Smith 1965). *A. texana* prefers areas near tree-bordered streams and surmounts its nests with craters ca. 38 cm in diameter and 13-15 cm high (Creighton 1950). Smith (1965) notes that a single nest may occupy 4,500 m^2 with 1,000 or more entrances (thus the name "town ant"). *A. mexicana* nests average 1-2/ha and can have a surface area of 8,000 m^2 (Mintzer 1979). With the help of a bulldozer, 911 chambers were found in a partially excavated *A. texana* nest more than 12 m deep (Smith 1965). Colonies may occupy the same location for 60-70 years (Smith 1965).

A. texana workers forage by day Sept-May, and by night April-Oct at temperatures ranging from 10-31°C (Waller 1986). The nests have distinct, vegetation-free foraging trails up to 600 m long (Smith 1965). The ants move the fungus gardens into chambers below 8 m. in winter (Moser 1963). Palatability, cuttability and season play a role in the plant material harvested by *A. texana* (Waller 1986). *Atta* workers will forage for many materials including seeds, caterpillar droppings, Spanish moss, corn, cornmeal, flour, rice, peas, wheat, oats, household foods, chicken feed, sugar, beans, ground coffee and chewing tobacco (Smith 1965). These ants can strip the leaves from a tree overnight (Truman et al. 1976). Although they do not sting, a large worker can draw blood with its bite (Smith 1965). Fortunately, the other leaf-cutting ant species live south of the United States. (See Mintzer and other leaf-cutting ant chapters in this volume).

WASMANNIA AUROPUNCTATA

This small ant (1.5-2.0 mm long), commonly called the little fire ant, is monomorphic and yellow. Females are the same color as workers but ca. 4 mm long. The 11-segmented antennae, bearing 2-segmented antennal clubs, can be folded into antennal scrobes on the sides of the head. The propodeum has a pair of spines.

The little fire ant is an aggressive, neotropical species found in south Florida (Spencer 1941) and California (Smith 1979).

Little fire ants do not build a nest but use any sheltered location, including exposed or covered soil, rotten wood and plant cavities (Smith 1965). These ants aggregate and have large, polygyne nests, which results in blurred colony boundaries and indistinct populations (Smith 1965). They nest deeper during droughts and move to trees during floods (Smith 1965). They tend aphids for honeydew and are predacious (Smith 1965).

In homes little fire ants are attracted to household fatty and oily foods like meats and peanut butter but not sweets (Fernald 1947). The ants have a painful sting but only sting when trapped between the body and clothing or some other object (Smith 1965).

Little fire ants are an agricultural pest, particularly of tree crops such as citrus, where they sting fruit pickers and drive out less aggressive ant species (Spencer 1941). Clark and his co-workers (1982) used baits to discover that *Wasmannia* workers rapidly recruited large numbers of workers, foraged 24 hours a day and showed no intraspecific aggression. Fortunately, as Smith pointed out in 1929, this ant is sensitive to cold temperatures and has yet to expand into non-subtropical areas of the United States (See Ulloa-Chacon and Cherix this volume).

LIOMETOPUM OCCIDENTALE

The velvety tree ant is easily identified by its two-tone coloration: a reddish brown thorax and somewhat darker and more brownish or blackish head and gaster. With some magnification, a covering of appressed hairs can be seen on the body, resulting in this species common name. Workers are polymorphic and 3-5 mm long; queens look similar but are larger.

Velvety tree ants are strictly a western U.S. genus, occurring no further east than Wyoming (Smith 1979).

Velvety tree ants nest in tree stumps and cavities, under tree bark, or under stones or mulch at tree bases (Truman et al. 1976; Ehmann 1982). Large nests, often located in oaks, cottonwoods or poplars, are sometimes lined with a paperlike substance (Truman et al. 1976; Ehmann 1982; Creighton 1950). These predacious ants tend honeydew-producing insects. They have a strong recruitment system, and can quickly assemble large numbers of workers (Ehmann 1982). No in-depth study has been made of this ant's biology.

These ants infiltrate clothing but do not always bite immediately. After they bite, they may inject a fluid into the wound, causing a "severe stinging sensation" which may persist for some time (Ehmann 1982). The ants are also a problem at outdoor events. When crushed, this ant has the same unpleasant odor as *T. sessile* (Ehmann 1982). It invades homes for shelter and warmth where it will feed on various foods (Ehmann 1982).

TETRAMORIUM CAESPITUM

The pavement ant, *T. caespitum*, is light brown to black and 2-3 mm long, with pale legs and antennae. Workers are monomorphic. There are parallel striae or lines on the head and thorax and two spines on the propodeum. The queen looks similar, but is larger (6 mm long).

Pavement ants are found from Ontario south to Tennessee, west to Nebraska, Missouri, Washington, Nevada and California. This species may have arrived with early colonists from Europe (Smith 1965), but Creighton (1950) argues that it is a native species.

T. caespitum nests are found under stones, next to buildings, in pavement cracks or in rotting wood (Truman et al. 1976; Smith 1965). Alates are present all year but mostly during June-July (Smith 1965). Colony founding is claustral; established queens lay 5-20 eggs/day and from egg to worker requires 43-63 days (Bruder and Gupta 1972).

Elsewhere the pavement ant is generally a rural species, but in the United

States it has been a major household ant in many large cities (Smith 1965; Mallis 1982). This species can nest in house walls, under floors and in insulation (Truman et al. 1976). It is a nuisance with slab-on-grade construction, invading homes through slab cracks or other natural openings (Truman et al. 1976). These ants move slowly, but are good foragers, especially for sweets and greases (Truman et al. 1976). This ant is almost omnivorous, and feeds on seeds, or girdles, scars, and scrapes the roots or stems of beets, cabbage, carrots, eggplant, lettuce, parsley, pepper, potatoes, radish, sugar beets, tomato, turnip, *Gaillardia, Coreopsis*, and aster (Smith 1965). It also tends Homoptera, mostly below ground (Smith 1965). It is not known if they bite or sting. They are host for two poultry tapeworm species (Smith 1965).

POGONOMYRMEX SPECIES

Commonly called harvester ants, *Pogonomyrmex* workers are large (5-10 mm long) and light to ferrugineous red or brown; the gaster may be dark brown to black. The queen is similarly colored, but ca. 12 mm long. Their most salient morphological characteristics are four fringes of long hairs, two on the underside of the head, and one on each mandible.

Cole's (1968) book on the taxonomy and distribution of United States *Pogonomyrmex* lists 22 species. He states that the harvester ant group "nearly blankets" the western United States' dry areas, with one species occurring in the southeast.

The large size and interesting habits of harvester ants have resulted in many studies. Nests vary in size, but are usually cone or crater shaped. *P. texanus* nests can be 1 to 10 meters in diameter (Barnes and Nerney 1957). *Pogonomyrmex* always nest in open sunny locations (Cole 1968). Indeed, experimental shading *P. badius* nests always resulted in nest relocation (Carlson and Gentry 1973). Nests of *P. occidentalis* average 141 chambers per colony, with tunnels as deep as 2 m. (Rogers and Lavigne 1974). Dean (1905) stated that *P. occidentalis* nests had at least 10,000 workers. Among 11 excavated *P. californicus* colonies, the mean worker number was 4500, with a maximum of ca. 6300 (Erickson 1972). At least two species move their nests up to 12 m from the original site after one or more years (Van Pelt 1976). Foraging trails may extend 200 feet (Barnes and Nerney 1957).

Harvester ants carry seeds back to their nests where they store and live on them (Cole 1968). This granivorous habit accounts for their common name. Harvester ants are also scavengers, particularly of arthropods.

Four *Pogonomyrmex* species cause such serious damage to rangelands because of their seed-harvesting activities, that they have ESA-approved common names: the Florida harvester ant (*P. badius*), the red harvester ant (*P. barbatus*), the California harvester ant (*P. californicus*) and the western harvester ant (*P. occidentalis*) (Cole 1968). Barnes and Nerney (1957) reported a seed loss of one pound/nest from alfalfa fields containing five *P. barbatus* colonies/acre (33 nests/acre have been reported)

Creighton (1950) believed *P. badius* had the most painful sting of any ant. Cole (1968) noted that sting effects are not localized but can travel through the lymph system, and susceptible individuals may experience anaphylactic shock. These comments are supported by Schmidt and Blum (1978) who found *P. badius* venom more toxic to mice than any other insect venom, and almost as toxic as that of the worst snakes (See Andersen and other harvester ant chapters in this volume).

CREMATOGASTER SPECIES

These ants, sometimes called acrobat ants, often hold their gasters over their heads or thoraxes when alarmed. The heart-shaped gaster is flattened dorsally and convex ventrally. Color varies even within species from yellow through yellowish- or reddish-brown to black (Creighton 1950). Workers are 2.5-3.0 mm long and monomorphic or slightly polymorphic. Queens are similarly colored but larger, about 7-8 mm long.

Crematogaster species are found in all the lower 48 states.

Little is known about *Crematogaster* biology. Colonies live in logs and stumps, under stones, in decayed standing trees, plant cavities or insect galls (Creighton 1950; Smith 1965). Many construct carton sheds of plant parts or soil over aphids, which they tend. The ant colony itself may live in these shelters (Creighton 1950).

Acrobat ants nest in houses, often in or around door and window frames (Smith 1965). *C. cerasi* may be the most common house-invading species in some northeastern states (Smith 1965). The ants can get into houses from tree limbs or power lines (Mallis 1982) and may strip the rubber insulation from phone wires (Smith 1965). Despite Smith's (1965) claims, Ehmann (1982) states that these ants rarely, if ever, attack sound wood. They are omnivorous, but prefer sweet or high protein foods (Smith 1965). The ants forage day and night, may bite if handled, and can give off a strong, unpleasant odor. Although they serve as tapeworm hosts and transmit Azalea flower spot disease (Smith 1965), Ehmann (1982) does not consider them a serious household pest.

PARATRECHINA LONGICORNIS

"Crazy ant" workers are monomorphic, 2.3 mm long and dark brown. Females look similar, but are larger. Both legs and antennae are very long and proportionally vary from usual ant morphology.

The crazy ant probably came from the orient, but was spread around the world by commerce (Smith 1965). It occurs in the southeast from South Carolina through Florida to Texas and in California (Smith 1979).

The crazy ant got its name from its jerky, erratic movement. It tends homoptera for honeydew, preys on other insects and eats seeds or other plant materials (Smith 1965). It nests in trash, plant cavities, soil or rotten wood (Smith 1965). Colonies contain up to 2,000 workers and 40 queens (Mallis 1982).

In northern states, the crazy ant seeks refuge indoors to survive cold winters (Smith 1965). It nests in small crevices and searches for food throughout a building (Truman et al. 1976). It feeds on meats, greases and many sweets as well as some crop seeds (lettuce, tobacco and others; Smith 1965). They have no sting and do not bite (See Zenner-Polania regarding a related species in this volume).

LASIUS ALIENUS AND *L. NEONIGER*

The "cornfield ants" are small (2-3.5 mm long), rotund, brown to blackish brown ants with no other prominent morphological characters. Queens are much larger: ca. 6-8 mm long.

Both *Lasius* species have extensive ranges. *L. alienus* occurs over most of the country, but is rare or absent in the southern Rockies, Great Basin and Gulf Coast states. *L. neoniger* ranges over most of the United States but not into the Pacific Northwest (Smith 1965).

These two *Lasius* species nest in different habitats. *L. alienus* nests in deep forests, rotting wood and under stones, while *L. neoniger* prefers grassy roadsides, croplands and lawns (Smith 1965). *L. alienus* does not construct distinct mounds, but *L. neoniger* nests have a characteristic ca. 5 cm diameter mound and are common in sidewalk cracks. The two species have been frequently confused in the literature. They probably have similar biologies (Smith 1965).

L. neoniger apparently causes more harm in cornfields due to its nest preferences. Forbes (1908) found that the ants store corn root aphid eggs over the winter and place the aphids on host plants in the spring. In homes, these ants nest in lower house regions, such as foundations or basements (Smith 1965). Smith noted that both species feed on sweet foods or meats. They cannot sting, and bites have not been reported. Numerous *L. neoniger* nest craters can be a nuisance on golf courses (Smith 1965).

Surprisingly, *L. neoniger* can successfully defend its territory against imported fire ants (Bhatkar et al. 1972). Apperson and Powell (1984) noted a negative correlation between numbers of imported fire ants and *L. neoniger* in pastures.

PHEIDOLE SPP.

Ant of this genus, called big-headed ants are dimorphic with normal workers and large-headed soldiers. Color varies from dark brown through reddish shades to light yellow. Workers (1.5-2.5 mm long) and soldiers (2.5-4.5 mm long) are smaller than the queens (4.5-6.5 mm long). The 12-segmented antennae bear distinct, 3-segmented clubs at the tips. The propodeum bears a pair of spines.

Big-headed ants range from the southern United States to as far north as New York (Smith 1965). This group encompasses at least 75 species in the United States and Canada (Naves 1985). Most species prefer the warmer and drier sections of the country (Mallis 1982).

Pheidole species nest in exposed soil, under cover or in rotting wood (Creighton 1950). Most species eat seeds, and utilize the large-headed soldiers' heavy jaws and muscles to husk them (Creighton 1950). Colonies usually contain 200-300 individuals, but sometimes as many as 3,000 (Creighton 1950).

Big-headed ants seldom nest indoors, but may invade homes to forage on meats, greases, and breads (Smith 1965). *Pheidole bicarinata vinelandica*, *P. floridana* and *P. dentata* are common house-infesting species (Smith 1965). *Pheidole floridana*, *P. morissi* and *P. moerens* have been collected in Florida homes. There are no records of bites or stings. At least two *Pheidole* species are hosts of various poultry tapeworms (Smith 1965) (see Andersen; and Reimer and Beardsley in this volume for more information on Pheidole).

PRENOLEPIS IMPARIS

This species is called the honey, false honey or the small honey ant. Workers are monomorphic, shiny, dark brown, and 2.5-4 mm long. Queens are larger, 7-8 mm long, and shiny, dark brown except for a more reddish thorax.

The honey ant is a native species that occurs throughout the United States (Smith 1965).

Honey ants are probably the most cold-adapted of all our ants. Wheeler (1930) observed foraging from 2 -13°C. In Florida, workers cease above-ground activity from April to November (Tschinkel 1987), but in the midwest, they are inactive only 1-2 months (Talbot 1943). They prefer moist, well-shaded soils, and build a circular crater of small soil pellets with a central entrance. The nest may be up to ca. 100 cm deep and have 6-40 chambers (Smith 1965). Tschinkel (1987) found nests in Florida 2.5-3.6 m deep, with all chambers off one vertical tunnel. Worker numbers varied from 560 to more than 10,000. Workers like honeydew and their gasters can become engorged with plant sap or honeydew, hence the common name. Lynch et al. (1980) noted that they were the dominant species in a Maryland hardwood forest study and were very aggressive at protein or fatty food baits.

Honey ant alates overwinter in the colony, have mating fights in March or April, and found colonies claustrally (Talbot 1943). Tschinkel (1987) found that Florida colonies are mostly polygynous--with up to six queens--but northern colonies are reportedly monogynous (Dennis 1941). Workers live at least two years, and Florida colonies can reach 7 to 9 years in age, or more (Tshinkel 1987). Tschinkel (1987) found that workers become engorged with fats and other nutrients, not sugars. He called them corpulents rather than repletes, a term associated with honeydew storage. The one brood raised per year is produced months after the ants retreat below ground. They are able to produce the new crop of workers using the corpulents' reserves.

Honey ants do not bite or sting, but occasionally invade homes, particularly feeding on sweet foods (Smith 1965). They damage flower buds, possibly weaken beehives, and can transmit at least one plant disease (Smith 1965).

CONOMYRMA SPP.

Pyramid or lion ants have a distinctive, cone-shaped or pyramidal dorsal surface of the propodeum which gave them the former common name, while their aggressive behavior gave rise to the latter. Workers are monomorphic, and vary from dark brown through reddish brown to a clear, light yellow. Intraspecies color variation is great. Workers are 1.5-2 mm long, while queens have similar coloration but are 5.5-7 mm long.

Conomyrma species range across the western and southern United States, occurring as far north as North Carolina (Smith 1979).

Pyramid ants prefer open places like pastures and meadows where they construct crater-like nests 4-8 cm in diameter (Smith 1965). Smith noted that colonies contain less than a few thousand individuals and that alates may overwinter in the nest. Workers have a strong butyric acid odor when crushed (Creighton 1950). Workers are excellent foragers, fond of honeydew and predacious (Smith 1965). Some species can forage during very hot weather (Creighton 1950).

Pyramid ants have economic importance as house ants, feeding on many foods, particularly sweet substances (Smith 1965). These ants cannot sting, but may bite. In California, *Conomyrma* species have been largely displaced as house pests by the Argentine ant (Mallis 1982). Their many craters in lawns are unsightly nuisances (Smith 1965).

ACANTHOMYOPS CLAVIGER AND *A. INTERJECTUS*

Commonly called the smaller yellow, larger yellow, perfumed yellow or citronella ant, workers of these species are reddish brown and about 3.5-4 mm long. Females are similar in color, larger (6-8 mm long) and have enlarged femurs.

Acanthomyops claviger ranges west to Kansas, but *A. interjectus* occurs further west and southwest (Idaho, New Mexico) (Smith 1979). The citronella nickname derives from the odor given off by these ants when they are alarmed or crushed. Citronellal and citral are produced in their large mandibular glands and probably function as an alarm pheromone (Wilson and Regnier 1971).

Creighton (1950) described yellow ants as "remarkably unobtrusive," leading a mostly subterranean life, while feeding largely on honeydew from aphids and other root-feeding insects. Nests are under logs and stones, at stump bases or, more rarely, in open ground (Wing 1968). Females probably are temporary social parasites on other ant species (Wing 1968). Although Smith (1965) reported that alates overwinter in the nest, Talbot (1963) observed *A. claviger* mating flights from Sept-Nov. Talbot (1963) found that 40,000 alates flew from one nest on 22 flights while another nest produced over 63,000 alates in 4 flights. *A. interjectus* flights occur during June-July, with alates apparently produced the previous fall or winter (Wing 1968). No information is available on other aspects of colony biology.
Yellow ants do not bite and cannot sting. However, reproductives swarm in large numbers in basements and around house foundations (Smith 1928), and are frequently mistaken for termites (Smith 1965). These ants occasionally invade homes but have never been seen eating household foods (Smith 1965; Truman et al. 1976).

CONCLUSION

The 17 genera and approximately 35 pest ant species discussed in this chapter represent only a fraction of the 600-plus North American ant species. Yet these ants have had considerable negative impact on mans activities in both rural and urban areas. We must not loss sight of the beneficial and neutral ant species as we attempt to control the pest species.

REFERENCES CITED

ADAMS, C.T. 1986. Agricultural and medical impact of the imported fire ants. In C.S. Lofgren and R.K. Vander Meer (eds.), Fire ants and leaf-cutting ants biology and management, Westview Press, Boulder.

APPERSON, C.S. and E.E. POWELL. 1984. Foraging activity of ants (Hymenoptera: Formicidae) in a pasture inhabited by the red imported fire ant (*Solenopsis invicta*). Fla. Entomol. 67: 383-392.

BANKS, W.A., C.T. ADAMS and C.S. LOFGREN. 1990. Economic loss of young citrus trees associated with foraging activity of the red imported fire ants, *Solenopsis invicta* Buren. In press.

BARNES, O.L. and N.J. NERNEY. 1957. The red harvester ant and how to subdue it. USDA Farmer's Bull. 1668: 11 p.

BARTELS, P.J. 1988. Reproductive caste inhibition by Argentine ant queens: New mechanisms of queen control. Insectes Soc. 35: 70-81.

BHATKAR, A.W.H. WHITCOMB, W.F. BUREN, P. CALLAHAN and T. CARLYSLE. 1972. Confrontation behavior between *Lasius neoniger* (Hym: Formicidae) and the imported fire ant. Environ. Entomol. 1: 274-279.

BOND, W. and P. SLINGSBY. 1984. Collapse of an ant-plant mutualism: the Argentine ant (*Iridomyrmex humilis* and myrmecochorus Proeteaceae. Ecology 65: 1031-1-37.

BRUDER, K.W. and A.P. GUPTA. 1972. Biology of the pavement ant, *Tetramorium caespitum* (Hymenoptera: Formicidae). Ann. Entomol. Soc. Amer. 65: 358-367.

BUREN, W.F. 1972. Revisonary studies on the taxonomy of the imported fire ants. J. Georgia Entomol. Soc. 7: 1-26.

BUYS, B. 1987. Competition for nectar between Argentine ants (*Iridomyrmex humilis*) and honeybees (*Apis mellifera*) on black ironbark (*Eucalyptus sideroxylon*). S. Afr. J. Zool. 22: 173-174.

CARLSON, D.M. and J.B. GENTRY. 1973. Effects of shading on the migratory behavior of the Florida harvester ant, *Pogonomyrmex badius*. Ecology 54: 452-453.

CLARK, D.B., C. GUAYASAMIN, O. PAZMINO, C. DONOSO and Y PAEZ DE VILLACIS. 1982. The tramp ant, *Wasmannia auropunctata*: Autecology and effects on ant diversity and distribution on Santa Cruz Island (Galapagos, Ecuador). Biotropica 14:196-207.

COLE, A.C. 1968. *Pogonomyrmex* harvester ants A study of the genus in North America. Knoxville: Univ. of Tenn. Press, 222 p.

CREIGHTON, W.S. 1950. The ants of North America. Bull. Mus. Comp. Zool., Cambridge. 585 p. Dean, G. A. 1905. The mound-building prairie ant. J. Kans. Acad.Sci. 19: 164-170.

DENNIS, C.A. 1941. Some notes on the nest of the ant, *Prenolepis imparis* Say. Ann. Entomol. Soc. Amer. 34: 83-86.

EDWARDS, J.P. 1986. The biology, economic importance, and control of the Pharaoh's ant, *Monomorium pharaonis* (L.). In S. B. Vinson (ed.), Economic Impact and Control of Social Insects, Praeger Scientific, New York.

EHMANN, N.R. 1982. Van Water and Rogers pest control library Vol. IV: Dooryard pests. 126 p.

ERICKSON, J.M. 1971. The displacement of native ant species by the introduced Argentine ant *Iridomyrmex humilis* Mayr. Psyche 78: 257-266.

ERICKSON, J.M. 1972. Mark-recapture techniques for population estimates of *Pogonomyrmex* ant colonies: an evaluation of the P^{32} technique. Ann. Entomol. Soc. Amer. 65: 57-61.

FERNALD, H.T. 1947. The little fire ant as a house pest. J.Econ. Entomol. 40: 428.

FORBES, S.A. 1908. Habits and behavior of the corn-field ant, *Lasius niger americanus*. Ill. Agric. Exp. Sta. Bull. 131: 31-44.

GORDON, D.M. 1984. The harvester ant (*Pogonomyrmex badius*) midden: refuse or boundary? Ecol. Entomol. 9: 403-412.

GRACE, J.K., D.L. WOOD and B.W. GRUNBAUM. 1986. Effect of Argentine ant contamination on ABO blood typing of human saliva samples. Bull. Etnomol. Soc. Amer. 32: 147-149.

HANEY, P.B., R.F. LUCK and D.S. MORENO 1987. Increases in densities of the citrus red mite, *Panonychus citri* (Acarina:Tetranychidae), in association with the Argentine ant, *Iridomyrmex humilis* (Hymenoptera: Formicidae), in southern California (USA) citrus. Entomophaga 32: 49-58.

KELLER, L. and L. PASSERA. 1988. Energy investment in gynes of the Argentine ant *Iridomyrmex humilis* (Mayr) in relation to the mode of colony founding in ants (Hymenoptera:Formicidae). Int. J. Invertebr. Reprod. Dev. 13: 31-38.

LOFGREN, C.S. 1986a. History of imported fire ants in the United States. In C.S. Lofgren and R.K. Vander Meer (eds.), Fire ants and leaf-cutting ants biology and management, Westview Press, Boulder.

LOFGREN, C.S. 1986b. The economic importance and control of imported fire ants in the United States. In S. B Vinson (ed.), Economic Impact and Control of Social Insects, Praeger Scientific, New York.

LYNCH, J.F., E.C. BALINSKY and S.G. VAIL. 1980. Foraging patterns in three sympatric forest ant species, *Prenolepis imparis, Paratrechina melanderi and Aphaenogaster rudis* (Hym.: Formicidae). Ecol. Entomol. 5: 353-372.

MALLIS, A. 1982. Handbook of Pest Control. Cleveland: Franzak and Foster Co. 1101 p.

MARKIN, G.P. 1970a. Foraging behavior of the Argentine ant in a California citrus grove. J. Econ. Entomol. 63: 740-742.

MARKIN, G.P. 1970b. The seasonal lifecycle of the Argentine ant, *Iridomyrmex humilis* (Hym: Formicidae) in southern California. Ann. Entomol. Soc. Amer. 63: 1238-1242.

McCOLLOCH, J.W. and W.P. HAYES. 1916. A preliminary report on the life economy of *Solenopsis molesta* Say. J. Econ. Entomol. 9: 23-28.

McNAMARA, C. 1945. A note on the swarming of *Solenopsis molesta* Say. Can. Entomol. 77: 40.

MINTZER, A. 1979. Foraging activity of the Mexican leafcutting ant, *Atta mexicana*, in a Sonoran Desert habitat (Hymenoptera, Formicidae). Insectes Soc. 26: 364-372.

MORENO, D.S., P.B. HANEY and R.F. LUCK. 1987. Chlorpyrifos and diazinon as barriers to Argentine ant (Hymenoptera:Formicidae) foraging on citrus trees. J. Econ. Entomol. 80: 208-214.

MOSER, J.C. 1963. Contents and structure of *Atta texana* nest in summer. Ann. Entomol. Soc. Amer. 56: 286-291.

NAVES, M.A. 1985. A monograph of the genus *Pheidole* in Florida (USA) (Hym: Formicidae). Insecta Mundi 1: 53-90.

NEWELL, W. and T.C. BARBER. 1913. The Argentine ant. USDA Bur. Entomol. Bull. 122: 98 p.

PASSERA, L., L. KELLER AND J.-P. SUZZONI. 1988. Control of brood male production in the Argentine ant *Iridomyrmex humilis* (Mayr). Insectes Soc. 35: 19-33.

RIBBLE, D.W. 1978. Visible trails of the Argentine ant. J. Kans. Entomol. Soc. 51: 796-797.

ROGERS, L.E. 1974. Foraging activity of the western harvester ant in the shortgrass plains ecosystem. Environ. Entomol. 3:420-424.

ROGERS, L.E. AND R.J. LAVIGNE. 1974. Environmental effects on western harvester ants in the shortgrass plains ecosystem. Environ. Entomol. 3: 994-997.

SAY, T. 1836. Descriptions of new species of North American Hymenoptera, and observations on some already described. Boston J. Nat. Sci. 1: 210-305.

SCHMIDT, J.O. and M.S. BLUM. 1978. Pharmacological and toxicological properties of harvester ant, *Pogonomyrmex badius*, venom. Toxicon 16: 645-652.

SMALLWOOD, J. 1982. Nest relocations in ants. Insectes Soc. 29: 65: 138-147.

SMITH, D.R. 1979. Formicoidea. p.1323-1867 In Catalog of Hymenoptera in America North of Mexico. Vol. 2. Washington, D.C.: Smithsonian Inst. Press. SMITH, M.R. 1928. The biology of *Tapinoma sessile* Say, an important house-infesting ant. Ann. Entomol. Soc. Amer. 21: 307-330.

SMITH, M.R. 1965. House-infesting ants of the eastern United States their recognition, biology and economic importance. USDA Tech. Bull. 1326:105p.

SPENCER, H. 1941. The small fire ant *Wasmannia* in citrus groves--a preliminary report. Fla. Entomol. 24: 6-14.

TALBOT, M. 1943. Population studies of the ant, *Prenolepsis imparis* Say. Ecology 24: 31-44.

TALBOT, M. 1963. Local distribution and flight activities of four species of ants of the genus *Acanthomyops* Mayr. Ecology 44: 549-557.

THOMPSON, C.R. 1988. A monograph of the thief ants, *Solenopsis (Diplorhoptrum)* of Florida. Univ. of Florida Dissertation, 114 p.

TRUMAN, L.C., G.W. BENNETT, and W.L. BUTTS. 1976. Scientific guide to pest control operations. Cleveland: Harvest Publishing Co. 276 p.

TSCHINKEL, W.R. 1987. Season life history and nest architecture of a winter-active ant, *Prenolepis imparis*. Insectes Soc. 34: 143-164.

URBANI, C.B. and P.B. KANNOWSKI. 1974. Patterns in the red imported fire ant settlement of a Louisiana pasture: some demographic parameters, interspecific competition and food sharing. Environ Entomol. 3: 755-760.

VANDER MEER, R.K., C.S. LOFGREN and F.M. ALVAREZ. 1985. Biochemical evidence for hybridization in fire ants. Fla. Entomol. 68: 501-506.

VAN PELT, A. 1976. Nest relocation in the ant *Pogonomyrmex barbatus*. Ann. Entomol. Soc. Amer. 69: 493.

VINSON, S.B. and L. GREENBERG. 1986. The biology, physiology, and ecology of imported fire ants. In S.B. Vinson (ed.), Economic Impact and Control of Social Insects, Praeger Publishers, New York.

VINSON, S.B. and A.A. SORENSEN. 1986. Imported fire ants: life history and impact. Texas Dept. of Agric. 28 p.

WALLER, D.A. 1986. The foraging ecology of *Atta texana* in Texas. In C.S. Lofgren and R.K. Vander Meer (eds.), Fire Ants and Leaf-Cutting Ants Biology and Management, Westview Press, Boulder.

WARD, P.S. 1987. Distribution of the introduced Argentine ant (*Iridomyrmex humilis*) in natural habitats of the lower Sacramento Valley (California, USA). Hilgardia 55: 1-16.

WHEELER, G.C. 1973. Ants of Deep Canyon. Deep Canyon Desert Research Center, Riverside: Univ. of California, 162 p.

WHEELER, W.M. 1930. The ant *Prenolepis imparis* Say. Ann. Entomol. Soc. Amer. 23: 1-126.

WHITFORD, W.G. and G. ETTERSHANK. 1975. Factors affecting foraging activity in Chihuahuan desert harvester ants. Env. Entomol. 4: 689-696.

WILSON, E.O. and F.E. REGNIER Jr. 1971. The evolution of the alarm-defense system in the formicine ants. Amer. Natur. 105: 279-289.

WING. M.W. 1968. Taxonomic revision of the nearctic genus *Acanthomyops* (Hymenoptera: Formicidae). Cornell Agric. Exp. Sta. Memoir 405: 173 p.

Systematics and Morphology

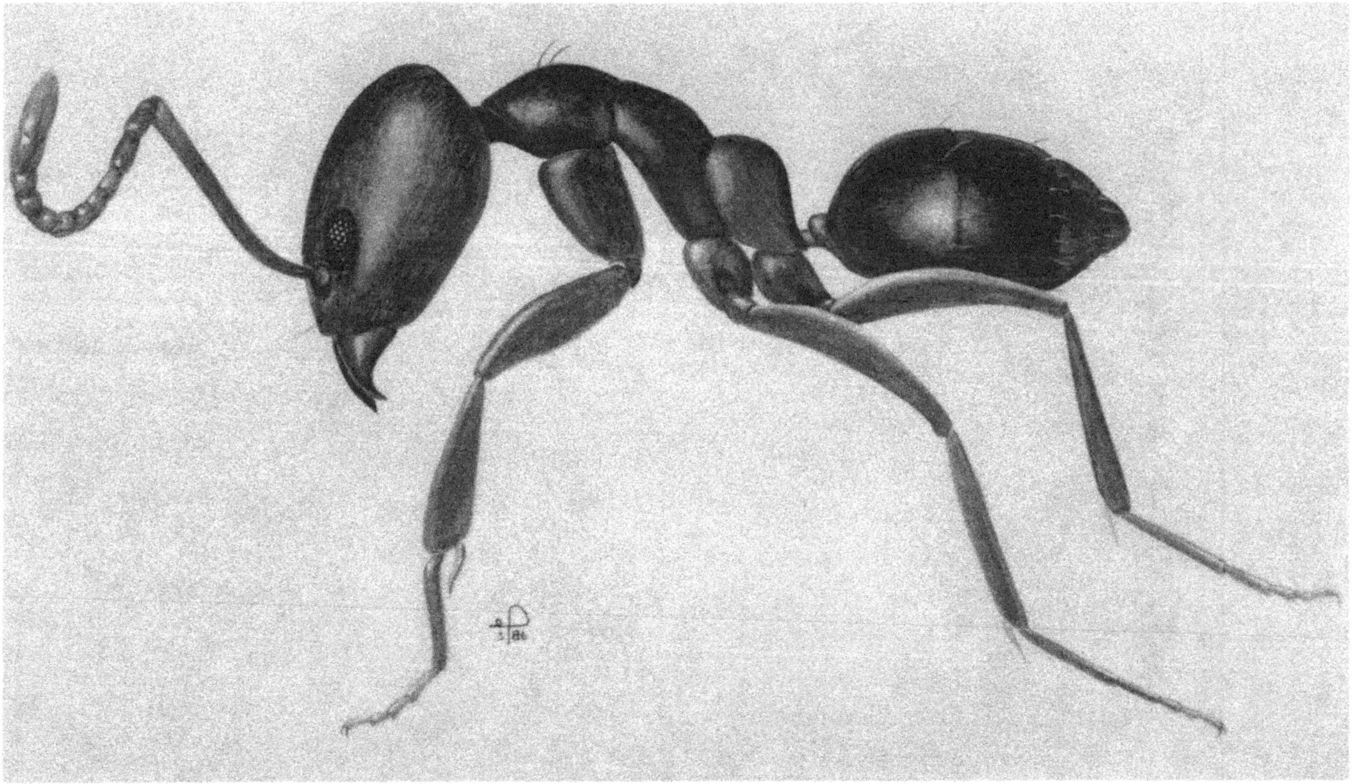

Iridomyrmex iniguus (x 28). Drawing by Eduardo Perez.

Overview

J.E. Lattke

There are an estimated 20,000 species of ants in approximately 300 genera. However, only about 8,000 of these have been named, thus we are ignorant of more than half of the probable number of species. Of the known genera only half have been the subject of a good taxonomic revision, despite the importance these insects play in relation with man (see this chapters in this volume) or in relation to natural environments (where they may represent 10-20% of the total animal biomass).

In recent years, ant systematics and taxonomy has been in a state of flux as concepts change, new are characters discovered, new techniques applied and more specimens collected and studied. Most of these changes have occurred since publication of Creighton's (1950) revision of the North American ant fauna. This book permits the separation of myrmecology from the penta- and quadrinomial system. The trinomen or subspecies concept was posteriorly dealt a crippling blow by Wilson and Brown (1953) so as to inhibit the formal description of ant subspecies.

Increased interest in methods and principles of systematics, along with the appearance of many taxonomic societies, theoretical journals and books (Hennig, 1950; Remane, 1956; Mayr, 1969; Sneath and Sokal, 1973; Wiley, 1981), has also affected ant systematics. Most subsequent work was pursued within the framework of evolutionary classification, but recent years have witnessed the ever growing influence of phylogenetic systematics (Bolton, 1987; Lutz, 1986; Agosti, 1989). The recognition of plate tectonics, and the realization of changing climate patterns during recent history have also influenced our understanding of phylogenies and geographic distributions.

Pismires (ants - of Germanic orgin) have numerous attributes that a taxonomist can study, but most work still relies on external morphology. Since this evidence is sometimes insufficient, other methods have come into use in systematic studies: biochemical, histological, genetic, cytological, geographic and ethological attributes. Vander Meer's contribution in this section offers biochemical, genetic, and behavioral evidence for the existence of a hybrid between *Solenopsis invicta* and *S. richteri*, and then examines the geographic distribution of the two species in their homeland. His conclusions exemplify how such evidence can clarify situations not discernible through external morphology alone, and also engender provoking questions on the role of geographic isolation and the treatment of hybrids. The usefulness of such characters is not only limited to interspecific problems, but also intraspecific differences such as chemical contrasts, and similarities between monogynous and polygynous colonies of *S. invicta* (Greenberg, this section).

A study of the morphology of *S. invicta*'s exocrine system (Billen, this section) reveals peculiarities apparently absent in most other pismires. The study goes beyond comparisons and also permits a view of relations between cell structure and function. Other usable attributes are karyotypes (Imai et. al., 1977), habitat temperature and humidity (Seifert, 1984b), and Malpighian Tubule counts (Brown, 1988). Even external morphology has not been fully exploited and the recognition of new characters are contributing to change (Bolton, 1988; Ward 1990). Comparative morphological studies can throw light on homology interpretations and help dictate the use of uniform names for morphological features. Such work is being carried out in other hymenopteran groups, and some ant problems could be solved through similar efforts.

Use of new techniques for analyzing attributes is another changing factor in ant systematics. Electrophoresis and chromatography are yielding useful data on alloenzyme variation, hydrocarbon profiles, and gene frequencies. Again, the contributions of Greenberg et al. and Vander Meer and Lofgren (this section) are examples of these developments. The use of electron microscopes (Billen, this section) and scanning electron microscopy (Taber et. al. 1987) have also proved useful. A more extensive use of statistics is another recent inclusion in ant systematics as is evident in Greenberg's treatment of hydrocarbon and venom profiles, or Seifert's (1984a) treatment of morphological data.

Introduction of new collecting techniques (litter sifting, berlese funnels, soil sample washing, and canopy fogging), especially in poorly known tropical areas have revealed an abundance of specimens previously thought rare and the appearance of new taxa. These novelties include genera that have broadened our view of the family, or in several cases have filled in blank spots between related taxa, causing synonymies or relations to become evident (Bolton, 1984; Lattke, 1990).

Even though ant systematics is undoubtably at a high point, a vast amount of labor remains. The higher classification has not been sorted yet and subfamilies still require definition. Internal anatomy is usually a more conservative condition and can be important at higher classification levels. Cruz-Landim (this section) presents data concerning the morphology of cephalic exocrine glands in a sample of 16 species from 6 subfamilies, and Caetano (this section) presents data on the morphology of the digestive tract and associated excretory organs of ants from different subfamilies. Such information shows a potential usefulness in establishing relationships between ant subfamilies. Some subfamilies badly need revision: Formicinae, Dolichoderinae and large parts of the Myrmicinae. The fungus-growing ants (Attini) are a large economically important group which need revision at the generic level and for each genus. The systematics of the regional fauna is presently unbalanced, with the Holarctic fauna much better known than those of other regions. Tropical ant fauna is poorly known and the rapid rate of habitat destruction in some critical areas makes sampling imperative now before it is too late. This same habitat alteration could promote the transformation of innocuous pismires into pest species. This book is proof that indepth research endeavours are mostly limited to the few species of immediate economic importance. Most studies involving modern (and usually expensive) techniques are produced in developed countries, reflecting the different social and economic contexts, and thus priorities, of scientific research. The natural history of many ants is still unknown and this kind of low-tech work should be encouraged in Third World countries. The absence of expensive apparatus does not make the work less valuable than that at other levels. Resident myrmecologists can carry out continuous research in tropical countries, as well as provide greater sampling diversity, than brief field trips by visiting colleagues.

In this section we have examples of the variety of attributes available for study and ways of examining them. Some can be treated by systematists themselves, while others need the assistance of specialists in other fields. We see a fusion of systematics with biology, and other auxiliary sciences. Our knowledge of formicid biology is taken from a small fraction of the estimated fauna and the more we know about them, the harder it is to stereotype and therefore classify ants.

REFERENCES

AGOSTI, D. 1989. Versuch einer phylogenetischen Wertung der Merkmale der Formicini, Revision der *Formica exsecta*-gruppe und List der Formicidae Europas. Ph.D. Thesis. Eidgenoessichen Technischen Hochschule Zuerich, Switzerland. 271 pp.

BROWN, JR., W.L. 1988. Data on Malighian Tubule numbers in ants. pp. 17-27 in Trager, J., ed., Advances in Myrmecology, E.J. Brill Publisher, New York, 551 pp.

CREIGHTON, W.S. 1950. The ants of North America. Bull. Mus. Comp. Zool., Harvard, 104:1-585

BOLTON, B. 1984. Diagnosis and relationships of the myrmicine ant genus *Ishakidrisgen.* n. Syst. Ent., 9:373-382

BOLTON, B. 1987. A review of the *Solenopsis* genus-group and a revision of Afrotropical *Monomorium* Mayr. Bull. Br. Mus. nat. Hist. (Ent.), 54:263-452

BOLTON, B. 1988. A new socially parasitic *Myrmica* with a reassessment of the genus. Syst. Ent., 13:1-11

HENNIG, W. 1950. Grundzuege einer Theorie der phylogenetischen Systematik. Deutscher Zentralverlag. Berlin. 370 pp.

IMAI, H.T., R. CROZIER, and R. TAYLOR. 1977. Karyoptype evolution in Australian ants. Chromosoma (Berl.), 59:341-93

LATTKE, J. 1990. Una nueva especie de *Pogonomyrmex* de bosque humedo tropical. Rev. Biol. Trop., (in press)

LUTZ, H. 1986. Eine neue Unterfamilie aus dem mittel-eozaenen Oelschiefer der Grube bei Darmstadt. Senckenbergiana Lethaea, 67:177-218

MAYR, E. 1969. Principles of Systematic Zoology. McGraw-Hill. N.Y. 428 pp.

REMANE, A. 1956. Die Grundlagen des natuerlichen Systems, der vergleichenden Anatomie und der Phylogenetik. Theoretische Morphologie und Systematik I. Akademische Verlagsgesellschaft. Leipzig. 364 pp.

SEIFERT, B. 1984a. Nachweis einer im Freiland aufgretretenen Bastardisierung von *Leptothorax nigriceps* Mayr und *Leptothorax unifasciatus* (Latr.) mittels eines multiplen Diskriminanzanalyse. Abh. Ber. Naturkundemuseum, Goerlitz, 58

SEIFERT, B. 1984b. Firm evidence for synonymy of *Myrmica rugulosoides* Forel, 1915 and *Myrmica scabrinodis* Nylander, 1846. Abh. Ber. Naturkundemuseum, Goerlitz, 58(6):1-10

SNEATH, P. and R. SOKAL. 1973. Numerical Taxonomy. The Principles and Practices of Numerical Classification. Freeman. San Francisco. 573 pp.

TABER, S., J. COKENDOLPHER, and O. FRANCKE. 1987. Scanning electron microscopic study of North American *Pogonomyrmex*. Proc. Ent. Soc. Wash., 89:512-526

WARD, P.S. 1990. The ant subfamily Pseudomyrmecinae (Hymenoptera:Formicidae): Generic revision and relationship to other formicids. Systematic Entomology, In Press.

WILEY, E. 1981. Phylogenetics: the Theory and Practice of Phylogenetic Systematics. Wiley. N.Y. 439 pp.

WILSON, E.O., and W.L. BROWN, JR. 1953. The subspecies concept. and its taxonomic application. Syst. Zool., 2:97-111

7

Chemotaxonomy Applied to Fire Ant Systematics in the United States and South America

R.K. Vander Meer and C.S. Lofgren

INTRODUCTION

Systematics

The genus, *Solenopsis* has undergone many changes over the last 60 years. It has been traditionally divided into several subgenera (Creighton 1930), which were synonymized by Ettershank (1966). In 1968, Baroni-Urbani separated *Diplorhoptrum*, small monomorphic species commonly known as thief ants, from *Solenopsis*. Within the genus *Solenopsis* there are several species complexes, based on morphology. The largest complex is *saevissima*. The *Solenopsis saevissima* complex has undergone many revisions, and is currently undergoing yet another (James Trager, personal communication). Creighton (1930) lumped 16 species into 8 subspecies or varieties. Wilson (1952) believed that most of the subspecies represented unstable hybrid variants and consolidated the complex into two species, *S. interrupta* and *S. saevissima*, and one subspecies, *S. saevissima* var. *richteri*. Ettershank (1966) preserved only *S. saevissima* and *S. interrupta*. Most recently Buren (1972) recognized Ettershank's two species, raised *S. richteri* (the black imported fire ant) and *S. quinquecuspis* to species rank and described two new species, *S. invicta* (the red imported fire ant) and *S. blumi*. There are probably many undescribed *saevissima* complex species in South America (Buren 1972; MacConnell et al. 1976); thus researchers anxiously await Trager's taxonomic revision.

Introduction into the United States

Black fire ants were accidently imported into the Mobile, Alabama area around 1918. Creighton (1930) identified them as *S. saevissima* var. *richteri*. Then, in the 1940's a red form of imported fire ant was also discovered around Mobile, Alabama; probably introduced in the 1930's (Lofgren 1986). The red and the black form were subsequently named *S. invicta* and *S. richteri*, respectively (Buren 1972). Little is known of the population dynamics that occurred between the two species during their rapid expansion over the last three to four decades (see Lofgren 1986 for fire ant distribution maps). However, in 1974 Buren et al. reported that *S. richteri* occupied only a small triangular area in northeastern Mississippi and northwestern Alabama, with an acute apex at Meridian, Mississippi (Buren et al. 1974). *S. invicta* was thought to occupy all other infested areas.

Biochemical Characters

Venom Alkaloids. The poison gland of *S. saevissima* complex workers contain *trans* 2-methyl-6-alkyl or alkenyl piperidines (MacConnell et al. 1971; MacConnell et al. 1976). The length of the alkyl or alkenyl group varies from C_{11} to C_{17} and may contain a single double bond. These compounds are easily referred to by stating their sidechain length and whether or not they have a double bond; e.g. *trans*-2-methyl-6-(*cis*-6-n-pentadecenyl)-piperidine can be simplified to *trans*-$C_{15:1}$. The venom alkaloid profiles of *S. invicta*, *S. richteri*, *S. geminata* and *S. xyloni* female sexuals are non-specific and consist primarily of *cis* and *trans* $C_{11:0}$ alkaloids (Brand et al. 1973). However, worker alkaloid patterns are species-specific (Brand et al. 1972; MacConnell et al. 1971) and have provided a useful tool for the study of a variety of fire ant problems (Vander Meer 1986a). MacConnell et al. (1976) analyzed the alkaloid profiles of 13 New World fire ant species (most undescribed) and concluded that they have taxonomic value.

Cuticular Hydrocarbons. Insect cuticular hydrocarbons are established taxonomic tools (Carlson 1988; Vander Meer 1986a). The cuticular hydrocarbons of *S. invicta* and *S. richteri* are species-specific (Lok et al. 1975) and composed of normal, methyl,and dimethyl branched compounds. The five major hydrocarbons of *S. invicta* are ubiquitous to the species, even being found in nest soil (Vander Meer et al. 1982; Vander Meer unpublished). *S. geminata* and *S. xyloni* (not members of the *saevissima* complex) also have species-specific cuticular hydrocarbon patterns, composed of normal and mono- and di-unsaturated compounds (Vander Meer unpublished). At the colony level, hydrocarbon patterns change quantitatively with time; however, they are qualitatively invariant (Vander Meer et al. 1989).

We will discuss the use of fire ant biochemical characters and associated behaviors to better understand the present distribution of *S. invicta* and *S. richteri* in the United States and to pinpoint the most probable South American source populations for these two invaders. In addition, we will identify and discuss some interesting taxonomic problems.

HYBRID FIRE ANTS IN THE UNITED STATES

Biochemical Evidence

Vander Meer et al. (1985) discovered hybridization between *S. invicta* and *S. richteri* when intermediate Dufour's gland chromatographic profiles were found from colonies identified morphologically as *S. richteri*. The two species are readily distinguished by their venom alkaloids (*S. richteri* does not produce the C_{15} alkaloids (Fig. 1) that dominate the *S. invicta* pattern), cuticular hydrocarbons (Fig. 1) and Dufour's gland profiles (Vander Meer and Lofgren 1989). Hybrids had alkaloid and hydrocarbon patterns intermediate to those of the parental species (see Vander Meer et al. 1985; Ross et al. 1987). The hybrids had escaped detection for several decades because

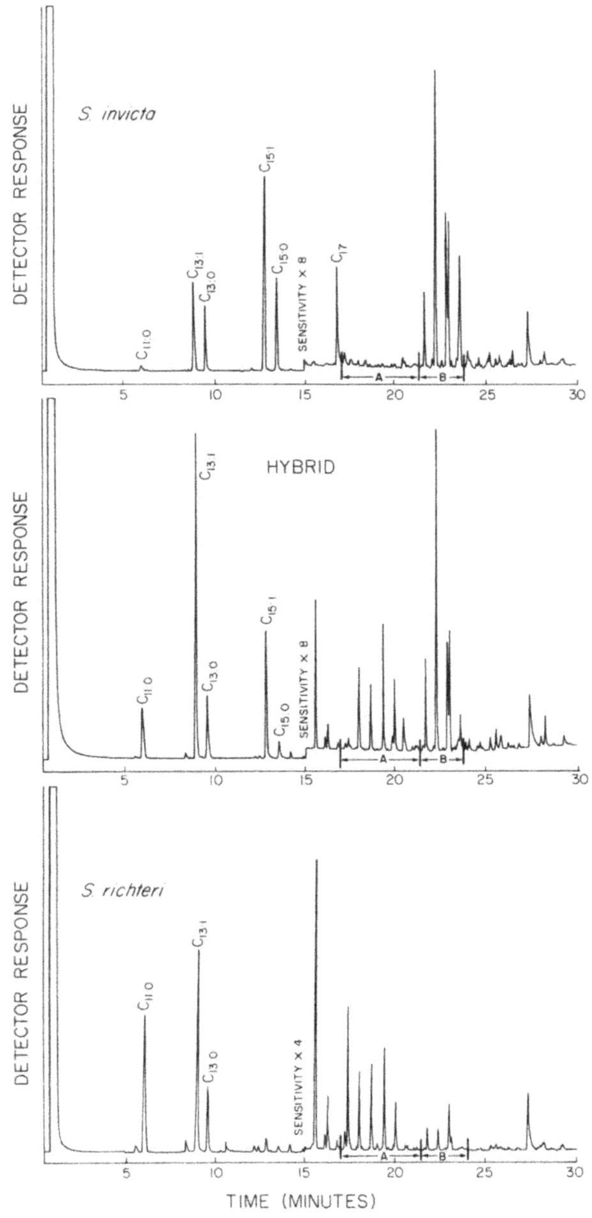

FIGURE 1. Gas chromatograph profiles of *Solenopsis invicta*, *Solenopsis richteri* and their hybrid. Characteristic *S. invicta* hydrocarbons = B; characteristic *S. richteri* hydrocarbons = A.

morphologically they look like *S. richteri*. Field observations were unlikely to detect hybridization (Vander Meer et al. 1985).

Genetic Evidence

Genetic evidence for hybridization based on gene frequency data was obtained by enzyme electrophoresis (Ross et al. 1987) of female sexuals. Four loci differentiated the two parental forms: *alpha-glycerophosphate dehydrogenase-1* (Agp-1, EC 1.1.1.8); *Octanol dehydrogenase* (Odh, EC 1. 1. 1. 73); *Esterase-2* and *Esterase-4* (Est-2,4; EC 3. 1. 1. 1). This methodology confirmed the occurrence of hybridization and found that recombinant genotypes predominated in the hybrid zone, indicating that F_1 hybrids are reproductively viable. Isozyme analyses of hybrid and parental workers were also diagnostic. *S. invicta* and *S. richteri* populations were fixed for specific electromorphs of *Glucose phosphate isomerase* (Pgi), whereas the hybrid showed segregation of both electromorph types (S. Narang, personal communication).

Biochemical and genetic methods were highly concordant in differentiating hybrid colonies from the two parental forms. Both methods revealed extensive gene introgression in a north-south study zone of about 120 Km in northeastern Mississippi (Ross et al.1987).

Behavioral Evidence

Recruitment Pheromone. The fire ant's complex recruitment strategy (Wilson 1962) is governed by a hierarchy of behaviors (Vander Meer 1987; Vander Meer 1986b). The Dufour's gland produces the recruitment pheromone (Wilson 1959). *S. invicta* workers are initially attracted to a trail (attractant pheromone; Vander Meer et al. 1988), then motivated to followed the trail (inducer pheromone; Vander Meer et al. in press), and finally follow it (orientation pheromone; Vander Meer et al. 1981). A bioassay that measured worker response to a point source of Dufour's gland extracts clearly differentiated between the two parental types (Vander Meer and Lofgren 1989). Bioassays using parental and hybrid worker Dufour's gland extracts demonstrated that hybrids could respond to either parental type. Similarly, parental workers responded well to hybrid Dufour's gland extracts (Vander Meer and Lofgren 1989). This provides definitive behavioral evidence for hybridization. As in the chromatographs of venom alkaloids and cuticular hydrocarbons, gas chromatographic examination of hybrid and parental Dufour's gland extracts clearly distinguished between the parents, with an intermediate hybrid pattern (Vander Meer and Lofgren 1989).

Nestmate Recognition. Nestmate recognition in *S. invicta* has been well studied (see Vander Meer this volume) and it is clear that both environmental and heritable cue sources are involved (Obin and Vander Meer 1988). However, the effect of environmental sources can be limited by rearing colonies in the laboratory under identical conditions (Obin 1986). The residual nestmate recognition response, measured in an aggression bioassay, is primarily due to heritable cues. Aggression bioassays showed no difference in the ability of *S. invicta*, *S. richteri*, or hybrid to recognize the other two populations. However, in intra-population bioassays, *S. invicta* and *S. richteri* both showed significantly reduced aggression responses, whereas the hybrid responded to other hybrid colonies as intensely as it did to its parents (Obin and Vander Meer 1989). This suggests greater genetic variability for heritable cues in the hybrid than in either parent. This goes along with the observed variability of genetically controlled cuticular hydrocarbons and venom alkaloids. Although these chemical classes have not been implicated as nestmate recognition cues, they represent a model for possible heritable cues. We found distinct and very consistent qualitative patterns for alkaloids and hydrocarbons of the two parent species (Vander Meer et al. 1985; Diffie et al. 1988; Vander Meer unpublished). However, for the hybrid there are patterns representing all possibilities between the two parents. Therefore, the expected high hybrid genetic variability, compared to its parents, is supported by alkaloid and hydrocarbon analyses and the results of nestmate recognition studies.

Extent of the Hybrid Population

Originally thought to be a localized phenomenon (Vander Meer et al. 1985; Ross et al. 1987), hybridization between *S. invicta* and *S. richteri* is now known to be extensive, with large reproductively viable hybrid populations throughout northeast Mississippi, northern Alabama and northwest Georgia (Vander Meer et al. 1985, Diffie et al. 1988). Hybrid populations in Alabama and Georgia have *S. invicta* at their southern border but do not have *S. richteri* at their northern border. The hybrid population is too extensive to have occurred by natural mating flights. Therefore, people may be largely responsible for the hybrids movement, just as in the case of the imported fire ant from the Mobile, Alabama area (Lofgren 1986).

A Look Back in Time

Is imported fire ant hybridization a recent event? Analysis of venom alkaloids from alcohol preserved specimens collected from Meridian to Starkville, Mississippi in 1964 (samples collected for E. O. Wilson) showed alkaloid patterns characteristic of both parental types and their hybrid (Vander Meer unpublished). The results demonstrated that hybridization occurred in and north of Meridian, Mississippi. In addition, analyses of samples collected from Mobile, Alabama in 1949 (E.O. Wilson) showed evidence of hybridization. Therefore, hybridization between *S. invicta* and *S. richteri* has been occurring wherever and whenever the two forms met.

A THEORETICAL MODEL

Initial introduction of both imported species occurred around the Mobile, Alabama area. However, their distributions are now disjunct, with *S. richteri* occupying only a small enclave in northeastern Mississippi, while *S. invicta* occurs in the southern half of the affected states, except for all of Florida. This corresponds to their distribution in South America, where *S. invicta* is in a tropical/sub-tropical area and *S. richteri* is in a temperate zone.

Initial reports (Wilson and Eads 1949; Wilson and Brown 1958) on the interaction of the two forms around Mobile, Alabama, indicate that the red form (*S. invicta*) competitively dominated the black form (*S. richteri*).

Competitiveness in ant populations is mediated to a large extent by their ability to discriminate the chemical cues associated with territory, recruitment and nestmate recognition (Hölldobler and Carlin 1987). In two of three recruitment pheromone behavioral sub-categories, attraction and orientation induction, there is asymmetric species-specificity favoring *S. invicta* over *S. richteri*. *S. richteri* does not respond to *S. invicta* Dufour's gland extracts in an olfactometer or orientation induction bioassay; however, *S. invicta* responds to *S. richteri* Dufour's gland extracts. Thus at least in this aspect of competition for territory and resources it appears that *S. invicta* has an advantage. Since we do not find hybrids south of an approximate East/West line at Meridian, Mississippi it must be concluded that the hybrid is competitively inferior to *S. invicta* in habitats south of this line. Chromatographic evidence indicates that the interface area around Meridian, Mississippi has been stable for at least 25 years, thus the hybrid and *S. invicta* have reached a standoff regarding competitiveness and environmental adaptability. On the northern end of the distribution, chromatographic evidence indicates that the hybrid is displacing a competitively inferior parent, *S. richteri*.

Many models have been developed to explain hybrid zones (see Vander Meer and Lofgren 1988). Moore's (1977) "bounded hybrid superiority" model is most useful. In this model an advantage exists for recombinant genotypes in habitats intermediate to those favored by the parental forms. This model is consistent with hybridization occurring wherever the two parental types meet and subsequent population equilibrium based on habitat suitability. This model also explains the persistence of hybrid populations in northern Alabama and Georgia. However, it does not deal with the apparent advance of the hybrid in northern Mississippi, displacing *S. richteri*. An asymmetric bounded hybrid superiority model would satisfy all known observations. The hybrid is bounded to the south by *S. invicta*, which is being held in check by the hybrid at a geographic point where an equilibrium has been reached balancing differences in competitiveness and environmental conditions. To the north, in Alabama and Georgia, the hybrid does not compete with either parent, and thus is asymmetrically bounded. However, in northern Mississippi the hybrid competes with both the environment and *S. richteri*. The observed northward displacement of *S. richteri* by the hybrid indicates that the hybrid may successfully out compete *S. richteri* through its entire range, eventually eliminating this parental form from the United States.

Buren (1972) reinstated *S. richteri* as a species and classified *S. invicta* as a new species based largely on the observation that no hybridization was found and there was little phenetic variability in the United States. The discovery of hybridization through chemical characters, coupled with the observation that hybrid morphology is almost identical to *S. richteri*, neutralizes the criteria used to specify *S. invicta*. Where the two forms have the opportunity to mate they are successful. Therefore, if they are "good" species, pre-mating isolation mechanisms must exist in their native South America.

THE FIRE ANT IN SOUTH AMERICA

The species-specific cuticular hydrocarbon and venom alkaloid patterns are largely invariant within the United States *S. invicta* and *S. richteri* populations. Thousands of samples from Brazil and Argentina were collected and analyzed by gas chromatography. One objective was to search for the specific chromatotypes that correspond to those found for *S. invicta* and *S. richteri* in the United States.

An exact match for *S. invicta* was found for specimens collected around Caceras, Brazil and south along the Paraguay River. No collections were made in Paraguay; however, the chromatotype resumed at the Paraguay/Argentina border and continued south along the Paraguay/Parana River to Sante Fe, Argentina. Different *Solenopsis* chromatotypes were found south of Sante Fe and to the east in a parallel direction along the Uruguay River. Therefore, the *S. invicta* chromatotype is found only along a narrow but very long corridor from Brazil to Argentina. This distribution approximates that given by Buren et al. (1974); although we extended its range further south in Argentina.

The United States *S. richteri* chromatotype was found only in a small lowland area of Argentina, approximately 100 km north of Buenos Aires, across the Parana River and close to the Uruguay border. These samples were identified morphologically as *S. richteri* (James Trager personal communication). Based on extensive surveys, the closest *S. invicta* and *S. richteri* come to each other is about 200 km. Other *Solenopsis* species exist between the two. No evidence for *S. invicta/S. richteri* hybrids was found. This data offers the simplest explanation for hybridization in the United States: in Argentina a geographic isolation barrier exists. Thus pre-mating isolation mechanisms were not selected for and when the geographic barrier was removed; i.e. accidental importation into the same area of the United States, hybridization occurred. Whether or not geographic isolation is enough to consider two populations different species is a good question, but one left to the taxonomists.

Other samples (identified morphologically as *S. richteri*; James Trager personal communication), collected from many areas of the Province of Buenos Aires, Argentina, including near the *S. richteri* type location, Buenos Aires (Buren 1972), had similar hydrocarbon profiles but differed dramatically in the alkaloid patterns. The alkaloid patterns were consistent and were composed primarily of *cis* and *trans* C_{11} alkaloids. This pattern characterizes *geminata* or *xyloni* complex species more than a *Saevissima* complex species. There was no evidence for mixing of the two *S. richteri* chromatotypes, even though in one instance colonies were sympatric. The two chromatotypes probably represent two cryptic species. Since the United States *S. richteri* do not match the *S. richteri* from the type location, the United States *S. richteri* need another name. In another scenario, if the two fire ant forms imported into the United States are ultimately considered a single species, then United States *S. richteri* and its corresponding chromatotype in Argentina can be synonimized to *S. invicta*.

CONCLUSION

Creighton (1930) recognized the difficulty of studying the *saevissima* complex by likening it to "entering a battle-field strewn with unexploded missiles" and that "there is a strong possibility that one of these taxonomic duds may, through tampering, bring the investigator to grief." These studies hoped to add another techniques to the taxonomic arsenal so that the battlefield may be a little less dangerous.

REFERENCES

BARONI-URBANI, C. 1968. Ueber die eigenartige Morphologie der mannlichen Genitalien des Genus Diplorhoptrum mayr (Hymenoptera:Formicidae) und die taxonomischen Schlussfolgerrungen. Z. Morph. Okol. tiere 63:63.

BRAND, J.M., M.S. BLUM, H.M. FALES, and J.G. MacCONNELL. 1972. Fire ant venoms: Comparative analyses of alkaloidal components. Toxicon 10: 259-271.

BRAND, J.M., M.S. BLUM and M.R. BARLIN. 1973. Fire ant venoms: Intraspecific and interspecific variation among castes and individuals. Toxicon 11: 325-331.

BUREN, W.F. 1972. Revisionary studies on the taxonomy of the imported fire ants. J. Ga. Entomol. Soc. 7: 1-26.

BUREN, W.F., G.E. ALLEN, W.H. WHITCOMB, F.E. LENNARTZ, and R. N. WILLIAMS. 1974. Zoogeography of the imported fire ants. J. N. Y. Entomol. Soc. 82:113-124.

CARLSON, D.A. 1988. Hydrocarbons for identification and phenetic comparisons: Cockroaches, honey bees and tsetse flies. Florida Entomol. 71:333-345.

CREIGHTON, W.S. 1930. The New World species of the genus *Solenopsis* (Hymenoptera: Formicidae). Proc. Am. Acad. Arts Sci. 66: 39-151, plates 1-8.

DIFFIE, S., VANDER MEER, R.K., and BASS, M.H. 1988. Discovery of hybrid fire ant populations in Georgia and Alabama. J. Entomological Sci. 23: 187-191.

ETTERSHANK, G. 1966. A generic revision of the World Myrmicinae related to *Solenopsis* and *Pheidologeton* (Hymenoptera:Formicidae). Aust. J. Zool. 14: 73-171.

HÖLLDOBLER, B. and CARLIN, N.F. 1987. Anonymity and specificity in the chemical communication signals of social insects. J. Comp. Physiol. A. 161: 567-581.

LOFGREN, C.S. 1986. History of imported fire ants in the United States. In Lofgren, C.S. and Vander Meer, R.K. (eds.), Fire Ants and Leaf Cutting Ants: Biology and Management. Westview Press, Boulder, CO, pp. 36-47.

LOK, J.B., E.W. CUPP and G.J. BLOMQUIST. 1975. Cuticular lipids of the imported fire ants, *Solenopsis invicta* and *richteri*. Insect. Biochem. 5: 821-829.

MacCONNELL, J.G., M.S. BLUM and H.M. FALES. 1971. The chemistry of fire ant venom. Tetrahedron 26: 1129-1139.

MacCONNELL, J.G., M.S. BLUM, W.F. BUREN, R.N. WILLIAMS, and H. M. FALES. 1976. Fire ant venom: Chemotaxonomic correlations with alkaloidal composition. Toxicon 14: 69-78.

MOORE, W.S. 1977. An evaluation of narrow hybrid zones in vertebrates. Quart. Rev. Biol. 52:263-277.

OBIN, M.S. 1986. Nestmate recognition cues in laboratory and field colonies of *Solenopsis invicta* Buren (Hymenoptera:Formicidae): Effect of environment and the role of cuticular hydrocarbons. J. Chem. Ecol. 12: 1965-1975.

OBIN, M.S. and R.K VANDER MEER. 1988. Sources of nestmate recognition cues in the imported fire ant, *Solenopsis invicta* Buren (Hymenoptera:Formicidae). Anim. Behav. 36: 1361-1370.

OBIN, M.S. and R.K. VANDER MEER. 1989. Between- and within-species recognition among imported fire ants and their hybrids (Hymenoptera:Formicidae): Application to hybrid zone dynamics. Ann. Entomol. Soc. Amer. 82: 649-652.

ROSS, G., R.K. VANDER MEER, D.J.C. FLETCHER and E.L. VARGO. 1987. Biochemical phenotypic and genetic studies of two introduced fire ants and their hybrid (Hymenoptera: Formicidae). Evolution 41: 280-293.

VANDER MEER, R.K. 1986a. Chemical taxonomy as a tool for separating *Solenopsis* spp., pp. 316-326. In: Fire Ants and Leaf Cutting Ants: Biology and Management. (C. S. Lofgren and Vander Meer, R. K., eds.), Boulder, CO, Westview Press, 435 p.

VANDER MEER, R.K. 1986b. The trail pheromone complex of *Solenopsis invicta* and *Solenopsis richteri*. In Lofgren, C. S. and Vander Meer, R. K. (eds.), Fire Ants and Leaf Cutting Ants: Biology and Management, Westview Press, Boulder, CO, pp. 201-210.

VANDER MEER, R.K. 1987. Fire ant foraging behavior: A multiplicity of strategies and mechanisms. In Eder, J. and Rembold, H. (eds.), Chemistry and Biology of Social Insects. Verlag J. Peperny, Munich, p. 541.

VANDER MEER, R.K. and C.S. LOFGREN. 1988. The use of chemical characters in defining populations of fire ants, *Solenopsis saevissima* complex, (Hymenoptera: Formicidae). Fla. Entomologist. 71: 323-332.

VANDER MEER, R.K. and C.S. LOFGREN. 1989. Biochemical and behavioral evidence for hybridization between fire ants, *Solenopsis invicta* and *Solenopsis richteri* (Hymenoptera:Formicidae). J. Chem. Ecol. 15: 1757-1765.

VANDER MEER, R.K., F.D. WILLIAMS and C.S. LOFGREN. (1981) Hydrocarbon components of the trail pheromone of the red imported fire ant, *Solenopsis invicta*. Tetrahedron Lett. 22: 1651-1654.

VANDER MEER, R.K., B.M. GLANCEY and C.S. LOFGREN. 1982. Biochemical changes in the crop, esophagus, and postpharyngeal gland of colony-founding red imported fire ant queens (*Solenopsis invicta*). Insect Biochem. 12: 123-127.

VANDER MEER, R.K., C.S. LOFGREN and F.M. ALVAREZ. 1985. Biochemical evidence for hybridization in fire ants. Florida Entomol. 68: 501-506.

VANDER MEER R. K., F.M. ALVAREZ and C.S. LOFGREN. 1988. Isolation of the recruitment pheromone of *Solenopsis invicta*. J. Chem. Ecol. 14: 825-838.

VANDER MEER, R.K., D. SALIWANCHIK and B. LAVINE. 1989. Temporal changes in colony cuticular hydrocarbon patterns of *Solenopsis invicta*: Implications for nestmate recognition. J. Chem. Ecol. 15: 2115-2125.

VANDER MEER, R.K., C.S. LOFGREN and F.A. ALVAREZ. In press. The orientation inducer pheromone of the fire ant *Solenopsis invicta.* Physiological Entomol.

WILSON, E.O. 1952. 0 complexo *Solenopsis saevissima* na America do Sul (Hymenoptera:Formicidae). Me. Inst. Oswaldo Cruz 50: 49-68.

WILSON, E.O. 1959. Source and possible nature of the odor trail of fire ants. Science 129: 643-644.

WILSON, E.O. 1962. Chemical communication among workers of the fire ant *Solenopsis saevissima* (Fr. Smith). 1. The organization of mass-foraging. 2. An information analysis of the odour trail. 3. The experimental induction of social responses. Anim. Behav. 10: 134-164.

WILSON, E.O. and J.H. EADS. 1949. A report on the imported fire ant, *Solenopsis saevissima* var. *richteri* Forel in Alabama. Ala. Dep. Conserv. Spec. Rep., 53 pp., 13 plates (Mimeographed).

WILSON, E.O. and BROWN, W.L. 1958. Recent changes in the introduced population of the fire ant, *Solenopsis saevissima* (Fr. Smith), Evolution 12: 211-218.

8

A Survey of the Glandular System of Fire Ants

J. Billen

INTRODUCTION

Ants of the genus *Solenopsis* are among the most thoroughly studied because of their enormous medical and agricultural impact on the environment (Lofgren and Adams 1982). It is surprising, however, to see how few reports deal with morphology and development of the exocrine glands, even though these few are well-known, forming the basis for social communication (Vander Meer 1983), and hence for most interindividual interactions. The existing contributions on glandular morphology refer to the cephalic glands, organs whose function is mostly digestive (Phillips and Vinson 1980; Petralia et al. 1980). Therefore, morphological data dealing with pheromone-producing glands are limited, in contrast to the many chemical and behavioral studies of the pheromonal secretions themselves.

For these reasons this paper will focus on the variety of exocrine glands in the fire ant *Solenopsis invicta* Buren, describing the morphological characteristics and ultrastructural organization of the various secretory tissues.

MATERIAL AND METHODS

Workers and queens were collected from both mono- and polygynous colonies of *S. invicta* in and around Gainesville (Florida), in June 1985. Dissected glands and cut body parts were fixed in cold two percent glutaraldehyde in Na-cacodylate buffer and postfixed in two percent osmium tetroxide. Dehydration was achieved through a graded acetone series, and preceded embedding in Araldite. Serial semi-thin sections for light microscopy were cut with a Reichert OmU3 microtome and stained with methylene blue and thionin. Thin sections obtained with a Reichert Ultracut microtome were double stained in the LKB 2168 Ultrostainer and viewed in a Philips EM 400 electron microscope.

RESULTS AND DISCUSSION

The exocrine system of *S. invicta* comprises nine main glands--some of which are paired or occur in multiple number--located all over the body (Fig. 1). As in other social Hymenoptera, they belong to either the epithelial gland

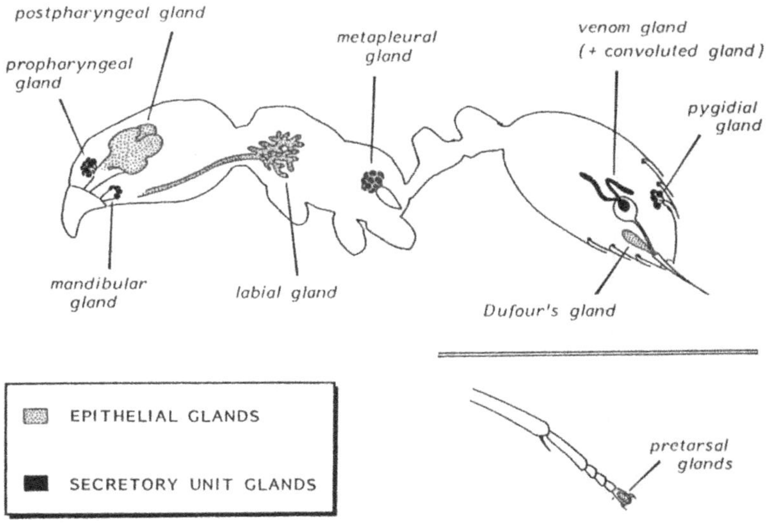

FIGURE 1. Survey of the location and appearance of the main exocrine glands in a fire ant worker.

or secretory unit types, according to their cellular organization (Billen 1987a).

Cephalic Glands

The pro- and postpharyngeal glands clearly illustrate the specific features of both glandular types. The propharyngeal gland consists of a left and right cluster of secretory cells with a corresponding duct cell for each cell that forms a narrow cuticular duct (internal diameter 0.4 μm). The several ducts of each cluster's secretory units (Fig. 2) open separately through the ipsilateral pharyngeal atrium. This pharyngeal opening site establishes that the older designation of "maxillary gland" is incorrect. The number of cells on each side varies from 15 in the smallest individuals to 47 in major workers, and 52 in queens (Phillips and Vinson 1980). Within each secretory cell the duct continues as a sinuous intracellular cuticular ductule, with fenestrated epicuticle, and is surrounded by mature microvilli (Fig. 3). Rounded nuclei are centrally located. The cytoplasm contains a well-developed granular endoplasmic reticulum (Fig. 4) and numerous membrane-bound and electron dense secretory vesicles (Fig. 3) which probably have a proteinaceous content. These morphological features are characteristic of cells that function to elaborate digestive enzymes, as has been demonstrated for the propharyngeal glands in other ant species (Ayre 1967).

The postpharyngeal gland, however, follows the epithelial gland organization, in which the glandular cells are arranged in a simple, monolayered epithelium. The secretory cells vary in height (10 and 15 μm) and have a microvillar apical plasmalemma adjacent to the cuticular lining (Fig. 5). The cytoplasm contains numerous mitochondria and a well-developed

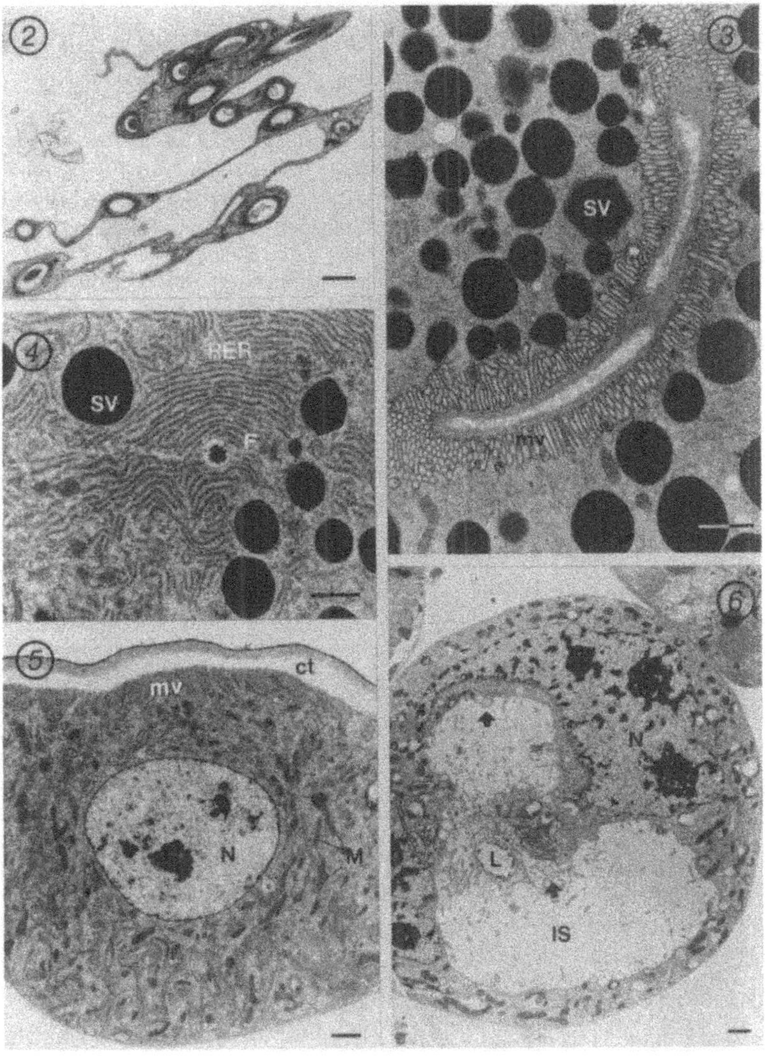

FIGURES 2-6: Ultrastructure of cephalic glands in *S. invicta* major workers (scale bar is 1 μm). 2) Section through propharyngeal gland duct cells. 3) Detail of end apparatus and surrounding secretory vesicles (SV) in a propharyngeal gland cell. mv = microvilli. 4) Cytoplasm of propharyngeal gland secretory cell with well-developed granular endoplasmic reticulum (RER) and electron-dense secretory vesicles (SV). 5) Postpharyngeal gland epithelium with apical microvilli (mv) and numerous mitochondria (M). ct = cuticle, N = nucleus. 6) Portion of labial gland tubule, showing the large interspace area (IS) with slender microvilli surrounding the central lumen (L). Arrows indicate exopinocytotic bleb formation. N = nucleus.

endoplasmic reticulum (smooth and tubular), extensions of which penetrate into the apical microvilli. This ultrastructural organization is consistent with the elaboration of non-proteinaceous, generally low molecular weight compounds (Noirot and Quennedey 1974), and supports the proposal that hydrocarbons are biosynthesized in the postpharyngeal gland (Vander Meer et al. 1982). A peculiar feature of the postpharyngeal gland in *Solenopsis* is its arrangement in a few flattened lobes (Phillips and Vinson 1980 for *S. invicta*; Attygalle et al. 1985, for *S. geminata)*. In all other ants studied so far, the gland has numerous finger-like tubular extensions. The function of the unusual anatomy in *Solenopsis* is unclear.

Each of the paired mandibular glands is small and comprises approximately five secretory unit cells. These open, through their corresponding ducts, into a common reservoir sac, which in turn opens at the inner base of the mandible. Although the gland cells are morphologically located on the outside of the reservoir, their invagination in it may give the impression they occur within the reservoir sac, as was reported by Phillips and Vinson (1980). Ultrastructural details for the mandibular gland in *Solenopsis* are unavailable, although a fine structure similar to other Formicidae can be expected. The products of the gland have been associated with alarm behavior in other ant species.

Located in the prothorax, the active part of the labial or salivary gland consists of a ramifying tubular system, from which a common median duct enters the head and leads to the base of the labium. Because of this opening site, the labial gland is considered a cephalic gland. The secretory cells in the tubules are epithelially arranged around a central cuticular duct, characterized by chitinous thickenings resembling the tracheal taenidia. Between the central cuticular duct and the apical plasmalemma of the secretory cells there is a wide interspace, traversed by numerous long and slender microvilli from the secretory cells. This interspace probably acts as a temporal storing area for the secretion (Fig. 6). Labial gland tubules have a diameter between 25 and 40 μm; the diameter of the large interspaces varies between 10 and 20 μm. According to Phillips and Vinson (1980), a well-defined reservoir is not present in *Solenopsis*. The fine structure of the glandular cells reveals irregularly shaped nuclei with dispersed chromatin, numerous mitochondria and free ribosomes. A well-organized granular endoplasmic reticulum, which occurs in the larval labial gland of *S. invicta* (Petralia et al. 1980), was not observed in the adult ants examined for the paper. Small electron-lucid vesicles are more frequent in the apical cell region and sometimes reach the large interspace area by exopinocytosis. No data are available on the labial gland secretion and its function in adult *S. invicta*. In the larvae, the gland produces digestive enzymes including proteases and amylases (Petralia et al. 1980).

Thoracic glands

True ant thoracic glands are restricted to the metapleural glands and the glands located in the legs. Both gland types in *Solenopsis* correspond to the general characteristics displayed in other ant groups (Billen 1987a).

Each paired metapleural gland consists of secretory cell clusters (20 to 30 individual cells) that open through accompanying duct cells into the sclerotized lateral atrium (Fig. 7), which in turn opens through a permanently open slit above the hindleg implantation. This peculiar anatomy, typical for the formicid metapleural gland, results in the continuous flow of materials produced by the gland. In some ant species, the gland products have antimicrobial activity and probably function in nest hygiene (Maschwitz et al. 1970). However, no antimicrobial activity has been demonstrated for the metapleural gland secretion of *S. invicta* (Vander Meer personal communication), and nest hygiene is most likely accomplished via the antimicrobial fire ant venom gland alkaloids. These compounds are dispersed in the colony as an aerosol (Obin and Vander Meer 1985).

Distinct epithelial gland sacs in the pretarsal segments of the six legs have been found in all Hymenoptera studied, including *S. invicta*. The gland is formed by an invagination with considerable thickening of the otherwise flattened epidermis from its most distal margin into the pretarsal segment (Fig. 8). The common occurrence of these glands in all legs of all species, and especially of both sexes, probably excludes a specialized pheromonal role. They may have a general mechanical function, providing adhesion to smooth surfaces, as illustrated for comparable glands in other insect orders (Hasenfuss 1977).

Abdominal glands

The pygidial gland, formed by two dorsolateral clusters of secretory unit glandular cells and their ducts, opens through the intersegmental membrane between the 6th and 7th tergite. In *Solenopsis*, the gland is small, each cluster having approximately 5 or fewer cells. The cells have a diameter

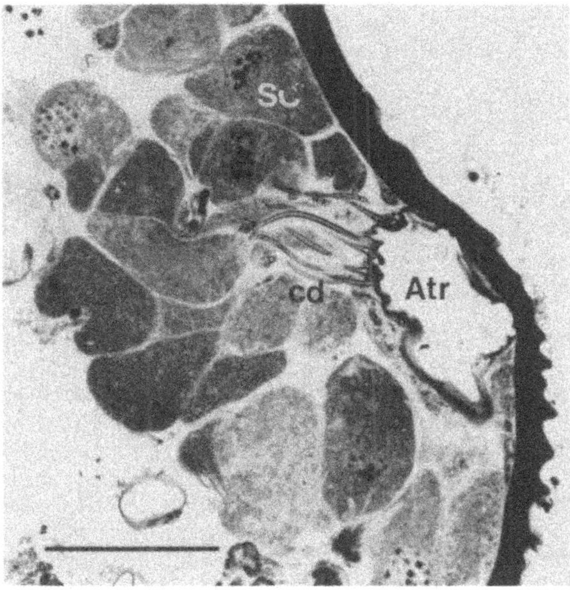

FIGURE 7. Semi-thin section through the metapleural gland of an *S. invicta* worker. Atr = atrium, cd = collecting ducts, SC = secretory cells.

between 25 and 35 µm, a more or less centrally-located rounded nucleus with dispersed chromatin. No information is available with regard to the function nor the composition of the fire ant pygidial gland secretion.

The Dufour gland, known as the source of the recruitment pheromone complex in fire ants (Vander Meer 1986), is a pear-shaped sac that opens through the sting base. The glandular cells form the epithelial lining of the sac and have a uniform thickness around 10 µm. As with all myrmicine ants, the ultrastructural organization lacks the typical modifications that characterize the Dufours gland in other ant subfamilies (Billen 1986). Mitochondria are randomly dispersed, and the apical and basal cell membranes display a simple topography without distinct invaginations. As for other Myrmicinae, the fine structure corresponds with the known production of hydrocarbons. The gland is surrounded by muscle fibers and some tracheoles.

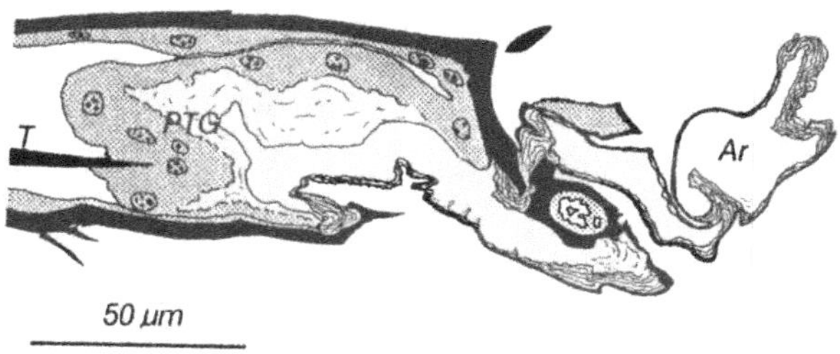

FIGURE 8. Longitudinal section drawing of the hindleg pretarsal gland (PTG). Ar = arolium, T = tendon.

The secretion of the venom gland in *Solenopsis* is the chemical basis of the defensive system and is notorious for the painful reactions it causes after an injection via the sting. The most characteristic chemical aspect of fire ant venom is its low proportion of protein and the dominant occurrence of piperidine alkaloids (Brand et al. 1972). This low protein content is clearly reflected at the ultrastructural level by the absence of granular endoplasmic reticulum in the cells of the secretory filaments. This is in contrast to the majority of the Formicidae. Instead, the glandular cells contain a few scattered ribosomes and vesicles of smooth endoplasmic reticulum, and an abundance of mitochondria (Fig. 9). Nuclei of the secretory cells are rounded and have a diameter of approximately 5 µm. Occupying a central position in the filaments as they surround the lumen, the nuclei of the duct cells have a more elongated shape and measure approximately 3 x 2 µm. The two venom gland filaments can reach a length of over 2 mm and are fragile; their diameters are 30 µm in workers and up to 70 µm in queens. Both free filaments merge into the convoluted gland, which forms the most voluminous part of the venom gland. It occurs as an extensive tissue invagination into the venom gland reservoir and can be considered a highly specialized transition zone between the free secretory filaments and the reservoir.

A convoluted gland is a common character of ant venom glands, but its development in *Solenopsis* is exceptional. Its mean diameter ranges from about 150 µm in minor to 250 µm in major workers, and over 300 µm in queens. A detailed anatomical and histological description of the fire ant convoluted gland is provided by Callahan et al. (1959). According to these authors, the convoluted gland is a continuation ("the main convoluted collecting duct") of the fused ducts from the two free filaments, with a ramifying system of medium and small collecting ductules (Figs. 10 and 11). The latter terminate in "what appear to be small reservoirs in the polygonal shaped cells making up the body of the convoluted poison gland" (Callahan et al. 1959). Based on work presented here, these polygonal cells have an intracellular end apparatus with surrounding microvilli, typical for the secretory unit type glands (Fig. 12). The fine structure of the convoluted gland cells resembles that of the free filament cells, having many mitochondria and few scattered free ribosomes (compare Figs. 9 and 12). They also contain numerous granular inclusions with a diameter of approximately 1 μm. Granular endoplasmic reticulum, again, is not present. Therefore, the convoluted gland cells represent a second and extensive secretory region of the fire ant venom gland. As for the free filaments, their ultrastructure illustrates the elaboration of non-proteinaceous molecules, which is the case with *Solenopsis* alkaloids. The precise contribution of the free filaments and of the convoluted gland in the secretory processes remains unknown.

The venom gland reservoir consists of a squamous epithelium, covered by a well-developed cuticular intima with a thickness around 1 μm and surrounded by loosely arranged muscle fibers (Fig. 13). The thick cuticular lining probably helps to prevent self-toxication from the stored venom secretion in the reservoir.

The secretion of both the venom gland and Dufour's gland is released through the sting, as is characteristic for the Formicidae (Billen 1987b). In their most posterior region, the slit-like ducts of both glands display a well-developed muscular apparatus. The Dufour gland duct is supplied with muscle fibers both dorsally and ventrally, while the venom gland duct only has a ventral muscular supply of two main muscle bundles (Fig. 14). The several sets of muscle fibers are responsible for an efficient and mutually independent discharge mechanism of both sting glands, and thus contradict the conclusion by Callahan et al. (1959) that the secretions from both glands must mix because Dufour's gland was described without its muscular apparatus. These authors mistook the existing muscle fibers for columnar cells which, together with the real flattened duct cells, were described as a stratified columnar epithelium.

CONCLUSION

Solenopsis invicta possesses a variety of exocrine glands, the occurrence and morphology of which generally corresponds to that of myrmicine ants, with a few peculiarities typical for fire ants. The fine structure of the various glandular tissues is consistent with the chemical composition of the secretory products, at least for glands with known chemical compositions. This is especially true of the venom gland, which produces the defensive substances as well as the queen attractant pheromone, and the Dufour's gland, source of the recruitment pheromone complex. For example, the venom gland, in contrast with other Formicidae, produces an alkaloidal secretion rather than proteinaceous compounds and lacks the otherwise typical granular

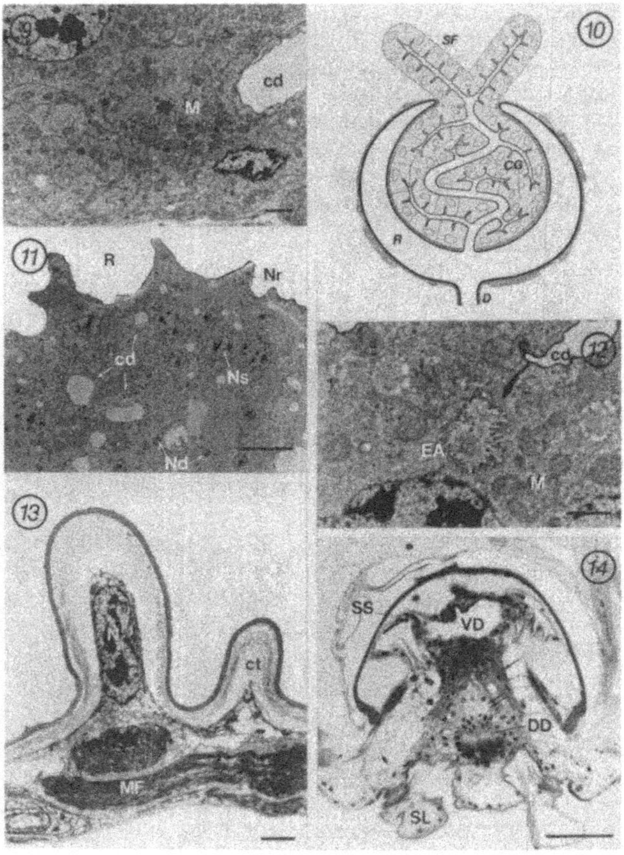

FIGURES 9-14: 9) Detail of secretory tissue in venom gland free filaments. cd = central collecting duct, M = mitochondria (*S. invicta* minor worker, scale bar 1 μm). 10) Schematic representation of the venom gland composition, combining data of Callahan et al. (1959) and own observations. CG = convoluted gland, D = duct, R = reservoir, SF = secretory filaments. 11) Part of the convoluted gland with differently sized collecting ducts (cd) and nuclei (Nd = duct cell nuclei, Nr = reservoir wall nuclei, Ns = secretory cell nuclei). R = reservoir, scale bar 10 μm. 12) Detail of convoluted gland showing contact between collecting ductule (cd) and secretory cell. EA = end apparatus, M = mitochondria (minor worker, scale bar 1 μm). 13) Venom gland reservoir wall showing relatively thick cuticular lining (ct) and surrounding muscle fibers (MF). N = nucleus (major worker, scale bar 1 μm). 14) Semi-thin cross section through the proximal region of the sting in a *S. invicta* queen. DD = Dufour gland duct, SS = sting shaft, SL = sting lancet, VD = venom gland duct. Single arrow point to muscular supply of venom gland duct, double arrow to muscles surrounding the Dufour gland duct.

endoplasmic reticulum. An abundance of mitochondria reflects the high energy demands and metabolic activity for the secretory cells in both the free filaments and the convoluted gland. The latter is definitely considered an additional secretory body; its strong development may be related to alkaloid synthesis.

For other glands, the chemical composition and behavioral function of the secretion remain unknown in *Solenopsis*. The lack of information about the mandibular and pygidial glands may partly be due to the small size of the glands in this genus. Their reduced size may pose a technical problem when attempting to collect detectable amounts of secretion. In other Formicidae (Attygalle and Morgan 1984), the clear pheromonal role of the mandibular gland in alarm behavior and the pygidial gland in recruitment behavior leaves an interesting, and so far unexplored, area of fire ant research.

ACKNOWLEDGEMENTS

I am grateful to Dr. E.D. Morgan and Dr. R.K. Vander Meer for their critical reading of this manuscript, and to E. Plaum for technical assistance. This research was supported by the Belgian National Fund for Scientific Research.

REFERENCES

ATTYGALLE, A.B. and E.D. MORGAN. 1984. Chemicals from the glands of ants. Chem. Soc. Reviews, 13: 245-278.

ATTYGALLE, A.B., J.P.J. BILLEN and E.D. MORGAN. 1985. The postpharyngeal gland of workers of *Solenopsis geminata* Hymenoptera:Formicidae. Actes Coll. Insectes Soc., 2: 75-86.

AYRE, G.L. 1967. The relationships between food and digestive enzymes in five species of ants Hymenoptera:Formicidae. Can. Ent., 99 : 408-411.

BILLEN, J.P.J. 1986. Comparative morphology and ultrastructure of the Dufour gland in ants Hymenoptera:Formicidae. Entomol. Gener., 11: 165-181.

BILLEN, J.P.J. 1987a. Morphology and ultrastructure of the exocrine glands in social Hymenoptera. In: Chemistry and Biology of Social Insects J. Eder and H. Rembold, Eds., J. Peperny Verlag, München, 81-84.

BILLEN, J.P.J. 1987b. New structural aspects of the Dufour's and venom glands in social insects. Naturwissenschaften, 74: 340-341.

BRAND, J.M., M.S. BLUM, H.M. FALES and J.G. MacCONNELL. 1972. Fire ant venoms: comparative analyses of alkaloidal components. Toxicon, 10: 259-271.

CALLAHAN, P.S., M.S. BLUM and J.R. WALKER. 1959. Morphology and histology of the poison glands and sting of the imported fire ant *Solenopsis saevissima v. richteri* Forel. Ann. Entomol. Soc. Am., 52: 573-590.

HASENFUSS, I. 1977. Die Herkunft der Adhäsionsflüssigkeit bei Insekten. Zoomorphologie, 87: 51-64.

LOFGREN, C.S. and C.T. ADAMS. 1982. Economic aspects of the imported fire ant in the United States. In: The Biology of Social Insects M.D. Breed, C.D. Michener and H.E. Evans, Eds., Westview Press, Boulder, Colorado, 124-128.

MASCHWITZ, U., K. KOOB and H. SCHILDKNECHT. 1970. Ein Beitrag zur Funktion der Metathorakaldrüse der Ameisen. J. Insect Physiol., 16: 387-404.

NOIROT, C. and A. QUENNEDEY. 1974. Fine Structure of insect epidermal glands. Ann. Rev. Entomol., 19: 61-80.

OBIN, M.S. and R.K. VANDER MEER. 1985. Gaster flagging by fire ants *Solenopsis* spp.: functional significance of venom dispersal behavior. J. Chem. Ecol., 11: 1757-1768.

PETRALIA, R.S., H.J. WILLIAMS and S.B. VINSON. 1982. The hindgut ultrastructure, and excretory products of larvae of the red imported fire ant, *Solenopsis invicta* Buren. Ins. Sociaux, 29: 332-345.

PHILLIPS, S.A. and S.B. VINSON. 1980. Comparative morphology of glands associated with the head among castes of the red imported fire ant, *Solenopsis invicta* Buren. J. Georgia Entomol. Soc., 15: 215-226.

VANDER MEER, R.K. 1986 - The trail pheromone complex of *Solenopsis invicta* and *Solenopsis richteri*. In: Fire Ants and Leaf-Cutting Ants, Biology and Management C.S. Lofgren and R.K. Vander Meer, Eds., Westview Press, Boulder and London, 201-210.

VANDER MEER, R.K., B.M. GLANCEY and C.S. LOFGREN. 1982. Biochemical changes in the crop, oesophagus and postpharyngeal gland of colony-founding red imported fire ant queens *Solenopsis invicta*. Insect Biochem., 12: 123-127.

WOJCIK, D.P. 1986. Bibliography of imported fire ants and their control: second supplement. Florida Entomol., 69: 394-415.

9

A Comparison of Venom and Hydrocarbon Profiles from Alates in Texas Monogyne and Polygyne Fire Ants, *Solenopsis invicta*

L. Greenberg, H.J. Williams and S.B. Vinson

INTRODUCTION

This preliminary study used gas chromatography to compare hydrocarbon and venom profiles in single and multiple queen forms of the fire ant, *Solenopsis invicta*. There are no obvious morphological differences between these forms, although a difference in the average size of workers is related to the number of queens (Greenberg et al. 1985). Behavioral differences are apparent, especially regarding the number of queens tolerated. The common hydrocarbon and venom components of the two forms appear similar and do not exhibit the kind of differences seen among Solenopsis species (for example, between *S. invicta* and *S. richteri*; Vander Meer et al. 1985; Vander Meer 1986). However, subtle differences in the hydrocarbons and venom profiles could discriminate between the two forms. In this study we qualitatively compared gas chromatograph peaks, using discriminant analysis and treating each peak as though it were a morphological character.

METHODS AND MATERIALS

Whole body washes of three alates from each selected colony (a pooled sample) were prepared by soaking the ants in hexane for 15 minutes. Samples from 25 monogynous (six collection sites) and 17 polygynous (six collection sites) laboratory colonies were taken. Figures 1 and 2 show the locations of these sites. One-three ul (later normalized to one ul) was injected onto a Hewlett-Packard 5890A chromatograph equipped with a Hewlett-Packard 7673A Automatic Sampler, flame ionization detectors and a SGE fused silica 25 m BP1 column with I.D.=0.22mm. Data were analyzed and visualized with a Hewlett-Packard 3393A Integrator. The gas chromatograph was held at 60°C for one minute, programmed at 15°C per minute to 280°C, where it was held for six minutes. The Helium carrier gas was set at 20 psi.

Fourteen peaks common to most samples were identified on the chromatograph. The identity of these peaks is not essential for this project, although most are known venom and hydrocarbon compounds (MacConnell et al. 1971; Brand 1978; Brand et al. 1973; MacConnell et al. 1971, 1976; Nelson et al. 1980; Lok et al. 1975; Vander Meer 1988). Figure 3 shows a typical plot with the peaks numbered. The areas of all the peaks were

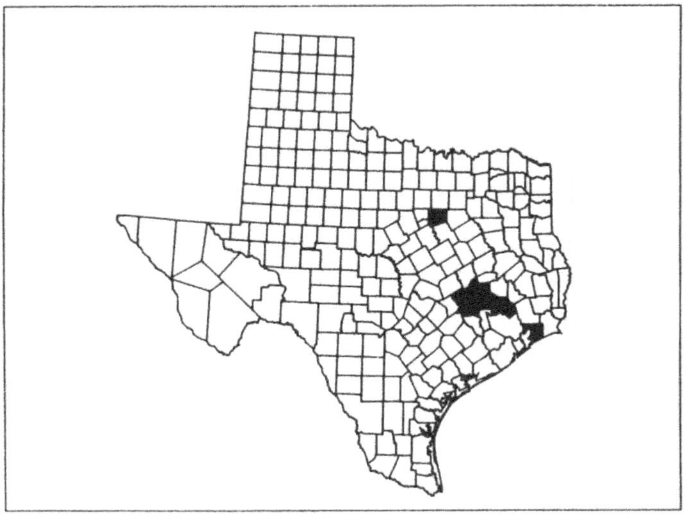

FIGURE 1. Map of Texas, showing counties (shaded) where samples were collected.

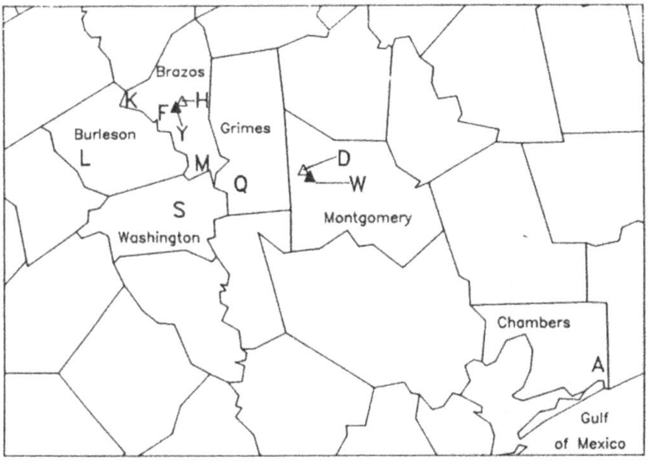

FIGURE 2. Enlarged map of southeastern Texas, showing collection sites by county. The northernmost county of Fig. 1 (Johnson) is not shown on this map. The large capital letters either are on the collection site or point to a symbol that shows the precise location. The map measures approximately 285 x 190 km.

summed and each peak expressed as a percentage of the total. An arcsine transformation of these percentages was followed by discriminant analysis using the first 13 peaks.

RESULTS

The output from the discriminant analysis (Table 1) shows the relative proportions of the 13 peaks for the monogyne and polygyne samples. The differences appear subtle. The univariate F-tests show a highly significant result only for peak 8, a minor component. The compounds' identity remains unknown at this time.

In contrast to the univariate statistics, the multivariate test is significant (Table 1, P=.002). Figure 4 plots the data using the factor scores from the discriminant analysis. Each factor score is a linear combination of the 13 variables that maximally discriminates between the monogyne and polygyne samples. The arbitrary Y-axis on this graph simply makes the points readable. The model predicts that colonies to the left of the arrow are monogyne, while those with factor scores to the right of the arrow should be polygyne. Lower case letters show monogyne colonies and capital letters indicate polygyne colonies. Figure 5 is a notch plot (Wilkinson, 1988) that shows the two populations' spread, medians and confidence limits.

Two more models were tested. The first one tested just the 5 alkaloid peaks (1-5 in Fig. 3). The multivariate F-test was not significant (P=0.49).

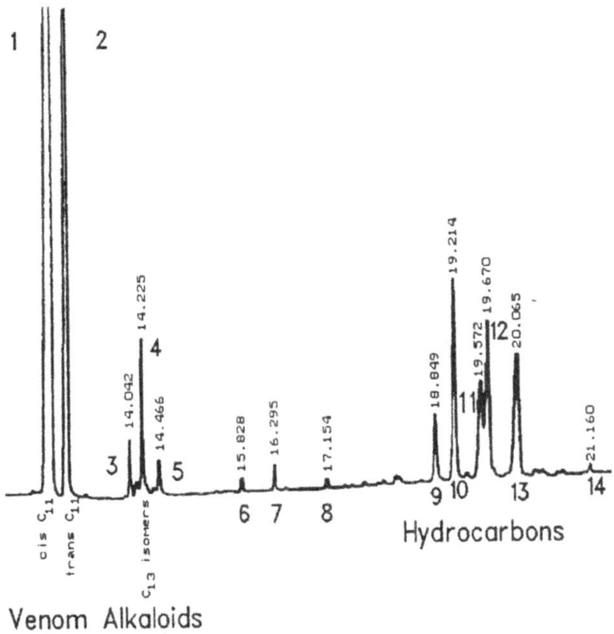

FIGURE 3. Chromatogram showing a typical plot. The peaks are numbered 1-14; the first 13 were used in the analysis. The numbers above each peak are retention times in minutes. Initial temperature = 60°C for 1 min followed by increments of 15 degrees per minute to 280°C, and 6 min at the final temperature. See Methods section for more details.

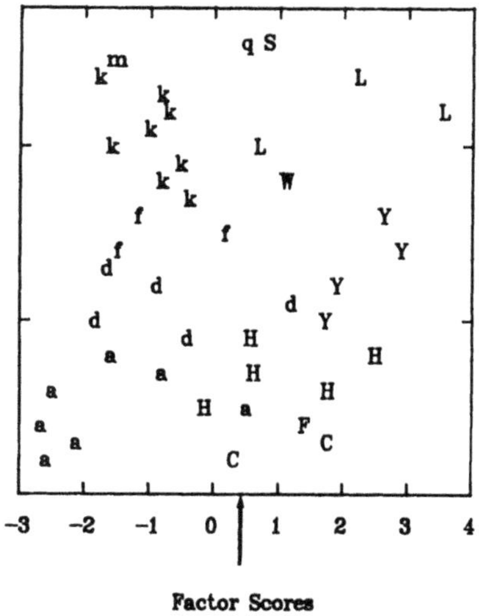

FIGURE 4. Plot showing collection sites and their factor scores. The letters correspond to those shown on the map of Fig. 2. The y-axis is an arbitrary scale chosen to make the points readable. The discriminant model predicts that colonies with factor scores to the right of the arrow are polygyne, while monogyne colonies are predicted to the left of the arrow. Actual (determined by laboratory observation) single-queen colonies are lower case, multiple-queen are upper case. The model correctly predicted the group membership in 37 out of 42 cases.

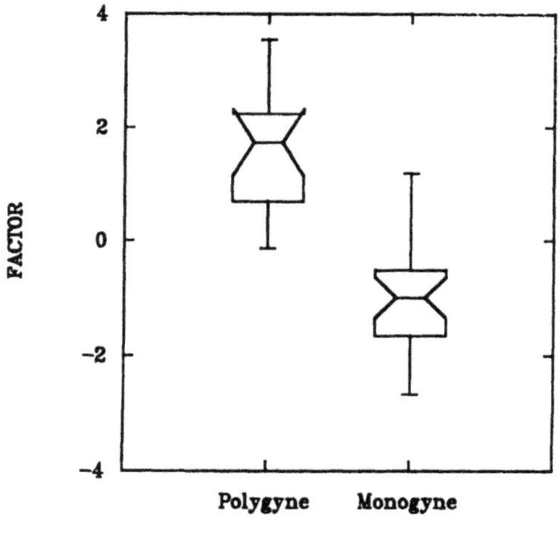

FIGURE 5. Notch plot of factor scores (Wilkinson 1988). The line through the narrow waist of each box is the sample median. Moving away from the medians, where the boxes first reach their full widths corresponds to the 95% confidence limits of the median. The lines extending from the boxes show the ranges.

TABLE 1. Statistical Analysis.

Peak Areas as a Proportion of Total Area (%)

Peak	Monogyne	Polygyne
1	52.733	50.794
2	26.622	26.017
3	6.747	7.882
4	12.115	13.377
5	6.469	7.167
6	0.829	2.141
7	0.515	1.346
8	0.607	2.849
9	5.889	7.574
10	8.792	8.070
11	6.548	5.755
12	6.552	8.575
13	6.688	6.965

UNIVARIATE F-TESTS

Peak	F	P
1	1.061	0.309
2	0.166	0.685
3	2.309	0.136
4	2.194	0.146
5	0.479	0.493
6	4.828	0.034
7	3.758	0.060
8	19.049	<0.001*
9	2.672	0.110
10	0.484	0.491
11	0.995	0.324
12	4.197	0.047
13	0.066	0.799

MULTIVARIATE TEST STATISTICS

All 13 Peaks in Model	Only Hydrocarbon Peaks 9-13 in Model
Wilks' Lambda = 0.368	Wilks' Lambda = 0.561
F-Statistic = 3.704	F-Statistic = 5.634
DF = 13, 28; Prob. = 0.002	DF = 5, 36; Prob. = 0.001
Canonical Correlation = 0.795	Canonical Correlation = 0.663

The second model analyzed just the 5 major hydrocarbon peaks (9-13 in Fig. 3). Its multivariate F-statistic was significant (Table 1, P=.001). The variability explained by the model with all variables included was 63.2%, while the comparable figure for only hydrocarbons was 44.0% (the squares of the canonical correlations).

DISCUSSION

It is tempting to do a stepwise discriminant analysis on these data to look for the most important variables. However, Wilkinson (1988) cautions that if a best subset of discriminators is selected from a large set, it will probably not replicate in a new sample. This is partly due to the problem of multiple comparisons and their influence on stated levels of significance. Therefore, this study tested only three possible models. Although peak 8 was the only one to show a significant difference by itself, the hydrocarbons by themselves (peaks 9-13) were also significantly different between monogyne and polygyne samples. The alkaloids, on the other hand, did not differ.

Figure 4 shows some interesting aspects of the results. The horizontal axis shows the similarity between samples. There is certainly a clumping of samples by collection sites (e.g., the k's), indicating some correlation between location and factor scores. However, the differences between single and multiple queen samples are even greater. All of the Foster samples (f's and F's) were collected in a single small field, historically mostly monogyne. A polygyne colony collected there clearly falls among the predicted polygyne samples. Likewise, the Cleburne samples (C's) are polygyne colonies most distant geographically from Hensel Park polygyne colonies (H's), yet overlap in factor scores. The most distinct groups were the monogyne colonies from Chambers County and the polygyne colonies from Burleson County (the a's and the L's).

Discriminant analysis makes a linear combination of variables to maximally discriminate between groups. The coefficients generated can predict group membership of new samples. The differences among the 42 samples may be due to chance events. The model's accuracy must be tested to see whether it can accurately predict group membership with new samples of ants collected at diverse locations. The eighth peak will also be identified to see whether it is a good character for separating the monogyne and polygyne forms. Finally, alates collected after mating flights will be checked to see whether their origins can be determined.

These results, along with other studies involving fire ant isozymes (Ross et al. 1985, 1987a, 1987b; Ross and Fletcher 1985; Dunton 1988) provide insight into the nature of the differences between monogyne and polygyne fire ants. This study shows good separation between the groups using extracts obtained with whole-body washes in hexane, although there is some degree of overlap and no wide gulf of separation between them. Ross et al. (1985) found no evidence for enzyme differences, but Dunton (1988) saw differences in banding patterns. If the results of this study are confirmed with additional samples from other locations, they will provide another criterion for discriminating between monogyne and polygyne colonies.

ACKNOWLEDGMENTS

We thank Ray Dunton for the ant samples from Chambers County. We are also very grateful to Sherry Ellison for technical support.

REFERENCES

BRAND, J.M., M.S. BLUM and H.H. ROSS. 1973 Biochemical evolution in fire ant venoms. Insect Biochem. 3: 45-51.

BRAND, J.M. 1978. Fire ant venom alkaloids: their contribution to chemosystematics and biochemical evolution. Biochem. Syst. Ecol. 6: 337-340.

DUNTON, R.F. 1988. Ecological genetics of monogynous and polygynous forms of the red imported fire ant, *Solenopsis invicta* Buren. Ph.D dissertation, Texas A&M University, College Station, TX 134 pp.

GREENBERG, L., D.J.C. FLETCHER and S.B. VINSON. 1985. Differences in worker size and mound distribution in monogynous and polygynous colonies of the fire ant *Solenopsis invicta* Buren. Journal of the Kansas Entomological Society 58: 9-18.

LOK, J.B., E.W. CUPP and G.J. BLOMQUIST. 1975. Cuticular lipids of the imported fire ants, *Solenopsis invicta* and *richteri*. Insect Biochem. 5: 821-829.

MACCONNELL, J.G., M.S. BLUM and H.M. FALES. 1971. The chemistry of fire ant venom. Tetrahedron 26: 1129-1139.

MACCONNELL, J.G., M.S. BLUM, W.F. BUREN, R.N. WILLIAMS and H.M. FALES. 1976. Fire ant venoms: Chemotaxonomic correlations with alkaloidal compositions. Toxicon 14: 69-78.

NELSON, D.R., C.L. FATLAND, R.W. HOWARD, C.A. MCDANIEL and G.J. BLOMQUIST. 1980. Re-analysis of the cuticular methyl alkanes of *Solenopsis invicta* and *Solenopsis richteri*. Insect Biochem. 10: 409-418.

ROSS, K.G. and D.J.C. FLETCHER. 1985. Comparative study of genetic and social structure in two forms of the fire ant *Solenopsis invicta* (Hymenoptera: Formicidae). Behavioral Ecol. Sociobiol. 17:349-356.

ROSS, K.G., D.J.C. FLETCHER and B. MAY. 1985. Enzyme polymorphisms in the fire ant, *Solenopsis invicta* (Hymenoptera: Formicidae). Biochemical Systematics and Ecology 13:29-33.

ROSS, K.G., R.K. VANDER MEER, D.J.C. FLETCHER, and E.L. VARGO. 1987a. Biochemical phenotypic and genetic studies of two introduced fire ants and their hybrid (Hymenoptera: Formicidae). Evolution 41:280-293.

ROSS, K.G., E. VARGO, and D.J.C. FLETCHER. 1987b. Comparative biochemical genetics of three fire ant species in North America, with special reference to the two social forms of *Solenopsis invicta* (Hymenoptera: Formicidae). Evolution 41: 979-990.

VANDER MEER, R.K., C.S. LOFGREN and F.M. ALVAREZ. 1985. Biochemical evidence for hybridization in fire ants. Florida Entomologist 68: 501-506.

VANDER MEER, R.K. 1986. Chemical taxonomy as a tool for separating *Solenopsis* spp. In: Fire ants and leaf-cutting ants: Biology and management, C.S. Lofgren and R.K. Vander Meer eds., Westview Press, Boulder, CO, pp. 316-326.

VANDER MEER, R.K. 1988. Behavioral and biochemical variation in the fire ant, *Solenopsis invicta*. In: Interindividual behavioral variability in social insects, R.L. Jeanne ed., Westview Press, Boulder, CO, pp. 223-255.

WILKINSON, L. 1988. Systat: The system for statistics. Systat, Inc., Evanston, IL., pp. 822.

10

Cephalic Exocrine Glands of Ants: A Morphological View

C. da Cruz-Landim

INTRODUCTION

The Hymenoptera have glandular structures embryologically linked to buccal pieces and known as the salivary gland system. Four pairs of glands form the system in ants: the labial or thoracic glands; the mandibular glands; the post-pharyngeal glands and the maxillary or hypopharyngeal glands. Most of these glands have their secretory portions located in the head, but the labial gland is located in the thorax, although their ducts open in the head labial segment. There is a large gap in time between the initial morphological studies of ant salivary glands and more recent work (Bordas, 1895; Janet, 1894; Otto, 1958; Forbes and MacFarlane, 1961; Toledo, 1967; Delage-Darchen, 1976; Brough, 1977; Phillips and Vinson, 1980; Gama e Cruz-Landim, 1982). These glands have different designations (Table 1), mostly related to the site of gland opening. Recently, study of these glands intensified and their origin (Gama, 1978), function (Delage-Darchen, 1976; Morgan et al, 1978; Brough, 1978; Vinson et al, 1980; Phillips and Vinson 1980; Attygale et al, 1985; Billen et al, 1987; Cammaerts et al, 1988) and ultrastructure (Petralia et al, 1980; Billen and Peusens, 1984; Gama, 1985; Billen, 1987) observed and described.

Although linked to the buccal parts and known as salivary glands, the secretions produced by these glands may not be used primarily in functions related to ingestion and digestion of food stuffs. Some, like the mandibular and post-pharyngeal glands, change and produce pheromones that ants use to establish social interactions in their colonies (Delage-Darchen, 1976; Morgan et al, 1978; Brough, 1978; Cammaerts et al, 1978; Cammaerts et al, 1988). Even without a complete understanding of the salivary glands' function, presence of sexual dimorphism, and sometimes caste polymorphism (Gama, 1978 and Toledo, 1967), is indicative of functional differences.

A comparative study of the glandular morphology in different Formicidae subfamilies, genera and species may aid in determining phylogenetic relationships in ants.

TABLE 1. Ants salivary glands system. A = Janet 1988; B = Emmert 1968; C = Almeida Toledo 1967; D = Gama 1978; E = Gama and Cruz-Landim 1982. Glands: max = Maxillary gland; ppg = Postpharyngeal; hpg = Hypopharyngeal; lab = Labial; sal = Salivary; salt = Salivary thoracic; man = Mandibular; iman = Intramandibular.

A	B	C	D	E
Formicidae	*Formica pratensis*	*Atta*	*Camponotus rufipes*	Formicinae
max	ppg	max	hpg	hpg
lab	lab	sal	lab or salt	lab or salt
man	man	man	man	man
-	max	--	--	--
ppg	ppg	ppg	ppg	ppg
--	--	iman	--	-

CEPHALIC GLANDS OF ANTS

Exocrine glands in the ant's head belong to the so-called salivary system and consist of mandibular glands, pharyngeal glands, hypopharyngeal glands and salivary or labial glands. The labial glands are the real salivary glands because they occur universally in the insects; they also originate from the larval salivary glands present in all holometabolous insects. Although considered here a cephalic gland, salivary glands have their secretory portion located in the thorax and only their excretory ducts enter the head through the occipital foramen to discharge the secretion in the labial segment.

The morphological features of these glands were studied in several ant species belonging to different subfamilies:

Poneroid Complex:

Subfamily Ponerinae	*Neoponera villosa*
	Odontomachus chelifer
	Pachicondyla striata
Subfamily Dorylinae	*Labidus praedador*
Subfamily Myrmicinae	*Atta capiguara*
	Cephalotes atratus
	Paracryptocerus pusillus
	Pheidole fallax
	Solenopsis saevissima
	Zacryptocenus clypeatus

Myrmecioid Complex:

Subfamily Pseudomyrmecinae	*Pseudomyrmex phyllophilus*
Subfamily Dolichoderinae	*Conomyrma brunnea*
Subfamily Formicidae	*Camponotus (Myrmothrix) abdominalis*
	Camponotus (Myrmothrix) rufipes
	Camponotus (Myrmofradris) arboreus
	Camponotus(Myrmeponis) sericeiventris

TABLE 2. Synopsis of the anatomic features of ants cephalic glands.

Subfamilies	Salivar thoracic	Mandibular	Pharyngeal	Hypopharyngeal
		PONEROID COMPLEX		
Ponerinae	glandular portion tubular or alveolar Actypic reservoir	Reservoir sac with long or short duct, and secretory cells localized	Poorly lobulated without peduncle	discharge point in sieve form (secretory cells discharging individually)
Dorylinae	glandular portion alveolar reservoir incipient	Idem	Most lobulated without peduncle	Idem
Myrmecinae	glandular portion tubular without reservoir	Reservoir without duct and secretory cells localized	Most lobulated with or without peduncle	Discharge through a short duct or sieve (Atta). Secretory cells discharge in excretory duct
		MYRMECOID COMPLEX		
Pseudomyrmecinae	glandular portion tubular. Reservoir in the junction of the collector ducts	Reservoir with short duct and secretory cells localized	Poorly lobulated without peduncle	discharge point in sieve
Dolichoderinae	glandular portion alveolar without reservoir	Idem	-	Idem
Formicinae	glandular portion tubular Reservoir in the lateral excretory duct	Reservoir with short duct and secretory cells non localized	Most lobulated without peduncle	discharge through duct sometimes branched

CEPHALIC GLAND ANATOMY

Table 2 is a synopsis of the results obtained from anatomic studies made in some ant subfamilies. Figure 1 shows the location of the glands in a hypothetical ant (Fig. 1 A, B), as well as the features of the glands in *Cephalotes atratus* (Fig. 1 C, D, E).

Labial/Thoracic Glands

The labial or salivary thoracic glands are formed by a pair of secretory parts located in the thorax lateral to the esophagus, and a system of canals that lead the secretion to the delivery point (Fig. 1 E).

The secretory portion may have alveolar or tubular forms. Figure 1 E shows a salivary gland with tubular secretory parts. The canal systems are formed by collector ductules that originate in and continue to the secretory tubules. The collector ductules fuse together and form the gland's excretory duct, one for each lateral glandular branch. The lateral excretory ducts join to form the final excretory duct that ends in the labium. Some species have a pair of glandular reservoirs which may be formed at the junction of the collector ductules or as an enlargement of the glandular lateral excretory duct (Table 2).

Mandibular Glands

The mandibular glands are linked to the mandible articular basis (Fig. 1 D). These glands are formed by glandular cells and a reservoir in the form of a bag with thin walls. The secretory cells may be grouped and located in the reservoir distal pole or distributed all around the bag (Table 2). The reservoir may be attached to the mandible directly or through a short or long excretory canal (Table 2).

Pharyngeal Glands

The post-pharyngeal glands are formed by two lateral and dorsal foliaceous evaginations of the pharyngeal epithelium in the passage of pharynx to esophagus (Fig. 1 C). The only differences among the species is the more or less lobulation of the glandular evaginations and their detachment from the pharynx wall. Sometimes, as seen in Figure 1 C, the evaginations are linked to the pharynx through a short peduncle, that in other cases is absent (Table 2).

Hypopharyngeal Glands

These glands are also known as maxillary glands due to their origin. They are masses of spherical secretory units attached to the hypopharyngeal plate in the infrabuccal pouch. Each secretory unit may be linked directly to the hypopharyngeal plate through an individual canalicule (Fig. 1 C). In this case, the plate became perforated by the canalicules openings, forming a sieve. In other species, the canalicules are connected to an excretory canal that leads the secretion to the infrabucal pouch (Table 2).

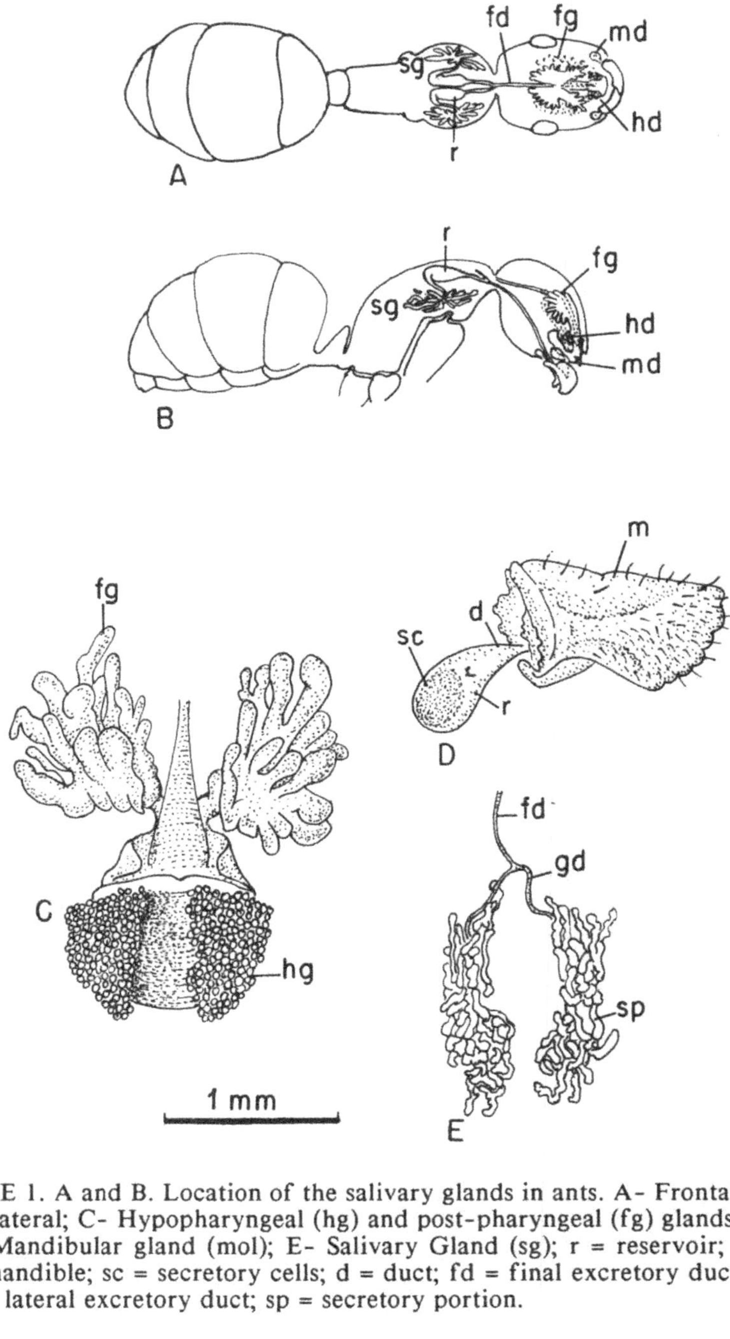

FIGURE 1. A and B. Location of the salivary glands in ants. A- Frontal; B- Lateral; C- Hypopharyngeal (hg) and post-pharyngeal (fg) glands; D- Mandibular gland (mol); E- Salivary Gland (sg); r = reservoir; m = mandible; sc = secretory cells; d = duct; fd = final excretory duct; gd = lateral excretory duct; sp = secretory portion.

TABLE 3. Synopsis of the histological features of ants cephalic glands.

Subfamilies	Salivar gland	Mandibular	Post-pharyngeal	Hypopharyngeal
		PONEROID COMPLEX		
Ponerinae	Central secretory cell surrounded by flat epithelial cells or secretory cells arranged as pseudotubules	Globular secretory cells connected to the reservoir by canalicules	Columnar epithelium lining an irregular lumen	Globular cells connected to the hypopharyngeal plate by individual canalicules
Dorylinae	Secretory cells forming pseudotubules	irregular secretory cells forming a compact mass	Idem	Idem
Myrmecinae	Secretory cells forming true tubules	secretory cells globular or forming as epithelium	Idem	globular cells connected to excretory duct or the hypophar. plate by canalicules
		MYRMECOID COMPLEX		
Dolichoderinae	Central secretory cells surrounded by epithelial cells	globular cells connected to the reservoir by individual canalicules	Idem	globular cells connected to the hypopharyngeal plate by canalicules
Pseudomyrmecinae	Secretory cells forming pseudotubules	Idem	Idem	Idem
Formicidae	Secretory cells forming true tubules	secretory cells distributed in the reservoir surface forming an epithelium	Idem	globular secretory cells connected to an excretory duct

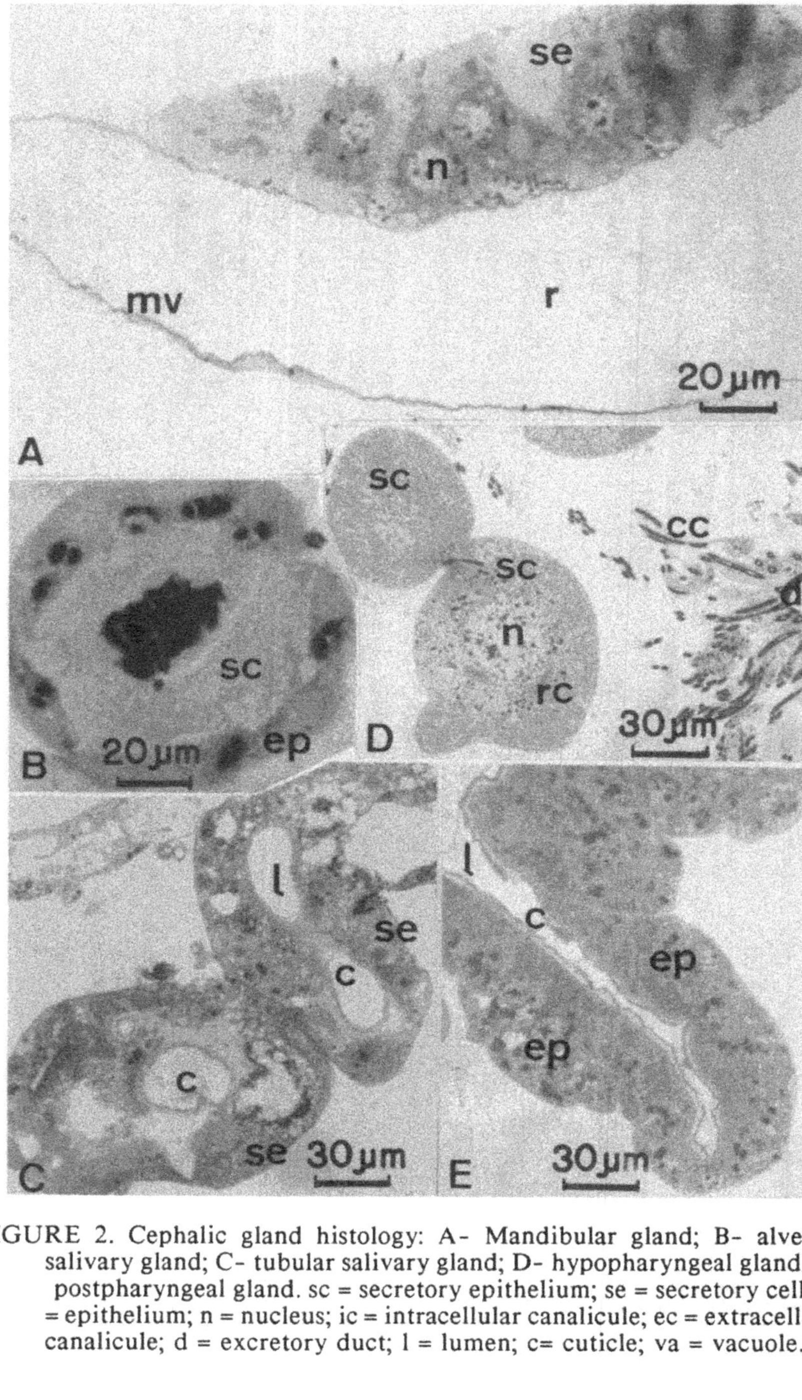

FIGURE 2. Cephalic gland histology: A- Mandibular gland; B- alveolar salivary gland; C- tubular salivary gland; D- hypopharyngeal gland; E- postpharyngeal gland. sc = secretory epithelium; se = secretory cell; ep = epithelium; n = nucleus; ic = intracellular canalicule; ec = extracellular canalicule; d = excretory duct; l = lumen; c= cuticle; va = vacuole.

CEPHALIC GLAND HISTOLOGY

All ant cephalic glands are of ectodermal origin. Only the salivary gland is present in the larva, although its morphology is very distinct from adults (Gama, 1978). The other glands form during the pupal phase.

The salivary gland originates from the larval salivary gland remnants. The post-pharyngeal and hypopharyngeal glands are evaginations of the pharynx. Mandibular glands form from epidermal invaginations at the mandible articulation point.

The tissues of the glandular lumens are of ectodermal origin, and are always lined by a cuticle.

Salivary Glands

The secretory portion of these glands may be composed of alveolus or tubules. The alveolus are formed by a large central secretory cell surrounded by small flat cells forming an epithelial lining (Fig. 2 B). This epithelium is continuous with the epithelial wall of the collector ductules. The secretion is delivered via a cuticular canal that runs between the secretory and surrounding epithelial cells.

The glands that anatomically appear as "tubular" may be histologically divided into two groups (Table 3). The tubular glands found in *Ectatomma* (Ponerinae), *Labidus* (Dorylinae) and *Pseudomyrmex* (Pseudomyrmicinae) are formed by a sequence of large secretory cells in the tubule axis surrounded by an epithelium of flat cells; i.e., pseudotubular organization. The collector ductule takes a sinuous course between the cells, giving a nodous aspect to these tubules. The true tubular glands, as from Formicinae, are formed by a secretory epithelium lining an axial lumen invested by a cuticle (Fig. 2 C).

The ducts of the salivary glands and the reservoir wall are formed by flat epithelial cells with apical cuticle reinforced by taenids.

Mandibular Glands

The mandibular gland reservoir wall is formed by a sheath of very flat epithelial cells with cuticular deposition in the luminal side (Fig. 2 A). In all cases, the secretory cells are individually connected to the reservoir through a canalicule. The canalicule is partly inside the secretory cell cytoplasm (collector canalicule) and partly extracellular (the excretory canalicule). The excretory is formed by one or more cells, different from the secretory cell. The canalicules, in or out of the cell, are lined by cuticle.

Post-Pharyngeal Glands

In all subfamilies the post-pharyngeal glands are formed by an epithelium of cuboidal or columnar cells with a cuticular lining in the lumen (Fig. 2 C).

Hypopharyngeal Glands

Histologically these glands are also very homogeneous in ants. In all species examined they are formed by globular secretory cells connected to the hypopharyngeal plate or to the excretory duct by individual canalicules (Fig. 2 D). As in the mandibular gland, the canalicule has an intracellular and an extracellular portion, both lined by cuticle. The extracellular canalicule are formed by proper cells (Fig. 2 D).

CEPHALIC GLANDS ULTRASTRUCTURE

Ultrastructural studies have been made only in *Camponotus rufipes*, an ant with a true tubular salivary gland and a racemose hypopharyngeal gland.

Salivary Gland

The secretory tubules on electron microscope examination appear formed by cells that have many basal membrane invaginations and apical microvilli; the cells may appear tall or flat depending upon the secretory stage. The secretion seems to accumulate between the secretory cells apices and cuticle (Fig 3 A) before passing into the lumen. The secretion has low electron density, so the subcuticular space appears empty, although the secretory cells detachment from cuticle and their flatness indicate presence of the secretion filing the subcuticular space.

The cells do not have Golgi, but show many mitochondria and a regularly developed rough endoplasmic reticulum (Fig. 3). Secretory granules are only rarely seen in these cells. When present they have median electron density (Fig. 3 B). These secretory cells have some vacuoles with clear content, and have junctions reinforced by septated desmosomes.

Mandibular Gland

The most prominent feature of mandibular gland cell ultrastructure is the intracellular canalicule and the mitochondria richness. The intracellular canalicule is formed by a cuticular tubule that travels through a cytoplasm groove formed by an invagination of the secretory cell surface membrane. The cell membrane projects microvilli to the canal (Fig. 4 A). Sometimes extensive secretion deposits are found around these canalicules (Fig 4 B). The secretion appears as electron dense granules (Fig. 4 B) whose content may be seen among the microvilli and within the cuticular canalicule (Fig. 4 A). These cells do not have rough endoplasmic reticulum, which suggests that the secretion has a lipid nature.

Post-Pharyngeal Gland

The cells that constitute the epithelium of this gland produce a thick cuticle that lines the glandular lumen. Below this cuticle, the cellular apical membrane shows many invaginations. The principle features of these cells are the presence of electron dense granules (Fig. 5 A) in the cytoplasm and a rough, poorly developed endoplasmic reticulum. The cellular junctions are reinforced by septated desmosomes (Fig. 5 A). Frequently fibrous (Fig. 5 A), electron dense material or amorphous (Fig. 5 B) median electron dense material is found in the lumen. The secretory function of the gland's cells, i.e., the glandular nature of this structure, is not accepted by all authors. Phillips and Vinson (1980) think they are digestive caeca.

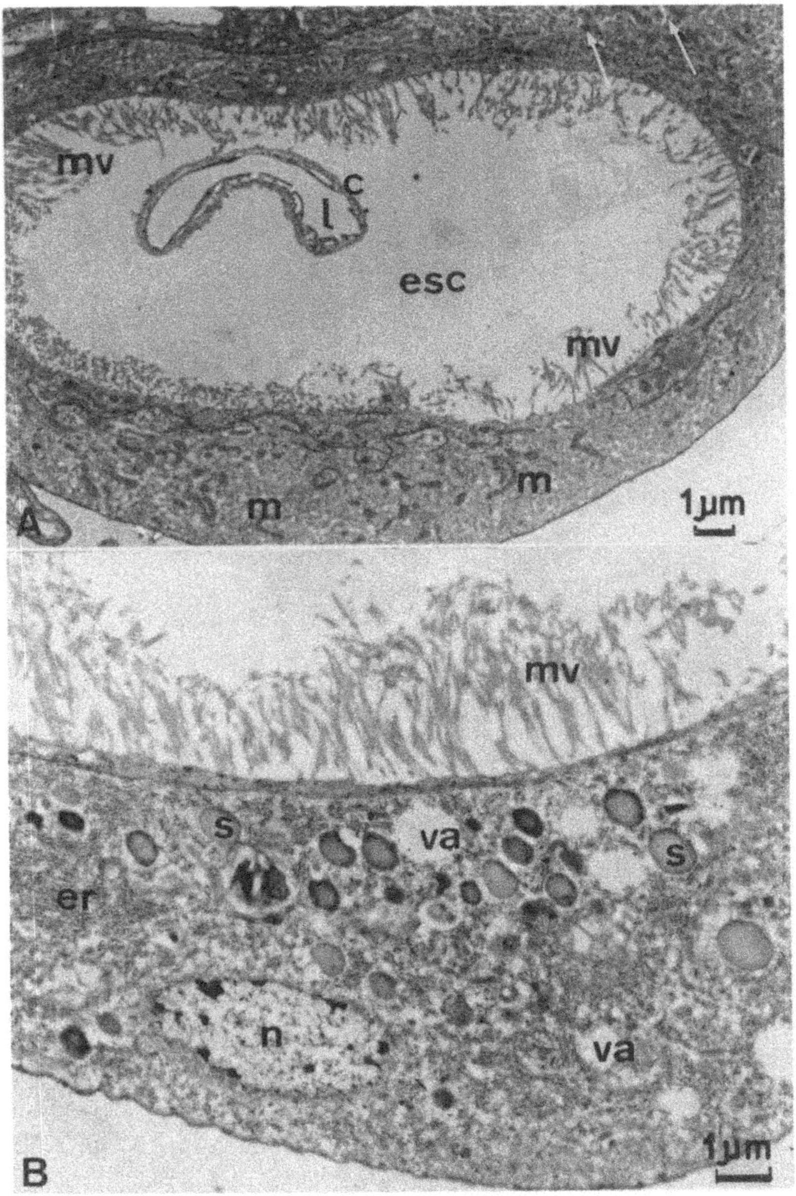

FIGURE 3. Salivary gland ultrastructure: A- cross section of a secretory tubule the arrow point to basal invaginations of the plasmic membrane; B- high magnification of the secretory cell. l = lumen; c = cuticle; ss = subcuticular space; mv = microvilli; s = secretion; er = endoplasmic rough reticulum; m = mitochondria.

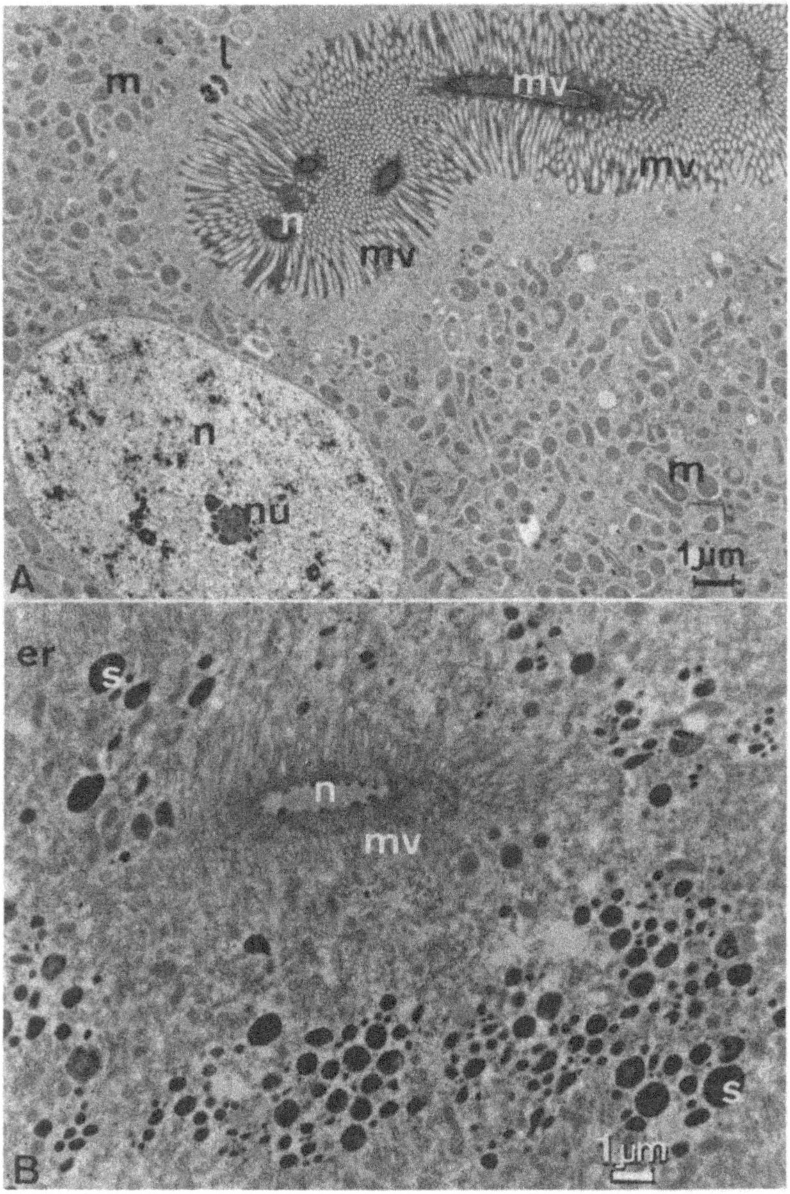

FIGURE 4. Mandibular gland ultrastructure: A- general aspect of the secretory cell; B- aspect of the secretion (s) around the intracellular canalicule (ic). mv = microvilli; m = mitochondria; n = nucleus; nu = nucleolus; er = rough endoplasmic reticulum; l = lysosome.

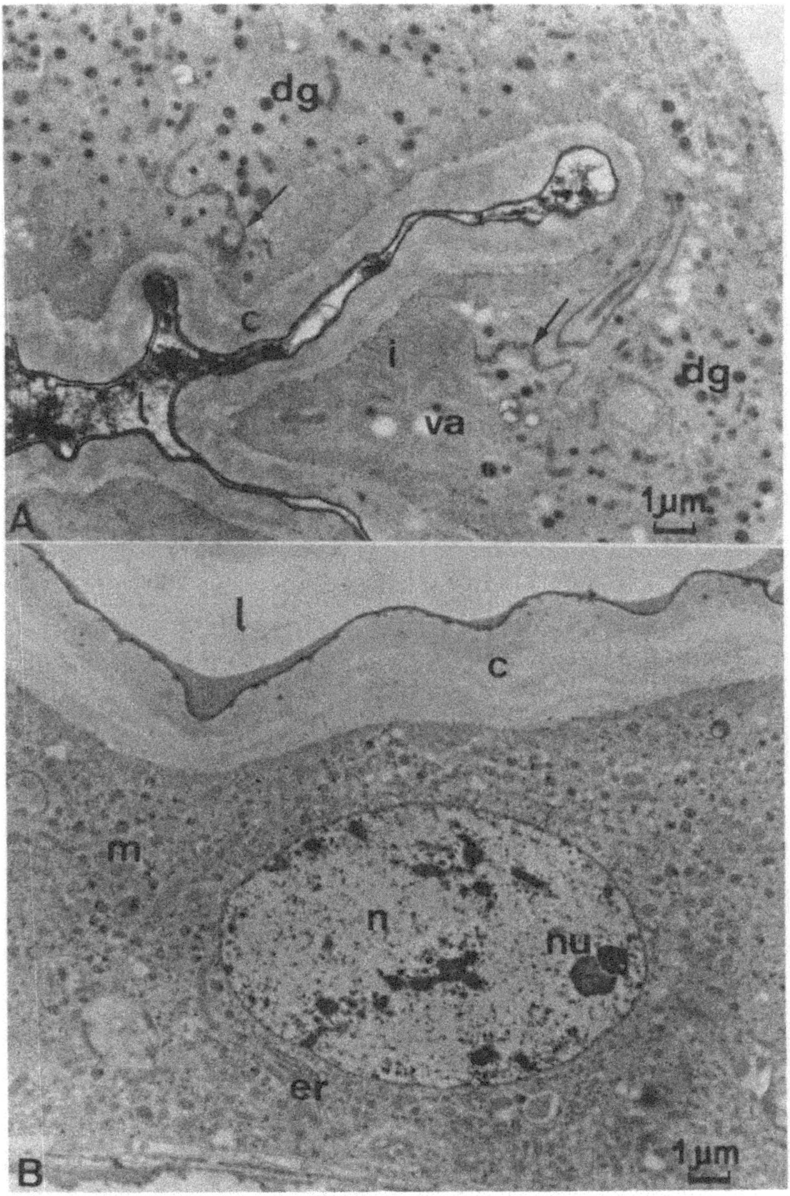

FIGURE 5. Postpharyngeal gland ultrastructure: A- general aspect of the epithelial cell. The arrow point to cellular junctions; B- high magnification of the epithelial cell. l = lumen with fibrous material in a and amorphous material in b; c = cuticle; dg = dense granule; va = vacuole; i = apical invaginations of plasmic membrane; n = nucleus; nu = nucleolus; er = rough endoplasmic reticulum.

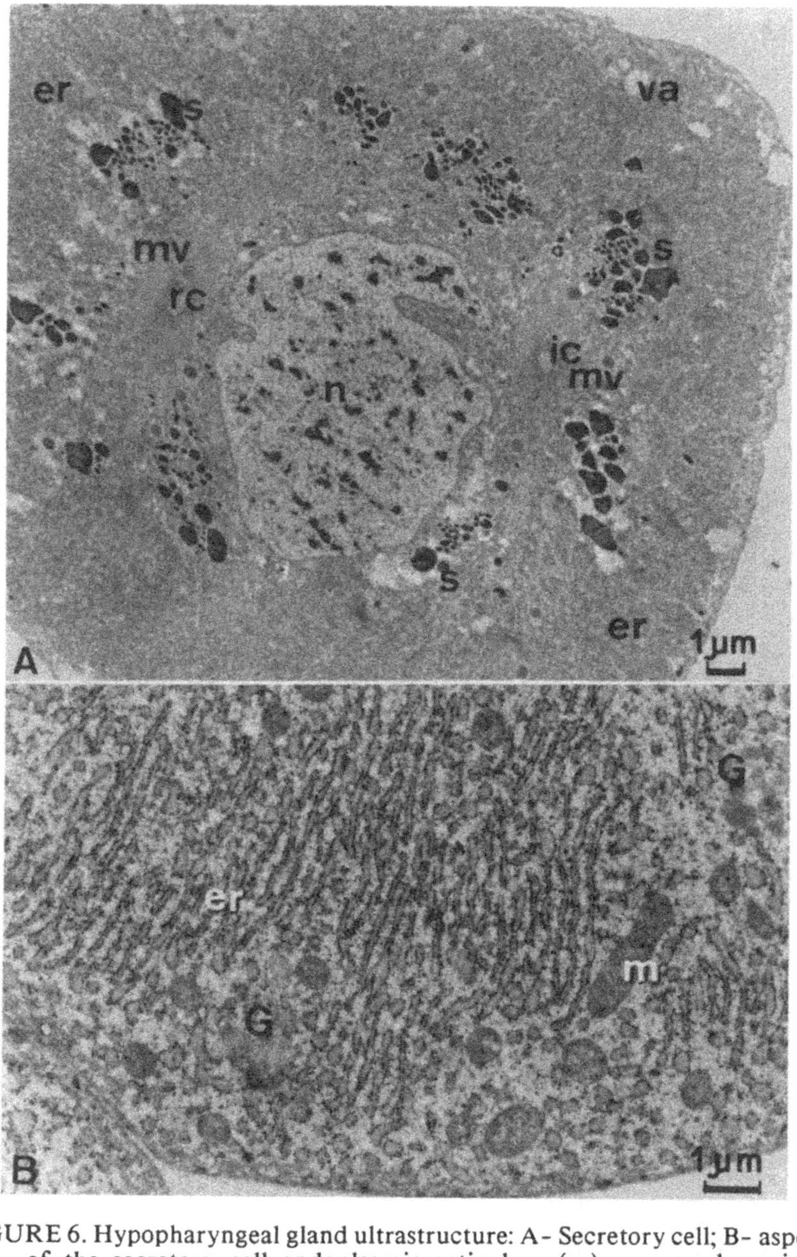

FIGURE 6. Hypopharyngeal gland ultrastructure: A- Secretory cell; B- aspect of the secretory cell endoplasmic reticulum (er). n = nucleus; ic = intracellular canalicule; s = secretion; er = rough endoplasmic reticulum; mv = microvilli; m = mitochondria; G = Golgi.

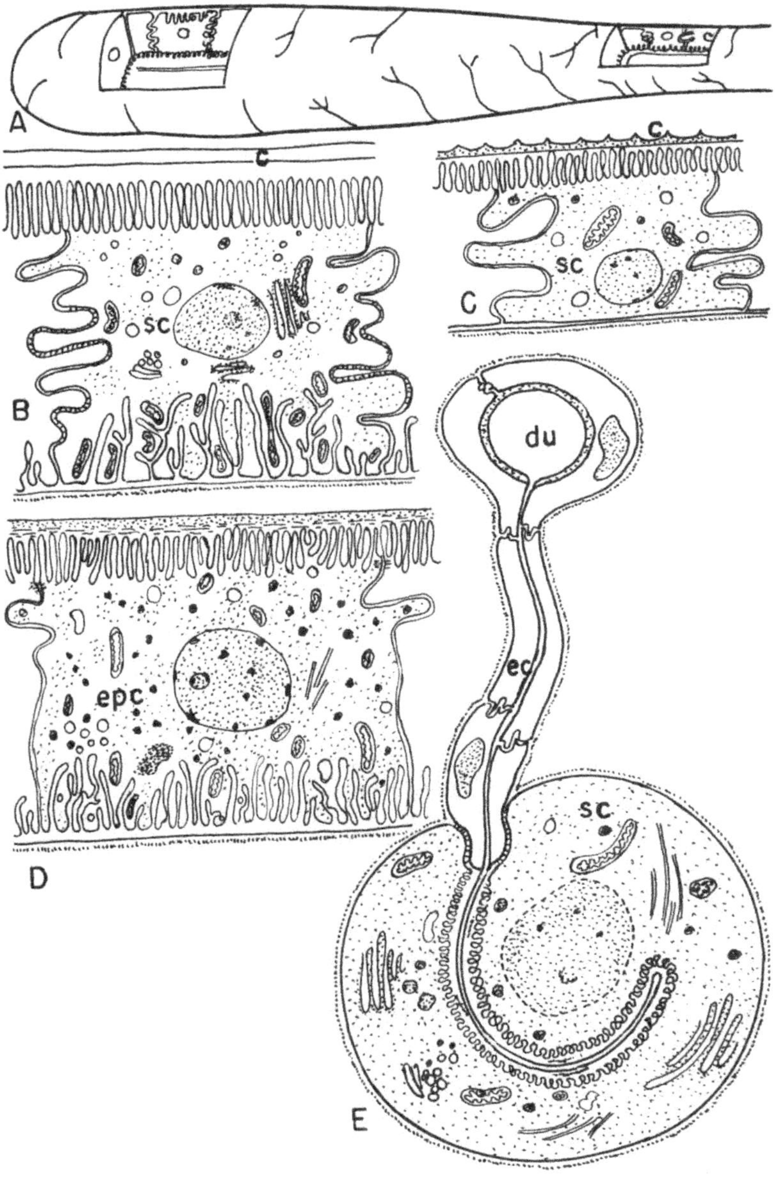

FIGURE 7. Schematic representation of cephalic glands ultrastructure: A- Salivary gland tubule; B- secretory cell (sc) of the tubule; C- epithelium of the salivary collector duct (sc); D- epithelial cell (ep) of the post-pharyngeal gland; E-hypopharyngeal gland. c= cuticle; ic = intracellular canalicule; ec = extracellular canalicule; du = excretory duct.

Hypopharyngeal Gland

The secretory cell of this gland shows an extensive rough endoplasmic reticulum (Fig. 6 A) whose cisternae are filled by a median electron dense material (Fig. 6 B) and several Golgi units. Their cytoplasm also has a cuticular canalicule located in a groove formed by the plasmic membrane invagination. This canal constitutes, as in the mandibular gland, the secretory pole of the cell. The secretion accumulates around it and is composed of electron dense irregular granule (Fig. 6 A). This canalicule also has a microvillar lining.

CONCLUSION

The general morphological organization of the ants' salivary glands system is almost the same as for other Hymenoptera (Cruz-Landim, 1967, 1968). This report attempted to show the morphological variations occurring in ants. Some morphological differences may be seen mainly at the anatomical level. The histology and ultrastructure of the glands are much related with the secretion type produced and, because of this, is much more uniform than the anatomy. Fig. 7 summarizes the ultrastructural features of the cephalic glands.

REFERENCES

ATTYGALLE, A.B., BILLEN, J.P.J. and MORGAN, E.D. 1985. The Postpharyngeal Gland of Workers of *Solenopsis geminata* (Hymenoptera: Formicidae). Actes Coll. Insectes Sociaux, 2 (Diepenbeek): 79-86.

BILLEN, J.P.J. and PEUSENS, G. 1984. Ultrastructure de la Glande Propharyngienne chez les Fournies Formicines (Hymenoptera, Formicidae). Actes Coll. Insectes Soc., 1: 121-129.

BILLEN, J.P.J., ATTYGALE, A.B., MORGAN, D. and OLLETT, D.G. 1987. The contents of the Dufour Gland of the Ant *Pogonomyrmex occidentalis*. Ed. Rembold, Chemistry and Biology of Social Insects. Verlag J. Peperny, München, pp. 426:427.

BILLEN, J. 1987. Morphology and Ultrastructure of the Exocrine Glands in Social Hymenoptera. Ed. Rembold, Chemistry and Biology of Social Insects, Verlag J. Peperny, München, pp. 81-84.

BORDAS, L. 1895. Glands salivaires des Hyménoptères. Annls. Sci. Nat., 19: 1-196.

BROUGH, J.E. 1977. The Morphology and Histology of the Mandibular Gland of an Australian Species of *Calomyrmex* (Hymenoptera: Formicidae). Zoomorphologie 87: 73-86.

BROUGH, J.E. 1978. The Multifunctional Role of the Mandibular Gland Secretion of an Australian Desert Ant *Calomyrmex* (Hymenoptera: Formicidae). Z. Tierpsychol, 46: 279-297.

CAMMAERTS, A.C., INWOOD, M.R., MORGAN, E.D., PARREY, K. and TYLER, R.D. 1978. Comparative study of the pheromones emitted by workers of the ants *Myrmica rubra* and *Myrmica scabrinodis* J. Insect Physiol., 24: 207-214.

CAMMAERTS, M.C., ATTYGALLE, A.B., VOSTROWSKY, O. and BESTMANN, H.J. 1988. Ethological Studies of the Mandibular Gland Secretion of the Ant *Manica rubida* (Formicidae: Myrmecidae). J. Insect Physiol., 34: 347-350.

CRUZ-LANDIM, C. 1967. Estudo comparativo de algumas glândulas das abelhas (Hymenoptera, Apoidea) e respectivas implicações evolutivas. Arq. Zoo., São Paulo, 15: 177-290.

CRUZ-LANDIM, C. 1968. Histoquímica e Ultrestrutura das Glândulas Salivares das Abelhas (Hymenoptyera, Apoidea). Arq. Zool São Paulo, 17: 113-166.

DELAGE-DARCHEN, B. 1976. Les Glandes Post-Pharyngiénnes des Fourmis. Connaissances Actuelles Sur Leur Structure, Leur Fonctionnement, Leur Rôle. Ann. biol. T. XV, Fasc. 1-2, 63-76.

EMMERT, W. 1968. Die Postembryonalentwicklung sebretorischer Koptdrüsen von *Formica pratensis* Retz. und *Apis mellifica* L. (Ins., Hym.). Z. *Morph* Tiere 63: 1-62.

FORBES, J. and MacFARLANE, A.M. 1961. The comparative anatomy of digestive glands in the female and male *Camponotus pensilvanicus* De Geer (Formicidae, Hymenoptera). J.N. York Ent. Soc., 69: 92-103.

GAMA, V. 1978. Desenvolvimento pos-embrionário das glândulas componentes do sistema salivar de *Camponotus (Myrmothrix) rufipes* (Fabricius, 1775) (Hymenoptera: Formicidae), Arq. Zool., São Paulo, 29: 133-183.

GAMA, V. 1985. O Sistema Salivar de *Camponotus (Myrmothrix) rufipes* (Fabricius, 1775) (Hymenoptera: Formicidae). Rev. Brasil. Biol., 45: 317-359.

GAMA, V. and CRUZ-LANDIM, C. 1982. Estudo Comparativo das Glândulas do Sistema Salivar de Formigas (Hymenoptera, Formicidae) Naturalia, São Paulo, 7: 145-165.

JANET, C. 1894. Sur le systeme glandulaire des fourmies. C.R. hebd. Seanc. Acad. Sci., 118: 989-992.

JANET, C. 1898. Systeme glandulaire tegumentaire de la *Myrmica rubra*. Observations diverses sur les fourmies. Éstudes sur les fourmies, les guêpes et les abeilles. Note 17. Paris: Carré et maud.

MORGAN, E.D., INWOOD, M.R. and CAMMAERTS, M.E. 1978. The Mandibular Gland Secretion of the Ant, *Myrmica scabrinodis*. Physiol., Entomol., 3: 107-114.

OTTO, von D. 1958. Uber die Homologreverhältnisse der Pharynse und Maxillardrüsen bei Formicidae und Apidae (Hymenoptera) Zool. Anz, 161: 216-226.

PETRALIA, R.S., SORENSEN, A.A. and VINSON, S.B. 1980. The Labial Gland System of Larvae of the Imported Fire Ant, *Solenopsis invicta*, Buren Cell Tissue Res., 206: 145-156.

PHILLIPS, S.A. and VINSON, S.B. 1979. Comparative anatomy of glands associated with the head among castes of the red imported fire ant, *Solenopsis invicta*, Buren. J. Ga. Ent. 15: 215-226.

PHILLIPS, S.A. and VINSON, S.B. 1980. Comparative morphology of Glands Associated with the Head Among Castes of the Red Imported Fire Ant, *Solenopsis invicta*. Buren. J. Georgia En. Soc., 15: 215-226.

TOLEDO, L.F.A. 1967. Histo-anatomia de glândulas de *Atta sexdense rubropilosa* Forel (Hymenoptera). Arq. Inst. Biol., 34: 321-329.
VINSON, S.B., PHILLIPS, S.A. and WILLIAMS, H.J. 1980. The Function of the Post-Pharyngeal Glands of the Red Imported Fire Ant, *Solenopsis invicta*. Buren. J. Insect Physiol., 26: 645-650.

11

Morphology of the Digestive Tract and Associated Excretory Organs of Ants

F.H. Caetano

INTRODUCTION

The digestive tract of ants generally show marked morphological similarity. Thus, individuals of different subfamilies with different diets show little variation in digestive tract morphology. However, the few changes detected have been used to characterize groups or even subfamilies (Caetano 1988).

DIGESTIVE TRACT MORPHOLOGY

From foregut to the hindgut, the following variations are observed:

Dilation of the Thoracic Portion of the Esophagus

This dilation occurs in *Myrmica, Leptothorax, Stenamma, Monomorium* and *Lasius* physogastric queens. The reproductive female's thorax is filled with wing muscles, which histolyze soon after the nuptial flight (Cruz-Landim and Silva de Moraes 1979). Concomitantly, ovariole development forces the contents of the full crop back to the esophagus where the esophagus can now expand into the void left by the histolyzed wing muscles. The result is the "thoracic crop" (Petersen-Braun and Buschinger 1975). This auxiliary crop was also detected in Ponerinae workers, although the process is due to the fusion of the IV abdominal segment's tergite and sclerite (for the support of sting muscles) (Caetano and Lage Filho 1982;Caetano et al. 1987; Caetano 1988). The "thoracic crop" has been found in other ants, though not to the extent cited above. Examples are *Acromyrmex* (Cristofoletti- Hellmeister and Caetano, personal communication) and *Dolichoderus attelaboides* (Caetano et al. In press).

Crop Size

This structure is extremely large in ants and is commonly dilated by a thick fluid. However, in the four *Atta* species Caetano (1984) studied, queens exhibited a rudimentary crop, while minima workers had a much larger crop than ants of the other castes despite being the smallest caste (Fig. 1). This

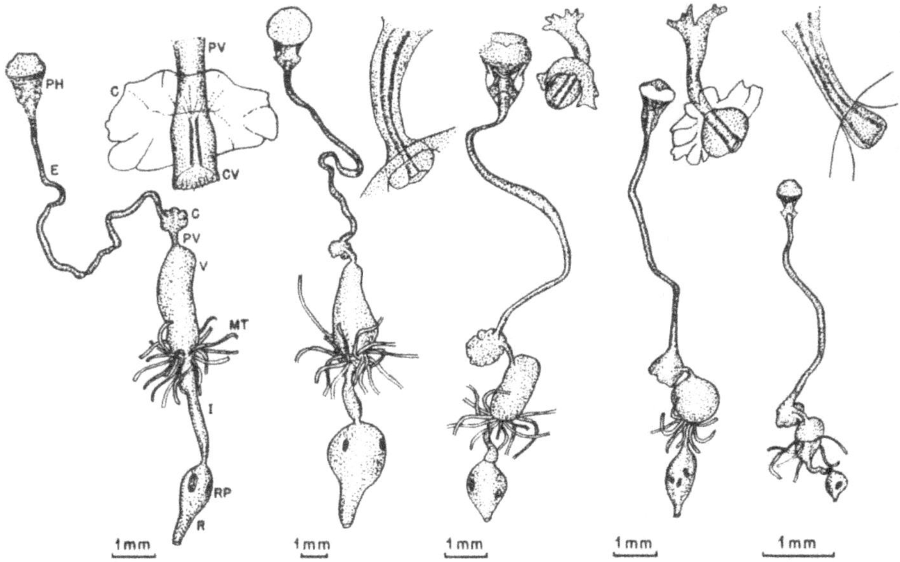

FIGURE 1. Schematic representation of the digestive tract of *Atta sexdens rubropilosa* (from left to right: minima worker, worker, soldier, male, and queen). PH = pharynx; E = esophagus; C = crop; PV = proventriculus; V = ventriculis; CV = cardiac valve; I = ileum; R = rectum; PVB = proventriculus bulb; MT = Malpighian tubules; RP = rectal pads.

may be related to the utilization of the crop to carry food targeted to colony young individuals and adults. This is the primary function of leaf-cutting ant minima workers (Mariconi 1970).

Proventriculus

This apparently insignificant region of extremely reduced size exhibits the most diverse morphological alterations. In some species, the proventriculus consists of a simple tube with no other definition. This type occurs in *Atta, Acromyrmex, Solenopsis, Ectatomma quadridens, Eciton* sp. and *Labidus praedator*. Others have a very elaborate bulb inside the crop. This structure occurs in representatives of the tribe Cephalotini (subfamily Myrmicinae);in the Pseudomyrmecinae; *Azteca bicolor, Tapinoma melanocephala* and *Dolichoderus attelaboides* (subfamily Dolichoderinae); *Odontomachus, Dinoponera gigantea* and *Paraponera clavata* (subfamily Ponerinae). In the subfamily Formicinae, the portion inside the crop is modified in the shape of long sepals. These sepals rest on a second dilation located in the neck of the proventriculus half way between the crop and the ventriculus (Eisner 1957; Caetano and Lage Filho 1982; Caetano 1984, 1988; Caetano et al. 1987; Caetano et al. In press). As to the proventriculus' function, those species without a bulb can retain food in the crop through a short petiole; the bulb's function is performed by the contraction of well developed circular muscles in this region. However, in a proventriculus with a bulb, crop contents may be passively held through the pressure of crop contents on the bulb. This retention of crop contents does not involve muscle action and therefore no energy is expended. This mechanism may be utilized by ants that need to hold food in the crop for long periods of time (Eisner 1957).

Contact of Proventriculus with Ventriculus

This contact usually occurs directly, but in members of the Ponerinae the contact area looks like a bud and is a dilation surrounding the neck of the proventriculus (Caetano and Lage Filho 1982; Caetano et al. 1987; Caetano 1988). Histologically, this is the point of transition from foregut to midgut (Caetano 1988).

Cardiac Valve

The cardiac valve (= stomodaeal valve) is part of the proventriculus and consists of a fold of the foregut epithelium which projects towards the inner part of the ventriculus lumen. The size of the cardiac valve varies, but in most ant species this structure is round and a little larger than the height of the ventricular epithelium. In Ponerinae, however, the cardiac valve is extremely long (Caetano and Lage Filho 1982; Caetano 1984; Caetano et al 1987). The cardiac valve pattern is identical to that observed in wasps by Caetano and Letizio-Machado (1982) and Caetano and Overal (1984).

Ventriculus

The ventriculus portion of the digestive tract does not show wide anatomical variations. It is more elongated in the larger species or castes and rounded in smaller species or castes (Caetano 1984, 1988). Fig. 1 clearly illustrates this situation in the various castes of *Atta sexdens rubropilosa*. The larger individuals (females, males and soldiers) have a more elongated ventriculus than smaller individuals (workers and minima workers).

Histologically, however, the ventriculus shows more or less specific modifications (Fig. 2); the presence of symbiotic microorganisms is easily seen (Caetano and Cruz-Landim 1983; Caetano 1988; Caetano et al. 1989 a,b).

Symbiont microorganisms occur in the ventriculus of Cephalotini ants (subfamily Myrmicinae) and of *Dolichoderus attelaboides* (subfamily Dolichoderinae). Two bacterial types have been detected in Cephalotini, one with two flagella (one at each end), and the other with a single flagellum at one end (Caetano 1989). One bacterial type and a non-septate fungus with dichotomous ramifications have been detected in *D. attelaboides* (Caetano et al. 1989 a,b). The function of these microorganisms in the ventriculus is unknown but they may participate in the process of food digestion (Fig. 3a and b).

In both cases, the ants' diet consists of plant exudates and freshly killed insects, and bird faeces in the case of Cephalotini. Laboratory studies show that both species clearly prefer honey diluted with water.

The epithelium of the ventriculus is formed by digestive cells and generative cells. The digestive cells normally show no organization, but in

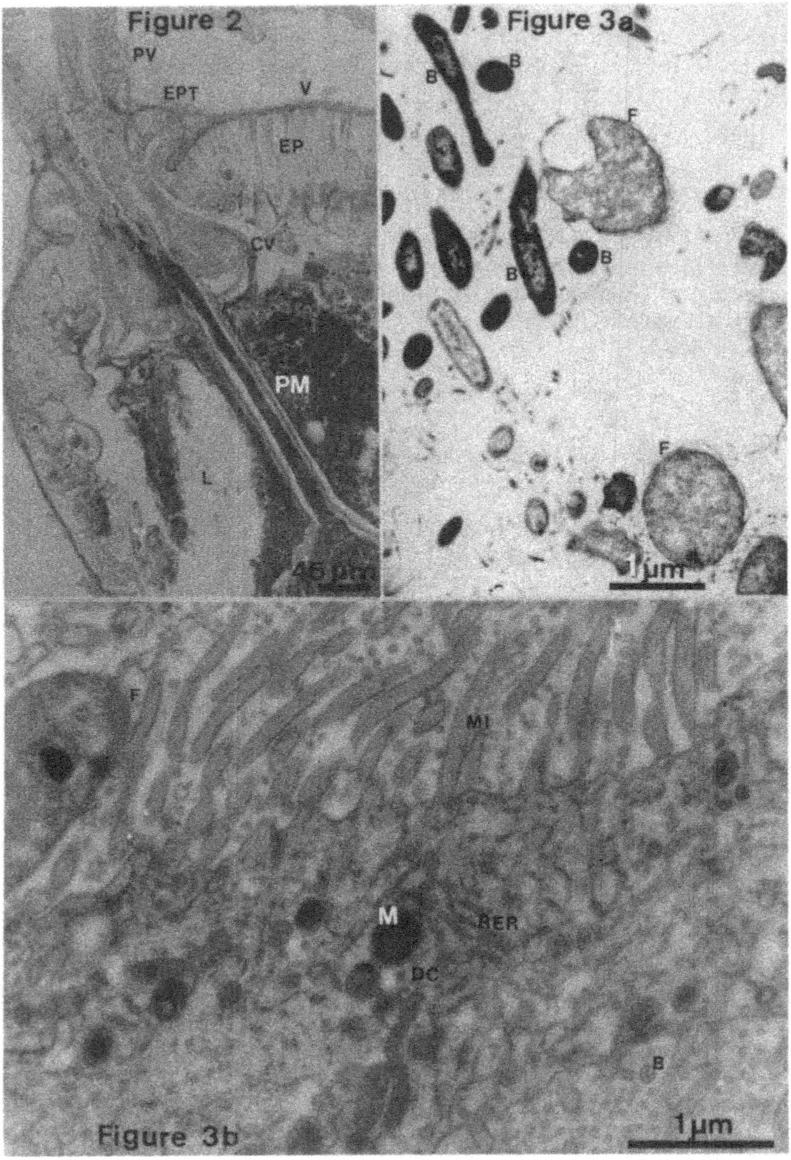

FIGURE 2-3. Anterior section of the ventriculus(V) (*Ectatomma quadridens*) showing the epithelial transition(EPT), Cardiac valve(CV) and peritrophic membrane(PM). PV=proventriculus; EP=epithelium; L=lumen. Figure 3. *Dolichoderus attelaboides*: [A] Cross section of fungus hyphae(F) and bacteria(B) from ventriculus. [B] Apical region detail from digestive cell (DC). Fungus(F) and bacteria(B) among microvilli(MI). M=mitochondria; RER=rough endoplasmatic reticulum.

Dinoponera gigantea and *Paraponera clavata* they are arranged in columns. The generative cells are located at the column bases. A similar arrangement exists in *Stenopelmatus* (Orthoptera) (Caetano et al. 1987).

Peritrophic Membrane

Even though this membrane occurs infrequently in ants, it may be found inside the ventriculus. When present, the peritrophic membrane of ants originates from foregut cells that form the cardiac valve. This membrane is seldom detected in common histological preparations, but is conspicuous in the subfamily Ponerinae (Caetano et al. 1987; Caetano 1988)(Fig. 2). However,in *Camponotus rufipes* this membrane can only be detected with a scanning microscope (Caetano and Hoffmeister 1987) (Figure 4). Of all these species, only *Dinoponera gigantea* showed faeces enclosed by the peritrophic membrane in the rectum. This membrane is believed to be common in ants, though it does not usually exert its primary function, i.e., enclosing the ingested food during the process of digestion/absorption.

Ileum

This is an undifferentiated segment in ants, but among the Cephalotini (subfamily Myrmicinae), the ileum's anteromedial portion consists of a pouch whose epithelium and muscles have peculiar characteristics and which contain a variety of bacterial types (about ten), as well as a filamentous fungus (Caetano and Cruz-Landim 1985 a; Caetano 1989). The function of these microorganisms appears related to processing food material not utilized by the ant with subsequent release of products used by the host insect.

Among Ponerinae and some Myrmicinae, the anterior portion of the ileum - immediately after the passage from the midgut to the hindgut - shows dilation which varies from species to species (Caetano and Lage Filho 1982; Caetano 1984, 1988; Caetano et al. 1987; Cristofoletti-Hellmeister and Caetano, personal communication).

The meaning of this dilation has only been explained for *Dinoponera gigantea, D. australis* and *Paraponera clavata*. In these species, the region also exhibits cuticular spines; their probable function may be to pull the peritrophic membrane, with its contents, toward the hindgut and also to perforate it, thus permitting more effective passage of the contents of the peritrophic membrane envelope into the ileal lumen. The entire process may be aided by the well-developed circular and longitudinal muscles that occur in this region (Caetano et al. 1987; Caetano and Hofmeister 1988) (Fig. 5).

These spines characterize two portions of the anterior ileal region: the first has single toothed cuticular spines, while the second has spines with up to ten teeth (Caetano and Hofmeister 1988).

ASSOCIATED EXCRETORY STRUCTURES

Two excretory structures are intimately associated with the digestive tract, i.e. Malpighian tubules and rectal pads.

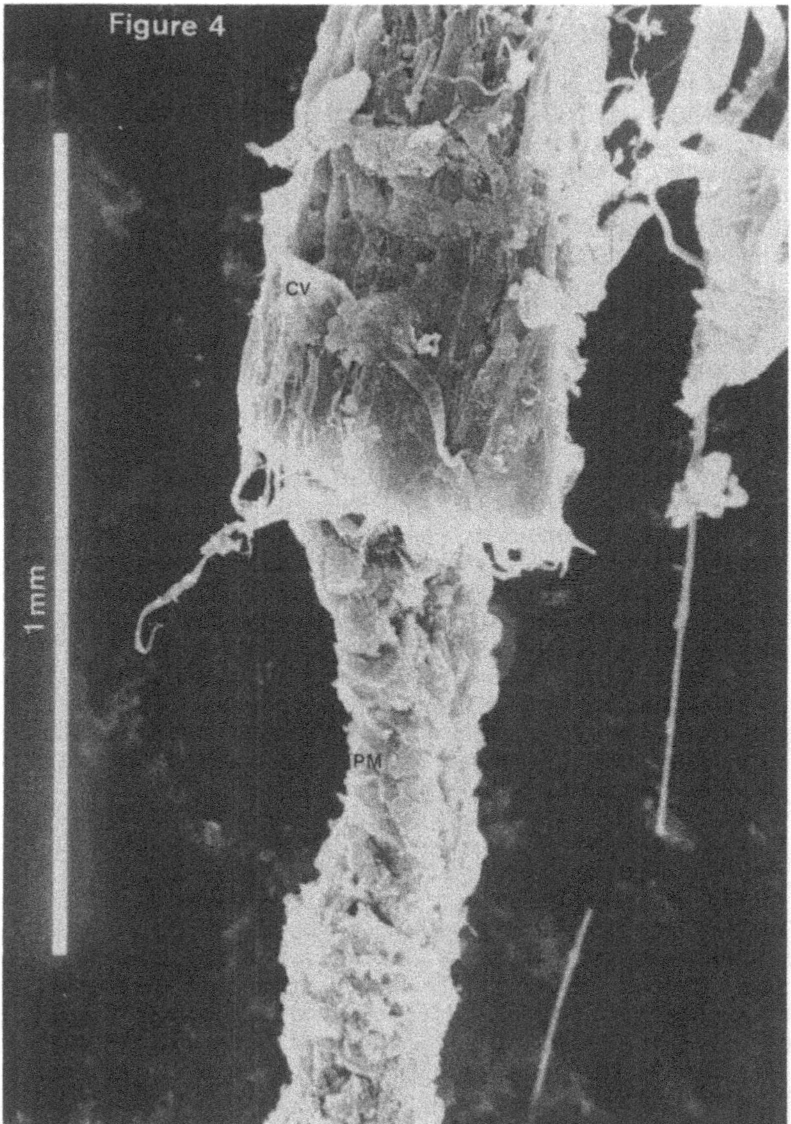

FIGURE 4. Scanning electron micrograph of cardiac valve (CV) and peritrophic membrane (PM) of *Camponotus rufipes*.

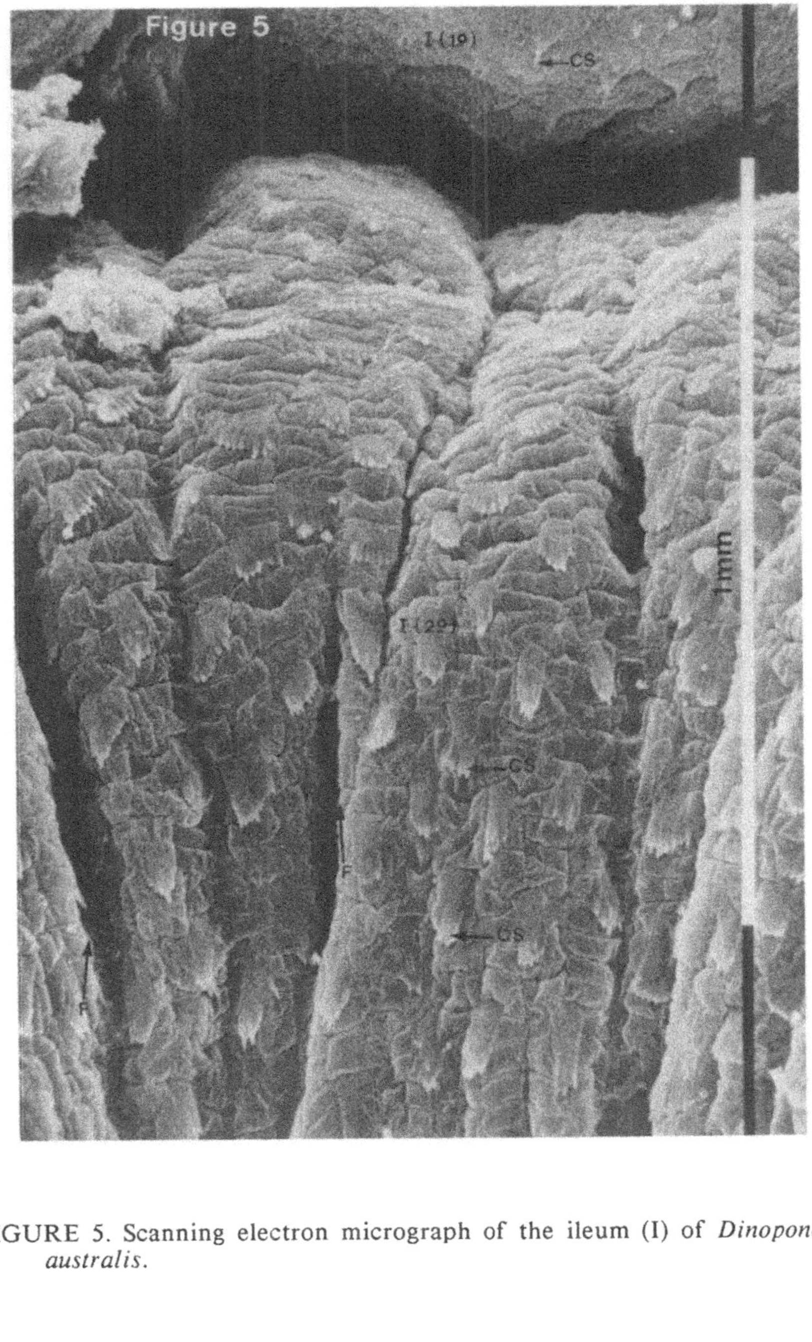

FIGURE 5. Scanning electron micrograph of the ileum (I) of *Dinoponera australis*.

Malpighian Tubules

These tubules are filiform structures with a free distal end in the body cavity between the gaster's organs. In *Solenopsis saevissima*, however, the distal end of these tubules adheres to the rectum wall (Caetano 1984). Each species has a constant number of Malpighian tubules, but the number varies with caste (Caetano and Cruz-Landim 1972; Caetano 1984). In *Atta*, the number of Malpighian tubules decreases from queens to minima workers.

In *Atta sexdens rubropilosa*, for example, the number of tubules for queens, males, soldiers, workers and minima workers is 34, 27, 18, 14 and 8, respectively (Caetano 1984). These structures are always multiples of two. The tubules originate from junction of the ventriculus and ileum, where they empty independently. Their primary function is to remove metabolites from hemolymph (Wigglesworth 1974).

Metabolite removal is aided by thin, spiraling muscle fibers present on the outer surface (Chapman 1975; Caetano and Cruz-Landim 1986). Contraction of these spiral fibers may promote a more rapid shift of luminal content from the distal to the proximal region and finally to the ileal lumen. It may also help in the movement of fluid from the body cavity.

The Malpighian tubules of *Bombus atratus* bees have a secretory function; the secreted material is composed of proteoglycans and acid glucosaminoglycans that will be used to build the cocoon (Mello, 1979). A similar function is assigned to the Malpighian tubules of the stingless bees *Plebeia droryana* and *Scaptotrigona postica* (Mello and Kerr 1984).

In *Dolichoderus attelaboides*, the Malpighian tubules of adult workers have cells characteristics of secretory cells, rich in vesicles of low electron density material. In the proximal portion of the tubule the lumen is filled with material identical to that present in the above cells. The composition of this substance is unknown, but seems to contain lipids, probably steroid hormones, since these cells are rich in smooth endoplasmic reticulum (Caetano et al. 1989) (Fig. 6A, B).

Rectal Pads

Formicidae rectal pads vary in shape, number and distribution on the rectal wall (Caetano 1984, 1988). Usually they are round and distributed in the median or anterior portion of the rectum, and more rarely in the posterior region. When this occurs, a total of six rectal pads are distributed two by two in the anterior, median and posterior regions of the rectum (Caetano and Lage Filho 1982; Caetano 1988).

The number of rectal pads is constant within the same subfamily. Two are present in Ecitoninae, six in Pseudomyrmicinae, Formicinae and Ponerinae, and three in Myrmicinae and Dolichoderinae (Caetano and Lage Filho 1982; Caetano 1984, 1988; Caetano et al. 1987). However, *Dolichoderus attelaboides* has six rectal pads, in disagreement with the number reported above for the subfamily Dolichoderinae. This species may be the least derived in the genus *Dolichoderus*, which in turn is the least derived in the subfamily. Thus, the genus may be phylogenetically very close to the subfamily Ponerinae, which has six rectal pads (Caetano et al. In press; Mackay, personal communication).

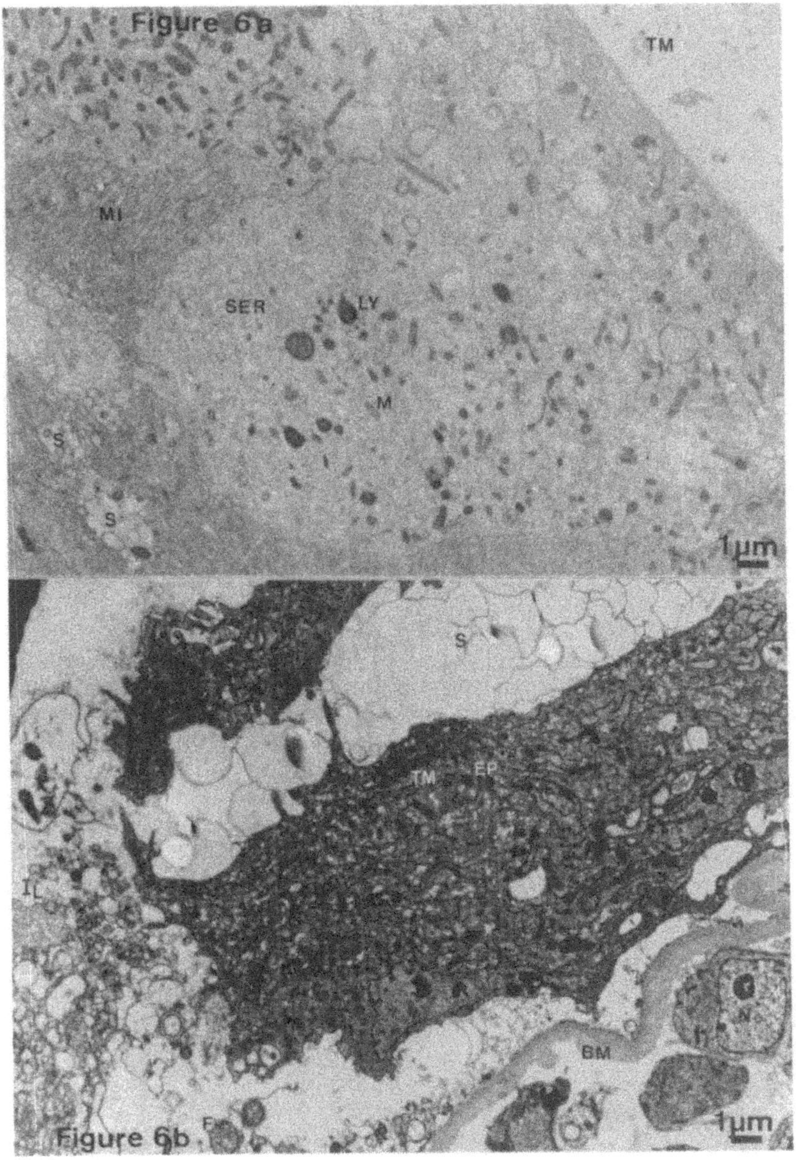

FIGURE 6. Malpighian tubules of *Dolichoderus attelaboides*. A. General view of the Malpighian tubule cells with microvilli (MI), Lysosomes (LY), few mitochondria (M) and the smooth endoplasmic reticulum (SER). Secretion (S) in the lumen. B. The contact of Malpighian cells with the ileum. S = Secretion; IL = ileum lumen; BM = Basement membrane.

The genus *Dinoponera* (subfamily Ponerinae) presents a strong chitin ring around the rectal pads. This ring is visible in morphological studies and known to occur only in this genus (Caetano et al. 1987).

REFERENCES

CAETANO, F.H., 1984. Morfologia comparada do trato digestivo de formigas da subfamilia Myrmicinae (Hymenoptera, Formicidae). Papéis Avulsos Zool., 35: 257-303.

CAETANO, F.H., 1988. Anatomia, histologia e histoquímica do sistema digestivo e excretor de operárias de formigas (Hymenoptera: Formicidae). Naturalia, 13: 129-174.

CAETANO, F.H., 1989. Endosymbiosis of ants with intestinal and salivary gland bacteria. Chap. 5, In Schwemmler, W. and Gassner, G., (eds.), Insect Endocytobiosis: morphology, physiology, genetics, evolution. C.R.C.Press, Boca Raton, 272pp.

CAETANO, F.H., M.I. CAMARGO-MATHIAS and W.L. OVERAL. 1986/1987. Anatomia e histologia comparada do trato digestivo de *Dinoponera gigantea e Paraponera clavata* (Formicidae, Ponerinae). Naturalia, 11/12: 125-134.

CAETANO, F.H., M.I. CAMARGO-MATHIAS, M.A. LIMA PIMENTEL, and M.E.M. TOMOTAKE. 1989. Tipos de microorganimos presentes no ventrículo de *Dolichoderus decollatus* (Formicidae: Dolichoderinae). Anais do XII Colóquio Soc. Bras. Micr. Elet., Caxambu.

CAETANO, F.H. and C.da. CRUZ-LANDIM. 1972. Morfologia comparada do trato digestivo de formigas do gênero *Atta* (Hymenoptera: Formicidae). Vol. Homenagem à W.E.Kerr, Rio Claro, SP, 22-26.

CAETANO, F.H. and C.da. CRUZ-LANDIM. 1983 - Ultra-estrutura das células colunares do ventrículo de *Camponotus arboreus* (Hymenoptera: Formicidae) e suas implicações funcionais. Naturalia, 8:91-100.

CAETANO, F.H. and C.da. CRUZ-LANDIM. 1985a. Presence of microorganisms in the alimentary canal of ants of the tribe Cephalotini (Myrmicinae): Location and relationship with intestinal structures. Naturalia, 10: 137.

CAETANO, F.H. and C. da. CRUZ-LANDIM. 1985b. Nota sobre a presença de fibras musculares nos túbulos de Malpighi de formigas *Neoponera villosa* (Hymenoptera: Ponerinae). X Colóquio Soc. Bras. Mic. Eletr., São Paulo.

CAETANO, F.H. and R.M. HOFMEISTER. 1987. Presença de membrana peritrófica em *Camponotus rufipes* (Hymenoptera: Formicidae). An. XI Colóquio Soc. Bras. Mic. Eletr., Caxambu.

CAETANO, F.H. and R.M. HOFMEISTER. 1988. Presence of cuticular spines in proctodeo of ants *Dinoponera australis* (Hymenoptera: Ponerinae). 4th International Congress of Cell Biology, Montreal.

CAETANO, F.H. and A.L. LAGE FILHO. 1982. Anatomia e histologia do trato digestivo de formigas do gênero *Odontomachus* (Hymenoptera: Ponerinae). Naturalia, 7: 125-134.

CAETANO, F.H. and V.L. LETIZIO-MACHADO. 1982. Morfologia e histologia do trato digestivo de *Protopolybia exigua exigua* (Hym., Vespidae) e suas estruturas excretoras anexas. Papéis Avulsos Zool., 34(25): 283-296.

CAETANO, F.H., M.A.L. PIMENTEL, M.E.M. TOMOTAKE. and M.I. CAMARGO- MATHIAS. 1989. Aspectos ultraestruturais do ventrículo de *Dolichoderus decollatus* com relação àpresença de microorganismos (Formicidae: Dolichoderinae). Anais XII Colóquio Soc. Bras. Mic,. Elet., Caxambu.

CAETANO, F.H. and W.L. OVERAL. 1984. Estudo do trato digestivo de vespas (Hymenoptera, Vespidae) e estruturas associadas. I. Morfologia. Rev. Bras. Ent. 28: 403-408.

CAETANO, F.H., M.E.M. TOMOTAKE, M.A.L. PIMENTEL and M.I. CAMARGO- MATHIAS. 1989. Presença de células secretoras nos túbulos de Malpighi de *Doliochoderus decollatus* (Formicidae: Dolichoderinae). An. XII Colóquio Soc. Bras. Mic. Elet., Caxambu.

CAETANO, F.H., M.E.M. TOMOTAKE, M.A.L. PIMENTEL and M.I. CAMARGO-MATHIAS. 1989. Morfologia interna de operárias de *Dolichoderus attelaboides* Fr. Smith, 1858 (Formicidae: Dolichoderinae). I. Trato digestivo e sistema excretor anexo. (in press).

CHAPMAN, R.F. 1975. The insects, structure and function. American Enselvier Publishing Company, Unc., New York, XII + 819pp.

CRUZ-LANDIM, C.da and R.L.M. SILVA DE MORAES. 1979. Electron microscopic study of flight muscle degeneration in queens of *Atta sexdens* (Hymenoptera: Formicidae). Cytobios, 24:13-23.

EISNER, T. 1957. A comparative morphological study of the proventriculus of ants (Hymenoptera: Formicidae). Bull. Mus. Comp. Zool. Harvard College, 116: 438-490.

MARICONI, F.A.M. 1970. As saúvas. Editora Agronomica Ceres, São Paulo. XI + 167pp.

MELLO, M.L.S. 1979. A mucous secretion in the Malpighian tubes of a Neotropical Bumblebee, *Bombus atratus* Franklin. Protoplasma, 99: 147-158.

MELLO, M.L.S. and W.E. KERR. 1984. Histochemistry on salivary gland and Malpighian tubule secretions crontributing to the Cocoon in *Plebeia droryana* and *Scaptotrigona postica* (Hymenoptera: Apoidea). Zool. Anz, Jena 213, 3/4, s. 177-189.

PETERSEN-BRAUN, M. and A. BÜSCHINGER. 1978. Entstehung und funktion eines thorakalen kroptfes bei Formiciden Königinnen. Insectes Sociaux, 22: 51-66.

ROEDER, K.D. 1953. Insect Physiology. John Wiley, Sons. Inc., New York., 1100pp.

WIGGLESWORTH, V.B. 1974. The principles of insect physiology. John Wiley & Sons. Inc., New York, 827pp.

Reproduction

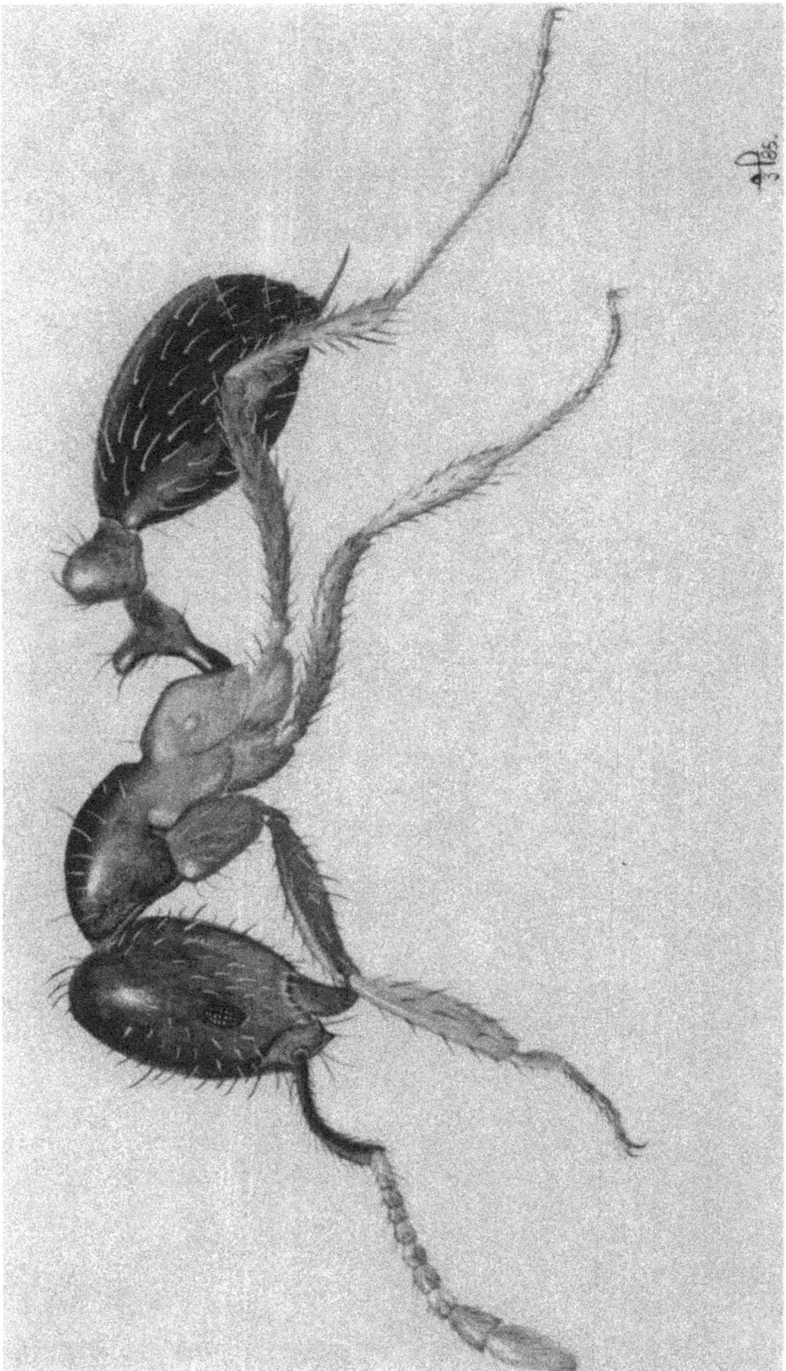

Solenopsis geminata (x 35). Drawing by Eduardo Perez.

Overview

L. Passera

Study of ant reproduction depends on the availability of biological material. Thus, major advances were made in the fifties from observations on the european red wood ant by Gösswald and Bier and others (see Schmidt, 1974); in the sixties and seventies from experiments on *Myrmica rubra* by Brian (Brian, 1979, 1989). Because of their abundance, easy rearing in standard units, economic importance and availability of funding, pest-ants are becoming good candidates for investigations concerning reproduction. As a consequence, current research focuses on pest ants, e.g. fire ants, leafcutting ants, Argentine ants or Pharaoh's ants. All these studies involve the major topics of colony founding, egg laying and social control.

COLONY FOUNDING

The mechanisms by which pest ants found new societies are no different from other ant species. Monogynous species, e.g. *Atta mexicana* or the monogynous form of *Solenopsis invicta*, employ independent colony founding, where newly mated queens excavate a cell after the nuptial flight and rear the first workers without help. In contrast, newly mated queens of polygynous species need worker cooperation, since they are unable to nurture the brood. This dependent colony founding is achieved by different means : (A) the queens may mate in their natal nest without taking flight (*Monomorium pharaonis*, *Iridomyrmex humilis*; or (B) They may be adopted by an alien colony after the nuptial flight, as suggested for polygyne *Solenopsis invicta* (Glancey and Lofgren, 1988); or (C) Newly mated queens may cooperatively found a colony, as for *Atta texana* , where queens excavate a cell together and cooperate in collective egg production. This is a rare example of primary polygyny (Mintzer, 1987 and this section). Perhaps this behavior facilitates pest ant establishment, since it also occurs in the pest ant, *Tapinoma simrothi* (Hanna and Hanna, in preparation).

Dependent and independent founding were shown to be linked to the amount of fat stored in gynes before the mating flight (Keller and Passera, 1989). In the former case, queens have a far lower relative fat content (g fat/g dry weight <40%) than in the latter (>40%). The fat body is presumeably used to provide energy to the foundress during the claustral period. It would be useful to estimate fat body content in gynes just prior to nuptial flights for both monogynous and polygynous fire ant colonies. Fletcher et al. (1980) and Willer and Fletcher (1986) showed that the two forms differ in weight. Monogynous colony inseminated queens weigh 18 to 23 mg, whereas their polygynous colony counterparts range from 10 to 12 mg. So the amount of fat body should be lower in polygynous queens than in monogynous queens. But the weight differences would be more important if the relative fat content were known. It would also be of interest to estimate the maturation period between emergence and mating flights. Differences between monogynous and polygynous founding behavior might be responsible for the amount of stored fat. These studies should provide insight into the origin and the future (founding) of these two intriguing queen forms.

During claustral founding in monogynous species it was shown that foundresses lose weight. In *Solenopsis invicta* (Williams, this section) the foundress weight (monogynous form) dropped to half their initial value during the first 30 days. These data are similar to those obtained in *Lasius flavus* (Passera, unpublished). In the time from landing after nuptial flight to emergence of the first workers, queens lost 49% of their initial weight. Surprisingly, weight loss in *Atta texana* and *mexicana* was lower (30-35%) (Mintzer, this section). Mintzer postulates that in leafcutting queen foundresses, a large part of female effort is devoted to the fungus culture, an estimated 40% of the weight loss. If confirmed, these foundresses invest only about 20% of their energy to rearing the first workers.

Another salient point from this section concerns the queen's weight loss. *Atta texana* foundresses continue to lose weight two or three months after the first workers emerge indicating that their energy investment does not end at this point in the cycle. In contrast, young fire ant queens gain weight immediately after minum workers emerge, about 30 days (cf Williams, this section). Facile colony founding in fire ants compared to other ant species may help explain the spreading process of this species.

There is a need for additional studies regarding the energy budget displayed by foundresses of various persuasions (e.g. dependent vs independent, haplometrosis vs pleometrosis, ...etc.). The estimation of energy investment in founding may contribute to a better understanding of the selected strategies.

OVIPOSITION

After colony founding, the queens oviposition rate determines colony growth. Data reported in this section for monogynous (Williams) and polygynous fire ant colonies (Vargo) agree that queen weight is highly correlated with the number of eggs laid. This correlation is generally true, as also demonstrated in monogynous/polygynous units of *Plagiolepis pygmaea* (Mercier et al. 1985) and *Iridomyrmex humilis* (Keller, 1988). This correlation provides a quick and safe way to estimate queen fecundity. Direct estimation of egg laying using oviposition tests may actually underestimate the true value, as demonstrated by Della Lucia (this section). In *Atta sexdens*, the number of eggs directly collected inside the nest by means of a container attached to the queen's abdomen is 1.5 times higher than for queens isolated in a test box. Other methods also have disadvantages; i.e. the number of eggs laid based on counts recorded from nests at regular time intervals also underestimates egg number, since workers consume an unknown number of eggs.

Egg viability is an important phenomenon, because it pertains to societal growth. On the other hand, trophic eggs are suspected to be important during founding, as they are fed to larvae and later in the ergonomic stage they interfere in caste determination control (Bartels, 1988). Trophic eggs are well documented in fire ants and leafcutting ants. Dealated fire ant virgin queens removed from the inhibitory influence of the inseminated queen laid large quantities of trophic eggs. Intriguingly, queens in monogynous colonies almost never laid nonviable eggs whereas their polygynous counterparts laid about 25% trophic eggs. The shift between viable and nonviable egg laying may be under JH control. In fact, virgin queens topically treated with JH laid an increased percentage of embryonated eggs (0% to 16%) (Voss et al., 1988). It would be interesting to see if JH treatment increases the number of trophic eggs laid by polygynous queens.

The fire ant would make an excellent candidate for study of the role morphogenesis plays in ants. Whereas gonadotropic JH effects are well documented in many non-social insects, data are scarce concerning ants. *Solenopsis invicta* queens treated topically with JH stopped laying eggs for two days (Robeau and Vinson, 1976), although, data obtained by Barker (1978) were not in agreement. Extirpation of virgin queen CA resulted in decreased egg production, which could be restored by topical JHA application. Whether JH has gonadotropic action in fire ants is still under discussion. Similarly, in *Plagiolepis pygmaea*, after topical application of JHA, the treated queens laid slightly more eggs than control queens (Passera and Suzzoni, 1974), whereas JH treatment of *Camponotus athiops* queens resulted in the cessation of egg laying (Suzzoni et al., 1986).

The fecundity of ant queens is dependent on the number of workers. There is a general relationship between queen and colony age and queen fecundity. Observations made by Keller and Passera (in press) supported such an interpretation in monogynous species e.g. *Camponotus herculeanus*, *Camponotus ligniperda* and *Lasius niger*. The fecundity of their queens after the mating flight was related to their age and the number of workers. Della Lucia (this section) confirmed this fact in *Atta sexdens*. Using garden volume as an index of colony size, Della Lucia showed that queens from large colonies laid more eggs than their counterparts from small colonies. But this result was not verified in the fire ant. Both Vargo and Williams (this section), Tschinkel (1988) and Vargo and Ross (1989) demonstrated that queen fecundity was not related to the number of workers, at least in medium sized societies. The apparent keystone of queen fecundity in fire ants is the number of last instar larvae. Another pest ant, *Monomorium pharaonis*, exhibited a similar mechanism (Borgensen 1989). Queens with large larvae produced three or four times as many eggs as queens without larvae; this increased fecundity seems to be due to liquid secretions licked by the queens from various larval parts.

SOCIAL CONTROL

The methods by which ants control and determine the production of sexuals have been under investigation for many years. Considerable progress has been made in discovering the role played by trophogenetic and neuroendocrine factors. Sexuals may well be reared as follows: A) dependent on the presence/absence of the queen, nurse workers can choose between a double larval feeding program, specifically over and underfeeding. B) in response to this nutritional switch, the larva's neuroendocrine system is stimulated - neuroendocrine hormones (JH and ecdysteroids) act during sensitive periods triggering a particular morphogenetic program (Passera and Suzzoni, in press). How workers know the queen is absent or present is unknown. Investigations reported in this section by Vargo present evidence of an inhibitory queen influence mediated by a pheromone. This is supported by the fact that corpses of functional mated queens inhibited the production of sexual larvae and furthermore resulted in the killing of previously existing sexualized larvae. This is the first time that a primer pheromone similar to the bee's queen substance has been observed in ants. Additional support for this type of pheromone action was recently found in the Argentine ant (Vargo and Passera, 1989 and in preparation). The daily addition of a freshly killed queen corpse to an experimental unit resulted in complete inhibition of female sexual production. This inhibitory affect was suppressed in large measure when freshly killed corpses were rinsed in pentane.

Finally, field studies of reproduction in pest ants, e.g. fire ants, leafcutting ants, Argentine ants or Pharaoh's ant, have a double effect. They increase the scope of fundamental knowledge and also present practical implications. Whatever means is selected to control pest-ants, researchers need to know the basic mechanisms which regulate the production of workers and sexuals.

REFERENCES CITED

BARKER J.F. 1978. Neuroendocrine regulation of oocyte maturation in the imported fire ant *Solenopsis invicta.* Gen. Comp. Endocrinol. 35: 234-237.

BARTELS P.J. 1988. Reproductive caste inhibition by Argentine ant queens: a new mechanism of queen control. Insectes Soc. 35: 70-81.

BORGENSEN L.W. 1989. A new aspect of the role of larvae in the Pharaoh's ant society. Insectes Soc. 36: 313-327.

BRIAN M.V. 1979. Social control over sex and caste in bees, wasps and ants. Biol. Rev. 55: 379-415.

BRIAN M.V. 1989. Social factors affecting queen fecundity in the ant *Myrmica rubra.* Physiol. Entomol. 14: 381-389.

FLETCHER D.J.C., M.S. BLUM, T.V. WHITT and N. TEMPLE. 1980. Monogyny and polygyny in the fire ant, *Solenopsis invicta.* Ann. Entomol. Soc. Amer. 73: 658-661.

GLANCEY B.M. and C.S. LOFGREN. 1988. Adoption of newly-mated queens: a mechanism for proliferation and perpetuation of polygynous red imported fire ants, *Solenopsis invicta* Buren. Fla. Entomol. 71: 581-587.

KELLER L. 1988. Evolutionary implications of polygyny in the Argentine ant, *Iridomyrmex humilis* (Mayr) (Hymenoptera : Formicidae) : an experimental study. Anim. Behav. 36: 159-165.

KELLER L. and L. PASSERA. 1989. Size and fat content of gynes in relation to the mode of colony founding in ants (Hymenoptera, Formicidae). Oecologia. 80: 236-240.

KELLER L. and L. PASSERA. Fecundity of ant queens in relation to their age and the mode of colony founding. Insectes soc. (in press);

MERCIER B., L. PASSERA and J.P. SUZZONI. 1985. Etude de la polygynie chez la fourmi *Plagiolepis pygmaea* Latr. (Hym. Formicidae). I. La fécondité des reines en condition expérimentale monogyne. Insectes Soc. 32: 335-348.

MINTZER A.C. 1987. Primary polygyny in the ant *Atta texana*: number and weight of females and colony foundation success in the laboratory. Insectes Soc. 34: 108-117.

PASSERA L. and J.P. SUZZONI J.P. 1974. Action d'un analogue de l'hormone juvenile sur la fécondité d'une fourmi (Hymenoptera, Formicidae). C.R. Acad. Sc. Paris, D. 279: 2079-2082.

PASSERA L. and J.P. SUZZONI. Role of morphogenetic hormones in caste polymorphism in ants. In "Morphogenetic Hormones of Arthropods" (A.P. Gupta, ed.). Rutgers University Press (in press).

ROBEAU R.M. and S.B. VINSON. 1976. Effects of juvenile hormone analogues of caste differentiation in the imported fire ant, *Solenopsis invicta*. J. Georgia Entomol. Soc. 11: 198-203.

SCHMIDT G.H. 1974. Steverung der Kastenbildung und Geschlechtsregulation im waldameisenstaat. In "Sozialpolymorphismus bei Insekten" (G.H. Schmidt, ed.). Wissensch. Verlag. MBH, Stuttgart, 404-512.

SUZZONI J.P., A. GRIMAL, and L. PASSERA. 1986. Modalités et cinétique du développement larvaire chez la fourmi *Camponotus æthiops* Latr. Bull. Soc. Zool. Fr. 111: 113-121.

TSCHINKEL W.R. 1988. Social control of egg-laying rate in queens of the fire ant, *Solenopsis invicta*. Physiol. Entomol. 13: 327-350.

VARGO E.L. and L. PASSERA, 1989. Etude de quelques facteurs impliqués dans la détermination de la caste royale chez la fourmi d'Argentine *Iridomyrmex humilis* Mayr. Actes coll. Insectes Sociaux. 5: 101-108.

VARGO E.L. and K.G. ROSS. 1989. Differential viability of eggs laid by queens in polygyne colonies of the fire ant, *Solenopsis invicta*. J. Insect Physiol. 35: 587-593.

VOSS S.H., J.F. McDONALD and C.H. KEITH. 1988. Production and abortive development of fire ant trophic eggs. In "Advances in Myrmecology" (J.C. Trager, ed.). Brill, Leiden, 517-534.

WILLER D.E.and D.J.C. FLETCHER. 1986. Differences in inhibitory capability among queens of the ant *Solenopsis invicta*. Physiol. Entomol. 11: 475-482.

12

Reproductive Strategies of the Fire Ant

A.P. Bhatkar

INTRODUCTION

Species that occur in monogyne and polygyne colonies are excellent subjects for reproductive strategy studies. Most solenopsidine ants (Tribe Solenopsidini, Subfamily Myrmicinae) occur in various grades of monogyne and polygyne forms in North America (Bhatkar 1989), oligogyny being a transient condition in both. One neotropical species, *Solenopsis invicta* Buren, is widespread in southern North America and discrete in its forage/nest habits (Bhatkar and Vinson 1987a,b). Comprehensively studied mono- and polydomous forms of *S. invicta* are dispersed syntopically in the south central and southeastern U.S.A. Polygyny is as much a characteristic of the polydomous form as monogyny of the monodomous form. Early records document possible oligogyny in *S. invicta* (Green 1952) and *S. geminata* (Banks et al. 1973), with sporadic oligogyny in *S. xyloni* (Summerlin 1976) and *S. richteri*. Since then, polygyny in *S. invicta* in Texas (Hung et al. 1974, Mirenda and Vinson 1982, Greenberg et al. 1985), Mississippi (Glancey 1975), Georgia (Fletcher et al. 1980, Fletcher 1983) and Florida (Glancey et al. 1987), and *S. geminata* in Mississippi (Adams et al. 1976) has been discovered in the southern U.S.A. The occurrence of mono- and polygyne forms should be expected throughout the neotropical distribution of the *S. saevissima*-complex and pantropical distribution of *S. geminata*-complex, analogous to that of the palearctic *Formica*-complexes (Gösswald 1951, Pamilo and Rosengren 1983). The findings on *S. invicta*'s reproductive strategies and relevant data on *S. geminata* from tropical Mexico during 1980-84 are summarized in this chapter.

MATERIALS AND METHODS

The reproductive investment of *S. invicta* colonies was based on the biomass of sexuals, trapped during nuptial flights using a portable flight cage (Bhatkar, unpubl.). Sexuals were collected from various localities in central and southern Texas during 1986 - 88. Air and mound temperatures (at 15 - 30 cm depth), RH and wind velocity were noted before and during the nuptial flights. Comparative data on red and black forms of *S. geminata* in north Florida and humid tropical Mexico were taken from the studies during

1975 - 84. The percent sterility in males was determined by dissecting individuals under a 40X stereoscope from earlier-characterized mounds and comparing their reproductive systems with those of normal males (Hung et al. 1974). Mounds were characterized as polygyne using Greenberg et al.'s (1985) worker morphometry criterion and if more than two fertile gynes were present. Thoracic width was also used to differentiate sterile and normal males. Colony founding and biomass studies were conducted on postnuptial and inseminated gynes, kept singly and in groups of 5 - 10, in castone-base petri dishes at 25 ± 3° C and 60 ± 5% RH. One sample chi-square and contingency tests were used for biomass analyses. F-tests on Rayleigh's statistic (R) for angular distribution of the nuptial flight times were used to evaluate significant differences between sets of observations (P < = 0.01) (Zar 1974).

RESULTS AND DISCUSSION

Reproduction or generation of offspring with the same potential as the parents is achieved by organisms through sexual and asexual means. In subsocial and social insects, asexual reproduction is a rule in order to utilize locally ephemeral patches of resources at a rapid rate (female-producing thelytokous parthenogenesis, such as that in aphids, parasitic Hymenoptera, etc.), and to create haploid male images of the self for sexual reproduction and dispersion (see Wilson 1971). These conditions seem to be a property of the tropical riparian forest ecosystems where social insects have their largest biomass (Erwin 1983). This reproductive strategy ensures the outbreeding of diploid females with haploid males (thus reducing the accumulation of deleterious genes), and create an outcrop of diploid female offsprings, a majority of which forfeit their reproductive success (no gonads) to that of the mother, and resort to a colonial life for "inclusive fitness" with their kin (Hamilton 1964).

The inseminated reproductive female (queen) in a mature colony produces both reproductive diploid females (alates, gynes) and haploid males. The production of alate offspring and their mating flights are synchronized with correct environmental conditions in order to maximize successful mating and colony foundation. The overall reproductive strategy involves, (1) energy investment in viable reproductives, (2) development of a successful mating system, (3) development of a successful dispersal mechanism, and (4) development of a successful colony founding mechanism.

Investment in Viable Reproductives

Investment in the sexes by monogyne *S. invicta* is somewhat higher (mean ± SD: female = 13.15 ± 1.75, male = 5.57 ± 1.08 mg), than its polygyne counterpart (female = 10.04 ± 1.75, male = 4.43 ± 0.74 mg); however, the overall difference on the same flight dates is insignificant (chi-square contingency, polygyne vs monogyne: = 0.77 for females, p = 0.68, and 0.63 for males, p = 0.68, df = 2, n = 50 - 200 males, 500 females) (Fig. 1). The biomass of landed sexuals on the same day but from different localities (female = 11.51 ± 1.98 - 14.34 ± 1.91 mg, male = 5.91 ± 0.84 mg) could not

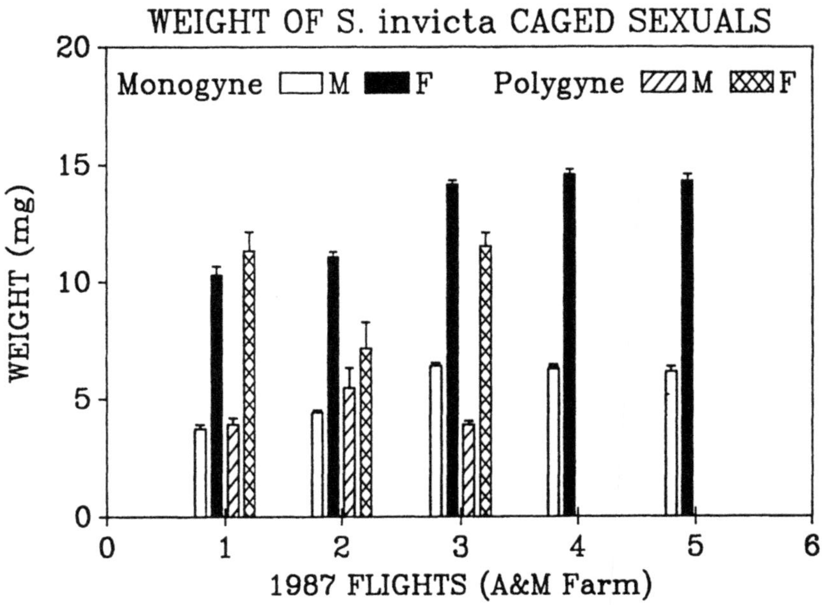

FIGURE 1. Weights of *Solenopsis invicta* caged sexuals from separate flights during March-September 1987 at the A&M Farm, Brazos County, Texas. M = males, F = females.

be differentiated from either monogyne or polygyne reproductives captured in flight traps. This is probably due to the mixed landing by both from different areas. Hypothetically, polygyny is pheromonal self-suppression of the multiple queen reproductive potential in *S. invicta* colonies (Fletcher and Blum 1981); reduced investment in reproductives may result. However, differential biomass may result from different amounts of nutrients fed to reproductive larvae or fewer nurses/larva in polygyne than in monogyne colonies. Thus, the balance of worker types (foragers and nurses) may affect the reproductive investment of a colony through resource availability. An investment similar to *S. invicta*'s can be found in monogyne and polygyne *S. geminata* (U.S. and tropical Mexico; monogyne female = 15.67 ± 1.02, male = 5.38 ± 0.87 mg, oligogyne and polygyne female = 12.50 ± 1.63, male = 5.03 ± 1.02 mg, n = 100-230). A darker variant of *S. geminata* was found to be oligo- (2-4 queens) or mostly monogyne in 133 colonies collected.

Development of a Successful Mating System

Fire ant mating systems are poorly known, due to the difficulty in observing aerial mating and failure to produce sexuals by artificially inseminated gynes (Greenberg, per. comm.). The gynes may mate singly (Hung et al. 1974, Ball and Vinson 1983) or at the most a single male may attempt to mate many times, as in the obligatorily polygyne *Monomorium minimum* (Bhatkar, unpubl.). The mechanics of mating is unknown; postnuptial landed males die within a few hours or days, often sperm exhausted. Multiple mating in social Hymenoptera has been documented but

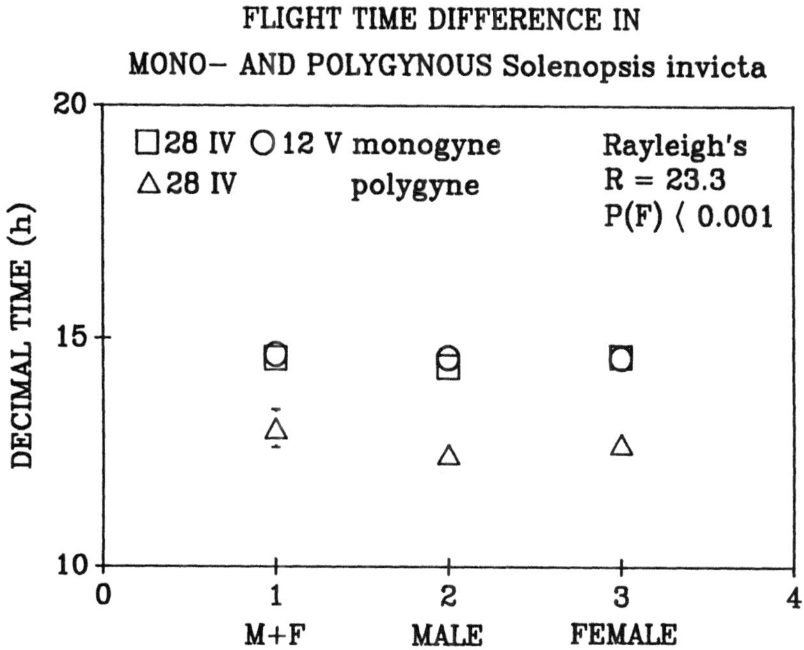

FIGURE 2. Flight time differences in mono- and polygyne *Solenopsis invicta* colonies on the same days of flights during 1987. M = males, F = females.

polygyne species are likely to mate singly rather than multiply (see Nonacs 1988). The sterile males flew into cages with the normal males and females, and were collected as they landed after nuptial flights, suggesting that they can fly. Whether they actually mate and/or compete with normal males is unknown, since the aspermic males tend to be feminized - diploid and somewhat larger in size than spermic males (thoracic width : head width, gynes = 0.90-1.03, spermic males = 1.33-1.36, aspermic males = 1.37-1.39, n = 50 - 100). Mating by monogyne males with polygyne females has been assumed, as mono- and polygyne colonies are suspected to be interbreeding populations (Ross et al. 1985). However, studies in Texas during the March-October mating season gave different results over the past three years. Either the two forms fly on different days, or their same-day flight times do not overlap; polygyne alates always fly 1 - 2 h earlier than do the monogyne (Fig. 2). This may suggest adaptation of the polygyne form to a temperate climate (Brazil-Argentine border region) and the monogyne form to a tropical zone (west central Brazil) in their homeland. The variable electromorph frequencies of a polymorphic flight muscle enzyme, alfa- glycerophosphate dehydrogenase, and its temperature sensitivity in mono- and polygyne forms of *S. invicta* may be responsible for the flight time differences (R.F. Dunton, A.P. Bhatkar, and S.B. Vinson unpublished results). This does not preclude the possibility that airborne sexuals of one form may mate with the other. Allozyme differences brought out recently by Dunton (1989) seem to support the reproductive isolation thesis rather than earlier findings (Ross et al. 1985),

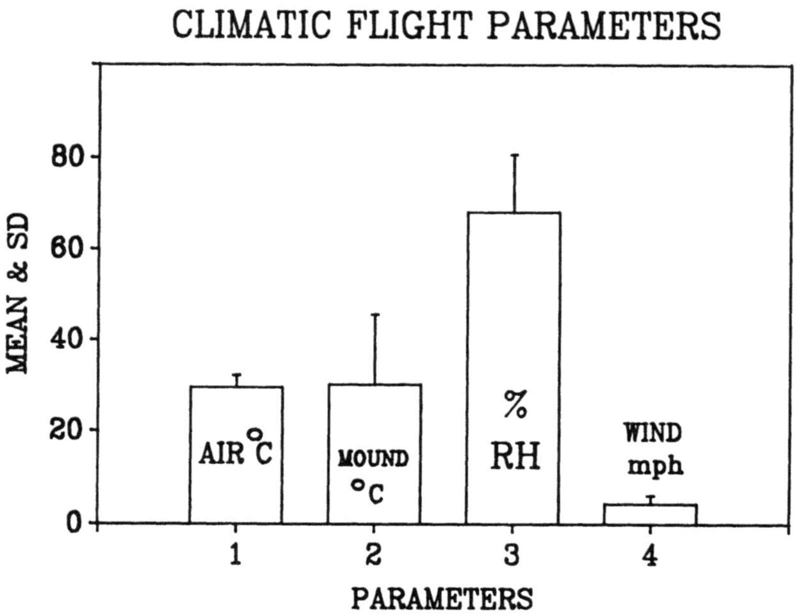

FIGURE 3. Climatic flight parameters during the nuptial flights at 3 Brazos County study sites during 1986-88.

at least for the Texas mono- and polygyne *S. invicta* populations.

Also, males flew at the same time as females in both populations, contrary to previous observations (Markin et al. 1971, Morrill 1974). Different flight times for the sexes seem to be a wasteful strategy unless mediated by a long-range sex pheromone or by intrademal mating. Local climatic conditions, such as 60 - 80% relative humidity, 25 - 35° C air temperature (more so than the mound temperature), 0 - 8 km/h (0 - 5 mph) wind velocity, few or no clouds trigger nuptial flights in alate-ready colonies (Fig. 3). A slight increase in wind gusts or a sudden drop in temperature induces the return of *S. invicta* and *geminata* alates to their nests, indicating that flights are tuned to local rather than long range dispersal. Increased aggressiveness and worker activity induces alates to fly from somewhat higher vegetation or mound topography. During the flight season, the humid tropico-temperate climates in the areas where fire ants seem to have evolved are more equitable with regard to these factors. The flights are easily induced artificially in an alate-ladened colony, suggesting they are meant for internidal rather than intranidal mating, outbreeding instead of inbreeding.

Sex Ratios

Even though the sex ratios in social insects are extraordinarily female-biased (worker castes being females also), an "effective", 1 gyne : 1 male deme level sex ratio, commonly occurs among monogyne *S. invicta* in Texas (5415 ± 2177 females: 5025 ± 3991 males, n = 12 flights), similar to a report from Florida (Morrill 1974). Most colonies specialize in the production of one or the other sex (95.82% of 160 mounds over 3 yrs), a phenomenon

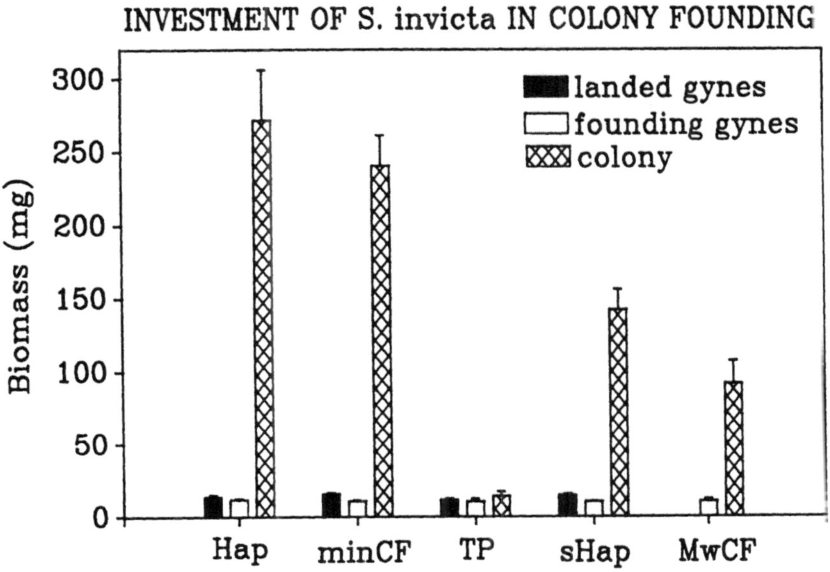

FIGURE 4. Investment of *Solenopsis invicta* during various colony founding conditions (means + SE of 7-10 replicates). Hap = single queen founding, minCF = cross-fostered with orphaned minims, TP = temporal polygyny, sHap = secondary single-queen condition, MwCF = cross-fostered with orphaned monogyne workers. Scale for the colony biomass is reduced to 50% for convenience. The colony biomasses, other than Hap and minCF, were significantly different ($p < 0.001$).

similar to that in *Formica* spp. (Pamilo and Rosengren 1983). An inseminated gyne may lay both diploid female- and haploid male-destined eggs, or diploid male-destined eggs (Ross and Fletcher 1986), possibly as a result of local inbreeding in a polygyne colony (Ross and Fletcher 1985). The capture of both sexes in the same flight cages and production of males by isolated queens from polygyne laboratory colonies indicated a high percentage of sterility (aspermicity) among male progeny. Haploid male progeny from dealated gynes are always normal (spermic). Thus, there is probably a reproductive division of labor in the fire ant colonies: uninseminated gynes dealate and remain in the colony, only to help produce males, while the inseminatedgynes produce females (Table 1). Readoption of several gynes that emerged from a nest and did not fly but dealated, was proven (unpublished) through the mark-recapture technique described earlier (Bhatkar and Vinson 1989). These recaptured gynes developed ovaries and produced males under laboratory conditions. The reproductive potential of queens may account for the sex ratio in a given deme, as evidenced through a large number of captured reproductives. Pamilo and Rosengren (1983) expected a male-biased sex ratio in polygyne *F. truncorum* and *F. exsecta* colonies due to "crowding," which may be regulated through winter persistence of the dealated females in a colony. The uncrowded nests of the same species gave a 1 : 1 sex ratio.

TABLE 1. Production of reproductives by isolated queens from 5 polygyne colonies collected from distant sites in Texas (Blacklands Res. Sta., A&M Farm, Bee Creek, Hopes Creek and Gulf Coast).

Queen rank*	Males (mean±SD)	Spermicity %	Females (mean±SD)
alfa	162.00 ±28.30	A 98.8	260.00 ±43.60
beta 1	31.40 ± 6.28	A 100.0	144.00 ±10.43
beta 2	6.00 ± 1.79	A 100.0	67.40 ± 6.18
beta 3	2.20 ± 0.40	A 100.0	53.00 ± 7.23
beta 4	1.20 ± 0.40	A 100.0	36.00 ± 7.26
beta 5	0.20 ± 0.40	A 100.0	3.00 ± 0.89
dealate 1	325.00 ±56.78	S 100.0	0.00
dealate 2	201.40 ±43.57	S 100.0	0.00
dealate 3	170.00 ±45.44	S 100.0	0.00
dealate 4	113.20 ±49.72	S 100.0	0.00
dealate 5	106.20 ±25.57	S 100.0	0.00

* Ranked according to alate productivity. S = spermic, A = aspermic.

The female-biased sex ratio (29.80 - 32.96 females : 1 male) in Texas polygyne *S. invicta* colonies may result from the pooled progeny of several nest queens and no contribution from the dealated females, or assassination of unwanted males. In those few mounds (3/7 flights during 1986-87), where males were produced, a near Fisherian 1 male : 6 female (243 - 667 : 1567 - 4089) ratio was observed.

Development of a Successful Dispersal Mechanism

Wind gusts often follow windless (wind velocity 0 - 8 km/h or 0 - 5 miles/h) post-rain sultry periods. Aerially-mated gynes drift with these gusts, suggesting one fire ant dispersal mechanism. Air currents and the gynes own wings (suited for downward gliding) may guide them to suitable adjacent habitats. The tropical flood plains where these species evolved may account for their attraction to reflecting surfaces, similar to the wet soil surface along rivers and marshes (Vinson and Greenberg 1986). Another mechanism of dispersal is through drifting of the floating post-nuptial reproductives (Morrill 1974) or colony fractions with the water currents. In riparian habitats, the fire ants adapted to flooding situations by forming a ball of workers, brood and queen/s that floats on the water surface until it makes landfall.

The mechanism of dispersal through colony fission usually occurs in polydomous polygyne colonies. Colony fission or "budding" by a single or a group of reproductives together with workers is a population dispersal strategy, unless associated with internidal mating or results from adoption of supernumerary inseminated gynes (potential queens). Colony fission occurs in the polygyne *F. rufa-* complex of the palearctic region (Rosengren and Pamilo 1983), *Iridomyrmex humilis, M. pharaonis, M. minimum* (Bhatkar, unpublished data) and other species (Wilson 1971). It would likely occur in polygyne *S. invicta* if internidal mating or adoption of postnuptial gynes were demonstrated. The order of spatial dispersal seems to be water currents, wind drifts and colony fission. However, the cross continental and regional spread of fire ants has occurred because of the accidental movement of postnuptial masses and fractioned colonies through man's transport of nursery stock and agricultural goods.

Development of a Successful Colony Founding Mechanism

Haplometrotic colony founding by a solitary inseminated gyne is thought to be the rule for *S. invicta*, although temporal pleometrosis or collective colony founding is apparent in the postnuptial gynes of this species, *S. geminata*, and other species of the fire ant complexes. Collective nest founding by 2-5 gynes, enhanced worker numbers and productivity in incipient colonies (Tschinkel and Howard 1983). However, productivity decreased significantly when the number of founding gynes was increased from 10 to 200 in 50 queen increments (Bhatkar, unpublished).

Despite the obvious advantage from the cumulative brood biomass of several egg-laying gynes, the situation invariably results in monogyne colony development within eight months of temporal pleometrosis. Supernumerary gynes are assassinated by other gynes, killed by newly emerged nanitic workers or left unattended. At the colony founding stage, brood robbing occurs, which for the successful queen significantly enhances colony growth (Stamps 1989).

Figure 4 compares the outcome of various plausible nest founding conditions. The most colony growth (indicated as the biomass of brood, minims and minors) during a four month period occurred in haplometrotically founded colonies (CF) (271.39 ± 69.51 mg), followed by colonies where 4 - 14 orphaned minims were cross-fostered by the nest founding gynes (minCF) (240.37 ± 40.98 mg). The difference was insignificant (chi-square = 1.88, df = 1, p = 0.17). Attempts to found primarily polygyne colonies by keeping the aggregates of 5 - 10 gynes for one to two months resulted in continuous mortality of gynes and poorly developed colonies (TP). Thus, pleometrotic colony founding is not a successful nest founding strategy. Due to the high incidence of initial oophagy and brood predation, colony growth is slow until the colonies become secondarily haplometrotic (sHap) and revitalize themselves. The predation of eggs and brood of the fostered nest-founding gynes by orphaned monogyne workers results in reduced colony growth (MwCF), which is superior to pleometrotic colony founding (TP)(chi-square brood wt, MwSF vs TP = 10.81, df = 1, p < 0.001) but not as efficient as minim worker- fostered nest founding (chi-square, minCF vs MwCF = 66.56, minCF vs sHap = 25.33, df = 1, p < 0.001). The initial predation decreased after the first month and the resulting colony mixed cross-fostered larger monogyne workers with smaller polygyne minims and minors. This condition may explain the occurrence of intermediate colonies observed by Greenberg et al. (1985) in the areas where mono- and polygyne colonies overlap at the A&M Farm; the solitary queens in monogyne colonies either die or colony swarms as a foraging-nesting strategy (Bhatkar and Vinson 1987a) leaving a part behind, and worker fostering may follow (Bhatkar, unpubl.). Ovipositing gynes tend to loose more weight in the presence of alien minims (minCF), workers (MwCF) or gynes (TP) than in the presence of their own offspring (Hap or sHap) (chi-square, 120 d, minCF = 0.86, p = 0.35; sHap = 0.79, p = 0.37) during the initial 15 - 30 d period (Bhatkar, unpubl.). However, no significant difference was found in the investment by nest founding queens after a 4-month colony development period (chi-square contingency = 0.30, df = 3, p > 0.95).

Adoption of a small proportion of released, newly-mated queens by a polygyne supercolony at Gainesville, Florida, has been documented (Glancey and Lofgren 1988). However, one to five gynes collected from Sulfur, Louisiana, edge of Lake Sommerville and College Station, Texas, introduced to the colonies originally collected from Brazos (the A&M Farm) and Travis (Bracon Ridge field reserve) counties, were killed within 48 hours. Thus, the origin of autoparasitic polygyne colonies remains enigmatic, but may occur similar to claustral nest founding *Monomorium solomonis*. Bolton (1986) delineated the advantages and disadvantages of a mating flight followed by group nest founding in *M. salomonis* species complex. Advantages were: protection against predation and death, colony founding in a resource-rich microhabitat, and fission with already existing workers. This was achieved at the cost of spatial dispersion and a potential for mass-destruction. The occurrence of intercolonial and interspecific trophallaxes in *Solenopsis* and *Formica* were suggested adaptive in social symbiotic colony formation (Bhatkar 1979, 1983). Comparative studies of obligatory or facultatively polygyne and monogyne species able to reproduce in captivity are necessary for an understanding of fire ant reproductive strategies.

A bivoltine strategy of mating, 1) internidal with close relatives and 2) extranidal or aerial with distant relatives, may commonly occur in a polydomous, polygyne supercolony. Adaptive significance of the inbred colonies has been discussed by Forsyth (1980). If gynes are the sisters of workers, they will be favored or disfavored in proportion to their ability to produce females (alfa or beta queens in Table 1), thus elevating the proportion of related females in a polygyne colony. Unfortunately, neither a specific genetic marker to "fingerprint" the heredity of each individual female exists nor has internidal mating been convincingly proven in fire ants. Part of the problem is the ants' hypogeal nesting habit. Some polygyne species show high intracolonial relatedness (Pamilo and Rosengren 1984), while others have low relatedness (Craig and Crozier 1979, Pamilo 1982, Pamilo and Rosengren 1984, Ross and Fletcher 1985). Once an individual-specific marker is discovered, the relatedness tests over several generations will be necessary to understand the regulation of a polygyne, polycalic *S. invicta* society.

ACKNOWLEDGMENTS

I thank Helga Sittertz-Bhatkar, Les Greenberg, Robert Vander Meer and Juan Alfredo Morales for reviewing the manuscript. This contribution is submitted through auspices of the Texas Agricultural Experiment Station (TA-24140), and is dedicated to Dr. S. B. Vinson on his 50th birthday for his contributions to basic biological research.

REFERENCES

ADAMS, C.T., W.A. BANKS and J.K. PLUMLEY. 1976. Polygyny in the tropical fire ant, *Solenopsis geminata* with notes on the imported fire ant, *Solenopsis invicta*. Fla. Entomol. 59: 411-415.

BALL, D.E. and S.B. VINSON. 1983. Mating in the fire ant *Solenopsis invicta* - evidence that alate mate only once. J. Georgia Entomol. Soc. 18: 287-291.
BANKS, W. A., J. K. PLUMLEY and D. M. HICKS. 1973. Polygyny in the colony of the fire ant *Solenopsis geminata*. Ann. Entomol. Soc. Amer. 66: 234-235.
BHATKAR, A.P. 1979. Trophallactic appeasement in ants from distant colonies. Folia Entomol. Mex. 41: 135-143.
BHATKAR. A.P. 1983. Interspecific trophallaxis in ants, its ecological and evolutionary significance, pp. 105-123. *In*, Social Insects in the Tropics (P. Jaisson, Ed.), Universite' Paris-Nord, Paris.
BHATKAR, A.P. 1989. Habitat relationship of the ants of Tallahassee Red Hills Region. Tall Timbers Research Sta. Bull. (in press).
BHATKAR, A.P. and S.B. VINSON. 1987a. Foraging in mono- and polydomous *Solenopsis invicta* Buren colonies, pp. 545-546. *In*, Biology and chemistry of social insects (J. Eder and H. Remboldt, Eds.), Verlag J. Peperny, Munich.
BHATKAR, A.P. and S.B. VINSON. 1987b. Colony limits in *Solenopsis invicta* Buren, pp. 599-600. *In*, Biology and chemistry of social insects (J. Eder and H. Remboldt, Eds.), Verlag J. Peperny, Munich.
BHATKAR, A.P. and S.B. VINSON. 1989. Mass-marking and recognition of marked workers in *Solenopsis invicta* Buren (Hymenoptera: Formicidae). Folia Ent. Mex. (in press).
BOLTON, B. 1986. Apterous females and shift of dispersal strategy in the *Monomorium salomonis*-group (Hymenoptera: Formicidae). J. Natur. Hist. 20: 267-272.
CRAIG, R. and R.H. CROZIER. 1979. Relatedness in the polygynous ant Myrmecia pilosula. Evolution 33: 335-341.
DUNTON, R.F. 1989. Ecological genetics of monogynous and polygynous forms of the red imported fire ant, *Solenopsis invicta* Buren. Ph.D. dissertation, Texas A&M University, College Station, Texas.
ERWIN, T.L. 1983. Tropical forest canopies: the last biotic frontier. Bull. Ent. Soc. Amer. 29: 14-19.
FLETCHER, D.J.C. 1983: Three newly-discovered polygynous populations of the fire ant, *Solenopsis invicta*, and their significance. J. Georgia Entomol. Soc. 18: 538-543.
FLETCHER, D.J.C. and M.S. BLUM. 1981. Pheromonal control of dealation and oogenesis in virgin queen fire ants. Science 219: 73-75.
FLETCHER, D.J.C., M.S. BLUM, T.V. WHITT and N. TAMPLE. 1980. Monogyny and polygyny in the fire ant, *Solenopsis invicta*. Ann. Entomol. Soc. Amer. 73: 655-661.
FORSYTH, A. 1980. Worker control of queen density in Hymenopteran societies. Amer. Natur. 116: 895-898.
GLANCEY, B.M. and C.S. LOFGREN. 1988. Adoption of newly-mated queens: a mechanism for proliferation and perpetuation of polygyne red imported fire ants, *Solenopsis invicta* Buren. Florida Entomol. 71: 581-587.
GLANCEY, B.M., J.C.E. NICKERSON, D. WOJCIK, J. TRAGER, W.A. BANKS and C.T. ADAMS. 1987. The increasing incidence of the polygyne form of the red imported fire ant, *Solenopsis invicta* (Hymenoptera: Formicidae), in Florida. Florida Entomol. 70: 400-402.

GLANCEY, B.M., C.E. STRINGER, C.H. CRAIG and P.M. BISHOP. 1975. An extraordinary case of polygyny in the red imported fire ant. Ann. Entomol. Soc. Amer. 68: 922.

GÖSSWALD, K. 1951. Die rote Waldameise im Dienste der Waldhygiene. Metta Kinau Verlag, Lüneberg.

GREEN, H.B. 1952. Biology and control of the imported fire ant in Mississippi. J. Econ. Entomol. 45: 593-597.

GREENBERG, L., D.J.C. FLETCHER and S.B. VINSON. 1985. Differences in worker size and mound distribution in monogynous and polygynous colonies of the fire ant *Solenopsis invicta* Buren. J. Kansas Entomol. Soc. 58: 9-18.

HAMILTON, W.D. 1964. The genetic evolution of social behavior. I. J. Theor. Biol. 7: 1-16, II. 17-52.

HUNG, A.F.C., S.B. VINSON and J.W. SUMMERLIN. 1974. Male sterility in the red imported fire ant, *Solenopsis invicta.* Ann. Entomol. Soc. Amer. 67: 909-1012.

MARKIN, G.P., J.H. DILLER, S.O. HILL, M.S. BLUM and H.P. HERMANN. 1971. Nuptial flights and flight ranges in the imported fire ant, *Solenopsis saevissima richteri.* J. Georgia Entomol. Soc., 6: 145-150.

MIRENDA, J.T. and S.B. VINSON. 1982. Single and multiple queen colonies of the imported fire ant in Texas. Southwest. Entomol. 7: 135-141.

MORRILL, W.L. 1974. Production and flights of alate red imported fire ants. Environ. Entomol. 3: 265-271.

NONACS, P. 1988. Queen number in colonies of social Hymenoptera as a kin-selected adaptation. Evolution 42: 566-580.

PAMILO, P. 1982. Genetic population structure in polygyne *Formica* ants. Heredity 48: 95-106.

PAMILO, P. and R. ROSENGREN. 1983. Sex ratio strategies in Formica ants. Oikos 40: 24-35.

PAMILO, P. and R. ROSENGREN. 1984. Evolution of nesting strategies of ants: genetic evidence from different populations types of *Formica* ants. Biol. J. Linn. Soc. London. 21: 331-348.

ROSENGREN, R. and P. PAMILO. 1983. The evolution of polygyny and polydomy in mound building Formica ants. Acta Entomol. Feenica 42: 65-77.

ROSS, K.G. and D.J.C. FLETCHER. 1986. Diploid male production - a significant colony mortality factor in the fire ant *Solenopsis invicta* (Hymenoptera: Formicidae). Behav. Ecol. Sociobiol. 19: 283-291.

ROSS, K.G., D.J.C. FLETCHER and B. MAY. 1985. Enzyme polymorphism in the fire ant, *Solenopsis invicta* (Hymenoptera: Formicidae). Biochem. System. Ecol. 13: 29-33.

STAMPS, W.T. 1989. Polygyne colony formation in the red imported fire ant, *Solenopsis invicta* Buren. M. S. thesis, Texas A&M University, College Station, Texas.

SUMMERLIN, J.W. 1976. Polygyny in the colony of the southern fire ant. Ann Entomol. Soc. Amer. 69: 54.

TSCHINKEL, W.R. and D.F. HOWARD. 1983. Colony founding by pleometrosis in the fire ant, *Solenopsis invicta.* Behav. Ecol. Sociobiol. 12: 103-13.

VINSON, S.B. and L. GREENBERG. 1986. The biology, physiology, and ecology of imported fire ants, pp. 193-226. *In*, Economic impact and control of social insects (S.B. Vinson, Ed.), Greenwood Press, Westport, CO.

WILSON, E.O. 1971. The insect societies. Belknap (Harvard Univ.) Press, Cambridge.

ZAR, J.H. 1974. Biostatistical analysis. Prentice-Hall, Engle wood Cliffs, NJ.

13

Oviposition and Growth of the Fire Ant *Solenopsis invicta*

D.F. Williams

INTRODUCTION

As monogyne ant colonies become larger, the queen's reproductive output must also increase (Wilson 1971). Oviposition and colony growth studies on the imported fire ant *Solenopsis invicta* have been reported by several investigators (Markin et al. 1973, Markin and Dillier 1971, Markin et al. 1972, Wood 1979, Green 1962, Wilson and Eads 1949 and more recently Porter and Tschinkel 1985,1986; Tschinkel 1988a). Tschinkel (1988b) reported on the social control of oviposition and indicated that queen weight, total follicles, follicles per ovariole, time spent feeding and the number of 4th instar larvae are positively correlated to *S. invicta* egg-laying rates. The purpose of this study was to examine how temperature affects *S. invicta*'s queen oviposition rates and colony growth. This paper will present an overview of this work.

OVIPOSITION RATE

Queens from 25 colonies were sacrificed each month (for eight months) to measure the effect of temperature, queen age, weight, and weight loss on the oviposition rate. The 200 queens were maintained in environmental chambers at one of five experimental temperatures in colonies held for growth studies. Before each colony was terminated for census, the oviposition rate and weight of the queen before and after oviposition were recorded. Oviposition tests isolated the queens on a damp cloth in a small weigh boat for a four-hr egg-laying period. We examined the cloth afterwards for eggs (Fletcher et al. 1980).

Temperature affected the number of eggs laid by *S. invicta* monogyne queens. Queens maintained at 27°C, 30°C, and 32°C oviposited more eggs than those at 24°C and 35°C. During eight months of tests, the mean number of eggs deposited varied less at 27°C than at 30°C; those at 30°C varied less than at 32°C. Thus, it can be concluded that temperature affects oviposition rate. Queen age had no direct effect on oviposition, although older queens are usually heavier than young queens, and heavier queens generally lay more eggs than lighter queens (Fig. 1). However, older lighter queens almost always laid fewer eggs than younger heavier ones. Tschinkel (1988b) also found a

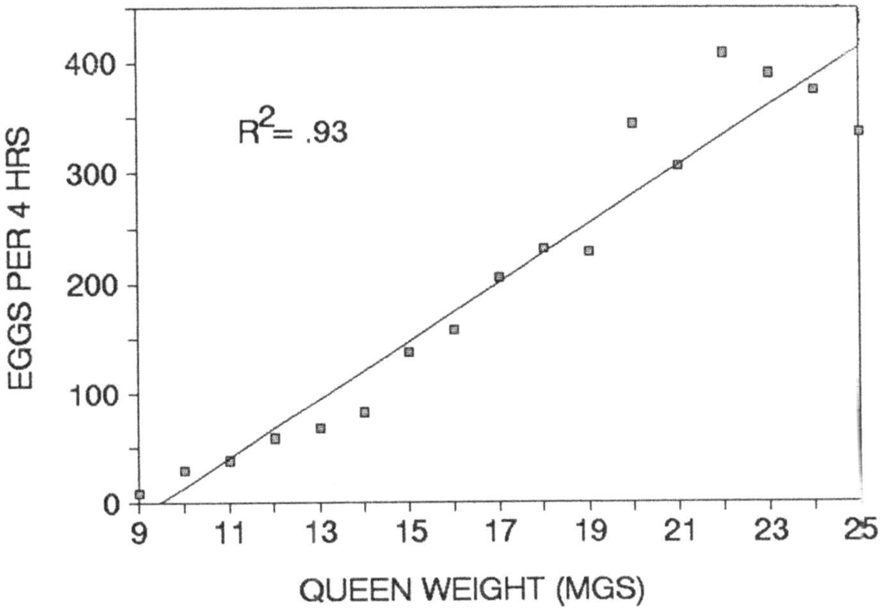

FIGURE 1. Correlation between body weight of the queen and the number of eggs laid.

positive correlation between queen weight and the egg-laying rate. In addition, queens that laid the most eggs lost the most weight during oviposition. Eggs accounted for half the weight loss during oviposition. The remaining weight loss could have been natural water or fluids and wastes expelled with the eggs.

In other studies, 100 colonies and their queens were maintained at 27°C. Food was available for only six hours twice a week. Another set of 100 colonies was fed continuously. The former group weighed less and laid fewer eggs, even though they were identical ages.

EGG WEIGHTS

Temperatures (27°, 30°, and 32°C) did not affect egg weight. However, the queen's age was an important factor in egg weights since older queens laid smaller eggs than younger queens. Tschinkel (1988a) also reported that younger queens laid larger eggs. We recorded the weights of 58,765 eggs weighed in clumps (ca. 189 eggs) from monogyne *S. invicta* queens; the mean wt. per egg was 0.0051mg (SD= ±.0025).

One hundred 80-day-old queens maintained at 30°C laid the most eggs in 24 hours, (mean ±SD) 2,748 ± 292 (Table 1). However, at 214 days and 243 days queens maintained at 27°C produced the largest number of eggs per 24 hour periods, (mean ±SD) 2,358 ± 494.1 and 2,082 ± 289.9, respectively. As

TABLE 1. Effects of age and temperature on 24-hr oviposition rates (based on 4-hr oviposition tests) and weight of queens of S. invicta. Five replicates per temperature per census day.

QUEEN AGE (CENSUS DAY)		MEAN AND SD OF EGGS LAID AND WEIGHT AT DEGREES C				
		24	27	30	32	35
90	EGGS	6±9.6	174±169.3	240±166.6	372±315.6	54±25.8
	WEIGHT	13±1.5	12±2.0	11±1.1	12±1.1	9±0.9
120	EGGS	168±68.9	804±206.2	600±381.7	808±447.3	
	WEIGHT	11±0.4	17±0.9	14±2.0	17±2.6	
151	EGGS	552±242.6	1980±204.2	2304±99.8	750±609.4	
	WEIGHT	15±1.4	19±0.7	20±0.8	17±1.5	
180	EGGS	594±170.5	1992±242.7	2748±292.3	2232±883.0	
	WEIGHT	15±1.5	20±1.0	22±0.9	18±2.6	
214	EGGS	558±348.9	2358±494.1	2082±474.1	1122±363.9	
	WEIGHT	16±3.2	23±2.0	20±2.3	21±1.9	
243	EGGS	834±226.4	2082±289.9	1734±413.7	1830±409.1	
	WEIGHT	18±0.6	23±1.6	21±1.8	19±2.2	

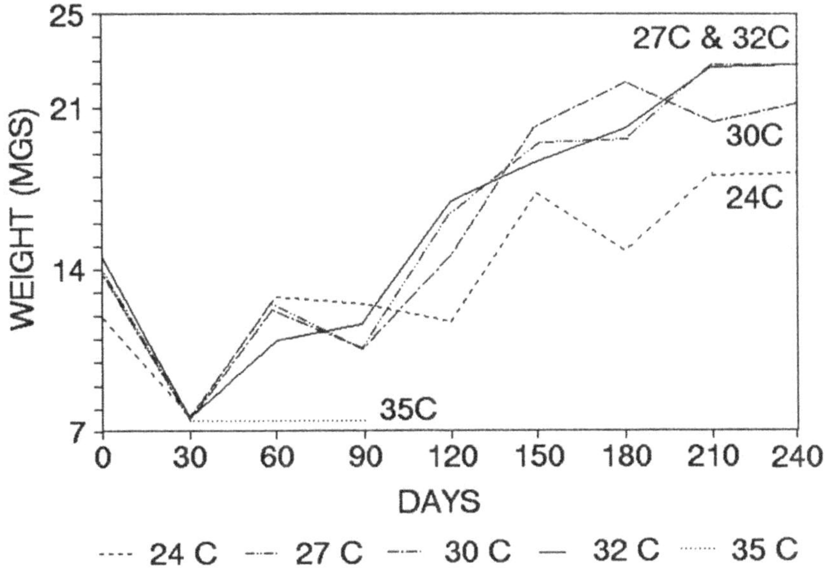

FIGURE 2. Changes in the body weights of queens maintained at constant temperatures over 243 days. Note decline in body weights during initial 30 days of colony founding regardless of temperature.

expected, queens held at the coldest temperature 24°C laid the fewest eggs over the entire study period. However, queen mortality was lower at this temperature than at any of the higher temperatures.

COLONY GROWTH STUDIES

Tschinkel (1988a) reported that colonies grew logistically, reaching half their size (110,000 workers) between 2 ½ and 3 ½ years and maximum size after 4-6 years. Porter (1988) reported *S. invicta* colony growth was highly temperature dependent and that post-founding colonies grew successfully between 24°C and 36°C with maximal rates at about 32° C.

Colony growth studies conducted at five constant temperatures (°C±SD): 24±0.4, 27±0.7, 30±0.6, 32±0.8, 35±0.6 for 8 months showed that the weights (mg) of the newly-mated queens dropped to half (ca. 7 mgs) their initial body weight during the first 30 days and then increased when minum workers appeared; the queen's weight returned to near the initial weight at 120 days (Fig. 2). By the 151st day, queens had increased their initial body weight by 50% at all temperatures except 24°C and 35°C. The greatest increases in colony growth rate (ca. 8 to 28-fold) occurred within the first 90 days at all temperatures except 24°C; at this temperature colonies grew in size between 90 and 120 days (Fig. 3). As expected, these colonies - maintained at the coldest temperature - 24°C grew the slowest.

After eight months, colonies maintained at 27°C had larger populations than those held at other temperatures (242,211 ± 42,099 total and 149,034 ±

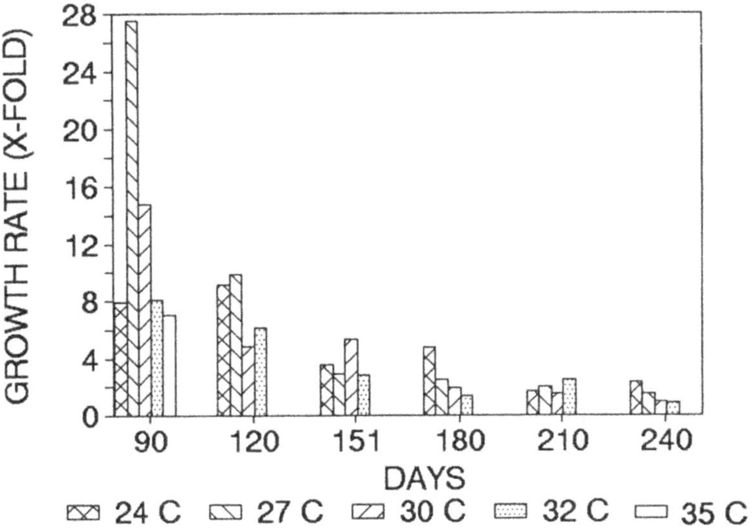

FIGURE 3. Growth rate (X-fold) in total colony size per 30 day intervals at constant temperatures during 8 months.

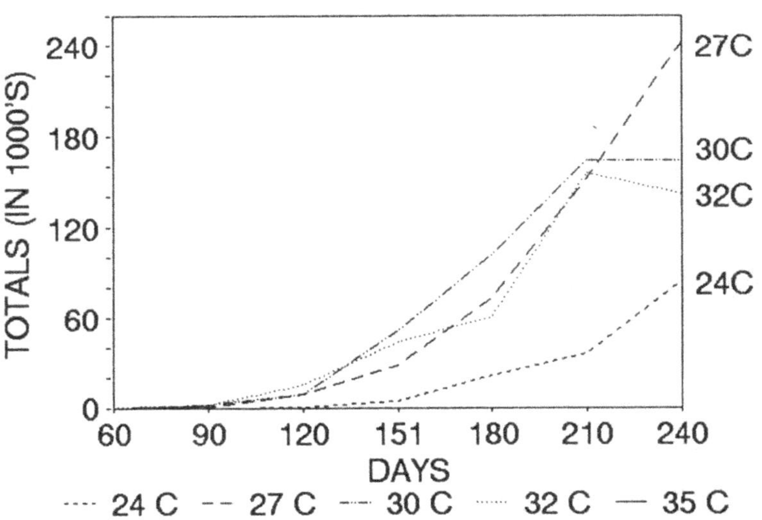

FIGURE 4. Effect of temperature on colony size in the fire ant *Solenopsis invicta*. Colonies maintained at 27°C produced larger numbers in all stages after 8 months.

31,535 workers; Fig. 4). Colonies maintained at 35°C produced workers at a faster rate than other colonies. Minum workers were present within three weeks in most colonies. However, many more workers died at 35°C, the queens constantly lost weight - never recovering their newly mated queen weight - and 91% of the colonies held at this temperature (the remaining 9 colonies were terminated for sampling) died by 90 days. Overall, queens held at 24°C and 27°C also had lower mortality rates than queens held at 30°C, 32°C, and 35°C, with the highest queen mortality at 35°C. Finally, after 8 months, the 27°C colonies contained the following mean numbers: workers 149,034 (61.5%), pupae 42,362 (17.5%), 4th instar larvae 17,613 (7.3%), and 1st-3rd instar larvae 33,203 (13.7%) with total colony size averaging 242,211 (SD = ±42099). These numbers are far greater than has previously been reported for 90 day colonies. For example, Markin et al. 1973 reported that the mean number of workers in seven-month-old field-collected colonies was 6,576 (1,850 - 19,400) while colonies 2 years old had ca. 25,000 workers. By three years, the fully mature colonies had produced 50,000 workers. At seven months, colonies in our growth studies averaged 106,934 workers (SD = ± 15,100) - more than twice the size of three year old fully mature colonies (Markin et al. 1973). One reason for the large differences is that the colonies in our studies were maintained under laboratory conditions at optimum temperatures and provided with unlimited food, factors which greatly aid colony growth. Whereas under natural conditions, seasonal effects and competition for finite food resources limit colony growth and therefore size. Tschinkel (1988a) checked the number of workers in three-six month old field colonies, (mean ± SD; 4,710 ± 3,950) and found their numbers equal to his five (mean ± SD; 1,926 ± 560) and six month-old colonies (mean ± SD; 8,076 ± 2,120). Again, these figures are small compared to those obtained in our studies. The reason is unknown but may be related to the differences in rearing procedures, temperature and /or diets used to maintain the laboratory colonies.

LOW TEMPERATURE GROWTH STUDY

Three month studies using newly-mated queens were also conducted, with the goal of collecting data on the effect of low temperatures (17, 20 and 24°C) on colony growth.

Previous studies indicated that fire ant queens could not develop and maintain colonies at temperatures below ca. 24°C. Our studies showed that queens produced colonies at 20°C. Forty percent of these colonies developed workers in 60 days, the rest within 90 days; only 20% of queens held at 17°C produced workers by the 90th day. However, all workers produced at 17°C died within a few days; therefore, no colonies were established.

CONCLUSION

In our studies, temperature, weight of the queen, and the weight loss by the queen during oviposition were factors positively associated with *S. invicta* queen's oviposition rate. This last factor may not be as important as the others since total loss may be a constant of egg number. Generally, queens laid more eggs at higher temperatures, heavier queens usually laid more eggs than lighter ones, and queens that lost the most weight laid the most eggs.

Age was unimportant except as it was related to weight. During oviposition, fifty percent of the queen's weight loss was due to the oviposited weight of the eggs.

Colony growth was temperature dependent with the maximum rate occurring at 27°C. The temperature window of growth producing complete colonies was 20 - 35°C, the lower limit being lower than has been reported in the past. At 32°C and 35°C, most growth curves dropped quickly and at 35°C colonies died within 90 days. Contributing to this was high worker and queen mortality. We believe as Porter (1988) indicated that the stress of high temperatures may have contributed to a decline in other functions such as egg-laying, brood care, and food flow through the colony. Queens lost less initial body weight at lower temperatures but by the time minum workers appeared in a colony, the queen's weight was reduced to half her initial weight. The greatest increase in colony growth (28-fold) occurred at 27°C within the first 90 days. The slowest growth and lowest queen mortality occurred at 24°C while the fastest growth but highest queen and colony mortality occurred at 35°C. Queens produced workers at 20°C and 17°C but all workers died within 1-2 days at 17°C. After eight months, the percentages (mean ± SD) for each stage in colonies maintained at all temperatures was 63.5% ± 3.0 workers, 18.0% ± 2.0 pupae, 6.4% ± 0.8 4th instar larvae, and 12.1% ± 1.9 1st - 3rd instar larvae.

The information obtained from these and previous studies have been used to develop a comprehensive computer model of the life cycle of *S. invicta.* The model simulates the major environmental variables such as temperature, precipitation, daylength, and colony food on the population dynamics of this ant in various habitats. The life cycle of the fire ant is incremented in the model into weekly age classes and simulations are run with weekly time steps. Our future investigations will improve and validate the model in the field. These studies will be directed towards survivorship, especially the influence of temperature on brood, effects of fluctuating temperatures on colony development and growth since colonies in nature are exposed to varying temperatures, the effects of humidity and photoperiod on colony development and growth, The effect of temperature and season on the production of alates, and the effect of temperature and precipitation on mating flights. Clearly, much work remains to be done on colony development, growth and population dynamics of the fire ant especially in the field.

ACKNOWLEDGMENTS

I am grateful to Gregory Knue, Bradford Lingo, and Lynne Bundy for their excellent technical assistance which made these long, tedious studies possible. Appreciation is also given to Sanford D. Porter, Dan Haile, and Dan Kline for advice and reviewing the manuscript.

REFERENCES CITED

FLETCHER, D. J. C., M. S. BLUM, T. V. WHITT, and N. TEMPLE. 1980. Monogyny and polygyny in the fire ant, *Solenopsis invicta.* Ann. Entomol. Soc. Amer. 73: 658-61.

GREEN, H.B. 1962. On the biology of the imported fire ant. J.Econ. Entomol. 55: 1003-4.

MARKIN, G.P., and J.H. DILLIER. 1971. The seasonal life cycle of the imported fire ant, *Solenopsis saevissima richteri*, on the Gulf coast of Mississippi. Ann. Entomol. Soc. Amer.64: 562-5.

MARKIN, G.P., H.L. COLLINS, and J.H. DILLIER. 1972. Colony founding by queens of the red imported fire ant, *Solenopsis invicta*.Ann. Entomol. Soc. Am. 65: 1053-8.

MARKIN, G.P., J.H. DILLIER, and H.L. COLLINS. 1973. Growth and development of colonies of the red imported fire ant, *Solenopsis invicta*. Ann. Entomol. Soc. Amer. 66: 803-8.

PORTER, S.D. 1988. Impact of temperature on growth and development of the ant *Solenopsis invicta*. J. Insect Physiol. 34: 1127-1133.

PORTER, S.D., and W.R. TSCHINKEL.1985. Fire ant polymorphism: the ergonomics of brood production. Beh. Ecol. Sociobiol. 16: 323-36.

PORTER, S.D., and W.R. TSCHINKEL. 1986. Adaptive value of nanitic workers in newly founded red imported fire ant colonies (Hymenoptera: Formicidae). Ann. Entomol. Soc. Amer. 79:723-6.

TSCHINKEL, W.R. 1988a. Colony growth and the ontogeny of worker polymorphism in the fire ant, *Solenopsis invicta*. Behav. Ecol. Sociobiol. 22: 103-15.

TSCHINKEL, W.R. 1988b. Social control of egg-laying rate in queens of the fire ant, *Solenopsis invicta*. Physiological Entomology. Physiological Entomology 13: 327-350.

WILSON, E.O. 1971. The insect societies. Belknap Press, Cambridge, Ma. 548p.

WILSON, E.O., and J.H. Eads. 1949. A report on the imported fire ant *Solenopsis saevissima var. richteri* Forel in Alabama. Ala. Dep. Conserv. Spec. Rep., 53 p., 12 plates (Mimeographed).

WOOD, L.A. 1979. Quantification and modification of worker size variation in young colonies of the fire ant *Solenopsis invicta*. M.S. thesis, Florida State University, Tallahassee, FL., 96 p.

14

Social Control of Reproduction in Fire Ant Colonies

E.L. Vargo

INTRODUCTION

The study of social control over reproduction occupies a key position in our attempt to understand the evolution and organization of insect societies. On one hand, colony-level selection can be expected to favor efficient cooperation of colony members to maximize the production of sexuals capable of founding new colonies (Oster and Wilson 1978). On the other, direct reproduction is generally dominated by only one or a few individuals in the colony, so that competition among nestmates for reproductive privileges is also a prominent feature of social life among the insects (West-Eberhard 1981; Fletcher and Ross 1985; Bourke 1988). The extent and means by which social factors regulate reproduction is determined by the interplay of these two evolutionary tendencies.

Social control of reproduction in the fire ant, *Solenopsis invicta*, has only recently been a focus of research, beginning with the discovery of pheromonal queen control over oogenesis in virgin queens (Fletcher and Blum 1981a). Progress has since been rapid, especially in the last two or three years, and a variety of reproductive processes in fire ant colonies are now known to be under social control.

CONTROL OVER OVIPOSITION

Alate Virgin Queens

Throughout most of its range in the southeastern U.S., *S. invicta* colonies produce sexuals in all but the coldest months (Markin and Dillier 1971). Both male and female alates reach sexual maturity in seven to 10 days (Lofgren et al. 1975); they then remain in the nest until favorable weather conditions bring about a mating flight. During this waiting period - which may last several weeks - female sexuals are prevented from ovipositing in the nest by means of a relatively nonvolatile primer pheromone produced by functional (egg-laying) queens (Fletcher and Blum 1981a, b). In addition to inhibiting oogenesis in virgin queens, this pheromone prevents histolysis of the flight muscles and dealation (shedding of the wings) which usually occur following disinhibition (Fletcher and Blum 1981a, 1983a). Study of this pheromone has been greatly facilitated by the development of a bioassay in which queen corpses can be tested for their ability to inhibit dealation in sexually mature virgin queens (Fletcher and Blum 1981b; Fletcher et al. 1983).

The inhibitory power of different queens, as assessed by this bioassay, bears a positive relationship to the degree of ovarian development (Willer and Fletcher 1986; Obin et al. 1988), suggesting that the quantity of pheromone produced by a queen is correlated to her level of fecundity. This is not a simple relationship, however, as certain types of queens are more or less inhibitory than others. For example, Willer and Fletcher (1986) found that virgin queens which had dealated and taken over reproductive function in queenless colonies for only one week were significantly more effective than "low weight" inseminated queens of monogyne (single-queen) colonies, even though their ovaries were less developed. Incidentally, the rapid production of this inhibitory pheromone by such virgin replacement queens is evidence of strong reproductive competition to become the principal egg-layer in these colonies.

As a first step toward locating the glandular source of the pheromone, Fletcher and Blum (1981b) tested parts of queen corpses for their inhibitory capability. Gasters alone were about as effective as whole corpses, whereas heads and thoraces together were only slightly inhibitory. These authors concluded that the pheromone is produced in the gaster, while the head and thorax become contaminated with it. This conclusion has been criticized by Vander Meer (1983) on the grounds that a queen attractant pheromone is produced in the venom gland located in the gaster (Glancey 1986). Thus, the gaster's strong inhibitory effectiveness may merely reflect the degree to which workers are attracted to this body region and thereby come into contact with substances on its surface, rather than indicate the site of production or storage of the inhibitory pheromone. Fletcher (1986a) has argued that as some gasters have remained inhibitory for a remarkably long period of time - more than three months - they must contain the source of the pheromone. In any event, dissemination of the nonvolatile inhibitory pheromone to virgin queens clearly depends on attractant pheromones, in so far as these mediate contact of the queen's body by workers. These two pheromones are, therefore, intimately related in the natural context, underscoring the complex nature of the pheromonal signals produced by the queen and the difficulties encountered when trying to sort them out experimentally.

Pheromonal inhibition of reproduction by virgin queens is remarkably effective, despite enormous colony size and complex nest structure. Mature monogyne colonies can contain 230,000 workers and tens of thousands of immature forms, all distributed over some 40 liters of nest (Markin et al. 1973). In addition, up to 5,000 alate queens may be present at any one time (Vargo and Fletcher 1987). Yet the queen produces enough pheromone to prevent these alates from producing males (see Ross and Fletcher 1985). Nevertheless, some female sexuals may oviposit if they remain in the nest for long periods. Fletcher and Blum (1983a) reported that despite having intact flight muscles, 68 of 300 (23%) female alates which overwintered in their parental monogyne nest had fully-developed ovaries and laid some eggs, although these eggs appeared to be nonviable.

The fact that flight muscle histolysis and dealation do not necessarily coincide with oviposition, raises the question of how many queen pheromones are involved in inhibiting reproductive activity of female sexuals. Fletcher

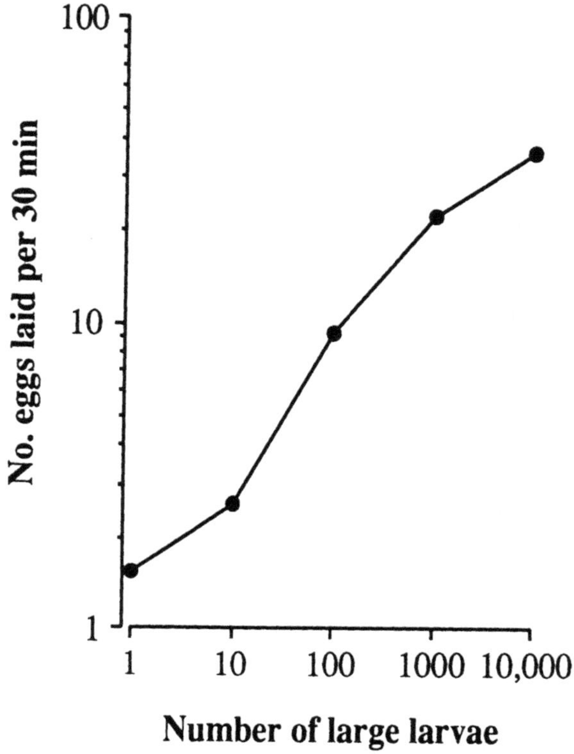

FIGURE 1. Effect of the number of fourth instar larvae on the egg-laying rate of queens. Each experimental colony contained about 3000 workers. n=4 in each case. (Redrawn with permission from Tschinkel 1988).

and Blum (1983a) have proposed that these different physiological events are controlled by a single pheromone which regulates the secretion and titer of juvenile hormone (JH). According to this hypothesis, oogenesis proceeds in the presence of low amounts of JH, while greater amounts are needed to effect histolysis of the flight muscles. These authors suggest that the pheromone does not inhibit JH production altogether, but reduces it to a degree permitting oogenesis through a continuous but low rate of vitellogenesis; a substantially higher titer is necessary to induce histolysis of the flight muscles. While this hypothesis is appealing, it has yet to be tested and the mode of action of this inhibitory pheromone remains uninvestigated.

Functional Queens

Monogyne Colonies. In a series of elegant experiments, Tschinkel (1988) investigated the role of workers and larvae in regulating queen oviposition in monogyne colonies. In ant colonies it is generally assumed that queens compete with larvae for food and worker attention, so that queen fecundity is correlated negatively with the number of larvae and positively with the number of workers (Tschinkel 1988). These experiments revealed a very different relationship in *S. invicta*. Surprisingly, the queen's oviposition rate was highly correlated with the number of fourth (last) instar larvae (Fig. 1), whereas it was unaffected by the number of workers, at least not within the range tested (300-6,000). As the number of fourth instar larvae depends directly on the queen's egg-laying rate, these are linked in a positive feedback loop, such that fecundity is pushed to higher and higher levels as eggs develop into fourth instar larvae. This larval stimulation of the queen's egg-laying rate declines with increasing larval numbers, as indicated by the logarithmic nature of this relationship (Fig. 1). Thus, at high larval numbers - above 10,000 or 15,000 - each additional larva has very little effect on fecundity. As a consequence, the declining effect of larval numbers on queen fecundity may help determine maximum colony size by setting the upper limit on the queen's oviposition rate.

The precise nature of larval influence on queen fecundity is unknown, but probably involves nutritional factors, most likely proteins. Fourth instar larvae are the only colony members capable of ingesting and digesting solid protein (Petralia and Vinson 1978; Petralia et al. 1980). In addition, the labial gland secretions of these larvae contain high concentrations of protein and amino acids which appear to be collected by workers (O'Neal and Markin 1973; Sorensen et al. 1983a, b). Material can be transferred from larvae to queens; Tschinkel (1988) observed that dye fed to larvae rapidly appeared in worker crops and then in the queen and her eggs. Thus Tschinkel speculated that fourth instar larvae may be a "digestive metabolic caste that processes protein for egg production by the queen". A similar stimulating effect of larvae on queen fecundity has recently been reported in another pest ant, *Monomorium pharaonis*, in which queens appear to feed directly on larval secretions (Bϕrgesen 1990).

Polygyne Colonies. In addition to the more common monogyne colonies, *S. invicta* forms polygyne (multiple-queen) colonies in both the U.S. and its native range in South America (Vargo and Fletcher 1987; K. G. Ross and J. C. Trager unpublished). In polygyne colonies the presence of other egg-laying queens strongly influences fecundity. Fletcher et al. (1980) were the first to report that queens from polygyne colonies are less physogastric and lay fewer eggs per individual than queens from monogyne colonies, suggesting a negative relationship between fecundity and the number of colony queens. Greenberg et al. (1985) found supporting evidence in laboratory colonies where the number of queens was negatively correlated with the number of eggs laid per queen but positively correlated with the total egg number.

The relationship between colony queen number and fecundity in field and standardized laboratory colonies was recently studied by Vargo and Fletcher (1989). Queen weight, a reliable measure of ovarian development and oviposition rate in this species (Fletcher and Blum 1983a; Willer and Fletcher 1986; Ross 1988; Tschinkel 1988, Vargo and Fletcher 1989), was used to assess the degree of fecundity. Although the fecundity of individual queens varied widely, the number of colony queens had a strong negative effect on mean fecundity in both field and laboratory colonies (Fig. 2). Uninseminated queens, which constitute a third or more of the functional queens in polygyne colonies (Lofgren and Williams 1984; Vargo and Fletcher 1987, 1989), were generally less fecund than their inseminated nestmates, weighing 5-25% less.

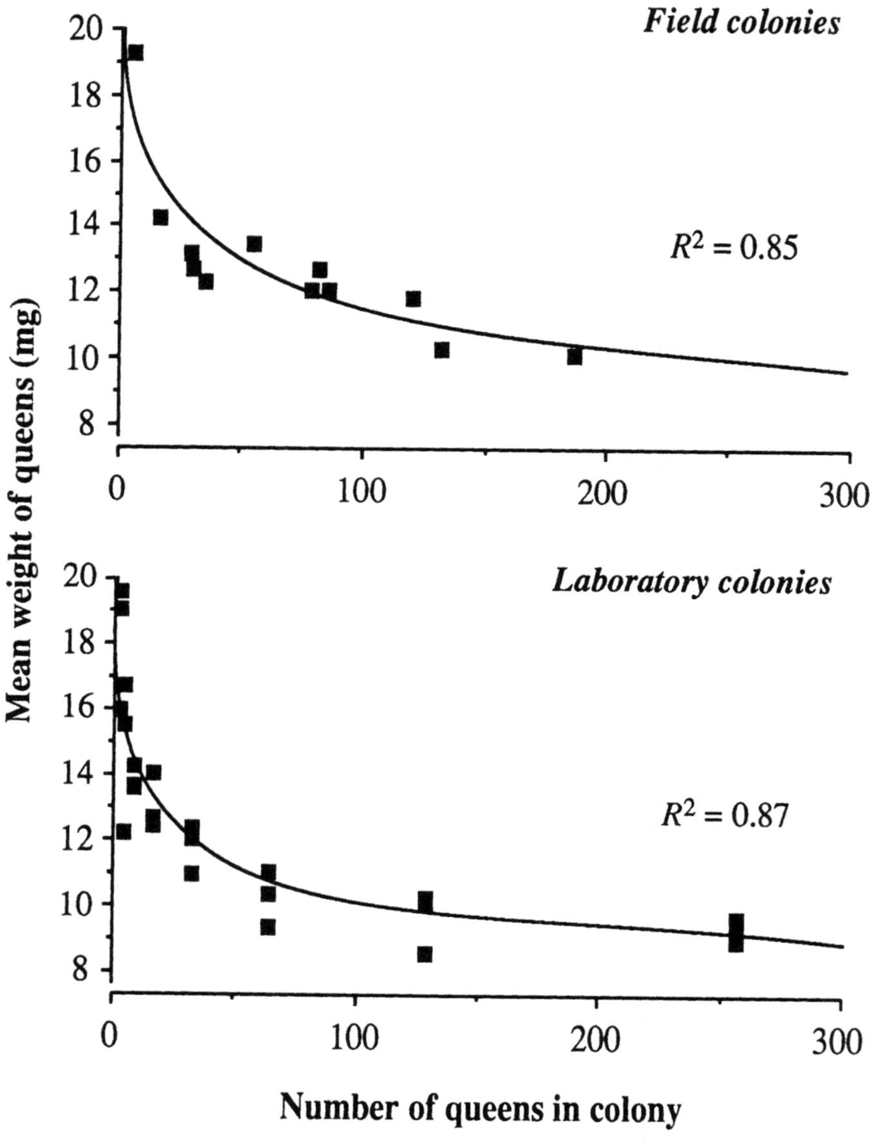

FIGURE 2. Relationship between queen number and mean queen weight in polygyne colonies of *S. invicta*. Queens placed in the standardized laboratory colonies were taken from a homogeneous pool and weighed after ca. nine weeks. (Adapted from Vargo and Fletcher 1989).

Colony size (total weight) was not significantly related to the mean weight of queens in either the field or laboratory colonies, suggesting that the number of workers or larvae bears little if any relationship to the fecundity of queens, at least within the range of colony size studied (12-265 g). While not specifically tested, it would seem that queen fecundity in polygyne colonies is not strongly influenced by the quantity of nutritional or other factors provided by larvae, in contrast to the role of such factors in monogyne colonies. Thus an intriguing possibility for a regulatory mechanism is mutual pheromonal inhibition of fecundity among nestmate queens in polygyne societies. As discussed above, functional *S. invicta* queens produce a pheromone that inhibits reproductive activity in virgin queens, presumably by suppressing JH production. Perhaps mutual inhibition of fecundity among cohabiting polygyne queens is also effected by pheromones which regulate JH production.

PRODUCTION OF VIABLE EGGS

Nearly all eggs laid by the mother queen in mature monogyne colonies are viable, i.e., they form embryos and can presumably develop into larvae (Fletcher and Blum 1983a; Voss and Blum 1988), though some may be consumed by fourth instar larvae before hatching (Petralia and Vinson 1978). However, nonviable (so called "trophic") eggs may be produced in monogyne colonies under certain social and physiological contexts. Newly-mated queens engaged in colony founding lay some eggs which fail to form embryos and presumably serve as an important source of food for the larvae of young colonies (Glancey et al. 1973; Cherix and Fletcher 1982; Voss and Blum 1988). In addition, virgin queens from monogyne colonies removed from the inhibitory influence of the mother queen lay almost exclusively nonviable eggs (Voss 1981; Fletcher and Blum 1983a). However, if such disinhibited virgin queens are maintained in the presence of large numbers of workers and brood, or treated topically with JH, they produce some embryonated eggs (Voss 1981; Fletcher and Blum 1983a), suggesting that nutritional factors as well as endogenous JH titers influence egg viability.

The viability of eggs produced by polygyne queens has been studied by Vargo and Ross (1989). Surprisingly, the proportion of viable eggs produced by functional queens from field colonies was highly variable, but inseminated queens produced a higher proportion than uninseminated queens (top half of Fig. 3). This variability among queens was unrelated to queen weight (fecundity), indicating that the oviposition rate of queens in polygyne colonies does not necessarily reflect their individual contribution to the production of colony members.

Interestingly, polygyne queens placed in monogyne conditions - kept individually with workers and brood in laboratory rearing units (Vargo and Ross 1989) - laid almost exclusively viable eggs (bottom half of Fig. 3), as does the mother queen in a natural monogyne colony. In contrast, the proportion of eggs laid by uninseminated queens held individually differed very little from those taken directly from the field. Experimental manipulation of the quantity of workers and brood had a strong effect on fecundity, reflecting the dependence of fecundity on fourth instar larvae under monogyne conditions (Tschinkel 1988), but the presence of larvae did not influence the ability of a queen's eggs to form embryos. This suggests that the lower viability of eggs produced by inseminated queens in polygyne colonies may result from mutual inhibition. In other species, queens are known to influence the viability of eggs produced by workers (see Vargo and Ross 1989); it is possible that fire ant queens also influence each other in this way.

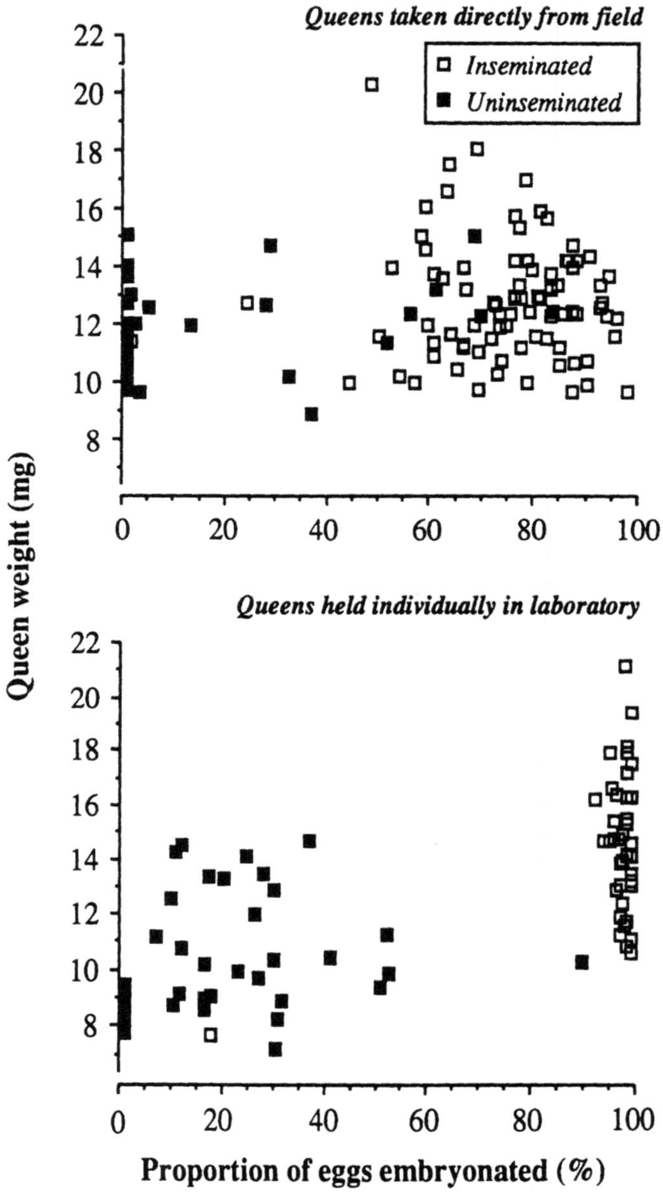

FIGURE 3. Proportion of viable eggs laid by polygyne queens of *S. invicta* in relation to insemination status and weight. Eggs were collected from individual queens, incubated with workers for 24-72 h, and then stained for chromatin (see Voss 1985). Eggs of queens held individually in the laboratory were sampled ca. six weeks after isolating the queens with nestmate workers and brood. (After Vargo and Ross 1989).

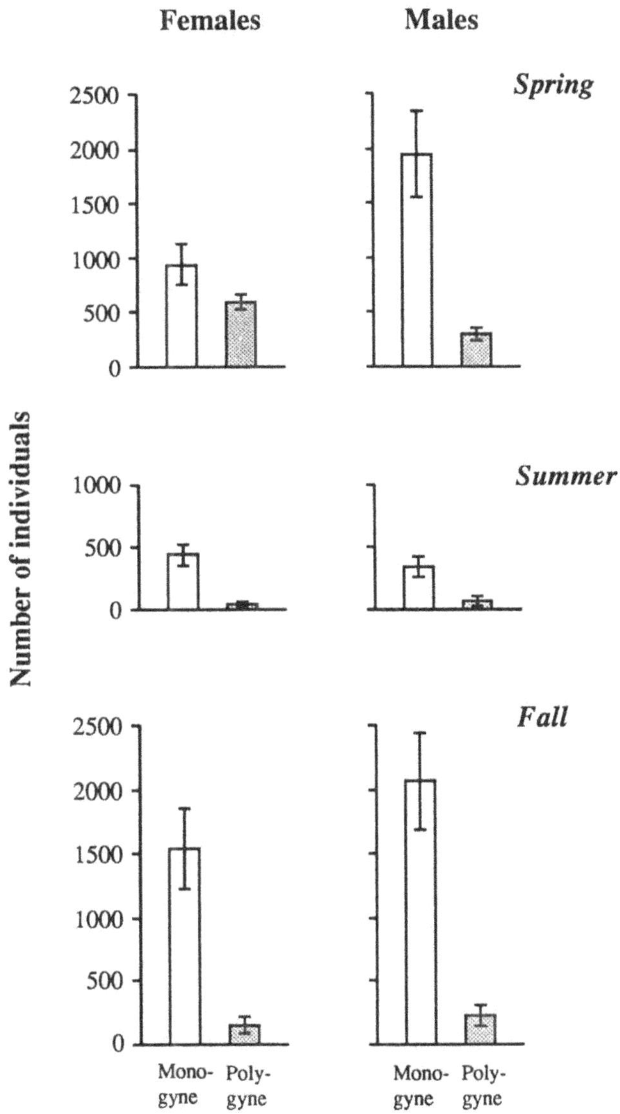

FIGURE 4. Numbers of sexuals (alates + pupae) present in monogyne and polygyne colonies of *S. invicta* from Georgia during three seasons of 1985. Bars=means±SE, n=25 in each case. Except for the number of females present in the spring sample, there were significantly fewer male and female sexuals present in the polygyne colonies during all three seasons (P<0.05, t-test). (Adapted from Vargo and Fletcher 1987).

CONTROL OVER THE PRODUCTION OF SEXUALS

Vargo and Fletcher (1987) discovered that polygyne colonies in the field produce fewer male and female sexuals than monogyne colonies during all three seasons when sexuals are actively produced (Fig. 4), suggesting a negative relationship between queen number and production of sexuals in fire ant colonies. This hypothesis was supported by laboratory studies in which halves of divided polygyne colonies containing many queens failed to produce sexuals, while monogyne or queenless halves produced hundreds of both male and female sexuals (Vargo and Fletcher 1986a). The inhibitory influence was demonstrated to be mediated by a queen pheromone (Vargo and Fletcher 1986b); the daily addition of 10 corpses of functional queens inhibited the production of sexuals in small experimental units (5 g) of workers and brood, whereas an equal number of corpses of reproductively inactive virgin queens had no inhibitory effect (Fig. 5). This inhibitory pheromone appears to act indirectly by affecting the behavior of workers toward larvae in a sex-specific manner. In response to the pheromone, workers presumably restrict the quantity and/or quality of food given to female larvae resulting in their development as workers; in contrast, they kill young male larvae, which have no such alternative line of development. In addition to preventing the sexualization of larvae, workers may also respond to the pheromone by killing late last instar larvae of both sexes after they have become sexualized (Vargo and Fletcher 1986b).

With the aim of providing a more sensitive test for the inhibitory pheromone, Vargo (1988) recently developed a bioassay in which individual queen corpses can be tested. With this procedure, the inhibitory activity of one queen equivalent or less can be detected in only two to three days, making it an exceptionally sensitive and rapid bioassay for a primer pheromone. Queen corpses - when placed in small cages which prevented direct contact by workers but through which volatile substances could readily pass - were not inhibitory. This suggests that this pheromone, like that inhibiting oogenesis and dealation, is relatively nonvolatile. Also, queen gasters were as effective as whole corpses, whereas heads and thoraces together were only slightly inhibitory. But as previously discussed, the effectiveness of gasters may relate to their attractiveness and not to the relative quantity of inhibitory pheromone they contain. Since the production of sexuals as well as oogenesis and dealation are under pheromonal inhibition in the fire ant, the same queen pheromone may mediate both responses. If so, this pheromone would be similar to the queen substance of the honey bee which, among other functions, inhibits both ovarian development in workers and the production of new queens (reviewed by Free 1987).

REGULATION OF QUEEN NUMBER

In themonogyne form of *S. invicta*, the presence of a single functional queen is usually enforced through execution of supernumary queens by workers. There are, however, two conditions under which temporary polygyny is known. Monogyne colonies of this species are founded in a manner typical of many ants; after the mating flight, queens seal themselves in subterranean nest chambers and rear the first brood from their body

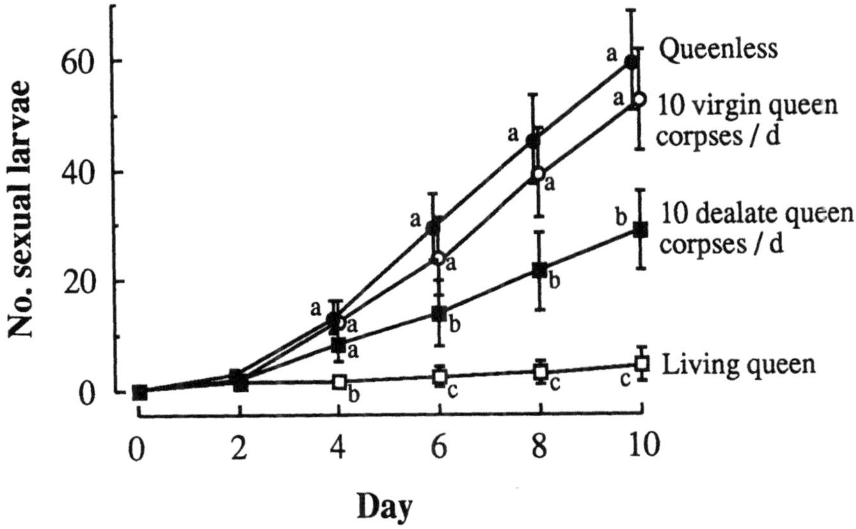

FIGURE 5. Numbers of sexual larvae produced in 5-g units of workers and brood. Shown are means ± SE. Treatments with different lower case letters at a particular day differed significantly ($P<0.05$, Newman-Keuls test, log transformation). $n = 20$ in each case. (From Vargo and Fletcher 1986b).

reserves (reviewed in Vinson and Greenberg 1986). In the field, newly-mated queens often associate in groups and cooperate in rearing the first brood. Such pleometrotically founded colonies always become monogyne when the first workers to emerge kill all but one queen (Markin et al. 1972; Tschinkel and Howard 1983; Fletcher 1986b).

In mature monogyne colonies, temporary polygyny may occur following removal or death of the mother queen. Once freed from her pheromonal inhibition, numerous virgin queens become reproductively active. However, the workers begin to execute them within 72 hours and continue until only one or two remain (Fletcher and Blum 1981a). How do workers choose among these virgin replacement queens? Fletcher and Blum (1983b) have proposed a multi-component hypothesis. They postulate that the quantity of pheromone produced by individual queens is positively correlated to fecundity, so that variable fecundity among coexisting queens forms the basis of a sort of pheromonal hierarchy. This hypothesis also holds that workers have a tolerance threshold for the level of some queen pheromone(s) in the colony and that they can distinguish individual queens according to the quantity of pheromone produced. Hence when the level of queen pheromone in the colony exceeds the tolerance threshold of workers, they begin to execute egg-laying queens based on their position within this hierarchy, eliminating the poorest producers first and thereby retaining the most productive (i.e., most fecund) queen.

Results of experiments performed by these authors support such a mechanism (Fletcher and Blum 1983b). When queenless groups of workers were given the choice between two foreign queens, one highly fecund and the other less so, they executed the queen of lower fecundity in 22 of 24 (92%) cases. As evidence of a positive correlation between fecundity and pheromone production, Fletcher and Blum (1983b) rely on the results of tests for the queen pheromone which inhibits oogenesis and dealation, though, as admitted by them, it is not known if this pheromonal signal is among those actually used by workers when they discriminate among queens under such conditions. The execution of foundress queens in pleometrotic associations may also be based on quantitative differences in pheromone production, but it has not been established whether the relative fecundity of queens is related to survival in such incipient colonies.

It has been suggested that in addition to the physogastric mother queen, monogyne colonies of *S. invicta* may contain one or more inseminated but reproductively inactive dealate queens. Tschinkel and Howard (1978) experimentally removed the mother queen from monogyne colonies in the field and reported that 27% of these nests possessed a functional inseminated queen when examined 8-10 weeks later. These authors hypothesized that such inseminated replacement queens were surviving foundresses from pleometrotic associations which had been reproductively inhibited by the previous queen. However, there has been no further evidence to support this hypothesis and it seems unlikely that monogyne colonies contain more than one inseminated queen. To my knowledge, not a single supernumerary dealate queen has been found among the thousands of monogyne colonies collected to date, and yet such queens should be present in more than one-quarter of these colonies if this hypothesis were true. The inseminated replacement queens found by Tschinkel and Howard may well have been adopted by the queenless colonies after a mating flight and/or the nests sampled after removal of the mother queen contained colonies other than those which were dequeened, as monogyne colonies are known to change nest sites often (Hays et al. 1982).

Since workers normally enforce monogyny in *S. invicta*, how is it that they tolerate multiple queens in some colonies? Two mutually compatible possibilities come to mind. First, polygyne queens may produce relatively little pheromone, so that their combined production does not exceed the pheromonal tolerance threshold of the workers. Second, workers of polygyne colonies may have an elevated tolerance threshold or even lack one altogether. The current evidence tends to exclude the first possibility. When workers from both polygyne and monogyne colonies were presented with 25 foreign inseminated queens collected from a single mating flight, monogyne workers executed all but one in six of eight (75%) cases, whereas polygyne workers accepted nine or more in seven of eight (87.5%) cases (Fletcher and Blum 1983b). Since these queens all came from a single pool, the greater tendency of polygyne workers to tolerate multiple-queens in this case cannot be explained by abnormally low levels of pheromone production. Rather, it seems that compared to workers from monogyne colonies, polygyne workers have an elevated tolerance threshold for the queen pheromone(s) which elicits execution behavior. The circumstances under which this pheromone threshold changes, thereby permitting the coexistence of multiple queens, remain for future studies to determine.

CONCLUDING REMARKS

Our understanding of social control over reproduction in fire ant colonies has come a long way since it first became a subject of serious study only nine years ago. While this knowledge is far from complete, more is presently known about social regulation of reproduction in *S. invicta* than perhaps any highly eusocial species besides the honey bee. Queen pheromones appear to play a central role, underscoring the importance of chemical signals in regulating colony function in highly eusocial insects (Blum 1974). The number of distinct fire ant queen pheromones remains unknown, but three different aspects of reproduction are clearly under pheromonal control: oviposition by virgin queens, production of male and female sexuals, and control of queen number in monogyne colonies. While none of these pheromones has been chemically identified, the existence of bioassays for two of the three and the large number of queens provided by the abundance of polygyne colonies of this ant give much hope for progress along these lines. Other promising areas for future research include the interaction between the exocrine and endocrine systems of queen fire ants as well as the regulation of oviposition and egg viability in polygyne colonies.

The study of social factors regulating reproduction in *S. invicta*, as in other social insect pests, has obvious practical implications. Deliberate disruption of any one of the reproductive processes, e.g., the application of synthetic queen pheromones to prevent the production of sexuals, could offer efficient and environmentally safe avenues for control. It is too early to say whether such methods might prove feasible in the control of fire ant colonies. Clear evaluation must await chemical identification of the queen inhibitory pheromones coupled with a deeper understanding of their behavioral and physiological consequences.

ACKNOWLEDGMENTS

I thank Kenneth Ross for his helpful comments on the manuscript. This review was written while supported by a grant from the National Science Foundation (INT-8702250).

LITERATURE CITED

BLUM, M.S. 1974. Pheromonal bases of social manifestations in insects, pp. 190-199. In M.C. Birch (ed.), Pheromones. North-Holland Publishing Co., New York.

BΦRGESEN, L.W. 1990. A new aspect of the role of larvae in the pharaoh's ant society (*Monomorium pharaonis* (L.), Formicidae, Myrmicinae): producer of fecundity-increasing substances to the queen. Insectes Soc. In Press.

BOURKE, A.F.G. 1988. Worker reproduction in the higher eusocial Hymenoptera. Q. Rev. Biol. 63: 291-311.

CHERIX, D. and D.J.C. FLETCHER. 1982. The eggs of founding queens of the imported fire ant, p. 249. In M. D. Breed, C. D. Michener and H. E. Evans (eds.), The biology of social insects, Westview Press, Boulder, Colorado. 419 pp.

FLETCHER, D.J.C. 1986a. Perspectives on some queen pheromones of social insects with special reference to the fire ant, *Solenopsis invicta*, pp. 184-191. In C.S. Lofgren and R.K. Vander Meer (eds.), Fire ants and leaf-cutting ants: biology and management. Westview Press, Boulder, Colorado. 435 pp.

FLETCHER, D.J.C. 1986b. Triple action of queen pheromones in the regulation of reproduction in fire ant (*Solenopsis invicta*) colonies. Adv. Invert. Repr. 4: 305-316.

FLETCHER, D.C.J. and M.S. BLUM. 1981a. Pheromonal control of dealation and oogenesis in virgin queen fire ants. Science 212: 73-75.

FLETCHER, D.C.J. and M.S. BLUM. 1981b. A bioassay technique for an inhibitory primer pheromone of the fire ant, *Solenopsis invicta* Buren. J. Ga. Entomol. Soc. 16: 352-356.

FLETCHER, D.C.J. and M.S. BLUM. 1983a. The inhibitory pheromone of queen fire ants: effects of disinhibition on dealation and oviposition by virgin queens. J. Comp. Physiol. A 153: 467-475.

FLETCHER, D.C.J. and M.S. BLUM. 1983b. Regulation of queen number by workers in colonies of social insects. Science 219: 312-314.

FLETCHER, D.C.J., M.S. BLUM, T.V. WHITT and N. TEMPLE. 1980. Monogyny and polygyny in the fire ant, *Solenopsis invicta*. Ann. Entomol. Soc. Am. 73: 658-661.

FLETCHER, D.C.J., D. CHERIX and M.S. BLUM. 1983. Some factors influencing dealation by virgin queen fire ants. Insectes Soc. 30: 443-454.

FLETCHER, D.J.C. and K.G. ROSS. 1985. Regulation of reproduction in eusocial Hymenoptera. Ann. Rev. Entomol. 30: 319-343.

FREE, J.B. 1987. Pheromones of social bees. Chapman and Hall, London. 218 pp.

GLANCEY, B.M. 1986. The queen recognition pheromone of *Solenopsis invicta*, pp. 223-230. In C.S. Lofgren and R.K. Vander Meer (eds.), Fire ants and leaf-cutting ants: biology and management. Westview Press, Boulder, Colorado. 435 pp.

GLANCEY, B.M., C.E. STRINGER and P.M. BISHOP. 1973. Trophic egg production in the imported fire ant, *Solenopsis invicta*. J. Ga. Entomol. Soc. 8: 217-220.

GREENBERG, L., D.J.C. FLETCHER and S.B. VINSON. 1985. Differences in worker size and mound distribution in monogynous and polygynous colonies of the fire ant, *Solenopsis invicta* Buren. J. Kans. Entomol. Soc. 58: 9-18.

HAYS, S.B., P.M. HORTON, J.A. BASS and D. STANLEY. 1982. Colony movement of imported fire ants. J. Ga. Entomol. Soc. 17: 266-274.

LOFGREN, C.S., W.A. BANKS and B.M. GLANCEY. 1975. Biology and control of imported fire ants. Ann. Rev. Entomol. 20: 1-29.

LOFGREN, C.S. and D.F. WILLIAMS. 1984. Polygynous colonies of the red imported fire ant, *Solenopsis invicta* (Hymenoptera: Formicidae) in Florida. Fla. Entomol. 67: 484-486.

MARKIN, G.P., H.L. COLLINS and J.H. DILLIER. 1972. Colony founding by queens of the red imported fire ant, *Solenopsis invicta*. Ann. Entomol. Soc. Am. 65: 1053-1058.

MARKIN, G.P., J.H. DILLIER and H.L. COLLINS. 1973. Growth and development of colonies of the red imported fire ant, *Solenopsis invicta*. Ann. Entomol. Soc. Am. 66: 803-808.

MARKIN, G.P. and J.H. DILLIER. 1971. The seasonal life cycle of the imported fire ant, *Solenopsis saevissima richteri*, on the Gulf Coast of Mississippi. Ann. Entomol. Soc. Am. 64: 562-565.

OBIN, M.S., B.M. GLANCEY, W.A. BANKS and R.K. VANDER MEER. 1988. Queen pheromone production and its physiological correlates in fire ant queens (Hymenoptera: Formicidae) treated with fenoxycarb. Ann. Entomol. Soc. Am. 81: 808-815.

O'NEAL, J. and G.P. MARKIN. 1973. Brood nutrition and parental relationships of the imported fire ant *Solenopsis invicta*. J. Ga. Entomol. Soc. 8: 294-303.

OSTER G.F. and WILSON E.O. 1978. Caste and ecology in the social insects. Princeton University Press, Princeton, New Jersey. 352 pp.

PETRALIA, R.S. and S.B. VINSON. 1978. Feeding in the larvae of the imported fire ant, *Solenopsis invicta*: behavior and morphological adaptations. Ann. Entomol. Soc. Am. 71: 643-648.

PETRALIA, R.S., A.A. SORENSEN and S.B. Vinson. 1980. The labial gland system of larvae of the imported fire ant, *Solenopsis invicta*: ultrastructure and enzyme analysis. Cell Tiss. Res. 206: 145-156.

ROSS, K.G. 1988. Differential reproduction in multiple-queen colonies of the fire ant *Solenopsis invicta*. Behav. Ecol. Sociobiol. 23: 341-355.

ROSS, K.G. and D.J.C. FLETCHER. 1985. Comparative study of genetic and social structure in two forms of the fire ant *Solenopsis invicta* (Hymenoptera: Formicidae). Behav. Ecol. Sociobiol. 17: 349-356.

SORENSEN, A.A., R.S. KAMAS and S.B. VINSON. 1983a. The influence of oral secretions from larvae on levels of proteinases in colony members of *Solenopsis invicta* Buren (Hymenoptera: Formicidae). J. Insect Physiol. 29: 163-168.

SORENSEN, A.A., T.M. BUSCH and S.B. VINSON. 1983b. Behaviour of worker subcastes in the fire ant, *Solenopsis invicta*, in response to proteinaceous food. Physiol. Entomol. 8: 83-92.

TSCHINKEL, W.R. 1988. Social control of egg-laying rate in queens of the fire ant, *Solenopsis invicta*. Physiol. Entomol. 13: 327-350.

TSCHINKEL, W.R. and D.F. HOWARD. 1978. Queen replacement in orphaned colonies of the fire ant, *Solenopsis invicta*. Behav. Ecol. Sociobiol. 3: 297-310.

TSCHINKEL, W.R. and D.F. HOWARD. 1983. Colony founding by pleometrosis in the fire ant, *Solenopsis invicta*. Behav. Ecol. Sociobiol. 12: 103-113.

VANDER MEER, R.K. 1983. Semiochemicals and the red imported fire ant (*Solenopsis invicta* Buren) (Hymenoptera:Formicidae). Fla. Ent. 66: 139-161.

VARGO, E.L. 1988. A bioassay for a primer pheromone of queen fire ants (*Solenopsis invicta*) which inhibits the production of sexuals. Insectes Soc. 35: 382-392.

VARGO, E.L. and D.J.C. FLETCHER. 1986a. Queen number and the production of sexuals in the fire ant, *Solenopsis invicta* (Hymenoptera: Formicidae). Behav. Ecol. Sociobiol. 19: 41-47.

VARGO, E.L. and D.J.C. FLETCHER. 1986b. Evidence of pheromonal queen control over the production of male and female sexuals in the fire ant, *Solenopsis invicta*. J. Comp. Physiol. A 159: 741-749.

VARGO, E.L. and D.J.C. FLETCHER. 1987. Effect of queen number on the production of sexuals in natural populations of the fire ant, *Solenopsis invicta*. Physiol. Entomol. 12: 109-116.

VARGO, E.L. and D.J.C. FLETCHER. 1989. On the relationship between queen number and fecundity in polygyne colonies of the fire ant, *Solenopsis invicta*. Physiol. Entomol. 14: 223-232.

VARGO, E.L. and K.G. ROSS. 1989. Differential viability of eggs laid by queens in polygyne colonies of the fire ant *Solenopsis invicta*. J. Insect Physiol. 35: 587-593.

VINSON, S.B. and L. GREENBERG. 1986. The biology, physiology and ecology of imported fire ants, pp. 193-226. In S. B. Vinson (ed.), Economic impact and control of social insects. Praeger, N.Y. 421 pp.

VOSS, S.H. 1981. Trophic egg production in virgin fire ant queens. J. Ga. Entomol. Soc. 16: 437-440.

VOSS, S.H. 1985. Rapid, simple DNA staining for fire ant eggs. J. Entomol. Sci. 20: 47-49.

VOSS, S.H. and M.S. BLUM. 1988. Trophic and embryonated egg production in founding colonies of the fire ant *Solenopsis invicta* (Hymenoptera: Formicidae). Sociobiol. 13: 271-278.

WEST-EBERHARD, M.J. 1981. Intragroup selection and the evolution of insect societies, pp. 3-17. In R. D. Alexander and D. W. Tinkle (eds.), Natural selection and social behavior. Chiron Press, New York. 532 pp.

WILLER, D.E. and D.J.C. FLETCHER. 1986. Differences in inhibitory capability among queens of the ant *Solenopsis invicta*. Physiol. Entomol. 11: 475-482.

15

Egg-laying in *Atta sexdens rubropilosa*, Under Laboratory Conditions

T.M.C. Della Lucia, E.F. Vilela, D.D.O. Moreira, J.M.S. Bento and N. Dos Anjos

INTRODUCTION

Oviposition of *Atta sexdens rubropilosa* generally occurs five to six days after colony foundation. At this time, the queen lays reproductive (normal) and trophic eggs (Autouri 1942; Mariconi 1970). The queen holds the eggs between her mandibles as they come out of her abdomen and eats them or places them on the developing fungus. As the population increases, foraging increases and the fungus grows; at the same time trophic egg production decreases, then ends. The queen lays only reproductive eggs, which are then collected by minor workers.

Egg-laying in social insects has been widely investigated in *Apis mellifera L.* (Gerig and Wille 1975; Harbo 1979); in *Coptotermes formosanus* Shiraki (Huang and Jung 1980); in *Myrmica rubra* L. (Hand 1983); in *Pheidole pallidula* (Nyl) (Passera and Suzzoni 1984) and in *Formica polyctena* Foerster (Winkler and Smith 1985).

A review of papers containing data on egg production by queens for several other ant and termite species was presented by Nielsen and Josens (1978).

Egg production and populations of larva, pupa and workers were studied within beginning colonies of *Atta capiguara* Goncalves (Pereira-da-Silva et al. 1986). However, literature on oviposition of leaf-cutting ants is scarce.

The objectives of this work were to determine the number of eggs laid by queens of *A. sexdens rubropilosa* under laboratory conditions; to relate the number of viable to non-viable eggs during each collecting period, and to investigate a possible relation between oviposition and the volume of the fungus garden.

MATERIALS AND METHODS

The queens used belonged to colonies maintained in the laboratory at 25 ± 2°C; 75 to 85% relative humidity and a photoperiod of 12 hours. Before initiation and during the tests, colonies were maintained in total darkness. Observations were made on queens removed from the nest and those maintained inside.

Eight queens were observed for oviposition outside the colony. Queens designated R2, R7 and R8 belonged to colonies which had fungus garden volumes of approximately 900 ml, while Rl belonged to a colony of 1100 ml. Queens R3, R4, R5 and R6 belonged to colonies with a 600 ml fungal garden, while R4 came from a colony of 500 ml.

Queens removed from their nests were placed individually in 110 ml plastic cups. A piece of cotton soaked in a 1:1 solution of distilled water and honey was placed near each queen and each cup covered with a perforated plastic lid. A cotton plug soaked in distilled water and placed over the lid maintained adequate moisture in the container. Eggs were collected every 12 hours for 96 consecutive hours. The queen was moved to a new cup containing fresh honey solution after every egg collection.

Four queens Rld, R2d, R3d and R4d--from colonies of 1100; 1000; 700 and 400 ml fungus gardens, respectively--were used to determine oviposition inside the colony.

For egg collection, each queen was placed within a polyethylene tubular container which had a hole in the lid which was split so that the queen could be positioned with her abdomen inside the container and her head and thorax free for feeding and grooming by the workers. Eggs were collected by changing the containers but not the lid and counted every 12 hours for 16 consecutive periods (a total of 192 hours).

RESULTS AND DISCUSSION

The number of viable eggs obtained during each 12 hour interval for queens outside the colony was highly variable (Table 1). However, cumulative values of egg laying over 96 consecutive hours showed interesting trends (Fig. 1). Despite the variation in the values (Table 1) the curves increased and never leveled off (Fig. 1). This could be an indication that the queens did not suffer stress. Non-viable eggs were also found, though infrequently. These eggs appeared as shriveled, jelly-like anomalies, normally in aggregates. Frequently they resembled ovules present in the abdomen of virgin queens. The number of these eggs varied from 0 to 10 per 12 hour interval for the eight queens.

TABLE 1. Number of viable eggs laid by *Atta sexdens rubropilosa* queens maintained outside the nest, determined in 12 hour-intervals.

	QUEENS							
HOURS	Rl	R2	R3	R4	R5	R6	R7	R8
12	196	154	31	8	6	20	85	116
24	276	143	105	33	57	52	104	194
36	300	179	37	8	72	46	88	260
48	257	119	57	20	48	12	94	202
60	213	68	64	29	17	38	120	166
72	268	94	70	22	23	38	114	138
84	205	126	81	27	62	48	119	88
96	171	144	76	17	59	40	117	174

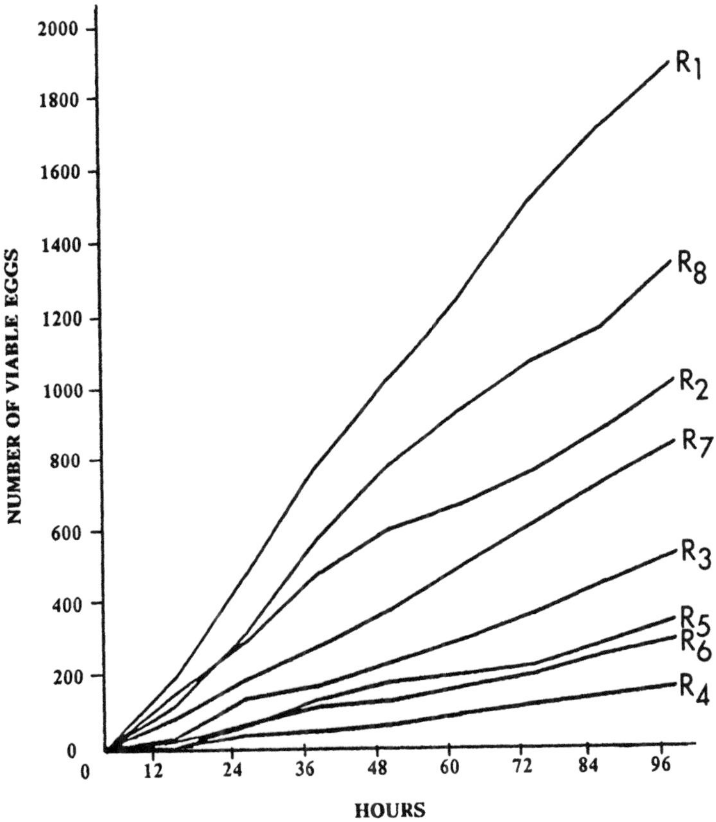

FIGURE 1. Cumulative numbers of viable eggs laid by eight queens of *Atta sexdens rubropilosa*, maintained inside the colony and counted at 12 hour intervals.

Both the average and total number of viable eggs was greater in queens from larger colonies (Rl, R2, R7 and R8) compared to those for queens from smaller colonies (R3, R4, R5 and R6 (Table 2). The rationale for these results may be associated with the reproductive potential of each queen and to the existence of smaller worker populations which care for the brood. In small colonies, the queen also helps to feed the larvae.

The average value of total number of eggs per queens, per 12 hours was 101.5 (viable and non-viable; eight eggs per hr per queen). From this total, 100.2 eggs per queen per 12 hours were viable (eight viable eggs per hour per queen). The queen laid one non-viable egg about every 10 hours.

Inside the colony, egg laying in 12 hour intervals was high but decreased (Table 3). Cumulative egg laying leveled off as the experiment progressed (Fig. 2). Egg-laying followed two distinct phases. The first was characterized by a nearly constant number of eggs laid at each interval, up to the eighth. The second phase was characterized by an abrupt reduction in egg production. This indicated that the queens were severely stressed.

TABLE 2. Mean values of total number of eggs and of viable and non-viable eggs per 12 hour interval per queen maintained outside the nest and related to the size of the fungus garden.

Queens	Garden Size (ml)	Total Eggs	Viable Eggs	Non-Viable Eggs
R1	1000	237.2	235.7	1.5
R2	900	130.4	128.4	2.0
R7	900	106.4	105.1	1.2
R8	900	168.6	167.2	1.4
AVERAGE		160.6	159.1	1.5
R3	600	66.7	65.1	1.6
R4	500	21.1	20.5	0.6
R5	600	44.1	43.0	1.1
R6	600	37.2	36.7	0.5
AVERAGE		43.2	41.3	1.0
TOTAL AVERAGE		101.5	100.2	1.3

TABLE 3. Number of viable eggs laid by queens of *Atta sexdens rubropilosa* maintained inside the nest, determined in 12 hour intervals.

	QUEENS			
HOURS	R1	R2	R3	R4
12	267	303	3	118
24	214	121	0	80
36	232	325	60	78
48	157	284	82	92
60	193	250	83	138
72	230	273	46	91
84	48	216	103	92
96	32	49	71	7
108	0	1	23	0
120	74	44	14	1
132	84	34	36	0
144	24	3	16	2
156	1	26	9	0
168	4	105	12	6
180	7	0	10	0
192	8	0	10	0

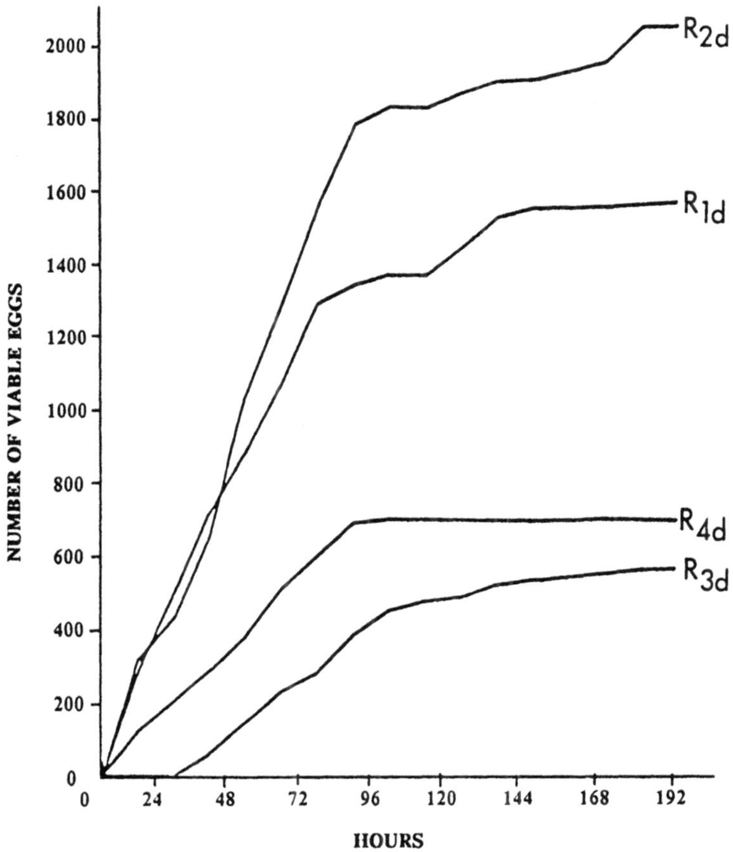

FIGURE 2. Cumulative numbers of viable eggs laid at 12-hour intervals by four *Atta sexdens rubropilosa* queens maintained inside the colony.

The handling and restraining of queens resulted in their death by the end of the 16th interval. During the experiment and before their death, the queens were fed and groomed in the head and thorax regions by the workers. Movement of parts free from the containers was also observed. The decreasing number of eggs laid may be due to stress (artificial conditions and handling) as well as a lack of stimuli to oviposit. According to Wilson (1974), the significance of "grooming" has not been completely elucidated. Grooming of the queen's abdomen may be necessary to stimulate oviposition, either through a pheromone or by simple physical contact. However, this hypothesis needs to be confirmed by further investigation.

In this experiment, queens from larger colonies also laid more eggs. The number of viable eggs for each 12-hour interval showed variation, both between queens and for the same queen. Queen R2d laid the most eggs while queen R3d laid the fewest. Interestingly, the increase of non-viable eggs into aggregates may reflect stress conditions. The greatest number of non-viable eggs was found inside the colony because of a longer observation period.

The average number of eggs and number of viable and non-viable eggs reflected both the oviposition phase and size of the fungus garden (Table 4).

TABLE 4. Mean values of total number of eggs and of viable and non-viable eggs per 12 hour interval per queen maintained inside the nest for the first and second phases of oviposition, relative to the size of the fungus garden.

Queens	Garden (ml)	Total No. Eggs 1st Phase	Total No. Eggs 2nd Phase	Viable Eggs 1st Phase	Viable Eggs 2nd Phase	Non-Viable Eggs 1st Phase	Non-Viable Eggs 2nd Phase
R1d	1000	178.1	34.5	171.6	25.2	9.0	9.2
R2d	1000	243.2	34.4	227.6	26.6	15.6	7.7
Average		210.7	34.4	199.6	25.9	12.3	8.5
R3d	700	57.4	19.7	56.0	15.0	1.4	4.7
R4d	400	93.5	2.2	77.0	1.1	6.5	1.1
Average		75.4	11.0	66.5	8.1	3.9	2.9
Total Average		143.1	22.7	133.1	17.0	8.1	5.7

In the first oviposition phase, the average was 143.1 eggs per queen per 12 hours, of which 133 were viable and 8.1 were non-viable. Therefore, approximately 12 viable eggs were laid for each non-viable egg. Among all queens the range was 6 to 20 viable eggs per hr. In the 2nd phase, these values reached only about 4 viable eggs laid per non-viable egg per queen.

Similar results were found with *Formica rufa.* (Nielsen and Jones 1978). According to Brian (1978), however, the rate of egg production under laboratory conditions differs from the rate in the field because there are several factors involved in oviposition, such as temperature and food restriction.

CONCLUSIONS

Great variations in the number of eggs laid by the queen and queens of *A. sexdens rubropilosa* under laboratory conditions were observed in this study. The average number of eggs laid by queens inside the colony was 1.5 times that for queens kept outside the colony. The proportion of viable to non-viable eggs was 80 to 1 outside the nest and 12 to 1 inside the nest. Queens from larger colonies had a greater egg-laying capacity than queens from smaller colonies. High mortality results if queens are handled more than 15 times at 12 hour intervals.

ACKNOWLEDGMENTS

We are grateful to Dr. Harold G. Fowler and Dr. Maria Jose Hebling-Beraldo for reading and criticizing this manuscript; to Eduardo Candido do Nascimento for the graphs and to CNPq (Brazil) for financial assistance.

REFERENCES

AUTUORI, M. 1942. Contribuicao para o conhecimento da sauva. (*Atta spp.* Hymenoptera - Formicidae). II - 0 sauveiro inicial (*Atta sexdens rubropilosa* Forel, 1908). Arq. Inst. Biol. 13: 137-148.

BRIAN, M.V. 1978. Production Ecology of Ants and Termites. Cambridge, Cambridge Press. 409 pp.

GERIG, L. and H. WILLE. 1976 Periodicity in the oviposition of queen mellifera L. (Hym., Apidae). Mitt. Schweiz. Entomol. Ges., 48: 91-97. 1975. Apud: Entomology Abstracts, 7: 180.

HAND, L. 1983. The effect of photoperiod by the red ant, *Myrmica rubra.* Entomol.Exp. Appl., 34: 169-173.

HARBO, J.R. 1986. Oviposition rates of instrumentally inseminated and naturally mated queen honey bees (Hymenoptera: Apidae). Ann. Entomol. Soc. Am., 79: 112-115.

HUANG, L.W. and H.W. JUNG. 1980. The influence of temperature and humidity on the initial colonies of *Coptotermes formosanus* Shiraki. Acta Entomol. Sin., 23: 32-36; Apud: Entomology Abstracts, 11: 131.

MARICONI, F.A.M. 1970. As sauvas. Agron. Ceres, SP. 167 pp.

NIELSEN, M.G. and G. JOSENS. 1978. Production by ants and termites, pp. 45-53. *In*: Brian, M. V. (ed.). Production Ecology of Ants and Termites. Cambridge, Cambridge University Press.

PASSERA, L. and J.P. SUZZONI. 1984. Caste-biased egg-laying in the ant *Pheidole pallidula* (Nyl): biological date related to the queens at the end of hibernation.Insectes Sociaux. 31: 155-170.

PEREIRA DA SILVA, V., L.C. FORTI, V.L. DE BIASI. and E.M. PEREIRA-DA-SILVA. 1986. Dinamica populacional das colonias de *Atta capiguara* Gonsalves, 1944 (Hymenoptera; Formicidae), pp. 58-59. Rio Claro, VII Encontro de Mirmecologia and do Estado de Sao Paulo, jun., Anais.

WILSON, E.O. 1974. The Insect Societies. Cambridge, Belknap Press. 548 pp.

WINKLER, I. and G.H. SCHMIDT. 1985. Influence of the egg dimorphism of the queen of *Formica polyctena* Foerster by environmental factors. Insectes Sociaux. 32: 86-95.

16

Foundress Female Weight and Cooperative Foundation in Atta Leaf-cutting Ants

A. Mintzer

INTRODUCTION

Colony foundation requires more complex behaviors and more intensive investment by the mother queen of *Atta* leafcutting ants than by any other ant species. The newly inseminated queen must establish and nurture a fungus culture, in addition to rearing her own offspring, without recourse to an external food source. It is not surprising that *Atta* queens are among the largest ants known, only exceeded by some doryline driver ant queens. This paper reviews certain dynamics of colony foundation in three *Atta* species specifically: female initial weight, weight loss and investment during laboratory colony foundation, and productivity, survivorship and success. Results of the lab work illuminate optimal alternative strategies for independent or cooperative colony foundation in *A. texana*, *A. mexicana*, and *A. cephalotes* in the field.

METHODS

Atta texana queens were collected in Washington Co., Texas after nuptial flights in 1984-86. These flights occurred just before dawn on days after heavy rainfalls in late April, May and early June. *Atta cephalotes* queens were collected from their claustral burrows on May 30, 1988 near Palenque, Chiapas, Mexico. Their flight had apparently occurred two or three days previously after early Summer rains, and the ants had no eggs or fungus cultures yet. Live, newly-mated *Atta mexicana* queens were received from collectors in Arizona, El Salvador, and two localities in Mexico. The flights took place after rainfalls in June and July.

The queens were weighed before being installed in clear plastic culture containers. Culture dishes--lined along the bottom with perlite granules or dental plaster premoistened with distilled water--were kept in a controlled environment of approximately 27-29 C and 65% relative humidity. The dishes remained closed until worker eclosure or imminent foundress female death. Dying females were removed for spermathecal examination. When worker eclosure was expected within 24 hours, dishes were opened and the female(s) weighed. Measurement of brood and fungus production took place at this time. The dish was covered with a lid fitted with a vertical plastic tube, initially plugged with a stopper. The number of offspring produced by each queen was determined 10 days after the first one emerged. When worker foraging began, young colonies received a diet of oatmeal and cornmeal.

RESULTS AND DISCUSSION

The initial weight of dealate females varied widely within and especially between *Atta* species. Females of *A. texana* and southern *A. mexicana* weighed only 300-400 mg, while females of *A. cephalotes* weighed about 500-600 mg. In Brazil, Pereira da Silva (1979) reports *A. capiguara* female weights ranging up to and exceeding 800 mg, and E. Vilela (pers comm.) reports *A. sexdens rubropilosa* weights from 400 to 500 mg. This corresponds closely with the largest ants ever collected in the United States, *A. mexicana* female sexuals from Arizona (unpublished data).

All females lost weight during colony foundation (typically 30-35%) before the first workers emerged in *A. texana* and *A. mexicana*. In these two species, the weight loss was better correlated than initial weight with production of fungus and offspring. Three surviving *A. cephalotes* queens had lost 17-39% of their initial weight when their first workers emerged. The first larval brood received eggs from the female and did not consume fungus. The fungus was cultured on fecal fluid produced by the female; about 40% of the female's weight loss in *A. texana* and *A. mexicana* was associated with fungus culture. In the lab, the initial fungus culture of all three species developed into a disk or mat of mycelium, with brood piled on top in the center. The surface area of the mycelial mat at the time of first worker emergence was used as a measure of fungal production. For independent females, this ranged from 3.0-5.5 cm^2 (mean 4.0 cm^2,n=23) in *A. texana*, and was 3.9 cm^2 for a successful *A. cephalotes* female.

In the laboratory, females of *A. cephalotes* and *A. texana* required 35-38 days to rear their first workers. Offspring numbers ten days after first emergence ranged from 15 to 48 for *A. texana* (mean 34, n=23), 3 to 37 for *A. mexicana* (mean 22,n=11), and 17 to 32 for *A. cephalotes* (n=3). First broods of *A. texana* and *A. mexicana* yielded workers with two to four mm body lengths (HW under 1.2 mm). However, first broods of *A. cephalotes* included some larger ants, up to 5.5 mm body length in one brood (HW over 1.5 mm).

Newly mated queens of all three species were able to establish colonies independently in the lab. *A. mexicana* females are very aggressive when placed together; children in El Salvador seek morbid entertainment by grouping these "Hormigas de Mayo" in containers and watching them slash and cut each other with their mandibles (L. Serrano, pers comm.). Available data indicate *A. cephalotes* colony-founding queens are also incompatible in the field and always start colonies alone. In contrast, *A. texana* foundress queens may form groups and cooperatively found colonies in the field or laboratory. These queens often land in local clusters which favor queen encounters and formation of dense clumps. Most of the dealates (ca. 75% inseminated in *A. texana* join groups, which can include more than one inseminated queen. In the laboratory, females may be paired at random without difficulty, and rare conflicts within groups never escalate to overt fighting.

Cooperative colony founding by *Atta texana* queens results in significantly higher survival rates and greater collective production of worker offspring and fungus. While they produced workers of the same small size in the same time period as single queens, groups of two or three females reared significantly more of them ($p<0.0001$, Mann Whitney U test). Measured fungal production was significantly greater in these groups than in single females ($p<0.0001$, Mann Whitney U test). Grouped females also lost less weight (unpaired $t=2.015$, $p<0.05$) than single females. Six months after the nuptial flight, surviving polygynous colonies had larger fungus gardens than surviving monogynous colonies.

Atta texana female groups are stable associations, lasting at least three years in the laboratory; in this time period, all worker sizes including majors are produced and the worker force grows beyond 10,000 individuals. In Louisiana, Moser (Moser and Lewis 1981) and others have uncovered established colonies with several inseminated females. *Atta texana* demonstrates a pattern of facultative primary polygyny which is uncommon in other ant species. Although detailed data is missing, Wilson (1985) characterizes *Atta* colony foundation as haplometrotic (=primary monogyny).

Why does cooperative behavior occur in *A. texana*, and why does it appear to be confined only to this species of leaf-cutting ant?

To survive, females without infrabuccal fungus pellets must join groups possessing viable fungus. Females without fungus have been collected from several species, including *A. texana*, but cooperative behavior has not been observed. Also, field groups have been collected in which nearly all females had pellets; lab rearing shows that the incidence of nonviable pellets is too low to account for observed cooperative behavior in *A. texana*. During colony foundation, queens do not groom each other, so there are no obvious hygienic benefits for cooperating ants.

Observations on other species suggest that female landing patterns also afford opportunities for postflight contact and group formation. Cooperation might allow digging groups to seal and complete the natal burrow more quickly, escaping some predation. However, female groups might also be more apparent to predators than single ants on the surface. No evidence indicates that *A. texana* faced more numerous surface predators than other *Atta* species, although the recent invasion of the tramp *Solenopsis invicta* has obscured our ability to assess this accurately.

While initial female weight was a poor predictor of fungus and worker offspring production, survivorship data show an advantage for larger *A. texana* females. Sixty-three percent of single females with initial weights over 400 mg survived until worker emergence; only 33% of females weighing 350 mg or less did so. Thirty-one successful females lost an additional 80 to 194 mg (mean 117 mg) in a two to three month period after the first workers emerged, indicating that continuing investment by the foundress(es) may be required even after worker foraging begins. Small single females may be particularly disadvantaged in this regard.

Three of four independent *A. cephalotes* females lived to oversee worker emergence. Comparing weight loss and productivity in *A. texana* and *A. cephalotes*, we found that the larger species did not rear more worker offspring or produce a larger fungus culture. In fact, two of the *cephalotes* females consumed most or all of their fungus cultures before workers emerged, possibly reflected in modest weight losses of 89 and 122 mg. The third ant produced a fungus culture and worker brood very close to mean values for single *texana* females, but lost more weight than the typical *texana* female in the process. This larger investment was reflected in the production of some larger workers in the *cephalotes* first brood. As reported by Wilson (1985), the larger workers are utilized for efficient leafcutting activity by the young colony, although it was noted that the four mm workers of *A. mexicana* and *A. texana* cut small pieces in quantity from mature leaves, as do *Trachymyrmex septentrionalis* workers of similar size.

The addition of a second female to the founding unit would probably have a less significant positive effect on survivorship and success in *A. cephalotes* than in the smaller *A. texana*. Given the automatic costs of group living (e.g. spread of pathogens, reproductive competition) which accompany female cooperation, the optimal strategy for large species such as *A. cephalotes*, *capiguara*, and *laevigata* may be shifted towards independence. The divergent colony founding strategies of *A. texana* and *A. mexicana* females in the same size range call for further comparative field studies in Mexico and Texas, to clarify as-yet unidentified crucial differences in these two species.

ACKNOWLEDGMENTS

I thank S.B. Vinson and the Texas Agricultural Experiment Station for their support of this project. I thank Luis Quiroz and Cuauhtemoc Deloya-Lopez for collecting *A. mexicana* in southern Mexico, and Leopoldo Serrano for collecting the same species in El Salvador, and shipping females by air freight to Texas. I also thank USDA-APHIS/PPQ (D. Knott and R. Pierce in particular) for import permission and assistance.

REFERENCES

MOSER, J. and J. LEWIS. 1981. Multiple nest queens of *Atta texana*. Turrialba 31: 256-257.

PEREIRA DA SILVA, V. 1979. Dinamica populacional, biomassa e estructura dos ninhos inicias de Atta capiguara Goncalves na regiao de Botucatu, SP. Ph.D. Thesis, UNESP Botucatu, Sao Paulo, Brasil.

WILSON, E.O. 1985. The sociogenesis of insect colonies. Science 228: 1489-1495.

Natural History and Biology

Atta laevigata soldier (x 5.5). Drawing by Eduardo Perez.

Overview

S.D. Porter

The world is full of ants. They crisscross virtually every square meter of land from the tropics to the subarctic. As a myrmecologist, I have difficulty stopping anywhere without shortly making their acquaintance.

Generally, ants are peaceable residents of natural communities. However, a few species have earned a reputation for "afflicting and tormenting man." Most of these species are exotics which have become especially abundant in heavily disturbed areas associated with human habitation. Practically everyone, it seems, has their own "war stories" about the depredations of one ant or another. On the positive side, research on pest species has provided much of what we know about the basic biology of ants. The following 13 chapters are replete with myrmecological data which should benefit both the naturalist and the pest control expert.

The first two chapters discuss the relationship between the fire ant, *Solenopsis invicta*, and native ant communities in North and South America. Camilo and Phillips evaluate the impact of imported fire ants at 12 sites in Central Texas. They report that "within the range of *S. invicta*, the diversity of the ant assemblages were adversely affected by increasing densities" of fire ants. These results are in contrast to those of Fowler, Bernardi and Romagnano in Brazil, where *S. invicta* is apparently a minor component of the native ant community. Fowler et al. conclude that, "*S. invicta* populations [in Brazil] are not limited by competitive interactions at food resources," and that *S. invicta*, "may be held in check by natural South American enemies."

The next seven chapters focus on various aspects of leaf-cutting ant biology. Weiss briefly describes the growth of a laboratory reared *Atta cephalotes* colony, from the founding queen to the colony's eventual death after 10 years. This colony produced more than 160 L of cumulative refuse and contained a maximum of 54 L of fungus. Cedeño and Leon document the growth and demography of *Atta laevigata* colonies in the field, from initial founding to colonies more than two years of age. They also suggest that young colonies may aggregate brood in their second fungus chamber because this, "liberates immatures from the action of secondary plant compounds," which are apparently detoxified in the upper or initial chamber. Claver evaluates several transect methods of estimating *Acromyrmex lobicornis* colony density and recommends that (when possible) transect widths should be, "based on distance estimates of widths derived from indefinite width transects."

The quantity of vegetation consumed by leaf-cutting ant colonies determines, to a large extent, the magnitude of their ecological and economic impacts. Fowler, Forti and Romagnano review and evaluate techniques used to estimate vegetation consumption of attine colonies in 32 published studies. The information they provide will facilitate more prudent use of those techniques in future research. Perera, González and Martínez describe activity patterns of *Atta insularis* in Cuba as a function of season and weather. They found that food retrieval only accounted for about 20% of surface activity and that the overall activity pattern did not change consistently with season. Jolivet assesses the effectiveness of ant-tended *Cecropia* trees in fending off *Atta* defoliation. He also hypothesizes that *Cecropia* species which are not tended by ants are defended by secondary chemicals which make them unacceptable for leaf-cutting ants. Waller and Moser discuss numerous symbionts associated with *Atta texana* colonies in Texas and Louisiana. Comparative studies of symbiotic communities in other attines would be very interesting from the standpoint of community ecology and coevolution. Such studies could also lay a foundation, as the authors suggest, for future "biological control programs using natural enemies."

The final four chapters of this section summarize our knowledge of the natural history and ecology of several lesser-studied groups of pest ants. These reviews will be a valuable resource for these groups. Hansen and Akre provide an excellent review of carpenter ant (*Camponotus*) biology including mating flights, colony founding and the structure of mature colonies. They also discuss foraging and nesting sites. Ulloa-Chacon and Cherix summarize previous studies of the little fire ant, *Wasmannia auropunctata* with special emphasis on nest structure, queen fecundity and sexual production. Zenner-Polanía details the biology and history of *Paratrechina fulva* in Colombia, where huge densities of this species can disrupt agriculture and greatly reduce biodiversity. Finally, Harada outlines the history, biology and economic importance of nine dolichoderine ant pests in the tribe Tapinomini, most notable of which is the Argentine ant, *Iridomyrmex humilis*. Harada also provides a key, distributional maps and several fine illustrations of the ants themselves.

This book covers a considerable amount of territory concerning pest ants, but I would like to mention several opportunities for future work. For example, information concerning the natural history and ecology of *Pheidole megacephala, Paratrechina longicornis* and *Anoplolepis* species is either limited or difficult to obtain. A considerable amount of research has been done with the Argentine ant, *I. humilis*; nevertheless, much of this information is scattered across six continents or buried in old bulletins. Similar complaints can be made about the Pharoah's ant, *Monomorium pharaonis*, and the tropical fire ant, *Solenopsis geminata*. Thorough literature reviews for each of these species would be very useful. Also, there is probably room for a carefully illustrated book on leaf-cutting ants written for the general scientific audience, a sequel to Weber's 1972 book.

In conclusion, I would like to highlight one important area for future research. I think we need to look more seriously at the possibility of biological control of pest ants. We know very little about the ecology of exotic species in their native habitats, specifically the biotic and abiotic factors which regulate populations of these pests. For instance, why do *P. fulva* populations in Colombia appear to decline after a decade or so (Zenner-

Polania, this section) and why do *Solenopsis* densities in Brazil appear to be much lower than they are in the United States (Fowler et al., this section). Certainly, in the case of *S. invicta*, pathogen and parasite levels are much higher in South America than in North America. Perhaps some pest ants attain high densities as a consequence of their release from natural control agents left behind in their homeland. The prospect of biological control of pest ants is still speculative; nevertheless, it offers the potential for long-term control that would be both cost-effective and environmentally sound.

17

Evolution of Ant Communities in Response to Invasion by the Fire Ant *Solenopsis invicta*

G.R. Camilo and S.A. Philips, Jr.

INTRODUCTION

Since the introduction of the red imported fire ant, *Solenopsis invicta* Buren, into the southern U.S.A., it has been considered a threat to the native ant fauna (Lofgren et al. 1975, Wilson et al. 1971). The rapid spread and successful establishment of *S. invicta* (Lofgren et al. 1975) has been attributed to its high reproductive potential (Glancey et al. 1976), its aggressive behavior (Bhatkar et al. 1972) and its efficient foraging and recruitment abilities (Jones 1985, Kidd et al. 1985, Phillips et al. 1986).

The occurrence of *S. invicta* in North America has been associated with the extinction or displacement of other fire ants. Wilson and Brown (1958) considered the substitution of the "dark phase" of the imported fire ant for the "light phase" as the evolution of character displacement due to a pleiotropic polygene interaction in the founding population of imported fire ants in Alabama. This color phase change is now interpreted as the substitution of one species of imported fire ant, *S. richteri* Forel, by another, *S. invicta* (Buren et al. 1974). Prior to Buren (1972), studies conducted on imported fire ants in North America inaccurately identified species because the taxonomy of the species complex was not well established (Baroni Urbani and Kannowski 1974).

After Buren's revision of imported fire ants in 1972, several papers were published dealing with competition and community disturbance by *S. invicta*. Baroni Urbani and Kannowski (1974) studied demographic parameters and interspecific competition of *S. invicta* in a Louisiana, U.S.A. pasture and found that only two other species were common: *Monomorium minimum* (Buckley) and *Paratrechina arenivaga* (Wheeler). These ants coexist with *S. invicta*, although they seemed to compete for the same food resources. Another species of fire ant present in the area, *Solenopsis xyloni* McCook, was found only in the shade beneath large trees and was probably already displaced to these areas by *S. invicta*, which prefers unshaded areas.

The replacement of other ants by *S. invicta* has been observed on several occasions. Using bait stations, Glancey et al. (1976) discovered that the once dominant *Iridomyrmex humilis* (Mayr) was displaced by *S. invicta*. *Solenopsis xyloni* and *Solenopsis geminata* (F.) were not found in their survey, although these species previously had been found in the study area.

The authors suggested that *S. invicta* displaced these species. Whitcomb et al. (1972) reported that in Florida soybean fields, *S. geminata* was one of the first species to disappear following invasion by *S. invicta*. They hypothesized that this rapid disappearance occurred because *S. invicta* "...is particularly aggressive to other ant species of its own trophic level." More recently, Tschinkel (1988) suggested that the outcome of competition between *S. invicta* and *S. geminata* is mediated by disturbance of the environment. Tschinkel (1986) described *S. invicta* as a weed species, because *S. invicta* is particularly adapted to ecologically disturbed habitats. In addition, Porter et al. (1988) quantified the competitive replacement of native fire ants by the red imported fire ant and found that polygyne *S. invicta* replaced S. geminata six fold. Their results suggest strong asymmetric competition between these ants and a restructuring of the ant community after invasion by *S. invicta*.

The objectives of our study were to determine (1) the composition of ant assemblages from heavily infested to uninfested *S. invicta* areas, (2) which factors may be controlling the diversity of ant communities and (3) the effect of *S. invicta* on the densities of other ant species.

MATERIALS AND METHODS

The study area was a transect paralleling Interstate Highway 10 in Kimble and Kerr Counties of central Texas, U.S.A. The transect extended from non-infested to heavily infested *S. invicta* areas and was composed of twelve plots (A-L), each 4.8 kilometers apart. Each plot consisted of 24 pitfall traps. A trap was a plastic cup (440 ml) embedded in the soil with the cup lip flush with the ground surface. A second plastic cup was inserted into the first and soil was piled around the cup in order to level the ground with the lip of the second cup. By so doing, cups could be set in place without having to excavate a new hole each time sampling occurred, thus minimizing the "digging-in" effect (Greenslade 1973). Ethylene glycol was used as the killing and preservative agent (Adis 1979). To prevent larger, unwanted arthropods and vertebrates from being trapped, asphalt shingles, 10 by 10 cm, were placed over each trap, and a space of approximately 1 cm was left between the trap and the cover. This procedure also prevented rainfall from flooding the traps. The transect was monitored every three days. The 72 hours allowed maximum capture of ants (Jansen and Metz 1979) and the time between monthly samplings provided a rest period during which no ants were removed from the habitat (Samways 1983). Two plots almost entirely infested with *S. invicta* (plots K and L) were eliminated from all analyses.

Communities were monitored once each month, from August 1986 through November 1986, and from March 1987 through October 1987. However, November 1986 and March 1987 were excluded from the analyses, along with the intervening months, to minimize variation resulting from seasonal fluctuations (Samways 1983). Voucher specimens were deposited in the entomology collection of the Department of Agronomy, Horticulture, and Entomology, Texas Tech University, Lubbock, Texas, U.S.A.

The importance value (IV) of each species was calculated for each plot as follows:

$IV_{ik} = X_{ik}\,(3N)^{-1}(FN)$; where X is the number of individuals of species x collected in a specific plot, i, at a specific sampling period, k. The number 3 accounts for the number of days the traps were in place, so the IV

TABLE 1. Species assemblages in central Texas, U.S.A., as detected by cluster analysis.Plots K and L were excluded from the analysis.

	Species Assemblage			
Species	One	Two	Three	Four
Pachycondyla harpax	+			
Labidus coecus	+			
Conomyrma flava	+			
Brachymyrmex depilis	+			
Camponotus discolor	+			
Tapinoma sessile	+			
Formica sp.	+			
Crematogaster laeviuscula	+			
Pogonomyrmex imberbiculus	+			
Crematogaster pilosa	+			
Leptothorax terrigena [a]	+			
Solenopsis aurea	+			
Leptothorax sp. [a]	+			
Solenopsis geminata	+			
Solenopsis (Diplorhoptrum) sp. 1 [a]	+			
Pheidole sp. 1	+			
Tetramorium spinosum	+			
Pogonomyrmex barbatus	+	+	+	
Pheidole tepicana	+	+	+	+
Paratrechina terricola / vividula [b]	+	+	+	+
Forelius pruinosus	+	+	+	
Leptogenys elongata	+			
Neivamyrmex nigrescens		+		
Camponotus sp.		+		
Forelius foetidus		+		
Atta texana		+		
Pheidole hyatti		+		
Monomorium pharaonis		+		
Solenopsis xyloni		+		
Monomorium minimum		+	+	+
Pheidole crassicornis tetra		+	+	+
Solenopsis (Diplorhoptrum) sp. 2			+	
Pheidole sp. 2			+	
Solenopsis invicta			+	+
Conomyrma insana			+	+

[a] Species detected as uncommon by PCA.
[b] Unable to differentiate between species.

will be given in days; N represents the number of traps in a plot and F the number of traps in which individuals of species x were found.

Cluster analysis performed using Procedure Cluster (SAS Institute Inc. 1985) identified distinctive assemblages. The clustering rule used was the centroid method (Pielou 1984) and the dendrogram generated by the Procedure Tree (SAS Institute Inc. 1985). A Kendall's correlation (Sokal and

Rohlf 1981) was performed to determine a response of species richness to variable densities of *S. invicta*.

Habitat disturbance was quantified by assigning ranked values to each habitat depending on the level and frequency of the disturbance. If mowed once a month, the habitat was assigned a rank of 10. If the habitat was undisturbed in the month between sampling periods it received a rank value of five, whereas a habitat never mowed was assigned a rank of one.

A principal components analysis (PCA) identified sources of variation in diversity among plots (Digby and Kempton 1987). The eigenvalue matrix for the species was generated by Procedure Cluster, and Procedure Princomp generated the values for the plots (SAS Institute Inc. 1985). The densities of *S. invicta* were excluded from the PCA to identify the effects of this species on the native ant fauna. Uncommon species were dropped from the analysis, but included in the list of their respective assemblages. The default stop of the PCA occurred when 99% of the variation was included in the ordination axes. Importance values of the species occurring with *S. invicta* were correlated (Sokal and Rohlf 1981) through time to *S. invicta* densities. This test will help identify which ants are increasing or decreasing in importance in response to *S. invicta* densities. Variation among plot diversity in the PCA was tested using Kendall's rank correlation for disturbance and Pearson's product-moment for temperature and densities of *S. invicta* (Sokal and Rohlf 1981).

RESULTS AND DISCUSSION

Seven thousand six hundred ninety-seven ants, representing five subfamilies and 35 species, were collected. Cluster analysis identified four distinct assemblages (Fig. 1). The first bifurcation in the dendrogram separates the site that occurs in an undisturbed habitat from those occurring in disturbed habitats. The second division separates those clusters that contain *S. invicta* from those that do not. The final division separates a cluster of sites with high *S. invicta* densities from one with low densities.

Density of *S. invicta* is negatively correlated with species richness and thus diversity (Kendall's $t = 0.62$; $n=27$; $P < 0.05$). Twenty one species of ants occur in the undisturbed, uninfested plot (Table 1), and 14 in the disturbed, uninfested plots. Comparison of the previous assemblages with the assemblages containing *S. invicta* (Table 1) suggests a pattern of species number reduction with increased densities of *S. invicta*: nine species in the assemblage with low densities of *S. invicta*, and five in the high density assemblage.

Species assemblage one (Table 1) was found in an undisturbed habitat (it had not been mowed or grazed for at least 8 years). It possessed the highest diversity and contained ecologically specialized forms such as *Pachycondyla harpax* (F.), *Labidus coecus* (Latreille), *Leptothorax terrigena* Wheeler, and *Tetramorium spinosum* (Wheeler). These species were found throughout most of the sampling periods (summer and fall 1986; spring, summer, and fall 1987). This assemblage might have differentiated from assemblage two, or from an assemblage with a similar community structure after a short ecological period, once mowing was stopped. Species assemblages two, three, and four (Table 1) were in disturbed habitats, i.e., mowed or grazed several times each year. Species assemblage two, characterized by high diversity, had

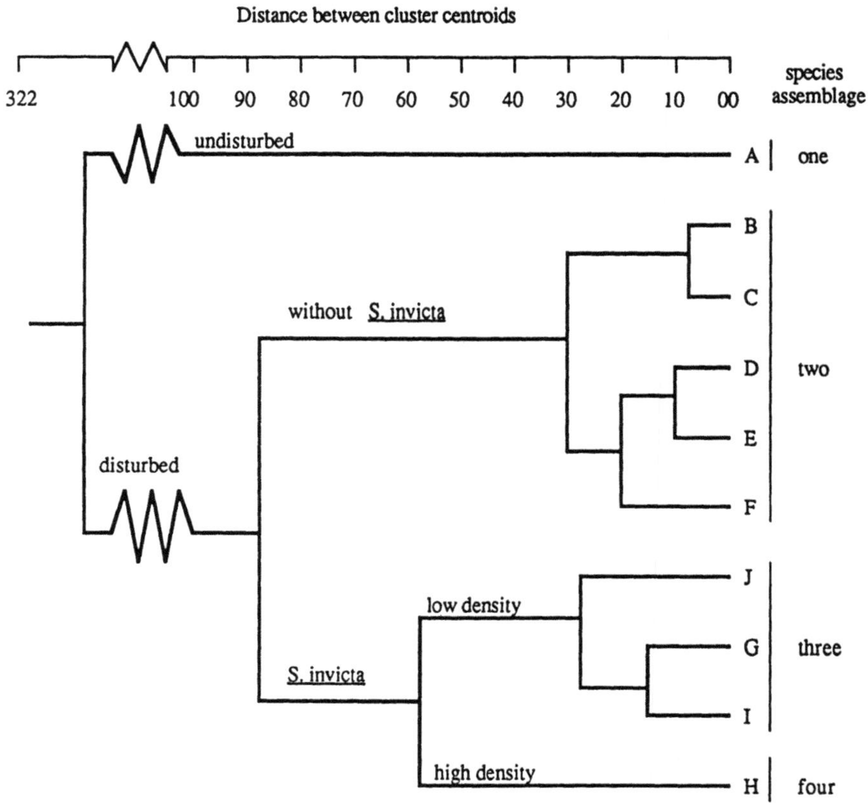

FIGURE 1. Dendrogram of ant assemblages located in Kerr and Kimble Counties, Texas, U.S.A. Plots K and L were excluded from the analysis.

some ecologically specialized species such as *Leptogenys elongata* (Buckley) and *Neivamyrmex nigrescens* (Cresson). However, these species were found only sporadically throughout the sampling periods. The major contrast noted between the first two assemblages may have resulted from the latter being disturbed several times each year, contrasted with the lack of disturbance of assemblage one. Assemblage one should not be considered a climax community because it is found in a patch habitat, the area originally being a grassland.

Species assemblages three and four were found within the range of *S. invicta*. Species assemblage three has low densities of *S. invicta* (< 2 foragers per day per trap). Most species were collected sporadically, with the exception of *M. minimum*. In this assemblage, a species of thief ant, *Solenopsis (Diplorhoptrum)* sp. 2, was captured repeatedly, even though this ant is a subterranean species. Tschinkel (1988) suggested that this subgenus may be

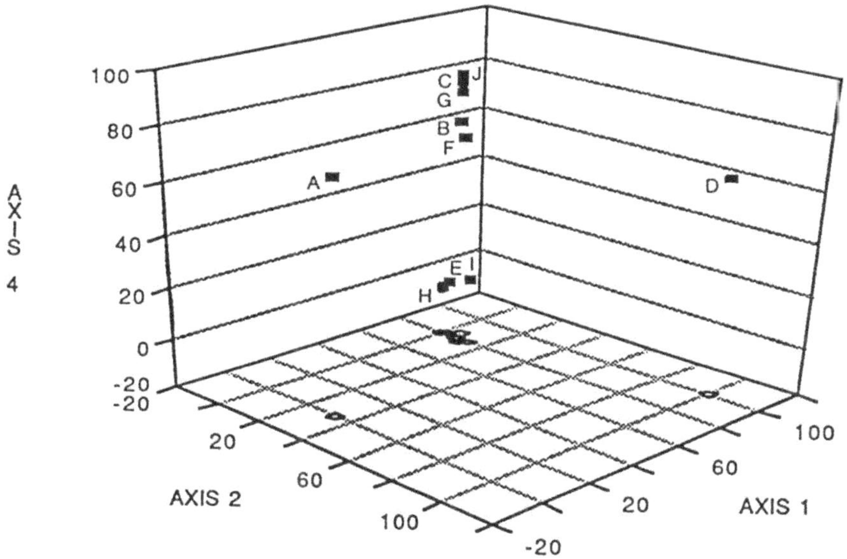

FIGURE 2. Principal components analysis of plots by species. Squares in the 1-2 axes plane are the geometrical projections of the plots in the ordination space. Axis 1 is correlated with disturbance, axis 2 with temperature, and axis 4 with density of *S. invicta.*

important in preventing the establishment of *S. invicta* because of their predacious habits.

Species assemblage four has a low species richness and a high density of *S. invicta* (> 200 foragers per day per trap). In this assemblage no species of the subgenus *Diplorhoptrum* were collected. Our observations do not refute Tschinkel's statements, but also do not constitute significant, quantitative support.

The PCA by species (Fig. 2) produced four aggregations, as did the previous analysis. Axis 1 accounts for most variation (Table 2) among plots in species composition and was correlated to disturbance of the habitat (Kendall's $t = 0.77$; $P < 0.01$). Plot A corresponds to species assemblage one (Fig. 1), and is isolated from the other plots along this axis. Variation along axis 1 is mostly due to the large species turnover (Table 1) between assemblage one and the other assemblages. Axis 2 is correlated to temperature despite measures taken to reduce this effect. The variation in axis 4 was highly correlated to densities of *S. invicta* ($r = 0.63$; $P < 0.01$). Among the significant axes, this axis accounts for the least variation (Table 2) because six of the ten plots did not contain *S. invicta*. Axis 3, even though it accounts for more variation than axis 4, is probably an undetected effect (i.e., herbicide or insecticide applications adjacent to the sampling plots) or a random effect.

A fifth aggregation of ants, plots K and L, was found in the study area.

TABLE 2. Principal components analysis scores of plots.

Plot	Axis 1	2	3	4
A	0.0	32.75	83.55	70.60
B	79.75	7.49	95.12	68.10
C	79.54	8.25	89.46	83.40
D	100.0	100.0	87.80	64.65
E	81.26	2.01	100.0	2.73
F	83.98	5.56	0.0	60.77
G	79.73	8.14	9.29	79.69
H	79.84	0.0	9.49	0.0
I	79.72	12.97	8.49	6.94
J	77.51	10.26	89.42	86.99
R^2	0.8849	0.9326	0.9791	0.9921
%	0.8849	0.0477	0.0465	0.0130

This aggregation consisted of *S. invicta* densities up to 1,000 ants per day per trap. Only one other species was found within this habitat: *M. minimum*. Plots K and L were located within a recreational area and the grounds mowed several times each month. In addition, park trash provided the fire ants with an excellent food resource.

Of the eight species associated with *S. invicta* (Table 3), two had significantly positive correlations with densities of *S. invicta* , and two were negative. The densities of *M. minimum* and *Conomyrma insana* (Buckley) were each positively correlated to *S. invicta* densities. These results suggest a steady restructuring of the community as observed by Porter et al. (1988), even several years after invasion by *S. invicta.* If the trend observed in importance values continues in the fire ant-infested plots, assemblage three might sometime become similar to assemblage four. Whether the observed structure of assemblage four is the most stable, or whether this assemblage is moving towards an even more stable assemblage like the recreation area infestation, is a question that should be addressed in future studies.

In summary, within the range of *S. invicta*, the diversity of the ant assemblages was adversely affected by increasing densities of this species. In addition, ants of the granivorous guild, such as *Pheidole tepicana* Pergande and *Pheidole crassicornis tetra* Wheeler, are being replaced faster than other ants in assemblages containing *S. invicta.* The granivorous ant, *Pogonomyrmex barbatus* (Smith) is not affected as negatively as ants of the genus *Pheidole.* Because no significant correlation was found between *P. barbatus* and *S. invicta*, this suggests that the two species can coexist. Ants of the genus *Pheidole* exhibit more specialized food habits than most of the other species occurring with *S. invicta.* Competition for seeds could be a reason for replacement of these seed-harvesting ants because workers of *S. invicta* are similar in size to those of *Pheidole*, but much smaller than *Pogonomyrmex.*

TABLE 3. Correlation of species importance values to densities of *S. invicta.*

Species	Coefficient of Correlation (r)
Conomyrma insana	+ 0.588**
Monomorium minimum	+ 0.489*
Forelius pruinosus	+ 0.424
Paratrechina terricola / vividula a	+ 0.079
Pogonomyrmex barbatus	- 0.111
Pheidole sp. 2	- 0.159
Pheidole	- 0.345*
Pheidole crassicornis tetra	- 0.584**

* P< 0.05; ** P< 0.01; [a] Unable to differentiate between species.

ACKNOWLEDGMENTS

We thank George C. and Jeanette N. Wheeler for identification of *Leptothorax terrigena* and James C. Trager for identification of *Solenopsis (Diplorhoptrum)* sp. 2. We thank Kathy Phillips and the graduate students of the Department of Entomology, Texas Tech University, for their assistance with the field work. We also thank Leland Chandler, Michael Willig, and James C. Cokendolpher for reviewing our manuscript. Finally, we thank the staff of the Texas Tech University Center at Junction for making those facilities available to us. This study constitutes Contribution no. T-4-249, College of Agricultural Sciences, Texas Tech University, Lubbock, Texas.

REFERENCES

DIS, J. 1979. Problems of interpreting arthropod sampling with pitfall traps. Zool. Anz. 202: 177-184.

BARONI URBANI, C. and P.B. KANNOWSKI. 1974. Patterns in the red imported fire ant settlement of a Louisiana pasture: Some demographic parameters, interspecific competition and food sharing. Environ. Entomol. 3: 755-760.

BHAKTAR, A., W.H. WHITCOMB, W.F. BUREN, P. CALLAHAN and T. CARLYSLE. 1972. Confrontation behavior between *Lasius neoniger* (Hymenoptera: Formicidae) and the imported fire ant. Environ. Entomol. 1: 274-279.

BUREN, W.F. 1972. Revisionary studies on the taxonomy of the imported fire ants. J. Georgia Entomol. Soc. 7: 1-26.

BUREN, W.F., G.E. ALLEN, W.H. WHITCOMB, F.E. LENNARTZ and R.N. WILLIAMS. 1974. Zoogeography of the imported fire ants. J. New York Entomol. Soc. 82: 113-124.

DIGBY, P.G.N. and R.A. KEMPTON. 1987. Multivariate analysis of ecologcal comunities. Chapman and Hall, London.

GLANCEY, B.M., D.P. WOJCIK, C.H. CRAIG and J.A. MITCHELL. 1976. Ants of Mobile county, AL, as monitored by bait transects. J. Georgia Entomol. Soc. 11: 191-197.

GREENSLADE, P.J.M. 1973. Sampling ants with pitfall traps: digging-in effects. Insect Soc. 20: 343-353.

JANSEN, M.J.W. and J.A.J. METZ. 1979. How many victims will a pitfall make? Acta Biotheor. 28: 98-122.

JONES, S.R. 1985. Foraging and recruitment abilities of *Solenopsis invicta* Buren, compared with other ant species indigenous to Texas. unpubl. M.S. thesis, Texas Tech University.

KIDD, K.A., C.S. APPERSON and L.A. NELSON. 1985. Recruitment of the red imported fire ant, *Solenopsis invicta* Buren, to soybean oil baits. Florida Entomol. 68: 253-261.

LOFGREN, C.S., W.A. BANKS and B.M. GLANCEY. 1975. Biology and control of imported fire ants. Ann. Rev. Entomol. 20: 1-30.

PHILLIPS, S.A., S.R. JONES and D.M. CLABORN. 1986. Temporal foraging patterns of *Solenopsis invicta* and native ants of central Texas. Chapter 10 in C.S. Lofgren and R.K. Vander Meer [eds.], Fire ants and leaf cutting ants: Biology and management. Westview Press, Boulder, CO.

PIELOU, E.C. 1984. The interpretation of ecological data, a primer on classification and ordination. John Wiley and Sons, New York.

PORTER, S.D., B. VAN EIMEREN, and L.E. GILBERT. 1988. Invasion of red imported fire ants (Hymenoptera: Formicidae): microgeography of competitive replacement. Ann. Entomol. Soc. Amer. 81: 913-918.

SAMWAYS, M.J. 1983. Community structure of ants (Hymenoptera: Formicidae) in a series of habitats associated with citrus. J. Appl. Ecol. 20: 833-847.

SAS INSTITUTE, INC. 1985. SAS user's guide: Statistics, Version 5 Edition. Cary, N.C.

SOKAL, R.R. and J.F. ROHLF. 1981. Biometry. W.H. Freeman, San Francisco.

TSCHINKEL, W.R. 1986. The ecological nature of the fire ant: some aspects of colony function and some unanswered questions. Chapter 7 in C.S. Lofgren and R.K. Vander Meer [eds.], Fire ants and leaf cutting ants: Biology and management. Westview Press, Boulder, CO.

TSCHINKEL, W.R. 1988. Distribution of the fire ant *Solenopsis invicta* and *S. geminata* (Hymenoptera: Formicidae) in northern Florida in relation to habitat disturbance. Ann. Entomol. Soc. Amer. 81: 76-81.

WHITCOMB, W.H., H.A. DENMARK, A.P. BHATKAR and G.L. GREEN. 1972. Preliminary studies on the ants of Florida soybean fields. Florida Entomol. 55: 129-142.

WILSON, E.O. and W.L. BROWN, JR. 1958. Recent changes in the introduced population of the fire ant Solenopsis saevissima (Fr. Smith). Evolution 12: 211-218.

WILSON, N.L., J.H. DILLIER and G.P. MARKIN. 1971. Foraging territories of imported fire ants. Ann. Entomol. Soc. Amer. 64: 660-665.

18

Community Structure and *Solenopsis invicta* in Sao Paulo

H.G. Fowler, J.V.E. Bernardi and L.F.T. di Romagnano

INTRODUCTION

Competition has been considered an important organizing force in structuring certain communities, and particularly, ant communities. However, it is often difficult to determine the intensity and pathways of competition in the field, as community-wide manipulation experiments are difficult to design and interpret. Few test procedures in the literature examine the roles of behavioral processes in community processes (Fowler 1988a; Phillips et al. 1986; Jones and Phillips 1987). For ants, the role of nutritional ecology coincides with behavior, and especially recruitment processes, to further define community structure (Fowler et al. 1988).

Recently, *Solenopsis invicta*'s range has been found to be more widespread in its native South America than originally reported (Buren, et al. 1974). *S. invicta* occurs outside of the Pantanal region of Brazil, in the states of Goias, Minas Gerais, and São Paulo, particularly in "cerrado" vegetation (Wojcik 1983, 1986). A previous study in the state of Mato Grosso (Fowler 1988a) indicated that behavior was important in understanding ant community structure, especially that of *S. invicta*.

The present report describes studies conducted in a derived cerrado vegetation, a "campo sujo," or an old-field pasture in the State of São Paulo, Brazil. *S. invicta* is a member of this community, although not a dominant member. We compare pitfall trapping with bait captures, using the procedure established by Fowler (1988b) to examine the importance of recruitment in community structure of ants, and we discuss the problems of comparing various indices of community structure, all of which give varying results. Based upon our bait studies, we compare the performance of *S. invicta* to other members of the community and discuss its low level of activity and density in its homeland. We also discuss the role of leaf-cutting ants (*Atta* spp. and *Acromyrmex* spp.) in Neotropical ant communities, as well as problems of analysis and interpretation of data due to their inclusion in the ant community.

MATERIALS AND METHODS

Studies were conducted in the municipality of Rio Claro, São Paulo, Brazil, in campo sujo vegetation belonging to the city. It was impossible to establish a grid of pitfall traps because indigents robbed the initial placement. Therefore, we randomly placed 45 pitfall traps at one end of the field, and another series of 49 pitfall traps at the opposite end to minimize losses to indigents. Pitfall traps consisted of plastic drinking cups, 6 cm in diameter, which were partially filled with a water and detergent solution when placed in the field. Bait stations were placed at the same locations when the pitfall traps were not in operation. Bait stations consisted of tuna fish placed into drinking straws cut into 1 cm sections. Both pitfall trapping and bait stations (N= 94 for both) were replicated during each of six monthly intervals (588 collections for both types of collections). In addition, 160 *Drosophila* baits were placed in the field at irregular intervals, at the same sampling points, to separate the role of individual food collection from recruitment to larger baits (Fowler 1988a).

We analyzed various indices derived from our collections. First, only the number of baits exploited by each species was recorded - our collection technique did not permit an analysis of the number of ants collected at these baits. A similar analyses of the pitfall traps (number of occurrences per species) was performed, as well as the number of ants collected in these traps. Using the total number of pitfall traps, and dividing this into the number of ants captured per species, an estimate of the general species-specific activity patterns for the whole habitat was obtained. An estimate of the site specific activity patterns was arrived at by dividing the number of captured ants by the number of pitfall traps in which the species occurred. This also allowed an estimation of the community's mosaic patterning.

For all species, we calculated the binomial probability of capturing ants at either large baits or in pitfall traps as a check of the methods. Non-parametric Spearman's correlations were performed between indices to examine the interrelations among the estimated indices. Finally, the results of our study were compared to a previous study - examining *S. invicta*'s role in relation to dominant ant species. We specifically looked at individually and group retrievable baits for four different Brazilian habitat types.

RESULTS

The results of our ant community survey are presented in Tables 1-3. A total of 48 species were discovered using all methods (Table 1), but this number varied depending on the collection technique (Tables 1 and 3). In five of six community indices, *Pheidole oxyops* was the dominant species. Other species of *Pheidole* (7 in all) were among the most common of the ants collected. *Paratrechina fulva* and *Iridomyrmex humilis* were also highly ranked in most of the community indices (Table 1), with species of *Conomyrma* also important by most community indices.

Our results suggest that approximately half (21/48) of the species detected were not equally registered in pitfall traps and at large-bait stations, especially species of leaf-cutters which do not frequent bait stations (Table 1), as well as the army ant, *Labidus praedator*. *S. invicta*, among others, appeared more frequently at bait stations than pitfall traps (Table 1).

TABLE 1. The ant fauna of the campo sujo vegetation studied, as well as several indices of community organization. Indices are followed by their respective species ranking in parenthesis.

Species	Frequency of bait occupation			Frequency of Individuals Collected[£]	Density of Activity[§]	
	Pitfalls	Large Baits	"Drosophila"		Habitat-wide	Local
Acromyrmex balzani	0.026 (12)	0 (48)*	0 (28)	0.031 (7)	0.303 (7)	5.45 (12)
Acromyrmex rugosus	0.008 (33)	0 (47)*	0 (29)	0.031 (12)	0.194 (12)	10.8 (3)
Acromyrmex niger	0.015 (17)	0 (46)*	0 (30)	0.055 (6)	0.541 (6)	16.78 (2)
Atta capiguara	0.031 (9)	0 (45)*	0 (31)	0.058 (5)	0.566 (5)	8.31 (5)
Atta sexdens	0.012 (24)	0 (44)*	0 (32)	0.024 (11)	0.237 (11)	8.8 (4)
Atta laevigata	0.010 (30)	0 (43)*	0 (33)	0.018 (14)	0.178 (14)	8.08 (6)
Cyphomyrmex	0.002 (44)	0 (42)	0 (34)	0.001 (44)	0.009 (44)	1.66 (35)
Trachymyrmex	0.014 (19)	0 (41)	0 (35)	0.010 (21)	0.102 (21)	3.35 (15)
Mycocepurus goeldi	0.003 (42)	0 (40)*	0 (36)	0.002 (41)	0.160 (41)	2.25 (27)
Pheidole oxyops	0.206 (1)	0.249 (1)	0.135 (1)	0.271 (1)	2.663 (1)	6.02 (10)
Pheidole radoszkowsii	0.051 (3)	0.050 (4)*	0.084 (4)	0.089 (2)	0.884 (2)	8.08 (7)
Pheidole sp. 1	0.019 (13)	0.009 (28)	0.090 (3)	0.025 (10)	0.247 (10)	6.0 (11)
Pheidole sp. 2	0.009 (31)	0.013 (23)	0.039 (9)	0.008 (24)	0.084 (24)	4.27 (14)
Pheidole sp. 3	0.016 (16)	0.014 (22)	0.026 (14)	0.009 (23)	0.093 (23)	2.6 (20)
Pheidole sp. 4	0.039 (6)	0.030 (9)	0.019 (15)	0.016 (17)	0.154 (17)	1.83 (32)
Pheidole sp. 5	0.014 (20)	0.018 (19)	0.032 (11)	0.015 (18)	0.151 (18)	4.94 (13)
Monomorium floricola	0.014 (21)	0.020 (16)	0.013 (20)	0.005 (29)	0.052 (29)	1.71 (13)
Crematogaster sp. 1	0.001 (46)	0.007 (31)*	0.019 (16)	0.003 (46)	0.004 (46)	2.0 (30)
Crematogaster sp. 2	0.008 (34)	0.009 (29)*	0.006 (22)	0.002 (36)	0.029 (36)	1.6 (38)
Crematogaster sp. 3	0.008 (35)	0.018 (18)*	0.006 (23)	0.011 (20)	0.109 (20)	6.1 (39)
Pogonomym. naegelii	0.012 (23)	0.016 (21)*	0.019 (17)	0.004 (33)	0.043 (33)	1.6 (39)
Diplorhoptrum sp. 1	0 (47)	0.002 (37)	0.006 (24)	0 (478	0 (47)	0 (47)
" " sp. 2	0 (48)	0.001 (38)	0 (37)	0 (48)	0 (48)	0 (48)

Continued.

Solenopsis invicta	0.008 (36)	0.037 (6)*	0 (38)	0.002 (40)	0.023 (40)	1.3 (43)
Solenopsis sp. 1	0.003 (43)	0.037 (7)*	0 (39)	0.001 (43)	0.011 (43)	1.5 (40)
Solenopsis sp. 2	0.006 (40)	0.013 (24)	0.032 (12)	0.002 (39)	0.027 (39)	2.1 (28)
Labidus praedator	0.007 (39)	0 (39)	0 (40)	0.063 (3)	0.625 (3)	43.6 (1)
Pseudomyrmex sp. 1	0.007 (38)	0.002 (36)	0.019 (18)	0.001 (42)	0.016 (42)	1.0 (46)
Pseudomyrmex sp. 2	0.002 (45)	0.003 (35)*	0.006 (25)	0.001 (45)	0.009 (45)	2.5 (24)
Pseudomyrmex sp. 3	0.001 (28)	0.007 (32)	0 (41)	0.002 (37)	0.029 (37)	1.23 (45)
Conomyrma brunnea	0.039 (7)	0.037 (8)	0.006 (25)	0.059 (4)	0.059 (4)	7.0 (8)
Conomyrma sp. 1	0.068 (2)	0.050 (5)	0.058 (6)	0.030 (9)	0.299 (9)	2.06 (29)
Conomyrma sp. 2	0.030 (10)	0.027 (12)	0.052 (7)	0.019 (13)	0.190 (13)	2.94 (17)
Conomyrma sp. 3	0.030 (11)	0.025 (13)	0 (42)	0.016 (15)	0.163 (15)	2.53 (22)
Iridomyrmex humilis	0.048 (4)	0.066 (2)*	0.097 (2)	0.030 (8)	0.303 (8)	2.914 (18)
Ectatomma planidens	0.009 (32)	0.006 (33)	0 (43)	0.003 (34)	0.032 (34)	1.63 (36)
Ectatomma quadridens	0.038 (8)	0.029 (10)*	0 (44)	0.014 (19)	0.138 (19)	1.67 (34)
E. tuberculatum	0.012 (25)	0.010 (26)	0 (45)	0.007 (25)	0.077 (25)	3.07 (16)
Odontomachus bauri	0.006 (41)	0.005 (34)	0 (46)	0.002 (38)	0.029 (38)	2.286 (26)
Pachycondyla	0.012 (26)	0.009 (30)*	0 (47)	0.004 (31)	0.047 (31)	1.857 (31)
Paratrechina fulva	0.045 (5)	0.058 (3)	0.071 (5)	0.016 (16)	0.158 (16)	1.63 (37)
Brachymyrmex sp. 1	0.014 (22)	0.017 (20)*	0.006 (27)	0.007 (26)	0.077 (26)	2.53 (21)
Brachymyrmex sp. 2	0.015 (18)	0.021 (15)	0.019 (19)	0.004 (30)	0.048 (30)	1.50 (41)
Camponotus blandes	0.017 (14)	0.028 (11)*	0.052 (8)	0.009 (22)	0.095 (22)	2.52 (23)
C. trapezoides	0.008 (37)	0.022 (14)*	0.039 (10)	0.004 (2)	0.045 (32)	2.50 (25)
Camponotus sp. 1	0.012 (27)	0.019 (17)	0.032 (13)	0.003 (35)	0.032 (35)	1.286 (44)
Camponotus sp. 2	0.017 (15)	0.011 (25)	0 (48)	0.005 (28)	0.054 (28)	1.429 (42)
Camponotus sp. 3	0.011 (29)	0.009 (27)	0.013 (21)	0.006 (27)	0.063 (27)	2.69 (19)

[£]Proportional representation with respect to the total number of ants collected during this study. [$]Habitat-wide activity density is the total number of ants per species divided by the total number of pitfall traps (N = 588); Local activity density is the number of ants collected per species divided by the number of pitfall traps in which that species was collected. *Significant differences between pitfall occurrences and large-bait occurrences based upon the binomial probability test.

TABLE 2. Spearman's rank correlation coefficients (top) and corresponding t-values (bottom) of community indices listed in Table 1.

	Frequency of occupation at			Frequency of	Activity density	
	pitfalls	large baits	small baits	individuals	habitat	local
	A	B	C	D	E	F
A	-------	0.477	0.136	0.790	0.790	0.331
B	3.68*	------	0.490	0.178	0.178	0.133
C	0.93	3.81*	------	0.279	0.279	0.148
D	8.74*	1.22	1.97*	------	0.780	0.730
E	8.74*	1.22	1.97*	8.47*	------	0.730
F	2.38*	1.26	1.01	8.47*	8.47*	-----

*t-values significant at the 95% probability level.

Community indices applied in this study were not concordant with respect to species ranking (Table 1), but most indices correlated significantly (Table 2), although correlation coefficients were not generally large. The number of pitfall traps occupied by a species indicated how many large baits were occupied by this species, but not the number of small baits (Table 2). This was highly correlated with how many individuals of a species were collected, as well as general activity in the habitat (occurrence)(Table 2). Surprisingly, local activity density was not as highly correlated with the number of pitfall traps occupied as it was with habitat-wide density (Table 2).

Comparing the community structure and relative performance of *S. invicta* in this study with other areas of Brazil (Table 3), *S. invicta* was never the dominant species in communities varying from 15 to 56 species. In all studies, almost every large and small bait was discovered, but *S. invicta* occupancy was always less than that of another ant species, despite variation of the dominant species (Table 3). In three of four cases, *S. invicta* fared better at exploiting larger food items than individually retrievable food items (Table 3), although the dominant species for these food items were not always the same in each of the study areas.

DISCUSSION

Our study focused on methods of evaluating community structure, as well as the role of *S. invicta* in Brazilian ant communities. Given the lack of computing facilities at our institution, a more rigorous analysis of the data, such as that employed by Phillips, et al. (1987), could not be performed. However, our results strongly suggest that simple indices may be informative in ascertaining the relative roles of ant species in communities. Our data on frequency of species occurrence at bait stations and the relative frequency of the number of individuals collected per species at pitfall traps has been reproduced many times before in ant community studies. However, our analysis of the number of pitfall traps capturing a certain species has not, to our knowledge, been used before, and, as with data from bait stations, can help assess heterogeneity in species distributions. Our activity density indices

TABLE 3. Comparative data on community structure and the relative importance of Solenopsis invicta in ant communities of the states of São Paulo (this study) and Mato Grosso (Fowler, 1988a).

Location and type of habitat sampled	Number of species	Bait Frequencies: found	by S. invicta	by dominant sp.	Dominant Species
Small baits (exploited by individual foragers)					
Mato Grosso					
corn-field	31	0.894	0.116	0.270	Conomyrma brunnea
lawn	31	0.926	0.032	0.268	Pheidole radoszkowskii
pasture	15	0.993	0.000	0.508	Pheidole radoszkowskii
São Paulo					
old-field pasture	27	0.932	0.000	0.135	Pheidole oxyops
Large baits (exploited through recruitment)					
Mato Grosso					
corn-field	30	0.986	0.193	0.203	Conomyrma sp.
lawn	38	0.986	0.016	0.170	Pheidole fallax
pasture	56	0.936	0.001	0.298	Pheidole radoszkowskii
São Paulo					
old-field pasture	46	0.982	0.008	0.206	Pheidole oxyops

on both the local and habitat-wide levels show that those species most frequently collected are not necessarily those with the highest local abundance; the dominant ant *Pheidole oxyops* is a good example. *P. oxyops* has an efficient recruitment system and its nests have also been shown to function as miniature pitfall traps (Fowler 1987), a behavior which further enhances its dominance in the community studied.

Species which greatly affect community indices due to their unique behaviors and ecologies must be considered. The army ant, *Labidus praedator*, and the leaf-cutting ants of the genera *Acromyrmex* and *Atta* are cases in point. These ants probably do not compete for the same food resources as the remainder of the ant community, although the competitive subcommunities they form are quite well structured (Fowler 1984). For the purposes of this research, which evaluated recruitment processes in community organization, the lack of activity at tuna-fish baits can be misleading as these ants do not exploit these resources, and are known to have a highly developed recruitment process (Fowler 1984).

Behavioral processes indicate *S. invicta* exploits large food resources more efficiently than would be expected from pitfall trapping in the community studied. Unlike in other situations, *S. invicta* did poorly at locating and exploiting individually retrievable food items, with even the remaining species of the complex *saevissima* group (Wilson 1952) faring much better. Thus, the relatively low index values given *S. invicta* may be due to sources other than competition for food resources, since *S. invicta* does well behaviorally. The strong dominance associated with *P. oxyops* in this study, and the mosaic pattern present in the habitat (based upon pitfall trap occurrences), suggest that they may be limited during colonization, or that large food items are too rare to be of significance. Similar results with other methods have been previously reported (Fowler 1988a). Why *S. invicta* fares so well in North America, even in areas recently invaded (Phillips, et al. 1987), remains unclear based upon its apparent low level of abundance in its homeland (Fowler 1988b).

Although much smaller than those previously studied by Fowler (1988a) or Bensen and Harada (1988), this community was much richer than communities reported for agro-ecosystems in the neighboring state of Minas Gerais (Della Lucia, et al. 1982). Although the communities studied by Benson and Harada (1988) do not contain *S. invicta*, they do contain close relatives (Wilson 1952), and may have been present in the community studied by Della Lucia, et al. (1982).

As pointed out by Wojcik (1983 1986), no apparent differences can be found between the behavior or ecology of *S. invicta* in North and South American populations, although the respective ant faunas are quite distinct. Possibly, other species are superior competitors in South America, but the available evidence does not support this contention (Fowler 1988a, 1988b, this paper). Diffuse competitive interactions may be more pronounced in South America (Fowler 1988a), or *S. invicta* and its relatives may be held in check by natural South American enemies (Wojcik 1986). It has been proposed that perhaps *S. invicta*, although known for its colonizing ability in the United States, is not as good a colonizer as other ant species in South America (Fowler 1988a), and thus enters the community at a disadvantage. In this study, we have examined a relatively undisturbed habitat and have reached conclusions similar to those which have been derived from more disturbed habitats (Fowler 1988a). Interestingly, no polygynous colonies of *S. invicta* or its relatives have been found during our studies, and swarming activity occurs at daily intervals different than that reported for the U.S. (unpublished). Whether these differences are due to the competitive environment remains to be tested.

In conclusion, we can reasonably state that *S. invicta* is a strong competitor for food in São Paulo, as well as in Mato Grosso (Fowler 1988a), doing better than expected in both localities (Fowler 1988b; Wojcik 1983). These studies suggest that *S. invicta* populations are not limited by competitive interactions at food resources, or at least at larger food resources, but fail to identify which factors are responsible for maintaining populations at low levels.

ACKNOWLEDGMENTS

We would like to acknowledge the assistance and encouragement of various colleagues during our studies, especially Marcos Antonio Pesquero and Luiza Carla Duarte. James Traeger identified reference samples of *S. invicta*, as well as many of the reference samples used for species identification during this study.

REFERENCES

BARONI-URBANI, C. and P.B. KANNOWSKI. 1974. Patterns of red imported fire ant settlement of a Louisiana pasture: some demographic parameters, interspecific competition and food sharing. Environ. Entomol., 3: 755-760.

BENSON, W.W. and A.Y. HARADA. 1988. Local diversity of tropical and temperate ant faunas (Hymenoptera: Formicidae). Acta Amazonica (in press).

BUREN, W.F., G.E. ALLEN, W.H. WHITCOMB, F.E. LENNARTZ and R.N. WILLIAMS. 1974. Zoogeography of the imported fire ant. J. N.Y. Entomol. Soc., 82: 113-124.

DELLA LUCIA, T.M.C., M.C. LOUREIRO, L. CHANDLER, J.A. FREIRE, J.D. GALVÃO and B. FERNANDES. 1982. Ordenação de comunidades de Formicidae em quatro agroecossistemas em Viçosa, Minas Gerais. Experientiae, 28: 67-94.

FOWLER, H.G. 1984. A organização das comunidades de formigas cortadeiras (*Atta* e *Acromyrmex*). An. Sem. Reg. Ecol., São Carlos, 4: 156-163.

FOWLER, H.G. 1987. The nest of *Pheidole oxyops* (Hymenoptera: Formicidae) as a pitfall trap. Pedobiologia, 30: 373-376.

FOWLER, H.G. 1988a. A organização das comunidades de formigas no Estado de Mato Grosso, Brasil. Actas X. Congr. Latinoamericano Zool., Valparaiso, Chile, Vol. III (in press).

FOWLER, H.G. 1988b. Hormigas como depredadoras de larvas y pupas de los curculionideos, *Contrachelus myrmicariae* and *Contrachelus psidii*: plagas de guyaba y jabuticaba. Turrialba, (in press).

FOWLER, H.G., L.C. FORTI, C.R.F. BRANDÃO, J.H.C. DELABIE and H.L. VASCONCELOS. 1988. Ecologia nutricional de formigas. *In* Panizza, A. and J.R.P. Parra. (eds.), Ecologia Nutricional de Insetos, Editora Manole, Sao Paulo, (in press).

JONES, S.R. and S.A. PHILLIPS, Jr. 1987. Aggressive and defensive propensities of *Solenopsis invicta* (Hymenoptera: Formicidae) and three indigenous ant species in Texas. Texas J. Sci., 39: 107-115.

PHILLIPS, S.A., S.R. JONES, and D.M. CLABORN. 1986. Temporal foraging patterns of Solenopsis invicta and native ants in central Texas. *In* Lofgren, C.S., and Vander Meer, R.K. (eds.) Fire ants and Leaf Cutting Ants: Biology and Management. Westview Press, Boulder, pp. 114-122.

PHILLIPS, S.A., Jr., W.M. ROGERS, D.B. WEBSTER and L. CHANDLER. 1987. Ordination analysis of ant faunae along the range of expansion front of the red imported fire ant in south-central Texas. Texas J. Agr. Nat. Res., 1: 5-10.

WILSON, E.O. 1952. 0 complexo *Solenopsis saevissima* na América do Sul (Hymenoptera: Formicidae). Mem. Inst. Oswaldo Cruz, 50: 49-68.

WOJCIK, D.P. 1983. Comparison of the ecology of red imported fire ants in North and South America. Fla. Entomol., 66: 101-111.

WOJCIK, D.P. 1986. Observations on the biology and ecology of fire ants in Brazil. *In* Lofgren, C.S. and Vander Meer, R.K. (eds.) Fire Ants and Leaf-Cutting Ants: Biology and Management. Westview Press, Boulder, pp. 88-103.

19

A Decade-long Study of an Attine Ant Colony

B.A. Weiss

INTRODUCTION

Attine ants in their natural habitat often establish the largest fixed colonies of any non-human species. Only the mobile army ants live in larger populations. However, laboratory studies of Attines have not produced colonies of great size. Weber (1976a, 1976b, 1976c) reported ten-year studies of *Sericomyrmex urichi*, *Atta cephalotes* and *Acromyrmex octospinosus* colonies in his laboratory. These colonies attained maximum fungus garden sizes of 2.25 L, 11.60 L, and 6.90 L respectively.

On July 16, 1976 in Panama, I excavated a newly founded colony identified by N.A. Weber (expedition leader) as *Atta cephalotes isthmicola*. The colony was expected to follow a growth course similar to those reported.

MATERIALS, METHODS AND RESULTS

The colony consisted of two garden chambers contained in two 1.25 L polystyrene boxes, each perforated with a hot needle to make several air holes and hand delivered by airplane to my laboratory at Drexel University. There the boxes were connected with 2.54 cm inside diameter acrylic tubing. A fine-screen covered chamber 51.5 cm long by 51.5 cm wide by 20 cm high was connected to the boxes. Substrate of red oak (*Quercus rubra*) leaves supplemented with oatmeal flakes was provided in the screened chamber at a rate that kept ahead of the colony growth.

The colony expanded slowly for the first two years and then grew rapidly, eventually reaching a maximum fungal garden size of 54 L in December, 1980. The colony sustained a garden size near 50 L for two years-from January, 1980 to January, 1982. Figure 1 depicts the growth of the fungus garden and associated refuse for the 114 months the colony lived in the laboratory.

During this period, the colony size reached 144 chambers arranged in seven layers, excluding the large box in which substrate was provided, arranged in seven layers. Figure 2 shows the size and arrangement of the colony.

The colony used substrate in quantities that necessitated freezing enough red oak leaves in the Fall to last until May. Figure 3 presents curves

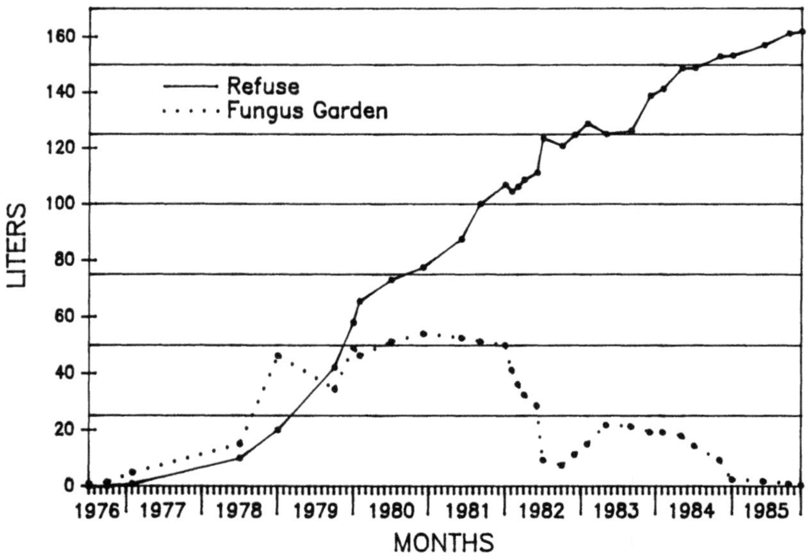

FIGURE 1. Growth of the fungus garden and associated refuse for the 114 months of the colony.

of the log of the monthly amounts and of the cumulative substrate provided during the life of the colony. The maximum monthly quantity of substrate taken was 2660 g in July, 1980. Total substrate for the entire life of the colony was 139,902 g.

The final weight of the colony was 22,825 g and the total volume of the 144 boxes were 180 L. Substrate weight greatly exceeded final colony weight because water in the substrate was transpired by the garden and generally lost through evaporation.

DISCUSSION

Laboratory use of large Attine colonies opens the possibility of studying the complex interconnecting array of garden and refuse chambers and their roles in the activity of the colony. During the life of this colony, workers were marked and traced to specific garden chambers from which they foraged (Weiss, 1983). Garden chambers were also observed in refuse chambers. The ants removed the refuse as the garden expanded. Garden temperature was critical. At 20°C the colony functioned well. If the temperature approached 30°C, the ants moved the brood to more ventilated areas out of the garden. The fungus began desiccating near 30 degrees celsius. A study of Attine biochemistry was also conducted (Ritter, et. al., 1982) and the ants were found to lack cholesterol in their bodies and brains.

In addition to the more natural habitat large colonies provide for research, such colonies could form sizable and dramatic displays for museum and zoological exhibits. Thus, public interest in Attines and the curiosity of future researchers might be stimulated.

FIGURE 2. Photograph showing the size and arrangement of the 144 1.25 L. boxes and substrate chamber in seven layers.

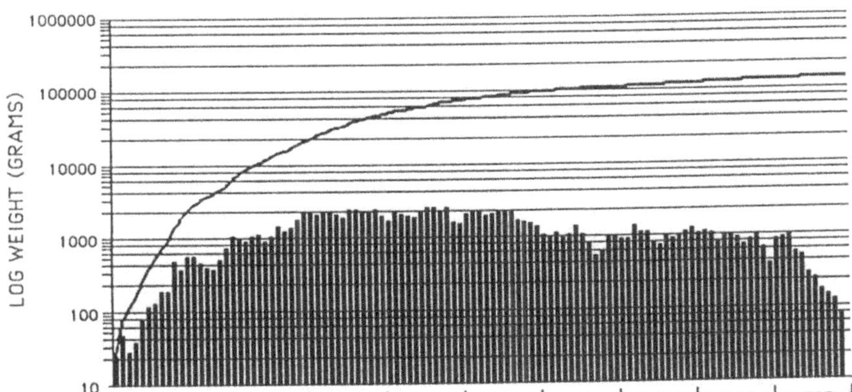

FIGURE 3. Curves of the log of the monthly amounts and of the cumulative substrate for the 114 months duration of the colony.

REFERENCES

RITTER, K.S., B.A WEISS, A.L. NORRBOM, and W.R. NES. 1982. Identification of 5,7-24- Methylene and Methylsterols in the brain and whole body of *Atta cephalotes isthmicola*. Comp. Biochem. Physiol. 71:345-349.

WEBER, N. A. 1976a. A Ten-year colony of *Sericomyrmex urichi* (Hymenoptera: Formicidae). Ann. Entomol. Soc. Amer. 69: 815-819.

WEBER, N. A. 1976b. A Ten-year Laboratory colony of *Atta cephalotes*. Ann, Entomol. Soc. Amer., 69: 825-829.

WEBER, N. A. 1976c. A Ten-year colony of *Acromyrmex octospinosus* (Hymenoptera: Formicidae). Proc. Entomol. Soc. Wash., 73: 294-292.

WEISS, B. A. 1983. Observations of Attines. In Hearing and Other Senses: Presentations in Honor of E. G. Wever, R. R. Fay and G. Gourevitch (Eds.), Amphora Press, Box N, Groton, Conn. 06340, USA, 113-122.

20

Development of the Ant-Fungus Relationship in *Atta laevigata*

A. Cedeno and M.J. Leon

INTRODUCTION

To understand the relationship between leaf-cutting ants and their mutualistic fungus, one must learn about their complicated caste and labor systems, their nest architecture, and the ants' interaction with anti-herbivore plant defenses.

This report traces the ontogeny of young nests of *Atta laevigata*, from colony foundation to an age of twenty-nine months. The study was made at the *Pinus caribaea* plantation of Uverito, Estado Monagas, Venezuela, where this leaf-cutter maintains an enormous population.

METHODS

Nests were marked at colony foundation and periodically selected at random for excavation. Capture of the whole population was assured by introducing a chloroform-impregnated wick of tissue paper into the nest gallery. Fungus gardens, together with their anesthetized ant populations, were dug out and fixed in 70% ethanol. External and internal architectural features were recorded.

In the laboratory, ants were counted and separated as immature (eggs, larvae, pupae) or adult forms. Pupal and adult head width (HW) was measured at its widest point.

RESULTS

An *Atta laevigata* queen initiates a fungus garden as a spongy, white sheet inside the only excavated chamber (13 to 35 cc capacity). She lays eggs on the fungal sheet and newly-hatched juveniles of several sizes live embedded in the fungus.

Pupae appear 28 days after colony foundation, but most of the immature population remained at the larval stage (see Table 1). The nests remained in a claustral phase.

Seventy-five days (2.5 months) after colony foundation, workers established contact with the external environment by opening the nest entrance and beginning to forage. The plant material they brought in was

TABLE 1. Immatures populations of three 28 days old cloistered nests. No adults have yet appeared.

Colony Code	N23	N28	N32
Chamber volume (cc)	12.5	30.6	35.0
N°Eggs	46	62	33
(%)	(36.8)	(31.8)	(24.5)
N° larvae	79	133	100
(%)	(63.2)	(68.2)	(74.1)
N° pupae	0	0	2
(%)			(1.5)
Total immatures	125	195	135

processed and incorporated into the surface of the fungus garden. From that time onward, the garden grew centrifugally, and eventually adopted a globular shape with immatures at the center. Table 2 shows the population's composition and its HW range. Chamber volume increased from 3 to 14 times during the first year. Based on a previous study made in an adult colony (León, 1986; Barral, 1987), HW was divided into three categories, each corresponding to one of three essential tasks: fungus cultivation (< 0.77 mm), excavation (0.77-1.18 mm), and foraging (> 1.18 mm). According to Table 2, most of the first adults belonged to the excavation and foraging categories. In the second batch of adults, represented by the pupae, the foraging component increased.

Table 3 shows the population characteristics of 4- to 12- month-old colonies. Fluctuations in total numbers and ratios of immatures to adults were notable. Chamber volume was variable, but did increase over time with an increase in size. Brood size could not predict colony age. Production of immatures in 11- and 12-month-old nests stopped in the dry season, during hydric stress (see colonies 35, 49, 1, 21, 28 in Table 3). From the fourth month onward, the dominant adult size was that of the excavation category. This coincides with the building of a deeper gallery leading to a future second chamber.

At the beginning of the second year, concurrent with the rainy season, a dramatic change in nest structure generally occurred. In some nests, a second chamber had already been excavated at greater depth and the fungus and immature population moved to the deeper chamber (see colonies 17, Pal, 27, 57 in Table 4). The initial chamber remained empty. In other nests, the single chamber was enlarged. Dimensions and population sizes of both chamber are shown in Table 4.

Once immature forms were placed in the deeper chamber, the initial chamber was enlarged and a new fungus garden cultivated there. This frees immatures from the action of secondary plant compounds, since the deeper fungal garden incorporated no fresh vegetation.

Pinus caribaea produces terpenoids α- and ß-pinene in abundance (Green, et al. 1974). These pinenes have potential antiinsect properties.

TABLE 2. Composition of the populations and size frequency distributions of adults from two 75 days old nests. Percentages in ( ).

PUPAE		ADULTS										
Colony	Chamber	Pop.	No.	No		Head Width (mm)				Head Width (mm)		
Code	Vol.	Size	Eggs	Larvae	No.	<0.77 :	0.77-1.18 :	>1.18	No.	<0.77 :	0.77-1.18:	>1.18
N6	175	255	106	32	27	5	9	13	90	16	37	37
			(41.6)	(12.5)	(10.6)	(18.5)	(33.3)	(48.1)	(35.3)	(17.8)	(41.1)	(41.1)
N10	108	147	38	40	27	7	8	12	42	12	15	15
			(25.9)	(27.2)	(18.4)	(25.9)	(29.6)	(44.5)	(28.6)	(28.6)	(35.7)	(35.7)

TABLE 3. Composition of the populations and size frequency distributions of adults from 12 nests during the first year of life. Percentages in ( ).

Colony	Age (months)	Chamber vol.(cc)	Pop size	N°	Adult Head With (mm)		
					<0.77	0.77-1.18	>1.18
Chip	4.2	360.0	107	83	14	44	25
				(77.6)	(16.9)	(53.0)	(30.1)
N13	5.1	165.0	103	40	6	27	7
				(38.8)	(15.0)	(67.5)	(17.5)
Can	6.7	198.0	59	43	13	12	18
				(72.9)	(30.2)	(27.9)	(41.9)
Mon	6.7	not measured	162	85	26	30	29
				(52.5)	(30.6)	(35.3)	(34.1)
Ptero	8.7	247.5	96	60	18	27	15
				(62.5)	(30.0)	(45.0)	(25.0)
13	11.2	720.0	425	307	90	140	77
				(72.4)	(29.3)	(45.6)	(25.1)
35	11.2	378.0	55	55	23	27	5
				(100.0)	(41.8)	(49.1)	(9.1)
49	11.2	392.0	102	102	25	44	33
				(100.0)	(24.5)	(43.1)	(32.4)
1	11.2	396.0	123	123	42	54	27
				(100.0)	(34.1)	(43.9)	(22.0)
21	12.1	210.0	24	24	15	9	0
				(100.0)	(62.5)	(37.5)	(0)
28	12.1	588.0	239	233	57	92	84
				(97.5)	(24.5)	(39.5)	(36.0)
59	12.1	392.0	392	206	69	93	44
				(52.5)	(33.5)	(45.1)	(21.4)

TABLE 4. Architectural changes and population sizes in 12 nests with only one fungal garden, 13-27 months old. Percentages in ( ); * = not measured; ** = not present yet.

		Chamber Vol. (cc)				Adult Head Width (mm)		
Colony	Age (months)	Initial	Deeper	Pop size	N°	<0.77	0.77-1.18	>1.18
7	12.7	364	*	574	261 (45.5)	88 (33.7)	122 (46.8)	51 (19.5)
17	12.7	270	180	186	90 (48.4)	27 (30.0)	47 (52.2)	16 (17.8)
51	12.7	150	*	244	87 (35.7)	18 (20.7)	39 (44.8)	30 (34.5)
36	14.5	768	**	915	610 (39.7)	242 (39.7)	235 (38.5)	133 (21.8)
53	16.3	810	**	567	351 (61.9)	137 (39.0)	155 (44.2)	59 (16.8)
Car	16.3	1170	**	466	272 (58.4)	65 (23.9)	140 (51.5)	67 (24.6)
60	17.3	200	**	50	20 (40.0)	1 (5.0)	9 (45.0)	10 (50.0)
Pal	17.3	576	48	417	252 (60.4)	60 (23.8)	138 (54.8)	54 (21.4)
55	18.8	*	**	1218	777 (63.8)	172 (22.1)	340 (43.8)	265 (34.1)
27	20.8	264	1300	1150	638 (55.5)	214 (33.5)	288 (45.1)	136 (21.4)
Aper	20.8	180	**	207	66 (31.9)	25 (37.9)	29 (44.0)	12 (18.1)
57	27.0	*	2016	2056	775 (37.7)	194 (25.0)	326 (42.1)	255 (32.9)

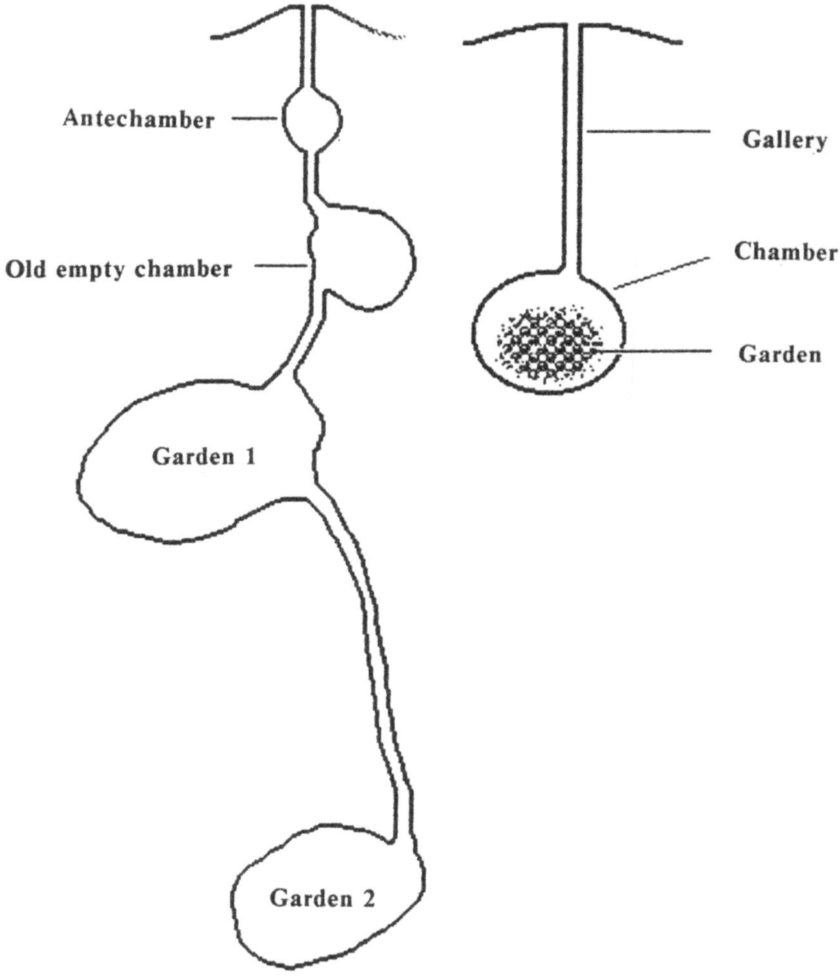

FIGURE 1. Schematic drawing of left - a two chamber nest and right- a one chamber nest.

The terpenoids will be used to explain the nature of the chemical interaction.

When preparing the fungus substrate and during the process of pine needle fragmentation, ant bites exposed the terpenoids present in pine needles. This could mean that terpenoids do not affect adults though immature forms may be more susceptible to their allelochemical influence.

To test this hypothesis, three single fungal garden nests and one two-fungal garden nest were dug out, the fungus separated into processed and

TABLE 5. The position of immature and adult ants in three single garden nests.

	Substrate	No. Immatures	No. Adults
Nest 1	Processed	157	27
	Fresh	7	123
Nest 2	Processed	160	61
	Fresh	0	126
Nest 3	Processed	58	134
	Fresh	38	15

fresh substrate fractions. Ants in each fraction were counted and classified as immature or adult forms. A gas chromatograph was used to analyze chloroform extracts of pine needles and portions on the fungal garden.

Figure 1 shows a schematic view of one- and two-fungal garden nests. Table 5 illustrates the population distribution in three one- fungus garden nests. Immatures tend to concentrate in the processed substrate of the garden, while adults may be anywhere.

Table 6 shows the population distribution in a nest with two-fungus garden, most recent being the initial garden (Garden 1), indicated by the presence of green material. It had only 3 immature forms, compared to 413 adults and the foundress queen. The deeper garden (Garden 2), where fresh substrate was absent, had a total number of 282 immatures and 266 adults.

Chromatographic analysis of terpenoids found no pinenes in the nests with only one-fungus garden. Table 7 reveals that α-pinene was found in pine needles but not in fungus fractions. ß-Pinene was present in needles and in Garden 1 but not in Garden 2.

TABLE 6. The number of immature and adult ants found in a nest with two gardens.

	No. Immatures	No. Adults
Garden 1 (younger)	3	413
Garden 2 (older)	282	266

TABLE 7. Chromatographic analysis of terpenoids in pine needles and in the two gardens of a nest.

	α-pinene	ß-pinene
Pine needles	+	+
Garden 1	-	+
Garden 2	-	-

DISCUSSION

Classical literature on leaf-cutting ants claims that the first emerging workers are the smallest - gardeners, or minima workers (Autori,1941; Weber, 1972). However, this report demonstrates that the first brood of *Atta laevigata* produces several sizes (Table 2), and the majority of first adults belong to the excavation and foraging categories. The size range discovered is the minimum required to perform the most essential tasks in the colony - namely, fungus cultivation, and the harvesting and processing of fresh vegetation. The same pattern has been found by Wilson (1983) for *Atta cephalotes*. From the fourth month onward, the dominant adult size belongs to the excavators. Individuals included in this category probably operate as a reserve force, performing other tasks and compensating for the absence of other size categories (Barral, 1987).

The cloistered queen chamber occupies the first level of nest architecture. Next, a gallery is opened to the outside and the chamber enlarged or retained as an antechamber. Then a deep gallery is built, leading to a future second chamber. Depth should attenuate the negative effects of high temperature and low humidity during the dry season.

While the nest remains in the one-chamber stage, immature forms are kept on the processed plant substrate. No pinenes were found in one-chamber nests. The low rate of green material incorporation may explain the absence of terpenoids in this gardens.

Removal of the fungus and the immature ant population to the deeper chamber liberates immatures from the action of secondary plant compounds, since only few pieces of fresh vegetation are incorporated into that garden. On the other hand, a new fungus garden cultivated in the old enlarged chamber has fresh green material placed within it. This garden contained ß-pinene but no α-pinene, even though pine needles have both. Either α-pinene is rapidly transformed by the fungus or it decays/evaporates quickly. ß-pinene is transformed more slowly, and does not affect adults (including the queen). Since both pinenes could be harmful to immature forms, the latter must be safeguarded until the potential toxic effect dissapears.

Terpenoid-containing plant species will be selected by ants as long as their fungus is capable of degrading the toxic compounds. *Atta cephalotes* discriminates against plant species that contain terpenoids harmful to its fungus (Hubbell, et al., 1983; Howard, 1987).

Substances potentially harmful to insects seem to affect larval growth. Leaf-cutting ants begin to defeat the plant's chemical defense strategy when adult workers process fresh material that contains allelochemicals. In addition, the fungus performs the chemical work of degrading secondary plant compounds. This work supports Cherrett (1980)'s hypothesis of dual (mutualistic) breakdown of plant defenses.

One could expect the exponential growth of the ant population to start from a two-fungus garden stage, provided that immature forms are now protected from allelochemical plant effects.

ACKNOWLEDGMENTS

We are grateful for field assistence given by José Barral and Jorge Zegarra. Jesús Alberto León, J.M. Cherrett and Alex Mintzer critized an early version of the manuscript. Jorge Zegarra revised the final version. CVG kindly furnished food and lodging at the Uverito camp. Funds were provided by CONICIT (S1- 1787) and Consejo de Desarrollo de la Universidad Central de Venezuela (C-03-1486).

REFERENCES CITED

AUTUORI, M. 1941. Contribuçao para o conhecimento da sauva (*Atta* spp.). I. Evoluçao do sauveiro (*Atta sexdens rubropilosa* Forel, 1908).Arquivos do Instituto de Biologia de Sao Paulo 12: 197-228.

BARRAL, J. 1987. Ontogenia, alometría y crecimiento de la colonia en *Atta laevigata*. Trabajo Especial de Grado. Universidad del Zulia. Maracaibo, Venezuela.

CHERRETT, J.M. 1980. Possible reasons for the mutualism between leaf-cutting ants (Hymenoptera, Formicidae) and their fungus. Revue de Biologie et d'Ecologie Mediterranéenne. 8: 113-122.

GREEN, C.L., B. KEEBLE and J. BURLEY. 1974. Gum turpentine analysis of some *Pinus kesiya, P. caribaea* and *P. oocarpa* provenances. Tropical Sciences 16: 195-206.

HOWARD, J.J. 1987. Leaf-cutting ant diet selection: The role of nutrients, water and secondary chemistry. Ecology 68: 503-515.

HUBBELL, S.P., D.F. WIEMER and A. ADEJARE. 1983. An antifungal terpenoid defends a neotropical tree (*Hymenaea*) against attack by fungus-growing ants (*Atta*). Oecologia (Berl) *60*: 321-327.

LEÓN, M.J. 1986. Castas de *Atta laevigata*. Trabajo Especial de Grado. Universidad Central de Venezuela. Caracas, Venezuela.

WEBER, N.A. 1972. Gardening ants, the attines. Memoir N° 92. The American Philosophical Society. Philadelphia.

WILSON, E.O. 1983. Cast and division of labor in leaf-cutter ants (Hymenoptera: Formicidae: *Atta*). IV. Colony ontogeny of *A. cephalotes*. Behav. Ecol. and Sociobiol. 14: 47-54.

21

Methods for Estimating the Population Density of Leaf-cutting Ant Colonies

S. Claver

INTRODUCTION

There are several ways to determine absolute nest density of leaf cutting ants (Table 1). However, no one discusses *why* one method was considered superior to another. Specifically, researchers have not evaluated the cost and precision of each methodology. Without these evaluations, the methods chosen may be inappropriate for the species and environment being studied. However, most researchers have shown a preference for transects in determining leaf-cutting ant densities, probably due to speed and flexibility (Table 1). Thus, an evaluation of variations on the transect method will prove valuable. The transect methods normally used for studying vertebrates may apply to sessile leaf-cutting ant colonies. This paper describes some transect methods and demonstrates their use in the field.

MATERIALS AND METHODS

For the field evaluation of transect methods, I chose to work with the leaf-cutting ant, *Acromyrmex lobicornis* (Emery). Common in Argentina's

TABLE 1. A selected listing of methods used for colony density studies of leaf-cutting ants.

Type	Source
<u>Quadrant</u>	
Small	Cherrett et al. 1974; Fowler 1977; ++
Large (aerial photos)	Bucher and Zuccardi 1967; Jonkman 1979;
Circular plot	Fowler and Haines 1983
<u>Transects</u>	
Fixed width	Rockwood 1973; Fowler 1983; this study
Indefinite width	this study
<u>Plotless</u>	
Nearest neighbor	Fowler 1984
Point quarter	not used

arid and semi-arid areas, this ant cuts broad-leaf vegetation (Kusnezov 1963). My study was staged in the MAB Nacuñan Ecological Reserve (Province of Mendoza) and adjacent areas used for cattle production. This flat area has sandy eolioc soils (Tanquilevich 1971) and vegetation comprised of *Proscopis flexuosa* open forest surrounded by other shrub and herbaceous strata (Roig 1970).

Acromyrmex lobicornis nests are characterized by low mounds (0.3 to 4.0 in diameter; x = 2.50 m.) with many mound openings. Foraging trails from the mound can be thirty meters in length. Spent fungal substrate is deposited on the soil surface at one end of the mound. The resulting dump - large and tabacco-colored - provides a clue to a colony's presence, as they are often hidden by the surrounding light-colored soil. Another identifying feature in some nests is a thatched dome made from plant stems mixed with soil. Nests usually occur beneath shrubs and often have a herbaceous cover.

Ac. lobicornis demonstrates a marked seasonal variation in foraging activity. Nocturnal patterns in summer become diurnal during the winter; intermediate patterns mark fall and spring. These changes are adaptive strategies to the arid environment (Claver unpublished).

Ac. lobicornis may move, but usually not over great distances. Moreover, nests may be inactive for months, without evidence of fresh soil excavations, waste deposition or foraging, and then become active again (Claver unpublished).

SAMPLING

I performed field studies over 400 ha. of land. Eighteen line transects, 250 m. long, were randomly run from the area's border towards the center. These transects were then reused for each of four sampling intervals corresponding to each season of the year. Every nest along each 50 m. sub-transect was recorded and mapped. Since nests were not always visible, I used additional evidences (Southwood 1978), in order of importance, (1) foraging trail activity, (2) fresh, spent-fungal substrate dumps, always associated with an active nest, (3) physical foraging trails, with or without freshly-cut vegetation fragments indicative of a nearby colony. The discovery of new nests was common due to previously discussed nest movement, nests becoming active, as well as the foundation of new nests.

I also recorded *all* nests encountered, the perpendicular distance from transect line to nest.

TRANSECT METHODS

Using the line transects as a base, I evaluated the following belt transects for efficiency: (1) fixed widths of 3, 5 and 10 m., and (2) indefinite widths using Hayne's (1949) and Eberhardt's (1968) methods. For each of the fixed width belt transects (1) I estimated colony densities for each census interval using the formula:

D = number of colonies recorded/(transect width) (transect length)

For the indefinite width belt transects (2), both methods were used for density estimates, as will be described separately. In Hayne's (1949) method, density estimates were performed using the formula:

$$D = ((A)(Z))/((Z)(Y)(x))$$

in which A = area; x = mean distance of all colonies to transect center line; Y = transect length; and Z = number of colonies recorded. This method is identical to Hoglund et al.'s (1967) - an indefinite width belt transect, because the mean distance of all colonies to the center transect line is determined *a posteriori*, when the transect width is fixed.

Eberhardt's (1968) indefinite width transect, like Hayne's (1949) method, depends upon the premise that the probability of detecting a nest from the center transect line falls as the distance between the two increases. This relation may be expressed as:

$$Px = 1 - (X/W)^k$$

in which X = colony distance from the center transect line; W = maximum distance from the center transect line at which a colony can be seen; and k = the slope of the curve.

The parameters k and W are estimated by linear regression of log(1-Px) on k(logW). To define the probability function, Px, distance classes from the center transect line are established, and using these distance classes, the colony distribution at each distance class can be determined. Regressions were evaluated by the F statistic, while goodness of fit of distributions were tested by the X^2 statistic. Thus, colony densities were estimated using the formula:

$$D = (Z(k+1)^2)/((4)(Y)(x)(k)(k+2)^2)$$

in which the parameters are the same as previously defined. D must be doubled because Eberhardt (1968) considered only one side of the transect (Seber 1973).

Absolute density estimates using the five types of transects, fixed belts of 3, 5 and 10 m. width, and Hayne's (1949) and Eberhardt (1968) indefinite width, were compared by one-way analysis of variance.

ANALYSIS OF EBERHARDT'S INDEFINITE WIDTH TRANSECT METHOD

As the precision of Eberhardt's (1968) method is proportional to the square root of the number of ant colonies observed during sampling, Eberhardt (1968) argued that all individuals observed should be recorded. Thus, the experimental design of indefinite width belt transects differs from that of fixed width belt transects. Fixed width belt transects require researchers to determine the sample area before sampling and to record all individuals in the area. Indefinite width transects determine this area after sampling.

To demonstrate how this method works, consider the basic calculations used to obtain parameter estimates for density estimations for the four census periods. For all census periods, the frequency distributions of observed

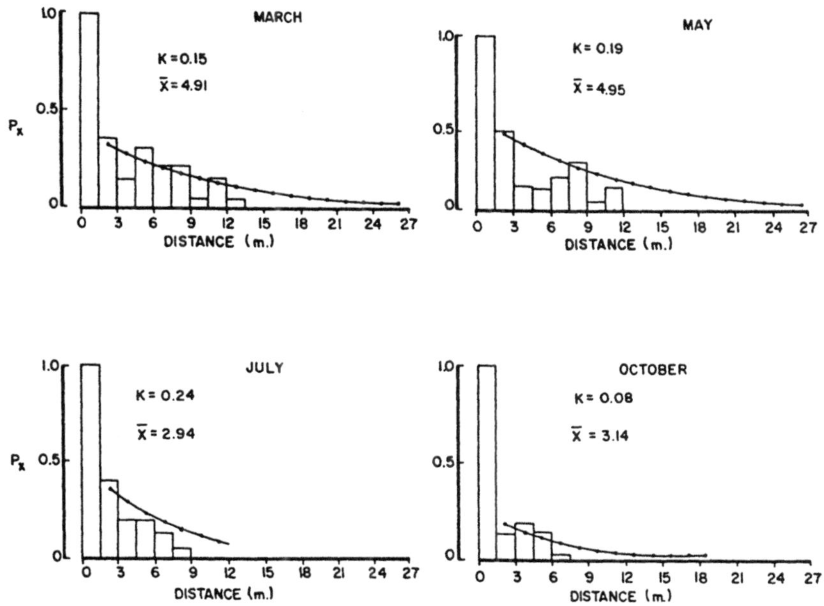

FIGURE 1. Eberhardt's (1968) indefinite width transect method applied to *Acromymex lobicornis* colonies in four sampling periods. The parameters k and x~ are also given. The theoretical distribution and fitted equation are shown. All sampling periods demonstrated a good fit of observed to expected frequencies, as well as good regression fits (Table 2).

distances from the center transect line varied among themselves (Fig. 1), which implies that mean distances (X) from the center transect line, and the slope of this relation, k, also varied. The frequency variations occurred because of the behavior and ecology of *Ac. lobicornis*, in addition to environmental changes (such as vegetative growth), which resulted in varying degrees of sampling error, as well as legitimate changes in nest densities. Nevertheless, Eberhardt's (1968) method fit all census data (Table 2), with k values homogeneous except for October, which varied significantly from other months (Table 2).

COMPARISON OF TRANSECT ESTIMATES

To compare transect methods, a researcher must have a knowledge of estimate precision and consistency among and between methods. Thus, highly variable methods could be discarded and other criteria used for final method selection. This analysis was performed with the transect estimates obtained during all sampling periods for all sampling methods (Table 3). The results can be used for method evaluation. Colony density estimates were not

TABLE 2. Parameter estimates and goodness of fit associated with fitting Eberhardt's (1968) indefinite width belt transect methods.

Month of Census	k ± S.E.	R^2	Regression F	P
March	0.15 ± 0.192	0.75	63.00	< 0.005
May	0.19 ± 0.008	0.21	5.13	< 0.005
July	0.24 ± 0.043	0.84	31.57	< 0.005
October	0.08 ± 0.020	0.59	17.50	< 0.005

TABLE 3. *Acromyrmex lobicornis* colony density estimates during four census by various fixed or indefinite transect methods in the region of the Ñacuñan MAB Ecological Reserve.

Transect type	Estimated colony density				
	March	May	July	October	X ± S.E.*
Fixed width					
3 m wide	14.8	18.5	14.1	22.2	17.41 ± 3.75a
5 m wide	10.7	12.9	9.7	14.7	12.00 ± 2.21 ab
10 m wide	7.1	8.0	6.0	9.3	7.61 ± 1.41 b
Indefinite width					
Mean distance	11.3	6.0	14.0	19.8	14.60 ± 3.66 ab
Eberhardt	11.5	9.3	10.6	33.7	16.62 ± 11.40 ab

* One way ANOVA: F =3.90, P < 0.05. Means followed by the same letter do not differ significantly (P < 0.05) using Tuckey's multiple range test.

significantly different, except for fixed belt transects with 3 versus 10 m widths (Table 3).

DISCUSSION

The information necessary for field surveys depends on the type of problem (Table 1). For some problems, presence-absence data is sufficient (Fowler and Claver 1989), while other problems may be resolved by relative, or scaled, abundance estimates. However, to answer precise questions on population management or related population dynamics, researchers must estimate density. The precision of absolute population estimates - which can be crudely gauged by examining the mean and variance of estimates produced by various methods - must be considered.

Researchers have relied heavily upon transect methods (Table 1) to obtain leaf-cutting ant colony estimates, while other methods have largely been ignored; the reasons for this were stated earlier. However, I could find no indication that researchers ever used quantitative criteria to establish transect width.

For fixed width belt transects, increasing the transect width results in a squared increase of the sampled area. There are fewer advantages to the transect method over the quadrant method as the transect width increases, since as the width increases, quadrants form and the quadrant method is preferred. One benefit of the transect method is sampling speed - a benefit which declines as transect width increases, along with search time.

For fixed width belt transects and quadrants, a principal operational problem concerns colonies that lie partially within the pre-defined areas. This operational problem never occurs in indefinite width transect methods, because researchers determine the dimensions of the sampled area after the sampling event and only the perpendicular distances from the center line transect to the colony are necessary. Because leaf-cutting ant colonies are sessile, researchers have no problem calculating flushing distances and angles from the center line transect, as occurs with work on mobile vertebrates. However, indefinite width transects are not free of sampling error. The principal source of error for the mean distance method is in the estimation of mean distance, which is a function of sample size (number of colonies recorded) and their spatial distribution patterns. Eberhardt's (1968) method has potential sources of error not found in other methods: 1) distance measures from the sighted colony to the center transect line, which should be minimal for studies of spatially fixed organisms; 2) defining the form of the function relating probability of sighting a colony to distance from the center transect line, based upon observed distance classes distribution; (3) estimating the slope of this function, k.

Thus, indefinite width belt transect methods convey additional precision, but with far greater possibility of estimation error. They are more appropriate at lower densities. This must be counter balanced with the simplicity and reduced assumptions of fixed width belt transects (Caughley 1977).

A comparison of *Ac. lobicornis* absolute density estimates obtained by each method indicates that there was no statistically significant difference (Table 3). Ideally, more than one method would be used to estimate absolute densities (Southwood 1978), but, in practice, a single method must be chosen. For *Ac. lobicornis* in Nancunan, the fixed width belt transects of 3 and 10 m widths could be discarded because they differed significantly from each other, but not from the other estimates. The 3 and 10 m wide belt transects produced consistently higher and lower absolute nest density estimates, respectively, than other methods. If one selects a method based upon coefficients of variation, the 5 m fixed width belt transect would be chosen.

Likewise, indefinite width belt transect methods, based on mean distance (Fig. 1; Table 3), produce absolute density estimates in accord with other methods, and agree well with the 5 m wide belt transect. Indefinite width transect methods would therefore be used in a first survey to define fixed width transect widths for posterior and, presumably, quick and precise surveys. It should be pointed out that the Eberhardt method produced an outlying density estimate (October), but otherwise performed well (Table 3).

For all belt or line transect methods, colony visibility is fundamental to method performance. Since visibility decreases exponentially with distance from the center transect line, both fixed and indefinite width belt transects have operational limitations, even if additional cues increase the probability of recording a nest. As the probability distributions of discovering a nest are unknown before sampling, I recommend that indefinite width transect methods be used for single or initial sampling surveys. Posterior surveys should then be run based on distance estimates of widths derived from indefinite width transects.

Any method will fail if the researcher is unfamiliar with an organism's ecology and behavior. In the case of leaf-cutting ants, a failure to recognize the signs of colony presence would have made the results incomplete. An understanding of *Ac. lobicornis*' seasonal and daily activity periods allows us to recommend that transects be conducted during spring or fall, when the colonies are diurnal. The absolute colony density study that considers such factors will be accurate and repeatable.

ACKNOWLEDGMENTS

I would like to express my appreciation to the following persons that helped me complete this work: Dr. Jorge Rabinovich for suggesting the use of various transect methods, Lic. Angela DiBlasi for statistical advice and assistance, Dr. Maura Kufner for enduring and sharing the joys of field desert research, Dr. R. Romderos and Ing. V. Roig for their assistance, Alberto Lombardi for kindly rendering the figure, and Dr. Harold Fowler for his valuable comments and support in the realization and completion of this work as well as for giving me additional confidence.

REFERENCES

BUCHER, E.H., and R.B. ZUCCARDI. 1967. Significación de los hormigueros de *Atta vollenweideri* Forel como alteradores del suelo en la provincia de Tucumán. Acta Zool. Lill. 23: 83- 95.

CAUGLEY, G. 1977. Analysis of vertebrate populations. John Wiley and Sons, New York. 234 pp.

CHERRETT, J.M., G.V. POLLARD and J.A. TURNER. 1974. Preliminary observations on *Acromyrmex landolti* (For.) and *Atta laevigata* (F. Smith) as pasture pests in Guyana. Trop. Agric. 51: 66-74.

EBERHARDT, L.L. 1968. A preliminary appraisal of line transects. J. Wildl. Mgmt. 32: 82-88.

FOWLER, H.G. 1977. Some factors influencing colony spacing and survival of the grass-cutting ant, *Acromyrmex landolti fracticornis* (Forel) (Hymenoptera: Formicidae), in Paraguay. Revta. Biol. Trop. 25: 89-99.

FOWLER, H.G. 1983. Distribution patterns of Paraguayan leaf-cutting ants (*Atta* and *Acromyrmex*) (Hymenoptera: Formicidae: Attini). Stud. Neotrop. Fauna Environ. 18: 121-138.

FOWLER, H.G. 1984. A organizaçao das comunidades de formigas cortadeiras (*Atta* e *Acromyrmex*). An. Sem. Reg. Ecol., Sao Carlos 4: 151-163.

FOWLER, H.G. and S. CLAVER. 1989. Leaf-cutting ant assemblies: effects of latitude, vegetation and behavior on species compositions. In C.R. Huxley (ed.), Interactions between ants and plants. Blackwell Scientific Publications, London. (in press).

FOWLER, H.G. and B.L. HAINES. 1983. Diversidad de especies de hormigas cortadoras y termitas de tumulo en cuanto a la sucesion vegetal en praderas paraguayas. pp. 187-201. In P. Jaisson (ed.) Social insects in the tropics. Vol. 259 pp.

FOWLER, H.G., V. PEREIRA DA SILVA, L.C. FORTI and B.N. SAES. 1986. Population dynamics of leaf-cutting ants: a brief review. pp. 123-145, In C.S. Lofgren and R.K. Vander Meer (eds.) Fire ant and leaf-cutting ants: biology and management. Westview Press, Boulder.

HAYNE, D.W. 1949. An examination of the strip census method for estimating animal number. J. Wildl. Mgmt. 13: 145-157.

HOGLUND, N., G. NILSSON and F. STALFELT. 1967. Analysis of a technique for estimating willow grouse (*Lagopus lagopus*) density. Trans. 8th Int. Congr. Game Biologists: 156-158.

JONKMAN, J.C.M. 1979. Population dynamics of leaf-cutting ant nests in Paraguayan pastures. Z. angew. Entomol. 87: 281- 293.

KUSNEZOV, N. 1963. Zoogeografia de las hormigas en sud-america. Acta Zool. Lill. 19: 25-186.

ROCKWOOD, L.L. 1973. Distribution, density and dispersion of two species of *Atta* (Hymenoptera: Formicidae) in Guanacaste Province, Costa Rica. J. Anim. Ecol. 42: 803-818.

ROIG, F.A. 1970. Flora y vegetacion de la Reserva Ecologica de Nacunan. Deserta 1: 25-232.

SEBER, G.A.F. 1973. The estimation of animal abundance. Griffin, London. 500 pp.

SOUTHWOOD, T.R.E. 1978. Ecological methods. 2nd ed. Methuen, London.

TANQUILEVICH, R. 1971. Suelos de la reserva de nacunan. Deserta 2: 131-206.

22

Methods for the Evaluation of Leaf-cutting Ant Harvest

H.G. Fowler, L.C. Forti and L.F.T. di Romagnano

INTRODUCTION

Leaf-cutting ants of the genera *Atta* and *Acromyrmex* are the major economic phytophagous pests of Neotropical agriculture (Cramer, 1967; Echandi, et al., 1972). These ants are responsible for an enormous amount of damage, yet information of how figures are obtained is scarce. Estimates based on pesticide costs and crop replacement require subjective interpretation. Are insecticides over-used? Are they effective? What criteria do control programs employ? No matter what the answers, economic data cannot estimate (1) colony vegetation input and requirements, (2) the impact of these ants on the energy flow and nutrient cycle and (3) the foraging activity of these ants to optimize control programs.

Various studies have been tried to estimate colony vegetation intake or energy flow. We provide a summary of these methods for the first time. We review the problems associated with these methods and the use of over-estimates in the literature. This review will provide researchers with a critical framework to evaluate previously published estimates, design their own research programs and provide government and private agencies with a critical overview of the ants harvesting behavior. To facilitate this discussion, each method employed to estimate leaf-cutting ant harvest will be treated separately and then discussed in an overview.

CONVERSION FACTOR METHOD

This method is based on the conversion of harvested vegetation into spent fungal substrate which is deposited in refuse chambers or piles in leaf-cutting ant colonies. The symbiotic ant fungi degrades the harvested vegetations' cellulose and the waste material is removed to discrete locations by the worker ants (Martin and Weber, 1969); thus, the weight difference between the vegetation input and the spent fungal output corresponding to the material passed along to the ants by the fungus. Autuori (1947) first used the reduction of biomass to estimate the vegetation intake of a 77 month old colony of *Atta sexdens rubropilosa.* Autuori excavated the colony, removed all spent fungal substrate from the refuse chambers, weighed the refuse and then converted this into the vegetation that entered the colony during the 77 months of colony life. A conversion factor of 12.4 was calculated based on the ratio of the entering vegetation biomass to that deposited in the refuse chambers. There are several short-comings to Autuori's method: (1) the laboratory experiment estimating the conversion factor lasted only two weeks; vegetation that enters a colony during two weeks is probably not deposited in the refuse chambers, especially since refuse deposition is cyclical rather than continuous (Weber, 1976); (2) the process Autuori used should have been based on dry-weight, not wet-weight; (3) Autuori did not consider the decomposition of the fungal substrate and assumed all substrate brought in would be utilized by the fungus. The present study found that spent fungal substrate (laboratory colonies) has a half-life of 45 days, which means that the conversion factor method will underestimate consumption if not corrected for decomposition.

Amante (1967d) used the conversion factor to estimate losses to leaf-cutting ants in *Eucalyptus* and *Pinus* plantations in São Paulo, Brazil (Table 1). Current evidence suggests that Autuori (1947) over-estimated the conversion factor (Table 1) (Schoereder, 1985); thus, losses were over-estimated by researchers such as Amante (1967d).

Autuori (1947) weighed the spent fungal refuse in 296 refuse chambers of the excavated colony and obtained 475.2 Kg. Using his conversion factor of 12.4, derived from laboratory studies, the colony should have consumed 5892.48 Kg of vegetation (475.2 Kg. refuse X 12.4 = 5892.4 Kg).

It is highly probable that Amante (1967a, 1967b, 1967c) used Autuori's (1947) conversion factor to estimate the amount of vegetation harvested by *Atta capiguara*, a grass-cutting ant, although there is no mention of how his estimates were obtained. A reanalysis of his data (Amante, 1972) is found at the end of this chapter (Table 2). Nakano, et al., (1981) also employed Autuori's (1947) conversion factor to estimate losses by *Atta* species.

Jonkman (1977a, 1977b) estimated a conversion factor for another grass-cutting ant, *Atta vollenweideri*, and discovered that this factor varied between 1.5 and 3.3, while it was 1.8 for the grass-cutting ant, *Acromyrmex landolti fracticornis* (Fowler and Robinson, 1979b). For another grass-cutting ant, *Atta capiguara*, Forti (1985) calculated a conversion factor of 1.9, but noted that this factor varied with respect to the applied methodologies. Jonkman (1980) judged that the amount of grass used by colonies of *Atta vollenweideri* was approximately 90.0 Kg/year/colony, while Forti (1985) postulated a vegetation input of 33.6 Kg/year/ colony for *Atta capiguara* (Table 1).

Schoereder (1985) re-calculated the conversion factor for *Atta sexdens rubropilosa* through laboratory experiments and found, on a dry-weight basis, a factor of 2.7, or approximately 7 for wet-weight. These numbers, as well as others, are given in Table 1.

A number of disadvantages and sources of potential error can be cited for this method:

<u>Disadvantages of the Conversion Factor Method</u>

(1)This method depends upon the herculean task of complete colony excavation, which requires extensive man-power. Colonies of *Atta* are large and it takes 4-5 laborers 30 days to completely excavate a colony (Forti, unpublished). It could be feasible to excavate *Acromyrmex* colonies due to their small size. Additionally, some species of *Atta* and *Acromyrmex* deposit their spent fungal substrate outside the nest, which would facilitate its use for these studies.

(2) Previous studies relied on a reduced number of samples, generally only one colony is excavated, and only some of its refuse chambers are collected. Given results on refuse decomposition, seasonal cycles of refuse deposition undoubtedly exist, but no studies have been performed to test this potential source of error.

(3) A neglected, but important source of error concerns spent fungal decomposition. It is impossible for the organic matter deposited in the first years of colony life to still exist as the colony grows to an age of seven to 10 years, since the refuse chambers have a rich fauna of decomposing food-web insects (Weber, 1972; Jonkman, 1977b; Autuori, 1952; Pereira-da-Silva, 1975). The present study's results indicate the refuse from spent fungal substrate would be recycled every 100 days for *Atta sexdens rubropilosa*, or approximately 3.5 times per year. Thus, a conservative estimate could be reached by multiplying the spent fungal substrate by the conversion factor by the annual (or monthly) renovation.

(4) Species of leaf-cutters, such as *Atta laevigata* which cut both broad-leafed plants and grasses, will probably have variable conversion factors, depending upon the type of vegetation harvested. It has already been shown that the conversion factor for grass-cutting ants is lower than for leaf-cutting ants. Thus it is important to know the percent of both vegetation types entering the nest.

(5) Given these shortcomings, it is likely that the conversion factor method, even if properly estimated, will result in a low estimate of vegetation harvest.

Advantages of the Conversion Factor Method

(1) However, this method can be applied in a relatively short period of time (the time necessary to excavate a colony, or approximately one month).

(2) Incorporating a decomposition constant, the estimates obtained by this method can be greatly improved.

FORAGING ACTIVITY METHOD

This method has been most popular in studies of leaf-cutting ants (Table 1). The earliest recorded usage of this method was by Hodgson (1955) to estimate the amount of vegetation harvested by colonies of *Atta cephalotes* (Table 1). The vast majority of studies using this method were performed over a short time-period, usually not more than a few months, and in some cases, were based on only one day of sampling (Table 1).

The foraging activity method consists of counting leaf fragments transported to the nest by workers during a fixed time period. These counts can be taken either visually or with photoelectric counters (Fowler and Robinson, 1979a; Forti, 1985). The mean number of fragments carried per time period is estimated, the loads carried by the ants are sampled to gauge mean load mass, generally on a dry-weight basis. The mean load mass is then

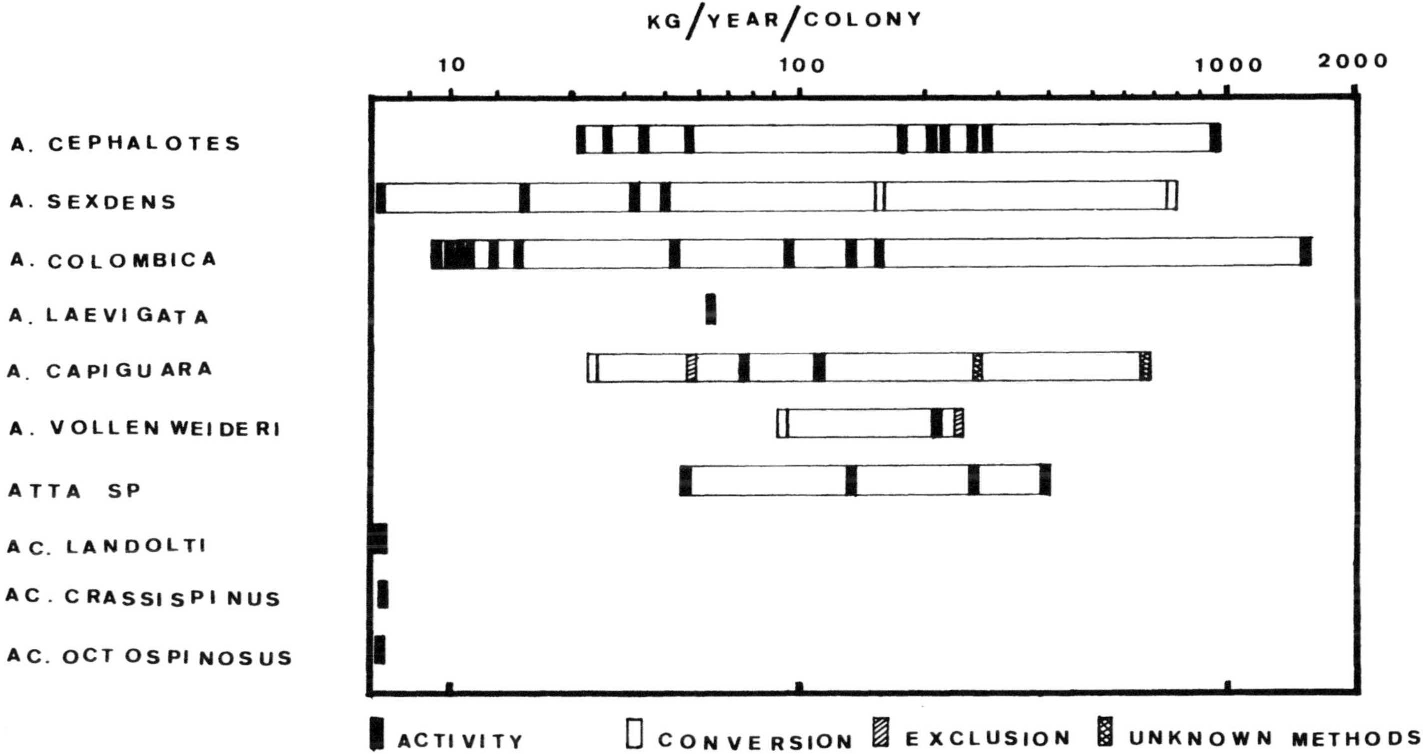

FIGURE 1. Variability in estimates of harvesting activity by species and by estimation method. See Table 1 for listing of the studies plotted.

multiplied by the mean time period load input to obtain estimates of the amount of vegetation entering the nest.

The vast majority of studies recorded in the literature have relied upon visual counts of foraging activity, but three studies used the more continuous records provided by photo-electric counters placed over the foraging trails (Dibley and Lewis, 1972; Fowler and Robinson, 1979a; Forti, 1985). Photo-electric counters have an efficiency of 40 to 120% (Dibley and Lewis, 1972) depending on the species of ant studied and the type of vegetation harvested

Visual counts of ant activity on foraging trails vary greatly over time. Forti (1985) found that trails of *Atta capiguara* change in leaf carrying activity from 2.5 to 83.8 fragments per minute, while Cherrett (1968) registered a range of 2 to 359 fragments/minute for *Atta cephalotes*. This method is further complicated by temporal variation of trail numbers; it is physically impossible to monitor more than one trail at a time (Fowler and Robinson, 1979a). Also, the location chosen on the trail as a counting point can influence the estimate of fragment input (Fowler and Robinson, 1979a). For both *Atta sexdens rubropilosa* (Fowler and Robinson, 1979a) and *Atta capiguara* (Forti, 1985) the number of trails used per colony may vary three times over the course of a year. Daily variability, and trail to trail variability, has been reported numerous times for leaf-input (Amante, 1972; Cherrett, 1968; Cherrett et al., 1974; Fowler and Robinson, 1979a; Robinson and Fowler, 1982; Vasconcelos, 1987).

As demonstrated in Table 1 and Fig. 1, variability is the most striking conclusion to be derived from these studies. This is the result of using different procedures for data acquisition, colony to colony variations, seasonal variations, and the duration of the study. Only for the last case can results be compared. Using Friedman's non-parametric analysis of variance, and grouping studies into less than 60 days, 60-100 days, and more than 100 days, and excluding Weber's (1976) laboratory study, a significant difference was found in the estimates of colony intake per year due to the length of study for *Atta colombica* ($H = 5.21$; $P < 0.05$), but not for *Atta cephalotes* ($H = 3.75$; $P > 0.05$) (Table 1), which are the only species with a sufficient amount of studies to permit an analysis. But the large variance in these estimates, does not permit a more critical analysis. As in all methods, the size of the colony (mound surface area) generally not goes unreported, a critical over-sight since vegetation harvest is a function of colony size (Fowler, et al., 1986a; 1986b).

Disadvantages of the Foraging Activity Method

(1) A large sample is required. As previously mentioned, the activity of foragers on trails varies, and the number of active trails changes during the year, as well as during the same time period.

(2) The yearly foraging pattern is marked by an auto-correlated error which does not permit the use of linear regression techniques to estimate yearly foraging patterns (Fowler, unpublished).

(3) Worker to worker variability exists with respect to load size carried (Fowler, 1979), further hindering an estimate of mean load size.

(4) If a photo-electric counter or visual count is used, errors will always be present, but no study has yet estimated these errors.

TABLE 1. Leaf-cutting estimates of Atta and Acromyrmex taken from the literature. To standardize the estimates, an activity period of 12 hours per colony (Fowler & Robinson, 1979a; Vasconcelos, 1987) and a yearly activity period of 270 days (Forti, unpublished) were used. See text for further details.

Leaf-cutting species	Source	Length of study (days)	Daily period studied	Vegetation harvested* g/hr/colony	Vegetation harvested* Kg/year/colony	Method employed
Atta	Harris, 1969[a]	60	0700-2200	80.0	259.2	Activity
cephalotes	Markham, 1966[a]	1	0000-2400	64.0	207.4	Activity
	Gara, 1970[a]	?	?	290.0	939.6	Activity
	Cherrett, 1972	70	0000-2400	13.1	42.4	Activity
	Hodgson, 1955	60	0700-1800	85.2	276.1	Activity
	Weber, 1976[b]	3285	0000-2400 (lab)	1.3	29.6	Activity
	Parsons, 1968	60	0530-1300	17.2	55.7	Activity
	Cherrett, 1968	60	0000-2400	10.9	35.3	Activity
	Blanton & Ewel, 1985	126	0000-2400	20.2	177.0	Activity
	Vasconcelos, 1987	210	2000-0030	64.3	208.3	Activity
	Rockwood, 1976	330	0000-2400	6.8	22.1	Activity
	Rockwood, 1976	330	0000-2400	3.5	11.2	Activity
	Rockwood, 1976	330	0000-2400	3.1	9.9	Activity

Continued.

TABLE 1. Continued.

Leaf-cutting species	Source	Length of study (days)	Daily period studied	Vegetation harvested* g/hr/colony	Vegetation harvested* Kg/year/colony	Method employed
Atta sexdens[c]	Prado, 1973	10	0000-2400 (lab)	0.61	1.98	Activity
	Autouri, 1947	2310	------------	47.7	154.5	Conversion
	Amante, 1967d	--	------------	227.5	738.0	Conversion
	Fowler & Robinson, 1979a	300	0000-2400	14.8	48.1	Activity
	Vasconcelos, 1985	210	2000-0030	12.3	39.9	Activity
	Shoereder, 1985	330	7 hrs	3.52	22.2	Activity
Acromyrmex octospinosus	Weber, 1977	360	0000-2400 (lab)	0.29	0.94	Activity
Acromyrmex landolti	Cherrett, et al., 1974	1	0000-2400	0.34	2.2	Activity
	Fowler & Robinson, 1975	90	0000-2400	0.23	1.5	Activity
	Robinson & Fowler, 1982	360	0000-2400	0.13	0.82	Activity
	Fowler (unpublished)	60	0000-2400	0.06	0.42	Activity

*When data were not provided, each leaf fragment was assigned a dry weight of 0.0092 g (leaf-cutters) or 0.0045 g (grass-cutters) (data estimates from Vasconcelos (1987) and Forti (1985)). [a]Original papers not seen. Data were taken from Lugo, et al. (1973). [b]Reanalysis of data.

Advantage of the Foraging Activity Method

(1) The method is comparatively cheap but requires man-power - a need which can be greatly reduced by the use of photoelectric counters. Given that long time periods are necessary to obtain acceptable estimates of activity, this method is also labor intensive.

EXCLUSION METHOD

This method was used only once by Robinson and Fowler (1982) to estimate harvest by Paraguayan grass-cutting ants. This method consists of establishing pasture areas (or crops) without established colonies and other areas with natural colony levels. If standard procedures are used, productivity in areas with ants can then be compared with productivity in areas without ants. The differences in these two values should reflect the amount of vegetation harvested by grass-cutting ants. Further combinations can be made with and without cattle to obtain estimates of ant-cattle-pasture interactions (Fowler, et al., 1986a).

Using this method, Robinson and Fowler (1982) concluded that *Atta capiguara* competes with cattle for pasture grass, but *Atta vollenweideri* does not. However, even if grass-cutting ants cut large quantities of vegetation, this does not mean they compete with cattle. However, inferences can be made from exclusion methods to evaluate their economic impact in pastures.

Although this method has only been used for grass-cutting ants, it would be highly productive in other cropping systems; even *Eucalyptus* and *Pinus* plantations over an appropriate (long) time period. For tree plantations, normal growth parameters could be taken, such as diameter at breast height (d.b.h.), to evaluate the impact of these ants on production.

Advantages of the Exclusion Method

(1) It is easily implemented for grass-cutting ants, without elevating operational costs.

(2) This is the most experimental of all methods as the researcher has complete control over the design.

Disadvantages of the Exclusion Method

(1) Maintenance and man-power costs may be high.

(2) To obtain a reasonable estimate of leaf-cutting ant harvest and competitive interactions with domestic animals, a large number of plots are needed. The heterogeneity of leaf-cutting ant activity makes this necessary.

(3) Other invertebrates and small mammals can enter the ant excluded plots and influence estimates.

(4) Plot estimates can evaluate the heterogeneity of leaf-cutting activity in both time and space.

FIXED TRANSECT METHOD

This method, currently under evaluation by H.G. Fowler, consists of establishing fixed transects, and marking all trees within a fixed width. Each tree is numbered, as well as each leaf of each tree. Allometric relationships

TABLE 2. A re-analysis of Amante's (1972) data, to check his previous harvest estimates of the grass-cutting ant, *Atta capiguara* (Amante, 1967a, 1967b, 1967c)

Yearly Trimester	Number of grass fragments collected	Grams dry weight of harvest per day	per trimester
First	7,928	158	14931
Second	12,780	256	24192
Third	10,669	213	20129
Fourth	7,355	147	13891

of tree growth and leaf-growth must be determined for each vegetative species. Periodic studies of these transects involve the analysis of each tree and each leaf. Previous leaf-cutting ant damage is factored out of leaf-growth by regressions based upon leaf allometric relationships. Missing leaves are hard to score unless evidence of stem cuts is present. This method also requires the observer to differentiate between leaf-cutting ant damage and other herbivore or pathogen damage.

Disadvantages of the Transect Method

(1) Vegetation height limits this type of study, which functions well for trees up to 3 m tall.
(2) A large number of repetitions are necessary (various transects), as foraging activity is not homogeneous in time or space.
(3) If the leaf is cut at the petiole's base, it is impossible to accurately differentiate leaf-cutting ant harvest or leaf abortion.
(4) A large amount of pre-test data is needed to obtain the appropriate allometric relations for each vegetative species studied.

Advantages of the Transect Method

(1) This method is useful in studies of the spatial and temporal habitat usage by leaf-cutting ants.
(2) Foraging activity can be estimated even if the colony itself is difficult to find.
(3) When colonies are located, harvesting activity can be correlated with colony density.

REMOTE SENSING METHODS

Aerial and satellite images have been used to study pulp-wood plantation losses caused by leaf-cutting ants (Moser, 1986; Anonymous, 1986) on a pilot scale. In 1983, the Brazilian Institute of Forestry Development (IBDF) analyzed Landsat satellite images, and estimated that leaf-cutting ant plus fire losses in Brazilian pulp-wood plantations (Mato Grosso do Sul) increased from 13% in 1980, to 34% in 1985, or a loss of approximately 230,000 ha (Anonymous, 1986). Moser (1986) estimated the damage inflicted by one colony of *Atta texana* on a *Pinus taeda* plantation through the use of aerial photographs.

Advantage of Remote Sensing Methods
(1) Large areas can be studied with ease.

Disadvantage of Remote Sensing Methods
(1) Costs of such methods are high, requiring specialized laboratories and highly-trained technicians to correctly interpret the data.
(2) Such methods only are valid when accompanied by extensive field surveys to check causes of losses.
(3) Satellite images or aerial photographs can only be taken at certain times of the year, and may not cover the area in question. A long record of such images over the same area is generally needed, but difficult, and expensive, to obtain.
(4) For aerial photographs, scales often vary greatly depending upon their original usage.

A RE-ANALYSIS OF THE DATA OF AMANTE (1972)

Amante estimated grass harvest by *Atta capiguara* in two publications (1967a, 1967b), the figures much larger than those obtained by Forti (1985) and Fowler and Robinson (1982) (Table 1, Fig. 1). How Amante obtained his estimates remains a partial mystery; the data used here is from his Ph. D. thesis (the only source available).

Using data from his tables (Amante, 1972), the number of fragments transported per minute during weekly observations was considered. A total of 24 observations were made during a one-minute period at the beginning of each hour of the day. From these data, the number of fragments transported per day was estimated and then multiplied by the mean mass of the fragments.

This re-analysis of Amante's (1972) data gave an estimate of 73.0 Kg of dry matter/year/colony (Tables 1 and 2), which indicates that his earlier estimates (Amante, 1967a, 1967b) were probably overestimated. Furthermore, our re-analysis of these data place the figures within the range of estimates by other researchers (Table 1).

DISCUSSION

Independent of the method, harvest estimates from leaf-cutting ant colonies vary significantly (Fig. 1). Given the lack of experimental standardization, it impossible to compare harvest, even by the same leaf-cutter species. While researching this paper, estimates were found for which the methods were unknown, such as those by Amante (1967a, 1967b); these estimates do not even coincide with the reanalysis of his available data (Amante, 1972).

Forti (1985) performed the only comparative test of methods by comparing the conversion factor method with the activity method for *Atta capiguara*. Without taking into account the decomposition of the fungal substrate, here observe that foraging activity produces estimates three times greater than estimates obtained by the conversion factor method (Table 1), but this is approximately the correction necessary to account for decomposition. Thus, conversion factor estimates probably underestimate true leaf-cutting ant harvest.

Incorrectly estimating the parameters involved in studies of harvest by leaf-cutting ants can promote posterior errors when used by other researchers. This point was proven when, based on Amante's data, only estimates of harvest obtained by Nakano, et al. (1981) were used to gauge pasture losses to grass-cutting ants. For forestry, Mendes Filho (1981) also based estimated losses on Amante (1967d), which as previously noted was a mal-application of Autuori's estimates (1947). Amante (1972) provides a good example of how extrapolating data can lead to costly errors. He used Autouri's 12.4 conversion factor for a broad-leaf cutter, then applied it to a grass-cutting ant — a subject with entirely different habits and conversion factors.

Data on occasional losses due to harvest by ants are best obtained by the experimental exclusion method. Using methodology similar to that proposed by Robinson and Fowler (1982), Precetti, et al. (1988) estimated losses produced by *Atta bisphaerica* in a São Paulo State sugar-cane field, and concluded that a mature colony produces a loss of 3.2 metric tons of sugar cane/ha in a field with a productivity of 60 metric tons/ha.

All methods have pros and cons. The most popular method, foraging activity, requires a large number of observations; however, this was generally not the case (Table 1), which may explain the variability of the data (Fig. 1). Of the methods which have been used, the conversion factor method is least recommended, unless supplementary studies are undertaken to quantify other involved parameters, such as decomposition. The transect method and remote sensing methods are best applied to ecosystem level studies; they are extremely expensive and time consuming for their wide applications. Of the available methods, depending upon the species involved, the exclusion method seems most appropriate for agricultural studies, and its use should be mandatory for harvest loss studies.

ACKNOWLEDGMENTS

Many people have contributed to our ideas and research, and we would like to thank Elena Fowler, Heraldo de Vasconcelos, and Virgilio Pereira-da-Silva for their valuable input and support. We acknowledge the support of UNESP during our research, and extend thanks to Jose Vicente E. Bernardi for his help with the decomposition studies.

REFERENCES CITED

AMANTE, E. 1967a. Saúva tira boi da pastagem. Coopercotia, 23: 38-40.

AMANTE, E. 1967b. A formiga saúva *Atta capiguara*, praga das pastagens. O Biologico, 33: 113-120.

AMANTE, E. 1967c. A saúva *Atta capiguara*, praga das pastagens. Instruções Práticas-DPA, No. 41: 1-12.

AMANTE, E. 1967d. Prejuizos causados pela formiga saúva em plantações de *Eucalyptus* e *Pinus* no Estado de São Paulo. Silvicult. S. Paulo, 6: 355-363.

AMANTE, E. 1972. Influência de alguns fatores microclimáticos sobre a formiga saúva, *Atta laevigata* (Fr. Smith, 1858), *Atta sexdens rubropilosa* Forel, 1908, e *Atta bispaerica* Forel, 1908, e *Atta capiguara* Gonçalves, 1944 (Hymenoptera, Formicidae), em formigueiros localizados no Estado de São Paulo. Ph.D. Dissertation, ESALQ-USP, Piracicaba, 175 p.

ANONYMOUS 1986. Marcha batica. Veja, 30 abril, 1986: 62.

AUTUORI, M. 1947. Contribuição para o conhecimento da saúva (*Atta spp.* IV). O sauveiro depois da primeira revoada (*Atta sexdens rubropilosa* Forel, 1908). Arch. Inst. Biol., 18: 39-70.

AUTUORI, M. 1952. Fauna das panelas de lixo do sauveiro (*Atta spp.*): Mydaidae - Diptera. Cienc. Cult., 4: 127.

BLANTON, C.M. and J.J. EWEL. 1985. Leaf-cutting ant herbivory in successional and agricultural tropical ecosystems. Ecology, 66: 861-869.

CHERRETT, J.M. 1968. The foraging behaviour of *Atta cephalotes* L. (Hymenoptera, Formicidae). I. Foraging patterns and species attacked in tropical rain forest. J. Anim. Ecol., 37: 387-403.

CHERRETT, J.M. 1972. Some factors involved in the selection of vegetable substrate by *Atta cephalotes* (L.) (Hymenoptera, Formicidae) in tropical rain forest. J. Anim. Ecol., 41: 647-660.

CHERRETT, J.M. 1986. The economic importance and control of leaf-cutting ants. In: Vinson, S.B. (ed.), Economic Impact and Control of Social Insects, Pragger Publishers, New York, pp. 165-192.

CHERRETT, J.M., G.V. Pollard and J.A. Turner. 1974. Preliminary observations on *Acromyrmex landolti* (For.) and *Atta laevigata* (Fr. Smith) as pasture pests in Guyana. Trop. Agric., 51: 69-74.

CRAMER, H.H. 1967. Plant protection and world crop production. Bayer Pflanzenschutz, Leverkusen, 524 p.

DIBLEY, G.C. and T. Lewis. 1972. An ant counter and its use in the field. Entomol. Exp. Appl., 15: 499-508.

ECHANDI, E., J.E. KNOKE, E.L. NIGH JR., M. SHENK and G.T. WEEKMAN. 1972. Crop production in Brazil, Uruguay, Bolivia, Ecuador, and the Dominican Republic. USAID Report, Univ. Calif. Berkley, 65 pp (mimeo).

EMMEL, T.C. 1967. Ecology and activity of leaf-cutting ants (*Atta sp.*). In: Advanced Biology: Insect Ecology in the Tropics, Winter, 1967, Organization for Tropical Studies, San Jose, Costa Rica, p. 125-131.

FORTI, L.C. 1985. Ecologia da saúva, *Atta capiguara* Gonçalves, 1944 (Hymenoptera, Formicidae) em pastagem. Ph.D. Dissertation, ESALQ-USP, Piracicaba, 234 p.

FOWLER, H.G. 1979. Environmental correlates of the foraging of *Acromyrmex crassispinus* Cienc. Cult., 31: 879-882.

FOWLER, H.G. and S.W. ROBINSON. 1975. Estimaciones acerca de la acción de *Acromyrmex landolti* Forel (Hymenoptera: Formicidae) sobre el pastoreo y la ganaderia en el Paraguay. Revta. Soc. Cient. Paraguay, 15: 64-71.

FOWLER, H.G. and S.W. ROBINSON. 1977. Foraging and grass selection by the grass-cutting ant, *Acromyrmex landolti fracticornis* (Forel) (Hymenoptera: Formicidae) in habitats of introduced forage grasses in Paraguay. Bul. entomol. Res., 67: 659-666.

FOWLER, H.G. and S.W. ROBINSON. 1979a. Foraging by *Atta sexdens* (Formicidae: Attini): seasonal patterns, caste and efficiency. Ecol. Entomol., 4: 239-247.

FOWLER, H.G. and S.W. ROBINSON. 1979b. Foraging ecology of the grass-cutting ant, *Acromyrmex landolti fracticornis* (Formicidae: Attini) in Paraguay. Intl. J. Ecol. Environ. Sci., 5: 29-37.

FOWLER, H.G., L.C. FORTI, V. PEREIRA-DA-SILVA and N.B. SAES. 1986a. Economics of grass-cutting ants. In: C.S. Lofgren and R.K. Vander Meer (eds.), Fire Ants and Leaf-Cutting ants: Biology and Management, Westview Press, Boulder, pp.18-35.

FOWLER, H.G., V. PEREIRA-DA-SILVA, L.C. FORTI and N.B. SAES. 1986b. Population dynamics of leaf-cutting ants: a brief review. In: C.S. Lofgren and R.K. Vander Meer (eds.), Fire Ants and LeafCutting Ants: Biology and Management, Westview Press, Boulder, pp. 123-145.

GARA, R.I. 1970. Report of forest entomology consultant (UNDP project 80). IICA, Turrialba, Costa Rica, 34 p (mimeo). Harris, L.D. 1969. A consideration of the nutrient "sumping" activities of leaf-cutter ants in the New World tropics. Research Report, In, Advanced Population Biology, July-August, Organization for Tropical Studies, San Jose, Costa Rica.

HODGSON, E.S. 1955. An ecological study of the behavior of the leaf-cutting ant, *Atta cephalotes*. Ecology, 36: 293-303.

JONKMAN, J.C.M. 1977a. Determination of the vegetative material intake and refuse production ratio in two species of grasscutting ants. Z. angew. Entomol., 84: 440-443.

JONKMAN, J.C.M. 1977b. Biology and ecology of the leaf-cutting ant, *Atta vollenweideri* Forel, 1898 (Hymenoptera: Formicidae) and its impact in Paraguayan pastures. Ph.D. Dissertation, Waenigenen, 132 p.

JONKMAN, J.C.M. 1980. Average vegetative requirement, colony size and estimated impact of *Atta vollenweideri* on cattle raising in Paraguay. Z. angew. Entomol., 89: 135-143.

LUGO, A.E., E.G. FARNWORTH, D. POOL, P. JEREZ and G. KAUFMAN. 1973. The impact of the leaf-cutter ant, *Atta colombica* on the energy flow of a tropical wet forest. Ecology 54: 1292-1301.

MARKHAM, C.P. 1966. The surface activity, behavior and energy influx rhythms of leaf-cutting ants, *Acromyrmex sp.* and *Atta cephalotes*. Research report. In, Tropical Biology: an ecological approach, Feb.-March, 1966, Organization for Tropical Studies, San Jose, Costa Rica.

MARTIN, M.M. and N.A. WEBER. 1969. The cellulose-utilizing capability of the fungus cultivated by the attine ant, *Atta colombica tonsipes* Santschi. Ann. Entomol. Soc. Am., 62: 1382-1387.

MENDES-FILHO, J.M. de A. 1981. Ação danosa de pragas desfolhadoras sobre as florestas de *Eucalyptus*. Circular Tecnica, IPEF, No. 131: 1-6.

MOSER, J.C. 1986. Estimating timber loss from a town ant colony with aerial photographs. South. J. Appl. Forestry, 10: 45-47.

NAKANO, O., S. SILVEIRA-NETO and R.A. ZUCCHI. 1981. Entomologia Econômica. USP, São Paulo, 314 p.

PARSONS, L.R. 1968. Aspects of leaf-cutter behavior: *Acromyrmex octospinosus* and *Atta cephalotes*. Research Paper. In, Tropical Biology: an Ecological Approach. July-August, Organization for Tropical Studies, San Jose, Costa Rica.

PEREIRA-DA-SILVA, V. 1975. Contribuição ao estudo das populações no Estado de São Paulo (Hymenoptera: Formicidae). Studia Entomol., 18: 201-205.

PRADO, L. 1973. Le comportement de decoupage des feuilles chez *Atta sexdens*. Ins. Soc., 20: 133-144.

PRECETTI, A.A.C.M., A.C.M. NASATO, G.J. BELTRAME, J.E. de OLIVEIRA and M. PALINI Jr. 1988. Perdas de produção em cana-de-açucar causadas pela saúva-mata-pasto, *Atta bisphaerica*. Parte I. Bol. Tec. Coopersucar, 42: 19-26.

ROBINSON, S.W. and H.G. FOWLER. 1982. Foraging and pest potential of Paraguayan grass-cutting ants (*Atta* and *Acromyrmex*) to the cattle industry in Paraguay. Z. angew. Entomol., 93: 42-54.

ROCKWOOD, L.L. 1976. Plant selection and foraging patterns in two species of leaf-cutting ants (*Atta*). Ecology, 57: 48-61.

SCHOEREDER, J.H. 1985. Aspectos ecologicos da saúva no cerrado. M.S. Thesis, Universidade de São Paulo, São Paulo, 102 p.

SHEPARD, J.D. 1985. Adjusting foraging effort to resources in adjacent colonies of the leaf-cutter ant, *Atta columbica*. Biotropica, 17: 245-252.

VASCONCELOS, H.L. de. 1987. Atividade forrageira, distribuição e fundação de colônias de saúvas (*Atta spp.*) em floresta da Amazônia Central. M.S. Thesis, INPA, Manaus, 62 p.

WEBER, N.A. 1972. Gardening ants: the attines. Mem. Am. Phil. Soc., 92: 1-146.

WEBER, N.A. 1976. A ten-year laboratory colony of *Atta cephalotes*. Ann. Entomol. Soc. Amer., 69: 825-829.

WEBER, N.A. 1977. A 10-year colony of *Acromyrmex octospinosus* (Hymenoptera: Formicidae). Proc. Entomol. Soc. Wash., 79: 284-292.

23

Seasonal Activity of *Atta insularis,* an Important Citrus Pest in Jaguey Grande, Cuba

L.F. Perez Perera, A. Gonzalez and J.F. Martinez

INTRODUCTION

Atta insularis Guérin is an insect whose presence in Cuba dates back to 1492 when the island was discovered. Years later, Fernández de Oviedo (1535) reported "damages" to isolated orange trees by "harmful ants." It was not until 1949 that Bruner and Valdés made the first biological observations of this species in the Agronomical Research Station of Santiago de las Vegas, Havana province. In the last 40 years, the results of different researchers (Alayo 1974, Pintera 1983, Pintera and Zorrilla 1981, Pérez et al 1977, 1979, 1981) have been reported regarding not only the biology but also emphasizing its control.

With the development and diversification of agriculture, new citrus projects have emerged on barren soils whose wild and native plant life have been wiped out, bringing about changes in the natural ecosystems and spreading *A. insularis*. This pest causes huge losses to newly planted citrus trees as well as to older ones (Fig. 1 and 2). This paper intends to modestly contribute to a better knowledge of the bio-ecology of this pest.

MATERIALS AND METHODS

Two *A. insularis* nests were selected in Lot T-25, Farm 1 from the Citrus Productive-Scientific Complex "Victoria de Girón," Matanzas province. These nests were in an eight year-old "Valencia Late" (*Citrus sinensis* [L.] Osbeck) orchard on "Sour" orange (*Citrus aurantium* L.) planted at 6 by 6 m distances. Cultural practices applied are according to the Technical Instructions from 1984.

The nests selected were under trees (Fig. 3). The workers of this species use leaves as substrate, which they incorporate into the fungus gardens they grow within the colonies (Fig. 4).

In 1985, selected nests were evaluated monthly to determine the number of workers transporting leaf portions; taking out earth from the interior of the galleries; introducing dry weeds (*Paspalum notatum* L.) or going loadless into the nests.

In each hour of the chosen days, five minute observations were made to determine the different activities employed by the ants in the nests. Work started at 7:00 a.m. and finished at the same time the next day.

FIGURE 1. Valencia Late orange trees affected by *Atta insularis*.

FIGURE 2. Close-up of Valencia Late orange shoot, showing a severe attack of *Atta insularis*.

FIGURE 3. Mound of *Atta insularis* under a Valencia late orange tree.

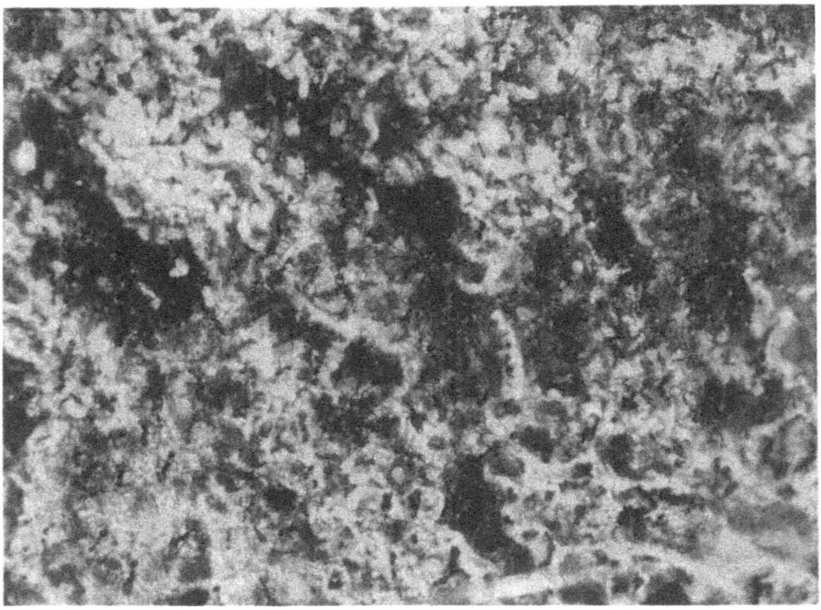

FIGURE 4. Fungal gardens grown by *Atta insularis* inside the colonies.

Trial data were statistically processed to determine the activity percentages of foraging ants in the two main Cuban seasons: drought and rains, for both colonies. The dry period is from October to March and the rainy season from April to September. Crop phenology was determined by visual observation, the beginning and the end of the massive and weak flush as well as flowering and phase of the developing fruits (Phenophases) were recorded.

Temperatures (maximum, minimum and mid °C), relative humidity (%) and rainfall (mm) were also recorded. These data were taken at the trial site during evaluation days and are shown in Fig. 9 along with Valencia orange phenophases.

RESULTS AND DISCUSSION

The workers of nest 1 had the following behaviors during the dry period.

From the total number of ants that came out, 21.5% went back to their nests with plant substrate (young leaves and albedo mainly); 27.1% took out earth from the interior of the openings; 2.6% contributed with other substrates, generally *P. notatum* (dry leaves); while the highest percentage was recorded for those walking loadless (48.9%). In many cases, these loadless ants accompanied those engaged in useful work (Fig. 5).

Fig. 6 shows worker behavior in a single colony during the rainy season. In this case, the ants carrying citrus plant material represented only 12.2% of the whole population.

Ants that took out earth to deepen openings and galleries, as well as those eliminated by rain water runoff, represented 34.5%. Dry weed supply was 4.8%, slightly above the one from the dry period. The percentage of loadless ants was similar to the former one, 48.6%.

In colony 2, different work patterns were obtained, since in the dry period 12.2% of the workers came back loaded with young Valencia late leaves and albedo from dropped fruits; 47.3% took out earth; 7.5% loaded dry leaves of *P. notatum* which were placed around the surface openings trying to cover the entrances a little. It happens more often when minimum temperatures are below 19°C, perhaps with the objective of regulating nest temperature. On the other hand, 33.0% of the ants carried nothing (Fig. 7). Fig. 8 shows *A. insularis* activity in colony 2 during the rainy season. In this colony, 11.0% of the active workers came back loaded with plant substrate; 38.4% took out earth; 7.6% carried *P. notatum* and the highest percentage belonged to those coming back loadless, 43.0%.

From the previous results, one can infer that *A. insularis*' seasonal behavior was different in both colonies, even when the conditions for development and culture were seemingly homogenous. Nevertheless, in the two cases studied, an apparently more favorable behavior for the insect was seen in the dry period, since foraging ants were more stimulated to incorporate green material into their nests during this time: maximum temperatures of 28 to 32°C, minimum of 18 to 21°C and mid of 19.5 to 23°C. Relative humidity ranged between 81 and 71% from January to April. In this period, massive flush, flowering and fruit development took place.

Late in April a low rain occurred (4.0 mm). The second massive

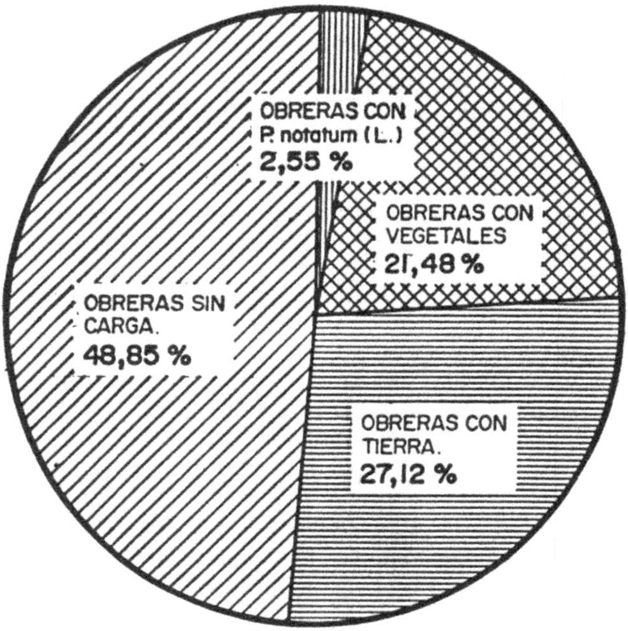

FIGURE 5. Dry season activity pattern of *Atta insularis* (Colony 1-field).

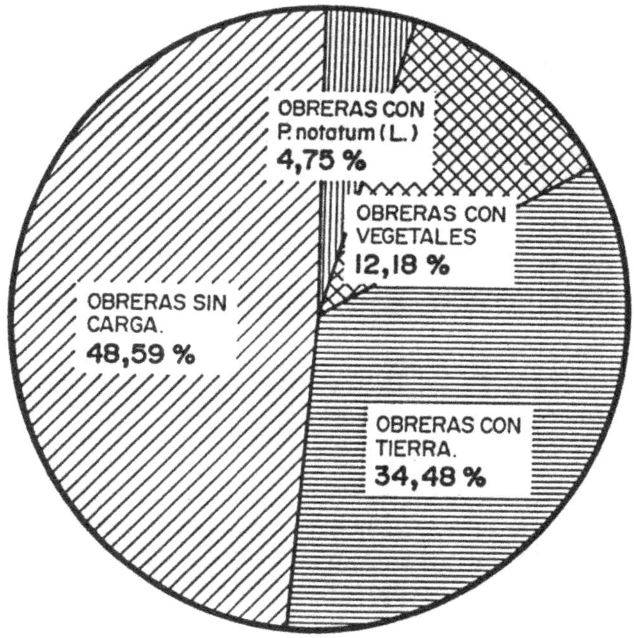

FIGURE 6. Rainy season activity pattern of *Atta insularis* (Colony 1-field).

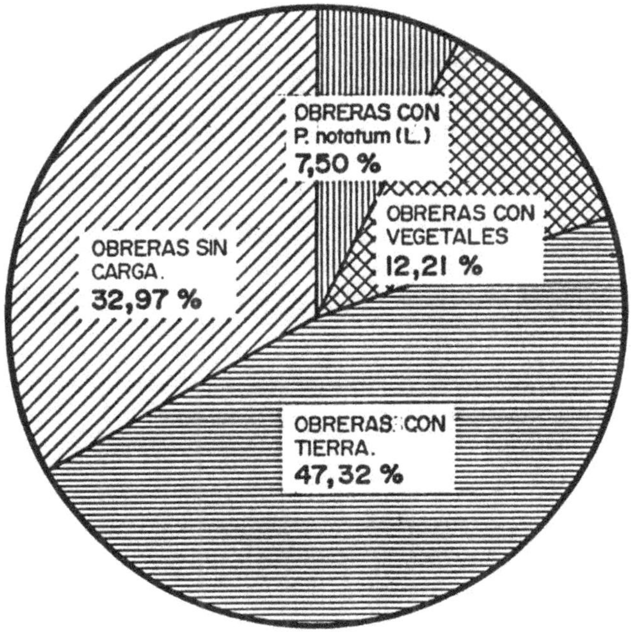

FIGURE 7. Dry season activity pattern of *Atta insularis* (Colony 2-field).

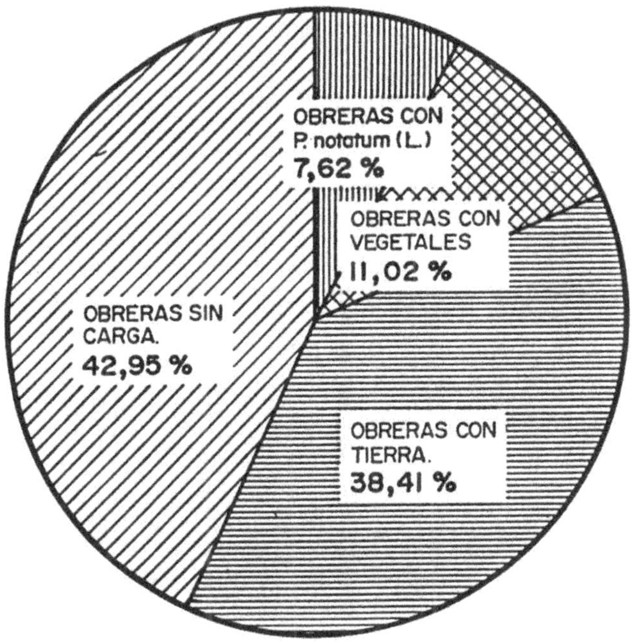

FIGURE 8. Rainy season activity pattren of *Atta insularis* (Colony 2-field).

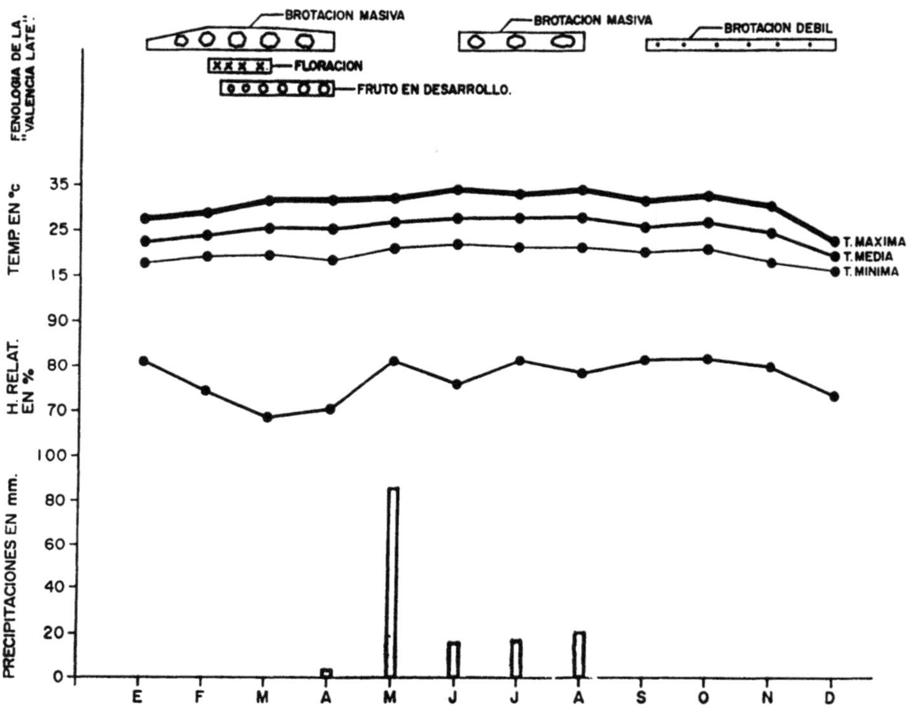

FIGURE 9. Bioclimatic factors present during the evaluation period.

vegetative period came about from June to August, accompanied by maximum temperatures of 34°C, minimum from 21 to 23°C and mid from 27.5 to 28.5°C. Relative humidity was 76.8% and in every month rains negatively influenced the outside activity of the workers, as did the maximum temperature of 34°C. Substrate availability was abundant in this stage, which makes us think that both abiotic factors have a negative influence upon the pest.

Finally, from September to December, a young leaf flush in Valencia orange took place, but it was weak, accompanied by maximum temperatures from 23 to 32°C; minimum from 16 to 20°C and mid from 19.5 to 26°C. Relative humidity was between 73 and 81%. Early in the year, there was no rain. In a situation like this, though the vegetative phase is not massive, it exists and climatic conditions are favorable to stimulate ant foraging activity (Fig. 9). Everything previously discussed allowed us to determine that there must be a perfect harmony among bioecological factors (pest, fungus culture, crop, climatic) for the insect to carry out these activities.

Of the four major activities made by *Atta insularis* workers outside the nests, the highest percentages were devoted to moving without a load and carrying earth. The lowest percentages were recorded for workers carrying *P. notatum* and plant substrate for the nests. Bucher and Montenegro (1974) reported variations in the foraging of *Acromyrmex heyeri* (For), *Acromyrmex lobicornis* (Emery), *Acromyrmex lundi* (Guérin) and *Acromyrmex striatus*

(Roger) during the rainy and dry periods in Argentina. Cherrett (1983) determined that *Atta mexicana* (F. Smith) picks up young leaves from attacked plants.

CONCLUSIONS

(1) Seasonal activity of *A. insularis* was different in both colonies throughout the dry and rainy period.
(2) Foraging ants tended to take more plant material into the colonies during the dry period due to favorable bioecological factors (availability of young shoots, slightly lower temperatures and no rains).
(3) Temperatures of 34°C, accompanied by rain, negatively influenced the ant's activity.
(4) Loadless workers and those taking earth outward had the highest frequency for all factors studied. Those incorporating plant material and *P. notatum* into the nests had the lowest frequencies.

RECOMMENDATIONS

Bearing in mind that the worker caste has the greatest activity in the dry period, poison baits should be applied in this period for them to be quickly taken to the nests. This keeps baits from getting lost due to adverse climatic conditions in the rainy period.

REFERENCES

ALAYO, D.P. 1974. Introducción al estudio de loas Hymenopteros de Cuba. Superfamilia Formicoidea S.B. No. 53, p. 20.

BUCHER, E.H. and R. MONTENEGRO. 1974. Hàbitos forrajeros de cuatro hormigas sinpàtridas del género Acromyrmex (Hymenoptera:Formicidae). Ecología. Asociación Argentina de Biología. 11: 47-53.

BRUNER, S.C. and V. BARRY. 1949. Observaciones sobre la biología de la bibijagua (Hymenoptera:Formicidae). Est. Exp. Agronómica Santiago de las Vegas. 19: 135-154.

CHERRETT, J.M. 1983. Resource conservation by the leaf-cutting ant Atta cephalotes in tropical rain forest. Ecology and Management. (Special Publication) Vol. 2. pp 253-263.

INRA. 1984. Instructivos técnicos para el cultivo de los citricos.

PÉREZ, F. and H. LEÓN. 1977. La bibijagua y su control químico en la Empresa de Cítricos "Victoria de Girón". Estación Experimental de Cítriocos, J. Grande, matanzas, Cuba. CIDA Vol. 2, pp 187-210.

PÉREZ, F. and H. LEÓN. 1981. Resultatos obtenidos en la lucha química contra la "bibijagua" *Atta insularis* Guérin y *Acromyrmex octospinosus* (Reich) (Formicidae, Attini) en la Empress de Cítricos "Victoria de Girón, Matanzas. Memorias 1er. Congreso Nacional de Inv. Cítricos y otros Frutales. Tomo II pp. 173-190.

PÉREZ, F. and N. ARIAS. 1981. Estudios biológicos y lucha contra la "bibijagua" en Cuba. Memorias 1er. Congreso Cacional de Cítricos y otros Frutales. Tomo II, pp. 193-205.

PÉREZ, F. and R. CASAMAYOR. 1979. Las hormigas cortadoras de hoja en Cuba. Frabricación de cebos envenenados para combatirlas. Attini newsletter, No. 6.
PINTERA, A. 1975. Obsevaciones preliminares sobre la biología de *Atta insularis* Guperin. Biblioteca ICIF, La Habana.
PINTERA, A. 1983. Selection of plants utilized by *Atta insularis* in Cuba (Hymenoptera:Formicidae). Acta Entomologica Bohemoslovaca. 80: 13-20.

24

Relative Protection of *Cecropia* Trees Against Leaf-cutting Ants in Tropical America

P. Jolivet

INTRODUCTION

Generally, myrmecophilic *Cecropia* trees are well-protected against leaf-cutting ants by their small inhabitants, the *Azteca* ants. *Atta*, which consume leaves, rarely invade these tree species. Myrmecophobic *Cecropia* are not attacked by leaf-cutters and do not harbor *Azteca*. A toxic substance produced by the *Cecropia* may repel ants, as in ant-less *Acacia*.

I have observed myrmecophilic plants during several stays in the tropics and been astonished to see how efficient the plant-ants were in defending them against defoliators. However, as one entomologist said, "No defense is sacrosanct." The ferocious *Iridomyrmex* of the epiphytic *Myrmecodia* in New Guinea tolerates weevils, leaf-beetles and Lycaenid caterpillars which produce limited damage to the plant's leaves. *Azteca* provide only incomplete protection for *Cecropia* trees in the American tropics since the ants tolerate a broad spectrum of defoliators, from mammals (*Bradypus tridactylus*, howler monkeys, bats) to birds and insects (Lepidoptera, belonging to at least eight families, Hymenoptera, Diptera, Coleoptera, Orthoptera, Hemiptera, Homoptera, Thysanoptera). Friebrig (1988) wrote, "When I compare the forms found in *Cecropia* with the Hexapods occurring on other plants, it seems that the [*Cecropia*] is more thoroughly infested than the majority of the woody plants of the region."

Friebrig's point of view is questionable. Although many insect defoliators are protected by such factors as toxicity, reflex bleeding, enteric discharge, ring defense behavior or even by trophallaxis, these strategies are never present together. For some reason one appears to exclude the others. Many birds, insects, spiders, even *Camponotus* ants, feed with total impunity on Mullerian bodies; but some of these animals can also be useful to the plant (e.g. in seed dispersal, pollination or predation).

In a recent paper (Jolivet 1987), I studied the biocoenosis of *Cecropia* trees. My findings indicated that myrmecophilous *Cecropia* trees suffer fewer attacks than surrounding trees, most attacks occurring in the tree's youth prior to the development of *Azteca* workers. Myrmecophilous *Cecropias* are fairly well-protected against the *Atta*; but surprisingly, so are the myrmecophobic species.

OBSERVATIONS

Wheeler's (1942) statement that "the Attini are not such a menace to the life or well-being of the plants" is untrue. The arguments supporting his contention have been that *Cecropia* leaves are not enticing or juicy, and that the tree might be protected from most of its enemies by the peculiar properties of its sap (Friebrig 1988).

On the contrary, *Cecropia* leaves are very attractive to *Atta* workers and if reported attacks are rare, it is due to the repelling activities of the small *Azteca* workers. When I joined the "Maraca Rainforest Project" last year in early June, I arrived just at the end of such an incident. (Maraca is a fluvial island situated in Brazilian Roraima, south of the Venezuelan frontier.) One *Cecropia paltata* tree, situated near a huge *Atta* nest, had suddenly been attacked by hordes of *Atta* workers; the invasion lasted just three days, the time needed to devour the tree's entire foliage. The *Azteca* ants withdrew within the stems in cowardly fashion but emerged after the raid; the leaves of the *Cecropia* rapidly grew again. The *Atta* were soon "punished." A burrowing animal, probably an armadillo, destroyed their nest in one night.

Since such *Atta* aggression against an "Ant-tree" is rather rare, we must conclude that *Azteca* protection is normally very efficient, at least against leaf-cutters. Baird (1967) also mentions two separate occasions in Central America when *Atta* cut all the leaves of *Cecropia* trees within 24 hours. Encounters between individuals of the attacking and defending species resulted in faster movement along the trails on the plant by the *Atta;* several *Atta* workers were thrown off by *Azteca* workers. However, such attacks by leaf-cutters remain very rare on *Azteca* occupied trees.

One question remains: Why are myrmecophobic *Cecropia* never attacked by *Atta* workers?

DISCUSSION

The total immunity of ant-less *Cecropia* remains a mystery. Among the 106 described *Cecropia* species, at least ten (9.43%) are myrmecophobic for either primary (not built to harbor ants) or secondary (could harbor ants but don't) reasons: *C. palmitisecta, C. auyantepuiana, C. santanderensis, C. kavanayensis* from mountainous areas, *C. hololeuca* from the eastern Brazilian plateau, *C. sciadophylla* from the Amazon, *C. pittieri* from Cocos Island, *C. monostachya* from Ecuador, *C. schreberiana* from the West Indies, while *C. peltata* from Jamaica is without ants and has lost its trichilia and mullerian bodies. Imported *Cecropia* in Singapore lose their trichilia or grow them anarchically. In the case of *C. peltata*, loss of trichilia and progressive reduction of Mullerian bodies is a secondary characteristic linked to the disappearance of *Azteca* and the reduction of selective pressure from defoliators. One can follow the regressive evolution from island to island--from Trinidad to Puerto Rico (Janzen 1973, Rickson 1977). *C. hololeuca*, sympatric with a myrmecophilous species belonging to the *C. adenopus* complex, has neither trichilia nor prostoma and Mullerian bodies and is common in eastern Brazil.

The case of *C. hololeuca* and *C. sciadophylla*, and Amazonian species, is difficult to interpret. *C. hololeuca* is theoretically protected from leaf-cutters by the white arachnoid hairs found on its spathes, stipules and the abaxial side of the lamina. That type of hair seems to repel sloths and some insects, but not all since leaf-beetles and caterpillars feed on the leaves. It should also be noted that the trunk and the petiole are brown, smooth and waxy, and theoretically prevent the *Atta* and other ants from climbing. However, ants do succeed in climbing the smooth stems of *Ricinus communis*.

The protection of the myrmecophobic *Cecropia* against ants must be chemical, similar to that conferred on non-stipulate *Acacia* in Central America by cyanogenic glucosides (Rehr et al. 1973).

To elucidate the mechanism, the following factors should be investigated: (1) Are there latex-bearing ducts in the vegetative parts of the antless *Cecropia*? (2) Are toxic chemicals stored in the leaves of the myrmecophobic *Cecropia* which are not found in the myrmecophilous ones? (A simple test could be made by feeding *Atta* ants with both types of trees.) and (3) Are there substances in antless *Cecropia* which either repulse leaf-cutter ants or have an antibiotic effect on the fungus gardens cultivated by the ants? In this respect, the properties could be compared with the effect on several plants such as *Hymenaea, Canavalia and Astronium.*

In conclusion, much remains to the investigated in the relationships between *Cecropia* trees and ants. Hopefully, that research can proceed rapidly to provide a better understanding of what prevents *Atta* ants from feeding on antless *Cecropia.*

ACKNOWLEDGMENTS

The author thanks the botanist of the Maraca Project, Mr. William Milliken, who first reported the invasion of the *Cecropia* tree. Also our best thanks go to the organizers of the Maraca Rainforest Project, including Dr. John Hemming, leader, Mr. Guttemberg Moreno de Oliveira and Steven Bowles, field leader, who enabled me to do research on Ant Plants in the Project area.

REFERENCES

ANDRADE, J.C., de. 1981. Biologia da *cecropia* lyratiloba Mig. var. nana Andr. & Car. na Restinga do Recreio dos Bandeirantes. Thesis, Rio. 71 pp.

BAIRD, J.N. 1967. Observations on the *Azteca-Cecropia* interaction. M.A. thesis, Univ. Kansas. 73 pp.

BEQUAERT, J. 1921-1922. Ants in their diverse relations to the Plant World. Bull. Am. Mus. Nat. Hist. 45: 501-504.

BERG, C.C. 1978. Cecropiaceae, a new family of the Urticales. Taxon. 27: 39-44.

BERG, C.C. 1978. Especies de *Cecropia* de Amazonia Brasileira Acta Amaz. 8: 149-182.

CARROLL, C.R. 1983. *Azteca*, in Janzen Costa Rican Nat. Hist.: 691-693.

FIEBRIG, K. 1909. *Cecropia peltata* und ihr Verhaltnis zu Azteca alfari, zu *Atta sexdens* un anderen Isekten. Biol. Centralbl. 29: 1-16.

JANZEN, D.H. 1969. Allelopathy by Myrmecophytes. The Ant *Azteca* as a Allelopathic agent of Cecropia. Ecology 50: 147-153.

JANZEN, D.H. 1973. Dissolution of Mutualism between *Cecropia* & its *Azteca* Ants. Biotropica 5: 15-28.

JANZEN, D.H., ed. 1983. Costa Rican Natural History. (Univ. Chicago Press). 815 pp.

JOLIVET, P. 1987. Les fourmis et les Plantes. Un example de coevolution. (Boubee & Cie., Paris). 254 pp., pls.

JOLIVET, P. 1987. Nouvelles observations sur les Plantes a Fourmis. Reflexions sur la Myrmecophilie. L'Entomologiste 43: 39-52.

JOLIVET, P. 1987. Remarques sur la Biovenose des *Cecropia* (Cecropiaceae) Biologie des Coelomera Chevrolet. Bull. Soc. Linn. Lyon 56: 255-276.

JOLIVET, P. 1988. A case of Parasitism of an Ant-*Cecropia* mutualism in Brazil. In: Advances in Myrmecology, J.C. Trager, ed. E.J. Brill Publisher, New York, 551 pp.

REHR, S.S., P.P. FEENY and D.H. JANZEN. 1973. Chemical Defence in Central American Non-Ant Acacias. J. Anim. Ecol. 42: 405-416.

RICKSON, F.R. 1977. Progressive loss of Ant-related traits of *Cecropia peltata* on selected Caribbean Islands. Amer. J. Bot. 64: 585-592.

VELASQUEZ, J. 1972. Contribucion al conocimiento de las especies del genero *Cecropia* L. (Moraceae) "yagrumos" de Venezuela. Acta Botanica Venezuelica 6: 25-64.

WHEELER, W.M. 1942. Studies of Neotropical Ant-Plants and their Ants. Bull. Mus. Comp. Zool. Harvard. 99: 3-362.

25

Invertebrate Enemies and Nest Associates of the Leaf-cutting Ant *Atta texana* (Buckley) (Formicidae, Attini)

D.A. Waller and J.C. Moser

INTRODUCTION

The complex nests of leaf-cutting ants (Formicidae; Attini) provide rich habitats for ant associates. These mounds consist of interconnected subterranean cavities that average 0.002 m^3 in volume (Walter et al. 1938) and extend to depths of 7 m (Moser 1963). Many cavities contain fungus gardens, cultivated on live plant material, frass and debris. Nests of some species include "detritus" chambers filled with discarded fungal garden, exhausted substrate and dead ants. Numerous invertebrate and vertebrate species associate with the ants, their fungus gardens or the detritus (Walter et al. 1938; Moser 1963; Weber 1972; Moser 1983), but little is known about their life cycles or their relationships with the ants. In this paper we review past literature and present observations on *Atta texana* parasites and nest associates in Texas and Louisiana (Table 1).

PARASITES

Leaf-cutting ants have few parasites. Nematodes (Nematoda) have been collected as internal parasites of *Acromyrmex octospinosus* (Weber 1945) and some species are common in deteriorating nests (Weber 1972; DAW). Mites (Acarina) also proliferate in many leaf-cutter colonies and are frequently phoretic on worker ants and alates (Weber 1972; JCM). On nest surfaces, phorid flies (Diptera, Phoridae) are important parasitoids of leaf-cutting ants (Weber 1945; Borgmeier 1926; 1931; 1963 Eible-Eibesfelt and Eible-Eibesfelt 1967; Feener and Moss 1989).

Phorid Flies

Over 20 phorid fly species attack *Atta* and *Acromyrmex* species; parasites and hosts are listed in Borgmeier (1931, 1963), Weber (1972) and Feener and Moss (1989). Females of these minute flies (approximately 1-2 mm body length) oviposit on adult ant workers by inserting their sclerotized ovipositors in the ant head (Feener and Moss, 1989) or thorax (Feener1987). Little is known about the immature stages, but larvae of the phorid,

TABLE 1. Species associated with *Atta texana* and their nests. Specimens were collected in Louisiana (L) or Texas (T).

Order (Family) Species	Reference []	Comments[1]
Isopoda (Platyarthridae)		
Platyarthrus scoblii	[Walter et al. 1938 (T)]	F
Chelonethida (Chernetidae)		
Allochernes n. sp.	[Walter et al. 1938 (T)]	D
Nematoda (nematode)	[JCM (L)]	D
Annelida		
Enchytraeidae sp.	[Moser 1963 (L)]	D
Chilopoda (Chilopoda sp).	[Moser 1963 (L)]	D, G; rare
Diplopoda (Diplopoda sp.)	[Moser 1963 (L)]	D, G; rare
Araneida (Agelenidae)		
Circuina arcuta = *C. robusta*	[Walter et al. 1938 (T)]	D
(Ctenizidae)		
Bothricocyrtum sp.	[Walter et al. 1938 (T)]	D
(Gnaposidae)		
Cesonia bilineata	[Walter et al. 1938 (T)]	F
(Linyphiidae)		
Grammonota pictilis	[Walter et al. 1938 (T)]	D
Masoncus sp.	[JCM (L)]	G; infrequent
(Lycosidae)		
Pirata spiniger	[JCM (L)]	G; infrequent
Acarina (Acaridae)		
Schwiebea n. sp.	[JCM]	D; phoretic on Geomysaprinus
(Tyroglyphid)	[Walter et al. 1938 (T)]	D; phoretic on *Gnathoncus*
Tyrophagus putrescentiae	[Moser 1963]	D; phoretic on Geomysaprinus
(Ascidae)		
Dendrolaelaps sp.	[JCM (L)]	In container with *A. texana* alates after mating flight.
Proctolaelaps n. sp.	[Moser 1967 (L)]	Phoretic on queen
(Digamasellidae)		
Protogamasellus primitivis	[JCM (L)]	D; lab colony
(Macrochelidae)		
Macrocheles spp. 1&2	[Walter et al. 1938 (T)] [Moser & Neff 1971 (L)]	D; phoretic on *Pholeomyia comans*; common
Macrochelidae n. gen. n. sp	[Walter et al. 1938 (T)] [JCM (L)]	D; common
(Tarsononemidae)		
Tarsonemus n. sp.	[Walter et al. 1938 (T)]	D; phoretic on *Gnathoncus*

TABLE 1. Continued.

(Trombidiidae)	
Microtrombidium sp. [Moser 1983 (L)]	F; infrequent
Trombidiid n. gen. n. sp. [Walter et al. 1938 (T)]	F
Typhlotrombidium tridentifer [Walter et al. 1938(T)]	F; common
(Uropodiae)	
Dinychus sp.(=Uraseius)[Moser 1963 (L)&Neff 1971]	D; phoretic on *Pholeomyia comans*,*P*.n.sp.nr. *leucogastra*
Oplitis attaae [Moser 1967] [Hirschmann 1972 (L)]	Phoretic on queen
Trichouropoda (=*Fuscuorpoda*) [Moser 1967 (L)] sp. nr. *marginata*	Phoretic on queen
(Vaigiidae)	
Vaigiidae sp. [JCM (L)]	D; lab colony
Entotrophi (Japygidae)	
Metajapyx n. sp. nr. *subterraneus* [Moser 1963 (L)]	G; infrequent
Thysanura	
Near *Atelura* [Walter et al. 1938 (T)]	F; common, *Grassiella* n. sp.?
Grassiella wheeleri [JCM (L)]	F; infrequent
Grassiella n. sp. [JCM (L)]	F; infrequent
Collembola (Entomobryidae)	
Pseudosinella violenta [Walter et al. 1938 (T)] [Moser 1963 (L)]	D, G; rarely found where ants are abundant; "cave species"
Dictyoptera (Blattidae)	
Arenivaga bolliana [Walter et al. 1938 (T)]	D, G
Arenivaga tonkawa [Walter et al. 1938 (T)]	D, G
Attaphila fungicola [Wheeler 1900, 1907 (T)] [Walter et al. 1938 (T)] [Moser 1964, 1967 (L)]	F; common
Orthoptera (Gryllacrididae)	
Ceuthophilus sp. nr. *umbratilis* [JCM (L)]	G; frequent
(Gryllidae)	
Myrmecophilia nebrascensis [Walter et al. 1938 (T)]	
Myrmecophilia n. sp. [Moser 1963 (L)]	G; infrequent
Hemiptera (*Tominotus unisetosus*) [Moser 1963 (L)] Froeschner (Cydnidae)	D; rare
Coleoptera (Carabidae)	
Tachyura dolosa [Moser 1963 (L)]	D; infrequent
(Histeridae)	
Acritus sp. [Walter et al. 1938 (T)] [Wenzel 1939]	D
Acritus attaphilitus [JCM (L)]	D, G; frequent
(type-Acritus sp. Boerne, TX,	Walter et al. 1938)

TABLE 1. Continued.

Geomysaprinus (=*Saprinus*) sp. [Moser 1963&1983(L)]	D; common
Gnathoncus sp. [Walter et al. 1938 (T)]	D *Geomysaprinus?*
Reninus salvini [Walter et al. 1938 (T)]	D
(Hydrophilidae)	
Oosternum attacomis [Spangler 1962 (L)]	D; frequent in active decomp.
(Ptiliidae)	
Acrotichus sp. or *Nephanes* sp. [JCM (L)]	D; rare
(Staphylinidae)	
Atheta sp. [Walter et al.1938 (T)] [JCM (L)]	D
Belonuchus sp. [JCM (L)]	D; common
Tinotus sp. [JCM (L)]	D; common
Gen. sp. 1 [JCM (L)]	D
Gen. sp. 2 [JCM (L)]	D
Gen. sp. 3 [JCM (L)]	D; common
Gen. sp. 4 [JCM (L)]	D
(Pselaphidae)	
Melba sp. [JCM (L)]	D; rare
Pselaphid n. gen. n. sp. [Walter et al. 1938 (T)]	D
(Elateridae)	
?*Conoderus xystus* [Walter et al. 1938 (T)] [Moser 1963 (L)]	D; infrequent
Elaterid adults and larvae [Walter et al. 1938 (T)]	D
(Scydmaenidae)	
Eumicrus sp. [Walter et al. 1938 (T)]	D; *Scydmaenus*?
Scydmaenus sp. [JCM (L)]	D, F; Infrequent
(Lycidae)	
Lycidae sp. [JCM (L)]	F; infrequent
(Alleculidae)	
Hymenorus discretus [Walter et al. 1938 (T)]	D; *L.opacicollis*?
Lobopoda opacicollis=L. aubruneata [Moser 1963 (L) [Campbell 1966 (L)]	D; common in active decomp.
(Scarabidae)	
Cotinus longitarsis [Walter et al. 1938 (T)]	D; common
Euparixia moseri [Woodruff & Cartwright 1967(L)] [Moser 1983 (L)]	F
Euphoriaspis aestuosa [Walter et al. 1938 (T)]	D
(Chrysomelidae)	
Megalostomis major [Walter et al. 1938 (T)]	G
Lepidoptera (Tineidae)	
Amydria confusella [Walter et al. 1938 (T)]	D
Diptera (Psychodidae)	
Phlebotomus n. sp. [Walter et al. 1938 (T)]	D; F
(Ceratopogonidae)	
Forcipomyia sp. [Walter et al. 1938 (T)]	D
(Bibionidae)	
Bibio femoratus [JCM (L)]	D; rare

TABLE 1. Continued.

(Sciaridae)	
Bradysia fenestralis [JCM (L)]	D; frequent
Sciaria varians [Walter et al. 1938 (T)]	D
(Phoridae)	
Apocephalus wallerae [Disney 1980 (T)]	S; frequent
Myrmosicarius texanus [Greene 1938 (L)] [Disney 1980 (T)]	S; frequent
(Fanniidae)	
Fannia moseri [Chillcott 1965 (L)] [Moser & Neff 1971] [Moser 1983]	D; frequent in active decomp.
Fannia sp. [Walter et al. 1938 (T)]	D
(Borboridae)	
Leptocerca sp. [JCM (L)]	?
(Ephydridae)	
Philygria fuscicornis [Walter et al. 1938 (T)]	D
(Milichiidae)	
Megaselia sp. [Walter et al. 1938 (T)]	D
Milichia sp. [Walter et al. 1938 (T)]	D; *P. texensis*?
Pholeomyia comans [Moser 1963 (L)]	D; common
Pholeomyia sp. nr. *leucogastra* [Moser&Neff 1971(L)]	D; infrequent
Pholeomyia texensis [Sabrosky 1959 (T)]	D; common
Hymenoptera (Chalcidae)	
Pheidoloxenus? sp. [JCM (L)]	S; in *Atta* head
(Cynipidae)	
Cynipidae sp. (2) [JCM (L)]	D; infrequent
(Diapriidae)	
Coptera (=*Psilus*) *pholeomyiae* [Moser&Neff 1971(L)] [Muesebeck 1980]	D;infrequent; pupal parasite of *P. comans*
Galesus n. sp. nr *punctiger* [Walter et al. 1938 (T)]	D; specimen is *C. pholeomyiae*
(Scoliidae)	
Scolia guttata [Walter et al. 1938 (T)]	S; entered nests; *Cotinus* parasite?
(Formicidae)	
Ponera trigona [JCM (L)]	D; rare
Neivamyrmex fallax [JCM (L)]	G; rare
Labidus coecus [JCM (L)]	G; rare
Apocrita	
Apocrita sp. [JCM (L)]	F

[1]D = detritus chamber, F = fungus garden or dormancy chamber, G = galleries and empty chambers, S = mound surface.

Apocephalus pergandie develop within the head capsules of *Camponotus pennsylvanicus* (Pergande 1900).

Two phorid species parasitize *A. texana* foragers. *Myrmosicarius texanus* occurs in Texas and Louisiana (Greene 1938; Borgmeier 1963; Disney 1980);

this species hovers over the ants at entrances and along trails and darts down to land on ant heads. In contrast, *Apocephalus wallerae* has been collected only in Texas and rarely hovers but instead runs up to *A. texana* foragers cutting leaves; females then dash between ant mandibles and appear to oviposit behind the occiput (DAW; Disney 1980). Although neither *M. texanus* nor *A. wallerae* has been reared from *Atta* workers, the sclerotized ovipositors of the females and the persistent oviposition behavior strongly suggests parasitism. In Louisiana, a large worker collected by JCM in 1962 had a phorid larva in one side of the head and a chalcidoid larva (*Pheidoloxenus*? sp.), in the other side.

Temporal and Seasonal Patterns. Both *M. texanus* and *A. wallerae* were common from April through November in Austin, Texas (Table 2). In one week in May 1980, DAW collected 162 *A. wallerae* at one ant entrance. Despite three years of careful searching, these phorids were not observed in the winter even though the ants foraged heavily at that time. Phorids did not appear to be active at night, while the ants foraged nocturnally during the summer and diurnally in the winter, and were sometimes active day and night during the spring and fall. Temporal patterns of ant activity may have been related to phorid abundance and/or weather. Figure 1 illustrates that nocturnal *A. texana* foraging and phorid activity occur during the warmest and wettest months in Austin.

Size Selectivity. Leaf-cutter foragers are continuously polymorphic, and *A. texana* workers on a trail can range in headwidth (HW; the measurement across the width of the head at the widest point) from less than 1 mm to over 3.6 mm (DAW). Because fly size is probably constrained by the dimensions of the ant host, phorids may selectively oviposit on hosts of a given size. In Trinidad, Weber (1972) noted large and small phorids attacking *Atta cephalotes* soldiers and workers, respectively, and large phorids pursued *A. cephalotes* soldiers and ignored workers in Vera Cruz, Mexico (DAW). To determine whether *M. texanus* selectively oviposited on *A. texana* workers according to size, DAW collected ants that had been attacked by *M. texanus* and compared HW with those of control ants (all unattacked ants) taken at the same nest entrance. There were usually too few oviposition attempts in a single sample to demonstrate significance, but when samples were combined for each month at a given entrance, the sizes of attacked ants were generally significantly larger than those of control ants (Table 3). Feener (1987) also found that the phorid *Pseudacteon crawfordi* selectively oviposits on larger workers of *Solenopsis geminata* and *S. xyloni*.

Parasitism Rates and Ant Defense. The phorid parasitism rate appears low. Although *M. texanus* females often hovered for up to an hour at an *A. texana* entrance, usually less than a dozen oviposition attempts were made during that time (DAW). A low rate of parasitism might be related to effective ant defensive behaviors. Ant workers fended off *M. texanus* ovipositions by rearing up on their hind legs with mandibles open and abdomens tucked forward, and *Atta* attacked by *A. wallerae* usually stopped cutting leaves. When flies of both species contacted ants during oviposition attempts, the workers vigorously cleaned their heads and mandibles with the forelegs.

Eible-Eibesfelt and Eible-Eibesfelt (1967) suggested that minim

TABLE 2. Months during which *Atta texana* parasitoids and nest associates were collected in Austin, TX, during 1979-1981.

Species	Month 1	2	3	4	5	6	7	8	9	10	11	12
Blattidae (Dictyoptera)												
A. fungicola				x								
Milichiidae (Diptera)												
P. texensis	x	x	x	x	x	x	x	x	x	x	x	
Phoridae (Diptera)												
A. wallerae				x	x	x	x	x	x	x	x	
M. texanus				x	x	x	x	x	x	x	x	
Minim hitchiking	x				x	x	x	x	x	x	x	x

TABLE 3. Mean (± SD) headwidths (mm) of ants attacked by the phorid, *M. texanus*, and unattacked control ants (sample sizes in parentheses) in Austin, TX.

Entrance	Date (1979)	Attacked ants	Control ants
1	June	1.55 ± 0.21 (23)	1.37 ± 0.33 (16)*
1	July	1.53 ± 0.32 (32)	1.28 ± 0.30 (48)***
2	July	1.45 ± 0.32 (21)	1.23 ± 0.29 (43)**
3	July	1.68 ± 0.35 (8)	1.58 ± 0.39 (52)NS
3	August	1.46 ± 0.33 (6)	1.36 ± 0.31 (55)NS
3	September	1.74 ± 0.35 (53)	1.50 ± 0.48 (189)***
3	October	1.76 ± 0.40 (39)	1.59 ± 0.37 (273)**

*P < .05, **P < .01, ***P < .001, NSnot significant, Student's t-test, 2 tailed.

hitchhiking - in which tiny minim leaf-cutting ant workers ride leaves carried by foragers - deters oviposition by phorids on the foragers. Recently, Feener and Moss (1989) demonstrated that *Atta colombica* foragers carrying hitchhikers on leaves have lower rates of parasitism than undefended ants. Minim hitchhiking probably also serves a nondefensive function, because it reduces the colony's energy expenditure by up to 10% when hitchhikers ride rather than walk to the nest (Feener and Moss, 1989). However, although *M. texanus* and *A. wallerae* were common in Texas, *A. texana* minims rarely rode leaves. When minim hitchhiking did occur, it appeared unrelated to phorid presence (DAW; Table 2).

Social Parasites

The workerless social parasite *Pseudoatta argentina* invades nests of *Acromyrmex lundi* (Weber, 1972), but little is known about this species. Weber (1972) suggested that the host and parasite are closely related.

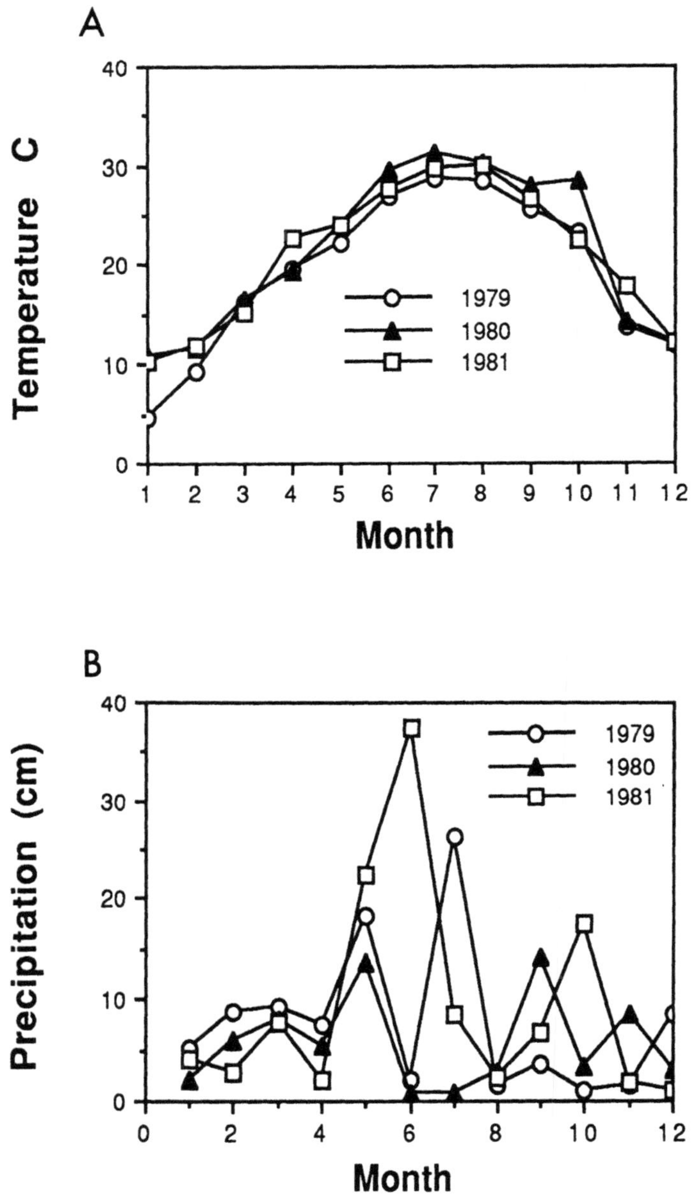

FIGURE 1. Mean average monthly temperature (A) and precipitation (B) in Austin, Texas from 1979-1981.

NEST ASSOCIATES

Moser (1983) briefly summarized the roles of the nest associates and natural enemies of *A. texana*. In general, these associates can be divided into broad habitat categories including fungus gardens, detritus, galleries and empty cavities, and dormant cavities which have characteristics of both the detritus and fungus gardens (Moser 1963). Some associates feed on the detritus (mostly remnants of fungus gardens) or fungi, others seem to be general feeders, and still others are predators or parasites. Internal detritus chambers occur in *A. texana* nests (Moser 1963), but the ants sometimes discard small quantities of garden material on the mound surface (DAW; JCM). In contrast, *Atta colombica tonsipes* and "*A. fervens*" (*A. mexicana*) deposit large amounts of waste outside the nest (Hinton 1935; Weber 1969). The common groups of associates (fungi, nematodes, mites, histerids, staphylinids and fly larvae) found in *A. texana* detritus cavities often bear striking similarities to those found in detritus habitats of other insects such as bark beetle galleries (Moser et al. 1971; Moser and Roton 1971). There are three stages of detritus decomposition: 1) recent (little, if any breakdown); the substrate looks and smells somewhat like cigarette tobacco and few associates are present, 2) active decomposition; maximum numbers of both associate species and individuals, 3) mature; where the substrate has changed to a "black humus," and fauna consists mostly of nematodes and earthworms. The biologies of species commonly found in detritus chambers, fungal gardens/galleries and empty cavities are discussed below:

Thysanura

Grassiella. *Grassiella wheeleri*, when present, seems to be the most common silverfish; JCM collected as many as 22 in galleries of one nest in Louisiana. However, it was not often found (JCM). Wheeler collected *G. wheeleri* from a number of ant species in Texas, including *Pachycondyla harpax* (P. Wygodzinsky, personal communication). *Grassiella* n. sp. was found in fungal gardens from two separate nests in Glenmora and Flatwoods, Louisiana (JCM). This species does not appear to be associated with nest galleries, and so may be ecologically distinct from *G. wheeleri*. Walter et al. (1938) recorded a species "near *Atelura*" (apparently a misspelling of *Attatelura*) common on walls of fungus garden chambers which may be *Grassiella* n. sp. The only other species of silverfish known to occur with *Atta* is *Attatelura geijskesi*, found with *A. sexdens* in Surinam (Wygodzinski 1942).

Dictyoptera

Attaphila. *Attaphila* is a small cockroach (to 3.5 mm long) that lives within leaf-cutting ant fungus gardens and disperses by riding leaf-cutter alates on the mating flight (Bolivar 1901; Moser 1964; DAW) (Fig. 2). Wheeler (1900) considered *Attaphila* to be the only insect that lives in close association with the ants. *Attaphila fungicola* follows the pheromone trails of *A. texana* and *Trachymyrmex* in the laboratory (Moser 1964), and *A. schuppi*. has been found on trails of *Acromyrmex* prob. *niger* (F. Smith) (Bolivar 1905). Within the nest, *A. fungicola* frequently rides upon soldiers (Wheeler 1900), while the related species *A. bergi* from Acromyrmex lundi nests in Argentina and

FIGURE 2. *Attaphila fungicola* phoretic on *A. texana* alate.

Uruguay apparently rides only alates (Bolivar 1901). Wheeler (1900) observed that *Attaphila*'s gut contents are whitish, and he initially suggested that the cockroaches feed on fungus, but he later believed that the roaches ingest ant secretions while riding upon the soldiers (Wheeler 1907). The ants generally ignore *Attaphila* in the nest (Wheeler 1900; DAW).

Both male and female of *A. fungicola* have been collected from Texan *A. texana* nests (Wheeler 1900; DAW), but only females have been collected from Louisiana (Moser 1964). In Louisiana there is one generation per year. JCM observed females and nymphs in the fungus garden cavities in Louisiana as early as March 24, and noted that the ratio of females to nymphs increased until after the first mating flights in late May when few, if any, of the large nymphs were seen. Small nymphs then appeared in early June, and nymphs predominated in nests until the following year. However, JCM has observed *A. fungicola* females in nests in July, September and January, suggesting that females can live over a year, and a female lived for one year in an artificial nest. During mound excavations in Austin, Texas, in June 1983, DAW recovered males (M), females (F) and nymphs (N) from fungus gardens in three separate mature nests. JCM has collected *Attaphila* oothecae in large numbers by capturing *Atta* alates on nest surfaces prior to the mating flights and placing them in jars. Oothecae were deposited at the bottoms of the jars within a few days. Information on the ovarioles and ooethecae of *A. fungicola* can be found in Roth (1968, 1971).

Wheeler (1900) noted that *Attaphila* adults and nymphs have never been collected with intact antennae and suggested that they are clipped by the ants tending their fungus gardens, although it is possible that the cockroaches practice antennal mutilation, as do many termite species (Hewitt et al. 1972).

Orthoptera

Ceuthophilus spp. These crickets are associated with the nest galleries, especially the upper galleries. Others are seen in tunnels, a type of gallery normally 0.3 m deep which may radiate as far as 300 m from nests. They occur in most nests, although not in great numbers, perhaps due to their large size. It is difficult to say how many species are involved because most specimens collected have been nymphs. In 1963, JCM collected females identified as members of a group that includes *C. umbratilis*, and *C. uhleri* complex nymphs. Unfortunately, males are required for a positive identification.

Myrmecophila n. sp. This tiny cricket was collected six times in nests in Louisiana by JCM under the following conditions:

Date	# found	Cavity type	Depth	Comments
8/60	1	Empty F.G.	3 m	Few workers present (Moser 1963)
4/61	1	Full F.G.	30 cm	Many workers, no brood
3/61	1	Gallery	8 cm	Nest dead; 2 *Ceuthophilus* present
3/61	2	Gallery	8 cm	Worker present
2/61	1	Gallery	15 cm	Worker present

It seems likely that *Myrmecophila* merely coexists with the ants and, from the dead nest observation, may be more dependent on the nest than on the ants. Walter et al. (1938) noted that *Atta* workers in Texas become annoyed when groomed by *Myrmecophila*, and the crickets nibbled at the chamber walls and runways encrusted with waxy ant secretions. Walter et al. (1938) listed this associate as *M. nebrascensis*, but A. B. Guerney identified the Louisiana specimens as a new species.

Hemiptera

Tominotus unisetosus. Five adults and several large and small nymphs of this cydnid bug were taken from a cavity containing mature detritus in Louisiana (Moser 1963). This species has only been found in Texas, Mexico, Guatemala and Costa Rica; we know little about its biology (Froeschner 1960). Cydnids feed on roots, but it is unclear how they sustain themselves in *Atta* nests.

Coleoptera

Geomysaprinus sp. Probably a predator, this histerid beetle is commonly found in detritus cavities. Moser (1963) cited it as *Saprinus*, and it may be the *Gnathoncus* sp. in Walter et al. (1938). The Louisiana material (Walter et al. 1938) is deposited at the Field Museum of Natural History.

Geomysaprinus sp. has been seen following foraging trails of *A. texana*

FIGURE 3. Adult histerid *Geomysparinus* n. sp. nr. *formicus* carried from the nest with refuse.

in Louisiana (JCM). Worker ants on the trail ignored the beetle. *Geomysaprinus* sp. followed an artificial trail made from major worker poison sacs dissolved in methylene chloride (JCM), but did not perform as well as the inquiline cockroach *Attaphila fungicola* (Moser and Blum 1963).

An *Atta* worker carried an adult *Geomysaprinus* from the nest (Fig. 3). Workers excavate dirt from mounds in October and November, but this rarely includes detritus and any associates. When a beetle was dropped on a previously removed refuse pile, the beetle immediately revived and ran off.

Staphylinidae. Rove beetles are common in detritus cavities (Moser 1963), but grouping these species produced a taxonomic quagmire. Recently, seven species were resolved (S. Ash, unpublished) from several hundred specimens collected from active detritus cavities in Louisiana (JCM). These include *Belonuchus* sp. (Staphylini), *Tinotus* sp. (Hoplandriini), *Atheta* sp. (Aleocharini) (Fig. 4) and four separate unknown species in the Aleocharini. All of these groups are probably predators and may feed on mites and/or nematodes. Kistner (1982) reviewed the relationships of staphylinid nest associates with their leaf-cutting ant hosts.

Scydmaenus sp. (Fig. 5). A total of eight adults of this small ant-like stone beetle were found in three nests. Two each occupied detritus cavities at

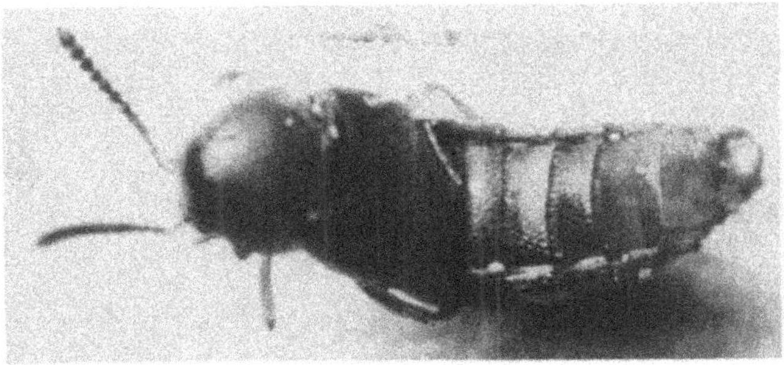

FIGURE 4. Staphylinid adult from detritus cavities.

FIGURE 5. Adult of *Scydmaenus* from detritus cavities.

depths of 2.5 and 2.75 m, one in a dormancy cavity at 3m, and three in a fungus garden at 4 m. All specimens are apparently the same species. The *Eumicrus* sp. listed by Walter et al. (1938) may be this species.

Lycidae sp. (Fig. 6). Small, probably first-instar larvae of this net-winged beetle were found in a central Louisiana detritus cavity (January 18, 1960), and large late-instar larvae were collected once in a dormancy cavity (February 6, 1959), and once in an active fungus garden with ants and brood (March 24, 1960). Arnett (1960) states that the larvae are carnivorous, so this species may eat *Atta* workers and/or their brood.

Lobopoda opacicollis. The known information for this large alleculid beetle is summarized by Campbell (1966). Larvae (Fig. 7) help decompose detritus. The *Hymenorus discretus* listed by Walter et al. (1938) may be this species. The presence of *L. opacicollis* in Mexico, Guatemala, and British Honduras suggests an association with the detritus of *A. cephalotes* and possibly *A.*

FIGURE 6. Lycid larvae from fungus gardens.

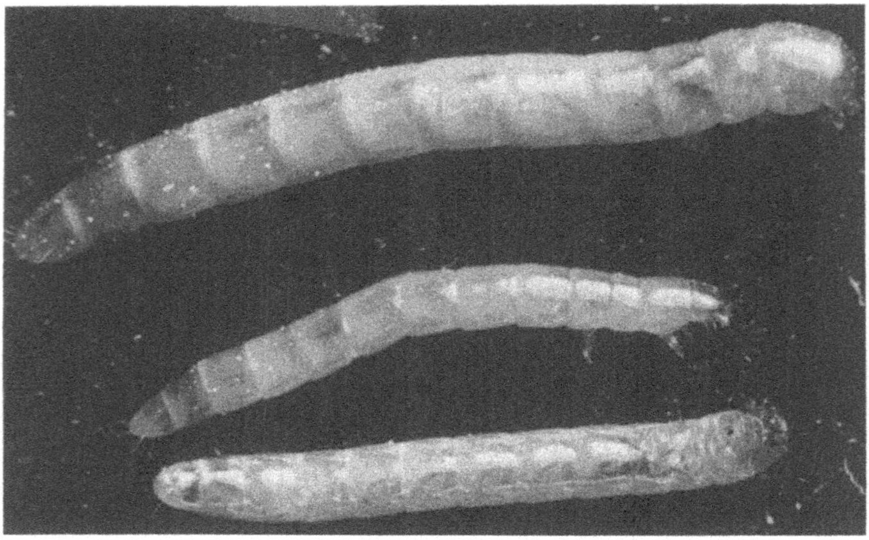

FIGURE 7. Larvae of various instars of *Lobopoda opacicollis* from detritus cavities.

mexicana. Adults of a closely related species, *L. granulata* (Campbell 1966) have been collected near a large *A. cephalotes* nest in Panama (Campbell 1971).

Euparixia moseri. This small scarab commonly inhabits *A. texana* fungus gardens, but is almost never in detritus cavities where "dung" beetles would be expected. Woodruff and Cartwright (1967) give a thorough review of five species that associate with *Atta* nests.

Diptera

Milichiidae. In Louisiana, the milichiid flies *Pholeomyia comans* and *P.* n. sp. nr *leucogastra* inhabit detritus cavities of *A. texana* where the larvae feed on nest wastes (Moser and Neff 1971). *P. comans* is probably the most common insect in the active detritus decomposition when all stages of this fly are present. *P.* n. sp. nr. *leucogastra* has been found only in galleries, but it carries a species of *Macrocheles* mite found in the detritus (JCM). *P. texensis* appears to replace *P. comans* in Texas (Sabrosky 1959; Waller 1980, 1984). Walter et al.'s (1938) report of *Milichia* sp. reared from nest refuse in Texas may be *P. texensis*. Females of *P. texensis* enter *Atta* nests either by hopping onto leaves carried by ants and riding them into the entrances (Waller 1980) or by waiting beside entrance holes and then dashing inside (DAW). Milichiids also rode leaves carried by *A. cephalotes* in Vera Cruz, Mexico (DAW), and Sabrosky (1959) reported that W.L. Brown and E.O. Wilson observed the milichiid *P. decorior* closely following a forager of the attine *Trachymyrmex septentrionalis* in Florida. *Pholeomyia texensis* rode leaves and was collected near *A. texana* entrances throughout the year (Table 2). *P. texensis* males, which have silvery abdomens, frequently occurred near *Atta* entrance holes, with mating observed on one mound in early June 1979 (Waller 1980). Sex ratios of *Pholeomyia* appear equal. Of 196 *P. comans* adults reared from nest detritus, 87 were males (JCM). Waller (1984) found that *P. texensis* females produce an average of 14.1 ± 5.3 SD females and 15.3 ± 7.7 males. Female *P. texensis* frequently had large red mites attached to their abdomens (DAW); Moser and Neff (1971) found three macrochelid mite species phoretic on both *P. comans* and *P.* n. sp. nr. *leucogastra*.

Hymenoptera

Coptera pholeomyia. This small diapriid wasp is an internal pupal parasitoid of *P. comans* (Moser and Neff 1971). The type series consists of females and a male collected in Louisiana by JCM, and a female collected in San Antonio, Texas, by Walter et al. (1938) who listed it as *Galesus* n. sp. nr. *punctiger* and noted that adults escaped by crawling into refuse rather than by flight.

Apocrita sp. This hymenopterous larva was found lying on a fungus garden containing a large number of workers and brood on May 15, 1964, in central Louisiana (JCM). B.D. Burks (personal communication) determined that it did not belong to the Formicoidea, Ichneumonoidea, Cynipoidea, Dryinidae, Bethylidae, Chalcidoidea or any of the Aculeata groups. Hence he concluded that it probably belonged to the Apocrita. Because many of the Apocrita are parasitic, it may parasitize the ant.

FUTURE RESEARCH

Much remains to be learned about the invertebrate enemies and other nest associates of leaf-cutting ants. Further study must determine the true nature of their relationship with the ants and assess their impact on leaf-cutter colony development and foraging success. Information on the parasite/predator load and insight into complex nest dynamics might also lead to the development of environmentally benign controls for leaf-cutters, since they are serious agricultural and silvicultural pests. Five research areas are particularly intriguing:

1) A comparison of natural enemies and nest associates from different populations of the same leaf-cutter species could provide information on leaf-cutter population dynamics and the speciation processes of their nest associates. For example, species compositions of milichiids and phorids associated with *A. texana* in central Texas and central Louisiana differ (Moser and Neff 1971; Waller 1980; Disney 1980), and the *Attaphila* cockroaches from these two populations have strikingly different sex ratios.

2) *Attaphila* diets should be examined to determine whether these cockroaches compete with the ants for fungus. A study of whether and how these insects escape detection by the host ants could lead to new insights about evolutionary patterns in myrmecophiles. Since the cockroaches seek out alates prior to dispersal, it may be possible to contaminate these insects with regicides and release them into the nest as a control measure.

3) A comparison of the parasite/predator loads of leaf-cutter species with internal and external detritus disposal could contribute to our understanding of differences in nest structure among leaf-cutter species. It could also help develop biological control programs using natural enemies such as nematodes, mites and phorid flies.

4) To consider phorids as potential control agents, we must thoroughly study the flies natural history, including host selection, fly developmental stages and dormancy patterns, and the behavioral and physiological defenses of ant hosts. Assessing the magnitude of phorid impact on leaf-cutter colony size, colony growth rates and foraging patterns will be important in understanding the constraints that limit the foraging success of these major herbivores. A detailed examination of phorid size variation in relation to host size-selectivity, phorid survival and fecundity will help explain the coevolutionary processes between parasites and hosts. Finally, a search for phorid parasitoids that attack leaf-cutter alates could result in a viable control.

5) We need to determine the "key" natural enemy (if any) regulating populations of the ant. For the southern pine beetle, which has a somewhat similar natural enemy complex, **parasites** were once thought to be the most important regulating factor. It is now apparent that **predators** provide important control. In order to develop coherent biological control programs for leaf-cutting ants, the natural enemies within nests (some of which are listed here for the first time) need to be investigated in depth, especially those such as *Microtombidium* that may threaten the queen.

ACKNOWLEDGMENTS

This paper is dedicated to Neal A. Weber for his outstanding work on the biology of attines. DAW's research was partially supported by a University of Texas fellowship and U.T. subvention funds. The use of the University of Texas Brackenridge Field Laboratory facilities and the assistance of its staff are gratefully acknowledged. We thank David H. Kistner for helpful comments on the manuscript, and we are grateful to the following individuals for species identifications and other assistance: James S. Ashe, Edward W. Baker, Thomas Bowman, Bernard D. Burks, Milton Campbell, Oscar L. Cartwright, Pedro Reyes Castillo, James G. Chillcott, R.H.L. Disney, Henry Dybas, Richard C. Froeschner, R.J. Gagen, Willis J. Gertsch, David Grimaldi, Ashley B. Gurney, Werner Hirschmann, T.H. Hubbell, Libbie Hyman, Daniel T. Jennings, David H. Kistner, Gerald W. Krantz, Merton C. Lane, Evert E. Lindquist, Lubomir Masner, Alex Mintzer, William B. Muchmore, Alfred F. Newman, Jr., Barry M. Oconnor, Charles L. Remington, Curtis W. Sabrosky, Robert M. Smiley, Leslie M. Smith, Marion R. Smith, Paul J. Spangler, William C. Stehr, Walter R. Suter, Paul H. Vercammen-Grandjean, Rupert L. Wenzel, W. W. Wirth, Robert E. Woodruff, and Pedro W. Wygodzinsky.

LITERATURE CITED

ARNETT, R.H. 1960. The beetles of the United States. The Catholic University of America Press. Washington, D.C. 1112 pp.

BOLIVAR, I. 1901. Un neuvo ortoptera mirmecofile *Attaphila bergi*. Com. Mus. Nac. Buenos Ayres 1: 331-336.

BOLIVAR, I. 1905. Les blattes myrmecophiles. Mitt. Schweiz. Entomol. Ges. 11: 134-141.

BORGMEIER, T. 1926. Phorideos novos ou pouco conhecidos do Brasil. Bol. Mus. Nac. 2: 39-52.

BORGMEIER, T. 1931. Sobre alguns phorideos que parasitma a sauva e outras formigas cortadeiras (Diptera, Phoridae). Archivos do Instituto Biologico 4: 209-228.

BORGMEIER, T. 1963. Revision of the North American phorid flies. Part 1 (Dipt. Phoridae). Studia Ent. 6: 1-256.

CAMPBELL, J.M. 1966. A revision of the genus *Lobopoda* (Coleoptera: Alleculidae) in North America and the West Indies. Ill. Biol. Monogr. 37, 203 pp.

CAMPBELL, J.M. 1971. A revision of the Alleuculidae (Coleoptera) of the West Indies. Mem. Entomol. Soc. Canada 81: 1-140.

CHILLCOTT, J.G. 1965. New species and stages of Nearctic *Fannia* R.D. (Diptera: Muscidae) associated with nests of Hymenoptera. Canadian Ent. 97: 640-647.

DISNEY, R.H.L. 1980. A new species of *Apocephalus* (Diptera: Phoridae) that parasitises *Atta texana* (Hymenoptera: Formicidae). Zeit. Ang. Zool. 67: 47-50.

EIBL-EIBESFELDT, I. AND E. EIBL-EIBESFELDT. 1967. Das Parasitenabwehren der Minima-Arbeiterinnen der Blattschneider-Ameise (*Atta cephalotes*). Zeit. fur Tierpsychol. 24: 278-281.

FEENER, D.H. Jr. 1987. Size-selective oviposition in *Pseudacteon crawfordi* (Diptera: Phoridae), a parasite of fire ants. Ann. Entomol. Soc. Am. 80: 148-151.

FEENER, D.H. JR. AND K.A.G. MOSS.1989. Defense against parasites by hitchhikers in leaf-cutting ants: A quantitative assessment. Behav. Ecol. and Sociobiol. In Press.

FROESCHNER, R.C. 1960. Cydnidae of the western hemisphere. Proc. U.S. Nat. Mus. 111: 337-680.

GREENE, C.T. 1938. A new genus and two new species of the dipterous family Phoridae. Proc. of the U.S. Mus. 85: 181-185.

HIRSCHMANN, W. 1972. Gangsystematik der Parasitiformes. Teil 103. Von J.C. Moser gesammelte Uropodiden aus Nordamerika und 2 neue *Oplitis* - Arten. Acarologie 17: 28-29, Abbildung 36-37.

HEWITT, P.H., J.A.L. WATSON, J.J.C. NEL and I. SCHOEMAN. 1972. Control of the change from group to pair behaviour by *Hodotermes mossambicus* reproductives. J. Insect Physiol. 18: 143-150.

HINTON, H.E. 1935. New Histeridae from the nests of ants of the genus *Atta* in Mexico (Coleoptera). Entomol. News 46: 50-54.

KISTNER, D.H. 1982. The social insects' bestiary, pp. 1-256. In H. R. Hermann (ed), Social Insects, vol 3. Academic Press, NY. 459 pp.

MOSER, J.C. 1963. Contents and structure of *Atta texana* nest in summer. Ann. Entomol Soc. Amer. 56: 286-291.

MOSER, J.C. 1964. Inquiline roach responds to trail-marking substance of leaf-cutting ants. Science 143: 148-149.

MOSER, J.C. 1967. Mating activities of *Atta texana* (Hymenoptera, Formicidae). Insect. Sociaux 14: 295-312.

MOSER, J.C. 1983. Town ant. In History, status and future needs for entomology research in southern forests. Proc. 10th Anniv. E. Texas Forest Entomol. Seminar. Oct. 6-7, 1983. Kurth Lake, Texas.

MOSER, J.C. and M. BLUM. 1963. Trail-marking substance of the Texas leaf-cutting ant; source and potency. Science 140: 1228.

MOSER, J.C. and S.E. NEFF. 1971. *Pholeomyia comans* (Diptera: Milichiidae) an associate of *Atta texana*: Larval anatomy and notes on biology. Zeit. Ang. Entomol. 69: 343-348.

MOSER, J.C. and L.M. ROTON. 1971. Mites associated with southern pine bark beetles in Allen Parish, Louisiana. Can. Entomol. 103: 1775-1798.

MOSER, J.D., R.C. THATCHER, and L.S. PICKARD. 1971. Relative abundance of southern pine beetle associates in east Texas. Ann. Entomol. Soc. Amer. 64: 72-77.

MUESEBECK, C.F.W. 1980. Nearactic parasitic wasps of genera *Psilus* Panzer and *Coptera* Say. USDA Tech. Bul. 1617. 71pp.

PERGANDE, T. 1900. The ant-decapitating fly. Proc. Entomol. Soc. Wash. 4: 497-502.

ROTH, L.M. 1968. Ovarioles of the Blattaria. Ann. Entomol. Soc. Amer. 61: 132-140.

ROTH, L.M. 1971. Additions to the oothecae, uricose glands, ovarioles, and tergal glands of Blatteria. Ann. Entomol. Soc. Amer. 64: 127-141.

SABROSKY, C.W. 1959. A revision of the genus *Pholeomyia* in North America (Diptera, Milichiidae). Ann. Entomol. Soc. Amer. 52: 316-331.

SPANGLER, P.J. 1962. A new species of the genus *Oosternum* and a key to the U.S. species (Coleoptera: Hydrophilidae). Proc. Biol. Soc. Washington 75: 97-100.

WALLER, D.A. 1980. Leaf-cutting ants and leaf-riding flies. Ecol. Entomol. 5: 305-306.

WALLER, D.A. 1984. *Pholeomyia texensis* Sabrosky (Diptera: Milichiidae): Laboratory notes on larval biology. Southwest. Nat. 29:356-357.

WALTER, E.V., L. SEATON and A.A. MATHEWSON. 1938. The Texas leaf-cutting ant and its control. USDA Circular No. 494: 1-18.

WEBER, N.A. 1945. The biology of fungus-growing ants. Part VIII. The Trinidad, B.W.I. species. Revista entomol. 16: 1-88.

WEBER, N.A.1969. Ecological relations of three *Atta* species in Panama. Ecology 50: 141-147.

WEBER, N.A. 1972. Gardening Ants, The Attines. Memoirs of the American Philosophical Society, Vol. 92, Philadelphia. 146pp.

WENZEL, R. L. 1939. A new genus and several new species of North American Histeridae. Ann. Entomol. Soc. Am. 32: 384-394.

WHEELER, W.M. 1900. A new myrmecophile from the mushroom gardens of the Texan leaf-cutting ant. Am. Nat. 34: 851-862.

WHEELER, W.M. 1907. Ants. Columbia University Press, New York. 663 pp.

WOODRUFF, R.E. and O.L. CARTWRIGHT. 1967. A review of the genus *Euparixia* with description of a new species from nests of leaf-cutting ants in Louisiana (Coleoptera: Scarabaeidae). Proc. US Nat. Mus. 123: 1-21.

WYGODZINSKY, P. 1942. Um novo genero e una nova especie de lepismatideo mirmecofilo (Thysanura). Rev. entomol. 13: 49-54.

26

Biology of Carpenter Ants

L.D. Hansen and R.D. Akre

INTRODUCTION

Componotus is the largest ant genus with about one thousand species worldwide (Brown 1973). However, only members in the subgenera *Componotus*, *Myrmentoma*, *Tanaemyrmex*, and *Myrmothrix* are pests of structures and other wood products. The carpenter ant's role as a structural pest in areas of the world other than North America is poorly documented. Indeed, only two European papers detail structural damage by carpenter ants (Butovitsch 1976; Benois 1972). Similarly, only in the last 10 years have carpenter ants been recognized as important North American structural pests. This has been mainly due to Creighton's (1950) misleading statements, which indicated that attacks by carpenter ants occur only on wood in various stages of decay. Information on the biology of *Componotus* is sparse; however, available data indicate that species causing structural damage are biologically similar. This chapter covers only the biology of North American and European wood-damaging species.

MATING FLIGHTS

In spring, alates produced the preceding summer and overwintered in the nest of a mature colony, leave to mate (Pricer 1908; Sanders 1964; Hölldobler and Maschwitz 1965; Benois 1972; Hansen and Akre 1985). A major swarm oçcurs between mid-April to mid-June for *C. modoc* Wheeler, *C. vicinus* Mayr (Hansen and Akre 1985), and *C. herculeanus* L. (Sanders 1964, 1970; Hölldobler and Maschwitz 1965). Mating flights of smaller numbers occur through June in most locations. The stimulus to fly is a function of complex environmental factors including photoperiod, temperature, humidity, and the number of hours of bright sunlight. Swarming occurs in the late afternoon of warm, sunny, humid days. The flight of both sexes is synchronized by a pheromone secreted from the male's mandibular glands as recorded for *C. herculeanus* and *C. ligniperda* Latreille (Hölldobler and Maschwitz 1965). This pheromone is not species-specific which accounts for more than one species swarming on the same day.

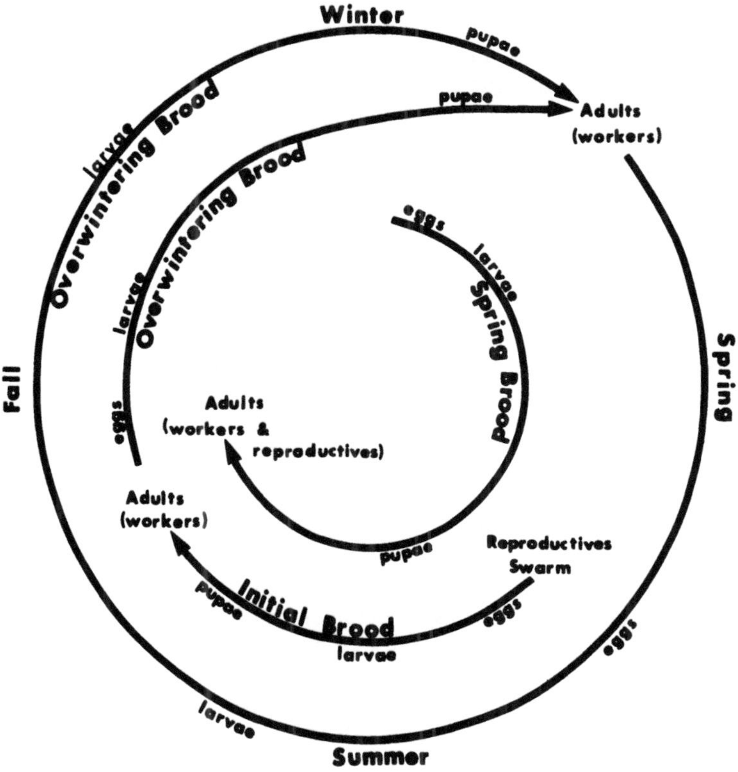

FIGURE 1. Life cycle of *Componotus*. The initiation of a colony occurs after the mating of alates that were produced the preceding summer. Initial brood develops into workers, and a second brood is produced which overwinters as larvae and completes development in early spring. The annual cycle of an established colony includes a spring brood which develops into workers and alates, and a second brood which overwinters as larvae and completes development in early spring.

INITIATION OF NEW COLONIES

Componotus queens found colonies singly and claustrally (Pricer 1908; Benois 1972; Hansen and Akre 1985). Following mating, males die and females seek a sheltered nest site. Eggs are laid in a clutch of 9 to 16 over a period of several days (Mintzer 1979; Hansen and Akre 1985). The eggs are stacked in clusters and tend to adhere to one another. If the queen is disturbed the entire cluster is moved to a more secure location. After two to three weeks, eggs hatch into legless, soft white larvae. The larvae are densely covered with hairs, the tips of which are curved into single sharp hooks. When larvae are stacked together, the hooked hairs interlock and clusters of larvae can easily be maneuvered by the queen or by workers. The queen obtains nourishment for herself and her first larvae by the dissolution and metabolism of her fat bodies and flight muscles. The larvae progress through several instars in the next two or three weeks until cocoons are spun and the larvae become prepupae. The prepupal-pupal period lasts another two to three weeks. Emerging adults cannot free themselves from the cocoons and must be assisted by the queen. In older colonies, workers assist newly-emerged adults from their cocoons. Workers in the first brood are minor and show little size variation (Mintzer 1979; Hansen and Akre 1985). One brood is reared the first summer with an average population of 4 to 25 workers (Pricer 1908; Mintzer 1979; Hansen and Akre 1985). Colony growth with a founding queen is slow. When the initial brood develops to the prepupal stage in late July and August, oviposition of the second brood commences. These eggs hatch and larvae develop before entering dormancy for the winter. Minor workers produced in the initial brood assist the queen in brood care, maintain and enlarge the nest site, and forage for food.

In the fall, colony activities decrease and all colony segments enter a dormant phase to overwinter. The queen resumes egg production in late winter or early spring, after larvae have broken diapause and developed to the prepupal stage (Fig. 1).

ESTABLISHED COLONIES

In colonies one year old or older, two periods of oviposition and larval development occur. The spring brood averages 55 to 70 days to develop from egg to worker. The second larval or summer brood enters diapause in the fall and only resumes development the following spring (Pricer 1908; Hölldobler 1961; Benois 1972; Mintzer 1979; Hansen and Akre 1985) (Fig. 1).

Componotus workers are polymorphic. Food supply determines worker size, with an optimum food supply producing larger forms with greater size variation (Smith 1942). Media and major workers are not produced until the third season in *C. pennsylvanicus*, *C. ferrugineus* (F.) (Pricer 1908), and *C. modoc* (Hansen and Akre 1985).

Mature colonies exist in partitioned populations with a parent colony and a number of satellite colonies (Sanders 1964; Kloft et al. 1965; Hansen and Akre 1985). Colonies maintain contact with one another by movement of workers and brood between the parent and satellite colonies along inter-connecting pathways which the ants build on and through the forest floor (Sanders 1964; Hansen and Akre 1985). The queen, some workers, eggs, and early instars are always located in the parent colony, established in an area of high humidity. Satellite colonies are located in drier, warmer areas--often in galleries excavated from sound wood. These colonies are composed of workers, mature larvae, pupae, and winged reproductives. The satellite colonies remain in place over the winter and contact among the partitioned segments of the colony is re-established in the spring.

The number of satellite colonies per parent colony varies. Sanders (1964) found up to eight for *C. herculeanus* and *C. noveboracensis* (Fitch) encompassing areas up to 830 m^2. Kloft et al. (1965) found a *C. herculeanus* colony living in 13 sites in an area of 130 m^2. On a number of occasions, we observed single colonies of *C. modoc* and *C. vicinus* inhabiting five sites. Ant numbers in *C. modoc* satellite colonies ranged from 200 to over 12,000 workers with larval and pupal populations up to 1,500. The large number of empty pupal cases (60,000+) recovered at one location indicates that satellite empty pupal cases (60,000+) recovered at one location indicates that satellite colony nesting sites are perennial (Hansen and Akre 1985).

Colony size has not been studied extensively and estimates vary. Sanders (1970) found colonies with worker populations of 12,240 for *C. herculeanus* and 8,900 for *C. noveboracensis*. Pricer (1908) reported *C. pennsylvanicus* colonies with 3,018 workers and *C. ferrugineus* with 2,631 workers. Size estimates for European colonies are not available, although Hölldobler and Maschwitz (1965) counted 6,300 winged reproductives of *C. herculeanus* leaving a West German nest in a medium size swarm during one hour. To produce this number of alates, the colony must have had an estimated 50,000 workers. In Spokane, Washington, 50,000 *C. vicinus* workers were counted, an estimated 50% of the total colony. From a small colony of *C. modoc* we recorded over 9,000 workers with no reproductives. Also, in a single satellite colony of *C. modoc* 12,000 workers were counted (Hansen and Akre 1985). Worker populations in single colonies of *C. modoc* and *C. vicinus* could easily number 80,000 to 100,000.

Estimates of the age of a colony when reproductives are produced vary. Pricer (1908) estimated the colony must be at least three to six years old or older. Laboratory rearings of inseminated queens and investigations of monogyne colony field nests of *C. modoc* and *C. vicinus* indicate extremely slow colony growth. The time required to produce reproductives probably should be extended to a minimum of ten years. However, in polygynous colonies where growth is accelerated reproductives may be produced sooner.

Colony founding in *Componotus* has generally been accepted as haplometrotic with primary monogyny because of inseminated queens' aggressiveness of inseminated queens toward one another. However, reports of polygyny in field-collected colonies of *C. herculeanus*, *C. ligniperda* (Hölldobler 1962), and *C. vicinus* (Hansen, unpublished) may indicate primary polygyny through pleometrosis or secondary polygyny through haplometrosis. Either of these methods would explain the abnormally large populations found in colonies of several species.

Laboratory tests with polygyne colonies of *C. modoc* (Hansen and Akre 1985), *C. vicinus* (Mintzer 1979), *C. herculeanus*, and *C. ligniperda* (Hölldobler 1962) indicate that if queens are isolated from one another, workers cooperate in brood care and foraging. In some instances, founding queens cooperate in colony initiation and tolerate one another for periods of a few weeks to two years (Mintzer 1979; Hansen and Akre 1985).

Another possible explanation for the large colonies in the West is the adoption of founding queens into satellite colonies or queenless parent colonies. Indeed, Hölldobler (1962) noted the adoption of new queens by queenless colonies in Germany, thereby extending the lifespan of an established colony.

Established colonies readily accept noninseminated queens. These queens may remain in the colony or return to a nest after the nuptial flight. They will produce eggs that develop into males and may be a major source of males for the colony (Hansen and Akre 1985).

TRAILS AND FORAGING

In nonforested areas, carpenter ants follow chemical and physical trails on the ground surface (Akre and Hansen this volume, Fig. 6). However, in forested areas carpenter ants construct extensive tunnel systems through soil litter (Sanders 1964; Hansen and Akre 1985). Tunnels are 1.5 to 3 cm diam and 1 cm to over a meter deep. These underground trails are lined with firmly packed wood chips and soil particles. They often follow tree roots where ants feed on aphid colonies of the genus *Cinara* (Sanders 1964). The total length of tunnels for a single *C. herculeanus* colony measured 185 m (Sanders 1970) and 200 m for *C. modoc* (David and Wood 1980).

A trail pheromone from the hindgut is used for orientation by *C. pennsylvanicus*, *C. americanus* Mayr (Barlin et al. 1976), and *C. modoc* (Hansen and Akre 1985). This pheromone is used in worker orientation between food sources and the nest, as well as to nesting sites for satellite colonies. *C. pennsylvanicus* workers intermittently touch the tips of their gasters on the substrate as they move between the food source and the nest (Barlin et al. 1976). Orientation to trails by *C. modoc* foragers is chiefly by hindgut pheromone during nocturnal hours since workers move at a steady pace with infrequent stops. During daylight hours, foragers orient chiefly to visual cues. Workers run short distances of 10 to 15 cm, then stop briefly before proceeding. If the view from the trail is altered, ants will stop trailing, cluster, and proceed only after some delay (Hansen and Akre 1985). In field tests, *C. modoc* oriented to landmarks near the ground but not to overhead cues such as tree tops or to polarized light (David and Wood 1980). Studies indicate carpenter ant trails are perennial; for the past nine years we have observed trails at identical locations.

Observations made on carpenter ant daily foraging patterns indicate the highest activity periods occur at night. A dramatic increase in numbers of foragers leaving *C. modoc* nests occurs at sunset, with peaks in foraging between 8 p.m. and 4 a.m. (Hansen and Akre 1985). In California, *C. modoc* foragers were observed on the trail all day from May to October, but in July and August were almost entirely nocturnal (David and Wood 1980). For *C. pennsylvanicus* in New Jersey, humidity and temperature were important parameters in June and August, whereas time of day was the dominant parameter in July (Fowler and Roberts 1980).

Carpenter ants feed on homopteran honeydew, plant exudates, and animal matter, particularly insects. They also scavenge on the carcasses of dead vertebrates. *Componotus* species are important predators of the jack pine budworm, western spruce budworm, forest tent caterpillar, and blackheaded budworm (Youngs and Campbell 1984).

The importance of trophallaxis as a means of distributing liquid nutrient among workers and from workers to larvae and the queen has long been recognized as a fundamental bond in the social life of ants. Whole food particles are transported to the nest in the mandibles, and liquid food transported in the crop. The gasters of returning foragers are greatly distended due to the large amounts of liquid carried in the crop. The quantity of food stored in the crop is controlled by the proventriculus, which regulates the amount of food that passes into the midgut. In *C. pennsylvanicus*, solid particles over 100 m diam are prevented from entering the crop by the filtering action of the mouth parts and the infrabuccal cavity. Particles from 10 to 100 m are progressively filtered as food is passed from worker to worker (Eisner and Happ 1962).

NESTING SITES

Carpenter ants burrow into wood to make nests. Wood is excavated by the mandibles and stored in nest areas, extruded through openings called windows, or packed into underground tunnels. Workers are constantly engaged in cutting galleries to accommodate the increasing brood in both parent and satellite colonies. Irregular in shape, galleries are generally excavated with the grain and follow softer portions of the wood (See Akre and Hansen this volume, Fig. 2). Some of the harder layers remain as walls separating the galleries. Openings are cut in these walls at intervals to allow movement of ants from one area to another. The nest walls are smooth and have a sand-papered appearance.

In natural areas carpenter ants nest in live or dead trees and in rotting logs and stumps. They enter live trees through cracks, scars, knot holes, and decayed or faulty places (See Akre and Hansen this volume, Fig. 4). Once inside the tree, ants remove decayed wood and extend galleries into adjacent sound wood. Reports that carpenter ants nest in soil and beneath rocks (Creighton 1950) are misleading in that these locations serve only as incubation sites for brood. The actual nest is found in decaying roots, buried timbers, or wood debris beneath the rocks or soil cover; these conditions aid thermoregulation.

Living trees provide most permanent nesting sites, but not all tree species are equally acceptable to the ants. In Europe and New Brunswick, Canada, carpenter ants prefer softwood species of *Abies* and *Picea* (Hölldobler 1962; Sanders 1964). In the Pacific Northwest, living trees of western red cedar and Douglas-fir are most often attacked by *C. modoc*, whereas wood debris in soil serves as a home to *C. vicinus*, and dead standing trees or fallen logs contain *C. herculeanus* (Hansen and Akre 1985). Parent colonies are confined to the base of the trees, with galleries extending into the roots and up the trunk (1 to 1.5 meters). Entrances to many of these sites are underground.

In urban areas, carpenter ant nests were found in 75% of shade trees sampled in New Jersey (Fowler and Parrish 1982). The greatest nesting usage was in silver maples, whereas white pine was the least used. Carpenter ants also nest in wooden structures and in other wood products, with a preference for moist wood that contains decay fungi, although sound wood is also mined.

LITERATURE CITED

BARLIN, M.R., M.S. BLUM, and J.M. BRAND. 1976. Species-specificity studies on the trail pheromone of the carpenter ant, *Camponotus pennsylvanicus* (Hymenoptera: Formicidae). J. Georgia Ent. Soc. 11: 162-164.

BENOIS, A. 1972. Etude ecologique de *Componotus* vagus Scop. (=pubescens Fab.) (Hymenoptera: Formicidae) dans la region d'Antibes: nidificacion et architecture des nids. Insectes Soc. 19: 111-129.

BROWN, W.L. JR. 1973. A comparison of the Hylean and Congo-West African rain forest ant faunas. *In* Meggers, B.J., E.S. Ayensu, and W.D. Duckworth (Eds). Tropical Forest Ecosystems in Africa and South America: A Comparative Review. Smithsonian, Washington, D. C. 350pp.

BUTOVITSCH, V. 1976. Uber Vorkommen und Schadwirkung der Robameisen *Camponotus herculeanus* und *C. ligniperda* in Gebauden in Schweden. Material Organ. 11: 161-170.

CREIGHTON, W.S. 1950. The Ants of North America. Bull. Mus. Comp. Zool. 104: 1-585.

DAVID, C.T. and D.L. WOOD. 1980. Orientation to trails by a carpenter ant, *Camponotus modoc* (Hymenoptera: Formicidae), in a giant sequoia forest. Can. Ent. 112: 993-1000.

EISNER, T. and G.M. HAPP. 1962. The infrabuccal pocket of a formicine ant: a social filtration device. Psyche 59: 47-60.

FOWLER, H.G. and M.D. PARRISH. 1982. Urban shade trees and carpenter ants. J. Arboricult. 8: 281-284.

FOWLER, H.G. and R.B. ROBERTS. 1980. Foraging behavior of the carpenter ant, *Camponotus pennsylvanicus*, (Hymenoptera: Formicidae) in New Jersey. J. Kansas Ent. Soc. 53: 295-304.

HANSEN, L.D. and R.D. AKRE. 1985. Biology of carpenter ants in Washington State (Hymenoptera: Formicidae: *Componotus*). Melanderia 43: 1-61.

HÖLLDOBLER, B. 1961. Temperaturunabhaengige rhythmische Erscheinungen bei Rossameisenkolonien (*Camponotus ligniperda* Latr. und *Camponotus herculeanus* L.). (Hym.Form.). Insectes Soc. 8: 13-22.

HÖLLDOBLER, B. 1962. Zur Frage der Oligogynie bei Camponotus ligniperda und *Camponotus herculeanus*. Zeit. Angew. Ent. 49: 337-352.

HÖLLDOBLER, B. and U. MASCHWITZ. 1965. Der Hochzeitschwarm der Rossameise *Camponotus herculeanus* L. (Hymenoptera: Formicidae). Zeit. Vergl. Physiol. 50: 551-568.

KLOFT, W., B. HÖLLDOBLER, and A. HAISCH. 1965. Traceruntersuchungen zur Abgrenzung von Nestarealen holzzerstorender Rossameisen *Camponotus herculeanus* L. und *C. ligniperda* Latr.). Ent. Exp. Appl. 8: 20-26.

MINTZER, A. 1979. Colony foundation and pleometrosis in *Camponotus* (Hymenoptera:Formicidae). Pan Pac. Ent. 55: 81-89.

PRICER, J.L. 1908. The life history of the carpenter ant. Biol. Bull. 14: 177-218.

SANDERS, C.J. 1964. The biology of carpenter ants in New Brunswick. Can. Ent. 96: 894-909.

SANDERS, C.J. 1970. The distribution of carpenter ant colonies in the spruce-fir forests of northwestern Ontario. Ecology 51: 865-873.

SMITH, F. 1942. Effect of reduced food supply upon the stature of *Componotus* ants (Hymen.:Formicidae). Ent. News 53: 133-135.

YOUNGS, L.C. and R.W. CAMPBELL. 1984. Ants preying on pupae of the western spruce budworm, *Choristoneura occidentalis* (Lepidoptera: Tortricidae), in eastern Oregon and western Montana. Can. Ent. 116: 1665-1669.

27

The Little Fire Ant *Wasmannia auropunctata* (R.) (Hymenoptera: Formicidae)

P. Ulloa-Chacon and D. Cherix

INTRODUCTION

The little fire ant, *Wasmannia auropunctata*, is a typical tramp species which has yet to be studied intensively. This paper summarizes the few research contributions to the biology, ecology and economic importance of this tiny ant, and some preliminary results obtained in our research program concerning reproductive biology of this polygyne species.

HISTORY

The little fire ant, which belongs to the subfamily Myrmicinae, is native to the American tropics and was described for the first time by Roger (1863) who, based on samples from the island of Cuba, named it *Tetramorium auropunctatum*. Forel (1884) found that the females exhibited certain morphological characters different from the genus *Tetramorium*. This led him to suggest that this species should be classified within the genus *Ochetomyrmex* (Mayr) (Forel 1886). Nevertheless, Forel (1887), placed it in the new subgenus *Xiphomyrmex (Tetramorium)*. Finally, Forel (1893), established the new genus, *Wasmannia*, comprised of *W. auropunctata* (Roger) and *W. sigmoidea* (Mayr).

Since Forel's time, *W. auropunctata* has been reported in Latin America and in the Caribbean Islands. According to Kempf (1972) six varieties have been described: *rugosa, australis, laevifrons, nigricans, obscura and brevispinosa.*

MORPHOLOGY

Within the genus *Wasmannia*, workers are monomorphic. They have an 11-segmented antenna with a 3-segmented club; the last segment is very large and longer than the two preceding segments combined. The clypeus is arched both longitudinally and transversely.

W. auropunctata has a long antennal scape reaching the occiput; it also has an antennal scrobe. The mandibles have 5 teeth, its maxillary palp are 3-segmented and the labial palp is 2-segmented. The large head is much broader than the thorax. The prothorax has angular humeri; and it lacks a promesonotal suture. The petiolar node in profile is subrectangular and higher than the postpetiole; the epinotal spines are prominent. The head and thorax are sculptured with large longitudinal and transversal rugules, interspersed with fine reticulations. Body hairs are long, erect and sparse. The ant is golden brown to brown, and reaches 1.2-1.5 mm in size. The sting is well developed.

The queen is very large (length 4.5-5.0 mm) compared to workers. The antenna, are like the worker's and the antennal scrobe, is well marked. The postpetiole node is much shorter than in the worker. The gaster is smooth and shiny. The first gastric tergite nearly covers the whole gaster in dorsal view; sculpture, pilosity and pubescence are as in the worker. The forewing has no closed discoidal cell. The queen is dark brown in color.

The male is about as long (4.2-4.5 mm) as the female, but with a more slender body. The antenna is 13-segmented. The gaster is elongate, with very conspicuous genitalia in which the external valves are curved and long, nearly two-fifths the length of the abdomen. Sculpture, pilosity and pubescence differ from those of the worker and queen. Color is more like that of the queen, except that the antenna, legs and external genitalia are yellowish. Forewing venation is as found in the queen.

ECONOMIC IMPORTANCE AND ECOLOGICAL IMPACT

W. auropunctata was accidentally introduced by man into several warm regions of the world. Additionally, in 1907 it was found in nursery plants at Kew Gardens (London) (Donistorphe 1915). Today, the ant is well-established in the United States, particularly in southern Florida (Smith 1929; Wheeler 1929; Spencer 1941), and in California (Nickerson 1983). Large numbers of the ants are also found in the Galapagos Islands (Silberglied 1972) and in New Caledonia (Fabres and Brown 1978). The species can be considered a pest in three different ways.

The Invasion of Cultivated Fields

This first point includes the symbiotic relationship with other pest species, such as Homoptera (aphids, white flies, scales), which excrete honeydew and in turn serve as an essential food source for the ants. At the same time, the ants protect these insects from their natural enemies (parasites and predators). This ant causes an ecological imbalance in phytophagous insect communities in Florida citrus crops (Spencer 1941); Puerto Rican coffee (Smith 1937); Cameroon cocoa (Bruneau de Miré 1969); Columbian coffee and cocoa (Posada et al. 1976); New Caledonian coffee, citric and ornamentals (Fabres and Brown 1978); and Brazilian cocoa (Delabie 1988).

Some emphasis should be placed on the stinging habits of the workers (Spencer 1941). In addition to its powerful sting, the mandibular secretion contains alkylpyrazines, compounds which may have both alarm and defensive functions (Howard et al. 1982).

Infestation of Human Houses

W. auropunctata as a house pest contaminates food; it is attracted to dirty and sweaty clothing and can infest bedrooms (Spencer 1941; Fernald 1947; Smith 1965).

Interspecific Competition with Local Fauna

Interspecific competition gradually eliminates native ant species and terrestrial invertebrates in the newly-colonized areas. In the Galapagos Islands, Clark et al. (1982) and Lubin (1984, 1985), found that the greatest density of other ant species occurred in dry areas which have the lowest density of the little fire ant. They suggest that the little fire ant's success as a colonizer can be attributed to its ability to nest in diverse habitats, its mass recruitment to food and its uninterrupted circadian activity.

Due to its importance as a destructive insect, some methods of crop control--such as destruction of nests, use of sticky strips (Spencer 1941) and chemical control with poisoned baits containing highly toxic products (Fernald 1947; Osburn 1948)--have been tested and implemented.

BIOLOGY AND ECOLOGY

Life History

Little is known about the life history of *W. auropunctata*. The morphology of immature stages and worker brood development was studied by Ulloa-Chacon and Cherix (1988). The egg is elliptical and whitish. It has an average size of 0.22 mm by 0.15 mm. The first instar larva measures 0.27-0.30 mm, and last instar 1.20 mm. The pupa is about 1.04 mm. In laboratory conditions (26°C and 60% relative humidity), the incubation period lasts from 8 to 10 days, the larval stage 16 to 18 days and the pupal stage 11 to 12 days. The entire development from egg to adult worker takes 37 days (min. 35 days, max. 40 days).

Wheeler and Wheeler (1954) described the worker and queen larvae.

Nest and Social Structure

W. auropunctata nest in a wide variety of substrata and in semi-arid to humid habitats (Spencer 1941; Kusnezov 1951). Nevertheless, it prefers high heat-humidity conditions for its foraging activity (Lubin 1985).

The nests, diffuse and inconspicuous, are composed of an aggregate of individuals (workers, queens and brood) (Clark et al. 1982). Several such nests are interconnected by worker trails forming a colony without boundaries. This type of social structure, called "unicolonial," is also common in other ant tramp species like *Monomorium pharaonis*, *Pheidole megacephala* and *Iridomyrmex humilis* (Hölldobler and Wilson 1977).

In *W. auropunctata*, nests are usually established at ground level, close to the surface. They are located around plant roots, under the rough bark of growing trees, in hollow branches or decaying logs, beneath dead leaves or stones, in preformed cavities (from borers), and under leaf bases of palms

(Spencer 1941; Kusnezov 1951; Fabres and Brown 1978; Torres 1984; Ulloa-Chacon and Cherix 1988).

In a Panamanian tropical forest composed of deciduous trees, Levings and Franks (1982) reported a density of 0.05 to 0.13 nests per m^2. Infested areas of the Galapagos Islands showed an increase of density of 0.75 to 1.75 aggregations per m^2 (Lubin, 1985). The highest density occurred in man-made habitats, like cultivated fields and pastures and ornamental plants (Torres 1984). In fruit crops we observed an average of 2.7±1.2 aggregations per m^2 (N=31 plots each 1 m^2).

The social structure of *W. auropunctata* was studied during the dry season at the Universidad del Valle experiment station in Cali, Colombia (altitude 970 m, average annual temperature: 24°C and 65-70% relative humidity). From 138 samples, we found three major kinds of aggregates within the unicolonial system.

- Queenright aggregates (54.4%), were composed of single or multiple queens with workers and brood (eggs, larvae and pupae).
- Worker-brood aggregates (40.0%), formed by workers and one or more immature stages (eggs only; eggs, larvae and pupae; pupae only), representing 40.0%.
- Worker aggregates (6%), formed by workers only (mainly young workers).

The number of queens per aggregate ranged from 1 to 16 with an average of 1.7±3.3 (N=44). The mean number of workers per queen ranged between 548.8±384.7 (N=48) in aggregates with a single queen to 259.3±192.3 (N=96) in aggregates with multiple queens. The mean number of workers per larva decreased significantly from 7.9±6.4 in aggregates with one or more queens to 3.9±3.6 in worker-brood aggregates ($t=3.06$, $p<0.01$, df59). Seasonal changes in composition and density of nests are under study.

Apparently new nests are formed by budding, since nuptial flights have never been observed (Spencer 1941; Lubin 1984).

Feeding Habits

The little fire ant is a polyphagous and opportunistic species. It feeds preferably on honeydew from sucking insects (Spencer 1941; Smith 1937; Clark et al. 1982), a great variety of prey (Arthropods, Gasteropods and Annelides), and vegetable material (seeds, flowers and leaves)(Clark et al. 1982; Torres 1984). They also eat dead insects (Spencer 1941; Kusnezov 1951; Torres 1984). In houses, the ants are attracted to fatty or oily material (Fernald 1947).

Workers are highly predaceous, and have been used as a predator of cocoa bugs (Bruneau de Miré 1969). They have also been observed taking floral and extra-floral nectar from plants (Lubin 1984; Meier 1985). Its importance as a factor in the evolution of plant-insect relationships has been studied by Schemske (1980) and Horvitz and Schemske (1984).

Reproductive Biology of Queens

Only *W. auropunctata* queens lay eggs; as workers are sterile (Ulloa-Chacon and Cherix 1988). There are, however, considerable differences in the oviposition rate of queens. These differences occur both

TABLE 1. Estimation of the influence of queen age on egg production in laboratory polygynous colonies (N=12). Total life span: 13.0 ± 1.8 months.

Age of queens (months)	Number of eggs/queen/day Mean	±	S. D.	t-test DF: 11 P< 0.001
2	6.0	±	1.1	
				8.2
5	12.0	±	2.5	
				10.5
10	4.7	±	1.4	

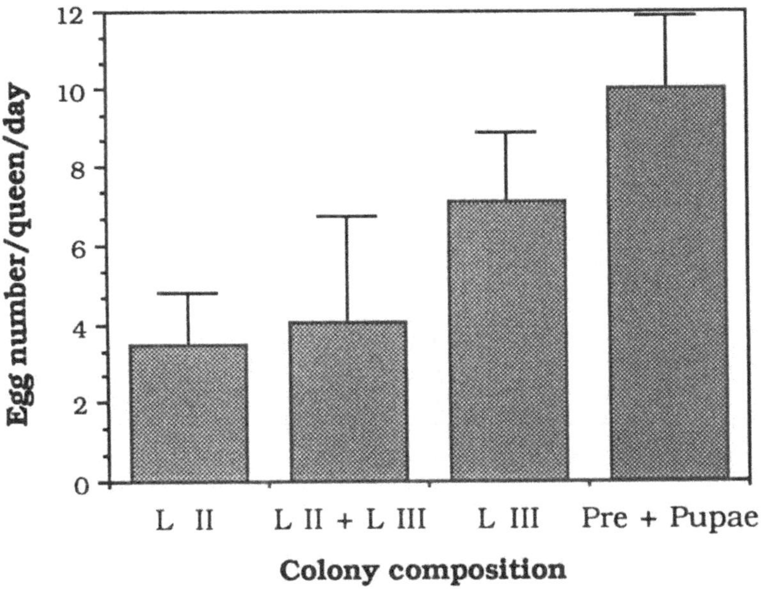

FIGURE 1. Influence of brood composition on egg production by *W. auropunctata* queens (L II = 2nd larval instar; L III = 3rd larval instar; Pre = prepupae).

among queens from the same population, and queens reared under identical laboratory conditions. Thus the total number of eggs laid per queen in twelve weeks ranged from 305 to 942 with an average of 599±193 (Ulloa-Chacon and Cherix 1988). This variability in queen fecundity may be due to genetic, individual and/or social factors. We found that queen age was an important influence on queen fecundity (Ulloa-Chacon and Cherix In Press). In polygyne laboratory colonies, egg production changes significantly during the queens' life-time (Table 1).

Queen fecundity is also influenced by colony composition (worker brood) (Ulloa-Chacon and Cherix In Press). Comparison between experimental monogyne colonies with worker brood (larvae only and pupae

TABLE 2. Brood production in monogynous and polygynous laboratory colonies containing queens of different fecundity rates. Colonies M= 1 queen; colonies T= 3 queens; colonies S= 6 queens and colonies N= 9 queens.

Colony No.	Average oviposition rate eggs/queen	eggs/colony (range)	Brood production during 90 days workers	sexuals
M-1	5.0	(1-5)	+	++
M-2	7.0	(6-10)	+	+
M-3	9.0	(6-10)	+	++
M-4	12.0	(11-15)	++	+a
M-5	14.0	(11-15)	+	+a
M-6	23.0	(16-25)	++	-
M-7	41.0	(> 25)	+++	-
T-1	2.3	(1-5)	+	++
T-2	7.7	(6-10)	+	+
T-3	13.7	(11-15)	++	+a
T-4	13.3	(11-15)	++	-
T-5	14.3	(11-15)	+	+a
T-6	16.0	(16-25)	++	-
T-7	21.3	(16-25)	++	-
S-1	3.2	(1-5)	+	++
S-2	11.7	(11-15)	++	-
S-3	14.3	(11-15)	++	-
S-4	22.0	(16-25)	+++	-
N-1	2.6	(1-5)	+	++
N-2	3.1	(1-5)	-	++

[a] Sexual larvae but no pupae were produced.

only) and broodless colonies revealed the following differences: the oviposition rate of queens was similar in broodless colonies and in colonies containing pupae only, whereas the oviposition rate was significantly lower in colonies with a majority of second instar larvae (Fig.1). This difference could result from food competition between queens and larvae, larval inhibition and/or oophagy.

Production of Sexuals

Under natural conditions, sexual forms are frequently reared in large worker-brood aggregates with a high worker/larva ratio (Ulloa-Chacon unpublished). Nevertheless, under certain laboratory conditions it was observed that sexual brood is produced both in queenless and queenright colonies. In experimental nests containing a single or several queens of different fecundity levels (established by counting the number of eggs per queen produced in 24 hours), gynes are reared only in colonies with a reduced average queen fecundity (Table 2). So, all colonies in which the average oviposition rate ranged between 1 and 10 eggs produced sexual brood. By contrast, all colonies in which the average oviposition rate was higher, 11-15;

16-25 or more than 25 eggs per colony, only workers are produced. In several colonies with an oviposition rate of 11-15 eggs, the sexual larvae produced did not survive to the pupal stage.

These preliminary observations suggest that the inhibition of alate production by "little fire ant" queens, as for the Pharaoh's ant (Edwards 1987), is dependent on queen fecundity. Extensive laboratory and field studies are required to determine what factors affect the production of sexuals in this species.

ACKNOWLEDGMENTS

We thank Dr. C. Baroni Urbani (Institute of Zoology, Basel), L. Keller (Museum of Zoology, Lausanne) and Dr. E. Vargo, University of Texas, for their critical comments on the manuscript, as well as, M. L. Baena and Hugo Ortiz, students of the Universidad del Valle (Cali, Colombia), for their valuable assistance in the field. P. Ulloa-Chacón is supported by a grant from the Universidad del Valle.

REFERENCES CITED

BRUNEAU de MIRÉ, P. 1969. Une fourmi utilisée au Cameroun dans la lutte contre les mirides du cacaoyer *Wasmannia auropunctata* (Roger). Café, Cacao, Thé 13: 209-212.

CLARK, D. B., C. GUAYASAMÍN, O. PAZMIÑO, C. DONOSO, and Y. PÁEZ DE VILLACÍS. 1982. The tramp ant *Wasmannia auropunctata* Autecology and effects on ant diversity and distribution on Santa Cruz Island, Galapagos. Biotropica 14: 196-207.

DELABIE, J. H. C. 1988. Ocorrência de *Wasmannia auropunctata* (Roger, 1863) (Hymenoptera, Formicidae, Myrmicinae) em cacauais na Bahia, Brasil. Rev. Theobroma 18: 29-37.

DONISTORPHE, J. K. 1915. British Ants Their life-history and classification. Plymouth William Brendon and Son. Limited. 379 pp.

EDWARDS, J. P. 1987. Caste regulation in the pharaoh's ant *Monomorium pharaonis* the influence of queens on the production of new sexual forms. Physiol. Entomol. 12: 31-39.

FABRES, G., and Jr. W. L. BROWN. 1978. The recent introduction of the pest ant *Wasmannia auropunctata* into New Caledonia. J. Aust. Entomol. Soc. 17: 139-142.

FERNALD, H. T. 1947. The little fire ant as a house pest. J. Econ. Entomol. 40: 428.

FOREL, A. 1884. Etudes Myrmécologiques en 1884, avec une description des organes sensoriels des antennes. Bull. Soc. Vaud. Sc. Nat. 20: 316-380.

FOREL, A. 1886. Espèces nouvelles de fourmis américaines. Ann. Soc. Ent. Belgique. 30: 38-49.

FOREL, A. 1887. Fourmis recoltées à Madagascar par le Dr. Conrad Keller. Mitt. Schweiz. Ent. Ges. 7: 381-389.

FOREL, A. 1893. Formicides de l'Antille St. Vincent, récoltées par M. H. H. Smith. Trans. Ent. Soc. London. 4: 333-418.

HÖLLDOBLER, B., and E. O. WILSON. 1977. The number of queens An important trait in ant evolution. Naturwissenschaften 64: 8-15.

HORVITZ, C. C., and D. W. SCHEMSKE. 1984. Effects of ants and an ant-tended herbivore on seed production of a neotropical herb. Ecology 65: 1369-1378.

HOWARD, D. F., M. S. BLUM, T. H. JONES and M. D. TOMALSKI. 1982. Behavioral responses to an alkylpyrazine from the mandibular gland of the ant *Wasmannia auropunctata*. Insectes Soc. 29: 369-374.

KEMPF, W. W. 1972. Catálogo abreviado das formigas da Regiïo Neotropical (Hymenoptera Formicidae). Studia Entomol. 15: 3-344.

KUSNEZOV, N. 1951. El género *Wasmannia* en la Argentina. Acta Zool. Lilloana 10: 173-182.

LEVINGS, S. C., and N. R. FRANKS. 1982. Patterns of nest dispersion in a tropical ground ant community. Ecology 63: 338-344.

LUBIN, Y. D. 1984. Changes in the native fauna of the Galapagos Islands following invasion by the little red fire ant *Wasmannia auropunctata*. Biol. J. Linn. Soc. 21: 229-242.

LUBIN, Y. D. 1985. Studies of the little fire ant, Wasmannia auropunctata, in a niño year, pp. 473-493. In El niño en las Islas Galápagos El evento de 1982-1983. Fundación Charles Darwin para las Islas Galápagos, Quito, Ecuador.

MEIER, R. E. 1985. Coexisting patterns and foraging behavior of ants on giant cacti of three Galapagos Islands, Ecuador. Experientia 41: 1228.

NICKERSON, J. C. 1983. The little fire ant, *Ochetomyrmex auropunctata* (Roger) (Hymenoptera Formicidae). Fla. Dep. Agric. Consumer Serv. Entomol. Circ. 248.

OSBURN, M. R. 1948. Comparison of DDT, chlordane and chlorinated camphene for control of the little fire ant. Fla. Entomol. 31: 11-15.

POSADA, L., I. Z. DE POLANÍA, I. S. DE ARÉVALO, A. SALDARRIAGA, F. GARCÍA, and R. CÁRDENAS. 1976. Lista de insectos dañinos y otras plagas en Colombia. Inst. Col. Agrop. Bol. Tec. 43 97 pp.

ROGER, J. 1863. Die neu aufgeführten Gattungen und Arten meines Formiciden-Verzeichnisses, nebst Ergänzungen einiger früher gegebenen Beschreibungen. Berl. Ent. Zeitschr. 7: 131-214.

SCHEMSKE, D. W. 1980. The evolutionary significance of extrafloral nectar production by *Costus woodsonii* (Zingiberaceae) An experimental analysis of ant protection. J. Ecology 68: 959-967.

SILBERGLIED, R. 1972. The little fire ant, *Wasmannia auropunctata*, a serious pest in the Galapagos Islands. Noticias de Galapagos 19/20: 13-15.

SMITH, M. R. 1929. Two introduced ants not previously known to occur in the United States. J. Econ. Entomol. 22: 241-243.

SMITH, M. R. 1937. The ants of Puerto Rico. J. Agric. Univ. Puerto Rico 20: 819-875.

SMITH, M. R. 1965. House-infesting ants of the eastern United States. USDA-ARS. Tech. Bull. 1326 105 pp.

SPENCER, H. 1941. The small fire ant *Wasmannia* in citrus groves-a preliminary report. Fla. Entomol. 24: 6-14.

TORRES, J.A. 1984. Niches and coexistance of ant communities in Puerto Rico repeated patterns. Biotropica 16: 284-295.

ULLOA-CHACÓN, P., and D. CHERIX. 1988. Quelques aspects de la biologie de *Wasmannia auropunctata* (Roger) (Hymenoptera Formicidae). Actes Coll. Insectes Soc. 4: 177-184.

ULLOA-CHACÓN, P., and D. CHERIX. 1989. Etude de quelques facteurs influençant la fécondité des reines de *Wasmannia auropunctata* (R.) (Hymenoptera, Formicidae). Actes Coll. Insectes Soc. In Press.
WHEELER, G. C. and J. WHEELER. 1954. The ant larvae of the Myrmicinae tribes Meranoplini, Ochetomyrmicini and Tetramoriini. Amer. Midland Nat. 52: 443-452.
WHEELER, W. M. 1929. Two neotropical ants established in the United States. Psyche 36: 89-90.

28

Biological Aspects of the "Hormiga Loca," *Paratrechina (Nylanderia) fulva* (Mayr), in Colombia

I. Zenner-Polania

INTRODUCTION

Although impossible to confirm, the research on *Paratrechina (Nylanderia) fulva* (Mayr) (Hymenoptera: Formicidae, Formicinae) indicates that this Brazilian ant was introduced almost simultaneously in 1969-1970 to Central Colombia's coffee growing area and to the Middle Magdalena Valley - a cattle breeding region - as a biological control agent against leaf-cutting ants and snakes. *P. fulva*, the crazy ant- or "Hormiga loca"- received its nickname because of its frenzied, apparently random movements. In Brazil, the ant is known as "formiga cuiabana" or "doceira" (Da Costa Lima 1936; Mariconi 1970). According to Bondar (in Brandao 1941), in 1912 it was used to control *Atta* spp. and snakes, but turned into a serious problem in orchards, due to associations with coccids.

Specimens collected in Colombia in 1971 were classified as *Paratrechina* sp. by Dr. D.R. Smith (SEL-PSI-USDA); later, the late Dr. W. Kempf, University of Brasilia, identified it as *Nylanderia fulva* (Mayr), noting that it is a household pest in southern Brazil and that its controlling effects on *Atta* spp. were unproven (ICA 1975). Muesebeck et al. (1951) placed *Nylanderia* as a subgenus of *Paratrechina*.

DISTRIBUTION AND ECONOMIC IMPORTANCE

The first official record of "hormiga loca" in Colombia was made in 1971 (Fig. 1), at Puerto Boyacá (Boyacá) where high populations disturbed cattle (ICA 1972). Ants blinded calves and contributed to the drying of rangeland grass. Lopez (1975) determined that in pastures the ant was associated with *Antonina* sp. (Homoptera: Pseudococcidae), a sap feeding insect responsible for the dry grass. Inquiries and conversations with farmers pointed to the ant's introduction "years ago" for control of abundant vipers and other poisonous snakes. In this area (150 meters above sea level, m.a.s.l.), populations have established in stable but low levels; the original ecological equilibrium has apparently been restored and the ant no longer causes problems. The high temperature, up to 40°C, is partly responsible for the low infestation.

A second outbreak occurred around Fusagasugá (Cundinamarca) at 1600

FIGURE 1. Distribution of *Paratrechina (Nylanderia) fulva* in Colombia.

m.a.s.l. in 1977, at a coffee plantation. The ants preyed on chickens, other birds and any domestic animal. The death of small animals was attributed to attacks by large numbers of worker ants, mainly due to asphyxia; larger animals were attacked around eyes, nasal fossae and hoofs (ICA 1977a). "Hormiga loca" was brought into this area in 1969 by a "progressive" Colombian farmer to control leaf-cutting ants and snakes. Its initial establishment was slow, but by 1974 it was again well known as a nuisance pest in households, coffee plantations and cattle ranges.

The next report of extremely high populations also came in 1977 from El Colegio (Cundinamarca) (Fig. 1) (ICA 1977b). The insect was accidently introduced into the area around 1975 from Fusagasugá hen dung shipments used as fertilizer. Besides attacking domestic animals, its association with Homoptera, on all types of vegetation, resulted in the formation of sooty mold *Capnodium* sp., a fungus which grows on excess honeydew secreted by these sap-feeding insects. The fungus disrupts the photosynthetic activity of the plants. However, the problem disappeared and the natural equilibrium was restored in this area ten years later (ICA 1985).

From this point on, *P. fulva* was introduced to new areas through soil and hen dung. The ant reached Bogotá in 1978 (Fig. 1), brought in potted ornamental plant material from Fusagasugá, giving the insect an adaptive range of 150 to 2600 m.a.s.l. (ICA 1978). At this altitude large populations have never been observed, but the insect is considered a common household pest during the rainy season.

The "hormiga loca" was transported within organic soil from El Colegio to Anapoima (Cundinamarca) in 1978 (Fig. 1), where after only one year it had invaded more than 10 ha of coffee plantations and cattle range (ICA 1979). Research was conducted in this region and control campaigns instituted. The problem abated until now the insect is only considered a secondary pest.

After years of silence, 1987 and 1988 again brought news about *P. fulva's* presence and activity. In three distant areas (Fig. 1) outbreaks occurred almost simultaneously. In El Espino (Boyacá) the ant affected domestic animals (ICA 1987); in Cimitarra (Santander) it again associated with Homoptera on cocoa, manihot and plantain; in cattle ranges and natural forest (ICA 1988a), and finally in Villeta (Cundinamarca) it attacked chickens and, indirectly, sugar cane due to its association with the pink sugar cane mealybug, *Saccharicoccus sacchari* (Cockerell) (ICA 1988b).

Besides its role as a direct pest of domestic animals, the ant's shear numbers make it a formidable household pest. The ant is also responsible for agricultural losses due to its association with sap feeding Homoptera and finally, although it does not bite nor sting humans, *P. fulva*'s presence on all types of plantations makes agricultural practices virtually impossible. Some small farms have been abandoned due to the presence of *P. fulva*.

P. fulva expands its territory slowly. At one coffee plantation, they advanced 100 meters a month, or only one kilometer a year. Rivers are the only geographical features which slow or stop their migration.

Some interesting conclusions can be drawn from these observations. People introduce the ant to new areas; to reach a peak abundance, the population has to occupy an area for at least two years and after that, a slow but continuous decline culminates in restoration of the natural equilibrium about 10 years later. *P. fulva* can develop within a wide range of altitudes (150 to 2600 m.a.s.l.) and temperatures (means of 13°C to 29°C), but thrives best at altitudes around 1500 m.a.s.l. and a mean temperature around 22°C.

DESCRIPTION OF THE DEVELOPMENTAL STAGES

Within "hormiga loca's" social organization the following forms exist and are described:

Eggs - Ovoid, milky white color; chorion smooth; mean width 0.18 mm; mean length 0.281 mm; oviposited in clusters of 17 to 25 eggs in the soil, in crevices, on or under rotten vegetation.

Larva - Hymenopteriform type, more or less cylindrical with both ends rounded; milky white color; the mean measures of a completely developed larva = length 1.84 mm, width 0.712 mm. The cephalic portion shows rudiments of sclerotized mandibles.

Pupa - Exarate type; same color as larva; mean length 2.12 mm and width 0.7 mm. Eyes brown color, quite conspicuous.

Workers - Typical of the subfamily Formicinae; color clear reddish-brown; large legs and antennae. Three very small oceli and compound eyes present; antennae 12-segmented; mandibles larger than half of the head with six teeth; mandibular and maxillary palps large; each abdominal segment with rows of brown setae. Body length without antennae from 2.17 to 2.41 mm.

Males - At first sight, this caste can be distinguished from workers by two pairs of well developed wings, a more robust thorax and a larger and more slender abdomen. The last abdominal segment shows the clasper in the form of a shovel; color dark brown, body covered with many large hairs. Legs and antennae large. Antennae 13-segmented; three large oceli and compound eyes; little developed mandibles, each with only one conspicuous tooth; both pairs of palps large. Length without antennae from 2.41 to 2.77 mm.

Queen - Color a little darker than workers and males; legs and antennae large. Antennae 12-segmented; compound eyes and three large oceli; mandibles with six teeth; palpi large. Meso- and metathorax show alar stumps. Abdomen as large as head and thorax together, twice as large as wide and with four dark brown stripes; each segment with a row of dark brown setae. Length without antennae from 5.06 to 5.3 mm.

NESTING CHARACTERISTICS

The "hormiga loca" does not construct elaborate nests but exploits any available humid site to raise its brood. The distribution of nests over 390 m^2 of coffee plantation is shown in Table 1. The plantation had good shade conditions and a heavy infestation during the rainy season. Most nests were found within the soil in crevices or cavities, either natural or constructed by other arthropods such as termites, spiders, white grubs, small beetles and annelids. The ants do not construct galleries or tunnels, but always use crevices around roots or stones, or sites previously occupied by other animals, which if not abandoned, they displace. Around farm houses, the main nesting sites are waste material, like garbage rubber implements and black plastic bags.

Two nest types can be distinguished. The first are transitory nests, which prevail during the rainy season and are found even in unprotected sites, such as under and on dead leaves. The egg-masses, larvae and pupae in a nest are moved almost daily to a new site, sometimes more frequently if

TABLE 1. Nesting sites of *Paratrechina fulva*, analysis of a 390 m^2.

Site of Permanent Nests	No.	Site of Transitory Nests	No.
Fallen banana stems	13	Soil crevices	324
Base of tree	13	Dead leaves	165
Base tree trunk	11	Base coffee tree	52
Root edge	39	Base coffee trunk	4
Under Rock	3	Base banana stem	8
Base citrus tree	1	Rotten root	1
		Grass root	1
		Empty snail shell	1
TOTAL	80		556

disturbed. Composition of these nests is quite variable. The mean number of workers and developmental forms are: 21 eggs, 84 larvae and 49 pupae, attended by only 46 workers. Female and male sexuals were absent.

The second nest type designated "permanent" - since they are always occupied by colonies - cover a large area, up to one square meter, and contain immatures to a depth of 40 cm below the surface. They are invariably found in well-drained, protected areas. During the dry season, only permanent nests are built. The content of these nests varied from 1 to 14 queens, 214 to 1034 eggs, 815 to 1314 larvae, 565 to 4539 pupae and 366 to 2302 workers within the nest site.

In areas completely colonized by *P. fulva*, permanent nests are prevalent with transitory ones scarce. In newly invaded plantations, the reverse is true and only 12.6% of the nests can be considered permanent. The population developing in the transitory nests forms the force which invades new areas, advancing in a column of workers.

The unbelievably large number of ants found in a recently established population is explained in part by the number of available nesting sites, estimated at over 16,.000 per hectare in coffee, cocoa or sugarcane plantations. From the control point of view, knowledge of where and how to look for nesting sites is critical.

FEEDING HABITS AND RELATIONSHIP WITH OTHER ARTHROPODS

Long term field observations discovered that *P. fulva*'s diet had two components - one liquid and one solid. The liquid intake consisted of honeydew secretions obtained from Homopteran insects - mainly Coccidae, Diaspididae, Pseudococcidae, Aleyrodidae and Aphididae - present under almost all tropical and subtropical conditions. The solid diet took the form of animal protein, from immature stages and adults of insects other than Homoptera, and parts of higher animals, like birds, calves, snakes and lizards (Zenner de Polania and Rulz Bolaños 1985). No morphological distinction could be made between workers foraging for liquid or solid food; on occasion, the ant's color was darker when they were in a predatory mode. If given a choice between liquid or solid bait, a worker would probe both, then decide

TABLE 2. Animal contents of a permanent *Paratrechina fulva* nest.

Food Source	Parts Found	Number
Lepidoptera Larvae	Head Capsule	18
Coleoptera		
Larvae	Head Capsule & Tarsi	28 & 3
Adults	Elytra	12
Isoptera Adults	Head	39
Hymenoptera - Ants	Head & Thorax	17 & 8
Arachnida	Legs	13
Birds	Bones & Eyes	5 & 2
Lizard (small)	Skeleton	1

on one. Workers can take either kind of food to the nest.

To obtain the honeydew, the ant drums the insect with her antennae and feeds on the droplets which she stores in the crop; the resulting distended abdomen can be three times the size of a flaccid one.

The above authors recorded 28 species from seven families of Homoptera associated with *P. fulva.* All can be detrimental to agriculture. Among them, the whitefly, *Aleurothrixus floccosus* (Maskell) and the green scale, *Coccus viridis* (Green) stand out because of their abundance, number of hosts and amount of damage. These two are also preferred by the ant.

Their symbiotic relationship was grouped into "protection" and "transport" categories. The ants protect their symbionts against predators such as larvae of coccinellids, syrphids, chrysopids and parasitoids (Eulophidae, Braconidae, Pteromalidae); furthermore, they construct protective soil particle covers over the colonies of several Pseudococcidae, among them *Saccharicoccus sacchari* on sugarcane and *Dysmicoccus brevipes* (Cockerell) on coffee berries. These insect pests always live in concealed parts of the host, such as roots, fruits or leaf-sheats; without this cover they cannot progress during the dry-season. The ants transport both adults and nymphs in their mandibles from infested plants to uninfested ones. For example, a *Psylla* sp. was brought to the shadow tree, *Inga* sp., in a newly invaded area. In all cases, the Homoptera were considered myrmecophiles (Zenner-Polanía and Ruiz-Bolaños 1985).

The typical types of insects and animal parts found in a nest is outlined in Table 2. The ant's general predatory habits are shown by the diversity of prey remains found in its nests. They are opportunistic and take whatever they find while foraging. Twenty-five or more workers may attack a potential prey, such as the larva of a noctuid moth. First the larva is immobilized, then the integument is pierced and the hemolymph sucked. New workers arrive and the larva is either taken to the nest or torn into large pieces and then transported.

P. fulva infested areas are marked by a noticeable reduction in white grubs, leaf-feeding Lepidoptera and termite colonies, when compared to non-infested areas. While the research was carried out over several years, no snakes were observed in infested regions; farmers confirmed their absence.

The diversity of American-tropical ant species is well documented. Within coffee plantations this diversity can easily be observed. As can be

TABLE 3. Relationship between *Paratrechina fulva* with other Formicidae.

Ant Species	Within	Within/Outside *Paratrechina fulva* territory	Outside
Pseudomyrmex simplex-a	X		
P. elongatus-a			X
Olygomyrmex sp.-b	X		
Crematogaster sp.-b		X	
Zacryptocerus sp.-b		X	X
Leptothorax echinatinodis-b	*X*		
Solenopsis sp.-b			X
Solenopsis geminata-b			X
Atta spp			X
Acromyrmex sp.			X
Ectatoma ruidum-b			X
Camponodus senex-b			X
Azteca spp.-b			
X			

a- Determined by P.S. Ward; b- determined by D.R. Smith.

deduced from Table 3, which shows the results of a long term ant survey in two contiguous areas, *P. fulva*'s major negative influence was on other ant species.

With the exception of *Olygomyrmex*, a subterranean ant which nests within nests of "Hormiga loca", specimens of the other three genera were only collected occasionally on coffee leaves. Equally rare were *Crematogaster* (never found) and *Zacryptocerus* (inhabits twigs of *Inga* spp.).

Within the genera displaced by *P. fulva*, five are considered economically important: three pest and two beneficial insects. *S. geminata*, feared in coffee growing areas for its painful stings, constructs its nests in the soil; they can easily be recognized by a small mound of loose soil. The leaf-cutting ants *Atta* and *Acromyrmex* are serious pests in and around coffee plantations. *Ectatomma ruidum* is a general predator, while Eberhard and Kafury (1974) state that *Azteca trigona* can protect its tree host against *Atta* spp. Finally, *Camponotus senex* provides reliable proof of the absence of *P. fulva* in a given area, since it is the first ant to disappear when the "hormiga loca" starts to invade a territory.

The farmer who brought the ant to Colombia was satisfied, but the introduced control agent became a much greater problem than the leaf-cutting ants and snakes combined.

LITERATURE CITED

DA COSTA LIMA, A. 1936. Terceiro catalogo dos insectos que vivem nas plantas do Brasil. Ministerio de Agricultura, Dept. Producao Vegetal. Escola Nal. da Agronomia. 381pp.

EBERHARD, G. and O. KAFURY. 1974. La ecología de la hormiga *Azteca trigona*, una posible defensa contra las arrieras. Mem. II Congreso Soc. Colombiana Entomol. Cali p. 33-37.

INSTITUTO COLOMBIANO AGROPECUARIO, Programa de Entomologia. 1972. Bogotá, (Colombia). Informe Anual de Progreso. Bogotá. 56pp.
INSTITUTO COLOMBIANO AGROPECUARIO, Programa de Entomología. 1975. Se comprueba identificación. Notas y Noticias Entomológicas (Colombia) Nov.Dic.:l.
INSTITUTO COLOMBIANO AGROPECUARIO, Programa de Entomología. 1977a. Invasión de hormigas. Notas y Noticias Entomológicas (Colombia). Sep.Oct. 59.
INSTITUTO COLOMBIANO AGROPECUARIO, Programa de Entomologia. 1977b. De nuevo las hormigas. Notas y Noticias Entomológicas (Colombia). Sep-Oct.:61.
INSTITUTO COLOMBIANO AGROPECUARIO, Programa de Entomologia. 1978. Hormiga loca. Notas y Noticias Entomológicas (Colombia) Marzo-Abril: 17.
INSTITUTO COLOMBIANO AGROPECUARIO, Programa de Entomologia. 1979. Falta de previción. Notas y Noticias Entomológicas (Colombia) Enero-Feb.: 1.
INSTITUTO COLOMBIANO AGROPECUARIO, Programa de Entomologia. 1985. Para dónde se fue? Notas y Noticias Entomológicas (Colombia) Enero-Feb.: 14.
INSTITUTO COLOMBIANO AGROPECUARIO, Programa de Entomologia. 1987. Amplia distribución. Notas y Noticias Entomológicas (Colombia) Jul-Agos: 45.
INSTITUTO COLOMBIANO AGROPECUARIO, Programa de Entomologia. 1988a. De nuevo en el Magdalena Medio. Notas y Noticias Entomológicas (Colombia) Jul.-Agosto: 45.
INSTITUTO COLOMBIANO AGROPECUARIO, Programa de Entomologia. 1988b. Se acentúa problema. Notas y Noticias Entomológicas (Colombia). Nov.Dic.:69.
LÓPEZ, A. 1975. Informe visita a algunas fincas ganaderas de Pueto Boyacá. ICA. Reg. 1. 4pp.
MARICONI, A. M. 1970. As saúvas. Ed. Agronómica Ceres, Sao Paulo. p.18
MUESEBECK, G. F. W., K. V. KROMBEIN and H. K. TOWNES. 1951. Hymenoptera of America North of Mexico. Synoptic Catalog. USDA, Agric. Monograph. No. 2. 848pp.
SOAREZ BRANDAO, J. S. 1941. A formiga cuiabana "versus" saúva. Sitios e Fazendas. Brasil. 6: 36-38.
ZENNER DE POLANIA, I. and N. RUÍZ-BOLAÑOS. 1985. Habitos alimenticios y relaciones simbióticas de la "hormiga loca" *Nylanderia fulva* con otros artrópodos. Rev. Colombiana Entomol. 11: 3-10.

29

Ant Pests of the Tapinomini Tribe

A.Y. Harada

INTRODUCTION

Several Tapinomini tribe ant species are important household and/or agriculture pests in many areas of the world. Nine genera occur in Neotropical regions (Kempf 1972), eight in Asia (Chapman and Capco 1951), two in Australia (Taylor 1985) and four in west Africa (Bolton 1973); at least two species have a wide zoogeographical distribution (Brown 1973; Harada 1988). The following nine species from four genera of this tribe infest houses and/or are pests of monocultures: *Azteca chartifex* var. *spiriti* Forel, *Azteca paraensis* var. *bondari* Borgmeier, *Conomyrma insana* (Buckley), *Iridomyrmex cordatus* (Fr. Smith), *Iridomyrmex humilis* (Mayr), *Iridomyrmex purpureus* (Fr. Smith), *Iridomyrmex pruinosus* (Roger), *Tapinoma melanocephalum* (Fabricius), *Tapinoma sessile* (Say).

We must clarify the taxonomy and understand these species' behavior and biology before we can solve the urban and agricultural problems caused by these ants.

Some Tapinomini pests are spread through commerce, and in some cases they are well adapted to the new environment and displace less aggressive native ants (Erickson 1971; Kock 1989). The most outstanding example is *I. humilis*, the most widespread Tapinomini pest. It originated in Argentina but now covers most tropical and temperate areas of the world.

I will give an overview of Tapinomini pest species, including historical and biological considerations, and their economic importance. A taxonomic key to the species with illustration is provided, as well as distribution maps (Figs. 4 and 5).

PEST SPECIES OF TAPINOMINI

Azteca chartifex var. *spiriti* Forel, 1912

History and Biology. *Azteca chartifex* var. *spiriti* was originally collected in humid primary forests where it nested in specific kinds of trees. It builds a "carton nest" with cellulose from fruit skins, and the nests are scalelike, conical or elongated, and brownish in color. The nests are 60 to 100 cm long, and weigh about 20 kg. They are attached to tree trunks or branches (Silva

1955). Because of this kind of nest they are called the "Caçarema" ant. Colony subnests are maintained by workers up to 10 meters from the main nest. The colonies, probably monogynous, contain thousands of individuals. The workers are polymorphic, agile, aggressive and forage on trails. When disturbed they fall down like rain from the big nests upon the aggressor who they bite. Omnivorous, they tend honeydew-excreting Homoptera which live on cocoa tree bark.

Economic importance. *Azteca chartifex* var. *spiriti* is a serious cocoa plantation pest. They hinder fruit production and corrode the fruit's skin causing decreased cocoa production. They may use the starch from the fruit's skin to build their nest and subnests (Silva 1955). The nests can impede flower production where they are attached, and because of their weight can bend or break cocoa tree branches. Workers contribute to the dissemination of the brown rottenness (podridäo parda) by carrying the fungal spores that cause this disease. They even facilitate the construction of moribund wasp nests next to their own nests, which make the harvest and cleaning of the cocoa plantation difficult. Also, the ants tend honeydew-excreting Homoptera, which may cause cocoa flowers to wither. But the ants also promote flower fertilization, increasing fruit production.

Azteca paraensis var. *bondari* Borgmeier, 1937

History and Biology. *Azteca paraensis* var. *bondari* has the common name "grafter ant" or "mangrove ant". Silva (1945) called this ant the "gardener ant."

Ule (1901) first collected *A. paraensis* in the Amazonas state (Brazil). In south Bahia state (Brazil), Zehntner first mentioned it in 1914. However, not until 1930 was this ant recognized as an important cocoa monoculture pest.

Azteca paraensis var. *bondari* is called the "mangrove ant" because it lives near the nests of mangrove wasps. The ants build sponge-like, circular-shaped nests (known as "garden nests") using a gum from cocoa tree sprouts and epiphyte seeds (Silva 1945). The epiphytes may be attached to branches of a variety of plants including cocoa tree (*Theobroma cacao* L.), "genipapo" tree (*Genipa americana* L.), "samuameira" (*Sterculea excelsa* Mart.) (Silva 1945).

The colonies are probably monogynous and contain thousands of individuals. They are omnivorous, but also tend honeydew-excreting Homoptera on cocoa and other plants.

Economic importance. *Azteca paraensis* var. *bondari* is the most important pest of cocoa cultivation in Bahia state, Brazil (Silva 1969). They cut and gnaw leaves, terminal sprouts, and sometimes fruits of cocoa trees. They also tend Homoptera (Membracidae, Coccidae, Aleyrodidae and Aphidae), which causes decreased cocoa production (Silva 1945, 1969). Sometimes they build nests in the same cocoa tree as *A. chartifex* var. *spiriti*, and disseminate a pathogenic fungi (Silva 1946). If the nests are large, they bend the tree's branches, making cocoa pruning and the fruit harvesting difficult.

Many studies have been done regarding eradication of this species because of the serious damage it causes to cocoa monocultures. This ant has been eliminated using poison liquids, and/or cutting down the nests (Silva

1957). From 1954 to 1960, nests were treated with pulverized BHC (Cruz 1957). However, this method proved ineffective. It endangered cocoa workers' security and the commodity itself. It was frequently necessary to destroy cocoa branches in order to burn the ant's nests (Silva 1969). A new method of BHC application was developed (Silva and Bastos 1965), which consisted of injecting 1% BHC solution into different points on "mangrove ant" nests. It is technically and economically efficient when compared to the classic method of dusting BHC over the nest (Silva 1969).

Conomyrma insana (Buckley, 1866)

History and Biology. Snelling (1973) reviews *Conomyrma insana*'s history. Snelling (1973) also synonymized seven North American *Conomyrma* species as *Conomyrma insana*. This species is called "pyramid ant" or "crazy ant," reflecting its specific name and habit of running erratically when disturbed.

Conomyrma insana builds nests in the soil, frequently in sunny, dry places such as pastures (Gregg 1963); rarely under objects. The nests have one opening, which leads to a shallow chamber about 5.5 m deep (Nickerson et al. 1975). The ants nest in various soil types: loam, sand, clay or limestone. Nests are frequently built near nests of the same or other species, e.g. *Conomyrma flavopecta* (Smith); the harvester ant *Pogonomyrmex occidentalis* (Cresson); and *P. barbatus* (Smith). *C. insana* colonies are monogynous (Berkelhamer 1984) and build multiple moderate-sized nests (Nickerson et al. 1975). They contain a few thousand individuals (Smith 1965). The reproductive castes can spend the winter as adults. The fast-moving workers forage using conspicuous trails. The workers invade houses from outside. Usually, they are predacious and carnivorous, but they also tend honeydew-excreting Homoptera. Also, they are predacious on the corn earworm, *Heliothis zea* (Boddie) and the fire ant *Solenopsis geminata* (Fabricius) (Smith 1965).

Economic Importance. *Conomyrma insana* are occasionally house-infesting ants. They build mounds and feed on many human foods, although they prefer sweets (Smith 1965).

Iridomyrmex cordatus (Fr. Smith, 1859)

History and Biology. Smith (1859), described *I. cordatus* as *Formica cordata* from Aru Island. It is native to the Solomon Islands lowland, and occurs sympatrically with *Oecophylla smaragdina* in 12 islands of this archipelago (Greenslade 1972).

I. cordatus is among the dominant ant species on Guadalcanal, Solomon Islands. It nests in bark fissures, old termitaria, swollen stems of epiphytes such as *Myrmecodia* and *Hydnophytum* (Rubiaceae), and under granular carton roofs on trees (Brown 1959). It does not usually nest on or under soil, but is frequently found in logs (Greenslade 1972). This polygynous species expands its territory by budding, rather than through mating flights (Greenslade 1971). In the Solomon Islands, it is a dominant and persistent species which takes a long time to expand its territory. It is almost unaffected by other ant populations, except for *Anoplolepis longipes* (Greenslade 1971). In Australia it inhabits advantageous epiphyte roots from the drier lowland, rather than the mid to high altitudes of the Papua New Guinea wetlands (Huxley 1982). This is an omnivorous species, but its main food is honeydew excreted by insects (Homoptera, Lycaenidae) (Greenslade 1972; Huxley 1982).

Economic Importance. *I. cordatus* probably contributes to decreased coconut production in the Solomon Islands, because it tends honeydew-excreting Homoptera (Greenslade 1972). On the beneficial side, it helps control *Amblypelta cocophaga* (coreid bug), the cause of coconut nutfall (Greenslade 1971).

Iridomyrmex humilis (Mayr, 1868)

History and Biology. The Argentine ant was first described by Mayr (1868) as *Hypoclinea humilis*, from specimens collected near Buenos Aires, Argentina. This species was introduced to North America 98 years ago via New Orleans on coffee shipped from Brazil (Newell 1908). Since then, *Iridomyrmex humilis* has spread throughout the American continent and caused both agricultural, urban and environmental problems (Barber 1916; Smith 1965; Ward 1987). However, in the southeastern United States *I. humilis* was largely displaced by another imported ant, *Solenopsis invicta*. Today, this species occurs throughout the Americas, Europe, South Africa, Asia and Australia, as well as many island groups (Markin 1970).

Although the wide distribution of the Argentine ant was through commercial shipments of plants and plant products, building materials, household goods, groceries, and other materials, this species has naturally dispersed to a lesser extent by heavy rains and floods.

Iridomyrmex humilis nests in exposed soil and soil under rotting wood or debris, bird nests, bee hives, hollow trees, and rubbish.

The omnivorous workers tend honeydew-excreting insects such as plant lice, scales and mealybugs (Smith 1965). They also, feed on floral and extrafloral nectaries of certain plants, gnaw into the buds of fruit trees, such as *Citrus* or even in ripened fruits such as figs (Smith 1979).

In temperate zones males and alate females usually occur during April and May. Mating probably takes place in the nests (Woodworth 1910; Skaife 1962); nuptial flights are rare. After mating, one or more inseminated females leave the mother nest with a group of workers and build a new colony. Females can not raise a colony by themselves. New colonies can be founded by fission or budding (Keller and Passera 1988). These ants form polydomous nests with a queen for each 500-1000 workers. Nest relocation depends on the season and soil condition, although exchange of individuals among nests in a given area is constant (Markin and McCoy 1968; Dechene 1970). During winter several small colonies aggregate to form one large colony that can again divide into small colonies in early spring or summer. Oviposition occurs all year. Outdoors, eggs are laid when the daily mean temperature is at least 18°C. The incubation period ranges from 12 to 55 days. The life cycle varies from 33 to 141 days with an average of 74 days (Newell 1908). Workers can walk on water (Borgmeier 1926).

This species has partially displaced *Pheidole megacephala* in Hawaii (Wilson and Taylor 1967), and Bermuda (Haskins and Haskins 1965). It also displaced the imported fire ant *Solenopsis saevissima richteri* Forel in Alabama (Wilson 1951), *S. geminata* F. in Hawaii (Fluker and Beardsley

1970), and *Pheidole californicus* Buckley, *P. gallipes* Wheeler and *Veromessor pergandei* Mayr in California (Erickson 1971). *I. humilis* and *Pheidole megacephala* are two species that actively extend their range and are capable of displacing long-established native species (Haskins and Haskins 1965; Erickson 1971; see Aron et al. this volume).

Economic importance. This species economic importance as a pest was first reported by Titus (1905). Since then, more attention has focused on the degree to which *I. humilis* invades urban and agricultural habitats (Erickson 1971).

The ability of *I. humilis* to nest in numerous habitats, produce many individual polydomous colonies, thrive on a variety of foods and cause displacement to other ant species explains its success as a competitive species. Also, the ant interferes with the biological control of scales and other Homoptera because of the workers foraging and honeydew gathering activities on *Citrus* (Ward 1987; Markin and McCoy 1968).

The Argentine ant is also one of the most troublesome home-infesting ants. It crawls into refrigerators, stoves, shelves, beds, clothing, over refuse, foodstuffs, filth, sputum, faeces, carrion, or sewage. Thus it can be an important disease vector (Ipinga-Regla, et al. 1984), including dysentery, typhoid fever and tuberculosis (Smith 1965). Its tending of honeydew-excreting insects causes serious damage to sugar cane, figs, citrus and many ornamental trees.

Argentine ant control has been attempted since it was recognized as an economic pest. In Argentina, carbon bisulphide, boiling water, petroleum, and other insecticides were applied under trails or nests (Gallardo 1912). In the United States, control of this pest started with fumigation of citrus trees. Different chemicals were applied, but chlorinated hydrocarbon insecticides were most effective (Gerhardt 1953) This is now the standard ant treatment on citrus trees (Musgrove and Carman 1965), except where the compounds are banned. Other chemicals such as chlorpyrifos, have successfully excluded *I. humilis* from citrus trees in California (Moreno et al. 1987; see Buys; Rust and Knight this volume).

Iridomyrmex pruinosus (Roger, 1863)

History and Biology. *Iridomyrmex pruinosus*, first placed in the *Tapinoma* genus, was moved to the *Iridomyrmex* genus because of its morphological characteristics. However, its taxonomic status is still confused. Snelling (In Hölldobler 1982) believes that this species should be moved to the *Forelius* genus.

Iridomyrmex pruinosus nests in areas of exposed soil with crater-like mounds or under cover of objects (Smith 1979), such as stones (Cole 1940), litter, and dead stumps (Van Pelt 1958). The nests are usually closed to other ant species (Hölldobler 1982). This species forms a unicolonial society with polydomous nests (Berkelhaner 1980, 1984) with three to seven queens (Hölldobler 1982). Nuptial flights occur in July, and females and males of the same colony can mate. Inseminated females return to mother colonies (Hölldobler 1982). The day after nuptial flights, many queens migrate, accompanied by workers and broods, and found new colonies (Hölldobler 1982). Immature forms occur in the nests all year (Van Pelt 1958). Colonies are small to moderate in size (Smith 1965). Workers are agile, fast and lay trails (Smith 1965). Workers of different nests do not show aggressive behavior toward members of same species (Hölldobler 1982). Foragers make characteristic trails even across sand and sometimes into vegetation. Workers reduce the foraging activity of competitors by chemical interference at food sources and nest entrances, particularly on *Myrmecocystus* when their nests are less than seven meters from each other. Foragers do not take large objects to the nests; each ant carries small pieces of dissected prey to the nest (Hölldobler 1982). Workers forage during the day and feed on dead insects, insect fragments, floral nectaries, termites, and tend honeydew-excreting Homoptera (Smith 1965; Hölldobler 1982).

Economic Importance. This species is a United States house pest (Gulf Coast States). Workers probably nest in houses, but usually enter from outdoors (Smith 1965). Workers prefer sweet foods, but will eat almost all the foods other ants eat (Smith 1965). This species is a possible intermediate host of tapeworm, *Raillietina tetragona* (Molin) (Smith 1965).

Iridomyrmex purpureus (Fr. Smith, 1858)

History and Biology. *I. purpureus* is composed of a group of eight complex color forms and is a common species of most Australian fauna (Halliday 1983). Of these, at least "*I. purpureus* strict" and the "blue" form can be considered sibling species (Halliday 1983).

The meat ant, *I. purpureus* is a terrestrial, nocturdiurnal, woodland (drier areas along watercourses) species (Taylor 1985). It nests in the ground in open areas, often near mature nests of the same species (Greaves and Hughes 1974; Hölldobler and Carlin 1985). They build mound nests in soil, and the number of nest mounds and nest entrance holes varies seasonally (Greenslade 1975b). Its distribution ranges from lowland to 600 m above sea level, where it nests in soil (Greenslade 1972). The species is omnivorous, and predacious, but they basically eat nectar and honeydew. In the Solomon Islands they tend 19 species of honeydew-excreting Homoptera. Mutual adaptation occurs between this ant and Homoptera (Greenslade 1972).

They make uni- or polycolonial societies (Greenslade 1975a) with flexible organization, extending from small individual nests to large, multi-nest colonies in which workers move freely between nests (Hölldobler and Carlin 1985). Colony foundation can be by haplometrosis, pleometrosis, colony budding, or by newly-mated queens founding nests near mature nests. Also, oligogynous colonies occur in this species by primary or secondary oligogyny (Hölldobler and Carlin 1985). Nuptial flights take place in October (Hölldobler and Carlin 1985) or March through April (Greaves and Hughes 1974), but alate forms overwinter inside nests (Hölldobler and Carlin 1985). Males from different colonies swarm on the same day, usually followed by the females. Females mate only once, starting in the air, and finishing on the ground (Hölldobler and Carlin 1985).

Greaves and Hughes (1974) found that claustral colony founding leads to monogynous colonies (Hölldobler and Wilson 1977). The species life cycle in the laboratory, takes 14-19 days, and first oviposition occurs 18 to 31 days after the nuptial flights (Hölldobler and Carlin 1985). Males are produced by eggs laid by queens (Halliday 1983).

Birds prey heavily on founding colonies (Hölldobler and Carlin 1985; Greaves 1973).

Meat ant workers delimit their territorial boundaries by exhibiting a characteristic confrontation behavior when they meet workers from other territories (Halliday 1983).

Economic Importance. The mutual adaptation of this species with 19 honeydew-excreting Homoptera species may contribute to decreased coconut production in the Solomon Islands (Greenslade 1972).

Tapinoma melanocephalum (Faricius, 1793)

History and Biology. It has been widely distributed by commerce. Its house habits have given it the name "house infesting ant" by English speaking people, "Hormiga boticaria" in Cuba (Wheeler 1913), and "albaricoque" in Costa Rica.

This ant has a variety of nesting habits. It nests in soil, beneath stones, rotten wood, decayed trees, under tree bark, in plant cavities, tufts of dead grass, plant stems, temporary cavities, beneath detritus and leaves, stumps, hollow twigs, and in urban environments: closets, instrument case lining, in piles of discarded clothing, and occasionally in greenhouses (Smith 1936, 1965; Kannowiski 1959). Colonies are polydomous with 100 to 1000 individuals. Colonies are typically broken into subunits that occupy different nest sites and exchange individuals along odor trails. According to Smith (1965), new colonies are founded when one or more inseminated females migrate with numerous workers. It is unknown if this species has nuptial flights. Members of different colonies are not antagonistic to each other. Worker movements are rapid and erratic. Although omnivorous, this species mainly feeds on sweets such as sugar, cakes, honeydew excreted from Homoptera, and dead and live insects. The workers excrete a substance which has a peculiar "Tapinoma odor".

Economic Importance. The "house infesting" ant is an important house pest because of its nesting and feeding habits. It can invade the house from outside or nest in the house. It can completely displace a resident *Tetramorium caespitum* population from a house (Brown 1964). This ant has propagated the grass root mealybug *Ripersia radicicola* Morrison on sugar cane roots. They also cause trouble by raiding insectaries (Smith 1936). On the beneficial side, this species preys on housefly, *Musca domestica* (L), eggs (Pimentel 1955) and on the bug, *Rhodnius prolixus* (Gómez-Nuñez 1971).

Tapinoma sessile Say, 1836

History and Biology. The odorous house ant uses diversified nesting sites that lack permanence. They can nest in soil, under stones, boards, dead leaves, rotten logs, stumps, in cavities, wasp and bird nests and houses. They can nest to heights of over 10,000 feet, and in a variety of environmental conditions (swampy or boggy sites, sandy areas along the sea coast, or higher locations). This species can survive a wide range of temperatures and humidity (Smith 1928).

In temperate regions, *Tapinoma sessile* colonies survive indoors during winter, with reproductive females, workers and brood. Workers are active throughout the year. If reproductive females are maintained at 70°F, they lay a few eggs which take about 26 days to hatch. However, outdoors the colony passes the winter with workers, reproductive females, and slowly growing larvae. Workers begin foraging activity in March. Reproductive females lay eggs from late April to early November, which coincides with brood development. From November to April, the larvae remain dormant, and mature to adults in April, thus requiring six to seven months to complete the life cycle. When eggs are laid from April to June adult development requires five to nine weeks, but when laid from June to September six to seven weeks are required. Alate females and males appear from mid June to early July. Nevertheless, males emerge a short time before females. Only mature colonies produce sexuals. Males inhabit the nest only for the time necessary to attain sexual maturity. However, mating can occur inside or outside nests (Smith 1928). Total brood development requires 34 to 83 days. Depending on the season, egg incubation takes 11 to 26 days; the larval stage 13 to 29 days; the prepupal stage two to three days and the pupal stage eight to 21 days (Smith 1928).

Usually, adult females and workers live for a few weeks, while the males live only a week to 10 days.

Normally, each *Tapinoma sessile* colony has many reproductive females (up to 200) producing brood. In the laboratory, each reproductive female can lay as many as 30 eggs a day from April to early October (Smith 1928). Queen size is indicative of their capacity to lay eggs (See Williams this volume and Vargo this volume). However, eggs laid by workers take longer than those laid by queens to develop to adults, with the exception of males (Smith 1928).

Colony size varies according to age and can range from a few hundred to many thousand individuals. Each colony has a large number of reproductive females and the sexual forms are only produced in older colonies (four to five years old; Smith 1928).

When the nest is disturbed workers run rapidly, emitting an odor like rotten coconut from their elevated abdomens. Workers of this species are good foragers and nurses. During both day and night, they forage on trails leading from their nest to the food supply. Forage areas include garbage cans, commodes, dirty linens, pantries, sinks, and refrigerators. They feed on many household foods: cooked and uncooked beef, fish, raw and fried liver, potatoes, dairy products and ripe fruit. However, the main food sources are honeydew, fresh organisms or fruit juices.

Economic Importance. "Odorous house ants" are aphidicolous and coccidicolous species and are associated with the occasional transmission of important plant diseases, and are known to disrupt the efficacy of biological control agents (Bobb 1965) and other parasitic insects (Kulman 1965), in addition to its propensity for household foods (Smith 1965).

This species has been controlled by topical application of chemicals such as dieldrin, diazinon, propoxur and chlordane, which are more efficient when applied for 48 hours (Wang and Brook 1970).

SUMMARY

The nine pest species of the Tapinomini tribe are omnivorous, make trails for foraging and recruitment. They may be polygynous and/or monogynous. The most widespread species, *I. humilis*, usually displaces native ant species. The house-infesting species usually nest inside houses, and mainly damage foodstuffs. However, *I. humilis* and *T. melanocephalum* can be a vector of human diseases and *T. sessile a* transmitter of plant diseases. Some prey on various economically important insects.

ACKNOWLEDGMENTS

I am grateful to Drs. Harold Fowler, William Overal and Pedrito Silva for reading and commenting on this manuscript. I also thank Dr. Guy Curturier for helping with some literature, and Mr. Artemio Silva for the drawings.

LITERATURE CITED

BARBER, E.R. 1916. The Argentine ant: distribution and control in the United States. U. S. Dept. Agr. Bull. 377:1-23.

BERKELHANER, R.C. 1980. Reproductive strategies in ants: a comparison of single-queened versus multiple-queened species in the subfamily Dolichoderinae (Hymenoptera: Formicidae). Dissertation, University of California, Berkeley.

BERKELHANER, R.C. 1984. An electrophoretic analysis of queen number in three species of Dolichoderinae ants. Insectes Sociaux 31:132-141.

BOBB, M.L. 1965. Insect parasite and predator studies in a declining sawfly population. Jour. Econ. Ent. 58:925-926.

BOLTON, B. 1973. The ant genera of west Africa: a synonymic synopsis with keys (Hymenoptera: Formicidae). Bull. Br. Mus. (Nat. Hist.), Entomol. 27:319-368.

BORGMEIER, T. 1926. Observaço~es biológicas sobre formigas brasileiras especialmente do Estado de Sa~o Paulo. Rev. Mus. Paulista 14:185-304, 4 figs.

BROWN, E.S. 1959. Immature nutfall of coconuts in the Solomon Islands. I. Distribution of nutfall in relation to that of certain species of ants. Bull. Ent. Res. 50:97-133.

BROWN, JR. W.L. 1964. Some tramp ants of old world origin collected in tropical Brazil. Ent. News 75:14-15.

BROWN, JR. W.L. 1973. A comparison of the hylean and Congo-west African rain forest ant faunas. pp:161-185. In Tropical ecosystems in Africa and South America: a comparative review. Smithsonian Institute Press, Washington, D. C. viii.

CHAPMAN, J.W. and S.R. CAPCO. 1951. Check list of the ants (Hymenoptera: Formicidae) of Asia. Inst. Sci. Tech. Manila. Philippines. 327pp.

COLE, A.C. 1940. A guide to the ants of the Great Smoky Mountains National Park, Tennessee. Amer. Mid. Nat. 24:1-88.

CRUZ, H.M. da. 1957. A assistência fitossanitaária na lavoura cacaueira da Bahia e seus resultados. pp:153-165. Annais IV. Reun. Fitos. Brazil, Min. Agr. P.D.S. V.

DECHENE, R. 1970. Studies of some behavioral patterns of *Iridomyrmex humilis* Mayr (Formicidae, Dolichoderinae). Wasm. Jour. Biol. 28:175-184.

ERICKSON, J.M. 1971. The displacement of native species by the introduced Argentine ant *Iridomyrmex humilis* Mayr. Psyche 78:257-266.

FLUKER, S.S. and J.W. BEARDSLEY. 1970. Sympatric association of three ants: *Iridomyrmex humilis*, *Pheidole megacephala* and *Anoplolepis longipes* in Hawaii. Ann. Ent. Soc. Amer. 63:1290-1296.

GALLARDO, A. 1912. Observaciones sobre una hormiga invasora *Iridomyrmex humilis* Mayr. Physis 3:133-138.

GERHARDT, P.D. 1953. Chlordane, dieldrin, and heptachlor for control of the Argentine ant in California citrus orchrnd. J. Econ. Entomol. 46:1063-1066.

GÓMEZ-NUÑEZ, J.C. 1971. *Tapinoma melanocephalum* as an inhibitor of *Rhodnius prolixus* populations. J. Med. Ent. 8:735-737.

GREAVES, T. 1973. Biological problems in the control of the meat ant *Iridomyrmex purpureus* (Smith) (Hymenoptera: Formicidae). J. Aust. Entomol. Soc. 12:284-288.

GREAVES, T. and R.D. HUGHES. 1974. The population biology of meat ant. J. Aust. Entomol. Soc. 13:329-351.

GREENSLADE, P.J.M. 1971. Interspecific competition and frequency changes among ants in Solomon Islands coconut plantations. J. Appl. Ecol. 8:323-350.

GREENSLADE, P.J.M. 1972. Comparative ecology of four tropical ant species. Insectes Sociaux 19:195-212.

GREENSLADE, P.J.M. 1975a. Dispersion and history of a population of the meat ant *Iridomyrmex purpureus* (Hymenoptera: Formicidae). Aust. J. Zool. 23:495-510.

GREENSLADE, P.J.M. 1975b. Short-term change in a population of the meat ant *Iridomyrmex purpureus* (Hymenoptera: Formicidae). Aust. J. Zool. 23:511-522.

GREENSLADE, P.J.M. and R.B. HALLIDAY. 1983. Colony dispersion and relationships of meat ants *Iridomyrmex purpureus* and allies in an arid locality in South Australia. Insectes Sociaux 30:82-99.

GREGG, R.E. 1963. The ants of Colorado with reference to their ecology, taxonomy, and geographic distribution. Univ. Colorado Press, Boulder. xiii + 792pp.

HALLIDAY, R.B. 1983. Social organization of meat ants *Iridomyrmex purpureus* analyzed by gel electrophoresis of enzymes. Insectes Sociaux 30:45-56.

HARADA, A.Y. 1988. Sistemática Filogenética da tribo Tapinomini Emery (Formicidae, Dolichoderinae) até o nível de gênero. Ph. D. Thesis, Univ. Paraná, Curitiba, PR. xiv + 176 pp.

HASKINS, C.P. and E.P. HASKINS. 1965. *Pheidole megacephala* and *Iridomyrmex humilis* in Bermuda-equilibrium or slow replacement? Ecology 46:736-740.

HÖLLDOBLER, B. 1982. Interference strategy of *Iridomyrmex pruinosum* (Hymenoptera: Formicidae) during foraging. Oecologia. 52:208-213.

HÖLLDOBLER, B. and N.F. CARLIN. 1985. Colony founding, queen dominance and oligogyny in the Australian meat ant *Iridomyrmex purpureus*. Behav. Ecol. Sociobiol. 18:45-58.

HÖLLDOBLER, B. and E.O. WILSON. 1977. The number of queens: an important trait in ant evolution. Naturw. 64:8-15.

HUXLEY, C.R. 1982. Ant-epiphytes of Australia. pp:63-73. In: Ant-plant interactions in Australia. R. C. Buckley ed. London. IX.

KELLER, L. and L. PASSERA. 1988. Energy investment in gynes of the Argentine ant *Iridomyrmex humilis* (Mayr) in relation to the mode of colony founding in ants (Hymenoptera: Formicidae). Int. Jour. Inv. Rep. Dev. 13:31-38.

KEMPF, W.W. 1972. Catálgo abreviado das formigas da Regiáo Neotropical (Hymenoptera:Formicidae). Studia Entomol. 15:3-344.

KOCK, A.E. 1989. A survey of the Argentine ant, *Iridomyrmex humilis* (Mayr), (Hymenoptera: Formicidae) in South African. J. Ent. Soc. Sth. Afr. 52:157-164.

KULMAN, R.M. 1965. Natural control of the eastern tent caterpillar and notes on its status as a forest pest. Jour. Econ. Entomol. 58:66-70.

MARKIN, G.P. 1970. The seasonal life cycle of the Argentine ant, *Iridomyrmex humilis* (Hymenoptera: Formicidae) in southern California. Ann. Ent. Soc. Amer. 63:1238-1242.

MARKIN, G.P. and C.W. McCOY. 1968. Occurrence of Nematode, *Diploscapter lycostoma*, in the pharyngeal glands of the Argentine ant, *Iridomyrmex humilis*. Ann. Ent. Soc. Amer. 61:505-509.

MORENO, D.S., P.B. HANEY and R.F. LUCK. 1987. Chloropyrifos and diazinon as barriers to Argentine ant (Hymenoptera: Formicidae) foraging on citrus trees. J. Econ. Entomol. 80:208-214.

MUSGROVE, C.H. and C.E. CARMAN. 1965. Argentine ant control on citrus in California with granular formulation of certain chlorinated hydrocarbons. J. Econ. Entomol. 58:428-435.

NEWELL, W. 1908. Notes on the habits of the Argentine or "New Orleans" ant, *Iridomyrmex humilis* Mayr. J. Econ. Ent. 1: 21-34.

NICKERSON, J.C., H.L. CROMROY., W.H. WHITCOMB and J.A. Cornell. 1975. Colony organization and queens numbers in two species of *Conomyrma*. Ann. Ent. Soc. Amer. 68:1083-1085.

PIMENTEL, D. 1955. relationship of ants to fly control in Puerto Rico. Jour. Econ. Entomol. 48:28-30.

SKAIFE, S.H. 1962. The study of ants. Scientific Club Edition, p.12. U.S. Dept. Agr. Tech. Bul. 798:1-44.

SILVA, P. 1945. "A formiga de enxêrto". Inst. cacau Bahia. Bol. 1:1-21.

SILVA, P. 1946. Formiga Caçarema versus formiga de enxêrto. Inst. cacau Bahia. 14:1-3.

SILVA, P. 1955. A formiga "Caçarema" e o cacaueiro. Inst. Biol. Bahia 2:1-13.

SILVA, P. 1957. Problemas entomológicos do cacaueiro com referência especial á Bahia. pp:59-72 VI reuniáo do Comité Técnico Interamericano de cacau.

SILVA, P. 1969. Novo método de aplicaçáo do BHC no combate à formiga de enxêrto, praga do cacaueiro. pp:291-295. In Memórias da segunda conferência internacional de pesquisas em cacau. Memórias. Sáo Paulo, Impres.

SILVA, P. and G.A.C. BASTOS. 1965. Polvilhadeira-injetora "PG" para combater formiga-de-enxêrto. Cacau Atualidades 2:20-23.

SMITH, D.R. 1979. Subfamily Formicoidea. In Catalog of Hymenoptera in America North of Mexico. Vol.II. pp:1323-1467 Apocrita (Aculeata). Krombein K.V. et. al. (eds.). Smithsonian Institution Press. Washington,D. C. 2209pp.

SMITH, M.R. 1928. The biology of *Tapinoma sessile* Say, an important house-infesting ant. Ann. Ent. Soc. Amer. 21: 307-329.

SMITH, M.R. 1936. The ants of Puerto Rico. Jour. Agr. Univ. Puerto Rico 20:819-875.

SMITH, M.R. 1965. House-infesting ants of the eastern United States. Their recognition, biology, and economic importance. USDA Tech. Bol. 1326:1-105.

SNELLING, R.R. 1973. The ants of genus *Conomyrma* in the United States (Hymenoptera: Formicidae). Nat. Hist. Mus. Los Angeles 238:1-6.

TAYLOR, R.W. 1985. Hymenoptera: Formicoidea. pp:1-149. In Zoological Catalogue of Australia. vol.II. D. W. Walton ed. Victoria. 381pp.

TITUS, E.S.G. 1905. Report on the New Orleans ant (*Iridomyrmex humilis* Mayr). U. S. Dept. Agr. Bureau Ent. Bull. 52:79-84.

VAN PELT, JR. A.F. 1958. The ecology of the ants of the Welaka Reserve, Florida (Hymenoptera: Formicidae). II. Annotated list. Amer. Mid. Nat. 59:1-57.

WANG, J.S. and F.S. BROOK. 1970. Toxological and biological studies of the odorous house ant, *Tapinoma sessile*. Jour. Econ. Ent. 63:1971-1973.

WARD, P.S. 1987. Distribution of the introduced Argentine ant (*Iridomyrmex humilis*) in natural habitats of the Lower Sacramento valley and its effects on the indigenous ant fauna. Hilgardia 55:1-16.

WHEELER, W.M. 1913. The ants of Cuba. Bull Mus. Comp. Zool., Harvard 54:475-505.

WILSON, E.O. 1951. Variation and adaptation in the imported fire ant. Evolution 5:68-79.

WILSON, E.O. and R.W. TAYLOR. 1967. An estimate of the potential evolutionary increase in species density in the Polynesian ant fauna. Evolution 21:1-10.

WOODWORTH, C.W. 1910. The control of Argentine ant. Univ. Calif. Agr. Exp. Sta. Bull. 207:53-82.

KEY TO PEST SPECIES OF TAPINOMINI, BASED ON WORKERS

1. Dorsal surface of propodeum in form of prominent tubercular elevation posteriorly. Maxillary palpi unusually long, the third segment longer than the combined lengths of the first and second segments. Ventral surface of the head with a weakly developed psamophore. Mesonotum in profile with a perceptible anterior angle to the mesopropodial suture (Fig. 1C) .... *Conomyrma insana* (Roger).
- Characters different from described above................ 2.
2. Petiolar node rudimentary or vestigial (Fig. 3) and hidden by the base of gaster when viewed from above.................. 3.
- Petiolar node developed, suberect to erect; in profile not hidden by the base of the gaster (Fig. 1, 2) .... 4.
3. Body color rather uniform, either a light brown or black. Length 2.4 - 3.25 mm (Fig. 3C) .... *Tapinoma sessile* (Say).
- Head and alitrunk black color and without pilosity, legs, antennae and mandibles usually light color. Length 1 - 1.5 mm (Fig. 3B) .... *Tapinoma melanocephalum* (Fabricius).
4. Apex of funiculi not thickened in appearance. Thorax broadest through anterior part of pronotum, but without pronounced shoulders. Mesothorax hunchback shaped and higher than pronotum...........5.
- Apex of funiculi somewhat thickened in appearance. Thorax broadest through anterior part of pronotum, and with rather pronounced shoulders. Mesothorax not hunchback shaped, as high as pronotum......6.
5. Maxillary and labial palpi five and three segmented, respectively. Tibia and scape without erect hairs. Body has sparse hairs. Make carton nests (Fig. 1B) .... *Azteca chartifex* var. *spiriti* Forel.
- Maxillary and labial palpi six and four segmented, respectively. Tibia, scape and body provided with long and erect hairs. Makes garden nests (Fig. 1A) .... *Azteca paraensis* Borgmeier.
6. Alitrunk arched with strong mesonotal declivity. Base of propodeum strongly swallow (Fig. 2) .... 7.
- Alitrunk arched with shallow mesonotal declivity. Base of propodeum with slightly shallow or without swallow (Fig. 3) .... 8.
7. Scape and tibia without erect hairs. Body uniform yellowish color. Length 2,2 - 2,6 mm (Fig. 2B) .... *Iridomyrmex humilis* Mayr.
- Scape and tibia with long and erect hairs. Head and alitrunk reddish color and gaster dark. Length 7 mm (Fig. 3A) .... *Iridomyrmex purpureus* (Fr. Smith).
8. Alitrunk in profile with a deep and long mesonotal declivity. Dorsum of alitrunk with long hairs. Length 2 mm (Fig. 2A) .... *Iridomyrmex cordatus* (Fr. Smith).
- Alitrunk in profile without deep and long mesonotal declivity. Dorsum of alitrunk without hairs. Length 3 mm (Fig. 2C) .... *Iridomyrmex pruinosus* (Roger).

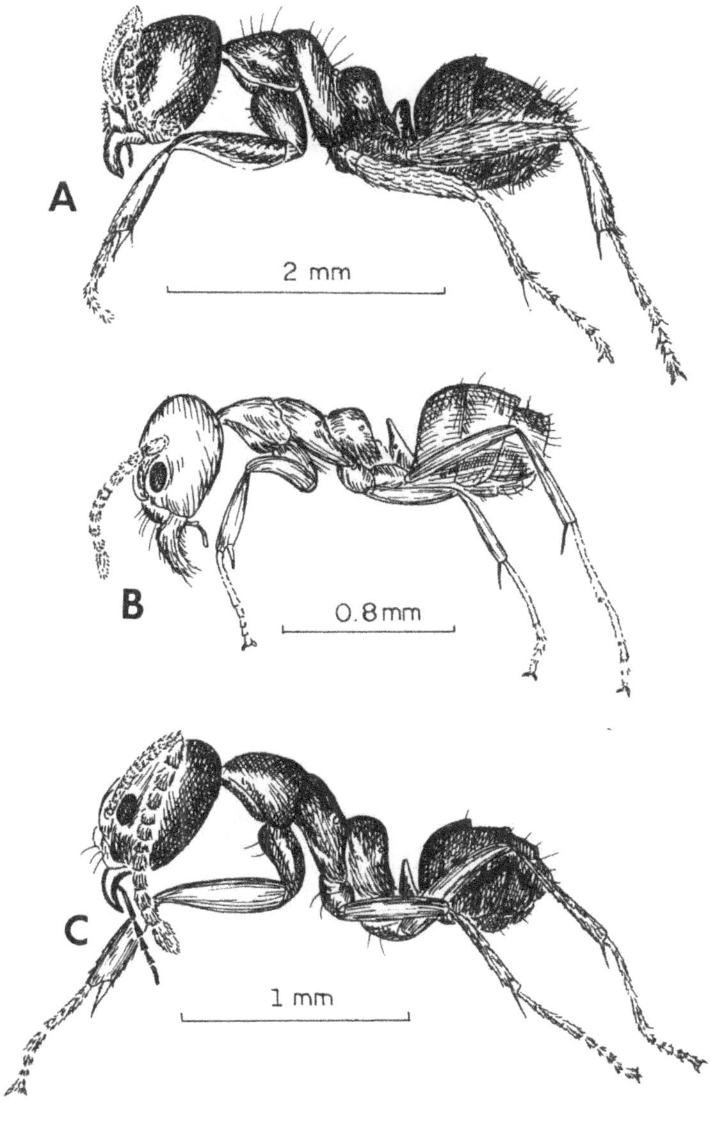

FIGURE 1. Lateral view of workers. A- *Azteca paraensis* var. *bondari*. B- *Azteca chartifex* var. *spirit*. C- *Conomyrma insana*.

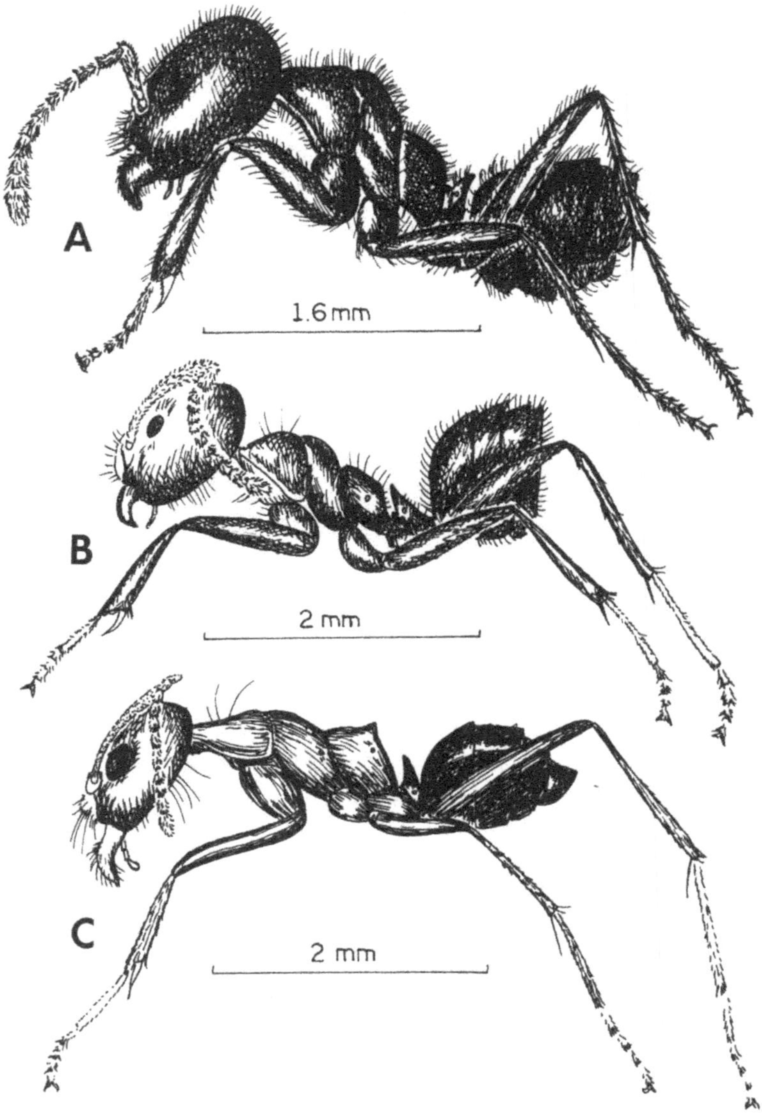

FIGURE 2. Lateral view of workers. A- *Iridomyrmex cordatus* (Fr. Smith). B-*Iridomyrmex humilis* (Mayr). C- *Iridomyrmex pruinosus* (Roger).

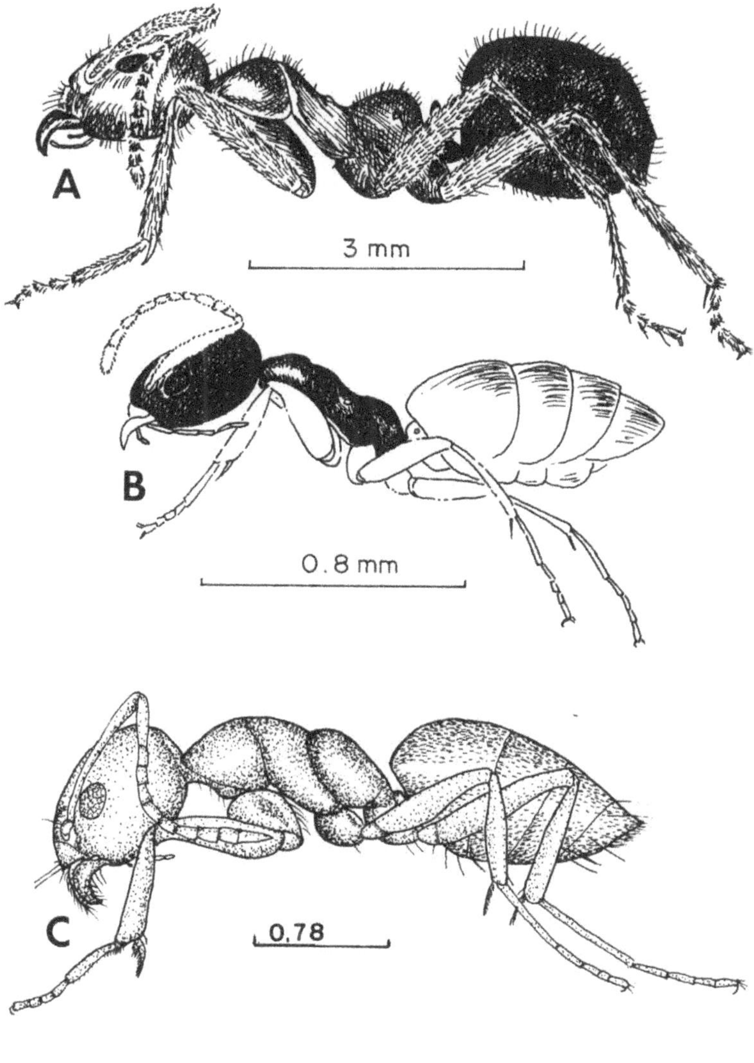

FIGURE 3. Lateral view of workers. A- *Iridomyrmex purpureus* (Fr. Smith). B- *Tapinoma melanocephalum* (Fabricius). C- *Tapinoma sessile* Say.

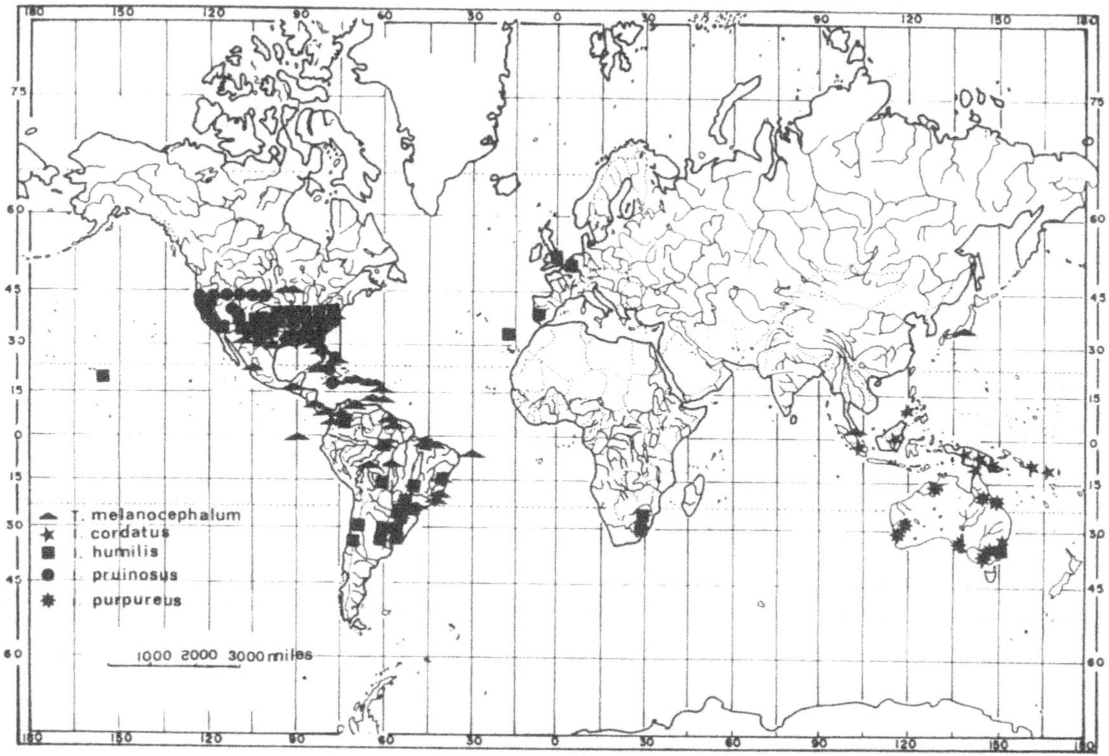

FIGURE 4. Map of geographic distribution of <u>I. cordatus</u>, <u>I. humilis</u>, <u>I. purpureus</u>, <u>I. pruinosus</u>, and <u>T. melanocephalum</u>.

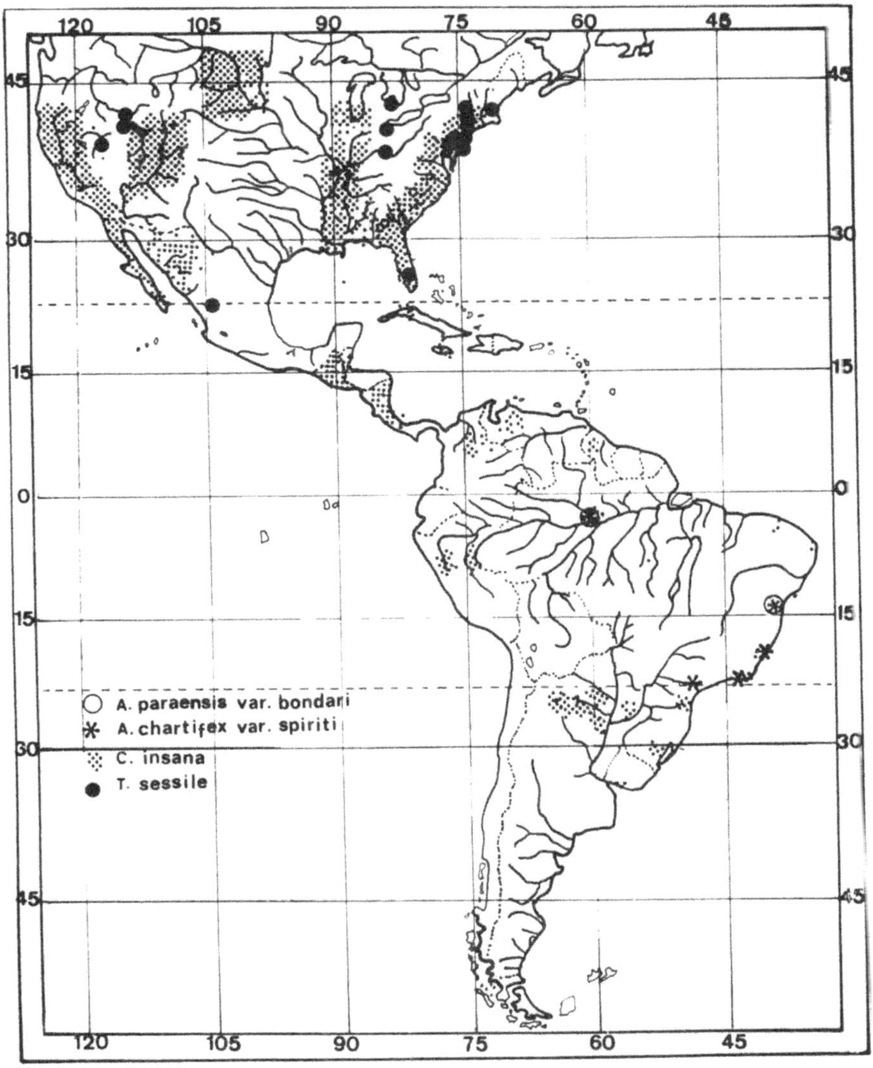

FIGURE 5. Map of geographic distribution of *A. chartifex* var. *spiriti*, *A. paraensis* var. *bondari*, *C.insana*, and *T. sessile*.

Behavioral and Chemical Ecology

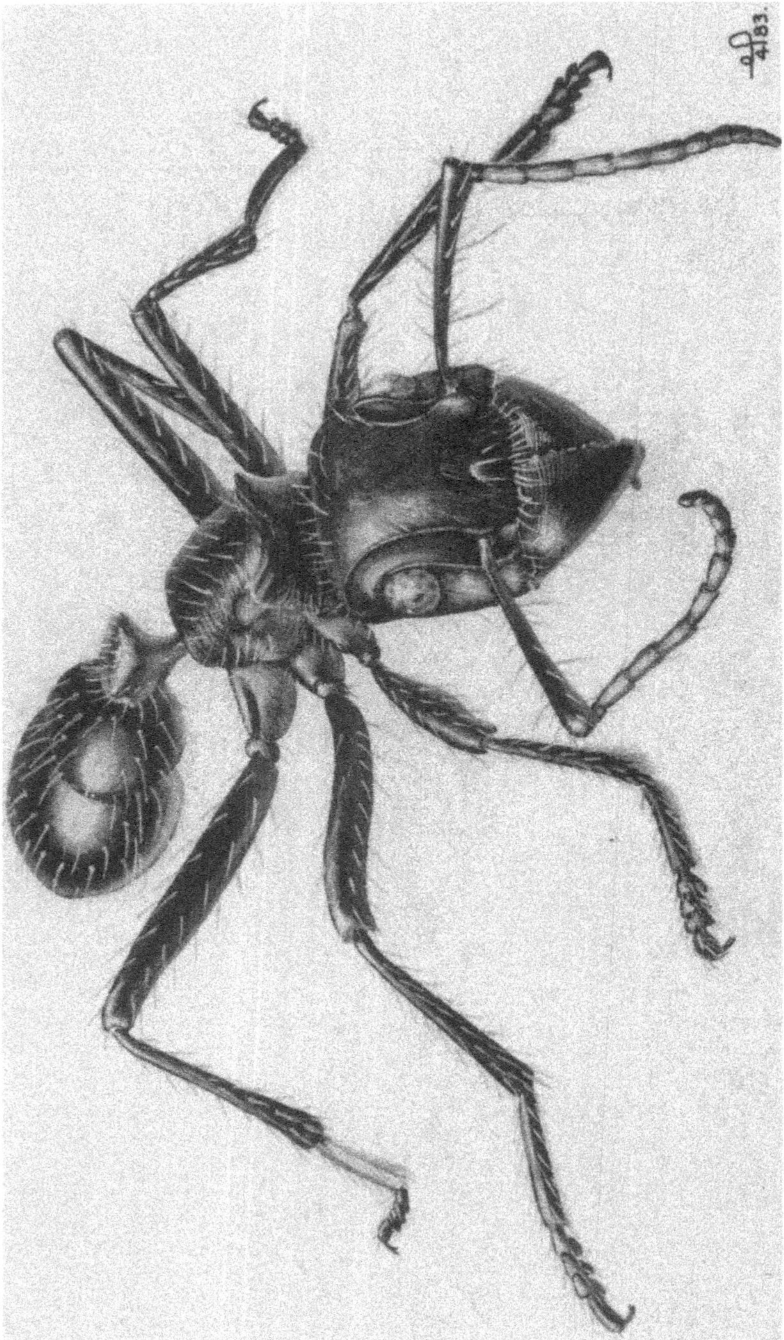

Paraponera clavata (x 8.4). Drawing by Eduardo Perez.

Overview

E.D. Morgan

The inexpert eye may find the activity of individual ants in a colony chaotic and purposeless. It requires patience or time-lapse photography to obtain a perspective view and show how the apparent chaos is coordinated and productive. Perhaps in the same way the individual contributions of researchers, each investigating a limited aspect of the ecology and behavior of these astonishing insects may appear uncoordinated and fragmentary. It requires the advantage of time to see how these investigations knit together, reinforce and illuminate each other. We shall read the contributions collected here with profit, and come back to re-read them many times before we fully appreciate all that they reveal.

Although colony mixing experiments with observations of tolerance or aggression were reported over a hundred years ago (McCook 1879; Lubbock 1882), we are only beginning to understand how ants distinguish nestmates from intruders. Saying that ants recognize a chemical signal is not enough. No doubt the signal is complex and contains much information. Part of the information probably comes from substances on the cuticle, but they alone do not always give a complete answer. Food and environment play a part, and different species react differently under the same experimental conditions. There have been few ideas to provide a new approach to this subject. Vander Meer, Obin and Morel have here tried an original approach by comparing aggressivity in monogyne and polygne groups of *Solenopsis*, and looking at the effect of genetic hybridization on aggressivity. Another approach to this problem of recognition is the study of the behavior of parasites, predators and inquilines. Wojcik surveys an impressively large list of arthropods which inhabit *Solenopsis* nests. There are many intriguing and unanswered questions: how do intruders enter the nest, how do they behave when first inside and how does that behavior change when the inhabitants become accustomed to the intruders?

The enormous economic importance of leaf-cutting ants ensure them a place of prominence in these studies. Salzemann and Jaffe provide a very interesting study on territorial behavior at a site occupied by *Atta laevigata*. Such detailed field experiments are rare and valuable, but only deepen the mystery of how these species distinguish friend and foe and are aware of when they are within their home territory, and where the boundary of that territory lies. The observations of these authors on the site over a long period are valuable in understanding the ever-changing pattern of colony growth, decline, territorial conflicts and new colony foundations. Their descriptions are bound to inspire new behavioral studies for the laboratory-bound investigator. One cannot help noting that the trees growing on nest mounds were never attacked. The suggestion of a "conservational grazing system" by Cherrett (1968) arose from his observations that palatable trees near an *Atta cephalotes* nest were not attacked. This concept has since been a familiar and attractive idea, though not without critics. Reed and Cherrett have a computer model of foraging which incorporates concepts formed using observations gathered from many sources. The results obtained from this model and the model's predictive value in showing that the resultant strategy will produce catastrophic damage in citrus monocultures, seem to be splendid vindications of Cherrett's concept. We must, however, await the response of its critics. Another computer modelling experiment is described here by Aron et al. (this section). It would be splendid if the two models could be fused together. We might then learn not only why *Atta cephalotes*' foraging pattern is successful, but how they achieve it.

The experiments of Guajara et al. (this volume) show that of the various senses and clues available to foraging *Acromyrmex*, the trail odor is the most important orientation clue. Visual and gravitational cues are secondary. This information is an appropriate introduction to the field experiments of Therrien and McNeil with another *Acromyrmex* species in Guadaloupe. *Ac. octospinosus* can create new branch trails to newly discovered food, regardless of whether the food is placed near or far from their trunk trails. Their ability to find food upwind or downwind is impressive. I would like to see Reed and Cherrett's model of foraging applied to Therrien and McNeil's observations of *Ac. octospinosus*' behavior.

A complication in understanding why leaf-cutting ants choose certain plants and not others, is that they are gathering food primarily for a fungus and only secondarily for themselves. What does palatability mean in such circumstances? Three groups of authors have addressed this question. Knapp et al. provide plentiful data on relative preferences of *Ac. octospinosus* for different plants, but our ignorance of what is in these plants makes interpretation of the information difficult. Vasconcelos and Fowler also attempt to understand *Atta* and *Acromyrmex* preferences for certain plants as fungal substrates in terms of palatability, physical deterrence of the plant and its nutrient value, but again we lack information about what is in these plants. It is unlikely that we will have complete chemical characterization of many target plants in the near or foreseeable future. Is it possible to produce a synthetic diet for leaf-cutting ants, from filter paper impregnated with suitable feeding stimulants and nutrients? Bueno et al. avoid some of the complications of this subject by focusing on plant substrate toxicity to the fungus. They draw attention to the deleterious effect of sesame (*Sesamum indicum*) leaves on the fungus garden. *Atta sexdens* readily collect leaves from this non-native plant.

Our knowledge of leaf-cutting ant pheromones extends over more than two decades, but it has not helped us move forward very far in understanding behavior. Howse has drawn together the available data on pheromones and made an admirable synthesis of it that should stimulate both chemists and behaviorists. In spite of the title of this section of the book, there is relatively little new chemistry described here. As an advocate of that craft, I believe that a greater contribution must come from chemists to produce that felicitous combination of ideas with biologists that will most quickly solve some of the problems discussed in these chapters.

The Argentine ant, *Iridomyrmex humilis*, is a tramp species at home in any suitable climate to which our human activity transports it. The vigor and aggressiveness of its colonies are remarkable, and its foraging patterns are a model of organization at this level. Aron, et al. have attempted a bold effort to model the self-organizing pattern of this activity. The construction of mathematical models that can test our assumptions about behavior is a recent advance, and the results here show a good correlation between the model and the way that *I. humilis* explores a virgin territory. Perhaps this and the other modeling efforts in this section will, in retrospect, be viewed as milestones.

Ants are the largest single group of eusocial insects, and although they share the qualities of sociality, they are remarkable for their diversity. 'So singular in each particular' said Shakespeare in another context. The contribution on *Monomorium pharaonis* by Hack and Vinson may be read best with the chapters on this species in Control section. The careful observations of Dejean on the way *Oecophylla longinoda* capture prey, is in marked contrast to the foraging strategy of *Atta* and *Acromyrmex*. Data on the pharmacological effect of any ant venom is scarce, it is therefore encouraging to have some information on the neuromuscular effects of the powerful venom from the giant *Paraponera clavata*. The alkaloidal venoms of *Solenopsis* and *Monomorium* (Escoubas and Blum) are looked at in a quite different way, measuring their insecticidal and antimicrobial activity. The broad spectrum of activity of these venoms prompts the unanswered questions - was there evolutionary pressure to produce antimicrobial activity (i.e. do the ants actively make use of this property) or is it an adventitious property of a venom developed for other purposes?

Ants have been the poor relation to Lepidoptera and Coleoptera in the great surge of research activity for insect pest control of recent decades. In proportion to the importance of social insects in terrestrial ecosystems, very little money has been devoted to understanding their behavior and ecology. At an Oxford meeting on Ant-Plant Interactions, It was stated that half the animal biomass in a tropical forest consists of ants (May 1989). Through the medium of meetings and books like this, the importance of ants in ecology is becoming better known, and through the investigations described here our knowledge of the biology and ethology of these fascinating objects of organization begin to grow.

REFERENCES CITED

CHERRETT, J.M. 1968. The foraging behavior of *Atta cephalotes* L. (Hymenoptera:Formicidae) 1. Foraging pattern and plant species attacked in tropical rain forest. J. Anim. Ecol. 37: 387-403.

LUBBOCK, J. 1882. Ants, Bees and Wasps. International Science Series, Vol. XL, Keegan Paul, Trench & Trubner, London, pp 119-152.

MAY, R.M. 1989. An inordinate fondness for ants. Nature, London. 341: 386-387.

McCOOK, H.C. 1879. Note on adoption of a queen ant. Proc. Acad. Nat. Sci. Philadelphia.

30

Nestmate Recognition in Fire Ants: Monogyne and Polygyne Populations

R.K. Vander Meer, M.S. Obin and L. Morel

INTRODUCTION

Nestmate recognition in ants is defined by the ability of workers to discriminate colony members from non-colony members among conspecifics. This is a well documented behavioral phenomenon (see Breed and Bennett 1987 for review). It is generally accepted that nestmate recognition is based on learned "chemical cues," a subset of overall colony odor (Vander Meer 1988). These cues can be derived from individuals, queens and/or workers ("discriminators") or from the environment (Hölldobler and Michener 1980). The recognition process has two elements: chemical cues on the intruding ant's surface and the resident worker's "neural imprint" or sensory template. If the intruder's cues match the template of the resident no aggression occurs; however, agonistic behaviors may ensue if they do not match.

S. invicta from the United States exist in both monogyne and polygyne forms. Polygyne *S. invicta* colonies were first reported by Glancey et al. (1973). Surveys indicate extensive polygyne populations in most of the 11 infested states of the south and southeast (Fletcher et al. 1980; Mirenda and Vinson 1982; Ross and Fletcher 1985; Glancey et al. 1987). Although the population dynamics of each form are poorly documented, monogyne and polygyne *S. invicta* occur in close proximity to each other. Hölldobler and Wilson (1977) predicted that the multiple matrilinies and patrilinies in a single polygyne nest would lead to reduced intraspecific aggression compared to a typical monogyne situation. For example, the highly polygyne *Monomorium pharaonis*, lives in "unicolonial populations," and does not display aggressive behavior between neighboring colonies (Petersen-Braun 1982). In addition, Hölldobler and Wilson (1977) predicted polygyne colony workers would be more aggressive at the interspecific level than monogyne workers, resulting in decreased ant species diversity.

This paper reviews studies aimed at elucidating the nestmate recognition mechanism of *S. invicta* in both monogyne and polygyne forms. Since both polygyne and monogyne *S. invicta* occur in the same species and the same geographic area, this ant makes a good model to test Hölldobler and Wilson's (1977) predictions. Recognition test comparisons for the two forms also provide insight into the potential practical application of the recognition process.

ENVIRONMENTAL VS HERITABLE CUES

Environmental

Heritable (Mintzer and Vinson 1985) and/or environmental (Jutsum et al. 1979) odors have been reported as ant nestmate recognition cues. Obin (1986) reared *S. invicta* workers from different colonies under identical laboratory conditions and found that they respond less aggressively to non-nestmate laboratory conspecifics than they do to field collected conspecifics. The almost 100 percent recognition of field collected non-nestmates by laboratory reared workers eliminates the possibility that laboratory rearing produced less aggressive workers.

The reduced aggression level observed among laboratory reared colonies supports the hypothesis that environmental odors play a role in *S. invicta* nestmate recognition (Obin 1986). Environmental odor sources include foods, soil and the atmosphere. These odors are passively absorbed unto the ant's lipophilic cuticle through simple contact, social interactions and trophallaxis. The continuously changing odors will be more or less uniformly distributed throughout the colony (Vander Meer 1988).

Heritable

The response of resident lab-reared workers to lab-reared intruders from monogyne colonies was significantly greater than the resident response to nestmates, despite over 18 months of uniform laboratory conditions (Obin 1986). By providing a uniform environment (lab-rearing), the effects of environmental cues can be nullified. The remaining aggression must be due to heritable cues. Although quantitative, the bioassay cannot determine the relative contributions of environmental and heritable cues to the recognition process (see next section).

THE HIERARCHY OF RECOGNITION CUES

Genetically correlated cues may be produced by any colony member; e.g., workers or the queen (Hölldobler and Michener 1980). The rapid distribution of cuticular compounds to other colony members has been demonstrated by: a) mixing of species-specific cuticular compounds in laboratory colonies with two different species (Errard and Jallon 1987; Vander Meer unpublished, b) rapid distribution of radiolabeled markers in fire ant colonies (Sorensen et al. 1985) and c) transfer of species-specific fire ant hydrocarbons to a myrmecophilous beetle (Vander Meer and Wojcik 1982). Transferable queen discriminators have been demonstrated for laboratory colonies of *Camponotus* spp. (Carlin and Hölldobler 1983).

Since aggression persists between non-nestmate *S. invicta* workers reared under identical laboratory conditions, heritable cues derived from either workers or queen are involved. Several experiments designed to test the hypothesis that queen discriminators influence nestmate recognition have been carried out. Addition of an alien queen to small, queenless subcolonies did not significantly increase aggression between recipient and non-recipient subcolonies derived from the same parental colony (Obin and Vander Meer 1988). In addition, parent colony response to intruders from queenless subcolonies was not significantly different from parent colony response to queenright nestmates; the response of parent and corresponding subcolony to non-kin was not different (Obin and Vander Meer 1989a). These data argue against the involvement of queen produced discriminators which, by elimination, leaves worker discriminators as the source of heritable nestmate recognition cues in *S. invicta.*

The aggression observed when colonies are held under uniform environmental conditions is investigative or challenging, rather than attack. This does not mean that environmental cues cause the most aggressive response in *S. invicta.* Until heritable cues can be nullified without affecting environmental cues, the relative contributions of the two cue sources cannot be determined; nor can any synergistic effects. Therefore, the question of cue hierarchy remains unanswered in *S. invicta.*

THE CHEMISTRY OF NESTMATE RECOGNITION

Researchers generally accept that nestmate recognition involves detection of specific chemicals on the cuticle (Wilson 1971). As already mentioned *S. invicta* utilizes environmental and heritable cues. Dietary effects (an environmental component) on nestmate recognition have been clearly demonstrated (Obin 1986, Obin and Vander Meer 1988, Obin and Vander Meer 1989c); however, the heritable part has been much more elusive. Cuticular hydrocarbons, potential heritable cues, have been associated with nestmate recognition in *Camponotus vagus* (Bonavita-Cougourdan et al. 1987) and with the integration and survival of a myrmecophilus beetle in fire ant nests (Vander Meer and Wojcik 1982). Undue excitement, regarding nestmate recognition, surrounds cuticular hydrocarbons as cues because they are readily analyzed and if studied using pattern recognition techniques, individual colonies are distinguishable. This may mean nothing to the ants.

In any event, the hydrocarbons are a part of the heritable component of colony odor and can be used as a general model for heritable nestmate recognition cues. For example they were recently used to investigate the transfer of colony odor, and therefore recognition cues, to a myrmecophilous wasp that parasitizes fire ants (Vander Meer et al. 1989a). While, *S. invicta* colony hydrocarbon patterns are distinctive from colony to colony for a given sampling time, they also change continuously within a colony over time (Vander Meer et al. 1989b). Assuming that all heritable cues are also dynamic rather than static, a worker must, during its lifetime, continually update its perception (template) of colony odor and nestmate recognition cues (Vander Meer et al. 1989b).

The experiment that unambiguously links cuticular chemistry with an aggression bioassay has yet to be done for any ant species.

POLYGYNE VS MONOGYNE RECOGNITION

Polygyne *S. invicta* colonies are polydomous, i.e. each colony occupies several mounds connected by subterranean foraging tunnels. Consequently, a colony's boundaries are not clearly defined. The average coefficient of relatedness within monogyne colonies is indistinguishable from the theoretical 0.75 expected in colonies with one queen inseminated by a single male. In contrast, within-nest coefficient of relatedness in polygyne populations is not significantly different from 0. This is due to the large number of different matrilinies in a single polygyne colony (Ross and Fletcher 1985).

Mirenda and Vinson (1982) found that, in the context of colony fusion, laboratory reared *S. invicta* polygyne nests tolerated neighboring mounds to a greater extent than monogyne mounds. However, we know overall discrimination is greatly diminished by the laboratory's uniform environment (Obin 1986, unpublished). To alleviate this problem we compared nestmate recognition in polygyne and monogyne *S. invicta* populations with aggression bioassays conducted on ants in their nest soil, soon after collection from the field (Morel et al. in press, unpublished).

No aggressive behavior was observed in intercolonial introductions, although there was a heightened awareness (rapid antennation) compared to intracolonial controls. This supports the contention that territoriality within polygyne populations is almost nonexistent and that all workers are treated as nestmates. Thus polygyne *S. invicta* populations fall under Hölldobler and Wilson's (1977) definition of a super- or unicolony.

Polygyne workers are also non-aggressive toward monogyne workers whether or not they are intruder or resident. But the monogyne workers were highly aggressive toward polygyne workers . However, *S. richteri* intruders were vigorously attacked by polygyne workers indicating that polygyne workers' potential aggressivity was not diminished.

Environmental cues can be dampened through uniform conditions, but apparently in natural *S. invicta* polygyne populations, both environmental and heritable cues are dampened. It is reasonable to assume that nestmate recognition cues would spread throughout a population by social interactions and unchallenged movement of workers. However, an area-wide uniform cue pattern would be unreasonable to expect. Instead we propose that since polygyne workers contact a large number of different cues, environmental and heritable, the observed lack of aggression is due to a much less discriminating template rather then uniform cue profile. The asymmetry in aggressive behavior of monogyne and polygyne *S. invicta* supports this hypothesis. Just as for monogyne workers, at any point in time each polygyne worker has a distinct cue pattern; however, the polygyne worker template is broadly defined compared to that of monogyne workers.

HYBRID VS PARENTS

Hybridization between *S. invicta* and *S. richteri* was first reported by Vander Meer et al. (1985). It was later found to be reproductively viable (Ross et al. 1987) and to occupy an extensive range in northeastern Mississippi, northern Alabama and northwestern Georgia (Diffie et al. 1988). Within this range it borders both parental types only in northeastern Mississippi. The most plausible explanation for this asymmetric distribution is hybrid superiority in habitats intermediate to those favoring parental forms (Moore's [1977] "bounded hybrid superiority"). Ross et al. (1987a,b) found greater heterozygosity values for the hybrid than parental forms, which supports hybrid superiority.

Competition for territory and resources is significantly mediated by competitor-enemy recognition (see Hölldobler and Carlin 1987 for review). Intraspecific and interspecific aggression bioassays were used to measure the ability of the two parental fire ant species and their hybrid to detect intruders (Obin and Vander Meer 1989b).

All aggression bioassays were done with laboratory colonies reared under identical conditions, thus the results were largely due to heritable nestmate recognition cues. Within the resolution of the bioassay, there were no significant differences for between population aggression scores; therefore, in the context of interspecific recognition no inferiority or superiority can be inferred.

Among the two parental forms, intraspecific aggression was significantly lower than interspecific; however, the scores for the hybrid were not significantly different. These data support the hypothesis that variability of heritable recognition cues is greater for hybrids than for either of the parents, as well as the suggestion (Ross et al. 1987) that fire ant hybrid zone stability may be a consequence of greater genetic diversity.

The intraspecific aggression scores suggest the following hierarchy of heterozygosity: *S. invicta* < *S. richteri* < hybrid (Obin and Vander Meer 1989b), which differs from the order determined by Ross et al. (1987) using isozymes. Thus, the data do not support an asymmetric " wave of advance" (see Barton and Hewitt 1985) in which the hybrid's movement is northward to a competitively inferior *S. richteri*. The extent that the aggression scores or for that matter isozymes reflect overall genetic heterozygosity is unknown, since each is looking at only a small part of the genome.

CONCLUSION

An understanding of the nestmate recognition process in fire ants has applications in many areas of study other than those discussed above. For example, differences in monogyne versus polygyne intraspecific aggression provide a facile way to determine if a nest is polygyne or not, without disruption of the mound (Morel et al. in press). The lack of intraspecific aggression in polygyne *S. invicta* gives a behavioral bioassay for determining if suspected South American *S. invicta* populations are the same as in the United States (Vander Meer and Patterson unpublished). Recognition of an intruder as non-nestmate initiates a host of non-chemical and semiochemical defenses, whether the intruder is another ant, parasitoid or parasite. Our understanding of nestmate recognition provided insights into myrmecophile integration mechanisms (Vander Meer and Wojcik 1982; Vander Meer et al. 1989a). Other areas of research for the future include the definition of recognition cue chemistry (their environmental and genetic bases) and the use of nestmate recognition differences to gain a better understanding of *S. invicta*'s transition from monogyne to polygyne.

REFERENCES

BARTON, N.H. and HEWITT, G.M. 1985. Analysis of hybrid zones. Ann. Rev. Ecol. Syst. 16;113-148.

BONAVITA-COUGOURDAN, A., CLEMENT, J. L., and LANGE, C. 1987. Nestmate recognition: Cuticular hydrocarbons and colony odor in the ant *Camponotus vagus* Scop. J. Entomol. Sci., 22: 1-10.

BREED, M., and BENNETT, B. (1987). Kin recognition in highly eusocial insects. In Fletcher, D. J. C., and Michener, C. D. (eds.), Kin recognition in animals. John Wiley, New York, pp.243-286.

CARLIN, N. F., and HÖLLDOBLER, B. 1983. Nestmate and kin recognition in interspecific mixed colonies of ants. Science, 222: 1027-1029.

DIFFIE, S., VANDER MEER, R. K., and BASS, M. H. 1988. Discovery of hybrid fire ant populations in Georgia and Alabama. J. Entomol. Sci. 23:187-191.

ERRARD, C. and JALLON, J. M. 1987. An investigation of the development of the chemical factors in ants intra-society recognition. In: Chemistry and Biology of Social Insects, J. Eder and H. Rembold (eds.), Verlag J. Peperny, Munchen, p. 478.

FLETCHER, D. J. C., BLUM, M. S., WHITT, T. V., and TEMPLE, N. (1980). Monogyny and polygyny in the fire ant, *Solenopsis invicta.* Ann. Ent. Soc. Amer. 73: 658-661.

GLANCEY, B. M., CRAIG, C. H., STRINGER, C. E., and BISHOP, P. M. (1973). Multiple fertile queens in colonies of the imported fire ant, *Solenopsis invicta.* J. Georgia Ent. Soc. 8: 327-328.

GLANCEY, B. M., NICKERSON, J. C. E., WOJCIK, D., TRAGER, J., BANKS, W. A., and ADAMS, C. T. (1987). The increasing incidence of the polygyne form of the red imported fire ant, *Solenopsis invicta* (Hymenoptera: Formicidae), in Florida. Florida Ent. 70: 400-402.

HÖLLDOBLER, B., and WILSON, E. O. (1977). The number of queens: An important trait in ant evolution. Naturwiss. 64: 8-15.

HÖLLDOBLER, B., and MICHENER, C. D. (1980). Mechanisms of identification nd discrimination in social Hymenoptera. In Markl, H. (ed.) Evolution of social behavior: Hypotheses and empirical tests. Verlag Chemie, Weinheim, pp. 433-439.

HÖLLDOBLER, B., and CARLIN. N. F. (1987). Anonymity and specificity in the chemical communication signals of social insects. J. Comp. Physiol. A 161: 567-581.

JUTSUM, A. R., SAUNDERS, T. S., and CHERRETT, J. M. 1979. Intraspecific aggression in the leaf-cutting ant *Acromyrmex octospinosus.* Anim. Behav., 27: 839-844.

MINTZER, A., and VINSON, S. B. 1985. Kinship and incompatibility between colonies of the acacia ant *Pseudomyrmex ferruginea.* Behav. Ecol. Sociobiol. 17: 75-78.

MIRENDA, J. T., and VINSON, S. B. (1982). Single and multiple queen colonies of imported fire ants in Texas. Southwest. Ent. 7: 135-141.

MOORE, W.S. 1977. An evaluation of narrow hybrid zones in vertebrates. Quart. Rev. Biol. 52:263-277.

MOREL, L., VANDER MEER, R.K. and LOFGREN, C.S. In press. Comparison of nestmate recognition between monogyne and polygyne populations of the fire ant, *Solenopsis invicta* Buren (Hymenoptera:Formicidae). Ann. Entomol. Soc. Am.

OBIN, M. S. (1986) Nestmate recognition cues in laboratory and field colonies of *Solenopsis invicta* Buren (Hymenoptera: Formicidae). J. Chem. Ecol. 12: 1965-1975.

OBIN, M. S. and VANDER MEER, R. K. (1988). Sources of nestmate recognition cues in the imported fire ant *Solenopsis invicta* Buren (Hymenoptera: Formicidae). Anim. Behav. 36:1361-1370.

OBIN, M.S. and VANDER MEER, R.K. 1989a. Nestmate recognition in fire ants (*Solenopsis invicta* Buren): Do queens label workers? Ethology 80:255-264.

OBIN, M.S. and VANDER MEER, R.K. 1989b. Between- and within-species recognition among imported fire ants and their hybrids (Hymenoptera: Formicidae): Application to hybrid zone dynamics. Ann. Entomol. Soc. Am. 82:649-652.

OBIN, M.S. and VANDER MEER, R.K. 1989c. Mechanism of template-label matching in fire ant, *Solenopsis invicta* Buren, nestmate recognition. Anim. Behav. 37: .

PETERSEN-BRAUN, M. (1982). Intraspezifishes Aggressionsverhalten bei der Pharaaomeise *Monomorium pharaonis* L. (Hymenoptera, Formicidae). Insectes Soc. 29: 25-33.

ROSS, K. G., and FLETCHER, D. J. C. (1985). Comparative study of the genetic and social structure in two forms of the fire ant *Solenopsis invicta* (Hymenoptera: Formicidae). Behav. Ecol. Sociobiol. 17: 349-356.

ROSS, G., R. K. VANDER MEER, D. J. C. FLETCHER, and E. L. VARGO. 1987a. Biochemical phenotypic and genetic studies of two introduced fire ants and their hybrid (Hymenoptera: Formicidae). Evolution 41: 280-293.

ROSS, K.G., VARGO, E.L. and FLETCHER, D.J.C. 1987b. Comparative biochemical genetics of three fire ant species in North America, with special reference to the two social forms of *Solenopsis invicta* (Hymenoptera: Formicidae). Evolution 41:979-990.

SORENSEN, A. A., FLETCHER, D. J. C., and VINSON, S. B. 1985. Distribution of inhibitory queen pheromone among virgin queens of the ant, *Solenopsis invicta*. Psyche 92: 57-69.

VANDER MEER, R. K. 1988. Behavioral and biochemical variation in the fire ant, *Solenopsis invicta*, In: Interindividual Behavioral Variation in Social Insects, R.L. Jeanne, (ed.). Westview Press, Boulder, CO, pp. 223-255.

VANDER MEER, R. K., and WOJCIK, D. P. 1982. Chemical mimicry in the myrmecophilous beetle *Myrmecaphodius excavaticolis*. Science, 218: 806-808.

VANDER MEER, R. K., LOFGREN, C. S., and ALVAREZ, F. M. (1985). Biochemical evidence for hybridization in fire ants. Florida Entomol. 68: 501-506.

VANDER MEER, R.K., JOUVENAZ, D.P. and WOJCIK, D.P. 1989a. Chemical mimicry in a parasitoid (Hymenoptera:Eucharitidae) of fire ants (Hymenoptera:Formicidae). J. Chem. Ecol. 15: 2247-2261.

VANDER MEER, R.K., SALIWANCHIK, D. and LAVINE, B. 1989b. Temporal changes in colony cuticular hydrocarbon patterns of *Solenopsis invicta*: Implications for nestmate recognition. J. Chem. Ecol. 15:2115-2125.

WILSON, E. O. 1971. The insect societies. Harvard Univ. Press, Cambridge, MA, 548 p.

31

Behavioral Interactions of Fire Ants and Their Parasites, Predators and Inquilines

D.P. Wojcik

INTRODUCTION

Arthropods that live with, or are associated with, ants are termed myrmecophiles (Wilson 1971, Kistner 1982). This relationship can be permanent or temporary, obligatory or facultative, integrated or nonintegrated. The biology and behavior of these relationships is inadequately known, with much of the information reported in the literature based on cursory observations or invalid interpretations (e.g. miscellaneous soil arthropods floated out of fire ant colonies, Collins and Markin 1971). References to myrmecophiles and their host relationships are numerous, widely scattered and often as difficult to find as the myrmecophiles.

The generally accepted classification of arthropod symbionts is based on a series of works by E. Wasmann, translated into English by Wheeler (1910). Wasmann devised 5 behavioral categories (1) synechthrans (persecuted guests), (2) synoeketes (indifferently tolerated guests), (3) symphiles (true guests), (4) ectoparasites and endoparasites and (5) trophobionts (provide secretions to the ants). While the categories are mutually exclusive, the myrmecophiles can often be placed in different categories at different stages in their life-cycles (Wilson 1971). Several other classification systems have been proposed (Kistner 1979), but none have become as generally accepted as the Wasmannian system. In most cases, the true relationships of the myrmecophiles are unknown, indeed many are still being discovered and their relationships with their hosts defined.

The mechanisms by which myrmecophiles integrate into the host societies have been the subject of many studies with much discussion of body shape, color, special morphological adaptations and appeasement substances (reviewed by Wilson 1971, Kistner 1979). Recently, some inquilines have been shown to use chemical mimicry to integrate themselves into the host colony. These organisms, several termitophilous Staphylinidae (Howard et al. 1980, 1982) and a myrmecophilous Scarabaeidae (Vander Meer and Wojcik 1982), acquire the cuticular hydrocarbons (part of the colony odor) of the host. Other studies by Vander Meer, Wojcik and Jouvenaz (unpublished) indicate this is a wide-spread mechanism in myrmecophiles of *Solenopsis* spp., including some discussed in this paper.

The majority of published papers often document only the occurrence

TABLE 1. Insect groups which have been reported associated with fire ants but their biology and behavior have not been studied.

ORDER Family	Reference
BLATTARIA	
Atticolidae	Roth and Willis 1960
COLEOPTERA	
Anticidae	Bruch 1917, Wasmann 1918b
Carabidae	Wasmann 1918b, Neece and Bartell 1981, Kistner 1982
Chrysomelidae	Wojcik et al. 1987a
Histeridae	Wasmann 1918a, 1918b, Bruch 1929, Helava et al. 1985, Wojcik et al. 1987a
Melandryidae	Neece and Bartell 1982
Scydmaenidae	Bruch 1917
Staphylinidae	Wasmann 1912, 1918a, 1925a, Bruch 1917, 1929, Travis 1941, Silveira Guido et al. 1965, Collins and Markin 1971, Kistner 1982, Neece and Bartell 1982, Banks et al. 1985
Tenebrionidae	Fall 1928, Silveira Guido et al. 1963 Summerlin 1978, Neece and Bartell 1981, 1982, Steiner 1982
DIPTERA	
Syrphidae	Borgmeier 1923, 1953
HEMIPTERA	
Cydnidae	Neece and Bartell 1981 Wasmann 1918a, Silveira Guido et al. 1965
HOMOPTERA	Borgmeier 1923, Travis 1941, Anon 1958, Wilson 1971, Bartoszeck 1976, Wojcik 1986
Cixiidae	Sheppard et al. 1979, Neece and Bartell 1982
HYMENOPTERA	
Bethylidae	Bruch 1917, Kieffer 1921, Wasmann 1925b, Ogloblin 1963
Ceraphronidae	Bruch 1929
Diapriidae	Kieffer 1921
Ichneumonidae	Wasmann 1912
STREPSIPTERA	
Myrmecolacidae	Ogloblin 1939, Teson and de Remes Lenicov 1979

of myrmecophiles with their hosts. In preparing this review, I have limited citations to those references which present or review biological or behavioral information on fire ant myrmecophiles. Groups which have been reported as associated wih fire ants, but where no biology or behavior have been reported are listed in Table 1. The fire ant names used in this paper are used in the sense of Trager (in press, A revision of the fire ants, *Solenopsis geminata* group (Hymenoptera: Formicidae, Myrmicinae).

DISCUSSION

Acarina

The mite fauna of fire ant nests has not been well studied. In only one case has a mite, a tarsonemid sp., been observed to feed on liquids from the mouthparts of a fire ant, *Solenopsis geminata* (F.) and occasionally feed on excreta at the ant's anus (Travis 1941). The mites stand erect, bobbing up and down on the ants, while tapping them with the first pair of legs. Some species belonging to known predaceous or parasitic groups were collected from fire ant nests (Hermann et al. 1970, Hunter and Costa 1971, Neece and Bartell 1982), but predation or parasitism was not observed. Most of the mites collected while on the ants bodies are assumed to be phoretic. Some have been collected only on workers (Anon 1958), others on female alates before (Hermann et al. 1970, Hirschmann 1972) or after mating flights (Hermann et al. 1970) and some on both ant castes (Hermann et al. 1970, Collins and Markin 1971, Wojcik unpublished). Occasionally the mites were so numerous that the female alates were unable to fly (Hermann et al. 1970). Some phoretic mite species are not host specific and have been collected on several genera of ants including *Solenopsis* (e.g. *Oplitis exopodi* Hunter and Farrier, *Oplitis virgilinus* Hunter and Farrier, Hunter and Farrier 1975, 1976). As might be expected since phoretic mites only use the ants for transportation, they are not always found in association with ants (e.g. *Oplitis* spp., Hunter and Farrier 1975, 1976).

Diplopoda

Several species of millipedes are considered fire ant myrmecophiles, although little is known of their biolcgy and behavior (Schubart 1950). *Calytodesmus schubarti* Causey is a facultative fire ant myrmecophile also found with other genera of ants in Florida and Georgia (Wojcik and Naves in press). This species scavenges from the nests of *S. geminata* and *S. invicta* Buren, feeding on organic debris that includes dead or injured fire ant adults and brood in the colonies. The millipedes are generally ignored by the ants, but can be killed and eaten by the ants in stress situations. The millipedes have a defensive secretion which usually stops ant aggressive behavior. Similar unidentified species have been found with several fire ant species in Brazil (Banks et al. 1985, Wojcik 1986, Wojcik et al. 1987a) and recently with *S. richteri* Forel in Argentina (Wojcik unpublished).

Coleoptera: Pselaphidae

A number of pselaphids are known myrmecophiles, ranging from well-integrated symphiles with well-developed trichomes to barely tolerated scavengers (Kistner 1982). Pselaphid beetles from 8 genera (8 species) have been collected from fire ant nests (Wojcik in preparation), but nothing is known of most species biology. *Eurhexius putzeyssi* Schauff., *Ctenisis fasciculata* Raffray and *Pselaphomorphus bruchi* Raffray (Bruch 1917) have been collected only in abandoned or semi-abandoned fire ant mounds. *Metopioxys gallardoi* Bruch is known from *S. richteri* nests in Argentina and Uruguay (Bruch 1917, San Martin 1968a). This species has been kept in laboratory colonies for up to 3 months, with the beetles often observed riding about the nest on the ants' backs (Silveira Guido et al. 1964, San Martin 1968a). The food habits of this species remain a mystery.

Fustiger elegans Raffray is considered a symphile and has been collected from nests of *S. richteri* and *Solenopsis* spp. in Argentina and Uruguay (Bruch 1930, San Martin 1968b, Wojcik unpublished). It is well integrated into the fire ant society with 572 beetles found in one mound (San Martin 1968b) and it survives during stress situations (Wojcik unpublished). San Martin (1968b) reported a laboratory lifespan for this species of up to 154 days. They have been observed moving about the colony with fire ant eggs in their mandibles (Silveira Guido et al. 1963, San Martin 1968b). These findings, coupled with the observations that the beetles usually stay in close proximity to or in the brood piles, are taken as evidence that this species preys on ant eggs (Silveira Guido et al. 1963, San Martin 1968b). This species was collected from *S. richteri* nests in Argentina and was observed preferentially staying in the brood piles in laboratory colonies (Wojcik unpublished). This species has trichomes which are highly attractive to the ants, even to the point of the ants rescuing live and dead individuals in disturbed laboratory situations (Wojcik unpublished). *F. elegans* can often be seen (in laboratory nests) climbing onto the ants and even riding around the nests on the ants (San Martin 1968b, Wojcik unpublished). Hexane washes of this species have shown the presence of the host ant cuticular hydrocarbons (Vander Meer and Wojcik unpublished) which implies chemical mimicry as one of its integrating mechanisms.

Coleoptera: Scarabaeidae

This large and diverse family has many myrmecophilous genera and species in 6 subfamilies (Kistner 1982). A mere 4 species in 4 genera have been associated with fire ants. Collection records are only available for 2 of the species: *Rhyssemus neglectus* Brown was collected from the nests of *S. invicta* and *S. xyloni* McCook (Summerlin 1978) and *Myrmecaphodius proseni* Martínez was collected from nests of *Solenopsis saevissima* (F. Smith) (Martínez 1952).

Euparia castanea Serville occupies the nests of *S. geminata* and *S. xyloni* from the southern United States to Costa Rica (Chalumeau and Howden 1984). Travis (1941) was the first to observe the ants gnawing on the beetles at the junction of prothorax and elytra. I have also seen this behavior, but have never found trichomes (Wojcik unpublished). I found that this beetle ate ant larvae, and fed by strigilation on ant larvae and dead ants (Wojcik 1975). The beetles also feed on other dead insects, indicating scavenging. The flight periods of this species are not correlated with the mating flights of its hosts (Wojcik et al. 1978); the beetles flew in all stages of ovarian development (Wojcik and Habeck 1977).

Chalumeau (1983) described the scarab beetle *Martinezia dutertrei* Chalumeau (= *Myrmecaphodius excavaticollis* Blanchard) from Cuba, giving its distribution as United States, Cuba, Brazil, Paraguay and Argentina. This species has been found with *S. geminata*, *S. xyloni*, *S. invicta* and *S. richteri* in the United States (Wojcik et al. 1977). *M. dutertrei* and 4 other *Martinezia* spp. were collected in Brazil, Bolivia, Paraguay, Argentina and Uruguay, but identifications of the host ants were not given (Chalumeau 1983). Beetles of this genus have been collected from *S. invicta*, *S. richteri* and *Solenopsis* spp. nests in South America (Banks et al. 1985, Wojcik 1986, Wojcik et al. 1987a, Wojcik unpublished). Wojcik (1975) demonstrated that this beetle is a predator feeding on ant larvae. It also fed as a scavenger on ants and other insects and by strigilation on dead ants. The ants usually treat this species as a symphile, even ignoring the beetles as they consumed living host larvae, but the ants will eat the beetles in stress situations (Wojcik 1975, unpublished). Ants are often observed to gnaw on the beetles at the junction of the prothorax and the elytra as observed on *E. castanea*, though no trichomes have been found (Wojcik unpublished). The beetles have the ability to acquire the host cuticular hydrocarbons even when switching hosts (Vander Meer and Wojcik 1982), allowing them to live with more than one ant host (Wojcik et al. 1977, Summerlin 1978). The flight periods of the beetle are not correlated with the mating flights of the host ants (Wojcik et al. 1978) and the beetles flew in all stages of ovarian development (Wojcik and Habeck 1977). This species was probably introduced into North America with one or more of its imported ant hosts (Lofgren et al. 1975). As its imported hosts, *S. invicta* and *S. richteri*, spread in the United States and displace the native fire ants, *S. geminata* and *S. xyloni*, *M. dutertrei* will probably become more common, while the native *E. castanea* will become rarer (Wojcik et al. 1977).

Coleoptera: Staphylinidae

Hundreds of staphylinid species from this widespread and diverse beetle family are known myrmecophiles (Kistner 1982). No information is available on 41 species in 23 genera in 4 subfamilies known from fire ant nests (Table 1). Only 4 genera, *Myrmecosaurus*, *Dinardopsis*, *Euthorax* and *Dinusella*, appear to have special morphological adaptations for living in fire ant nests (Kistner 1982). Out of the 22+ species in these genera, behavioral information is available only for *Myrmecosaurus ferrugineus* Bruch. Wojcik (1980) observed this non-predaceous symphile soliciting food and being fed by the host *S. invicta*. It is well integrated into the fire ant society and was probably introduced into North America with one of its imported hosts (Lofgren et al. 1975). Hexane washes of *Myrmecosaurus* spp. (collected from *S. richteri* and *Solenopsis* spp. nests in Argentina) reveal that they acquire the hosts' cuticular hydrocarbons (Vander Meer and Wojcik unpublished).

The genus *Oxypoda* does not have special morphological adaptations for the myrmecophilous existence. Six *Oxypoda* sp. specimens (identified by J.H. Frank, Univ. Florida), collected from a *S. geminata* nest in Gainesville, Fl, were observed in laboratory nests for up to 8 days (Wojcik unpublished). The beetles were treated as synechthrans (persecuted guests) with the host ants pursuing and attacking the beetles continuously. The beetles constantly presented their abdominal tips to the ants' heads, an act which neutralized the ants aggressive behavior. Similar behavior has been reported in other staphylinid genera (e.g. 2 *Pella* spp., Kistner and Blum 1971), and there are many reports of defensive secretions in staphylinids (Kistner 1982). The *Oxypoda* sp., often hiding in the brood pile, licked (strigilated) the ant pupae and appeared in several instances, to grab the pupae with their mouthparts, trying to tear the pupal cuticle without success.

Diptera: Phoridae

Most Phoridae are of an unknown status, including most of the genera associated with fire ants (Kistner 1982, Wojcik in preparation). Biological information exists only for some *Pseudacteon* species from North and South America. The ant head is commonly the oviposition site (*P. obtusus* Borgmeier, Williams and Banks 1987; *P. crawfordii* Coquillett, Coquillett 1907, Wheeler 1910 and Smith 1928), although Feener (1987) provided detailed observations of *P. crawfordii* ovipositing on the host's prothorax at the junction of the dorsal and ventral sclerites. In only a few instances have eggs (*P. obtusus*; Williams and Banks 1987), or pupae (unknown genus, species; Wojcik et al. 1987b), been observed on or in the host. Generally when the flies attack the ants, they hover over foraging trails (*P. obtusus*, William and Banks 1987; *Pseudacteon* spp., Wojcik unpublished), over nests during mating flights (*P. spatulatus* (Malloch), *P. crawfordii*, Smith 1928), or over disturbed nests (*Pseudacteon* spp., Williams 1980; Wojcik unpublished). These situations are probably similar as far as the flies are concerned. The flies dart onto the ants, and quickly attempt to deposit an egg (*P. obtusus*, Williams and Banks 1987) or insert the ovipositor in the appropriate place (*P. crawfordii*, Feener 1987). The ants sometimes appear stunned after oviposition (*P. spatulatus* and *P. crawfordii*, Smith 1928; *Pseudacteon* spp., Williams 1980). The ants vigorously defend themselves against the flies (*P. spatulatus*, *P. crawfordii*, Smith 1928; *P. crawfordii*, Feener 1987; *P. obtusus*, Williams and Banks 1987; *Pseudacteon* spp., Williams et al. 1973; Wojcik unpublished), often chasing the flies away before oviposition (*Pseudacteon* spp., Williams 1980).

In fire ants, the major workers are preferentially attacked while the minors are usually ignored (*P. crawfordii*, Feener 1987; *P. obtusus*, Williams and Banks 1987). The flies probably oviposit selectively on major workers because they represent a larger, more stable and longer-lived food source (Feener 1981, 1987, Williams and Banks 1987). Alate ants, while representing a larger food source, are much less common in colonies and are rarely exposed to parasitism. Only one instance of phorid parasitism of alate ants is known (Wojcik et al. 1987b).

A different scenario of parasitism has been reported for *Apodicrania termilophila* (Borgmeier), a fly found parasitizing several *Solenopsis* spp. in South America (Williams 1980). The adult flies walked around on the disturbed nest and the ants ignored them. Immature fly larvae are endoparasitic in the ant larvae while mature fly larvae and pupae are found in the nest being tended by the ants (Williams 1980).

Commoptera solenopsidis Brues, a wingless species, was discovered running around inside nests of *S. geminata* (Brues 1901). Unidentified wingless phorids have been found in *S. richteri* nests in Argentina (Wojcik unpublished). These phorids, like the ones described by Brues (1901), run jerkily among the ants, never contacting the ants.

Hemiptera

Many free-living Hemiptera prey on ants (Kistner 1982), with *Acanthaspis concinnula* Stal reported as a specific predator of *S. geminata* in Ceylon (Mühlenberg and Maschwitz 1976). The bugs lie in wait for an ant to pass by, then pierce the ants with their mouthparts and inject poison. Once subdued, the ant is sucked dry. The carcass, added to the debris on the bugs back, acts as camouflage and an attractant for more ants. This relationship needs more study because *S. geminata* is not native to Ceylon, but was probably introduced in colonial times.

Two species of true myrmecophiles are known from fire ant nests in South America. *Anommatocoris coleopteratus* (Kormilev), family Tingidae, is known from *S. richteri* and *S. saevissima* nests in Argentina and Uruguay, but nothing is known of its biology (San Martin 1966b). *Neoblissus parasitaster* Bergroth, family Lygaeidae, is a well-known associate found in *S. richteri* and *S. saevissima* nests in Argentina and Uruguay (San Martin 1966a). These bugs can be extremely numerous in fire ant mounds. Bruch (in San Martin 1966a) reported thousands in one nest and San Martin (1966a) commonly captured several hundred from one nest. I have collected over 100 from a single nest sample several times (Wojcik unpublished). The bugs are treated as symphiles by the ants and only killed in laboratory-created stress situations (San Martin 1966a, Wojcik unpublished). The bugs are aided in ecdysis by the ants (San Martin 1966a). The feeding relationship of bug to ant is unclear at this time. Although they commonly accumulate in brood piles, the ants sometimes chase them away (San Martin 1966a, Wojcik unpublished). This species has a distinctive strong defensive odor which it uses on the ants when they become too aggressive (Wojcik unpublished). San Martin (1966a) thinks the bugs feed on the abundant plant roots, particularly grasses, found in the ant mounds.

Hymenoptera: Eucharitidae

All eucharitid wasps are considered parasites of ants (Kistner 1982). Fire ant parasites are known only from the genus *Orasema* which also includes species parasitic to other ant genera. The following synthesis includes the known information about the fire ant parasites *Orasema crassa* De Santis, *Orasema aenea* Gahan, *Orasema rapo* Walker (Williams 1980) and *Orasema* spp. (Silveira Guido et al. 1964, Williams 1980, Wojcik unpublished). The females lay their eggs in plant tissue away from the hosts; apparently the wasps are not restricted to specific plants or plant parts (Parker 1942, Tochetto 1942). *Orasema* spp. have a highly sclerotized scimitar-shaped ovipositor (Heraty 1985) and insert their eggs singly into incisions in the leaf surfaces (Clausen 1940).

After hatching, planidia (first instar larvae) wait for an appropriate ant host on the plant surface. The legless planidia (Heraty and Darling 1984) climb on to the ant, and ride phoretically back to the ant nest, attached by means of an anal sucker. In the ant nest, the planidia leave the adult ant and transfer to the ant brood (larvae and pupae), probably via the nurse ants during feeding. Upon locating an ant larvae, the planidia burrow beneath the larvae's cuticle (Wheeler and Wheeler 1937, Heraty and Wojcik unpublished). The parasite only developes after the ant larva reaches the late prepupal stage

(Wheeler 1907, Clausen 1941). Then the parasite emerges from the host and feeds as an ectoparasite on the host (at least in some cases, Wojcik unpublished) after the host pupates. In other cases (Wheeler and Wheeler 1937), the parasite feeds at least partially as an endoparasite. This feeding results in a characteristically deformed pupae called a phthisergate (Wheeler 1907). Indirect evidence (Wojcik unpublished) indicates that the *Orasema* spp. which parasitize *Solenopsis* spp. in Mato Grosso, Brazil may require more than one host pupae. This is suggested by the mature parasite larvae's size and the variable sizes of the phthisergates. In those instances where the phthisergates survive the feeding by the wasp larvae, they are unable to complete development and eventually die (Wheeler 1910, Wojcik unpublished).

After the mature wasp larva terminates feeding, it leaves the host and is kept with the ant brood, where it is cared for by the worker ants (Wojcik unpublished). The parasite pupates with the assistance of the worker ants (just as they assist the ant larvae) (Silveira Guido et al. 1964, Wojcik unpublished). During the pupal period, the parasite pupae are mixed in with the host pupae, and are cared for like ant pupae (Silveira Guido et al. 1964, Williams 1980, Wojcik unpublished). Worker ants assist eclosion by the parasites in the same manner as the ants assist their own pupae (Silveira Guido et al. 1964, Wojcik unpublished). Fed and groomed by the worker ants, the adult parasites are integrated completely into the ant society (Silveira Guido et al. 1964, Williams 1980, Wojcik unpublished). When a nest is disturbed, the pupal and adult parasites are rescued by the ants in preference to their own brood (Wojcik unpublished). The manner in which the *Orasema* spp. adults leave the nest has not been observed, but in other eucharitid genera the ants either ignore the parasites when they are leaving the nest (Clausen 1923, 1941) or they actively remove them from the nest (Ayre 1962). The males hover over the ant nest, or rest on surrounding vegetation (Williams 1980, Wojcik and Jouvenaz unpublished). The adult parasites must be sexually mature when they leave the nest because mating takes place immediately (Williams 1980). Studies of cuticular hydrocarbons have shown that *Orasema* sp. larvae, pupae and adults possess only *Solenopsis* sp. host cuticular hydrocarbons while in the ant nest; upon leaving the host nest, the adult *Orasema* sp. acquire species-specific cuticular hydrocarbons (Jouvenaz et al. 1988, Vander Meer et al. 1989).

I should note that oviposition by *Orasema* spp. in plant tissue can sometimes cause cosmetic economic damage to bananas and other fruits (Tocchetto 1942, Nicolini 1950, Roberts 1958). In one case, damage to bananas by *Orasema* sp. populations required insecticidal bait treatment of the host ants to alleviate the problem (Ostmark and Evers 1976).

Hymenoptera: Formicidae

Social parasitism of ants by other ants runs the gamut from temporary parasitism by a newly-mated queen to slave-making to obligatory parasitism (Buschinger 1986). Temporary parasitism has been suggested for several *Solenopsis* spp. which produce a worker caste (Williams 1980), but nothing is really known of their biology. The true parasitic ants do not produce a worker caste (Wilson 1971). The workerless *Solenopsis* (*Labauchena) daguerrei* Santschi does not execute the host *S. richteri* (Silveira Guido, et al. 1973) as has been erroneously reported (Bruch 1930, Askew 1971). Instead, it attaches itself to the host queen's neck and steals the food the workers offer their mother queen (Silveira Guido et al. 1968, 1973). Some evidence suggests the parasite queen is preferentially fed before the host queen, and even the parasite immatures are more attractive to the workers than their own brood (Silveira-Guido et al. 1965, 1973). Eventually the host queen is deprived of so much food that her egg production declines and the colony's vigor and size decreases, particularly when more than one parasite queen is present (up to 10 have been found on one host queen, Silveira-Guido et al. 1965, 1973). The host queen eventually dies resulting in the demise of the colony. *S. richteri* colonies have been found with over 3,300 parasite adults (males and females) and immatures (70% of the collected colony) (Silveira-Guido et al. 1965, 1973). The incidence of parasites in active fire ant mounds was as high as 31% in some of the areas sampled (Silveira-Guido et al. 1965, 1973). The method of invasion and integration into the host colony is not understood. This species of parasite is a prime candidate for biological control of fire ants in the United States.

Lepidoptera

The many species of Lycaenidae and Riodinidae, associated with ants, are considered trophobionts (Wilson 1971, Kistner 1982), although the late-larval instars of some species prey on ant brood (Kistner 1982). To date, only 2 riodinid species are known to associate with fire ants. *S. geminata* (among other species of ants) tended *Eurybia elvina* Stichel in Mexico (Horvitz et al. 1987). This ant frequently constructed soil shelters over late instar larvae on the host plant. *Hameris epulus signatus* Stichel is known from nests of *S. richteri* from Argentina and Uruguay (Bruch 1926, Silveira Guido et al. 1964). The larvae possess excretory organs that secrete fluids upon which the ants feed (Bruch 1926). According to Bruch (1926), the larvae left the ant nests at night to feed on their preferred host plant, *Vicia graminea* Sm. (Leguminoseae). The larvae could be predaceous since cuticular fragments of ant larvae have been found in the caterpillar digestive tracts (A. Oglobin in Silveira Guido et al. 1964). The caterpillars, which pupate in the ant nests, and the pupae are not molested by the ants in laboratory nests (Bruch 1926, Silveira Guido et al. 1964, Wojcik unpublished).

Orthoptera

Only crickets from the genus *Myrmecophilia* (monogeneric family Myrmecophilidae) are known myrmecophiles (Kistner 1982). The species of this genus have multiple ant hosts including *Solenopsis* (Kistner 1982). *Myrmecophilia nebrascensis* Lugger has been collected from *S. invicta* and *S. xyloni* nests in Texas (Neece and Bartell 1981, 1982); the biology and behavior of this species was summarized by Wheeler (1910). The cricket is a synoekete which apparently feeds by strigilation on the surface secretions of its ant hosts and may steal food from them. Another species, probably *M. pergandei* Bruner, has been collected from *S. geminata* nests (Travis 1941, Wojcik unpublished) and *S. invicta* nests (Wojcik unpublished) in the southeastern United States. The behavior of these crickets in *S. invicta* laboratory nests was similar to that reported for other species. The crickets fed on or licked the surfaces of the ants and fled whenever approached by them. The crickets did not follow artificial or natural *S. invicta* trails and were eventually killed by the ants (Wojcik unpublished).

Thysanura

Thysanurans were among the first myrmecophiles studied and are found with many genera of ants (Kistner 1982). They steal food during trophallaxis, feed on body fluids of booty, or are strigilators. Many are found with more than one host, e.g. *Grassiella praestans* Silvestri has been collected from *Atta* spp. and *Solenopsis* spp. nests in Brazil (Eidmann 1937) and *S. geminata* nests in Mexico (Wasmann 1912). *Grassiella* sp., *Mirolepisma* sp. and *Prolepismina* sp. have been collected from several *Solenopsis* spp. nests in the United States (Collins and Markin 1971, Neece and Bartell 1981, 1982), but the only reported behavior is that they flee from the ants (Collins and Markin 1971). I have observed a species of thysanuran collected from Florida and Georgia *S. invicta* nests (Wojcik unpublished), which is probably the same species collected by Collins and Markin (1971) and Neece and Bartell (1981, 1982). The thysanurans displayed rapid movement in the laboratory colony, avoiding the ants. When the opportunity arose the thysanurans appeared to strigilate the cuticle of the ant workers and brood without attempting to pierce the cuticle. In other cases the thysanurans seized the brood and carried them away from the brood pile. The ant pupae were eaten by the thysanurans. The ants acted aggressively towards the thysanurans, killing them when possible. Thysanurans can be abundant in fire ant nests in the United States and Brazil, but are difficult to capture alive and uninjured particularly when the fire ant colonies are floated out in the laboratory (Wojcik unpublished).

CONCLUSION

The true relationship of most myrmecophiles with their ant hosts remains unknown. Much of the older literature on myrmecophiles needs to be critically revaluated in light of recent knowledge and updated taxonomy. There appears to be more variation in biology and behavior among related genera and species than has been assumed in the past. Therefore, many previous generalizations are invalid. The Wasmannian classification system for myrmecophiles (and other social insect symbionts) is imperfect, as shown in past discussions. However, it is convenient to use and when combined with categories such as predator or parasite, the combined classification system can be very informative. Additional studies on the biology and behavior of myrmecophiles will supply new information on the complex relationships between ants and their inquilines and provide the basis for a better understanding of these relationships.

Although we have little data on the effects of fire ant myrmecophiles on colony longevity, it is reasonable to assume that these organisms reduce colony fitness by usurping resources. Myrmecophiles such as *Solenopsis* (*Labauchena*) spp. can cause colony mortality, albeit slowly. Myrmecophiles, that eat or parasitize fire ant immatures (e.g. some scarab beetles, phorids and Thysanura), cause the malformation and death of immatures (e.g. *Orasema* spp.), or utilize their food resources (e.g. some mites and staphylinid beetles), produce additional stress on fire ant colonies. This combination of additive mortalities and draining of resources must effect fire ant colony vigor and perhaps survival. The abundance and effects of a biological control agent in its homeland are not necessarily indicative of its utility when introduced into a new environment (Doutt and Debach 1964). Therefore, each myrmecophile must be evaluated for effectiveness in the new environment before it can be discarded from an IPM program for fire ant control.

LITERATURE CITED

ANON. 1958. Observations on the biology of the imported fire ant. USDA, ARS 33-49, 21 p.

ASKEW, R.R. 1971. Parasitic insects. American Elsevier Publ. Co., Inc., New York. 316 p.

AYRE, G.L. 1962. *Pseudometagea schwarzii* (Asm.) (Eucharitidae: Hymenoptera), a parasite of *Lasius neoniger* Emery (Formicidae: Hymenoptera). Can. J. Zool. 40: 157-64.

BANKS, W.A., D.P. JOUVENAZ, D.P. WOJCIK and C.S. LOFGREN. 1985. Observations on fire ants, *Solenopsis* spp., in Mato Grosso, Brazil. Sociobiology 11: 143-52.

BARTOSZECK, A.B. 1976. Afideos de laranjeira (*Citrus sinensis* Osb.) e minoseira (*Citrus reticulata* B.), seus predadores e parasitas. Acta Biol. Parana, Curitiba 5: 15-48.

BORGMEIER, T. 1923. Novos phorideos brasileiros. Bol. Mus. Nac. (Sao Paulo) 1: 51-59.

BORGMEIER, T. 1953. Syrphidenlarven in Ameisennestern. Naturwissenschaften 40: 36.

BRUCH, C. 1917. Nuevas capturas de insectos mirmecófilos. Physis 3: 458-65.

BRUCH, C. 1926. Orugas mirmecofilas de *Hameris epulus signatus* - Stich. Rev. Soc. Entomol. Argent. 1(1): 1-9.

BRUCH, C. 1929. Neue myrmekophile Histeriden und Verzeichnis der aus Argentinien bekannten Ameisengaste. Zool. Anz., Wasmann-Festband 82: 421-37.

BRUCH, C. 1930. Notas preliminares acera de *Labauchena daguerrei* Santschi. Rev. Soc. Entomol. Argent. 3: 73-80, plates I, II.

BRUES, C.T. 1901. Two new myrmecophilous genera of aberrant Phoridae from Texas. Amer. Natur. 35: 337-56.

BUSCHINGER, A. 1986. Evolution of social parasitism in ants. Trends Ecol. Evolut. 1: 155-60.

CHALUMEAU, F. 1983. *Batesiana* et *Martinezia*, nouveax genres d'Eupariini (Coleoptera: Scarabaeidae: Aphodiinae) du nouveau monde. Bull. Mensuel Soc. Linn. Lyon 52: 142-53.

CHALUMEAU, F. and H.F. HOWDEN. 1984. Le genre *Euparia* (Coleoptera: Scarabaeoidea: Aphodiinae) Saint-Fargeau and Serville, 1828. Mitt. Schweiz. Entomol. Ges. 57: 85-94.

CLAUSEN, C.P. 1923. The biology of *Schizaspidia tenuicornis* Ashm., a eucharid parasite of *Camponotus*. Ann. Entomol. Soc. Am. 16: 195-217, plates XIV-XV.

CLAUSEN, C.P. 1940. The oviposition habits of the Eucharidae (Hymenoptera). J. Washington Acad. Sci. 30: 504-16.

CLAUSEN, C.P. 1941. The habits of the Eucharidae. Psyche 48: 57-69.

COLLINS, H.L. and G.P. MARKIN. 1971. Inquilines and other arthropods collected from nests of the imported fire ant, *Solenopsis saevissima richteri*. Ann. Entomol. Soc. Am. 64: 1376-80.

COQUILLETT, D.W. 1907. A new phorid genus with horny ovipositor. Can. Entomol. 39: 207-8.

DOUTT, R.L. and P. DEBACH. 1964. Some biological control concepts and questions, p. 118-42. In P. Debauch (ed.), Biological control of insect pests and weeds. Reinhold Publ. Co., NY. 844 P.

EIDMANN, H. 1937. Die gäste und Gastverhältnisse der Blattschneiderameise, *Atta sexdens* L. Z. Morphol. Oekol. Tiere 32: 391-462.

FALL, H.C. 1928. *Alaudes*. Pan-Pac. Entomol. 4: 145-150.

FEENER, D.H., Jr. 1981. Competition between ant species: outcome controlled by parasitic flies. Science 214: 815-7.

FEENER, D.H., Jr. 1987. Size-selective oviposition in *Pseudacteon crawfordi* (Diptera: Phoridae), a parasite of fire ants. Ann. Entomol. Soc. Am. 80: 148-51.

HELAVA, J.V.T., H.F. HOWDEN and A.J. RITCHIE. 1985. A review of the new world genera of the myrmecophilous and termitophilous subfamily Hetaeriinae (Coleoptera: Histeridae). Sociobiology 10: 127-382.

HERATY, J.M. 1985. A revision of the nearctic Eucharitinae (Hymenoptera: Chalcidoidea: Eucharitidae). Proc. Entomol. Soc. Ontario 116: 61-103.

HERATY, J.M. and D.C. DARLING. 1984. Comparative morphology of the planidial larvae of the Eucharitidae and Perilampidae (Hymenoptera: Chalcidoidea). System. Entomol. 9: 309-28.

HERMANN, H.R., M.S. BLUM and A.N. HUNT. 1970 (1971). Myrmecophilous arthropods associated with the imported fire ant, *Solenopsis saevissima* (Hymenoptera: Formicidae). Proc. La. Acad. Sci. 33: 13-8.

HIRSCHMANN, W. 1972. Gangsystematik der Parasitiformes. Teil 103. Von J.C. Moser gesammelte Uropodiden aus Nordamerika und 2 neue *Oplitis*-arten. Acarologie 17: 28-9. Abbildung 36-7.

HORVITZ, C.C., C. TURNBULL and D.J. HARVEY. 1987. Biology of immature *Eurybia elvina* (Lepidoptera: Riodinidae), a myrmecophilous metalmark butterfly. Ann. Entomol. Soc. Am. 80: 513-9.

HOWARD, R.W., C.A. MCDANIEL and G.J. BLOMQUIST. 1980. Chemical mimicry as an integrating mechanism: cuticular hydrocarbons of a termitophile and its host. Science 210: 431-3.

HOWARD, R.W., C.A. MCDANIEL and G.J. BLOMQUIST. 1982. Chemical mimicry as an integrating mechanism for three termitophiles associated with *Reticulitermes virginicus* (Banks). Psyche 89: 157-67.

HUNTER, J.E., III. and M.H. FARRIER. 1975. Mites of the genus *Oplitis* Berlese (Acarina: Uropodidae) associated with ants (Hymenoptera: Formicidae) in the southeastern United States. Part I. Acarologia 17: 595-623.

HUNTER, J.E., III. and M.H. FARRIER. 1976. Mites of the genus *Oplitis* Berlese (Acarina: Uropodidae) associated with ants (Hymenoptera: Formicidae) in the southeastern United States. Part II. Acarologia 18: 20-50.

HUNTER, P.E. and M. COSTA. 1971. Description of *Gymnolaelaps shealsi* n. sp. (Acarina: Mesostigmata) associated with the imported fire ant. J. Ga. Entomol. Soc. 6: 51-3.

JOUVENAZ, D.P., R.K. VANDER MEER and D.P. WOJCIK. 1988. Chemical mimicry in a parasitoid, *Orasema* sp., (Hymenoptera: Eucharitidae) of fire ants. Proc. 18th Intern. Congr. Entomol., Abstr. p. 243.

KIEFFER, J.J. 1921. Proctotrypides hôtes des fourmis en Argentine. An. Soc. Cientif. Argent. 91: 36-41.

KISTNER, D.H. 1979. Social and evolutionary significance of social insect symbionts, p. 339-413. In H.R. Hermann (ed.), Social insects. Academic Press, New York. Vol. 1. 437 p.

KISTNER, D.H. 1982. The social insects' bestiary, p. 1-244. In H.R. Hermann (ed.), Social insects. Academic Press, N.Y. Vol. 3. 459 p.

KISTNER, D.H. and M.S. BLUM. 1971. Alarm pheromone of *Lasius (Dendrolasius) spathepus* (Hymenoptera: Formicidae) and its possible mimicry by two species of *Pella* (Coleoptera: Staphylinidae). Ann. Entomol. Soc. Am. 64: 589-94.

LOFGREN, C.S., W.A. BANKS and B.M. GLANCEY. 1975. Biology and control of imported fire ants. Annu. Rev. Entomol. 20: 1-30.

MARTINEZ, A. 1952. Scarabaeidae nuevos o poco conocidos, III (Coleoptera). Publ. Misión Estud. Patol. Region. Argentina 23: 53-118.

MULENBERG, M. and U. MASCHWITZ. 1976. E1937/1975 *Acanthaspis bistillata* und *Acanthaspis concinnula* (Reduviidae) Maskierung (Freilandaufnahmen). Encycl. Cinematogr. 1975: 3-12.

NEECE, K.C. and D.P. BARTELL. 1981. Insects associated with *Solenopsis* spp. in southeastern Texas. Southwest. Entomol. 6: 307-11.

NEECE, K.C. and D.P. BARTELL. 1982. A faunistic survey of the organisms associated with ants of Western Texas. Grad. Stud. Texas Tech. Univ., Lubbock, TX. 25, 36 p.

NICOLINI, J.C. 1950. La avispita costurera y la tuberculosis del olivo. Rev. Agron., Porto Alegre (35-36): 20.

OGLOBLIN, A.A. 1939. The Strepsiptera parasites of ants. Verh. 7th Intern. Kongr. Entomol. 2: 1277-84.

OGLOBLIN, A.A. 1963. Especies nuevas del genero *Pseudoisobrachium* Kieff. (Fam. Bethylidae, Hymenoptera). Rev. Soc. Entomol. Argentina 26: 133-8.

OSTMARK, H.E. and C. EVERS. 1976. Manipulation of ant populations to control *Orasema costaricensis* in bananas. Paper presented at the XV Intern. Congr. Entomol., Washington, D.C. 23 August 1976.

PARKER, H.L. 1942. Oviposition habits and early stages of *Orasema* sp. Proc. Entomol. Soc. Washington 44: 142-5.

ROBERTS, F.S. 1958. Insects affecting banana production in central America. Proc. X Intern. Congr. Entomol. 3: 411-5.

ROTH, L.M. and E.R. WILLIS. 1960. The biotic associations of cockroaches. Smithson. Misc. Collect. 141, 470 p.

SAN MARTIN, P.R. 1966a. Nota sobre *Neoblissus parasitaster* Bergroth, 1903 (Blissinae, Lygaeidae, Hemiptera). Rev. Bras. Biol. 26: 247-51.

SAN MARTIN, P.R. 1966b. Nota sobre *Anommatocoris coleopteratus* Kormilev, 1955 (Vianaidinae, Tingidae, Hemiptera). Rev. Bras. Biol. 26: 327-8.

SAN MARTIN, P.R. 1968a. Nuevo hallazgo de *Metopioxys gallardoi* Bruch, 1917 (Coleoptera, Pselaphidae, Metopiinae). Rev. Bras. Biol. 28: 27-8.

SAN MARTIN, P.R. 1968b. Notas sobre *Fustiger elegans* Raffray (Coleoptera, Pselaphidae) en el Uruguay y la Argentina. Physis 28: 59-64.

SCHUBART, O. 1950. Ameisen und Diplopoden in ihren gegenseitigen Beziehungen. Rev. Entomol. 21: 615-22.

SHEPPARD, C., P.B. Martin and F.W. Mead. 1979. A planthopper (Homoptera: Cixiidae) associated with red imported fire ant (Hymenoptera: Formicidae) mounds. J. Ga. Entomol. Soc. 14: 140-4.

SILVEIRA-GUIDO, A., J. CARBONELL and C. CRISCI. 1973. Animals associated with the *Solenopsis* (fire ants) complex, with special reference to *Labauchena daguerrei*. Proc. Tall Timbers Conf. Ecol. Anim. Control Habitat Manage. 4: 41-52.

SILVEIRA-GUIDO, A., J. CARBONELL-BRUHN, C. CRISCI-PISANO and J. BRIOZZO-BELTRAME. 1968. Investigations on natural enemies of ants. Second report. Entomol. Sect., Centro de investigación en Fruitcultura, Horticultura y Vitivinicultura, Minist. de Granaderia y Agric. Montevideo, Uruguay. 22 p.

SILVEIRA-GUIDO, A., C. CRISCI-PISANO, P. SAN MARTIN and J. CARBONELL-BRUHN. 1965. Investigations on the biology and biological control of the fire ant, *Solenopsis saevissima richteri* Forel in Uruguay. Final report. Departmento de Sanidad Vegetal, Facultad de Agronomia, University of the Republic. Montevideo, Uruguay. 70 p., 36 plates.

SILVEIRA-GUIDO, A., P. SAN MARTIN, C. CRISCI-PISANO and J. CARBONELL-BRUHN. 1963, 1964. Investigations on the biology and biological control of the fire ant, *Solenopsis saevissima richteri* Forel in Uruguay. Second and third reports. Departmento de Sanidad Vegetal, Facultad de Agronomia, University of the Republic. Montevideo, Uruguay. 78 p., 67 p.

SMITH, M.R. 1928. *Plastophora crawfordi* Coq. and *Plastophora spatulata* Malloch (Diptera: Phoridae), parasitic on *Solenopsis geminata* Fabr. Proc. Entomol. Soc. Washington 30: 105-8.

STEINER, W.E., Jr. 1982. *Poecilocrypticus formicophilus* Gebien, a south american beetle established in the United States (Coleoptera: Tenebrionidae). Proc. Entomol. Soc. Washington. 84: 232-9.

SUMMERLIN, J.W. 1978. Beetles of the genera *Myrmecaphodius*, *Rhyssemus* and *Blapstinus* in Texas fire ant nests. Southwest. Entomol. 3: 27-9.

TESON, A. and A.M.M. de REMES LENICOV. 1979. Estrepsipteros parasitoides de hymenopteros (Insecta - Strepsiptera). Rev. Soc. Entomol. Argent. 3: 115-22.

TOCCHETTO, A. 1942. Bicho costureiro. Rev. Agron., Porto Alegre 6: 587-8.

TRAVIS, B.V. 1941. Notes on the biology of the fire ant *Solenopsis geminata* (F.) in Florida and Georgia. Fla. Entomol. 24: 15-22.

VANDER MEER, R.K., D.P. JOUVENAZ and D.P. WOJCIK. 1989. Chemical mimicry in a parasitoid (Hymenoptera: Eucharitidae) of fire ants (Hymenoptera: Formicidae). J. Chem. Ecol. 15:2247-61.

VANDER MEER, R.K. and D.P. WOJCIK. 1982. Chemical mimicry in the myrmecophilous beetle *Myrmecaphodius excavaticollis*. Science 218: 806-808.

WASMANN, E. 1912. Neue Beiträge zur Kenntnis der Termitophilen und Myrmecophilen. Z. Wiss. Zool. 101: 70-115, Taf. 5-7.

WASMANN, E. 1918a. Ueber *Solenopsis geminata saevissima* Sm. und ihre gäste. Entomol. Blätter 14: 69-76.

WASMANN, E. 1918b. Übersicht der myrmecophilen Paederinengattung *Myrmecosaurus* Wasm. (Staphylinidae). Entomol. Blätter 14: 210-4.

WASMANN, E. 1925a. Drei neue myrmecophile Staphyliniden (Col.). Deuts. Entomol. Zeit. 1: 1-10.

WASMANN, E. 1925b. Die Ameisenmimikry. Ein exakter Beitrag zum Mimikryproblem und zur Theorie de Anpassung. Abh. Theort. Biol. 19: 1-164.

WHEELER, G.C. and E.H. WHEELER. 1937. New hymenopterous parasites of ants (Chalcidoidea: Eucharidae). Ann. Entomol. Soc. Am. 30: 163-73, plates I, II.

WHEELER, W.M. 1907. The polymorphism of ants, with an account of some singular abnormalities due to parasitism. Bull. Amer. Mus. Natur. Hist. 23: 1-93, plates I-VI.

WHEELER, W.M. 1910. Ants their structure development and behavior. Columbia Univ. Press, New York. 663 p.

WILLIAMS, D.F. and W.A. BANKS. 1987. *Pseudacteon obtusus* (Diptera: Phoridae) attacking *Solenopsis invicta* (Hymenoptera: Formicidae) in Brazil. Psyche 94: 9-13.

WILLIAMS, R.N. 1980. Insect natural enemies of fire ants in South America with several new records. Proc. Tall Timbers Conf. Ecol. Anim. Control Habitat Manage. 7: 123-34.

WILLIAMS, R.N., J.R. PANAIA, D. GALLO and W.H. WHITCOMB. 1973. Fire ants attacked by phorid flies. Fla. Entomol. 56: 259-62.

WILSON, E.O. 1971. The insect societies. Belknap Press, Cambridge, Ma. 548p.

WOJCIK, D.P. 1975. Biology of *Myrmecaphodius excavaticollis* (Blanchard) and *Euparia castanea* Serville (Coleoptera: Scarabaeidae) and their relationships to *Solenopsis* spp. (Hymenoptera: Formicidae). Ph.D. dissertation, University of Florida, Gainesville, FL. 74 p. [Dissert. Abstr. Int. B 36: 5962]

WOJCIK, D.P. 1980. Fire ant myrmecophiles: behavior of *Myrmecosaurus ferrugineus* Bruch (Coleoptera: Staphylinidae) with comments on its abundance. Sociobiology 5: 63-8.

WOJCIK, D.P. 1986. Observations on the biology and ecology of fire ants in Brazil, p. 88-103. In C.S. Lofgren, R.K. Vander Meer (ed.), Fire ants and leaf-cutting ants, Biology and management. Westview Press, Boulder, CO, 435 p.

WOJCIK, D.P., W.A. BANKS and D.H. HABECK. 1978. Fire ant Myrmecophiles: Flight periods of *Myrmecaphodius excavaticollis* (Blanchard) and *Euparia castanea* Serville (Coleoptera: Scarabaeidae). Coleopt. Bull. 32: 59-64.

WOJCIK, D.P., W.A. BANKS, D.M. HICKS and J.W. SUMMERLIN. 1977. Fire ant myrmecophiles: new hosts and distribution of *Myrmecaphodius excavaticollis* (Blanchard) and *Euparia castanea* Serville (Coleoptera: Scarabaeidae). Coleopt. Bull. 31: 329-34.

WOJCIK, D.P. and D.H. HABECK. 1977. Fire ant Myrmecophiles: Breeding period and ovariole number in *Myrmecaphodius excavaticollis* (Blanchard) and *Euparia castanea* Serville (Coleoptera: Scarabaeidae). Coleopt. Bull. 31: 335-8.

WOJCIK, D.P., D.P. JOUVENAZ, W.A. BANKS and A.C. PEREIRA. 1987a. Biological control agents of fire ants in Brazil, p. 627-628. In J. Eder, H. Rembold (ed.), Chemistry and biology of social insects. Verlag J. Peperny, München. 757 p.

WOJCIK, D.P., D.P. JOUVENAZ and C.S. LOFGREN. 1987b. First report of a parasitic fly (Diptera: Phoridae) from a red imported fire ant (*Solenopsis invicta*) alate female (Hymenoptera: Formicidae). Fla. Entomol. 70: 181-2.

WOJCIK, D.P. and M.A. NAVES. *Calytodesmus schubarti* (Diplopoda: Stylodesmidae: Polydesmida) as a facultative myrmecophile. Sociobiology (in Press)

32

Territorial Ecology of the Leaf-cutting Ant, *Atta laevigata*

A. Salzemann and K. Jaffe

INTRODUCTION

Organisms use the space available to them in diverse ways. Thus, it should come as no surprise that "social animals" share in this diversity. In what areas do these differences appear? Unfortunately, compared to the amount of data compiled on solitary or subsocial animals, the territorial behavior of social animals has yet to be extensively studied. Ants are no exception, though some important laboratory studies have been completed.

Hölldobler and Wilson (1980) showed that the shape of the territory of three different ant species strongly depended on each species' foraging strategy: the weaver ant (*Oecophylla longinoda*) permanently occupies an "absolute territory" ranging up to 1600 m^2 including no more than 17 trees. A colony specific territorial pheromone acts as a repellent to foreign conspecifics, protecting the territory against invasions, even in the absence of residents. The harvester ants (*Pogonomyrmex barbatus* and *P. rugosus*) also have stable territories, but their territory is defined by trunk trails used by workers to collect seeds. Thus, Hölldobler and Wilson claim that the territory only consists of the trail system and the ant nests. Trails from different colonies never cross; rather, they tend to be oriented toward opposite directions. Accidental convergence of trails from different colonies causes aggressive confrontations lasting until the trails no longer intersect. Thus, the layout of a colony's trail system depends on that of its neighbor. The honey ant (*Myrmecocystus mimicus*) has a variable territory, which changes depending on the foraging activity and availability of termites. Hölldobler and Wilson (1980) claim that termites are a changing resource which make a stable territory inefficient and contact between workers from different colonies more frequent. This justifies the existence of ritualized confrontations that eliminate worker injury. Such a system also allows different colonies to exploit the same resource or use the same geographical area, thanks to separate spatio-temporal foraging patterns. If an area or resource is used simultaneously, neighboring colonies battle until one gives up. The winning colony takes over the loser's nest, killing the queen and taking larvae, pupae and repleates to their nest.

Most of what is known of ant territorial behavior is based on descriptions from laboratory studies (Hölldobler and Wilson 1976, 1977;

Cammaerts et al. 1977; Fresneau 1980; Jaffe and Puche 1984) but little is known about territorial behavior in the field (Wilson et al. 1971; Skinner 1980; Hölldobler 1981; Hölldobler 1983). The present work is part of a series which will exhaustively study territorial behavior of *Atta* colonies in the field.

Atta laevigata, the most common *Atta* species in Venezuela and Colombia, is normally found in savannas or disturbed habitats, where it forms gigantic nests, possibly the largest *Attini* colonies. Little is known about its territorial behavior, whereas for other *Attini*, territorial behavior of laboratory colonies has been repeatedly observed (Jaffe et al. 1979; Vilela 1983; Hölldobler and Wilson 1986). Despite these studies, not much can be inferred about territorial behavior in a natural environment, since the habitat offered to captive colonies is not comparable to the nest areas available in the field. The present work studies the dynamics and territorial strategy of *Atta laevigata* colonies in the field. In the process, some insight into the significance of *Atta* colony territorial behavior should be gained.

MATERIALS AND METHODS

The study was performed on four neighboring *Atta laevigata* colonies in commercial pine tree plantations (*Pinus caribea*) at Estado Monagas, Venezuela. The site had 15 year-old pines and showed an *A. laevigata* density of over 30 mature nests per hectare. To study the topographic distribution of the territorial mark, a series of test were performed along a trunk trail. An arena consisting of a bottomless plastic cylinder (9 cm diameter, 8 cm high) was placed on the ground either at a nest mound where workers were excavating earth, at the beginning of an active trail (0-1 m from a nest mound), or at the middle or end of the trail (cutting area). The resident workers were taken from the site where the experiment was performed and were chosen from among the workers walking from the nest to the foraging area. The intruders were collected from among the workers exiting their nests, at the beginning of the trail. The experiment introduced either the resident or the intruder to the arena, followed five minutes later by the other ant. After introducing the second ant to the arena, the behavior of both ants was observed for five minutes.

Each time, the cylinder was placed on a different site, avoiding bias produced by a possible site-dependency. The cylinder was covered with Fluon (ICI), avoiding the escape of test ants or the passage of ants from the outside.

Controls consisted of confrontations between ants of two neighboring colonies on virgin territory. The cylinder was placed outside trails in sites where no ant activity was observed.

The four test nests were observed for 19 months. The experimental area (1500 m^2) was drawn on a chart and the trunk trails, mounds, etc. replotted approximately every four months; i.e. in March 1987 (middle of the dry season), June 1987 (beginning of the rain period), November 1987 (end of the wet season) and March, April and September of 1988. Such maps allowed a study of the nests' temporal dynamics and the ants' foraging trails.

The identification of each nest mound was performed progressively by observing chronological appearance, connection with common active trails, synchronous foraging activities with the nearest central nest, the results of experimental confrontations of workers from different mounds and observation of spontaneous conflicts.

RESULTS

Geographical distribution of the territorial mark

At nest mounds and the beginning of the trail, the tests showed a clear dominance of the resident over an intruder (Table I), regardless of which ant first entered the arena. The presence of a territorial mark is also shown when the resident was observed transporting the intruder. At the middle of an active trail, the resident worker demonstrated reduced aggressiveness if entering the arena after the intruder, but was always the only biting ant. No transport of the intruder was observed in this case. At the end of a trail, the first ant to enter the arena always had an advantage in aggressive encounters, suggesting the absence of a territorial mark. As expected, confrontations on virgin territory revealed an advantage for the first ant entering the area, independent of its origin. Transport of the vanquished was never observed in this case.

TABLE 1. Frequencies of aggressive behavior shown (in % of tests performed, n= 40) by the first (1) to the second (2) introduced workers when either the intruder (I) was introduced into the arena with a 5 min delay respect to the resident (R) or vice-versa (R-I or I-R, respectively).

	R-I	I-R	X^2	
Active nest mound without trails or foraging activity:				
Bitings 1-2	55	5	<0.01	(n= 20)
Transport of I	5	5	---	
Beginning of the trail:				
Bitings 1-2	57,5	10	<0.001	
Transport of I	10	10	>0.05	
Middle of the trail:				
Bitings 1-2	30	0	<0.001	
Transport of I	0	0	----	
End of the trail:				
Bitings 1-2	12,5	7,5	>0.05	
Transport of I	0	0	----	
Virgin territory (workers from societies A and B):	A-B	B-A		
Bitings 1-2	15	12,5	>0.05	
Transport	0	0	----	

Figure 1 shows the topographic layout of nests, mounds and pines during the observation period, whereas Figure 2 shows the annual dynamics of some important features. The following observations merit special comment:

MARCH 1987: Twenty three per cent of the pine trees on the study site were connected to foraging mounds through trails. Worker activity was essentially nocturnal.

JUNE 1987: The foraging activity stopped and the trails were neglected. Only 4% of the trees were connected to clean but apparently inactive trails. The reduction in nest mounds was significant, ranging between -26 % to -66% per nest, compared to the number of mounds counted in March. Ants' activity concentrated on excavations below some of the central nest area mounds. On June the 7th, at sun rise, nests displayed intense preswarming activity (another had been observed 10 days before). Sexual forms emerged only from nests two and three but intense activity was also observed on nests one and four. In all cases, workers and soldiers were found outside the mounds near the nest center walking in all directions. At 10 a.m., the nuptial flights ended. Ill-formed females which did not fly and males resting on the ground were exterminated, even by their nestmates. After the flight, inseminated queens dug a vertical gallery, 20 to 30 cm deep, ending in a circular chamber, five to 10 cm in diameter. During this time, workers from nests which did not produce sexual forms were walking in all directions, killing any queen they met. Young nests were carefully excavated and queens, if present, killed. The next day, small mounds with characteristic pellets of sand indicated the success of 65 founding queens.

NOVEMBER 1987: Thirty two percent of the trees were connected with active clean trails and an increase in the number of foraging mounds was observed (+145%). Ants were collecting leaves, particularly during the night and the early morning. Nest three had 71 m of active trails. This strong activity helped to discover incipient colonies: only 15 newly-founded nests were now present, 77% of the founding queens had disappeared and some of the empty nests were now colonized by *Camponotus* sp. and a solitary wasp.

MARCH 1988: We recorded the maximum foraging activity (Figure 2). A total of 175 m of active trails were measured at this point. These were active during day and night. Workers stopped foraging only during the warmest hours of the day (between 11 a.m. and 7 p.m.). Only three mature nests survived the dry season and more than 95% of the incipient colonies had disappeared. The young nests were active at night until early in the morning. Nest four died at the beginning of the year and was regularly explored by workers of neighboring nest three which emerged from the dead heap with pellets of sand which they carried to their nest. Conflicts were observed between two neighboring mounds of nests one and two which formed a battle front since a foraging mounds of nest two started to emerge in the foraging territory of nest one.

APRIL 1988: Nest three showed a decrease of its foraging activity and nest two benefited using some of its neighbor's trails (Figure 1). Conflicts between

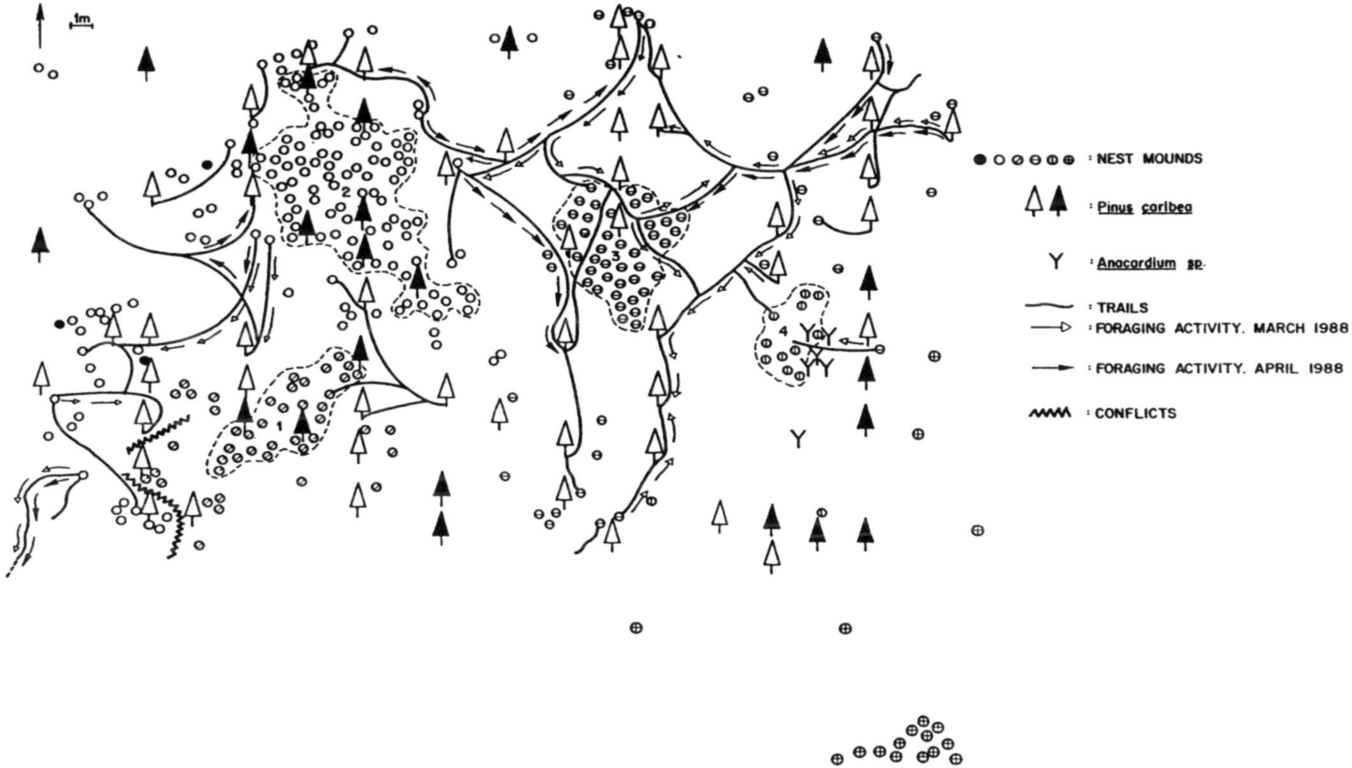

FIGURE 1. Topographic layout of the four Atta laevigata nests. Dark mounds represent young nests less than one year old. Trails are drawn to show nest activities at two different dates. Black pine trees are never attacked during the study period.

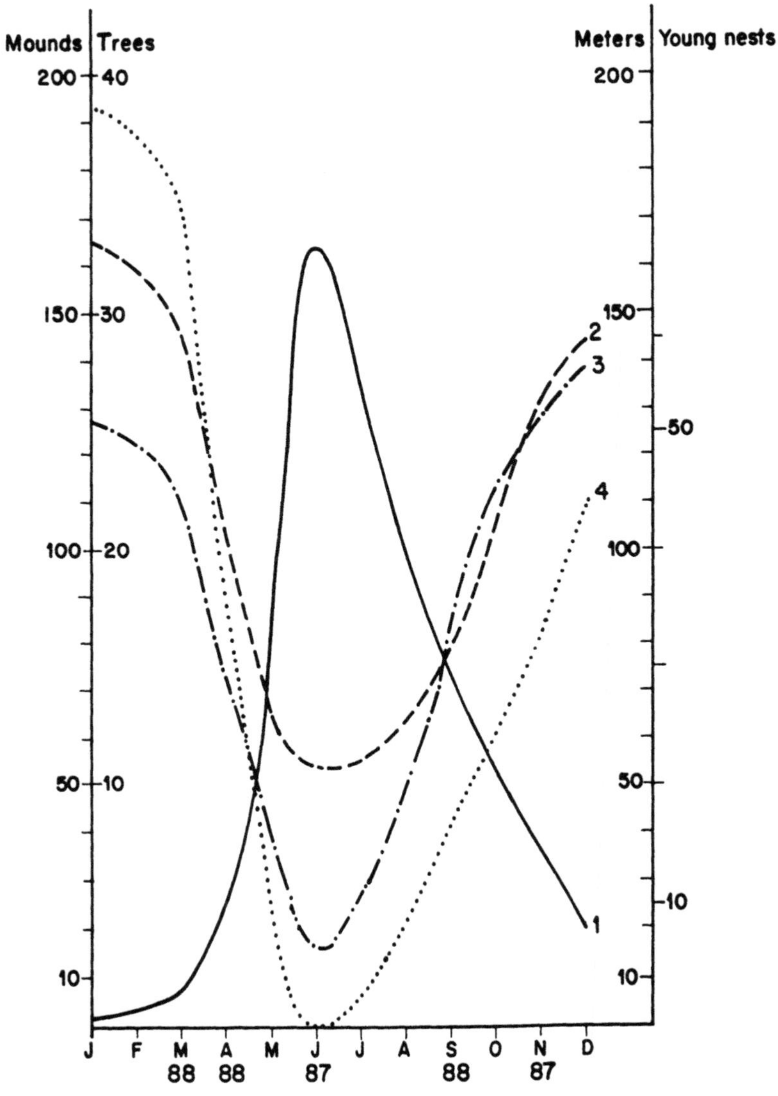

FIGURE 2. (1) the intensity of reproductive activity (represented by the number of young nests less than one year old in the site); (2) temporal fluctuations of the number of foraging nests; (3) number of pine trees connected with trails; and (4) length of active trails.

nests one and two continued underground. Fighting workers and cadavers emerged from foraging mounds of nest one suggesting that the galleries of both societies probably crossed. This conflict was observed for one month, taking place on the site with the highest foraging mound density. This fact was probably related to the low activity of nest one after the mating flights of June 87.

SEPTEMBER 1988: Fifty nine young nests resulted from the 1988 flights; only one nest survived from those founded in 1987. Mounds of nest four were collapsed. Nests two and three showed foraging activity, nest three had 75% of the 41 m of active trails leading to 25% of trees joined through trails to the nest.

In addition to these observations, it is interesting to note that each nest maintained a constant proportion of foraging mounds during the entire observation period, despite changes in their absolute number (Table II).

Forty-eight trees of the 70 trees at the experimental site were regularly foraged and represented 78% of the trees in the inter-nest-heap area. Ants foraged only 25% of pine trees situated on the nest heaps, a percentage significantly less than expected from random attacks ($p< 0.05$, X2 test= 3,98). The trees on which foraging activity was observed more than three separate times had a diameter (1.5 m above the ground) smaller by 15% than trees which were never attacked (means of the diameters were: 19 cm, and 22,5 cm, $N= 18$; $p < 0.05$, Student's t-test).

The highest reproductive activity was observed at the beginning of the rainy season (June in our study site). Foraging activity decreased at this time but increased between November and March.

Laboratory and field observations showed that 24 hours after the nuptial flight, each queen closed her nest entrance and initiated egg-laying three to four days later. *A. laevigata* queens stayed in their chamber, eating trophic eggs and part of the fungus. The first workers appeared seven to eight weeks later; they stayed with their mother about one week before they opened the nest and started to forage. At this stage, the workers closed their nest following a short period of foraging activity.

DISCUSSION

This study demonstrates the existence of *Atta laevigata* colony-specific territories in the field; i.e., defended areas used exclusively by a colony. The colony territory consists of the trail system and the ant nest, similar to territory described for *Pogonomyrmex* (Hölldobler and Lumsden 1980), but in *Atta* spatio- temporal patterns of different colonies allow trail-sharing.

Territories seemed stable, with few observed conflicts between colonies. However, one inter-colony conflict was observed. Autuori (1941) first reported spontaneous wars between *Atta* societies. Our study confirmed his observations and found that such battles may take place on the boundaries of two foraging territories, above and below ground, and may last for over four weeks, affecting the foraging activity of the colonies involved.

The four societies studied followed the same seasonal rhythm and the expansion of territory resulted in the reduction of that of its neighbors. This relative stability in foraging areas and number of mounds between neighboring nests suggests an established dominance of leader nests which

TABLE 2. Number of foraging mounds and annual distribution of mounds per nest (in % of the total mounds on the study site). Nest four died at the end of 1987 and was not included.

	MAR 87	JUN 87	NOV 87	MAR 88	SEP 88
TOTAL MOUNDS (absolute value)	70	51	117	147	78
NEST Nr 1 (%)	16	22	19	12	12
NEST Nr 2 (%)	50	45	47	53	57
NEST Nr 3 (%)	34	33	34	35	31

occupy the more spacious territories and control the expansion of neighboring nests, possibly younger or less combative. This limitation forces neighboring societies to adopt a spatio-temporal foraging strategy that allows them to share in time trails of a common territory. This system is advantageous because it avoids fatal conflicts between conspecific nests; it also allows societies to make better use of available space.

Trees on the nest heap were usually not attacked, a phenomenon also observed in *Atta laevigata* and *Atta cephalotes* (see Salzemann and Jaffé; and Cherrett this volume). The unforaged trees on the inter-nest area suggest the existence of a boundary between nest territories, a neutral "no ants land," which prevents massive and inopportune encounters between workers of neighboring colonies. By contrast, two to four mounds of the same nest may forage on the same tree. This shows the possible benefits to workers of a single pheromone being used for both orientation and mound specification (Hölldobler and Wilson 1986). As suggested elsewhere (Salzemann and Jaffe, in preparation), this pheromone seems to be identical to the territorial pheromone; i.e. the same secretion has two complementary functions.

The killing of inseminated queens before their newly-excavated nests are closed has been reported for *Atta cephalotes* (Cherrett *in* Rockwood 1973), *A. colombica* (Rockwood 1973), *A. capiguara* (Fowler et al. 1984), *A. laevigata* (Jaffe 1986) and *A. bisphaerica* (Fowler 1987). The search for newly-mated queens and incipient colonies were preferential in the direction of the last foraging area used by each adult nest. Thus, this area can be considered as a territory in agreement with the definition of Noble (1939, in Krebs and Davies 1978), i.e., an actively defended area for the colony's exclusive use. The elimination of competitors controls population density, as had also been reported for *A. bisphaerica* (Fowler 1987). Young queens dig their nests in areas without vegetation, often next to old trails or mounds of adult nests. Fowler (1987) also observed high colonization rates of the young *A. bisphaerica* queens in sites already occupied by mature nests of the same species. These behaviors, together with the low dispersal ability of *Atta* queens, may explain why an *Atta* nest occupies the same site for several decades. A nest that recently died will soon be occupied again.

ACKNOWLEDGMENTS

We thank C.V.G. Proforca for supporting the research. We thank especially N. Agrizonis and R. Tirado of the plantation project of Uverito for technical assistance and A.Ribbi-Jaffe from Conicit for editorial help.

REFERENCES

AUTUORI, M. 1941. Contribuicao para o conhecimento da sauva (*Atta* spp. Hymenotera Formicidae), I- Evolucao do sauveiro (*Atta sexdens rubropilosa* Forel 1908). Arq. Inst. Biol. S. Paulo 12, 197-228.

CAMMAERTS, M.C., E.D. MORGAN and R. TYLER. 1977. Territorial marking in the ant *Myrmica rubra L.* (Formicidae). Biol. Behav. 2, 263-272.

FOWLER, H.G. 1987. Colonization patterns of leaf-cutting ant, *Atta bisphaerica* Forel: evidence for population regulation. J. Appl. Ent. 104, 102-105.

FOWLER, H.G., S.W. ROBINSON and J. DIEHL. 1984. Effect of mature colony density on colonization and initial colony survivorship in *Atta capiguara*, a leaf cutting ant. Biotropica 16-1, 51-54.

FRESNEAU, D. 1980. Fermeture des societes et marquage territorial chez les fourmis Ponerines du genre *Neoponera*. Biol. Ecol. Med. VII-3, 205-206.

HÖLLDOBLER, B. 1981. Foraging and spatiotemporal territories in the honey ant *Myrmecocystus mimicus* Wheeler (Hymenoptera: Formicidae). Behav. Ecol. Sociobiol. 9, 301-314.

HÖLLDOBLER, B. 1983. Territorial behavior in the green tree ant (*Oecophylla smaragdina*). Biotropica, 15-4, 241-250.

HÖLLDOBLER, B. and C.J. LUMSDEN. 1980. Territorial strategies in ants. Science 210, 732-739.

HÖLLDOBLER, B. and E.O. WILSON. 1976. Weaver ants: social establishment and maintenance of territory. Science 195, 900-902.

HÖLLDOBLER, B. and E.O. WILSON. 1977. Colony specific territorial pheromone in the african weaver ant *Oecophylla longinoda* (Latreille). Proc. Nat. Acad. Sci. U.S.A., 74-5, 2072-2075.

HÖLLDOBLER, B. and E. O. WILSON. 1986. Nest area exploration and recognition in leaf cutter ants (*Atta cephalotes*). J. Insect. Physiol. 32-2, 143-150.

JAFFE, K. 1986. Control of *Atta* and *Acromyrmex* spp. in pine tree plantations in the Venezuelan llanos. In Fire Ants and Leaf-Cutting Ants, Biology and Managment, C.S. Lofgren and R.K. Vander Meer Edit. 409-416.

JAFFE, K. and H. PUCHE. 1984. Colony specific territorial marking with metapleural gland secretion in the ant *Solenopsis geminata* (Fabr.). J. Insect. Physiol. 30-4, 265-270.

JAFFE, K., M. BAZIRE-BENAZET and P.E. HOWSE. 1979. An integumentary pheromone secreting gland in *Atta* sp.: territorial marking with a colony-specific pheromone in *Atta cephalotes*. J. Insect. Physiol. 25, 833-839.

KREBS, J.R. and N.B. DAVIES. 1978. Behavioural ecology, an evolutionary approach. Blackwell Scientific Publications, 494 pp.

ROCKWOOD, L.L. 1973. Distribution, density and dispersion of two species of *Atta* (Hymenoptera: Formicidae) in Guanacaste Province, Costa Rica. J. Anim. Ecol. 42, 803-817.

SKINNER, G.J. 1980. Territory, trail structure and activity patterns in the wood-ant *Formica rufa* (Hymenoptera: Formicidae) in limestone woodland in north-west England. J. Anim. Ecol. 49, 381-394.

VILELA, E.F. 1983. Behaviour and control of leaf cutting ants (Hymenoptera, Attini). Ph.D. Thesis, University of Southampton, 209pp.

WILSON, N.L., J.H. DILLIER and G.P. MARKIN. 1971. Foraging territories of imported fire ants. Ann. Ent. Soc. Am. 64, 660-665.

33

Foraging Strategies and Vegetation Exploitation in the Leaf-cutting Ant *Atta cephalotes* (L.)— A Preliminary Simulation Model

J. Reed and J.M. Cherrett

INTRODUCTION

When foraging for leaves in tropical rain forest, leaf-cutting ants such as *Atta cephalotes* adopt a foraging pattern which has long puzzled naturalists; Eidmann (1935) remarked that, contrary to expectation, the ants do not always forage first on vegetation immediately surrounding the nest. Cherrett (1968) highlighted the problem, citing the plant *Terminalia amazonica*, which *A. cephalotes* was cutting 65.4m from its nest in Guyana's tropical rain forest. However, an intact specimen of the same species actually overhung the entrance holes of the nest and in 100 instances at least 6.2 specimens of the same species were nearer to the nest than the one being cut. More recently, Rockwood and Hubbell (1987) calculated that *A. cephalotes* colonies in Costa Rica could improve the quality of the leaves cut by 40% and reduce the foraging distance by 20% just by visiting the trees in their foraging area which had the highest quality leaves.

At first glance this apparent inefficiency contradicts optimal foraging theory. Cherrett (1968) suggested that it provided evidence of a "conservational grazing system," which spread grazing pressure over the whole foraging area and thus minimized over-exploitation of resources near the nest. Subsequently, Cherrett (1983) claimed that this system also minimized the over-exploitation of preferred plant species, so that both the quantity and quality of vegetation around the nest was conserved throughout the 20-year lifespan of these large, sedentary colonies. He suggested that the replacement of forest with citrus monocultures disrupted the system, producing massive tree death. This interpretation has proved controversial, and was challenged by Fowler and Stiles (1980), Shepherd (1982) and Rockwood and Hubbell (1987) on the grounds that simpler explanations can be provided for the observed phenomena.

To date, the arguments have been largely intuitive. This paper is a preliminary attempt to use a simulation model to assess the possibility that the foraging pattern of *Atta cephalotes* colonies permits a long-term maximization of energy input by preventing favored tree species near the nest from dying because of repeated defoliation. The simulation, written in Pascal and run on a Dec System 10, occupied about 50Kb memory when running. A one year execution took approximately 20s CPU time. A brief description of the model's more essential features follows.

THE MODEL

The Matrix of Trees

A circular arena set up around a central nest was divided into a grid of squares or units, each 3m x 3m. This foraging territory had a radius of 45 units in which trees were randomly placed with a minimum spacing of 3m (one tree per unit). Other tree densities could be used. The trees could vary in size (1 to 5 tree units, representing the number of days required to defoliate their palatable parts); in palatability to the ants (1 being highly palatable, and 5 inedible); and in the seasonal pattern with which they produced young leaves. Rockwood (1976) observed that of the 50% to 70% of tropical rain forest species cut by *Atta*, only young leaves were palatable for half while the rest retained palatability at any age. Four patterns of changing palatability were used following Langman and Jenik (1974): (i) continuous evergreen (constant palatability throughout the 4 seasons), (ii) periodic evergreen (leaves fall after flushing; no deciduous period), (iii) deciduous, (one season with no leaves), (iv) leaf exchanging (leaf fall and flushing are concurrent).

Tree Response to Defoliation

During an attack, a tree may be partially or fully defoliated depending on how long foragers remain on one tree before leaving it in favor of a newly-discovered, more palatable one. Little is known about the response of tropical rain forest trees to insect defoliation. Rockwood (1973) recorded a response to artificial defoliation ranging from no reflushing to refoliation after 2 weeks. Rockwood (1975) also observed the death of *Cordia deamii* from repeated defoliation by *Atta. Trema micrantha* and *Cestrum laevigatum* can be killed (Leston 1978); most mature citrus trees in an orchard observed by Cherrett and Jutsum (1983) were killed after 7 years of *Atta* attack; and Mendes Filho (1979) claimed that eucalyptus in Brazil died after the third successive defoliation. In the model, defoliation records are kept for each tree, and at the end of each quarter, defoliated trees reflush, with the restriction that a tree may only reflush a maximum of twice in any one year. A tree dies only if it is totally defoliated for the maximum of six times over a 3 year period, which we regard as a reasonably conservative estimate.

The Patterns of Ant Foraging

Ants forage daily. The scouting effort (in matrix units searched) and the maximum daily harvest can be stipulated. Four possible foraging strategies were modeled:

Concentric search. For a given scouting effort, all palatable trees closest to the nest were ranked by their palatability and the nearest most palatable trees exploited by one size unit each. This strategy simulated reconnaissance by individual scouts, each starting from the nest. If no palatable trees were

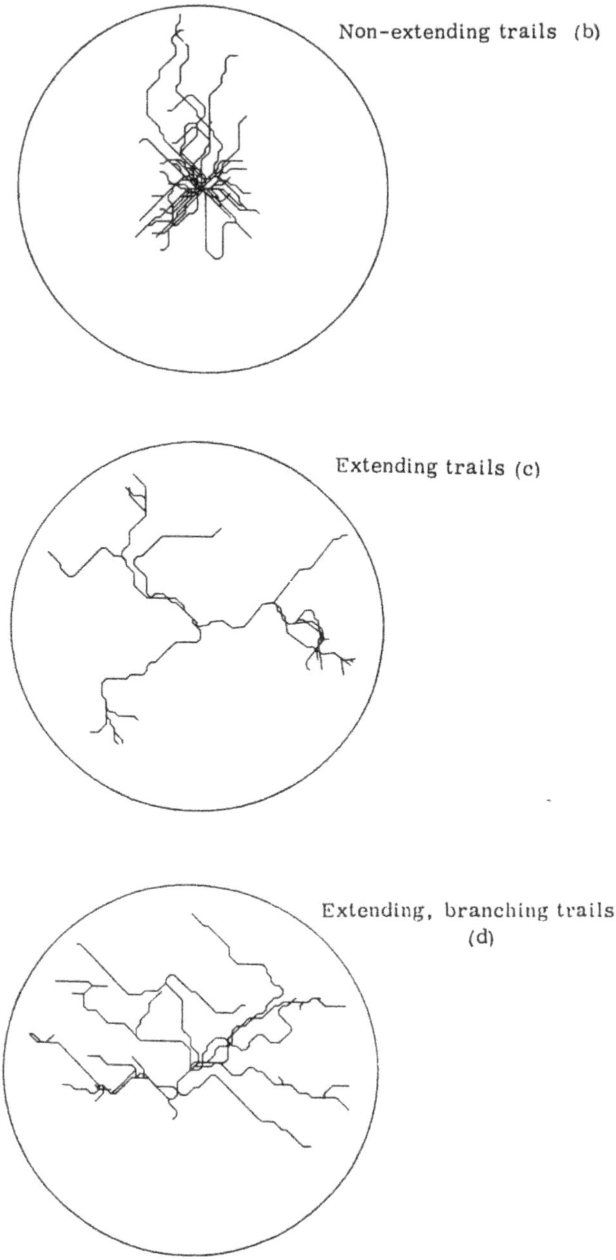

FIGURE 1. Trail patterns developed by three foraging strategies.

located within the specified area on any one day, the scouting effort was extended and the ants walked straight through the depleted zone to continue foraging. At the end of each quarter, when trees reflushed, all scouting began again from the nest. An individual search strategy for invertebrate prey was described in *Formica fusca* by Brian (1965). The other three strategies employed trunk trails.

Non-extending trail. The trail, initiated from the nest, followed in one of eight directions chosen at random. Each area unit a trail passed through and the units on either side were searched for palatable trees. When located, a palatable tree was defoliated, the trail abandoned and a new one initiated. If not, the trail extended by another area unit with a probability p of turning 45° to the present direction, and 1-2p of continuing straight ahead. A restriction was imposed that any subsequent change in direction could not be more than 90° to the original direction from the nest, otherwise unrealistic loops back to the nest and trail recrossing resulted, which have not been observed in nature. The value of p determined the sinuosity of the trail, and was eventually set at 0.25 as this gave a ratio of 1.33 for the distance along the trail to the straight line distance. This value is close to that obtained by Cherrett (1968) for *A. cephalotes*, and by Bowers and Porter (1981) for *A. colombica*. Each day ended when the prescribed scouting effort had been expended or the maximum permitted daily harvest obtained. Fig. 1b gives examples of the trails obtained.

Extending trail. Trail development occurred in the same way as in (2), but the scouting effort was first used to locate as many palatable trees as possible. These were ranked as in (1), and exploited by one size unit each until the permitted daily harvest had been obtained. These trees were included in the next day's ranking, enabling further exploitation of remaining resources. The following day, the trail and an area unit to each side, were re-searched for trees up to the most distant tree attacked on the previous day. Any unused scouting effort further extended the trail. If no palatable tree was encountered during one day's search, or if the trail reached 45 area units in length (the edge of the foraging territory), the trail was abandoned and the ants initiated a new one (Fig. 1c).

Extending, branching trail. In addition to (3), the ants created branches originating from the main trail at the points where trees were being harvested. These trail branches were extended and searched as in the non-branching trail but if no palatable tree was located during a day's search, then the ants abandoned the branch (Fig. 1d). This strategy most closely resembles that used by *A. cephalotes*.

In the simulations, only one trail from a nest was allowed at a time. Although unrealistic, it allowed these strategies to be compared while simplifying the model. The random elements in the model mean that each run provides a unique result. Consequently, all simulations were replicated five times, and results are expressed as means ± one standard error.

Energy Considerations

The gross benefits of foraging were estimated as the annual input of either tree size units harvested, or of nutrient units. The latter was defined as the product of tree size units and their mean quality (palatability). The net benefits of foraging (the energy value of leaf material brought in, relative to the cost of transporting it to the nest) were more difficult to calculate, and did not include the costs of scouting, recruitment, trail clearing and maintenance. Briefly, energy input took account of the energy content of each leaf fragment (an overestimate, since expelled refuse is still high in energy, and the fungus respires), and the likely amount of sap utilized as an energy source by workers (Littledyke and Cherrett, 1976; Quinlan and Cherrett, 1979). The cost per unit distance of transporting a leaf fragment was estimated by summing the basic metabolic and net carrying costs, and allowing for the fact that during a 24h period, only a proportion of returning foragers carry leaf fragments. Because there are still too many uncertainties to give a proper estimate of net energy inputs to the colony, the highest and lowest likely ratios of the cost to the energy input in J m^{-1} of trail length was calculated. These ranged from 0.0016 to 0.0161 - the "best" and "worst" case values.

Parameter Values

Three types of vegetation were created, a monoculture of uniform size class one and three unit palatability to represent a citrus plantation; a variable forest consisting of five tree types with both mean size and palatability at three units, and a highly variable forest with the same mean values, but using all five size categories, palatabilities two to four, and all seasonal types. Tree types should not be confused with tree species, as any one type could contain many species. Trees were placed at three densities, 535, 356 and 178ha^{-1} the highest simulating the density of tropical rain forest trees > 10cm diameter at breast height, and the lowest a citrus plantation. Scouting effort was set at four levels of 15, 30, 45 and 60 area units searched per day, although no field data to validate this are available. The maximum number of foraging sites attacked per trail per night was set at four, a value which included 90% of all observations made by Cherrett (1968).

RESULTS

The Effects of Tree Density and Scouting Effort on Tree Mortality

Increasing the number of units searched each day by scouts increased the percentage of trees killed at all tree densities up to a threshold search effort. Thereafter, percentage mortality fell (Table 1). This was due to (1) more frequent withdrawal from defoliating a tree when a more palatable alternative became available, and (2) less frequent revisiting of areas near the nest, as the trails reached further outward into unexploited resources. For similar reasons, increasing tree density (or a lower percentage of inedible trees, or ones defended by other ant species) generally lowered the percentage killed, although this was not true for the lowest scouting effort where trails lasted for only a short time. Percentage tree mortality was always higher in

TABLE 1. The percentage of trees killed in the whole territory, (and in the inner 80%) by an extending, branching trail foraging strategy, in a variable forest after 20 years. Four levels of scouting activity and 3 tree densities were investigated.

	No. units searched day -1			
Tree density ha-1	15	30	45	60
178	1.8 (8.7)	5.6 (22.1)	8.9 (33.9)	8.6 (26.1)
356	2.6 (11.3)	4.8 (18.2)	4.1 (13.3)	2.7 (6.5)
535	2.5 (10.3)	3.3(11.4)	1.9 (4.7)	1.4 (2.5)

the inner 20% of the foraging area, but at high tree densities and scouting effort, mortality was reduced to 1.4% overall, and 2.5% in the inner region.

The Effects of Foraging Strategy

The performance of the four possible foraging strategies (a--concentric search; b--non-extending trail; c--extending trail; d--extending, branching trail) were compared for two types of vegetation, the high density variable forest (rain forest), and the low density monoculture (citrus orchard), using a scouting effort of 60 area units per day.

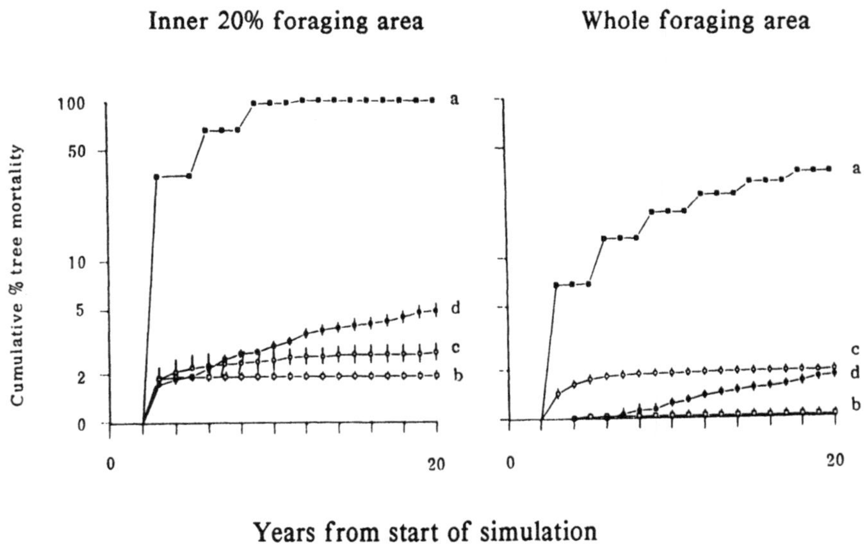

FIGURE 2. Effects of four ant foraging strategies on tree mortality in simulated tropical rain forest (means ± 1 S.E., N=5). a = Concentric search; b = Non-extending trails; c = Extending trails; d = Extending, branching trails.

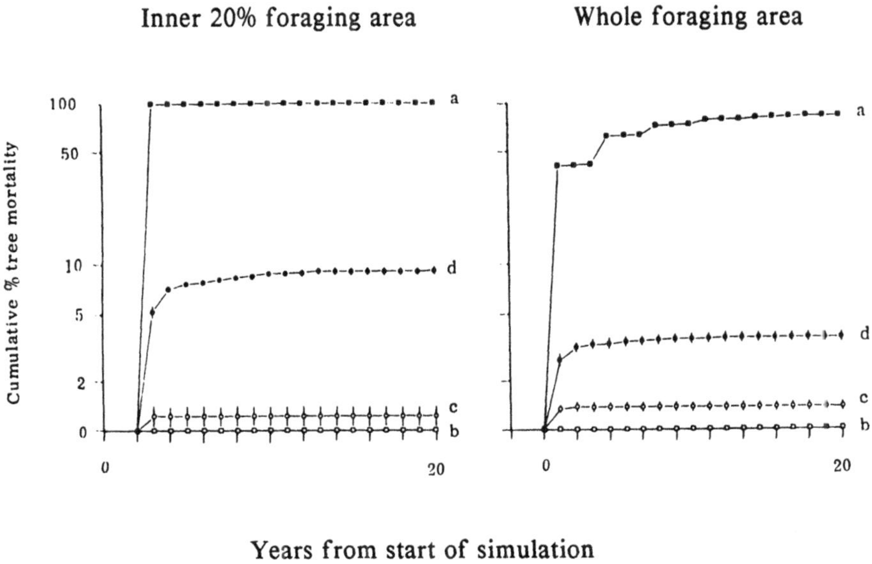

FIGURE 3. Effects of four ant foraging strategies on tree mortality in simulated citrus orchard (means ± 1 S.E., N=5). a = Concentric search; b = Non-extending trails; c = Extending trails; d = Extending, branching trails.

Tree mortality. The cumulative percentage tree mortalities for both the inner 20% and the whole foraging area in rain forest were highest for foraging strategy (a), but realistically low for all three trail- making strategies. The concentric search produced mortality in a three year cycle, as the same trees (those most palatable and closest to the nest) were attacked repeatedly until killed; all trees in the inner 20% died within seven years, and 40% over the whole territory within 20 years (Fig.2.). In the low density citrus plantation, however, the most realistic extending branch trail (d) produced about twice the percentage mortality, while the concentric search killed all the inner trees in three years, and again doubled the overall percentage mortality (Fig.3).

Gross leaf harvest. Success in bringing back leaf material to the nest, not tree mortality, is the parameter of importance to ant colonies. This can be measured as the number of tree size units harvested or as nutrient units input, and the performance of the four foraging strategies in rain forest is compared in Fig.4 using both measures. Initially, strategy (a) produced a gross input comparable with those of the extending trail strategies (c) and (d), but over the 20 year period, this fell to about 65% of the initial value, while (c) and (d) remained nearly constant at a level only slightly below the maximum allowed input of 1440 units per annum. The non-extending trail (b) produced a gross input over the first few days which was as high as the other strategies, but by the end of the first year (the first point recorded in Fig.4), the input was low, falling still farther in subsequent years. The harvest as measured in

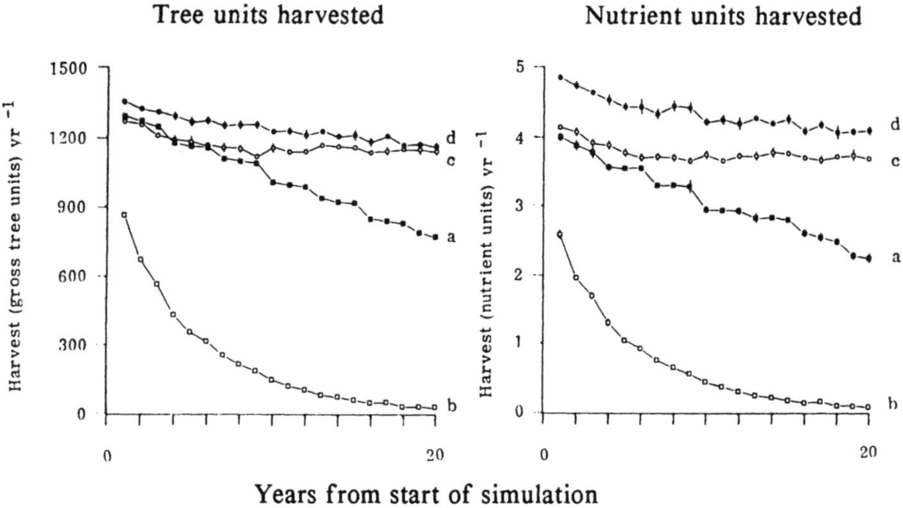

FIGURE 4. Effects of four ant foraging strategies on gross leaf harvest in simulated tropical rain forest (means ± 1 S.E., N=5). a = Concentric search; b = Non-extending trails; c = Extending trails; d = Extending, branching trails.

nutrient units shows similar trends, although the distinctions between strategies (c) and (d) are now clearer, with no overlapping confidence intervals. Strategy (d) was particularly effective at selecting the most palatable (nutritious) substrate.

By contrast, in the low density citrus plantation, the concentric search strategy (a) proved the most efficient over the 20 year period. In this monoculture, tree and nutrient units are identical. (Fig. 5). The trail-making strategies did badly and produced virtually no return after the tenth year.

Net tree harvest. However, success in bringing back leaf material is not the ultimate measure of a foraging strategy's optimality because energy is spent walking out, cutting and collecting the leaves, and bringing them back to the nest. Accordingly, the net energy return or "profit" to the colony needs to be estimated. In the "best case," where energy capture by the ant colony is highest with a cost to energy input per meter of trail length of 0.0016, the performance of the four strategies in rain forest is very similar to that for gross leaf harvest (Fig.6). However, the "worst case" ratio of only 0.016 did produce changes in the relative efficiencies of the strategies. Initially, the concentric search pattern (a) was the most efficient, but this decreased until after about 10 years, it was less efficient than the extending trails (c) and (d). The latter maintained a fairly constant net input over 20 years, as the mean distance of attacked trees from the nest did not change with time. The extending branching trail again proved to be the most efficient in the long run.

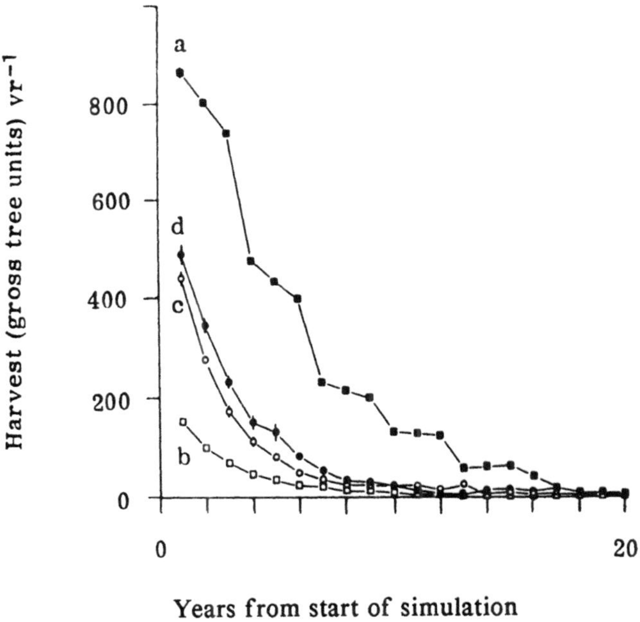

FIGURE 5. Effects of four ant foraging strategies on gross tree units harvested in a simulated citrus orchard (means ± 1 S.E., N=5). a = Concentric search; b = Non-extending trails; c = Extending trails; d = Extending, branching trails.

Finally, simulations were conducted on the highly variable mixed forest containing 60 types of trees. Tree harvest, mean palatability and tree mortality were all similar to those with five tree types in a variable forest. The only difference was that a greater number of trees (200 rather than 100) were abandoned by the ants each year before being completely defoliated, in favor of newly discovered more palatable trees. Tree mortality was low in both cases.

DISCUSSION

Constructing a preliminary model of this sort always serves to highlight the parameters which need to be measured in the field, but for which no data are available. Undoubtedly, the estimates that have been made of the energetic costs and benefits of foraging at different distances from the nest are crude and allow substantial margin for interpretation. Lighton et al. (1987) suggest that such calculations may be premature. However, they do permit a re-examination of Cherrett's (1968, 1983) controversial suggestion that the ants may be operating a conservational grazing system, which prevents deterioration of resources immediately around the nest during its 20 or so years of life.

The model does indeed confirm that the foraging strategy of branched,

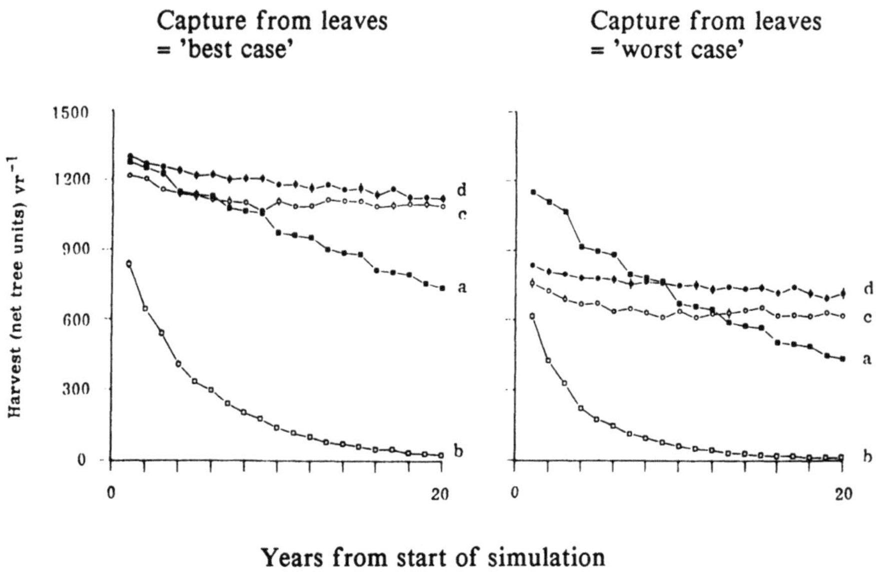

FIGURE 6. Effects of four ant foraging strategies on net tree units harvested in a simulated tropical rain forest (means ± 1 S.E., N=5)rd. a = Concentric search; b = Non-extending trails; c = Extending trails; d = Extending, branching trails.

persistent trails adopted by *Atta cephalotes* in tropical rain forest conserves the resource base, thus giving a consistent, maximal gross and net yield over the likely 20 year life span of the nest. It does this by spreading the foraging effort evenly over the foraged territory and thus minimizing tree mortality. As sexuals in this species may not be produced until five or six years after colony founding, colony fitness in terms of the total number of sexual offspring produced during its life must be greatly influenced by foraging success in the last ten or more years of life. It is interesting to note that net harvest in the worst case calculations (Fig. 6) is initially best using a concentric search strategy (the one which intuitively seems the most efficient), but after about ten years, this has so damaged the resource base, that the apparently less efficient extending, branching trail strategy becomes the more efficient.

Of the four possible foraging strategies simulated, extending branching trails (d) provides optimal foraging in the long run, judged over the whole colony life span, and Rockwood and Hubbell (1987) were right to stress that the ants' behavior follows predictions of optimal foraging. However, conserving resources, by minimizing tree mortality near the nest remains an important element of that optimality. As Rockwood and Hubbell (1987) state, foragers are unaware of the total pool of resources available, and so scout only near their extending trunk trails. As Shepherd (1982) remarks, "Leaf cutter foraging is better understood as a long term optimization that effectively exploits resources over the lifetime of the colony...," although it is not clear why he adds "....than as 'prudent' predation that 'husbands' resources." No one has suggested that the ants are "prudent," in any meaningful sense of the word, and if resources remain viable during a 20 year colony life span, then the mechanism which accomplishes this is effectively conserving the trees in perpetuity.

Rockwood and Hubbell (1987) propose a simpler hypothesis for the observed foraging patterns: (1) That nearer undefoliated specimens of a given plant species have been discovered by scout ants but rejected because they are less palatable as they contain more repellents, antifeedants and fungistatic substances. This has yet to be demonstrated and, as the colony may be attacking a wide variety of other species at the same time, it implies that at any one point in time, intraspecific variability in defensive chemicals is greater than interspecific variability; (2) That the use of cleared trunk trails reduces travel time to the extent that exploiting distant food sources may require less energy than exploiting nearer ones, if no cleared trail leads to them. However, discovering new trees to exploit necessarily involves scouting over uncleared forest litter whether the search begins from the nest, or from the side or end of an existing trail. In addition, long cleared trails require continual maintenance from litter fall, and wind and animal disturbance of the surrounding litter; (3) That foragers are unaware of the total pool of resources available, and so in ignorance miss nearby suitable trees. The latter interpretation can be accepted without reservation as it is a prime tenet of this study's simulation. It is less clear how it accords with Rockwood and Hubbell's other view (1) that nearby trees are found, but rejected on the grounds of palatability. Clearly, more research is needed before these issues can be resolved.

Although it can be argued that the extending, branching trail system which evolved for the ants' native habitat of tropical rain forest is an optimal one, the present simulation confirms the farmers' experiences--that it produces catastrophically damaging results in low density monocultures of citrus trees (Fig.3). Even from the ants' point of view, a concentric search pattern would produce a better harvest than the system actually employed, but as Cherrett (1983) has pointed out "... when the strategy evolved, ancestral *A. cephalotes* never entertained the idea that their tropical rain forest would be cut down and replaced by farms."

ACKNOWLEDGMENTS

The work was carried out as part of a NERC studentship held by J. Reed, and we are grateful for the comments of two anonymous referees.

REFERENCES

BOWERS, M.A. and S.D. PORTER. 1981. Effects of foraging distance on water content of substrates harvested by *Atta colombica* (Guerin). Ecology 62: 273-275.

BRIAN, M.V. 1965. Social insect populations. Academic Press, London.

CHERRETT, J.M. 1968. The foraging behavior of *Atta cephalotes* L. (Hymenoptera, Formicidae). 1. Foraging pattern and plant species attacked in tropical rain forest. J.Anim. Ecol. 37: 387-403.

CHERRETT, J.M. 1983. Resource conservation by the leaf-cutting ant *Atta cephalotes* in tropical rain forest. pp. 253-263. *In* S.L. Sutton, T.C. Whitmore and A.C. Chadwick (eds.), Tropical Rain Forest: Ecology and Management. Blackwell Scientific Publications, Oxford.

CHERRETT, J.M. and A.R. JUTSUM. 1983. The effects of some ant species, especially *Atta cephalotes* (L.), *Acromyrmex octospinosus* (Reich) and Azteca sp. (Hym. Form.) on citrus growing in Trinidad, pp. 155-163. *In* P. Jaisson (ed.), Social Insects in the Tropics. Proc. First Int. Symp. Int Union Study Soc. Ins. and the Soc. Mex. Ent. 2, Universite Paris-Nord.

EIDMANN, H. 1935. Zur Kenntnis der Blattschneiderameise *Atta sexdens* insbesondere ihrer Oekologie. Z. angew. Ent. 22: 385-436.

FOWLER, H.G. and E.W. STILES. 1980. Conservative resource management by leaf-cutting ants? The role of foraging territories and trails, and environmental patchiness. Sociobiology 5: 25-41.

LANGMAN, K.A. and J. JENIK. 1974. Tropical forest and its environment. Longman, London.

LESTON, D. 1978. A neotropical ant mosaic. Ann. Entomol. Soc. Amer. 71: 649-653.

LIGHTON, J.R.B, G.A. BARTHOLOMEW and D.H. FEENER JR. 1987. Energetics of locomotion and load carriage and a model of the energy cost of foraging in the leaf-cutting ant *Atta colombica* Guer. Physiol. Zool. 60: 524-537.

LITTLEDYKE, M. and J.M. CHERRETT. 1976. Direct ingestion of plant sap from cut leaves by the leaf-cutting ants *Atta cephalotes* (L.) and *Acromyrmex octospinosus* (Reich)(Formicidae, Attini). Bull. Entomol. Res. 66: 205-217.

MENDES FILHO, J.M. 1979. Technicas de combate as formigas. Piracicaba, I.P.E.F. Circular Technica 75: 14pp.

QUINLAN, R.J. and J.M. CHERRETT. 1979. The role of fungus in the diet of the leaf-cutting ant *Atta cephalotes* (L.). Ecol. Entomol. 4: 151-160.

ROCKWOOD, L.L. 1973. The effect of defoliation on seed production of six Costa Rican tree species. Ecology 54: 1363-1369.

ROCKWOOD, L.L. 1975. The effects of seasonality on foraging in two species of leaf-cutting ants (*Atta*) in Guanacaste Province, Costa Rica. Biotropica 7: 176-93.

ROCKWOOD, L.L. 1976. Plant selection and foraging patterns in two species of leaf-cutting ant (*Atta*). Ecology 57: 48-61.

ROCKWOOD, L.L. and S.P. HUBBELL. 1987. Host-plant selection, diet diversity, and optimal foraging in a tropical leafcutting ant. Oecologia (Berlin) 74: 55-61.

SHEPHERD, J.D. 1982. Trunk trails and the searching strategy of a leaf-cutter ant, *Atta colombica.* Behav. Ecol. and Sociobiol. 11: 77-84.

34

Senses Used by *Acromyrmex subterraneus molestans* During Homing Orientation, Under Laboratory Conditions

M.S. Guajara, E.F. Vilela and K. Jaffe

INTRODUCTION

The orientation of ants has been studied for a long time. Santschi (1911) reported on the homing orientation of *Cataglyphis bicolor* and *Messor barbatus* using the sun compass. Many ant species use visual clues for orientation. Workers of *Lasius niger* rely on visual cues as well as light polarization (Carthy, 1951). Rosengreen (1971) demonstrated that route fidelity occurs in *Formica* due to a detailed memory of visual cues along the trail. Holldobler (1976) found that the use of chemical trail and visual memory allows accurate homing by *Pogonomyrmex* species. Holldobler (1980) also demonstrated that workers of *Paltothyreus tarsatus* are incapable of sky orientation. They employ canopy orientation instead, a more apropriate method for a species that lives in tropical forest. Vilela et al. (1987) demonstrated that *Atta cephalotes* and *Atta laevigata* use different homing cues in the following order: 1) presence of chemical trail, 2) visual cues or layout of trail, 3) odor cues on the trail and 4) gravitational cues. Vilela (1983) suggested a different hierarchy for the orientation of *Acromyrmex octospinosus*, with the odor trail having a more important role in homing.

There have been few studies on the biology and behavior of the highly polymorphic *Acromyrmex* spp. This study was conducted to investigate the senses used in homing orientation by *Acromyrmex subterraneus molestans workers* under laboratory conditions.

MATERIAL AND METHODS

Four single queen colonies of *A. subterraneus molestans* were maintained in laboratory at 25 ± 2°C, 70 to 80% R.U. in glass containers (Fig. 1). The bioassays were conducted in labyrinths connected to the colony nest, as in Fig. 1.

Observations began at 8 a.m., just two hours into the light period and continued until noon. Darkness was stimulated by covering the labyrinth arms with aluminum foil and by use of red light in the room. A new labyrinth was used for each bioassay. The following combinations were tested:
(1) new labyrinth not chemically marked, under light (OL) or dark (ND)

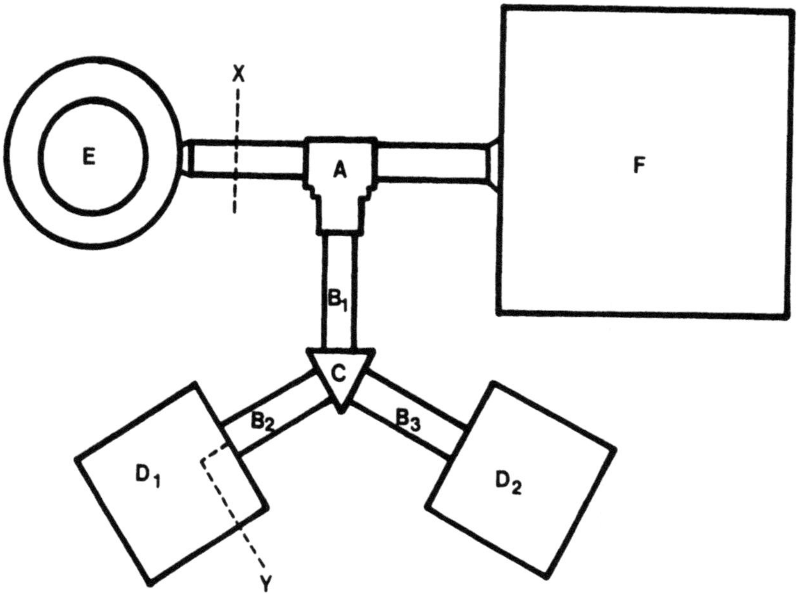

FIGURE 1. *A. subterraneus molestans* nest in the laboratory with a labyrinth connected to it: fungus garden (E); foraging arena (F); "T" connection (A); labyrinth arms (B); cardboard connection (C); plastic receptacles for ant imprisonment (D) and Y to X = ant route observed (ca. 30 cm).

conditions; (2) old labyrinth chemically marked, under light (OL) or dark (OD) conditions; (3) arms reversed 2:1 (Fig. 2) after chemical marking, under light (2 - 1, L) or dark (2 - 1, D) conditions; and (4) arms reversed 2: 3 (Fig. 3), after chemical marking, under light (2 - 3, L) or dark (2 -3, D) conditions.

Observations were initiated after 30 min of ant foraging activity, when the labyrinths were considered chemically marked. After arm reconnection, the time it took the first worker, collector or non-collector, to move from Y to X (Fig. 1) was measured. The amount of time the ant took to orientate itself to each experimental situation was considered a function of the situation's difficulty.

Workers transporting plant material to the nest were collectors, whereas all the other homing ants were non-collectors. Only B2 provided foraging, which left the container connected to B3 always empty for collector experiments. Forage was provided in both labyrinth arms for trials with non-collectors.

RESULTS AND DISCUSSION

Orientation cues used

The results of the bioassays OL, NL, OD and ND (Tables 1 and 2) demonstrated that the workers took significantly more time to walk through the labyrinth in the absence of a chemical trail but even in these conditions, the ants found their way back to the nest, showing that other orientation cues might be involved in homing. Collector workers displayed a higher degree of homing difficulty than the control (OL) in all other experimental situations.

TABLE 1. Comparison between the mean time spent by collector workers of A. subterraneus molestans in the different biossays (see text for abbreviations). Values compared using Mann-Whitney test, n = 20.

	BIOASSAYS	P ≤ 0.05	U
1	OL x NL	*	0
2	OD x ND	*	13.5
3	OLx OD	*	105.5
4	NL x ND	*	79.0
5	2 - 1, L x OL	*	6.5
6	2 - 1, D x OD	*	97.0
7	2 - 1, L x 2 - 1, D	ns	195.0
8	2 - 3, L x OL	*	6.0
9	2 - 3, D x OD	*	80.0
10	2 - 3, L x 2 - 3, D	*	177.0

TABLE 2. Comparison between the mean time spent by non-collector workers of A. subterraneus molestans in the different biossays (see text for abbreviations). Values compared using Mann-Whitney test, n = 20.

	BIOASSAYS	P ≤ 0.05	U
1	OL x NL	*	1.0
2	OD x ND	*	0
3	OL x OD	ns	157.0
4	NL x ND	*	106.5
5	2 - 1, L x OL	*	8.0
6	2 - 1, D x OD	*	20.5
7	2 - 1, L x 2 - 1, D	ns	155.0
8	2 - 3, L x OL	ns	145.5
9	2 - 3, D x OD	*	112.0
10	2 - 3, L x 2 - 3, D	ns	142.5

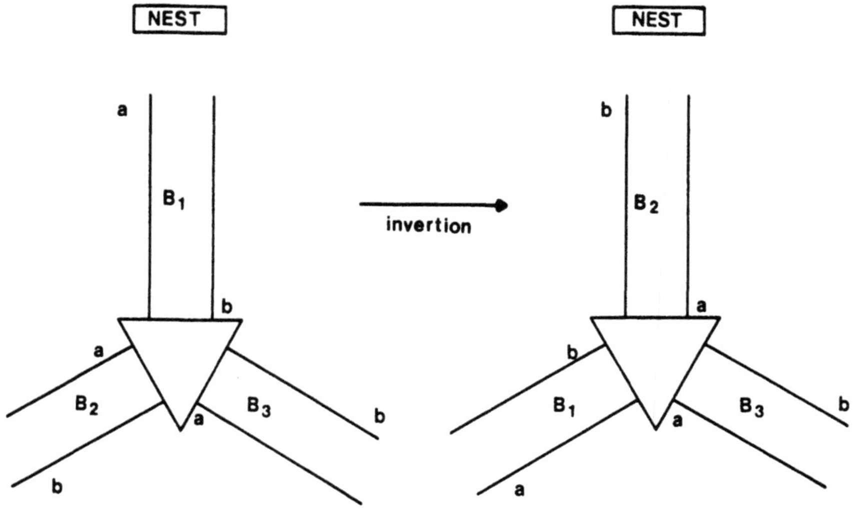

FIGURE 2. Labyrinth arm positions before and after inversion in bioassay 2-1.

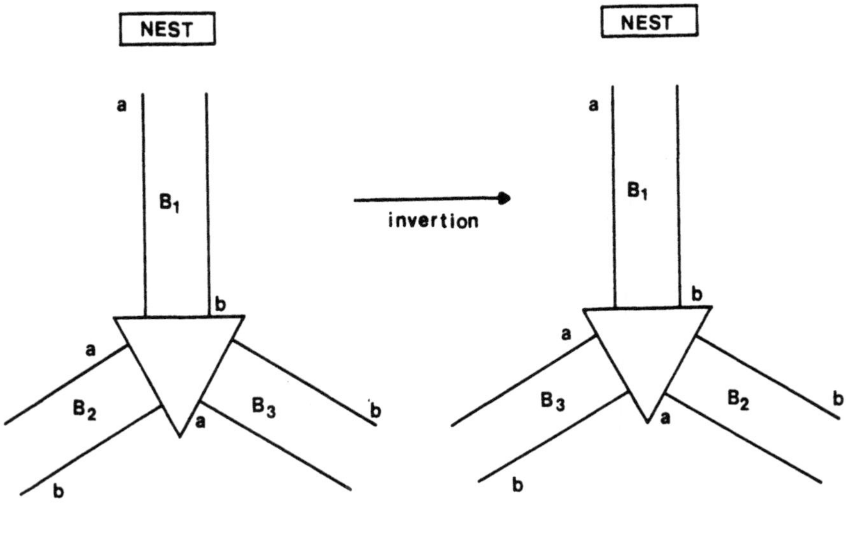

FIGURE 3. Labyrinth arm positions before and after invertion in bioassay 2 - 3.

When the trail was reversed 2 - 1, L and 2 - 1, D the workers took more time in homing compared with the control (OL and OD experiments). This suggests the existence of asymetrical information within the labyrinth, polarizing the trail and thus interfering with the worker's ability to orientate themselves. The presence of polarization in *Acromyrmex* trail had already been suggested by Vilela (1983), although his results were inconclusive. This study, however, supports the hypothesis that *Acromyrmex* workers are more dependent on the chemical trail than *Atta* workers. Trail polarization might be due to the presence of specific chemical marks along the trail or to odor concentration gradient. Further studies may elucidate this matter.

The bioassays OL vs. OD and NL vs. ND indicated that collector workers are better at homing in light than darkness. The time spent by these workers increased considerably in the absence of light (Table 1, OL vs. OD). However, non-collectors showed the same ability to orientate under both conditions (Table 2, OL vs. OD). In the absence of a chemical trail and of light (ND), both collectors and non-collectors had great difficulty orientating themselves (Table 3). Despite insignificant differences in the bioassay results NL vs. ND (Table 1 and 2), collectors had much more difficulty initiating homing; only 14 out of 20 trials could be evaluated within the established observation time (up to 10 min).

The comparison between time spent by collector and non-collector workers to cross the labyrinth after inversion (2 - 1, L) revealed that collectors perform better. This suggests that differences exist in the utilization of visual cues by both categories of workers. It is certain that the chemical trail is the main cue for the workers. Visual cues, however, come second only for the collectors. For the non-collectors, a chemical trail is immediately followed by utilization of odor differences in the trail. The activity rhythm of this species and the daily activities of the colonies should be investigated.

TABLE 3. Mean time (± s.d.) spent by workers of *A. subterraneus molestans* in different bioassays (see text for abbreviations). Values compared using Mann-Whitney test, n = 20.

BIOASSAYS	P < 0.05	U	X ± s.d.(sec) COLLECTORS	NON-COLLECTORS
OL	*	109.0	34.6 ± 9.3	24.4 ± 8.9
NL	*	109.5	173.2 ± 85.6*	110.5 ± 56.5
OD	*	73.5	60.9 ± 36.2	31.5 ± 12.5
ND	*	84.0	255.0 ± 97.1	163.8 ± 53.2
2 - 1, L	*	124.0	94.3 ± 42.4	153.7 ± 90.8
2 - 1, D '	ns	194.0	93.6 ± 37.4	123.8 ± 93.9
2 - 3, L	*	33.0	105.4 ± 65.7	36.4 ± 18.9
2 - 3, D	+	34.5	107.5 ± 51.3	54.4 ± 21.6

* n = 14.

Orientation of collector and non-collector workers

Collectors had much more difficulty in orientation after reversal of the arms (2 -3L and 2 - 3, D), when they had to walk through the arm used exclusively by non-collecting workers (Table 3). The mechanisms involved in this phenomenon remain a mystery. During the study, a higher number of individuals worked at activities other than collecting plant material, although this number was not quantified. It can be assumed that among the non-collectors some individuals might be responsible for the replacement of active individuals. Polymorphism, as in *Atta*, could be responsible for the differences found between the two worker categories studied. However, a great variation in worker size was corroborated, though this variation has yet to be adequately quantified. Studies on the divison of labor in *A. subterraneus molestans* should also be conducted.

ACKNOWLEDGMENTS

We want to thank Dr. Harold Fowler and Dr. Terezinha Della Lucia for reading and criticizing the manuscripts. We are also in debt to FINEP and CAPES (Brazil) for financial support and to Clara E. Dias for typing the manuscript.

REFERENCES

CARTHY, J.D. 1951. The orientation of two allied species of British ant 1. Visual direction finding in *Lasius niger*. Behaviour, 3: 275-303.

HÖLLDOBLER, B. 1976 Recruitment behaviour, home range orientation and territoriality in harvester ants, *Pogonomyrmex*. Behav. Ecol. Sociob., 1: 3-44.

HÖLLDOBLER, B. 1980. Canopy orientation: a new kind of orientation in ants. Science, 210: 86-88.

ROSENGREEN, R. 1971. Route fidelity, visual memory and recruitment behaviour in foraging wood ants of the genus *Formica*. Acta Zool. Fennica., 133: 1-106.

SANTSCHI, F. 1911. Observations et remarques critiques sur le mecanisme de l'orientation chez les fourmis. Revue Suisse de Zoologie, 19: 303-338.

VILELA, E.F. 1983. Behaviour and control of leaf-cutting ants (Hymenoptera: Attini). University of Southampton (Tese Ph.D.).

VILELA, E.F., K. JAFFE and P.E. HOWSE. 1987. Orientation in leaf-cutting ants (Formicidae:Attini). Anim. Behav., 35:1443-1453.

35

The Discovery of New Resources and Subsequent Trail Formation by *Acromyrmex octospinosus* in Guadeloupe

P. Therrien and J.N. McNeil

INTRODUCTION

Little is known about the importance of established trunk trails in the discovery of new food sources by foraging leaf-cutting ants. Cherrett (1968) and Stradling (1978) studied the foraging behavior of *Atta cephalotes*, and concluded that the majority of scouting activity takes place along the length, or at the extremity of trunk trails. This conclusion is supported by subsequent studies on *Atta sexdens* (Fowler and Robinson 1979) and *Atta colombica* (Shepherd 1982), which suggest that scouting is trail centered, a strategy termed "area-restricted search" (Fowler and Stiles 1980, Shepherd 1982).

Contrary to the above *Atta* species, which form large colonies (Weber 1972) with well maintained trail systems (Fowler 1978), *Acromyrmex octospinosus* colonies are smaller (Weber 1972, Lewis 1975, Pereira-da-Silva *et al.* 1981, Shepherd 1982) and frequently have only one trunk trail (Therrien 1986). Observations of 10 colonies in Guadeloupe also showed that during a 16 month period, the azimuthal orientation of the trunk trails rarely changed (Therrien 1986). Under an "area-restricted search" strategy, new resources in areas of the territory not accessed by the trail are predicted to have a lower probability of being discovered and exploited by *Ac. octospinosus*, thereby leading to an uneven coverage of the territory. However, in preliminary studies short-lived pheromone trails were seen leading to resources far from the trunk trail (Therrien 1986), something unlikely to arise with the "area-restricted search" strategy. This paper reports results of a field experiment designed to test the "area-restricted search" model for *Ac. octospinosus* in Guadeloupe.

MATERIAL AND METHODS

The experiment was carried out in 1984 at the Lemesle area of Grande-Terre, Guadeloupe (F.W.I.) (Therrien *et al.* 1986). To determine the effect of the trunk trail on resource discovery and trail formation, banana pieces (3 x 4 cm) were placed, one piece every meter, along both sides of the first 12 m of the only trunk trail of an *Ac. octospinosus* colony. The pieces were alternately placed 50 and 100 cm from the trail; on the first day those at uneven distances were at 50 cm while those at even distances were at 100 cm.

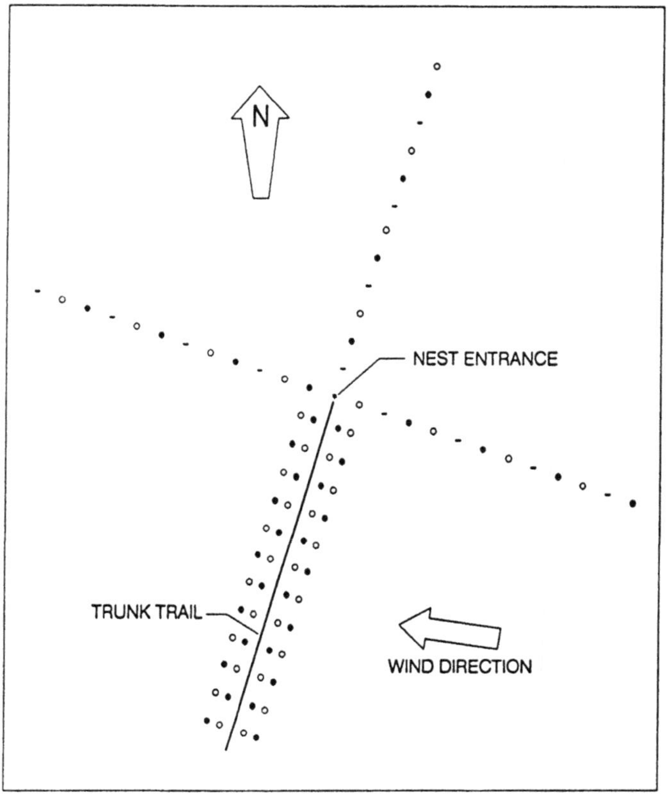

FIGURE 1. Experimental layout for examining search behaviour of *Ac. octospinosus* at Lemesle, Guadeloupe. Symbols indicate locations of banana pieces on different days of observation: black circle = first day, white circle = second day and dash = third day.

The positions were reversed on the second day (Fig. 1). At the same time pieces were placed along three 12 m transects, at 90, 180 and 270° relative to the trunk trail. Banana pieces were placed 3m apart and the positions changed relative to the nest (Fig. 1). A trial required two days along the trunk trail and three days along the other transects. We made four replicates for the trunk trail and three for the other transects.

The ant's daily foraging window at this site was usually from 5:30 AM to 7:30 PM (Therrien 1986). Continuous observations were made from 8:00 to 11:00 AM and food sources were removed at the end of each daily observation period. Two people were needed to take observations: one along the trail and the other for the three transects. A maximum of 20 seconds was needed to check all pieces on the territory. The wind direction relative to the trunk trail was noted at the beginning of each observation period, since ants can orient upwind to food sources (Littledyke and Cherrett 1978). During the experiment the wind was always from the south-east (Fig. 1).

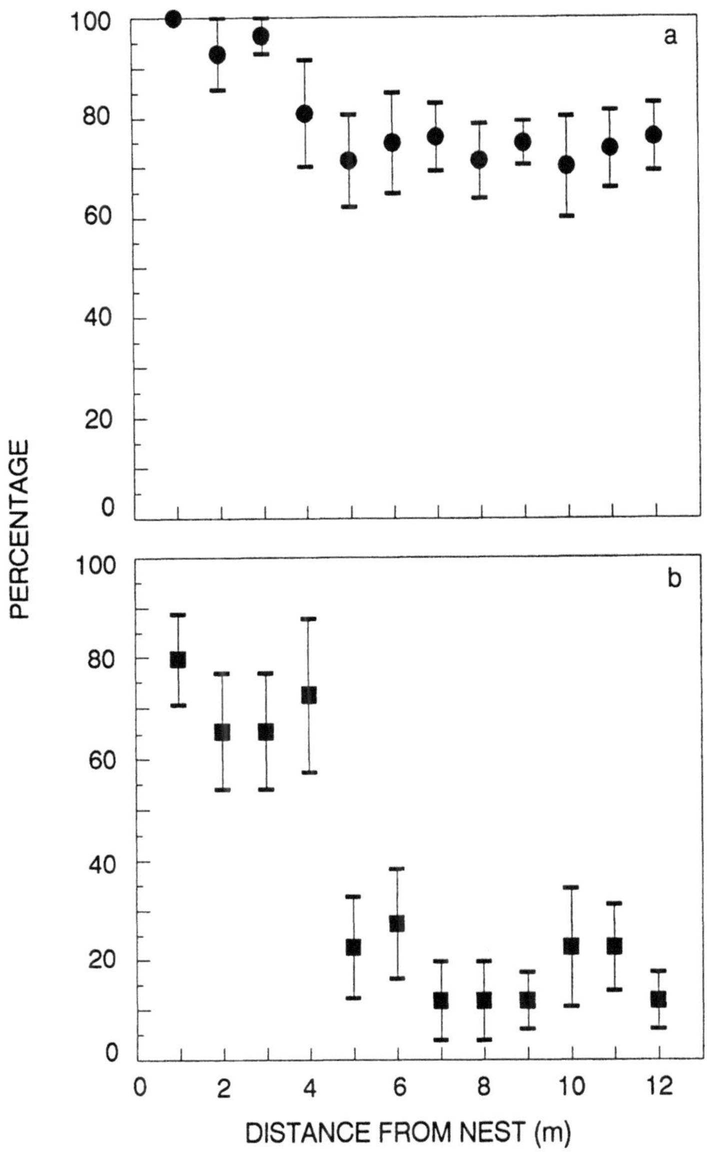

FIGURE 2. Effect of distance from nest on the percentage of banana pieces discovered (a) and on the percentage of trail formation (b) by *Ac. octospinosus*. Vertical bars indicate the SE. For all means N = 7.

Trail formation was assumed if five foraging workers were simultaneously present on a given banana piece. This was an accurate indicator, as it was always followed by constant and regular foraging on the food source. Time data were $\log_e$ transformed before analysis of variance. In the text, back transformed means of time data are presented with 95% CI in brackets. Percentage data needed no transformation and means ± SE are given. All inferences presented were computed using the TYPE III SS (Freund *et al.* 1986).

RESULTS

The overall probability of discovery of any banana piece was 80% (240/300) and was not affected by (i) the location of the piece on any of the four transects ($F = 0.58$; 3, 71 dF; $P > 0.05$), or (ii) by the distance from the trail (50 vs 100 cm) for pieces on the transect along the foraging trail ($F = 0.23$; 1, 71 dF; $P > 0.05$). However, on all four transects, the probability of discovery was affected by distance from the nest entrance ($F = 8.45$; 1,71 dF; $P < 0.05$), being high within the first 3-4 m from the nest and decreasing sharply at distances beyond this (Fig. 2a).

The average time to first discovery was 9.8 min [7.8; 12.27] (N = 76) for pieces upwind of the trail, 12.3 min [10.3; 14.7] (N = 89) for pieces along the three transects away from the trail and 16.9 min [13.3; 21.5] (N = 75) for pieces downwind of the trail. The time to first discovery increased with distance from the nest entrance for pieces upwind of the trail (Fig. 3a) (F regression = 10.37; 1 dF; $P < 0.05$; t-test for slope = 0: $t = 3.22$, $P < 0.05$), varying from an average of 3.4 min [2.3; 5.0] (N = 8) at 1 m to 25.8 min [9.2; 72.2] (N = 5) at 12 m; but it remained essentially constant with distance from the nest entrance for pieces downwind along the trail (Fig. 3b) (F regression = 0.04; 1 dF; $P > 0.05$; t-test for slope = 0: $t = 0.21$, $P > 0.05$) and those on transects away from the trail (Fig. 3c) (F regression = 2.98; 1 dF; $P > 0.05$; t-test for slope = 0: $t = 1.73$, $P > 0.05$). Along the trail, distance from the trail (50 or 100 cm) had no significant effect on time of first discovery when taken alone ($F = 2.39$; 1, 226 dF, $P > 0.05$) or in interaction with position relative to the trail (upwind or downwind) ($F = 0.56$, 1, 226 dF, $P > 0.05$).

Trails were formed to 35.3 % (106/300) of all pieces placed on the territory, or 44.2 % of those discovered by foragers (106/240). As was the case for discovery, trail formation was not influenced by the orientation of the transects with respect to the foraging trail ($F = 0.32$; 3, 65 dF; $P > 0.05$), or by the distance (50 or 100 cm) from the foraging trail ($F = 1.19$; 1, 65 dF; $P > 0.05$). It was influenced, however, by distance from the nest entrance ($F = 28.6$; 1, 65; $P < 0.05$) (Fig. 2b).

Trail formation downwind from the foraging trail took twice as much time (46.5 min [37.8; 57.3], N = 23) as it took upwind (21.3 min [15.8; 28.8], N = 43) or along the three transects away from the trail (21.3 min [17.1; 26.6], N = 40) ($F = 20.91$; 1, 104 dF; $P < 0.05$ for downwind mean compared to mean obtained by combining observations for pieces upwind and those away from the trail). Distance from the nest did not affect time to trail formation ($F = 0.24$; 1, 86 dF; $P > 0.05$). As for time to discovery, time to trail formation along the trunk trail was not influenced by distance from the trail (50 or 100 cm) taken alone ($F = 0.18$; 1, 86 dF; $P > 0.05$) or in interaction with position relative to wind (upwind or downwind) ($F = 0.63$, 1, 86 dF, $P > 0.05$). The time for trail formation was positively correlated with time to first discovery of pieces both upwind (Fig. 4a; $P < 0.05$, N = 43) and along transects away from the trail (Fig. 4c; $P < 0.05$, N=40), which is not surprising considering that pieces must be discovered before recruitment can occur.

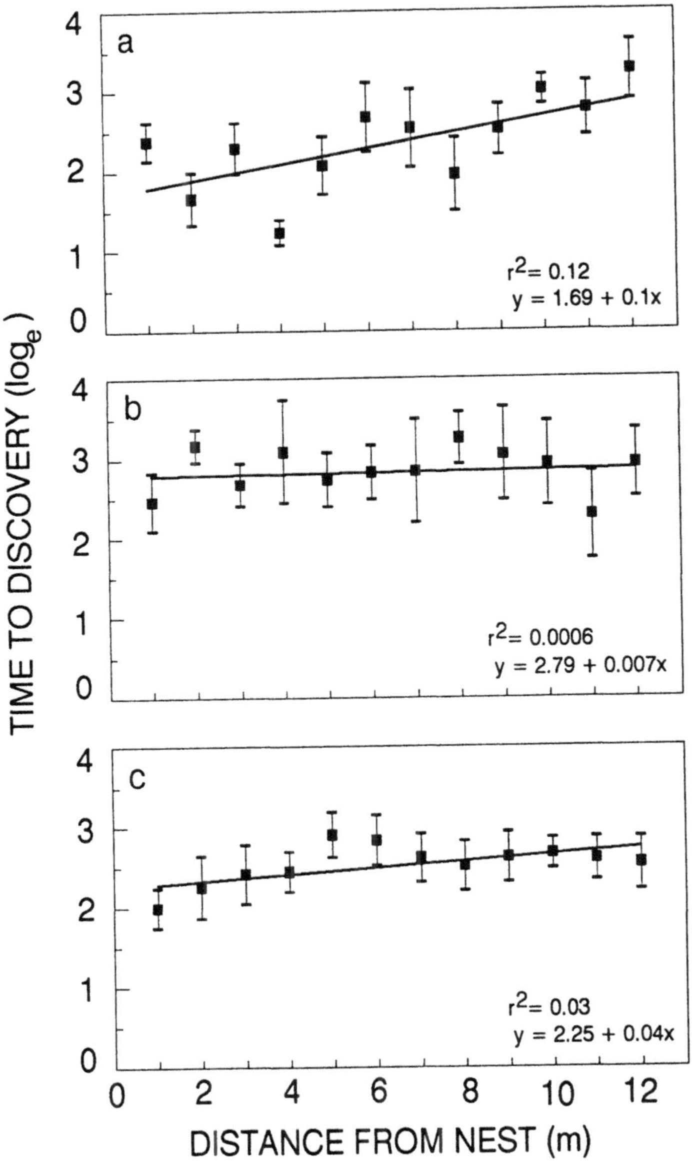

FIGURE 3. Effect of distance from nest on the time to discovery of newly introduced banana pieces by foragers of *Ac. octospinosus*. Squares indicate the mean and vertical lines represent the SE. N varied from 4 to 9. (a) Upwind along the trail, (b) Downwind along the trail and (c) Away from the foraging trail.

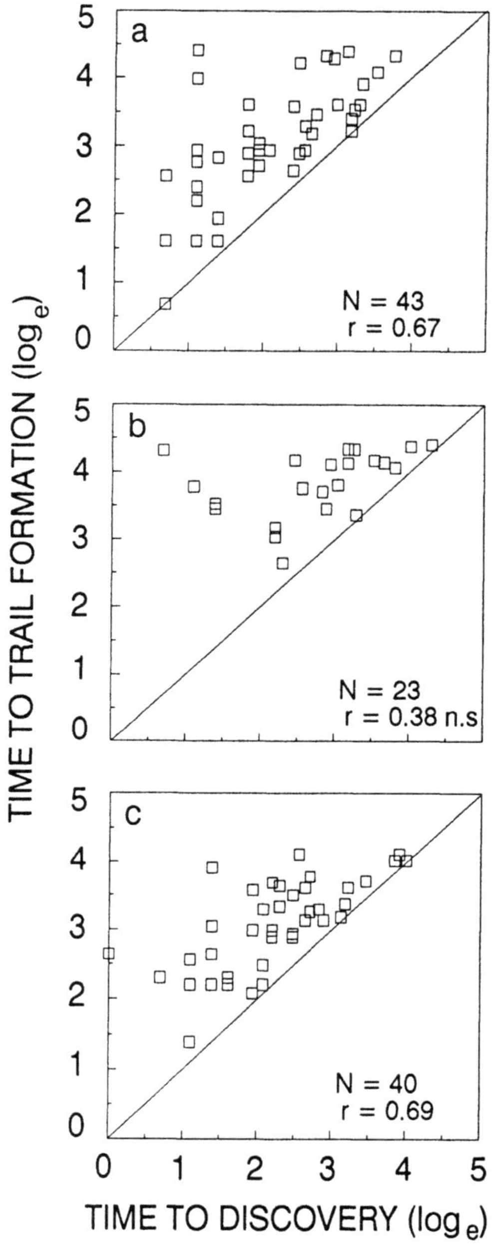

FIGURE 4. Relation between time to discovery of a new food source (banana) and the time to trail formation to the same resource by foragers of *Ac. octospinosus*. (a) Upwind along the trail, (b) Downwind along the trail and (c) Away from the foraging trail.

However, the time to trail formation was not significantly correlated with time to discovery of pieces downwind of the trail (Fig. 4b; $P > 0.05$, $N=23$).

DISCUSSION

Ac. octospinosus is very opportunistic in the area studied, exploiting transient resources, such as flowers or fruits of many plant species, as soon as they appear in the territory, as well as more permanent resources such as leaves (Therrien 1986). Three lines of evidence suggest that new food discovery and trail formation are not trail centered for *Ac. octospinosus* in Guadeloupe. First, workers found the food sources on transects away from the trail as often as comparable food sources near the foraging trail. Second, the probability of new trail formation to food on transects away from the trail was similar to that of food near the trail. Third, banana pieces at 50 and 100 cm on either the upwind or downwind side of the trail had the same probability of being discovered or exploited. This differed from data on *A. colombica*, where distance from the trail (25 or 50 cm) affected the probability of food discovery (Shepard 1982).

While the probabilities of discovery and trail formation were not affected by the food's position with respect to the trunk trail, in all cases probabilities decreased with increasing distance from the nest. This suggests that scouts radiate out randomly from the nest entrance, favoring no specific direction to search the territory. Trunk trails access a very small part of a colony's territory. Shepherd (1982) estimated that only 2-5% of the territory of the *A. colombica* colonies he studied was within 10 m of a trunk trail. Considering that the occurrence of transient resources favored by *Ac. octospinosus* is unpredictable both in time and in space, a scouting effort centered on the nest would be more adaptive than an "area-restricted search" as foragers could equally cover all parts of the territory.

The presence of a nest-centered scouting effort does not negate the importance of the trunk trail in the foraging strategy of *Ac. octospinosus* colonies. Trunk trails lead to permanent resources (Carroll and Janzen 1973, Fowler and Stiles 1980), that are exploited continuously throughout the year. However, these resources are more heavily exploited when the availability of transient ones is low (Therrien 1986).

The reduction in probabilities of both discovery and trail formation beyond 4 m from the nest entrance is intriguing. Therrien (1986) has shown that only large foragers move far from the nest; smaller ones remain close to the nest. It is possible that 4 m represents the limit beyond which smaller foragers will not go, and that forager density remains relatively constant within a 4 m radius around the nest. However, beyond 4 m, the relative density of foragers would decline, not only due to a decrease in absolute forager numbers, but also because the territory surface area increases with distance from the nest.

Wind direction did not influence the probability of new food sources being discovered. However, it did affect the time taken to discover resources upwind of the foraging trail, which increased with distance from the nest. No such effect was seen on transects away from the trunk trail or downwind along the trail. Littledyke and Cherrett (1978) showed that *Ac. octospinosus* can orient upwind to a variety of odor stimuli and, in our experiment, wind borne banana odor would attract foragers already susceptible to leaving the trail. Fowler and Robinson (1979) showed that there is a natural forager leakage from *A. sexdens* trails, with 25% leaving in the first 5 m of a 25 m trail. If a similar situation is true for *Ac. octospinosus*, more foragers would be on the trail close to the nest. They could respond to wind borne odors and more rapidly discover resources close to the nest rather than those farther away. Downwind and away from the trail, ants presumably received no olfactory cues over any appreciable distance and thus food was discovered at random, as foragers "bumped" into them.

The position of food with respect to wind direction was also important in trail formation. At all distances along the trail, it took less time after discovery to create a trail to banana pieces when they were upwind than when they were downwind. One possible explanation is that foragers recruited by scouts from banana pieces downwind subsequently responded to the odor and moved towards the upwind banana pieces. This hypothesis is supported by (i) our results on trail formation away from the foraging trail, where, in the absence of "upwind odor effects", mean time to trail formation was not significantly different than that observed to resources upwind along the trail, and (ii) by the absence of a significant correlation between time of discovery and trail formation for downwind banana pieces (Fig. 4).

Our results show that *Ac. octospinosus* in Guadeloupe does not use the "area restricted search" strategy proposed for three *Atta* species, but instead uses a search strategy centered on the nest. Much of the recent foraging theory assumes that a species' current observed behaviors have been shaped under pressures of local conditions by natural selection (Hassell and Southwood 1978, Kareiva 1983). Viewed this way, the mixture of pheromone trails, physical trails and search strategy centered on the nest, as employed by *Ac. octospinosus* in Guadeloupe, can be interpreted as an environmental adaptation where transient and permanent resources often coexist. However, to fully elucidate the foraging strategies of leaf-cutter ants, comparative research is needed for many species and under a wide range of ecological conditions.

LITERATURE CITED

CARROLL, C.R. and D.H. JANZEN. 1973. Ecology of foraging by ants. Ann. Rev. Ecol. Syst. 4: 231-257.

CHERRETT, J.M. 1968. The foraging behaviour of *Atta cephalotes* L. (Hymenoptera, Formicidae) I. Foraging pattern and plant species attacked in tropical rain forest. J. Anim. Ecol. 37: 387-403.

FOWLER, H.G. 1978. Foraging trails of leaf-cutting ants. N-Y Entomol. Soc. 86: 132-136.

FOWLER, H.G. and S.W. ROBINSON. 1979. Foraging by *Atta sexdens* (Formicidae: Attini): seasonal patterns, caste and efficiency. Ecol. Entomol. 4: 239-247.

FOWLER, H.G. and E.W. STILES. 1980. Conservative resource management by leaf-cutting ants? The role of foraging territories and trails, and environmental patchiness. Sociobiology 5: 25-41.

FREUND, R.J., R.C. LITTELL and P.C. SPECTOR. 1986. SAS system for linear models. 1986 edition. SAS Institute Inc. Cary. 210 pp.

HASSELL, M.P. and T.R.E. SOUTHWOOD. 1978. Foraging strategies of insects. Ann. Rev. Ecol. Syst. 9: 75-98.

KAREIVA, P. 1983. Influence of vegetation texture on herbivore populations: resource concentration and herbivore movement. pp 259-289. *In* Variable plants and herbivores in natural and managed systems. R.F. Denno and M.S. McClure (eds.). Academic Press, New York.

LEWIS, T. 1975. Colony size, density and distribution of the leaf-cutting ant, *Acromyrmex octospinosus* (Reich) in cultivated fields. Trans. R. ent. Soc. Lond. 127: 51-64.

LITTLEDYKE, M. and J.M. CHERRETT. 1978. Olfactory responses of the leaf-cutting ants *Atta cephalotes* (L.) and *Acromyrmex octospinosus* (Reich) (Hymenoptera: Formicidae) in the laboratory. Bull. ent. Res. 68: 273-282.

PEREIRA-DA-SILVA, V., L.C. FORTI and L.G. CARDOSO. 1981. Dinâmica populacional e caracterizaçao dos ninhos de *Acromyrmex coronatus* (Fabricius, 1804) (Hymenoptera, Formicidae). Rev. bras. Ent. 25: 87-93.

SHEPHERD, J.D. 1982. Trunk trails and the searching strategy of a leaf-cutter ant, *Atta colombica*. Behav. Ecol. Sociobiol. 11: 77-84.

STRADLING, D.J. 1978. The influence of size on foraging in the ant *Atta cephalotes* and the effect of some plant defence mechanisms. J. Anim. Ecol. 47: 173-188.

THERRIEN, P. 1986. Behavioral ecology of the leaf-cutting ant, *Acromyrmex octospinosus* (Reich), in Guadeloupe, F.W.I. The University of British Columbia, Ph. D. Thesis. 166 pp.

THERRIEN, P., J.N. McNEIL, W.G. WELLINGTON and G. FEBVAY. 1986. Ecological studies of the leaf-cutting ant, *Acromyrmex octospinosus* in Guadeloupe. pp 172-183. *In* Fire ants and leaf-cutting ants. Biology and management. C. S. Lofgren and R. K. Vander Meer (eds). Westview Press, Boulder.

WEBER, N.A. 1972. Gardening ants, the Attines. Mem. Am. Phil. Soc. Philadelphia. 92: 1-146.

36

Factors Controlling Foraging Patterns in the Leaf-cutting Ant *Acromyrmex octospinosus* (Reich)

J.J. Knapp, P.E. Howse and A. Kermarrec

INTRODUCTION

Many plant species escape attack from leaf-cutting ants (Cherrett 1968a, 1968b, 1972, Rockwood 1976, Pintera 1983). Host plant selection may be influenced by repellent chemicals in the plants (Cherrett 1972, Hubbell *et al.* 1983, Howard and Wiemer 1986), physical characteristics of the leaves (Stradling 1978, Waller 1982a and 1982b), the proximity of plants to established trails (Shepherd 1982), the nutritional (Cherrett and Seaforth 1970, Littledyke and Cherrett 1975, 1978, Mudd *et al.* 1978) and water content of leaves (Cherrett 1972, Bowers and Porter 1981) and resident ant defense of plants against leaf-cutting ant attack (Jutsum *et al.* 1981).

Generally, ants forage on material near to their trails. As a particular trail develops, the majority of scouting behavior occurs along the trail and at the end of it (Cherrett 1968a). Rockwood (1973) noted that while ants on one trail continue to cut one plant species for long periods of time, other trails switch after a relatively short time from one plant species to another. Cutting generally dropped to low levels 60-80 meters from the nest, but very common trees near the nest were ignored. Foraging preferences were thus clearly independent of abundance and distance of individual trees from the nest (Rockwood 1976, 1978, Fowler 1982, Shepherd 1985).

In order to explain this pattern of foraging, Cherrett (1968a, 1983) proposed that *Atta cephalotes* in its natural environment has adopted an optimal conservational foraging strategy. He argued that fitness will be maintained by conserving the quality and quantity of vegetation in the vicinity of the nest over the life of the colony (approx. 20 years.) By scouting mainly from the ends of existing trails rather than from the nest itself, grazing pressure is to some extent evened out over the whole foraging area. However, the costs of retrieval are greatly increased, although it reduced the danger of trees near the nest being killed. Cherrett argues that it is this strategy, which suits the species so well in the rainforest, that results in total devastation when the forest is replaced by citrus orchards.

Rockwood (1976) observed that while *Atta* colonies did not concentrate their activity on those trees closest to the nest, any palatable trees in this area had a high probability of being visited. They suffered neither more nor less defoliation than plant species visited 60 meters from the nest. He concluded that the foraging pattern could not be called truly conservative. Fowler and Stiles (1980) have argued that attine foraging patterns reflect the response of workers to the patchiness of their palatable resources in the environment. Some evidence suggests that leaf-cutting ants are able to make subtle distinctions in leaf quality, not only between species, but also between individual plants within a species and between leaves from different parts of the same plant (Hubbell and Wiemer 1983). Ants prefer certain species at certain times of year, and young leaves and flowers of many species are more heavily cut than older leaves (Rockwood 1973, 1976). The lack of correlation between ant attack and the abundance and distance of plants from the colony suggests that plants are selected on the basis of their vulnerability to attack and for their quality as food for the ants, the fungus or both (Howard and Wiemer 1986).

Leaf mechanical defenses (e.g.: toughness, trichome density, copious or latex sap) may prevent cutting altogether or reduce the cutting rate to the point where it is unprofitable to attempt to cut (Cherrett 1968b, Howard and Wiemer 1986). Larger workers more easily cut tough leaves than small workers (Wilson 1980, Waller 1982a, 1986, Seri *et al.* 1986). Latex adheres to the mandibles, rendering them useless (Stradling 1978).

Some researchers suggest that the nutritional quality of leaves will influence selection. Some species preferentially select substrates coated with mono- and disaccharides in the laboratory (Cherrett and Seaforth 1970, Littledyke and Cherrett 1975, 1978, Mudd *et al.* 1978). Amino acids were also shown to attract some species of leaf-cutting ant (Cherrett and Seaforth 1970). Field preferences are related to the mineral composition of the leaves, for example, *Atta cephalotes* preferred leaves with higher concentrations of nitrogen and potassium (Berish 1986). In addition, some ant species select plant material of higher moisture content (Cherrett 1972). It is also possible that the further the distance from the nest, the higher the moisture content of the leaves selected (Bowers and Porter 1981).

The importance of plant secondary compounds in attine host plant selection has become more evident. Many plants contain chemicals toxic to insects and microorganisms (Feeny 1976, Rhodes and Coates 1976). Some evidence now suggests that the selection of fungal substrate is in part based on plant chemistry (Martin 1970, Cherrett and Seaforth 1970, Cherrett 1972, Barrer and Cherrett 1972, Littledyke and Cherrett 1975, 1978, Stradling 1978, Mudd *et al.* 1978, Rockwood and Glander 1979, Cherrett 1980, Hubbell and Wiemer 1983, Hubbell *et al.* 1983, 1984, Howard and Wiemer 1986). Many secondary compounds are deleterious to fungal development (Stossell 1970, Levin 1971, Whittaker and Feeny 1971). Indeed two plant compounds--a phytohaemagglutinin from Jackbean (*Canavalia enisformis*) and carophyllene epoxide from *Hymenaea courbaril*--are toxic to the attine fungus (Fowler and Robinson 1979, Fowler 1981, Mullenax 1979, Hubbell *et al.* 1983). Some researchers believe that attines tend to avoid plants containing antifungal compounds (Mudd and Bateman 1979, Hubbell and Rockwood 1983, Hubbell *et al.* 1984).

Many plants contain terpenoid compounds which are deterrent or repellent to leaf-cutting ants (Wiemer and Ales 1981, Chen *et al.* 1983, Chen *et al.* 1984, Hubbell *et al.* 1984, Hurbert and Wiemer 1985, Ojkunade and Wiemer 1985a, 1985b, Howard and Wiemer 1986). Certain varieties of yam and citrus that are not attacked (Pollard *et al.* 1983, Febvay *et al.* 1985) may contain chemical defenses. Resistance of the yam, *Diosorcea cayenesis*, was thought to be based on saponins (Febvay *et al.* 1985) but this theory has not been positively confirmed in subsequent bioassays (Rahbe *et al.* 1988).

Several leaf-cutting ant species seasonally avoid many of their regular hosts (Fowler and Robinson 1979, Fowler 1981, Waller 1982a, Waller 1986). Others remain palatable throughout the year. Ants may require different nutrients at different seasons depending on brood, temperature or moisture conditions in the nest (Lewis *et al.* 1974, Rockwood 1973, 1976). Seasonal rejection may result from fluctuations in plant levels of nutrients, water, or secondary compounds. These fluctuations may reflect natural seasonal variations or they may be induced by environmental stresses, including fungal or herbivore attack (Haukioja and Niemela 1977).

The present study investigates whether laboratory colonies of *Acromyrmex octospinosus*, the only attine species on the island of Guadeloupe (F.W.I), showed a scale of preference for plants occurring on the island and whether this preference could be explained in laboratory bioassays in terms of repellency or feeding deterrents.

FORAGING ACTIVITY

Methods and Materials

To monitor the ants' behavior and control their leaf intake, laboratory nests were used. This ensured that the test plant species was presented alone, and not in competition with other species. A range of plant material from each test species was offered to the ants. Activity on the foraging trail was measured and used to determine the plant's relative acceptability. Throughout an experiment, the ants' behavior was observed and notes made.

Ac. octospinosus nests, collected from the field in plastic containers, were transported to the laboratory (Fig. 1). The plastic transporting lid was replaced with a glass lid to enable the observation of the nest. The container

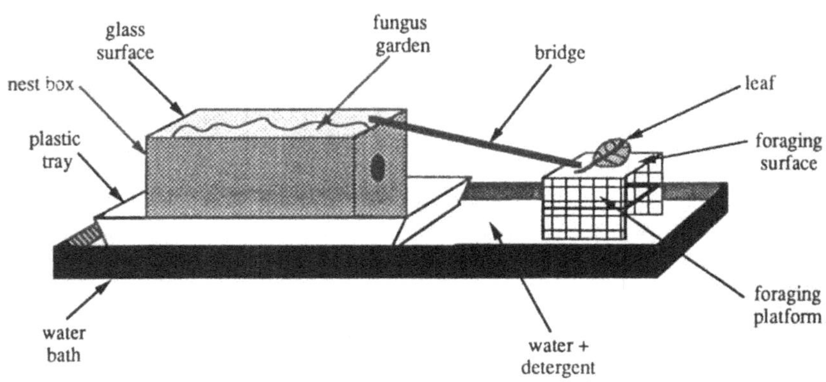

FIGURE 1. Laboratory nest of the leaf-cutting ant, *Ac. octospinosus*.

had one entry hole. The new nests were not used for experimental purposes until the ants had restructured themselves and reestablished a fungus garden of between 0.5 and 1.0 liters. All the experimental nests contained queens. The nests were housed in a specially designed ant room and maintained at 25°-30°C and 70-80% relative humidity. They experienced the natural/dark cycle (i.e. LD 12:12) of the tropics. The nests were fed regularly with a range of plant material, but not with the test species (except for hibiscus).

Guadeloupe has a restricted but mixed flora with many pantropical introduced species as well as purely neotropical species. The areas of *Ac. octospinosus* infestation are predominantly agricultural regions. Cultivation varies from sugar cane plantations to the more traditional "jardin creole". There are also many trees, particularly fruit trees and domestic gardens with a variety of ornamental plants. Some of the land is also given over as pasture.

Fresh plant material was collected from the field, placed in polythene bags to reduce desiccation and transported back to the laboratory. If available, leaves still attached to their stems and branches were selected. Either leaves of a range of ages or only leaves of approximately the same age were chosen. Whenever possible both flowers and fruits were collected.

In the laboratory, plant material was placed on the foraging platform. If a particular species of plant was unavailable locally and could not be collected on consecutive days, it was stored in a sealed polythene bag in the refrigerator overnight. Although not ideal, this method of storage preserved enough material to complete an experiment.

Three nests were prepared for each experiment. Twenty-four hours prior to the actual run, the nests were fed with fresh samples of the test species. This was done so that any initial "explorative" cutting could not be misinterpreted as true palatability. Reports (Febvay *et al.* 1985, Rahbe *et al.* 1988) describe *Ac. octospinosus* "switching off" responsiveness to certain plant species after an initial cutting period. By allowing the ants to sample each test species for a period of twenty-four hours, this problem might be avoided. However, observations were made at the time of this preliminary feeding to see if indeed the ants did cut the material provided.

Following the twenty-four hour period any plant material remaining on the platform was noted, removed and discarded. Fresh test material was then placed on the foraging platforms. Attempts were made to give each of the three nests approximately the same quantity and type of material. The platform could only be reached by crossing a bridge ca 40 cm. long (Fig. 1). This arrangement allowed ants on the trail to be counted with relative ease. A record was made of the type of material provided (leaves, flowers, stems, etc) and which was cut first.

After a period of at least 15 minutes and not more than 30 minutes, trail activity was measured. This was done by recording the number of ants on the trail during a three minute period. The figure was broken down into three parts: (1) The number of ants traveling away from the nest towards the food source; (2) The number of ants returning to the nest *unladen*; (3) The number of ants returning to the nest *laden*.

Occasionally, when recruitment was very slow, observations continued after 30 minutes. Long response latencies could have been due to deterrent properties of the plant material, but they could also have been influenced by the ants' circadian activity rhythms. If, after some time, no cutting occurred, the latter possibility was tested by placing some highly acceptable species

TABLE 1. The 43 plant species tested for acceptability (%=proportion laden ants on trail).

No.	Plant species	Common name	%	Type
1	*Ricinus communis*	Castor oil plant	32.4	Herb
2	*Hibiscus hibiscus*	Hibiscus	31.8	Shrub
3	*Artocarpus altilis*	Bread-fruit tree	30.6	Tree (F)
4	*Manihot esculenta*	Cassava	27.4	Veg (H)
5	*Dioscorea alta*	AsiaticYam	27.0	Veg (V)
6	*Melia azedarach*	Chinaberry	26.8	Tree
7	*Citrus sinensis*	Sweet Orange	26.7	Tree (F)
8	*Annona muricata*	Sour sop	24.6	Tree (F)
9	*Passiflora edulis*	Passion fruit	24.4	Vine
10	*Spondias dulcis*	Golden apple	22.9	Tree
11	*Passiflora foetida*	Love-in-a-mist	20.7	Herb
12	*Persea americana*	Avocado	19.4	Tree (F)
13	*Colocasia esculenta*	White eddo	19.2	Herb
14	*Carica papaya*	Papaya	18.9	Tree (F)
15	*Vigna unguiculata*	Black-eye pea	18.8	Veg(S)
16	*Nerium oleander*	Oleander	18.5	Shrub
17	*Ascelpias curassavica*	Blood flower	16.6	Herb
18	*Cajanus cajan*	Pigeon pea	16.3	Veg(S)
19	*Psidium guajava*	Guava	15.5	Tree (F)
20	*Centalla sp.*	–	15.4	Herb
21	*Ipomoea batatas*	Sweet potato	12.9	Veg (V)
22	*Annona squamosa*	Custard apple	9.9	Tree (F)
23	*Lantana camara aculeata*	Red sage	8.1	Shrub
24	*Passiflora laurifolia*	Water lemon	7.0	Vine
25	*Daucus carota*	Carrot	7.0	Veg (H)
26	*Veronia albicaulis*	–	5.7	Herb
27	*Musa acuminata*	Banana	5.3	Tree (F)
28	*Cucurbita moschata*	Winter squash	5.3	Veg (V)
29	*Xanthosoma jaquini*	Tannia	4.6	Veg (H)
30	*Eupatorium sp.*	–	3.6	Herb
31	*Triphasia trifolia*	Lemon grass	2.8	Herb
32	*Capsicum sinense*	Pepper	2.5	Shrub
33	*Momordica charantia*	Balsam Pear	1.9	Herb
34	*Dioscorea cayenesis*	Yellow yam	1.6	Veg (V)
35	*Peperomia pellucida*	Man-to-man	1.5	Herb
36	*Lantana involucrata*	Rock sage	1.0	Shrub
37	*Eupatorium odoratum*	Christmas bush	0.6	Herb
38	*Curcuma domestica*	Safron	0.6	Herb
39	*Mangifera indica*	Mango	0	Tree (F)
40	*Pogostemon heyneanus*	–	0	Herb
41	*Allamanda cathartica*	Allamanda	0	Shrub
42	*Passiflora quadrangularis*	Giant granadilla	0	Vine
43	*Eryngium foetidum*	Fit feed	0	Herb

(such as hibiscus) on the platform and observing if recruitment continued to be low. If so, the test was discontinued.

Observations were made throughout the experiments on the ants' behavior (ie. aggression, antennae waving, cleaning, licking the surface, drinking sap from cut stems etc.) and the fate of the test plant material (i.e. incorporated in fungus garden, discarded en route to the nest, taken to rubbish pile etc.).

Results

The proportion of ants carrying leaves, in relation to the total of all ants using the trail, was calculated for each species by taking the percentages from all three nests and calculating the average. These percentages provide an approximate index to the selection of each species and overcame the problem of comparing nests with different background levels of activity. Although levels of activity differed in the three nests, the acceptability rankings were generally similar. Table 1 shows the plants arranged in descending order of acceptability, as measured by the numbers of laden ants returning to the nest.

In total, 43 species were tested. The proportion of laden ants on the trails ranged from 0 to 32%. The average was 12.8% (standard error 1.6). The median was 11%. These results indicate a scale of preference, with a plant's acceptability being more than a simple yes/no decision.

Each species in Table 1 was assigned to one of the following categories:

Herb: Herbaceous plant usually not cultivated, although some used for culinary or medicinal purposed (eg: Citronella).

Shrub: Woody plants sometimes used for ornamental purposes (eg: hibiscus, oleander).

Vine: Plants that use others for support and have elongate stems, can be herbaceous or woody, also called lianas (eg: passion vines, sweet potatoes).

Veg(etable): Commonly cultivated plants. They may be any of the above three types (as indicated by the letter following in brackets).

Tree: Usually woody plants which climb upwards to form a canopy. These species which are responsible for producing important local fruit crops are specified by the letter (F).

There are no signs that any one habit is more or less vulnerable to attack by *Ac. octospinosus*. The 43 plant species were separated into 4 ranks of "palatability" or acceptability (pick-up and transport of ant plant pieces) as shown in Table 2.

Table 3a, 3b, 3c and 3d describe the observations made during the course of the preparation feed and during the test run itself. The various plant parts (leaves, flowers, stems, fruits, tendrils etc.) provided for the preparation feed and the order in which they were cut were noted. The time period which elapsed between putting the leaf material in place and the first cut was recorded during the test run. Additional general observations were made and are also recorded in these tables.

Flowers, when present were usually cut first, the only exceptions were *Annona squamosa* (22) where flowers were cut second and *Peperomia pellucida* (35) where neither flowers nor leaves were cut. In contrast the

TABLE 2. Shows the 43 species of plant divided into four ranks of acceptability.

Rank	Acceptability	Plant No.	%	Mean %
1	Highly acceptable	1-11	≥ 20%	26.3
2	Moderately acceptable	12-21	≥10%	17.8
3	Low acceptability	22-28	≥5%	6.9
4	Unacceptable	29-43	<5%	1.4

flowers of *Allamanda cathartica* (41) were the only part of the plant cut during the test, although leaves were also cut during the preparation feed. This change in behavioral response to plant material, which is usually long-lasting, can be called delayed rejection.

Delayed rejection of the above type was observed with several plant species (eg: *Allamanda cathartica, Dioscorea cayenesis*) which were readily foraged by *Ac. octospinosus* during the preparation feed but 24 hours later during the test run were totally rejected. Other subtle behavioral differences were observed which led to the categorization of four different behavioral responses, including two kinds of delayed rejection.

(1) Harvesting: All or most of the plant material is cut and transported back to the nest.
(2) Delayed rejection (type A): In the initial presentation, the colony cut and transport the leaves but in all subsequent presentations rejects the plant material. This involved some kind of information storage by the colony.
(3) Delayed rejection (type (A): Initially the ants will forage but numbers on the trail are often low. As time continues, foraging wanes and eventually ceases altogether. This response occurs each time a nest is presented with the species and so is unaffected by experience.
(4) Immediate rejection (type A): The ants harvest none of the plant material. Initially partial or whole cuts may be made on the leaves but fragments are not transported to the nest. When cut, leaf discs are abandoned or occasionally transported to the rubbish pile (as with *Lantana involucrata trifolia*).
(5) Immediate rejection (type B): There is no cutting and no harvesting.

In summary, *Ac. octospinosus* sometimes rejects certain plant species after a latent period, but others are rejected almost immediately. The latent period may be short, a matter of a few hours, or longer (at least 24 hours).

Checks were carried out on colonies which exhibited behavioral responses of categories 2, 3 or 4 where cutting was slow or non-existent, to determine whether the low level of activity was general to the colony. Hibiscus leaves were placed on the foraging platform at the end of the test period (max. 5 hours). In each case, activity on the trail increased dramatically and foraging quickly commenced. This suggests that the previous lack of activity may have been due to the quality of the test species and not to some internal mechanism. Certain other parameters can also give a guide to the relative acceptability of one species over another. These include the speed with which ants harvest a certain quantity of material: this also measures the level of recruitment to a particular food source. However, high recruitment levels do not always correlate with high levels of cutting and transportation back to the nest.

Plant species which elicited behavior responses of types 2, 3, 4 and 5 are listed below.

2. Type A delayed rejection	3. Type B delayed rejection	4.+ 5 Immediate rejection
Allamanda cathartica	*Eupatorium odoratum*	*Eryngium foetidum*
Dioscorea cayenesis	*Momordica charantia*	*Eupatorium sp.*
Cucurbita moschata	*Xanthosoma jaquini*	*Triphasia trifolia*
Annona squamosa	*Capsicum sinensde*	*Curcuma domestica*
Mangifera indica	*Lantana camara a.*	*Peperomia pellucida*
Ipomoea batatas	*Musa acuminata*	*Pogostemon heyneanus*
	Veronia albicaulis	*Lantana involucrata*
	Passiflora laurifolia	*Passiflora quadrangularis*

Some plants were cut and then discarded or taken to the rubbish pile. Others were not cut at all but still had large numbers of ants exploring the surfaces of the leaves and sometimes drinking sap from cut stems. Interpretations of acceptability based solely on levels of recruitment can thus be misleading.

The time to the first cut (Table 3) is the time in minutes which elapsed from the start of the test run until the first cut by a foraging ant. It should be noted that this time represents the time taken to cut *leaves* rather than flowers, fruits etc. For each rank the mean time was calculated:

Rank 1 (Harvesting) Mean time to first cut = 8.33 min.
Rank 2 (Harvesting) Mean time to first cut = 30.33 min
Rank 3 (Harvesting) Mean time to first cut = 60.85 min
Rank 4 (Harvesting) Mean time to first cut = 123.33 min

These figures show that, as acceptability declines, the length of time taken to the first ant cut increases. While this may be generally true, there are some dramatic exceptions. These are most evident in the group of plant species which exhibit the type B delayed rejection response. For instance *Lantana camara aculeata* (23) was first cut after only 6 minutes. This is a Rank 3 species yet its cutting time compares more favorably with the Rank 1 species. It is therefore dangerous to draw too many conclusions from these time periods.

It is interesting to compare the results obtained from members of the particular plant family or members of the same plant species. In some cases, the response of *Ac. octospinosus* to all the members of the group was similar.

TABLE 3a. Laboratory observations of the foraging behavior of *A. octospinosus* of plant species of rank 1.

No.	Plant Species	Order cut	Time to 1st cut (mins)	Additional Observations
1	*Ricinus communis*	Leaves 1st Stems 2nd	4-8	Rapid recruitment, all plant material given consumed
2	*Hibiscus hibiscus*	Flowers 1st Leaves 2nd stems 3rd	2-4	Very rapid recruitment, all parts cut quickly
3	*Artocarpus altilis*	Only leaves provided	6-16	Active exploration of leaf surfaces before cutting begins
4	*Manihot esculenta*	Only leaves provided	4-5	Rapid recruitment
5	*Dioscorea alta*	Leaves 1st Stems 2nd	4-6	Rapid recruitment
6	*Melia azedarach*	Flowers 1st Leaves 2nd Stems 3rd	1-4	Extremely rapid recruitment very active
7	*Citrus sinensis*	Only leaves provided	2-10	Active exploration of leaf surfaces, drinking sap from cut stems
8	*Annona muricata*	Only leaves provided	12-18	Slower to recruit, selects younger leaves first
9	*Passiflora edulis*	Tendrils 1st Leaves 2nd Stems 3rd	8-14	Initial period of exploring followed by cutting, all consumed
10	*Spondias dulcis*	Leaves 1st Stems 2nd	10-13	Rapid recruitment
11	*Passiflora foetida*	Tendrils 1st Leaves 2nd Fruit 3rd	9-12	High level of recruitment especially exploring fruit

In others, there was a distinct difference between the members. Four members of the family Passifloreaceae were tested. *P. edulis* (9) and *P. foetida* (11) were highly acceptable species in Rank 1. The latter is, however, found towards the lower end of the scale and could be considered borderline between ranks 1 and 2. *P. laurifolia* (24) had a low acceptability and was placed in Rank 3. A fourth species, *P. quadrangularis* (42) was immediately rejected (rank 4). Different species within the same genera can thus differ markedly in their acceptability. Could physical differences account for this rejection? Or are there more subtle distinctions in leaf chemistry which make one species acceptable but not another closely related species.

Other species from the same families are also found in different ranks. *Dioscorea alta* (Family: Dioscoreaceae) in rank 1, *D. cayenensis* in rank 4;

TABLE 3b. Laboratory observations of the foraging behavior of *Ac. octospinosus* of plant species of rank 2.

No.	Plant Species	Order cut	Time to 1st cut (mins)	Additional Observations
12	*Persea americana*	Leaves 1st Stems 2nd	14-27	Cuts younger leaves first, Initial activity declines after a period
13	*Colocasia esculenta*	Leaves 1st Stems 2nd	16-25	Initial recruitment wanes and continues at lower level of activity
14	*Carica papaya*	Only leaves provided	22-27	Long period of exploration of leaf surfaces
15	*Vigna unguiculata*	Stems 1st Leaves 2nd	26-41	Stems are soft not woody and are easily cut
16	*Nerium oleander*	Flowers 1st Leaves 2nd Stems 3rd	23-31	Rapid recruitment to flowers, leaves take much longer
17	*Ascelpias curassavica*	Flower 1st, Fruit 2nd Leaves+stems	21-28	All parts of plant eventually attacked including the roots,
18	*Cajanus cajan*	Stems 1st Leaves 2nd	23-50	Initial activity is high but this is reduced , could be due to leaves drying up
19	*Psidium guajava*	Leaves 1st Stems 2nd	35-41	Younger leaves cut first, only a few mature leaves cut.
20	*Centalla sp.*	Stem 1st Leaves 2nd	47-75	Succulent stems, drinking sap from cut leaves and surfaces.
21	*Ipomoea batatas*	Leaves 1st Stems 2nd	20-38	Rapid recruitment initially this slows down and activity reduces during cutting.

Annona muricata (Family: Annonaceae) in rank 1, *A. squamosa* in rank 3; and *Nerium oleander* (Family: Apocynaceae) in rank 2, *Allamanda cathartica* in rank 4.

Some species in the same family had similar acceptabilities. For example, members of the Compositae; *Eupatorium odoratum* (37), E. sp. (31) and *Veronia albicaulis* (26), rank low. Members of the Leguminosae family, *Vigna unguiculata* (16) and Cajanus cajan, were in intermediate ranks.

It can be concluded that the factor which determines plant acceptability is often genus specific and even specific to a particular species. Some members of the genus have managed to escape attack while others remain vulnerable. Possibly the experiences of the individual colony enter into this, because it is never possible to work with a completely naive colony.

TABLE 3c. Laboratory observations of the foraging behavior of *Ac. octospinosus* of plant species of rank 3.

No.	Plant Species	Order cut	Time to 1st cut (mins)	Additional Observations
22	*Annona Squamosa*	Leaves 1st Flowers 2nd Stems 3rd	27-34	Very active trail but slow to cut.
23	*Lantana camara aculeata*	Flowers 1st leaves, fruit Stems	5-8	High activity including some initial cutting then slows down
24	*Passiflora laurifolia*	Only leaves provided	112-130	Very slow to cut, little material taken eventually stops altogether
25	*Daucus carota*	Leaves } Stems } 1st	11-30	Cuts whole plant but as soon as it becomes desiccated stops altogether.
26	*Veronia albicaulis*	Leaves 1st Stems 2nd	78-112	Leaf cutting declined and eventually stopped
27	*Musa acuminata*	Only leaves provided	128-147	Very slow recruitment, exploration of leaf surfaces
28	*Cucurbita moschata*	Only leaves provided	22-27	Initial cutting declines and eventually stops

CHOICE EXPERIMENTS

The preceding experiments show that the recruitment level to a food source depends on the quality (eg. lack of deterrents) of the material offered. Not only did the number of ants using the trails increase for certain plant species, but more significantly so did the proportion of ants cutting and transporting leaves back to the nest. However, these experiments were somewhat crude and such factors as leaf texture and cuticle hardness could have accounted for some of the observed differences. In order to eliminate these possibilities and produce a more accurate assessment of plant preferences, a computerized experiment was undertaken.

In this experiment the ants were given a choice between a standard, *Hibiscus hibiscus* and various test species. *H. hibiscus* was chosen as the standard because fresh material was readily available in large quantities and, according to preliminary results, it is a highly acceptable species (Table 2).

Hibiscus was tested against a variety of locally grown plants (in particular those cultivated in jardins creoles) and also against several fruit trees. Those species which appeared to have a low acceptability were examined to see if rejection occurred upon first contact or after a period of exposure. Such acquired rejection has been observed in *Ac. octospinosus* feeding on yams (Rahbe *et al.* 1988).

The computerized system used in these experiments was developed by Febvay *et al.*(1986). The system recorded continuous, rapid behavior in large

TABLE 3d. Laboratory observations of the foraging behavior of *A. octospinosus* of plant species of rank 3.

No.	Plant Species	Order cut	Time to 1st cut (mins)	Additional Observations
29	*Xanthosoma jaquini*	Only leaves provided	111-167	Low activity, very little cutting or exploration
30	*Eupatorium sp.*	Leaves 1st Stems 0	247-291	Very low activity, a few "taster" cuts only
31	*Triphasia trifolia*	Only leaves provided	53-69	After a period a few pieces cut but taken to rubbish not to the nest
32	*Capsicum sinense*	Only leaves provided	41-48	A few pieces cut then all activity on leaves ceased
33	*Momordica charantia*	Stems 1st leaves 2nd	20-59	A few pieces cut then all activity on leaves ceased
34	*Dioscorea cayenesis*	Only leaves provided	112-128	Exploration of leaf surfaces but few cuts made and no transport
35	*Peperomia pellucida*	Flowers 0 Leaves 0	176-244	No parts cut not even the flowers, a lot of exploring on leaf surfaces.
36	*Lantana involucrata*	Flowers 1st Leaves 2nd Stem 0	86-205	A few pieces cut but taken to rubbish pile not into the nest
37	*Eupatorium odoratum*	Leaves 1st Stems 0	43-48	A few pieces cut then all activity on leaves ceased
38	*Curcuma domestica*	Only leaves provided	52-65	A few pieces cut then all activity on leaves ceased
39	*Mangifera indica*	Only leaves provided	0	Cuts made in leaf but not completed, no transport
40	*Pogostemon heyneanus*	Only leaves provided	0	Cuts made in leaf but not completed, no transport
41	*Allamanda cathartica*	Flowers 1st Leaves 0	0	Flowers taken very quickly but leaves remain uncut
42	*Passiflora quadrangularis*	Only leaves provided	0	Active exploration, test cuts made but no transportation
43	*Eryngium foetidum*	Leaves 0 Stems 0	0	Some pieces cut but abandoned others 1/2 cut

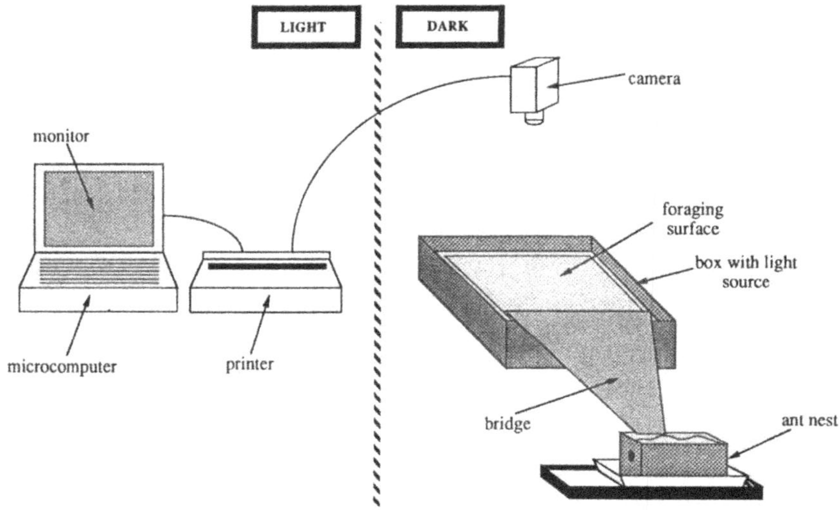

FIGURE 2. Schematic representation of the system used to record foraging behavior in *Ac. octospinosus* (after Febvay et al. 1986).

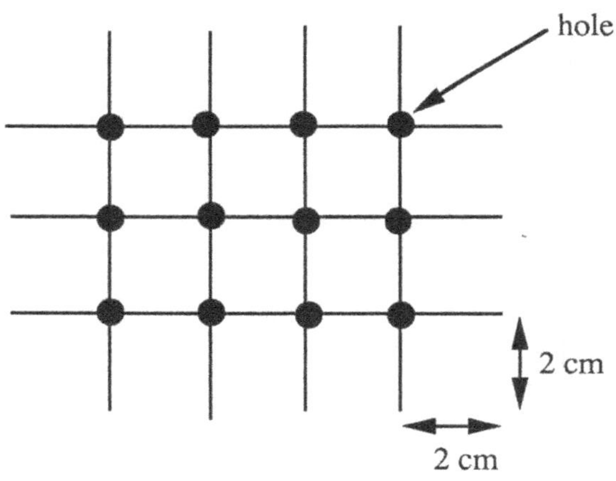

FIGURE 3. Grid pattern of holes punched through black card to create an array of light points (after Febvay et al. 1986).

numbers of small insects. Normally such recording is difficult and exhausting, producing results plagued with operator bias and inaccuracies caused by fatigue. The Febvay system eliminates most of these problems. Basically it is a microcomputer (PC) coupled to a video camera which records the behavior of the ants as they forage on leaf discs cut from the various plant species (see Figure 2).

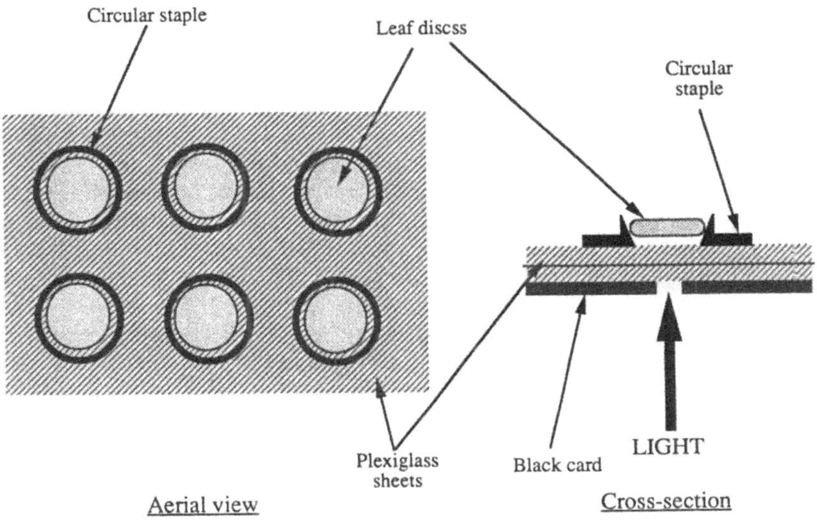

FIGURE 4. Aerial view and cross section of the foraging platform showing the arrangement of the circular staples on the plexiglass surface and their positioning above the holes in the black card (after Febvay et al. 1986).

Materials and Methods

The video camera, positioned directly above a foraging platform, recorded the selection and removal of leaf material. The platform itself was a modified version of that described by Hubbell and Wiemer (1983) and consisted of a light box containing three neon tubes. The lid of the box was made of a sheet of thick black card covered with a sheet of plexiglass. The card was marked with a grid forming 25 x 25 squares of 2 cm^2. Holes of 1.5 mm were then punched through the plexiglass at all points on the grid (see Fig. 3). This resulted in an array of holes with light shining through from underneath. Using an additional card to block out the holes not required, arrays of different sizes could be created eg: 20 x 20, 12 x 12 etc. At this point, the image sent to the computer and visible on the monitor screen consisted of small white spots arranged in a grid pattern. The number of spots depended on how many of the holes were covered by card. For each run the grid size was selected and the card secured in place with tape. The foraging surface itself consisted of a detectable sheet of transparent glass. Its upper surface was covered in circular metal staples (see Fig. 4). These were fixed 2 cm apart, directly above the holes. The circular staples served two functions: first, to mark where the leaf discs should be placed and secondly, to prevent the leaf discs from becoming dislodged once in position. The staples did not interfere with pick-up of leaf material by the ants.

When all the leaf discs were in position, the light from below was blocked out. As a result, the image on the monitor screen appeared totally black. When a leaf was removed, a spot appeared on the screen. The different types of leaf discs were placed on the foraging surface according to a computer-generated random distribution pattern

The purpose of this experiment was to follow in time the pick up of leaf discs by monitoring the appearance of light spots. Complete darkness was necessary for the experiment to run correctly, but *Ac. octospinosus* forages equally well in the light or dark.

The software program "Attinetho V2.0 Juin 1985" (Febvay *et al.* 1986) ran the experiment. It recorded chronologically, the order, time and type of each leaf disc picked up. At the end of each run the data was saved, filed and then made available for statistical analysis and graphic representation.

Leaf discs (approx. 6 mm diameter) of the plant species to be tested were cut from fresh leaves using a hole punch. The leaves selected for each species were all approximately the same size and age. Holes were cut along the margins of the leaf, care being taken not to cut across major veins. Once cut, the discs were placed in labelled petri dishes to avoid desiccation. With a minimum of delay, the discs were then placed on to the clean forage surface using a pair of clean forceps. The discs from each plant species were carefully positioned on the appropriate spots according to the random pattern. The laden sheet was then placed on top of the light box in position with all the light holes covered over. The experiment began when the bridge was placed across the gap from the nest to the foraging platform and the computer was

TABLE 4. Results from two-way choice tests between hibiscus and various test species using the computerized bioassay. ***= significant ($p<1\%$); NS = not significant; + = greater than hibiscus; - = less than hibiscus; O = no difference. Chi squared with one degree of freedom.

	Plant Species	Common name	Chi- square	Significance
	Terminalia catappa	Indian almond	33.10	* * *
+	*Nerium oleander*	Oleander	17.81	* * *
	Artocarpus altilis	Bread fruit tree	16.26	* * *
	Passiflora quadrangularis	Giant granadilla	36.0	* * *
	Allamanda cathartica	Golden trumpet	30.42	* * *
	Musa acuminata	Banana	24.17	* * *
	Cucurbita moschata	Winter squash	24.17	* * *
-	*Mangifera indica*	Mango	21.42	* * *
	Annona squamosa	Custard apple	18.68	* * *
	Psidium guajava	Guava	13.9	* * *
	Ipomoea batatas	Sweet potato	9.32	* * *
	Persea americana	Avocado	6.95	* * *
O	*Passiflora laurifolia*	Water lemon	0.75	NS
	Citrus grandis	Pomelo	0.035	NS

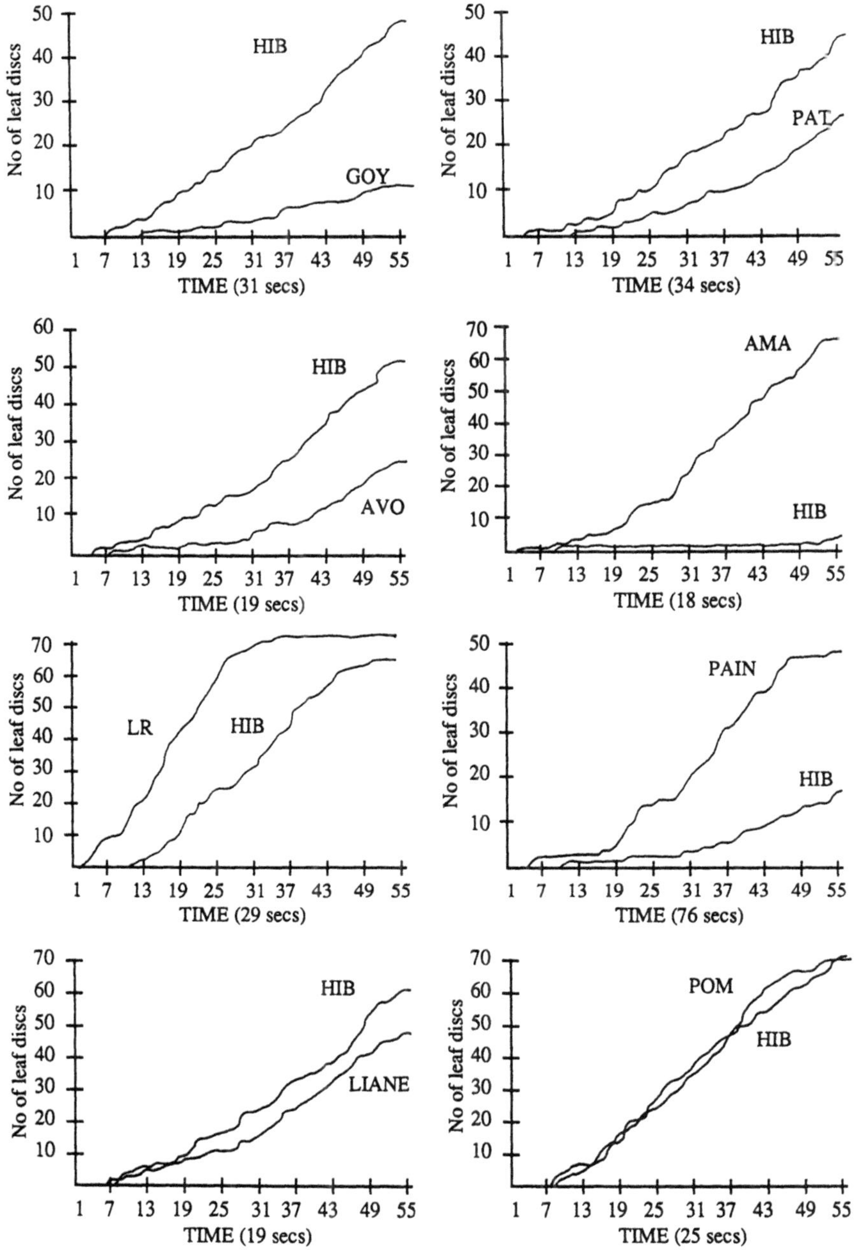

FIGURE 5a. Cummulative pick-up rates of leaf discs in two-choice experiments. Time in arbitrary periods of seconds, noted on each abscissa. HIB = *Hibiscus hibiscus*, *GOY* = *Psidium guajava*, *PAT* = *Ipomoea batatas*, AVO = *Persa americana*, AMA = *Terminalia catappa*, LR = *Nerium oleander*, PAIN = *Artocarpus altilis*, LIANE = *Passiflora laurifolia*, POM = *Citrus maxima* (pomelo).

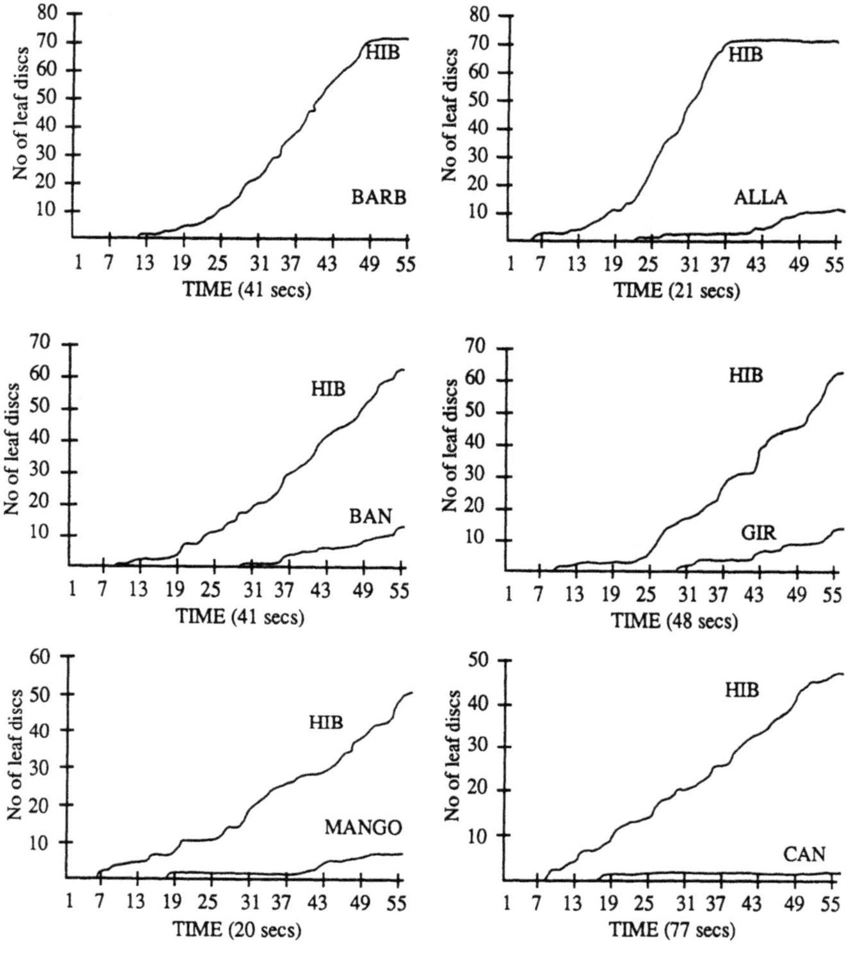

FIGURE 5b. Cummulative pick-up rates of leaf discs in two-choice experiments. Time in arbitrary periods of seconds, noted on each abscissa. HIB = *Hibiscus hibiscus*, *BARB* = *Passiflora quadrangularis*, ALLA = *Allamanda cathartica*, BAN = *Musa acuminata*, GIR = *Cucurbita moschata*, MANGO = *Mangifera indica*, CAN = *Annona squamosa*.

activated.

The run was stopped when either all of one species' discs had been picked up by the foraging ants or after a certain time period had elapsed. The progress of the experiment could be monitored on the screen throughout each run (the computer and monitor were located outside the darkened experimental chamber). Once a run was completed, the nest and trail areas were examined to confirm that the leaf discs had indeed been taken into the nest and not discarded *en route* to the nest or deposited in the rubbish pile.

Several measures were taken to minimize the possibility of ant escapes during experimental runs. The nest platform itself was isolated in a water bath containing a solution of mild detergent. The upper rim of the light box was made 8 cm high and coated with talcum powder to prevent escapes from the experimental arena.

For each plant species tested three runs were carried out. Each run was performed using a different nest. In this way the effects of individual nest preferences--if any--could be accounted for. In each case, the nest had been exposed to the standard and test species for at least 24 hours prior to the run. An exception was made for species in which delayed rejection was being investigated. Here, runs were first carried out on naive nests and subsequently at certain time intervals after food had been given.

Results

The results of runs using hibiscus as the standard species are summarized in Table 4. Chi-squared analysis was performed on the data generated by each run. This showed whether or not the difference in pick up between the two plant species was significant, the null hypothesis being that there should be no preference shown for either species.

The fourteen-test plant species tested can be divided into three groups based on the foraging response of *Ac. octospinosus*. These were: those species the foragers preferred to hibiscus (+), those species the foragers did *not* prefer (-) and those over which no significant preference was shown (0). The higher Chi-squared value, the greater the degree of preference or non-preference. There are three species which were preferentially foraged over hibiscus: *Terminalia catappa* (Indian almond), *Nerium oleander* (oleander) and *Artocarpus altilis* (breadfruit tree). Two species for which *Ac. octospinosus* showed no significant preference were: *Passiflora laurifolia* (water lemon) and *Citrus grandis* (pomelo). The remaining nine species were all rejected in favor of hibiscus. The degree of preference of hibiscus over the nine plant species makes it possible to rank their acceptability.

The graphs in Fig. 5 represent the response of *Ac. octospinosus* during the actual test runs, and show the rate of leaf pick-up. Time is measured in units of seconds, the size of each unit is shown in brackets on each graph.

The graph of *Terminalia catappa* clearly shows the high degree of preference the foraging ants demonstrated for the test species over hibiscus. Foraging on *T. catappa* began quickly and only after all its leaf discs had been removed was hibiscus selected. The ants searched out all possible *T. catappa* discs before resorting to hibiscus. By contrast, *Nerium oleander* and *Artocarpus altilis*, while both being significantly preferred, were not selected to the exclusion of hibiscus. Pick up of oleander and breadfruit tree occurred at a faster rate than for hibiscus and in each case there was a delay before hibiscus discs were removed. However this delay did not, as in *T. catappa*, last until all discs of the test species had been removed. Approximately 25% of each test species were removed before hibiscus pick-up began.

The ants completely rejected giant granadilla, *Passiflora quadrangularis* in favor of hibiscus: foragers did not pick up a single giant granadilla leaf disc during the course of the run. Giant granadilla was not even taken after the hibiscus was gone. In these experiments, almost every physical factor was eliminated by presenting the nests with material that had already been cut in the form of leaf discs. Since the upper and lower leaf surfaces of giant granadilla were smooth, i.e. absence of trichomes or other surface structures, any deterrence must be chemically-based.

Golden trumpet, *Allamanda cathartica*, was also rejected in favor of hibiscus. In this case, a small degree of foraging occurred once all the hibiscus discs had been removed, after which the level remained static, suggesting that the initial pick ups were "tasters" and that golden trumpet was deterrent.

Foraging on banana, *Musa acuminata*, began only after approximately 35% of hibiscus leaf discs had been removed by *Ac. octospinosus*. The rate of banana pick up was slow and did not significantly increase after the removal of hibiscus discs. Thus, a preference for hibiscus was not the only factor limiting banana pick up; the plant itself had negative qualities. The same was true for winter squash and mango, in which foraging on the test species began only after 40% of the hibiscus leaves had been removed. The pick up rates were much slower compared with those for hibiscus and pick up rates did not improve once all hibiscus discs had been removed. This would have been expected if hibiscus had been the only factor limiting pick up of these two species. As with banana, these results suggest that a chemical factor in the leaves themselves is responsible.

When presented with a choice between custard apple, *Annona squamosa* and hibiscus, the ants picked up hibiscus exclusively. However once the supply of hibiscus was exhausted the ants began to forage on custard apple discs, though this later increase is not shown on the graph. The pick up rate was not as fast as that for hibiscus.

When presented with a choice between guava, *Psidium guajava* and hibiscus, the ants preferred hibiscus. The pick up rate for hibiscus was high and although pick up of guava discs started without a long delay, it was much slower. Similar results were obtained for sweet potato, *Ipomoea batatas* and avocado, *Persea americana*, although for these somewhat higher pick up rates suggested that they were more acceptable to the colony than those previously described.

Finally the remaining two test species, water lemon, *Passifolora laurifolia* and pomelo, *Citrus grandis*, had approximately the same pick up rate as hibiscus. From their graphs, it is clear there was no significant difference between the starting times for either the standard or test species. This suggests that they have an acceptability similar to hibiscus.

From all this data a hierarchy can be constructed, based upon the calculation of means and chi-square values (which were homogenous for the three replicates). (1) Species preferable to hibiscus, which are thus at the top of the hierarchy. (2) Species not significantly different from hibiscus. (3) Species rejected in favor of hibiscus, assigned lower ranks depending on the degree of rejection as determined by their chi-squared values.

Chi-squared values alone can mislead. These were calculated after a certain number of leaf discs, but not all, had been removed. The fate of the remaining discs is not taken into account in Chi-squared analysis. The ants' behavior once hibiscus discs had been removed and foragers faced only with the test species is important. If the test species continues to be ignored, as in giant granadilla, or if it then becomes acceptable, as in custard apple, this will have important bearings on a species' acceptability. The rate of pick up, as

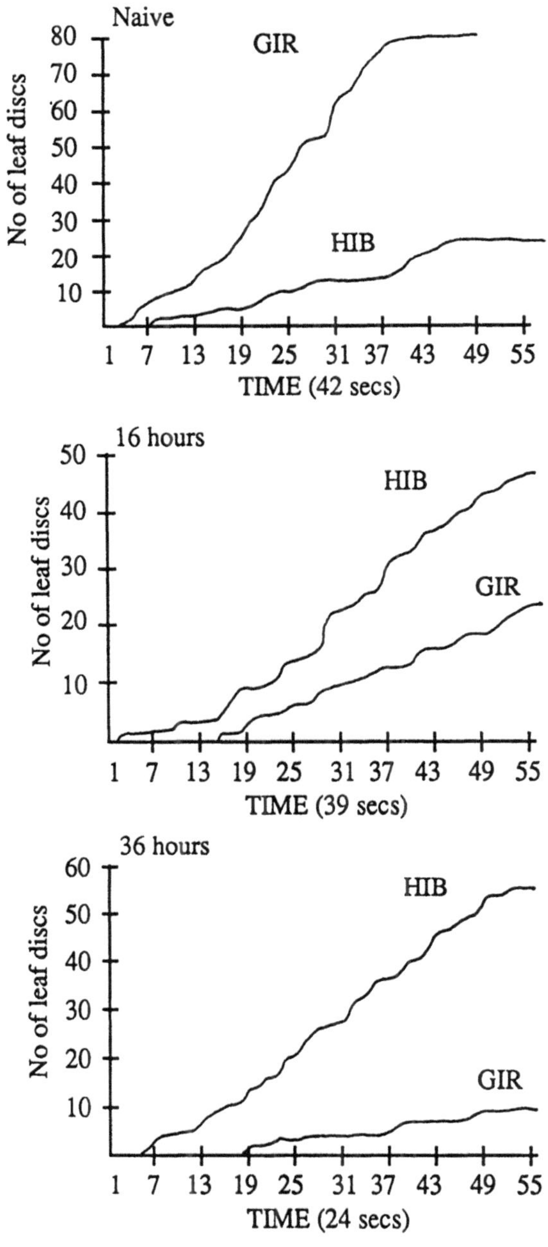

FIGURE 6. Cummulative pick-up rates of leaf discs in two-choice experiments showing type A delayed rejection of *Cucurbita moschata* (GIR). For other details see Figure 5.

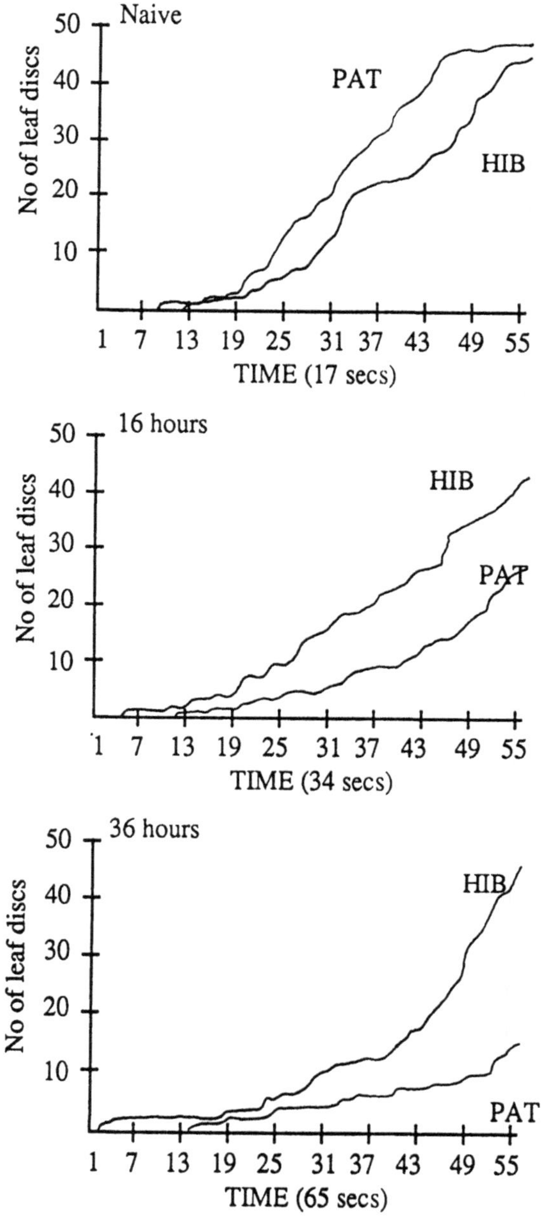

FIGURE 7. Cummulative pick-up rates of leaf discs in two-choice experiments showing type A delayed rejection of *Ipomoea batatas* (PAT). For other details see Figure 5.

represented by the slope of the curves in Fig. 5, provides the key to determining whether a species is genuinely acceptable or whether its acceptability is masked by the ants' preference for hibiscus.

Delayed rejection

Type A delayed rejection means a behavioral response which leads to an initially acceptable plant species becoming unacceptable for a relatively long period following an initial exposure. In order to test if indeed this was occurring, naive nests had to be used, ie. nests which had not been previously exposed to the test plant species. Naive nests were provided with test plant material with hibiscus as the standard.

Three plant species were examined in this way: sweet potato; *Ipomoea batatas*, winter squash; *Cucurbita moschata* and custard apple *Annona squamosa.* (see Figures 6, 7 and 8). Initially, the test species were highly acceptable, but as time passed the ants foraged less and less on the test species and turned instead to hibiscus. Eventually the level of test species foraged in each case was extremely low.

Figure 7 shows how foraging on sweet potato changes over a period of 36 hours. The naive nest is seen to have a slight preference for sweet potato (with a chi-squared = 2.864 at $p <10\%$). After 16 hours, however, this situation had reversed and hibiscus was the preferred species (with a chi-squared = 3.947 at $p <1.0\%$). After 36 hours the level of sweet potato foraging had fallen even more (with a chi-squared = 14.659 at $p < 0.1\%$).

This pattern was repeated with winter squash (see Fig. 6), although within species the switch was even more pronounced. From the graph it is clear that the naive nest prefers winter squash over hibiscus (a chi-squared = 23.439 at $p < 0.1\%$). After 16 hours winter squash foraging had dropped and hibiscus was then preferred (with a chi-squared = 8.647 at $p < 1.0\%$). After 36 hours winter squash foraging dropped to a very low level (a chi-squared = 21.428 at $p < 0.1\%$).

The final test was performed with custard apple (Fig. 8), which had a rapid acquired response. After only 16 hours, the level of foraging fell to almost zero. The naive nest accepted custard apple, even preferring it to hibiscus (with a chi-squared = 8.647 at $p < 1.0\%$). After 16 hours, the situation had completely reversed and custard apple was effectively ignored (a chi-squared = 33.108 at $p < 0.1\%$).

It is clear from such results that delayed rejection does indeed occur and can be long-lasting. Rejection of these species continued for the 4 to 5 day period of observation. Apart from occasional cuts, no recovery to previous levels occurred.

DISCUSSION

A number of authors (eg: Littledyke and Cherrett 1976; Hubbell *et al.* 1983; Powell and Stradling 1986) developed the theory that the catholicity of feeding in leaf-cutting ants follows from the ability of the fungus to detoxify plant allelochemicals. What selectivity exists is probably due to deterrents that act on individual foragers. If this is so, the chemical deterrents may be detected by the ants as token stimuli announcing the presence of anti-fungal compounds deleterious to the entire colony. Another possibility is that the

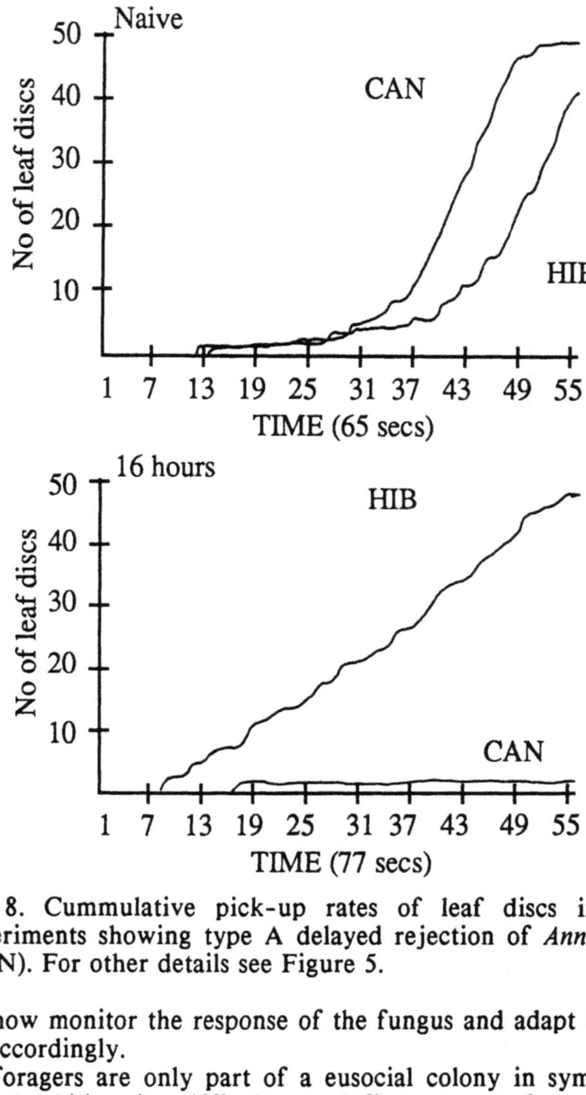

FIGURE 8. Cummulative pick-up rates of leaf discs in two-choice experiments showing type A delayed rejection of *Annona squamosa* (CAN). For other details see Figure 5.

ants somehow monitor the response of the fungus and adapt their foraging behavior accordingly.

As foragers are only part of a eusocial colony in symbiosis with a fungus, palatability is difficult to define. Apart from the obvious subjectivity, it is necessary to decide whether "palatability" applies to the fungus, a particular caste or size-class of workers, or to the whole "superorganism" including the fungus. There is also the problem of what units it should be measured in. Waller (1982a), for example, defines it as the rate of cutting measured as the number of laden ants on the trial per minute. In the present work, the more objective term "acceptability" is defined in terms of pick-up and transport in laboratory tests.

The results here demonstrate that *Ac. octospinosus* colonies exhibit four different categories of rejection responses to leaf-material. The leaves of some plant species are inspected but not cut (eg: *Allamanda, Peperomia*) suggesting some form of odor-mediated rejection (type B immediate rejection). In other species (eg: *Mangifera indica*), cutting begins but is not completed suggesting the involvement of aversive gustatory stimuli (type A immediate rejection). In yet others, recruitment ceases with time, but the rejection is short-lived and recruitment resumes on representation after a short interval (eg: with *Lantana* and *Passiflora laurifolia*). This has been designated type B delayed rejection. In type A delayed rejection, which is clearly an acquired response, the rejection is persistent (as with *Ipomoea* and *Annona squamosa*).

Type B delayed rejection is characterized by an initial very low level of trail activity which gradually declines to zero, but foraging resumes each time the plant is re-offered. In type A delayed rejection, ants from a naive nest forage initially at very high rates, but then, after about 12 hours, suddenly slow down their transport rate and often take no further material. Rahbe *et al* (1988) found that *Ac. octospinosus* was still inhibited from feeding on yellow yam after a two week period without contact with the plant, but found evidence of recovery beginning 38 days after the last contact -- which may have been when a new population of foragers began to eclose. A likely explanation for this type of response is that the foraging workers become negatively conditioned to the plant. Drinking of the sap (which commonly occurs: Littledyke and Cherrett (1976) provide negative reinforcement for withdrawal from an innocuous conditional stimulus coming from the plant material. A more complicated hypothesis would be that a signal from the stressed fungus is detected by the minims, which then inhibit foraging workers from recruiting to that plant. The difficulty here lies in explaining why concurrent recruitment to other plants is not also inhibited.

Conditioning processes are well-known in social insects, and have been especially well-documented in honeybees by Menzel (1969), Koltermann (1973a and b) and others. These authors showed that honeybees rapidly associated feeding at a particular source with scents or visual stimuli presented there and that the insects had an hierarchical learning ability. Floral scents such as those of rosemary acted as more effective conditional stimuli than scents of non-floral origin, such as caprylic acid, and scents from flowers in the main geographical range of *Apis mellifera* were preferred to those from outside that range. Changing the quality of the food reinforcement (unconditional stimulus) could change the scale of preferences. Thus honeybees have a scale of preferences based upon innate levels of sensitivity, combined with their individual foraging experience. Learning can occur on one or a few trials and reversal learning is easily demonstrated.

Brower (1984) used conditioning theory to explain predation selectivity on butterflies, and hence the evolution of mimicry; his avoidance learning theory can be transferred *mutatis mutandis* to leaf-cutting ant foraging behavior. Brower divided defensive chemicals into two classes. Class I chemicals have toxic effects when ingested, and can act as negative reinforcers for the innocuous class II chemicals which act only on olfactory or gustatory sensory receptors (although it should be noted that a sufficiently odorous toxin can be both a class I and a class II chemical). Class II chemicals may elicit *distal avoidance*, which is advantageous for the predator as it need not come into contact with the prey or *proximal avoidance* in which it must taste (or perhaps smell) on contact before showing a conditioned avoidance. Depending on the mode of action of toxins, they may have immediate effects on the autonomic nervous system which result in violent illness and one-trial learning, or illness that occurs some hours after digestion has taken place (i.e. the so-called Garcia effect, in which the CS-UCS interval is greatly extended), leading to avoidance of stimuli associated with the last consumed food. In predatory birds, Class I chemicals associated with Class II emetics potentiate distal avoidance on subsequent retesting.

If the hypothesis of avoidance conditioning is used to explain feeding selectivity in leaf-cutting ants, the various responses to plants shown in Table 3 can be explained. Type B immediate rejection suggests distal avoidance of a class II chemical as a result of previous conditioning. Type A immediate rejection appears to be proximal avoidance. Type B delayed rejection may have a variety of explanations but must involve only weakly active Class I chemicals which produce delayed illness and avoidance conditioning that is relatively resistant to extinction.

It follows that immediate rejection--frequently observed in either laboratory and field tests--may only partially (if at all) reflect the innate tendencies, but may be an artefact of the foragers' past experience, in which case, as the literature demonstrates, different authors may come to differing conclusions about the "palatability" of a given plant species. Significantly, the laboratory colonies examined showed type A delayed rejection only to plants unlikely to be encountered in their field environment. A further expected complication is the effect resulting from generalization: Do toxic neotropical plants have Batesian mimics using the same class II chemicals?

REFERENCES

BARBER, P. M. and J. M. CHERRETT. 1972. Some factors affectingthe site and pattern of leaf-cutting activity in the ant *Atta cephalotes*. J. Ent. (A) 47: 15-27.

BERISH, C. W. 1986. Leaf-cutting ants select nitrogen rich forage. Am. Midl. Nat., 115: 268-276.

BOWERS, M. A. and S. D. PORTER. 1981. Effect of foraging distance on water content of substrates harvested by *Atta columbica*. Ecology, 62: 273-275.

BROWER, L. P. 1984. Chemical defence in butterflies. *In*: R. Vane-Wright (ed.) Academic Press, London.

CHEN, T. K., D. C. ALES, N. C. BAENZIGER, and D. F. WIEMER. 1983. Ant repellent triterpenoids from *Cordia alloidora*. J. Org. Chem. 48: 3525-3531.

CHEN, T. K., D. F. WIEMER, and J. J. HOWARD, 1984. A volatile leaf-cutter repellent from *Astronium graveolens*., Naturwissenschaften, 71: 97-98.

CHERRETT, J. M. 1968a. The foraging behavior of *Atta cephalotes* 1. Foraging pattern and plant species attacked in tropical rain forest. Anim. Ecol., 37: 387-403.

CHERRETT, J. M. 1968b. Some factors involved in the selection of vegetable substrate by *Atta cephalotes* (L.) (Hymenoptera: Formicidae) in tropical rain forest. J. anim. Ecol. 41: 647-660.

CHERRETT, J. M. 1972. Chemical aspects of plant attack by leaf-cutting ants. *In* J. Harborne (ed.), Phytochemical Ecology, Academic Press, London.

CHERRETT, J. M. 1980. Possible reasons for the mutualism between leaf-cutting ants (Hymenoptera: Formicidae) and their fungus. Biol.-Ecol. Med. 7: 113-122.

CHERRETT, J. M. 1983. Resource conservation by the leaf-cutting ant *Atta cephalotes* in tropical rainforest. In: Tropical Rain Forest: Ecology and Management (ed.). Sutton, Whitmore and Chadwick, Blackwell, London.

CHERRETT, J.M. and C. E. SEAFORTH. 1970. Phytochemical arrestants for the leaf-cutting ants, *Atta cephalotes* and *Acromyrmex octospinosus*, with some notes on the ants' response. Bull. Ent. Res. 59: 615-625.

FEBVAY, G., P. BOURGEOIS, and A. KERMARREC. 1985. Antifeedants for the Attine Ant, *Acromyrmex octospinsosus* (Reich) (Hymenoptera - Formicidae), in Yam (Dioscoreaccae) leaves. Agronomie, 5: 439-444.

FEBVAY, G., Y. RAHBE and A. KERMARREC. 1986. Analyse par digitalisation d'images due comportement d'affouragement d'une fourmi attine: application aux tests de choix. Agronomie, 1986, 6: 743-749.

FEENY, P. 1976. Plant Apparency and chemical defense. Recent Adv. Phytochem, 10: 1-40.

FOWLER, H. G. 1981. Subtropical seasonality and the foraging activity of a grass-cutting ant *Acromyrmex landolti fracticornis* (Formicidae: Attini). Ciencia e Cultura 32: 252-256.

FOWLER, H. G. 1982. Habitat effort on fungal substrate selection by a leaf-cutting ant. N.Y. Ent. Soc. XC(2), 64-69.

FOWLER, H. G., S. W. ROBINSON. 1979. Foraging by *Atta sexdens* (Formicidae: Attini) seasonal patterns, caste and efficiency. Eco. 4: 239-247.

FOWLER, H. G., E. W. STILES. 1980. Conservative Resource Management by leaf-cutting ants? The role of foraging territories and trails, environmental patchiness. Sociobiology, 5: 25-41.

HAUKIOJA, E. and P. NIEMELA. 1977. Retarded growth of a geometrid larvae after mechanical damage to leaves of its host tree. Annal. Zool. Fennici, 14: 48-52.

HOWARD, J. J. and D.F. WIEMER. 1986. Chemical Ecology of Host-Plant selection by the leaf-cutting ant, *Atta cephalotes*, pp 26o0-273. *In* Fire Ants and Leaf-Cutting Ants: Biology and Management. Westview Press, Colorado, USA.

HUBBELL, S.P. and D.F. WIEMER. 1983. Host plant selection by an Attine ant. *In* P. Jaisson (ed.). Social Insects in the tropics. Universite Paris-Nord, p. 133-154.

HUBBELL, S.P., D.F. WIEMER and A. ADEJARE. 1983. An anti-fungal terpenoid defends a neotropical tree (Hymenea) against attack by fungus-growing ants. Oecologia, 60: 321-327.

HUBBELL, S. P. and L.L. ROCKWOOD. 1983. A hypothesis concerning the foraging strategy of a tropical leaf-cutting ant. Biotropica.

HUBBELL, S. P., J.J. HOWARD and D. F. WIEMER. 1984. Chemical leaf repellency to an Attine ant: Seasonal distribution among potential host species. Ecology, 65: 1067-1076.

HURBERT, T. D. and D. F. WIEMER. 1985. Ant-repellent terpenoids from *Melampodium divericatum*. Phytochemistry, 24: 1197-1198.

JUTSUM, A.R., J.M. CHERRETT and M. FISHER. 1981. Interactions between the fauna of citrus trees in Trinidad and the ants *Atta cephalotes* and *Azteca sp.* J. Appl. Ecol. 18:187-195.

KOLTERMANN, R. 1973a. Retroaktive Hemmung nach sukzessiver Informationseingabe bei *Apis mellifera* und *Apis cerana* (Apidae). J. comp. Physiol. 84: 299-311.

KOLTERMANN, R. 1973b. Rassen bzw. artspezifische Duftbewertung bei der Honigbiene und ockologische Adaptation. J. comp. Physiol. 85: 327-361.

LEVIN, D. A. 1971. Plant Phenolics: an ecological perspective. Am. Nat. 105: 157-181.

LEWIS, T., G.V. POLLARD and G.C. DIBLEY. 1974. Micro-environmental factors affecting diel patterns of foraging in the leaf-cutting ant *Atta cephalotes* (L.)(Formicidae:Attini). 43: 143-153.

LITTLEDYKE, M. and J. M. CHERRETT. 1975. Variability in the selection of substrate by the leaf-cutting ants *Atta cephalotes* (L.) and *Acromyrmex octospinosus.* (Reich) (Formicidae:Attini). Bull. Ent. Res. 65: 33-47.

LITTLEDYKE, M. and J.M. CHERRETT. 1976. Direct ingestion of plant sap from cut leaves by the leaf-cutting ants *Atta cephalotes* (L.) and *Acromyrmex octospinosus* (Reich) (Formicidae:Attine). Bull. Ent. Res. 66: 205-217.

LITTLEDYKE, M. and J.M. CHERRETT. 1978. Defence mechanisms in young and old leaves against cutting by the leaf-cutting ants *Atta cephalotes* and *Acromyrmex octospinosus.* Bull. Ent. Res. 68: 263-271.

MARTIN, M.M. 1970. The biochemical basis of the fungus-attine ant symbiosis. Science, 169: 16-20.

MENZEL, R. 1969. Das Gadaechtnis der Honigbiene fuer Spektralfarben. II. Umlernen und Mehrfachlernen. Z. vergl. Physiol., 63: 290-309.

MUDD, A. and G.L. BATEMAN. 1979. Rates of growth of the food fungus of the leaf-cutting ant *Atta cephalotes* (L.)(Hymenoptera:Formicidae) on different substrates gathered by ants. Bull. Ent. Res. 69: 141-148.

MUDD, A., D. J. PEREGRINE, and J. M. CHERRETT. 1978. The chemical basis for the use of citrus pulp as a fungus garden substrate by the leaf-cutting ants *Atta cephalotes* and *Acromyrmex octospinosus.* Bull. Ent. Res. 68: 673-685.

MULLENAX, C.H. 1979. The use of Jackbean (*Canavalia ensiformis*) as a biological control for leaf-cutting ants (*Atta* spp). Biotropica 11: 308.

OJKUNADE, A. L. and D. F. WIEMER, 1985a. Ant repellent sesquiterpene lactones from *Eupatorium quadrangulare.* Phytochemistry, 24: 1199-1201.

OJKUNADE, A. L. and D. F. WIENER, 1985b. Jacquinonic Acid: An ant repellent triterpenoid from *Jacquinia pungens.* Phytochemistry, 24: 1203-1205.

PINTERA, A. 1983. Selectivity of plants utilized by *Atta insularis* in Cuba. (Hymenoptera: Formicidae). Acta. Ent. Bohemoslov. 80: 13-20.

POLLARD, G. V., R. RILEY, and E. WATTIE. 1983. Preliminary Investigations on the selection of citrus sp. by the leaf-cutting ant *Acromyrmex octospinosus* (Reich)(Formicidae: Attini). Tropical Agriculture, 60: 282-285.

POWELL, R. J. and D. J. STRADLING. 1986. Factors influencing the growth of *Attamyces bromificatus*, a symbiont of attine ants. Trans. Br. mycol. Soc., 87: 205-213.

RAHBE, Y., G. FEBVAY and A. KERMARREC. 1988. Foraging activity of the attine ant *Acromyrmex octospinosus* (Reich)(Hymenoptera: Formicidae) on resistant and susceptible yam varieties. Bull. ent. Res. 78: 329-337.

RHODES, D. F. and COATES. 1976. Toward a general theory of plant anti-herbivore chemistry. Recent Adv. Phytochem. 10: 168-213.

ROCKWOOD, L. L. 1973. Distribution, density and dispersion of two species of leaf-cutting ants (*Atta*) in Guanacaste Province, Costa Rica. Biotropica., 7: 176-193.

ROCKWOOD, L. L. 1976. Plant selection and foraging patterns in two species of leaf-cutting ants (*Atta*). Ecology, 57: 48-61.

ROCKWOOD, L. L. 1978. Foraging patterns and plant selection in Costa Rican leaf-cutting ants. J.N.Y. Entomol. Soc. 85: 222-233.

ROCKWOOD, L. L. and K. E. GLANDER. 1979. Howling Monkeys and Leaf-cutting ants: Comparative foraging in a Tropical Deciduous forest. Biotropica, 11: 1-10.

SERI, G. RUDOLPH and C. LOUDON. 1986. Load size selection by foraging ants (*Atta cephalotes*). Ecol. Entom. 11: 401-410.

SHEPHERD, J. D. 1982. Trunk trails and the searching strategy of a leaf-cutter ant *Atta colombica*. Ecol. Behav. Sociobiol. 11: 77-84.

SHEPHERD, J. D. 1985. Adjusting foraging effort to resources in adjacent colonies of the leaf-cutter ant, *Atta colombica*. Biotropica, 17: 245-252.

STOSSEL, A. 1970. Antifungal compounds produced by higher plants. Recent Adv. Phytochem, 1: 143-180.

STRADLING, D. J. 1978. The influence of size on foraging in the ant, *Atta cephalotes* and the effect of some plant defence mechanisms. J. Anim. Ecol. 47: 173-178.

WALLER, D. A. 1982a. Leaf-cutting Ants and avoided plants: Defenses against *Atta texana* attack. Oecologia, 52: 400-403.

WALLER, D. A. 1982b. Leaf-cutting ants and live oak: the role of leaf toughness in seasonal and intraspecific host choice. Ent. Exp. App. 32: 146-150.

WALLER, D. A. 1986. The Foraging Ecology of *Atta texana* in Texas. pp. 46-158. *In* Fire Ants and Leaf Cutting Ants: Biology and Management. C.S. Lofgren, R.K. Vander Meer, (eds.). Westview Press, Boulder, Colorado.

WHITTAKER, R. H. and P. P. FEENY. 1971. Alleochemics: Chemical Interactions between species. Science 153: 757-770.

WIEMER, D. F. and D. S. ALES. 1981. Lasidiol Angelate: Ant repellent sesquiterpenoid from *Lasiantheae fruiticosa*. J. Org. Chem. 46: 5449-5450.

WILSON, E.O. 1980. Caste and Division of Labor in Leaf-cutter ants (Hymenoptera: Formicidae: *Atta*). II. The ergonomic optimization of leaf-cutting. Behav. Ecol. Sociobiol. 7: 157-165.

37

Foraging and Fungal Substrate Selection by Leaf-cutting Ants

H.L. Vasconcelos and H.G. Fowler

INTRODUCTION

Endemic to the neotropics, *Atta* and *Acromyrmex* leaf-cutting ants defoliate many natural and agricultural ecosystems (Haines 1975: Fowler et al. 1986). Several authors have suggested their ecological success is linked to the fungus they cultivate. This fungus grows on a vegetable substrate gathered by the ants and is the ant larvae's sole source of food (Quinlan and Cherrett 1979).

Since Cherrett's (1968) pioneering work on *Atta cephalotes*, other researchers have developed an interest in leaf-cutting ant foraging patterns and leaf selection. Field studies have been conducted on other leaf-cutter species in diverse habitats, in addition to laboratory studies which deal with the chemistry of substrate selection. Research was primarily stimulated by the ants' economic importance as a pest and secondarily by theoretical interest in plant-herbivore interactions. This paper summarizes the published data and delineates the major factors which determine foraging patterns and fungal substrate selection by leaf-cutting ant colonies.

FUNGAL SUBSTRATE: GENERAL PATTERNS

Based on available information, colonies include about half of the available plant species as fungal substrates, but the proportion ranges from five to 77% (Table 1). Despite this apparent catholic selection pattern, leaf-cutting ants do demonstrate strong preferences. Plant resource species are not equally exploited, and only a few of these comprise the bulk of gathered vegetal substrate, as mirrored by the index for even distribution of harvested plant species (Table 1). The index varied between 0.11 and 0.34 (Table 1) for the various species studied. A value of 1.0 indicates equal exploitation of resources; therefore, the actual values demonstrate the heterogeneity of resource exploitation. Indeed, variation in the relative contribution of plant species to the fungal substrate is sometimes greater within a leaf-cutting ant taxa in different habitats, than between taxa inhabiting similar habitats.

What factors contribute to so many resource species being selected as fungal substrates by leaf-cutting ants? What factors contribute to the marked differential utilization of leaf resources?

TABLE 1. Leaf-diet of various leaf-cutting ants, the number of plant species exploited, and the equitability of utilization.

Species and Habitat	Location	Period of study (mo)	Plant spp. attacked (% of available)	Equitability in species utilization*	Source
Atta capiguara					
pasture	Brazil	12	9-13(?)	0.12-0.24	Forti (1985)
Atta cephalotes					
second growth	Costa Rica	5	17(5)	0.11	Blanton & Ewel (1985)
dry forest	Costa Rica	1	30(51)	0.33	Rockwood & Hubble (1987)
dry forest	Costa Rica	11	?(49-77)	0.37-0.61**	Rockwood (1976)
wet forest	Guyana	2	36(50)	0.29	Cherrett (1968)
wet forest	Brazil	13	103(44)	0.34	Vasconcelos (1989)
Atta colombica					
dry forest	Costa Rica	11	?(49-67)	0.58-0.62**	Rockwood (1976)
wet forest (old)	Colombia	12	130(65)	0.24	Shepherd (1985)
wet forest (new)	Colombia	12	103(75)	0.11	Shepherd (1985)
Atta insularis					
wet forest	Cuba	12	52(52)	0.21	Pintera (1983)
Atta mexicana					
desert scrub	Mexico	4	4-5(?)	0.23-0.86	Mintzer (1979)
Atta sexdens					
wet forest	Brazil	13	16(7)	0.25	Vasconcelos (1989)
dry forest	Paraguay	11	53(67)	0.36	Fowler (unpublished)

*Ratio between the observed and maximum possible diet diversity. Diet diversities are calculated using the inverse of Simpson's (1949) index. **Overestimate: estimates based solely on the more utilized plant species

THE ROLE OF PLANT ABUNDANCE

Optimal foraging theory (reviewed by Pyke 1984) predicts that the inclusion of a given food in an animal's diet does not depend upon its abundance, but rather on the absolute abundance of the "higher ranked" food items. Thus, where higher ranked food items are abundant, the diet should be more specialized. Data on leaf-cutting ant foraging patterns support this theory. Blanton and Ewel (1985) observed that *A. cephalotes* harvest measurable amounts of only 17 plant species in their rich (332 available plant species), though cassava-dominated, experimental ecosystems. Cassava leaves are highly attractive to leaf-cutting ants. In fact, in Brazil *A. sexdens* is called "sauva da mandioca". In Blanton and Ewel's (1985) study, cassava was the most frequently and extensively attacked plant species, accounting for more than 76% of the total leaf area cut. This foraging pattern contrasts with that observed in tropical forests, where resources are distributed more evenly. In tropical forests, more plant species are included in the leaves harvested by *A. cephalotes* and each species' relative importance is smaller (Fig. 1). Interestingly, similar plant species selection patterns are found between *A. cephalotes* in cassava dominated habitats, *A. capiguara*, a grass-cutting specialist which inhabits homogeneous habitats (Fig. 1) and with *A. colombica*. Shepard (1985) compared the foraging patterns of two adjacent colonies whose territories differed by the inclusion of high quality plant species. The colony, located in the young forest, had 46 percent more of three desirable plant resource species, and, consequently, a much lower dietary diversity than the colony located in the older forest.

THE ROLE OF PLANT DEFENSES

Plant defenses against leaf-cutting ant attack include several chemical compounds (allelochemicals) as well as physical traits which inhibit cutting and/or fungal growth. Recent extensive work on the chemical aspects of plant selection by leaf-cutting ants revealed that some secondary plant compounds are potent ant repellents (Hubbell et al. 1983; Howard 1987: Hubert et al. 1987). The occurrence of such compounds is widespread in nature. Hubbell et al. (1984) surveyed 42 randomly selected plant species for ant repellency and found that three-quarters of these exhibited significantly repellent non-polar extracts, while half exhibited significantly repellent polar extracts. Nevertheless, leaf preference field studies demonstrated that avoidance was principally correlated with non-polar compounds (Hubbell et al. 1984).

Many plant species demonstrate seasonal repellency and attractivity patterns (Fowler and Stiles 1980). Studies by Howard (1987) confirmed that non-polar compounds are the most important chemical factors in determining seasonal plant palatability to leaf-cutting ants. Of the few non-polar compounds characterized, the vast majority are terpenoids (Wiemer and Ales 1981; Chen et al. 1983; Hubbell et al. 1983; Hubert et al. 1987). Furthermore, terpenoids have potent anti-fungal properties (Hubbell et al. 1983). Cyanogens and saponins also act as strong leaf-cutting ant deterrents (Kawanishi and Raffauf 1986; Febvay and Kermarrec 1986; Howard 1987). However, these secondary plant compounds are less widespread than the non-polar compounds.

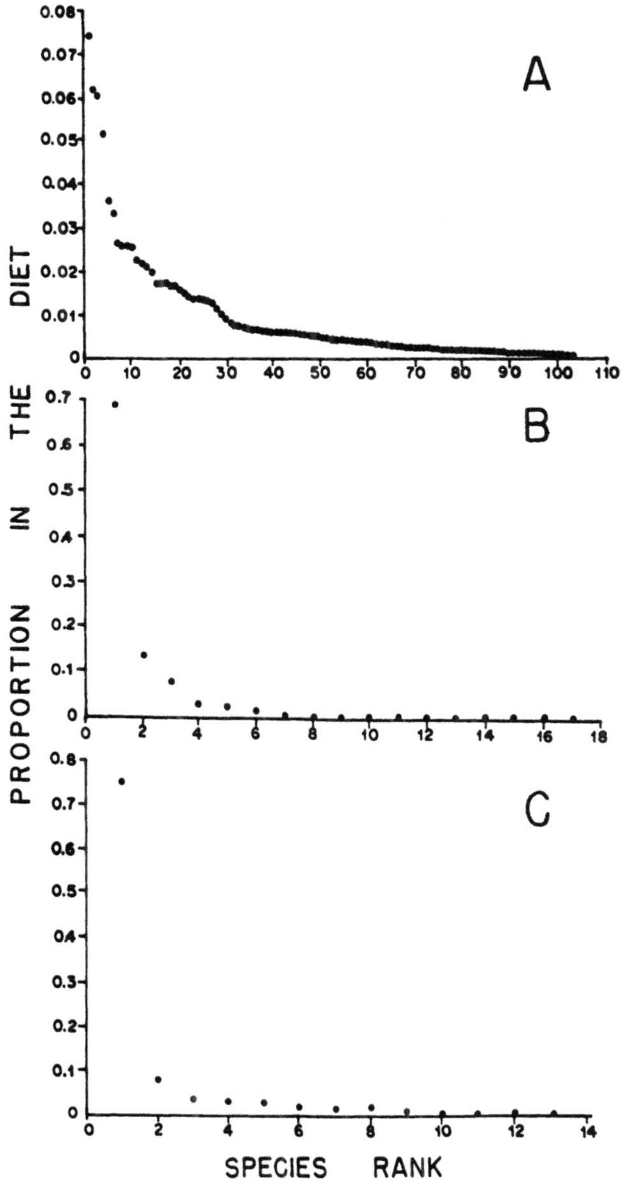

FIGURE 1. A: Relative contribution of various plant species to the leaf-diet of *Atta cephalotes* in a tropical wet forest (Vasconcelos, unpublished). B: Same for *A. cephalotes* in a diverse, but dominated by high ranking plant species (Blanton and Ewel 1985). C: Same for *Atta capiguara* in a homogeneous habitat, a pasture (Forti 1985). Plant species ranked according to the total amount harvested.

Evidence for plant physical defenses against leaf-cutting ant attack is more circumstantial than experimental. In general, foragers select tender vegetal material (Cherrett 1972; Fowler and Robinson 1977; Fowler and Stiles 1980; Waller 1982a, 1982b). However, these results are difficult to interpret because separating chemical and physical factors in an experimental design is difficult. The production of gluey or copious saps, especially latex, are also thought to be important ant deterrents (Stradling 1978), although rubber trees are attacked in force by leaf-cutters on plantations. In these cases, leaf sap may gum up the workers' mandibles or other appendages during cutting. Nevertheless, many lactiferous species are highly ranked in the plant substrates harvested by leaf-cutting ant colonies (Hubbell et al. 1984; Howard 1987; Vasconcelos, unpublished), though they are often selected only seasonally or are restricted to senescent leaves. Available evidence suggests that these defenses are not as efficient as plant secondary chemicals.

Finally, many aggressive ant species with arboreal habits, such as *Azteca*, *Dolichoderus*, and *Pseudomyrmex* species, exclude leaf-cutter foragers from their host-plants (Janzen 1967; Jutsum et al. 1981). As field studies progress, especially in tropical forests, the role of other ant species in modifying leaf-cutter foraging patterns will undoubtedly enable us to explain more of the observed variation than in the past.

THE ROLE OF PLANT NUTRIENTS

The nitrogen content of leaves is an important factor in plant choice among herbivores (Mattson 1980). However, studies of leaf-cutting ant foraging patterns have not revealed any influence of plant nitrogen content on plant selection by these ants (Hubbell and Wiemer 1983; Howard 1987). The independence of plant nitrogen content may reflect the ability of the ants' symbiotic fungus to synthesize amino acids and protein from dilute nitrogen sources. If so, it would not be unique in nature. Many herbivorous animals track their nitrogen deficiencies with the help of endo- or ecto-symbiontic microorganisms (Mattson 1980). In contrast, the carbohydrate content of plants affects palatability. The addition of sucrose solutions to intact leaves results in a dramatic increase in cutting rate (Barret and Cherrett 1972; Waller 1982b).

Information on fungal metabolism is scarce, and thus we cannot evaluate the roles of nutrients. Both nitrogen and carbohydrates, studied extensively in herbivores that directly use the ingested plant material, may have little or no validity for leaf-cutting ants, because we do not know what nutrients the symbiotic fungus requires to grow (Fowler et al. 1989). We may be overlooking nutrients by focusing on the most preferred plant species, rather than evaluating the contributions to the fungal growth of lower-ranked plant species (Fowler et al. 1989).

Studies on the effects of leaf water content have produced conflicting results. Cherrett (1972a), Fowler and Robinson (1979) and Bowers and Porter (1981) found a positive correlation between mean water content and plant palatability, whereas Howard (1987) found none and Blanton and Ewel (1985) found a negative relationship. This apparent contradiction was probably related to the different dietary status of the colonies studied. In addition, water content covaries with respect to other plant chemicals. The influence of water content on plant selection is probably greater for water-stressed colonies, as suggested by Bower and Porter's (1981) study at the end of Panama's dry season. Many laboratory colonies freely drink water; laboratory conditions are often drier than the soil environment, thus adding further credibility to this interpretation.

Changes in nitrogen content, and, more importantly, in allelochemicals occur under intensified plant water stress (Mattson 1980). In cassava leaves, for example, hydrocyanic acid content drops when plants enter water stress (Gondwe 1974). These correlations may explain why some plant species only serve as fungal resources after leaves senesce.

SAMPLING

Plant selection by leaf-cutting ants is constrained by several factors, with allelochemicals apparently exercising the stronger constraint. Many plant species produce allelochemicals only during certain seasons (Hubbell et al. 1984), and many plant species are only attacked when flushing new leaves (Cherrett 1972a; Rockwood 1976; Fowler and Stiles 1980; Shepard 1982; Vasconcelos, 1989). Resource availability for leaf-cutting ant colonies, therefore, has a strong temporal as well as spatial component. To track these spatio-temporal changes in resource availability, colonies should function as statisticians (Pyke 1984). To best use their resource base in the foraging territory, colonies should collect and store information while foraging, and then use this information to make decisions (Pyke 1984). To achieve this goal, sampling is the optimal foraging procedure (Oster and Heinrich 1976; Pyke 1984). Indeed, a plot of the cumulative distribution of plant species attacked by leaf-cutting ant colonies (Fig. 2) produces patterns similar to the "species-sample curve" widely used by ecologists to infer limits of associations or communities.

For herbivorous mammals, sampling increases food variety and permits preference changes depending on a food's properties (Westbody 1974, 1978). Leaf-cutting ants share this behavior (Fowler et al. 1989). For example, laboratory colonies of leaf-cutting ants demonstrate marked fluctuations in the ranking of different fungal substrates, as well as of different chemical fractions of a single substrate type (Cherrett and Seaforth 1970; Cherrett 1972b). The apparent preference of leaf-cutting ants for "novel" substrates (Cherrett 1972b) may be a consequence of this sampling procedure.

TRUNK TRAILS AND RESOURCE MONITORING

During the course of evolution, leaf-cutting ants have probably faced two major problems while perfecting an efficient foraging strategy: (1) most plant species only produce palatable patches of vegetation seasonally (Fowler and Stiles 1980), and (2) the production of these resources continues over a long time scale, possibly encompassing the life of the colony. Thus, colonies must continually gather information about the current availability, or stock, of resources and then store this critical information for future use. These problems were apparently solved by the development of foraging trails, which are both chemically and physically marked. The construction and maintenance of these trails, also called trunk trails (Holldobler 1976; Shepard 1982), characterizes the majority of leaf-cutting ant taxa, with the exception of some *Acromyrmex* species which depend upon ephemeral or homogeneously

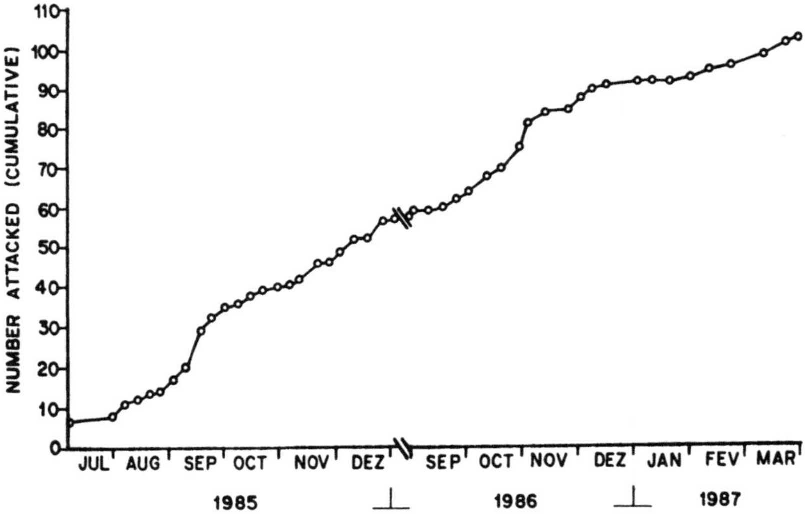

FIGURE 2. Cumulative distribution of the number of plant species attacked by an *Atta cephalotes* colony in a tropical wet forest during 13 months (Vasconcelos, unpublished).

dispersed plant resources (Fowler 1978). Although a few researchers have demonstrated the energetic advantages of walking over uncluttered foraging trails (Shepard 1982; Lighton et al. 1987), as well as their role in reducing territorial conflicts with neighboring colonies (Fowler and Stiles 1980; Hölldobler and Lumsden 1980), we believe that these trails are primarily used to locate and monitor plant resources. In most cases, resources are spatially predictable for a period which exceeds the average lifespan of a worker. Visual signals transmitted by physical trails have prolonged activity; e.g. after harvest stops on a trail, foraging may resume several months to a year later (Shepard 1982; Vasconcelos 1989). This system permits not only a rapid familiarization of naive workers with the most productive or palatable portions of the colony's foraging territory, but also increases the chances of finding previously exploited plant resources when these are once again palatable (Fowler and Stiles 1980).

CONCLUDING REMARKS

We have briefly summarized those factors which may affect plant selection by leaf-cutting ants. However, the data base of information needed to elucidate factors involved in leaf-cutter ant selection of fungal substrate is scanty. Even the long-standing debate over whether or not leaf-cutting ant colonies practice conservative resource management within their foraging territories (Cherrett 1968, 1984; Rockwood 1976; Fowler and Stiles 1980; Rockwood and Hubbell 1987) can be solved by examining leaf-cutting colonies as statisticians, with foraging trails biasing sampling choices; these are further fine-tuned by the plant species' chemical and physical factors.

However, the uniformity of the relative contribution of only a few of the plant species present in the diet of leaf-cutting ants (Table 1) suggests that further field and laboratory studies coupled with detailed pattern analysis may finally permit us to understand the patterns of resource choice.

ACKNOWLEDGMENTS

We thank Malcom Cherrett for sharing his ideas and views of leaf-cutting ecology with us, and for support of our continuing research we gratefully acknowledge the support of the Conselho Nacional de Desenvolvimento Cientlfico e Tecnôlogico (CNPq) of Brazil, the World Wildlife Foundation, and the FUNDUNESP. Nozor Pinto kindly penned the figures.

REFERENCES

BARRER, P.M. and J.M. CHERRETT. 1972. Some factors affecting the site and pattern of leaf-cutting activity in the ant, *Atta cephalotes*. J. Ent. (A). 47: 15-27.

BLANTON, C.M. and J. EWEL. 1985. Leaf-cutting ant herbivory in successional and agricultural tropical ecosystems. Ecology 66: 861-869.

BOWER, M.A. and S.D. PORTER. 1981. Effect of foraging distance on water content of substrates harvested by *Atta colombica* (Guerin). Ecology 62: 273-275.

CHEN, T.K., D.C. ALES, N.C. BAENZIGER and D.F. WIEMER. 1983. Ant-repellent triterpenoids from *Cordia allidora*. J. Org. Chem. 48: 3525-3531.

CHERRETT, J.M. 1968. The foraging behaviour of *Atta cephalotes* L. (Hymenoptera: Formicidae). 1. Foraging pattern and plant species attacked in tropical rain forest. J. Anim. Ecol. 37: 387-403.

CHERRETT, J.M. 1972a. Somem factors involved in the selection vegetable substrate by *Atta cephalotes* L. (Hymenoptera: Formicidae) in tropical rain forest. J. Anim. Ecol. 41: 647-660.

CHERRETT, J.M. 1972b. Chemical aspects of plant attack by leaf-cutting ants. pp. 13-24. *In* J.B. Harbone (ed.), Phytochemical Ecology, Academic Press, London.

CHERRETT, J.M. 1984. Resource conservation by the leaf-cutting ant, *Atta cephalotes*, in tropical rain forest. pp. 253-265. In S.L. Sutton and T.C. Whitmore (eds.), Tropical Rain Forest: Ecology and Management. Blackwell Scientific Publications, Oxford.

CHERRETT, J.M. and C.E. SEAFORTH. 1970. Phytochemical arrestants for the leaf-cutting ants, *Atta cephalotes* (L.) and *Acromyrmex octospinosus* (Reich) with some notes on the ants' response. Bull. Ent. Res. 59: 615-625.

FEBVAY, G. and A. KERMARREC. 1986. Prevention of feeding by *Acromyrmex octospinosus* with antifeedants from yams. pp. 247-259. *In* C.S. Lofgren and R.K. Vander Meer (eds.), Fire Ants and Leaf-Cutting Ants: Biology and Management. Westview Press, Boulder.

FORTI, L.C. 1985. Ecologia da atividade forrageira da saúva *Atta capiguara* Gonçalves, 1944 (Hymenoptera, Formicidae) e seu impacto em pastagem. Ph.D. Thesis, Esc. Sup. Agric. L. Queiroz, Piracicaba, Sao Paulo. 235 pp.

FOWLER, H.G. 1978. Foraging trails of leaf-cutting ants. J. N.Y. Entomol. Soc. 86: 132-136.

FOWLER, H.G. and S.W. ROBINSON. 1977. Foraging and grass-selection by the grass-cutting ant, *Acromyrmex landolti fracticornis* (Forel) (Hymenoptera: Formicidae) in habitats of introduced forage grasses in Paraguay. Bull. Ent. Res. 67: 659-666.

FOWLER, H.G. and S.W. ROBINSON. 1979. Foraging ecology of the grass-cutting ant, *Acromyrmex landolti fracticornis* (Formicidae: Attini). Int. J. Ecol. Environ. Sci. 5: 29-37.

FOWLER, H.G. and E.W. STILES. 1980. Conservative resource management by leaf-cutting ants? The role of foraging territories and trails, and environmental patchiness. Sociobiology 5: 25-41.

Fowler, H.G., L.C. Forti, V. PEREIRA-DA-SILVA and N.B. SAES. 1986. Economics of grass-cutting ants. pp. 18-35. *In* C.S. Lofgren and R.K. Vander Meer (eds.), Fire Ants and Leaf- Cutting Ants: Biology and Management. Westview Press, Boulder.

FOWLER, H.G., L.C. FORTI, C.R.F. BRANDÃO, J.C. DELABIE and H.L. VASCONCELOS. 1989. Ecologia nutricional de formigas. *In* A.P. Panizzi and J.R.P. Parra (eds.), Ecologia Nutricional de Insetos. Editora Manole, Sao Paulo.

GOMDWE, A.T.D. 1974. Studies on the hydrocyanic acid contents of some local varieties of cassava (*Manihot esculenta* Crantz) and some traditional cassava food products. East Afric. Agric. Forest J. 40: 161-167.

HAINES, B. 1975. Impact of leaf-cutting ants on vegetatioin development at Barro Colorado Island. pp. 99-111. *In*, F.B. Golley and E. Medina (eds.), Tropical Ecological Systems. Springer Verlag, New York.

HÖLLDOBLER, B. 1976. Recruitment behavior, home range orientation and territorality in harvester ants *Pogonomyrmex* spp. Behav. Ecol. Sociobiol. 1: 3-44.

HÖLLDOBLER, B. and C.J. LUMSDEN. 1980. Territorial strategies in ants. Science 210: 732-739.

HOWARD, J.J. 1987. Leaf-cutting ant diet selection: the role of nutrients, water and secondary chemistry. Ecology 68: 503- 515.

HUBBELL, S.P., J.J. HOWARD and D.F. WIEMER. 1984. Chemical leaf repellency to an attine ant: seasonal distribution among potential host plant species. Ecology 65: 1067-1076.

HUBBELL, S.P. and D.F. WIEMER. 1983. Host plant selection by an attine ant. pp. 133-153. *In* P. Jaisson (ed.), Social Insects in the Tropics. Vol. 2. Universite' Paris Presses, Paris.

HUBBELL, S.P., D.F. WIEMER and A. ADEJARE. 1983. An antifungal terpenoid defends a neotropical tree (*Hymnanea*) against attack by fungus-growing ants (*Atta*). Oecologia 60: 321- 327.

HUBERT, T.D., A.L. OKUNADE and D.F. WIEMER. 1987. Quadrangolide, a heliangolide from *Eupatorium quadranqualate*. Phytochemistry 26: 1751-1753.

JANZEN, D.H. 1967. Interaction of the bull's horn acacia (*Acacia cornigera*) with an ant inhabitant (*Pseudomyrmex ferruginea*) in eastern Mexico. Kansas Univ. Sci. Bull. 47: 315-558.

JUTSUM, A.R., J.M. CHERRETT and M. FISHER. 1981. Interaction between the fauna of citrus trees in Trinidad and the ants *Atta cephalotes* and *Azteca* sp. J. Appl. Ecol. 18: 187-195.

KAWANISHI, K. and R.F. RAFFAUF. 1986. *Caryocar microcarpum*: an ant repellent and fish poison of the northwest Amazon. J. Nat. Prod. 49: 1167-1168.

LIGHTON, J.R.B., G.A. BARTHOLOMEW and D.H. FEENER, JR. 1987. Energetics of locomotion and load carriage in the leaf- cutting ant, *Atta colombica* Guer. Physiol. Zool. 60: 524- 537.

MATTSON, W.J., Jr. 1980. Herbivory in relatiion to plant nitrogen content. Ann. Rev. Ecol. Syst. 11: 119-161.

MINTZER, A. 1979. Foraging activity of the Mexican leaf-cutting ant, *Atta mexicana* (F. Smith), in a Sonoran Desert habitat (Hymenoptera: Formicidae). Ins. Soc. 26: 364-372.

OSTER, G.F. and B. HEINRICH. 1976. Why do bumble bees major? Ecol. Monogr. 46: 129-133. Pintera, A. 1983. Selection of plants utilized by *Atta insularis* in Cuba. Acta Entomol. Bohemoslov. 80: 13-20.

PYKE, G.H. 1984. Optimal foraging theory: a critical review. Ann. Rev. Ecol. Syst. 15: 523-575.

QUINLAN, R.J. and J.M. CHERRETT. 1979. The role of fungus in the diet of the leaf-cutting ant, *Atta cephalotes* (L.). Ecol. Ent. 4: 151-160.

ROCKWOOD, L.L. 1976. Plant selection and foraging patterns in two species of leaf-cutting ants (*Atta*). Ecology 57: 48-61.

ROCKWOOD, L.L. and S.P. HUBBELL. 1987. Host plant selection, diet diversity, and optimal foraging in a tropical leaf- cutting ant. Oecologia 74: 55-61.

SHEPHERD, J.D. 1985. Adjusting foraging effort to resources in adjacent colonies of the leaf-cutter ant, *Atta colombica*. Biotropica 17: 245-252.

SIMPSON, E.H. 1949. Measurement of diversity. Nature 163: 668.

STRADLING, D.J. 1978. The influence of size on foraging in the ant, *Atta cephalotes*, and the effect of some plant defense mechanisms. J. Anim. Ecol. 47: 173-188.

VASCONCELOS, H.L. 1989.Foraging activity of twospeciesofleaf- cutting ants (*Atta*) in a primary forest of Central Amazon. Ins. Soc. (in press).

WALLER, D.A. 1982a. Leaf-cutting ants and avoided plants: defence against *Atta texana* attack. Oecologia 52: 400-403.

WALLER, D.A. 1982b. Leaf-cutting ants and live oak: the role of leaf toughness in seasonal and intraspecific host choice. Ent. Exp. Appl. 32: 146-150.

WESTBODY, M. 1974. An analysis of diet selection by large generalist herbivores. Am. Nat. 108: 290-304.

WESTBODY, M. 1978. What are the biological bases of varied diets? Am. Nat. 112: 627-631.

WIEMER, D.F. and D.C. ALES. 1981. Lasidiol angelate: ant repellent sesquiterpenoid from *Lasianthea fructiosa*. J. Org. Chem. 46: 5449-5450.

38

Toxic Effect of Plants on Leaf-cutting Ants and Their Symbiotic Fungus

O.C. Bueno, M.J.A. Hebling-Beraldo, O. Aulino da Silva, F. Pagnocca, J.B. Fernandez and P.C. Vieira

INTRODUCTION

Many plant species escape attack by leaf-cutting ants. Several factors play an important role in whether or not ants cut a plant's leaves. In addition to physical factors, the presence of secondary compounds toxic to ants and/or their symbiotic fungus is of fundamental importance. Depending on what substance is present, the plant may act as a repellent, a deterrent, or as a toxicant.

Researchers looking for alternative control methods are interested in the plant substances toxic to leaf-cutting ants. In this paper, we present information about the toxicity of plants immune or susceptible to attack by leaf-cutting ants, as well as specific data about the deleterious effect of *Sesamum indicum* (Pedaliaceae) and *Virola sebifera* (Myristicaceae) on *Atta sexdens rubropilosa* Forel, 1908.

TOXICITY

"Plant toxicity" is the deleterious effect that a given plant may have on other plants or on animals due to the release of the chemical substances. These substances may be uniformly distributed within the plant and thus present in every organ (roots, stem, branches, leaves, fruits), or may be found only in certain organs. In many species, the fruits are the main reservoirs of toxic material. Abiotic factors such as climate, altitude and type of soil may interfere with production of these substances. Some species contain varying amounts of toxic substances throughout their life cycle, whereas others are toxic only near or during the flowering phase. Some ecologists and evolutionists believe that during the flowering phenophase, when metabolism accelerates and the plant has flowers, certain substances may be produced that repel some animals or even induce intoxication through ingestion or contact. Signalling is an important feature of insect-plant relationships. If the emitter of the signal benefits, i.e. an ant repellent emitted by a plant, then the compound is called an allomone. If the signal is detrimental to the emitter, i.e. a plant producing an ant phagostimulant, then the chemical is called a kairomone.

Cohabiting toxic and nontoxic plants may be chemically similar and represent a complex problem for herbivores that need to differentiate them. Understanding the mechanisms of chemical interaction should aid in interpreting the evolution of plant-insect interactions and perhaps lead to control of several crop pests. This is important because attempts to biologically control leaf-cutting ants have failed "in vivo" because the ants and their symbiont fungus possess several protection and defense mechanisms against biohazards (Kermarrec et al. 1986).

PREFERENTIAL FEEDING

The relationship between plants and leaf-cutting ants is complex and difficult to interpret based on currently accepted theories about insect-plant interrelationships (Littledyke and Cherrett 1975). According to them, Belt (1874) first suspected leaf-cutting ants used leaves as a substrate for their fungus and other researchers later determined that the ants fed on cultivated fungus. The discovery that adult ants also ingest the sap of cut leaves (Barrer and Cherrett 1972; Littledyke and Cherrett 1976) cast doubts on the hypothesis that these insects were monophagous. Indeed, because they cut leaves from different plant species, leaf-cutting ants are considered generalist herbivores (Rockwood 1977). However, leaf-cutting ants prefer certain plant species, some of which are particularly susceptible to attack (especially introduced and/or cultivated plants), whereas others are almost immune.

Why ants prefer some plants over others has yet to be fully clarified. Hubbell and Wiemer (1983) summarized a series of factors that may play a role in this selectivity: (1) the presence of secondary compounds toxic to ants and/or their fungus, (2) presence of secondary compounds that may reduce ant or fungus ability to digest the plant (e.g., tannins), (3) the plant's nutritive value for the fungus, (4) amount of nitrogen (protein), carbohydrates and lipids, (5) moisture content, (6) leaf toughness and physical plant defense mechanisms, such as trichome density and presence of latex. The same investigators propose other factors that may affect foraging: (1) the trail system developed, (2) the ants' previous experience with certain plants, (3) higher energy expenditure (e.g., plant distance), (4) competition with rival colonies, (5) presence aggressive ant species. Other factors may be equally important: (1) seasonal changes in the plant's chemistry and (2) different phases in the colonies' foraging activity. Today, few plants are known to be toxic to ants. Even though we know the effects of certain plant insecticides such as nicotine, rotenone and pyrethrin, little information is available about other plant toxins that may interfere with leaf-cutting ant activity.

REPELLENTS, DETERRENTS AND ANTIFEEDENTS

The toxicity of a chemical compound always depends on the animal and on the mode of absorption/excretion. Thus, the alkaloid solanine occurs in all varieties of domestic potatoes, but the concentration is so low that potatoes are rarely dangerous (Harborne 1988).

Recent studies found repellent, deterrent and antifeedant secondary chemical compounds in protected plants native to the neotropical region (Table 1). The table is indicative of the few plant species and leaf-cutting ants studied thus far. In general, the bioassays used have provided information for identifying compounds that act as ant repellents or deterrents

TABLE 1. Secondary compounds from plants that are repellent to the leaf-cutting ants.

Species	Compounds	Activity	Leaf-cutting ants	References
Lasiantheae fruticosa	lasidiol angelate (sesquiterpenoid)	repellent	A. cephalotes	Wiemer and Ales (1981)
Cordia alliodora	3-α-hydroxyolean-12-en-27-oic acid and oxidized derivatives	repellent	A. cephalotes	Chen et al. (1983)
Hymenaea courbaril	caryophyllene epoxide	repellent (anti-fungal)	A. cephalotes	Hubbell and Wiemer (1983)
Astronium graveolens	trans-β-ocimene (volatile)	repellent (fungicide)	A. cephalotes	Chen et al. (1984)
Melampodium divaricatum	caryophyllene oxide sphatulenol, guaianol and kolavenol	repellent	A. cephalotes	Hubert and Wiemer (1985)
Eupatorium quadrangularae	sesquiterpene lactones	repellent	A. cephalotes	Okunade and Wiemer (1985a)
Jacquinia pungens	jacquinonic acid	repellent	A. cephalotes	Okunade and Wiemer (1985b)
Vismea baccifera	caryophyllene epoxide, kolavenol and nerolidol	deterrent	A. cephalotes	Cancilla (1985 apud Howard et al. 1988)
Dioscorea cayenensis	saponin group	antifeedant	A. octospinosus	Febvay et al. (1985)

or as antifeedants. Neither crude extracts nor isolated active compounds of these plants have been widely studied for their effects on ant colony development (Febvay et al. 1985), the fungus "in vitro," isolated ants (Hubbell et al. 1983; Howard et al. 1988) and foraging behavior (Rahbe et al. 1988).

We have never observed leaf-cutting ants attacking *Virola sebifera*, a plant distributed throughout Brazil. Bueno et al. (1989a) confirmed this fact in laboratory preference tests. Crude extracts of *V. sebifera* obtained with chloroform and dichloromethane inhibited the development of the symbiont fungus "in vitro" (Pagnocca, F.C., personal communication), confirming that one the important factor in plant selection is toxicity to the ants' symbiont fungus. Fractionation of *V. sebifera* leaf extracts (Fraga 1987) revealed dibenzylbutyrolactone and furofuranic lignans. Sesamine was detected among the latter and caused marked inhibition of the symbiont fungus in tests carried out under the same conditions as for the crude assay tests (Pagnocca et al. 1987b). According to Gangadharam et al. (1963) and Ramaswy (1957 and Budowski 1964), sesamine displays bactericidal activity, but its action on the fungus cultivated by leaf-cutting ants was previously unknown.

Sesamine occurs seasonally in *V. sebifera* leaves (Fraga 1987), while other, more consistently present lignans have also been studied (Fernandes, J.B., personal communication) for possible deterrent and/or fungicidal activities. Lignans have antitumor, antimitotic and antiviral activity and some are also toxic to fungi, insects and vertebrates (MacRae and Towers 1984).

In general, leaf-cutting ants must recognize plants deleterious to them or their fungus; no delayed effects on the nests of the ants have been observed. However, sesame (*Sesamum indicum*), an exotic plant introduced to America 200 years ago (Nayar 1976) and highly preferred by *Atta sexdens rubropilosa* (Souza 1965; Bueno et al. 1989 a) has been used empirically to control leaf-cutting ants in the field (Barbiellini 1926; Borges 1926; Barreto 1930 and 1936). In an attempt to determine how useful sesame can be in control of leaf-cutting ants, Gonçalves (1944) conducted a laboratory study which treated ant nests with sesame leaves. He observed no deleterious effects. However, Bueno et al. (1989 b), using the Hebling-Beraldo et al. (1985) method to maintain ant nests with sesame leaves in the laboratory, noted an ant and fungal mass decrease until complete nest extinction. During the 60 days of treatment, ant behavior changed, as did the moisture level inside the fungal chambers and the consistency and general aspect of refuse. The effect of sesame leaves may be caused by toxic secondary metabolites ingested with leaf matter or indirectly through the symbiont fungus. A second possibility may be the direct action of these metabolites on the fungus, thus depriving adult ants, and particularly larvae, of food. The simultaneous occurrence of both possibilities cannot be ruled out.

Measurement of oxygen consumption by ants removed from nests previously offered sesame leaves indicated an increased respiratory rate after the second week of treatment (Hebling-Beraldo et al. 1989). However, when ant groups were maintained on an artificial diet containing crude sesame leaf extracts prepared with chloroform and methanol, the ants did not die. To test possible anti-fungal action, the fungus was isolated from *Atta sexdens rubropilosa* nests and cultured in medium containing different concentrations of crude sesame leaf extracts prepared with chloroform, methanol and water. Both the chloroform and methanol extracts partially inhibited fungal growth starting at concentrations equivalent to 30 mg dry leaf weight/ml culture medium (Pagnocca et al. 1987a). Inhibition increased with extract concentration and 90% or more reduction in fungal growth was observed when a 60 mg dry leaf weight/ml medium was used. Other sesame leaf extracts prepared with material obtained at different times of year or from plants in different stages of development showed similar results, indicating no relationship between inhibition and season or plant age (Pagnocca, F.C., personal communication). Attempts to extract sesamine from sesame leaves proved unsuccessful, which was not unexpected since Budowski (1964) reported that sesamine is synthesized in the seed oil. Fractionation of organic extracts prepared from sesame leaves, a process our laboratory is developing, should allow the isolation of compounds responsible for fungal inhibition. Fungicides probably helped *Canavalia ensiformis* leaves control *Atta* nests (Mullenax 1979).

CONCLUSION

The apparent contradiction between the cutting and transportation of sesame leaf pieces to the nest and their negative effect on *Atta sexdens rubropilosa* and its symbiont fungus may lead to a better understanding of fungus-ant symbiosis and other peculiarities of the leaf-cutting ant-plant interaction. In addition, this information opens new alternatives for controlling these insects, either by large-scale synthesis of toxic secondary metabolites through bait and other vehicles, or by the joint use of these plants with cultivated crops.

ACKNOWLEDGMENTS

We gratefully acknowledge the support of the Conselho Nacional de Desenvolvimento Científico e Tecnológico, Fundaçao Banco do Brasil and Fundaçao para o Desenvolvimento da Universidade Estadual Paulista.

LITERATURE CITED

BARBIELLINI, A. 1926. A saúva e o gergelim. Chac. Qui. 33: 532.

BORGES, A. 1926. Ainda o gergelim e as saúvas. Chac. Qui. 33: 538-539.

BARRER, P.M. and J.M. CHERRETT. 1972. Some factors affecting the site and pattern of leaf-cutting activity in the ant *Atta cephalotes* L. J. Entomol. 47: 15-27.

BARRETO, U.P. 1930. Ainda o gergelim como sauvicida. Chac. Qui. 41: 252.

BARRETO, U.P. 1936. O gergelim e a saúva. Chac. Qui. 54: 338-339

BUDOWSKI, P. 1964. Recent research on sesamin, sasamolin, and related compounds. J. Am. Oil Chem. Soc. 41: 280-285.

BUENO, O.C., M.J.A. HEBLING-BERALDO, S.L.R. CASTRO, O.A. DA SILVA and F.C. PAGNOCCA. 1989a. Testes de preferencia de *Atta sexdens rubropilosa* por folhas de *Sesamum indicum* e *Virola sebifera*. Anais do IX Encontro de Mirmecologia, 24-28, Viçosa - Minas Gerais (Brasil).

BUENO, O.C., M.J.A. HEBLING-BERALDO, O. AULINO DA SILVA and A.M.C. MATENHAUER. 1989b. Effects of the sesame (*Sesamum indicum* L.) on the nest development of *Atta sexdens rubropilosa* Forel. Insect Soc. In press.

CHEN, T.K., D.C. ALES, N.C. BAENZIGER and D.F. WIEMER. 1983. Ant-repellent triterpenoids from *Cordia alliodora*. J. Org. Chem. 48: 3525-3531.

CHEN, T.K., D.F. WIEMER and J.J. HOWARD. 1984. A volatile leafcutter ant repellent from *Astronium graveolens* Naturwissenschaften, 71: 97-98.

FEBVAY, G., P. BOURGEOIS and A. KERMARREC. 1985. Antiappétants pour la fourmis attine, *Acromyrmex octospinosus* (Reich) (Hymenoptera, Formicidae), chez certaines especies d'ignames (Dioscoreaceae) cultivées aux Antilles. Agronomie, 5: 439-444.

FRAGA, R.L. 1987. Lignanas das folhas de *Virola sebifera* Aubl. (Myristicaceae), variaçao sazonal e transformaçoes em análogos de podofilotoxina e esteganacina. M. S. Thesis, Univ. Federal de Sao Carlos - Sao Paulo (Brasil). 118pp.

GONÇALVES, C.R. 1944. O gergelim no combate a saúva. Bol. Fitossanitário, 1: 19-27.

HARBORNE, J.B. 1988. Introduction to ecological biochemistry. Third Edition. Academic Press Limited, London, 356pp.

HEBLING-BERALDO, M.J.A., O.C. BUENO, O.A. SILVA and A.M.C. MATENHAUER. 1985. Relaçao entre consumo foliar e crescimento do fungo em formigueiros iniciais de *Atta sexdens rubropilosa* Forel, 1908 (Hymenoptera, Formicidae). Revta. Bras. Ent. 29: 449-452.

HEBLING-BERALDO, M.J.A., O.C. BUENO, R.E. ALMEIDA and O. AULINO DA SILVA. 1989. Influência do tratamento com folhas de *Sesamum indicum* sobre o metabolismo respiratório de *Atta sexdens rubropilosa*, Forel 1908. (Hymenoptera, Formicidae). An. Soc. Entomol. Brasil. In Press.

HOWARD, J.J., J. CAZIN, JR. and D.F. WIEMER. 1988. Toxicity of terpenoid deterrents to the leafcutting ant Atta cephalotes and its mutualistic fungus. J. Chem. Ecol. 14: 59-69.

HUBBELL, S.P. and D.F. WIEMER. 1983. Host plant selection by attine ant, pp. 133-154. *In* P. Jaisson (ed.), Social Insects in the tropics, vol. 2. University of Paris Press, Paris. 252pp.

HUBBELL, S.P., D.F. WIEMER and A. ADEJARE, 1983. An antifungal terpenoid defends a neotropical tree (*Hymenaea*) against attack by a fungus-growing ant (*Atta*). Oecologia, 60: 321-327.

HUBERT, T.D. and D.F. WIEMER. 1985. Ant-repellent terpenoids from Melampodium divaricatum. Phytochemistry, 24: 1197-1198.

KERMARREC, A., M. DECHARME and G. FEBVAY. 1986. Leaf-cutting ant symbiotic fungi: A synthesis of recent research, pp. 231-246. *In* C.S.Lofgren and R.K.Vander Meer (eds), Fire ants and leaf-cutting ants: Biology and management. Westview Press, Boulder. 435pp.

LITTLEDYKE, M. and J.M. CHERRETT. 1975. Variability in the selection of substrate by the leaf-cutting ants *Atta cephalotes* (L.) and *Acromyrmex octospinosus* (Reich) (Formicidae, Attini). Bull. ent. Res. 65: 33-47.

LITTLEDYKE, M. and J.M. CHERRETT. 1976. Direct ingestion of plant sap from cut leaves by leaf-cutting ants *Atta cephalotes* (L.) and *Acromyrmex octospinosus* (Reich) (Formicidae, Attini). Bull. ent. Res. 66: 205-217.

MACRAE, W.D. and G.H.N. TOWERS. 1984. Biological activities of lignans. Phytochemistry, 23: 1207-1220.

MULLENAX, C.H. 1979. The use of jackbean (*Canavalia ensiformis*) as a biological control, for leaf-cutting ants (*Atta* spp). Biotropica, 11: 313-314.

NAYAR, N.M. 1976. Sesame, *Sesamum indicum* (Pedaliaceae). pp. 231- 233. *In* N.W.Simmonds (ed.), Evolution of crop plants. Longman Group Limited, London. 339pp.

OKUNADE, A.L. and D.F. WIEMER. 1985a. Ant-repellent sesquiterpene lactones from *Eupatorium quadrangularae*. Phytochemistry, 24: 1199-1201.

OKUNADE, A.L. and D.F. WIEMER. 1985b. Jacquinonic acid, an ant-repellent triterpenoid from *Jacquinia pungeus*. Phytochemistry, 24: 1203-1205.

PAGNOCCA, F.C., O.A. SILVA, O.C. BUENO and M.J.A. HEBLING-BERALDO. 1987a. Inibiçao do crescimento do fungo cultivado por *Atta sexdens rubropilosa* Forel, 1908 (Hymenoptera, Formicidae), em presença de extratos foliares de *Sesamum indicum*. Paper presented at the XI Congr. Brasil. Entomol., Campinas - Sao Paulo (Brasil). 12 July 1987.

PAGNOCCA, F.C., O.A. SILVA, O.C. BUENO, M.J.A. HEBLING-BERALDO, J.B. FERNANDES and P.C. VIEIRA. 1987b. Efeito da sesamina sobre o desenvolvimento do fungo cultivado por *Atta sexdens rubropilosa* Forel, 1908. Paper presented at the XIV Congr. Brasil. Microbiol., Viçosa - Minas Gerais (Brasil). 8 August 1987.

RAHBE, Y., G. FEBVAY and A. KERMARREC. 1988. Foraging activity of the attine ant *Acromyrmex octospinosus* (Reich) (Hymenoptera: Formicidae) on resistant and susceptible yam varieties. Bull. ent. Res. 78: 329-337

ROCKWOOD, L.L. 1977. Foraging patterns and plant selection in Costa Rica leaf cutting ants. N. Y. Entomol. Soc. LXXXV: 222-233.

SOUZA, F.L. 1965. Plantas preferidas pela saúva. Divulgaçao Agronômica, Rio de Janeiro, 14: 23-29.

WIEMER, D.F. and D.C. ALES. 1981. Lasidiol angelate: ant-repellent sesquiterpenoid from *Lasiantheae fucticosa*. J. Org. Chem. 46: 5449-5450.

39

Pheromonal Control of Behavior in Leaf-cutting Ants

P.E. Howse

INTRODUCTION

"The ceaseless, toiling hosts impress one with their power, and one asks - what forests can stand before such invaders? How is it that vegetation is not eaten off the face of the earth?"
--Belt, 1874; "The Naturalist in Nicaragua"

One answer to this question (see Knapp et al. this volume) is that defenses reside in secondary plant chemicals and the physical characteristics of leaves which oblige the ants to switch between different plant hosts. The ability of leaf-cutting ants to exploit vegetation selectively over a wide area - which can be over 6000 m^2 in *Atta sexdens* in Paraguay (Fowler and Robinson 1979) - implies a sensitive communication system capable of controlling rapid recruitment to new alternative feeding sites on the basis of food quality. Figure 1 shows the main sources of leaf-cutting ant semiochemicals. The more that is learned about the chemistry of these substances, the more questions arise about their role in behavior, the ways in which chemical signals are integrated with other sensory information, and the importance of their environmental context.

ORIENTATION

Interesting parallels exist between the discovery of the identity of the silkworm sex pheromones by Butenandt et al. (1959) and the identification of *Atta texana's* pheromone by Tumlinson et al. (1971). In both cases, the insects' low response thresholds suggest that a minute quantity of each could siphon the species out of existence.

The detection threshold of the ants for the main trail pheromone component, methyl-4-methylpyrrole-2-carboxylate (M4MP2C), was 0.08 pg cm^{-1} so that 0.33 mg should theoretically be sufficient to lay a detectable trail around the earth (Tumlinson et al. 1971). Successful mimicry of Lepidopteran or leaf-cutting ant pheromones has not been possible because (a) insects are able to switch between different cues in different circumstances, (b) insects normally have upper and lower thresholds to pheromones, so they can be repellent at high concentrations, (c) physiological factors must be appropriate

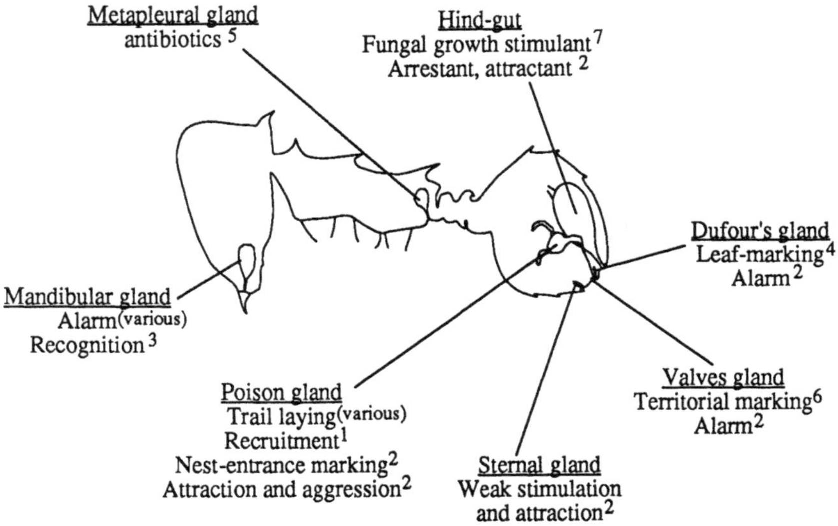

FIGURE 1. Location and possible role of some exocrine glands of a typical leaf-cutting ant, according to various authors. (1) Jaffé and Howse (1979); (2) Hölldobler and Wilson (1986); (3) Jaffé (1983); (4) Bradshaw et al. (1986); (5) Schildknecht and Koob (19700; (6) Jaffé et al. (1979); (7) Weber (1966).

and, finally, (d) other components of the pheromone blend may be crucial to the full behavior response.

M4MP2C trails are followed by other attines, notably *Atta cephalotes* (Riley et al. 1974) and *Acromyrmex octospinosus* (Robinson et al. 1974) but not by *A. sexdens*; *A. cephalotes* and *Ac. octospinosus*, however, discriminate between each other's trails. This is not surprising in view of the differences in poison gland secretion among the various species. Cross et al. (1979) found the major component to be 3-ethyl-2,5-dimethylpyrazine in *A. sexdens* although M4MP2C is present along with methyl and ethyl phenylacetates. *Ac. octospinosus*, however, has M4MP2C as its principal component with 3-ethyl-2,5-dimethylpyrazine and 2-ethyl-3,5-dimethylpyrazine.

The inference from these analyses is that the pheromone blend determines the specificity of trail following, although the reported bioassays did not test this hypothesis. As most ant trail substance bioassays only measure persistence of following, investigation of other behavioral roles for poison gland constituents is limited.

The possibility that a leaf-cutting ant trail contains a variety of chemical cues is suggested by the work of Vilela et al. (1987), which also illustrates the ants' ability to exploit information sources for orientation. Leaf-carrying workers of *A. cephalotes*, *A. laevigata* and *Ac. octospinosus* compensated for the rotation of a bridge connecting a food table and the nest when it was turned through 180° while they traversed it, and found their way back to the nest. In the dark, approximately half the ants failed to compensate for the turn, except for *Ac. octospinosus* workers, only 25% of which regained the nest. This simple experiment suggests that the ants normally combine chemical and visual cues, but in the absence of the latter *Ac. octospinosus* workers can detect the polarity of the trail. When the bridge was constructed of two parallel runways, ants could, in theory, learn one of two possible turn sequences, depending on the runway chosen. When such a bridge connected foraging table and nest, workers of all three species took significantly longer to reach the nest in the dark after a 180° rotation than after a 360° control rotation. Further experiments using a symmetrical Y maze showed that gravitational cues may be used to detect the nest direction and also demonstrated that in completely darkened conditions *A. cephalotes* workers could select the arm carrying a pre-laid trail back to the nest even after the maze had been rotated 120°, irrespective of the arm into which they were released. Thus, in addition to visual, kinaesthetic and gravitation cues, the ants can also--to a greater or lesser extent depending on the species - make use of polarized chemical cues on the trail. Trails may differ both quantitatively and qualitatively. The trail may be tapered in one direction. It may fluctuate in intensity at intervals, giving a repetitive asymmetry; it may fluctuate in intensity at strategic points or it may fluctuate in intensity irregularly so the ants can learn the topography. One factor that may contribute to trail polarity is fade-out time. Therrien et al. (1986) showed that significant fade-out of *Ac. octospinosus* trails in the field occurs after 10-12 h and this process may be accelerated by the action of ultraviolet light. A trail must therefore be renewed throughout its length or it will develop patchiness.

In qualitative terms one trail may differ from another by virtue of the differing minor components of individual workers that use each trail. Such minor components, or the secretion of a second gland, could also modulate the chemical trail. Ants can discriminate between a trail leading to food in laboratory colonies and a trail leading to the rubbish dump (Jaffe 1983). There are, therefore, a number of possible roles for minor components of the poison gland secretion and more discriminating assays are needed to test for other behavioral effects. The apparent ability of foragers to utilize a variety of information sources while orientating poses severe problems in the experimental design of bioassays. Vilela et al. (1987) believe that they use cues in a particular hierarchy: odor-trail - visual cues - spatial layout - odor differences on the trail - gravity. The ants' ability to switch between these sources of information, implying specific learning abilities, means not only that the workers have back-up systems when ambiguity arises, but they can orient to and from the nest at night, in both vertical and horizontal planes, an ability that cannot be explained in terms of simple trail-following.

Another function of the poison-gland secretion is to stimulate recruitment. Mass recruitment to a foraging trail in *A. cephalotes* laboratory colonies can be achieved by the contents of a single worker poison gland (Jaffe and Howse, 1979). A trail of 4-methylpyrrole-2-carboxylate has the same effect except the ants follow this more hesitantly. The relative food quality can be communicated by the pheromone concentration. Jaffe and Howse (1979) believe the foragers respond to the density of the ants at the food source or on the trail by modulating the amount of pheromone they add to the trail.

LEAF MARKING

Barrer and Cherrett (1972) established that *A. cephalotes* workers cutting leaves attracted other foragers to the immediate vicinity. They found no evidence of pheromone marking but attributed the effect to the release of chemicals from the cut leaf and to the geometry of the cut leaves, which made further cutting easier. Bradshaw et al. (1986) later found that *A. cephalotes* workers marked the leaf discs they cut in two ways. First the tip of the gaster was streaked against the leaf surface and then was applied one to five times to the edge of the leaf disc while held between the mandibles. When marked and unmarked (control) leaf segments were placed alongside foraging trails the former attracted ants from a radius of up to 2 cm and were picked up 3-4 times more quickly. After storage for 300s, the attraction of marked leaves faded. Its origin was traced to the Dufour's gland, the secretion of which significantly reduced pick-up time and increased carrying time. (Z)-9-nonadecene mimicked the effect. When applied to filter paper discs, significantly higher levels of subsequent marking resulted than with other hydrocarbons in the active Dufour's gland fraction. n-Tridecane had similar effects, notably in stimulating marking of treated paper discs, but the effect was highly dose-dependent.

The significance of this behavior may lie in its relation to chain transport or "bucket brigade" foraging. Hubbel et al. (1980) reported that when *A. cephalotes* scouts discovered a new food source in the field they returned to the main trail, where 83% then dropped their leaf fragments or transferred them directly to other ants on the first meter of the trail. Experiments with a square of cardboard previously marked and then placed on the main trail showed that the deposition of fragments was a response to the chemical trail only. The authors concluded that this behavior allows rapid exploitation of a new food source; foragers rely on their short-term memory to relocate the food source and reinforce the connection to the main trail. After this has occurred, the foragers take their leaf particles all the way to the nest. If they did this initially, the return journey might take up to four hours, by which time the new recruitment trail would have evaporated.

Transfer of cut leaf discs also occurs in *A. cephalotes*, in which foragers cutting in trees let their cuttings fall to the ground where they are transported back to the nest by a separate cohort of "carriers." This is ergonomically efficient because it avoids the necessity of journeys up and down the trunk. Hubbell et al. (1980) calculate that the two-forager method is less costly than the one-forager method when

$$\sim C/T < n - 1/m,$$

where C is the cost of cutting a fragment, T the cost of the trip up and down the tree, n the number of fragments and m the number of additional fragments cut for each n found. For cutting sites of a given height, therefore, it follows that there will be selection pressure to increase the probability of pick-up and transport of cut leaf particles. In *A. sexdens rubropilosa*, three foraging stages have been reported (Fowler and Robinson, 1979). One subset of foragers cuts in trees; a second searches out the fallen cuttings, cuts them into smaller pieces and puts them on the main trail. The third subset takes the leaf fragments back to the nest. The total foraging efficiency is 49%, but 99% of material is harvested once deposited on the trail.

Leaf-marking also occurs in *Ac. octospinosus* (Howse, unpublished) in which transfer of leaf discs on the trail has not been reported - although dropped discs are usually recovered. In laboratory colonies, marked leaf fragments are picked up sooner and transported more rapidly to the nest than unmarked control discs. Furthermore, foragers can discriminate leaf fragments marked by members of their own colony from those of another colony.

TERRITORIAL MARKING

Atta cephalotes workers explore new territory in laboratory colonies and mark them with an often visible secretion daubed on to the substrate from the tip of the abdomen. On unmarked territory they adopt a characteristic posture with the gaster lifted, which has been called abdomen raising behavior (ARB) by Jaffe et al. (1979). The frequency of ARB is least on territory marked by the same colony and greatest on substrates marked by other species; thus it can be used as an assay.

The origin of the colony-specific marker was traced to the secretion of the valves gland of the sting. This secretion is active for about one hour after deposition on paper. It contains proteins (Bazire-Benazet and Zylberberg, 1979) and adheres to cuticle: possibly the proteinaceous fraction retards the evaporation of the volatile components.

Poison gland secretion elicits frequent ARB when freshly applied (Jaffe et al. 1979), as does synthetic pyrrole (4MP2C). However, the effect of the synthetic trail pheromone in producing ARB is effectively eliminated when added to a previously marked paper.

Support for the territorial pheromone hypothesis comes from its defensive role. Vilela and Howse (1986) isolated workers of various attine species on previously marked papers. On their own territory *Atta laevigata* or *A. cephalotes* workers were significantly more aggressive, showed more threat responses to intruders and spent more time on marked substrates than on unmarked ones.

Hölldobler and Wilson (1986) also investigated territorial marking in *A. cephalotes*, but came to different conclusions. They studied the behavior of the ants in the immediate vicinity of the nest exit or in an arena, the entrance to which was considered the nest exit. Marking with poison gland secretion served as a colony-specific "nest exit pheromone" orientating workers towards the opening and stimulating trail laying, leaf-cutting and leaf retrieval. Ants became more aggressive after they had been on the same territory for a day or more, but the authors failed to find any evidence for a territorial pheromone. They could not reproduce the effects of the valves gland secretion but found instead that it produced typical alarm behavior.

These results are not entirely inconsistent with those of Jaffe et al. (1979) or Vilela and Howse (1986). In those experiments, nests were housed in plastic boxes or jars placed in the center of a table 53 x 38 cm. Hölldobler and Wilson (1986) did most of their bioassays immediately adjacent to the nest exit, having observed that most of the marking took place within 5 cm of the exit. The other authors carried out bioassays with paper substrates placed alongside a foraging trail which lead from the nest to food placed on the nest table. Thus, the ants may have behaved differently at the nest entrance from the way they do alongside foraging trails. A simple observation confirmed this idea. In laboratory colonies a piece of spent fungus is taken from a foraging site to the nest with the ant following the foraging trail. Only at the nest entrance is the decision then made to transport it to the rubbish pile (Jaffe, 1983, and personal observations).

A tentative synthesis of these data is as follows. The nest entrance is marked with poison gland secretion by abdominal dipping and typical trail-laying which intergrade into one another (Hölldobler and Wilson 1986). This stimulates further trail-laying and recruitment to food. In areas of potential trail development or leaf-cutting, abdomen dipping leads to deposition of a marker pheromone which does not originate from the poison gland. Although Hölldobler and Wilson (1986) repudiate the view that the valves gland is ineffective in reducing ARB and stimulates attack, the latter response may result from the bioassay method, in which the contents of one gland are presented on a 5 x 5 cm square of absorbent paper. This represents an abnormally high concentration, raised still further by the rapid release of the volatiles from paper substrate. Hölldobler and Wilson (*op. cit.*) note that the valves gland secretion caused no aggression when distributed as 0.6 gland over 15 x 21 cm.

Later, Jaffe (1984) used another bioassay for territorial marking, in which two ants from different colonies were put onto a marked filter paper and the initiator of the attack and the first ant to leave the substrate were noted. Again, evidence was found for territorial marking in *A. cephalotes, A. laevigata* and *Ac. octospinosus landolti*. This resulted in colony-specific recognition of marked papers, and a separate series of bioassays showed this was not due to visual cues.

ALARM PHEROMONES

Recognition

While species or colony recognition depends on chemical markers deposited at the nest entrance, their chemical nature remains unknown. However, they increase the ants' aggressive tendencies on their own territory, possibly by facilitating the discrimination of aliens on an odor basis. Jutsum et al. (1979) showed that odors from the diet play a role in individual recognition. For example, *Ac. octospinosus* colonies given the same food in the laboratory show less aggression towards each other than colonies given different food. Workers of *A. sexdens, A. cephalotes* and *Ac. octospinosus* all attack aliens from the same or other species, although intraspecies recognition takes place at 0-8 mm and interspecies recognition at 5-15 mm. Jaffe (1983) showed that dead *Ac. octospinosus* ants were more quickly removed from a nest table of *A. cephalotes* than dead *A. cephalotes* workers. Freeze-dried workers of either species were removed most quickly, suggesting that species recognition depends upon volatile chemicals. Further experiments located the recognition signals mainly on the head and thorax. Freeze-dried ants treated with extracts of heads took significantly longer to be rejected than those treated with thorax or gaster extracts. Species recognition may therefore depend on a spectrum of compounds forming part of the mandibular alarm gland secretion. Colony recognition in *Ac. octospinosus* may be related to differing ratios of two major mandibular gland compounds.

Defense

The multifunctional role of "alarm pheromones" has been reviewed by Bradshaw and Howse (1984). This role depends on the multicomponent nature of the secretions. Moser et al. (1968) identified 4-methyl-3-heptanone and 2-heptanone from the worker mandibular gland of *A. texana*. In a simple bioassay, in which they introduced an airstream containing pheromone into a glass vessel containing the ants, they concluded that 4-methyl-3-heptanone released attraction in quantities of 3×10^9 molecules cm^{-2}, and "alarm" with fast running and open mandibles at ten times this concentration. 2-heptanone was 1000 times less effective.

Blum et al. (1968) confirmed the presence of 4-methyl-3-heptanone in a series of *Atta* species: *texana*, *biosphaerica*, *capiguara*, *colombica*, *robusta* and *sexdens*. Subsequently, Crewe and Blum (1972) examined the alarm pheromones of certain species of *Cyphomyrmex*, *Trachymymex* and *Acromyrmex*, identifying mixtures of 3-octonal, 3-octanone, 4-methyl-3-heptanone and 4-methyl-3-heptanol; they believed the patterns of occurrence had phylogenetic significance. However, the value of such information in phylogeny has diminished because more detailed analysis shows mandibular gland pheromones to be more complex than originally thought. Riley et al. (1974) who found the S(+) 4-methyl-3-heptanone enantiomer to be active for *A. cephalotes* and *A. sexdens*, also found 3-octanone, 3-octanol and methyl-3-heptanone and 2-heptanone in both species. Indeed, Schildknecht (1976) identified seventeen compounds from *A. sexdens rubropilosa* mandibular gland secretions, including monoterpenes, such as Beta-pinene, citral and farnesol, a lactone and various straight- and branched-chain aliphatic compounds. The discovery of citral in *A. sexdens* dates back to Butenandt et al. (1959) and structural confirmation by Blum et al. (1969) who also found geraniol and nerol. The discovery of extensive ranges of minor components, suggests that species differences are variations on a common keyboard of compounds.

Bradshaw and Howse (1984) and Bradshaw (1981) promoted the concept that certain alarm pheromone components of mandibular gland secretions diffuse from a release point at different rates, creating a changing qualitative gradient of active spaces. Attraction and "alarm" to 4-methyl-3-heptanone occur at different thresholds in *A. texana* (Moser et al. 1968) and a sequence of attraction and alarm can be obtained even in reaction to a single compound (Bradshaw and Howse, 1984). Complex behavioral sequences are possible if other components of the secretion have subtle distinguishable effects, this has yet to be investigated for *Atta* species. In *Ac. octospinosus*, the two principal components, 3-octanone and 3-octanol, have different effects on behavior (Howse, unpublished). The former induces rapid running and recruitment of workers from the surrounding area, while the latter stimulates ants to lay odor trails focusing on the source, and biting at the source. Crewe and Blum (1972) recorded a specific ratio of 128:1 for 3-octanone: 3-octanol in *Ac. octospinosus*, but Howse and Febvay (in prep.) found ratios of less than 10:1 (Guadaloupe FWI). The lowest ratios (less than 2:1) and the largest quantities were found in the largest workers, while the smallest amounts and highest ratios (8:1) were found in the smaller workers. It can therefore be suggested that large workers have a greater ability to recruit others to the general area of a disturbance, which correlates with their more defensive role.

PHEROMONES IN LEAF-CUTTING ANT CONTROL

Various trials have been undertaken to investigate the potential of trail pheromones for improving leaf-cutting ant poison bait. Robinson and Cherrett (1978) found that the discovery of paper discs by *A. cephalotes*, *A. sexdens* and *Ac. octospinosus* laboratory colonies was increased by the addition of M4MP2C. Field tests in Paraguay and Brazil (Robinson et al. 1982) in which the synthetic pyrrole was added to soybean, citrus pulp or paper disc baits showed no increase in the pick-up of treated baits compared with untreated baits (except for one trial with *Ac. octospinosus*). Thus the addition of the pheromone did not significantly enhance the attractive effect of food odors. Cross et al. (1979) also carried out field experiments in Paraguay, adding M4MP2C, 3E25DMP, methyl phenyl acetate and ethyl phenyl acetate, all components of the trail pheromone. These were put on to vermiculite baits in various concentrations. No enhanced pick-up was obtained even when florasil was added to the baits to provide a slow release substrate for the volatile pheromone components. Trials were also carried out with various concentrations of (Z)-9-nonadecene, believed to be the principal component of the leaf-marking pheromone (Bradshaw et al. 1986), but again no significant increase in pick-up of impregnated baits was found in the field.

Although we have been able to show that certain aspects of leaf-cutting ant behavior are controlled by glandular secretions, and identified the main components of these secretions, we are left with the paradox that it has so far been impossible to enhance pick-up and transport of baits in the field by the addition of such components. Four main areas need additional research.

(1) Pheromones are generally effective only within a limited range of concentrations. This has been repeatedly noted in bioassays on trail pheromone components and applies also to the leaf-marking pheromone (Bradshaw et al. 1986). Controlled release of volatiles from baits is therefore required. It has been achieved, for example, by adding pheromone to soybean oil (Robinson et al. 1982). The bait formulation must also be able to protect pheromones from degradation by sunlight.

(2) While the multicomponent nature of pheromones has been well-researched chemically, the multifunctionality has not, and few behaviorally discriminating bioassays have been carried out. Not only may minor components synergize the action of major components, but they may also control separate links in a sequence. Pheromones from two or more distinct glandular sources may also combine to increase the multicomponent nature of a secretion.

(3) The ability of ants to switch their attention between different sensory stimuli can confuse an understanding of pheromone action. This ability shows itself in trail following and in recruitment to food sources (see Knapp et al. this volume). In alarm-defense behavior, visual and acoustic cues may also play important roles, as well as territorial marking and other factors.

(4) To achieve maximal effectiveness, an artificial bait must stimulate

a sequence of behavior starting with attraction from a distance, pick-up, marking, transport, transfer at the nest entrance (or earlier), and acceptance within the nest for incorporation in the fungus garden, and recruitment of more foragers. The future objective lies in potentiating discovery of baits and stimulating recruitment processes to baits. Incorporating major components of the trail pheromone in broadcast particles provides a short-range attractant but clearly does not trigger part of the normal harvesting sequence. Not all of the necessary stimuli are presented to the ants to elicit the full sequence of behavior.

REFERENCES

BARRER, P. M. and J.M. CHERRETT. 1972. Some factors affecting the site and pattern of leaf-cutting in the ant *Atta cephalotes* L. J. Ent.(A) 47: 15-27.

BAZIRE-BENAZET, M. and L. ZYLERBERG. 1979. An integumentary gland secreting a territorial pheromone in *Atta* spp.: detailed structure and histochemistry. J. Insect Physiol. 25: 751-756.

BLUM, M. S., F. PADOVANI and E. AMANTE. 1968. Alkones and terpenes in the mandibular glands of *Atta* species. Comp. Biochem. Physiol. 26: 291-299.

BRADSHAW, J. W. S. 1981. The physiochemical transmission of two components of a multiple chemical signal in the African weaver ant, *Oecophylla longinoda*. Anim. Behav. 29: 581-585.

BRADSHAW, J. W. S. and P. E. HOWSE. 1984. Sociochemicals of ants, pp. 429-473. *In* W. J. Bell and R. T. Carde (eds.), Chemical ecology of insects. Chapman Hall, London. 524 pp.

BRADSHAW, J. W. S., P. E. HOWSE and R. BAKER. 1986. A novel autostimulatory pheromone regulating transport of leaves in *Atta cephalotes*. Anim. Behav.34: 234-240.

BUTENANDT, A., R. BECKMAN, D. STAMM and E. HECKER. 1959. Ueber den Sexual Lockstoffe des Seidenspinners *Bombyx mori*. Reinarstellung und Konstitution. Z. Naturforsch. 14b: 283- 284.

BUTENANDT A., B. LINZEN and M. LINDAUER. 1959. Ueber einen Duftstoff aus der Mandibeldruese der Blattschneiderameise *Atta sexdens rubropilosa* Forel. Arch. Anat. microsc. Morph. exp. 48: 13-19.

CREWE R. M. and M. S. BLUM. 1972. Alarm pheromones of the Attini, their phylogenetic significance. J. Insect Physiol. 18: 31-42.

CROSS J. H., R. C. BYLER, U. RAVID, R. M. SILVERSTEIN, S. W. ROBINSON, P. M. BAKER, J. SABINO DE OLIVEIRA, A. R. JUTSUM, and J. M. CHERRETT. 1979. The major component of the trail pheromone of the leaf-cutting ant, *Atta sexdens rubropilosa* Forel: 3-Ethyl-2,5-dimethylpyrazine. J. Chem. Ecol. 5: 187-203.

FOWLER, H. G. and S. W. ROBINSON. 1979. Foraging by *Atta sexdens* (Formicidae: Attini): seasonal patterns, caste and efficiency. Ecol. Entom. 4: 239-247.

HÖLLDOBLER, B. and E. O. WILSON. 1986. Nest area exploration and recognition in leafcutter ants (*Atta cephalotes*). J. Insect Physiol. 32: 143-150.

HUBBEL, S. P., L. K. JOHNSON, E. STANISLAV and B. WILSON. 1980. Foraging by bucket-brigade in leaf-cutter ants. Biotropica 12: 210-213.

JAFFE, K. 1984. Nestmate recognition and terratorial marking in *Solenopsis geminata* and in some Attini. *In* Fire ants and leaf-cutting ants, biology and management. C. S. Lofgren and R. K. Vander Meer (eds.) Westview Press. Boulder, pp. 261-222.

JAFFE, K. 1983. Chemical communication among workers of leaf-cutting ants. *In* P. Jaisson (ed.), Social Insects in the Tropics Vol. 2. Univ. Paris-Nord, France pp. 165-180.

JAFFE, K. and P. E. HOWSE. 1979. The mass recruitment system of the leaf-cutting ant *Atta cephalotes*. Anim. Behav. 27:930-939.

JAFFE, K., M. BAZIRE-BENAZET and P. E. HOWSE. 1979. An integumentary pheromone secreting gland in *Atta* sp. Territorial marking with a colony-specificity pheromone in *Atta cephalotes*. J. Insect Physiol. 25: 833-839.

JUTSUM, A. R., T. S. SAUNDERS and J. M. CHERRETT. 1979. Intraspecific aggression in the leaf-cutting ant *Acromyrmex octospinosus*, Anim. Behav. 27: 839-844.

MOSER, J. C.,R. C. BROWNLEE and R. SILVERSTEIN. 1968. Alarm pheromones of the ant: *Atta texana*. J. Insect Physiol. 14:529-535.

RILEY, R. G., R. M. SILVERSTEIN, B. CARROLL and R. CARROLL. 1974. Methyl 4-Methyl pyrrole-2-Carboxyate a volatile trail pheromone from the leaf-cutting ant *Atta cephalotes*. J. Insect Physiol. 20: 651-664.

ROBINSON, S. W., J. C. MOSER, M. S. BLUM and E. AMANTE. 1974. Laboratory investigations of the trail-following responses of four species of leaf-cutting ants with note on the specificity of a trail pheromone of *Atta texana* (Buckley). Insectes Sociaux, 21: 87-94.

ROBINSON, S. W. and J. M. CHERRETT. 1978. The possible use of methyl-4-methylpyrrole-2-carboxylate, an ant trail pheromone, as a component of an improved bait for leaf-cutting ants. Bull. Entomol. Res. 68: 519-529.

ROBINSON, S. W., A. R. JUTSUM, J. M. CHERRETT and R. J. QUINLAN. 1982. Field evaluation of methyl-4-methylpyrrole-2-carboxylate, an ant trail pheromone, as a component of baits for leaf-cutting ants (Hymenoptera: Formicidae) control. Bull. Entomol. Res. 72: 345-356.

SCHILDKNECHT, H. 1976. Chemical ecology - a chapter of modern natural products chemistry. Angew. Chem. Int. Ed. Eng. 151: 214-222.

SCHILDKNECHT, H. and K. KOOB. 1970. Plant bioregulators in the metathoracic glands of myrmicine ants. Angew. Chem. Int. Ed. Eng. 9: 173.

THERRIEN, P. J. M. MCNEIL, W. G. WELLINGTON and G. FEBVAY. 1984. Ecological studies of the leaf-cutting ant, *Acromyrmex octospinosus*, in Guadeloupe. *In* Fire ants and leaf-cutting ants, biology and management. C. S. Lofgren and R. K. Vander Meer (eds.) Westview Press, Boulder, pp. 172-183.

TUMLINSON, J. H. R. M. SILVERSTEIN, J. C. MOSER, R. G. BROWNLEE, and J. M. RUTH. 1971. Identification of the trail pheromone of a leaf-cutting ant *Atta texana*. Nature, 234 (5328): 348-349.

VILELA, E. F. and P. E. HOWSE. 1984. Territoriality in leaf-cutting ants, *Atta* spp. *In* Fire ants and leaf-cutting ants, biology and managemant. C. S. Lofgren and R. K. Vander Meer (eds.). Westview Press, Boulder, pp. 157-171.

VILELA, E. F. and P. E. HOWSE. 1988. Pheromone performance as an attactive component in baits for the control of the leaf-cutting ant *Atta sexdens rubropilosa* Forel, 1908 (Hymenoptera: Formicidae). An. Soc. Entomol. Brasil. Ano 17: 107-124.

VILELA, E. F., K. JAFFE and P. E. HOWSE. 1987. Orientation in leaf-cutting ants (Formicidae: Attini). Anim. Behav. 35: 1443-1453.

40

Self-organizing Spatial Patterns in the Argentine Ant *Iridomyrmex humilis* (Mayr)

S. Aron, J.M. Pasteels, S. Goss and J.L. Deneubourg

INTRODUCTION

Two conspicuous characteristics of the Argentine ant, *Iridomyrmex humilis*, are its highly agonistic and opportunistic behaviors. Originating in South America, this species spread to most temperate and sub-tropical zones of Australia, South Africa, North America, and Europe. Once installed, *I. humilis* eliminates native ants through continuous warfare and exploits a maximum range of food sources, destroying buds, blooms and fruits of crops such as sugar cane, coffee and lemons (Newell, 1908; Crowell, 1968; Markin, 1970; Fluker and Beardsley, 1970; Erickson, 1971; Bond and Slingsby, 1984; Ward, 1987). In addition, the ants attend and protect numerous scale insects and plant lice from natural predators, leading to large population increases of Coccidae and Aphididae and failure of natural control (De Bach et al. 1951; Samways et al. 1982; Dreistadt et al. 1986; Haney et al. 1987). The Argentine ant is considered a major economic pest.

Their ecological success depends on at least three factors: (1) they tolerate a wide range of habitats as long as there is sufficient moisture in logs, under rocks or in the ground; (2) they establish polygynic and polydomic societies consisting of many thousands of individuals. Colonies are parcelled into sub-colonies (through seasonal "divisional migration" - Skaife, 1955; Markin, 1970; Benois, 1973) with nests established near food sources and permanent contact between sub-colonies maintained by means of trunk trails. Moreover, these nests are not permanent structures and may fuse with adjacent nests. Entire populations frequently move from area to area in response to unfavorable environmental conditions (food-source depletion, flooding, desiccation,...); (3) they perform massive and rapid recruitment mediated by odor trails, allowing them to respond quickly to local disturbances or the discovery of food.

Recent observations of *I. humilis* showed that exploration of new territories proceeds according to a well-defined behavior pattern. An ant column extends from the nest, then fans out at its distal extremity into a front of workers exploring individually in random directions. Whereas exploratory recruitment generally leads to the dispersion of ants on the new terrain, *I. humilis*' exploratory activity is a collective process, and resembles army ant hunting raids, albeit on a smaller scale (Aron et al. 1989; Deneubourg et al. 1989a).

This paper will investigate how *I. humilis* exploits its territory at the colony level. This will involve territorial marking, the formation of exploratory trails, the discovery and exploitation of food sources, the formation of trails linking neighboring colonies and the selection of the shortest trails. These collective spatial patterns can be generated by simple interactions between a large number of identical explorers, through their trail laying and following behavior. This illustrates how complex structures may be based on self-organization processes.

EXPLORATORY BEHAVIOR

In order to study exploratory recruitment, a bridge was placed between the nest container and a chemically unmarked sand-filled arena (80 x 80 cm). The access of Argentine ants colonies to such unfamiliar terrain elicits an exponential outward surge of ants associated with collective exploratory behavior. The first workers entering the new arena individually explore, in random directions, the area closest to the nest. As this front progresses into the arena, a trail is formed which connects the front to the nest (Fig. 1a). In general, the larger the colony size, the faster the pattern formation, and the stronger the trail. For colonies above 1500 workers, the formation of more than one exploratory trail may occur. The rate of trail formation-progression is about two to three m/hour for a 600 ant colony. Some days later, the trail is still present, but no longer so well marked, with only a small number of foragers seen in the arena.

During recruitment, explorers returning to the nest move rapidly among quiescent nestmates, vigorously antennating them, and then return to the new

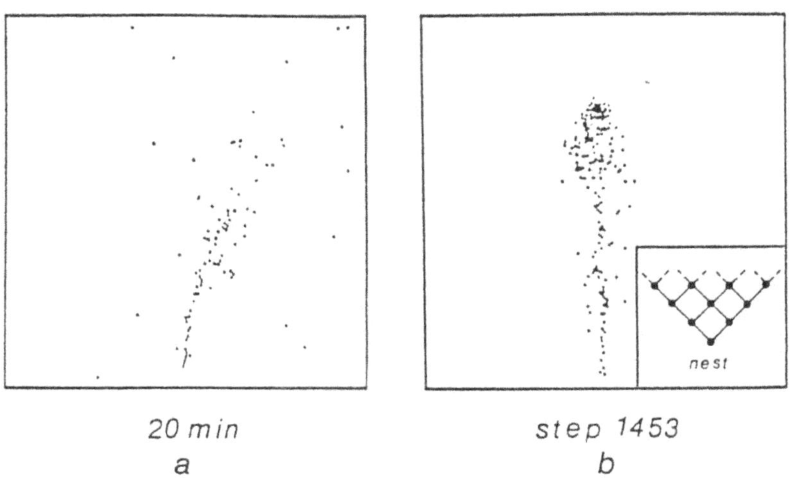

FIGURE 1. A) Collective exploratory pattern on a 80x80 cm. sand-filled arena after 20 min, colony of 600 workers. B) Idem, generated by Monte-Carlo simulation (see text). (from Deneubourg et al. 1989).

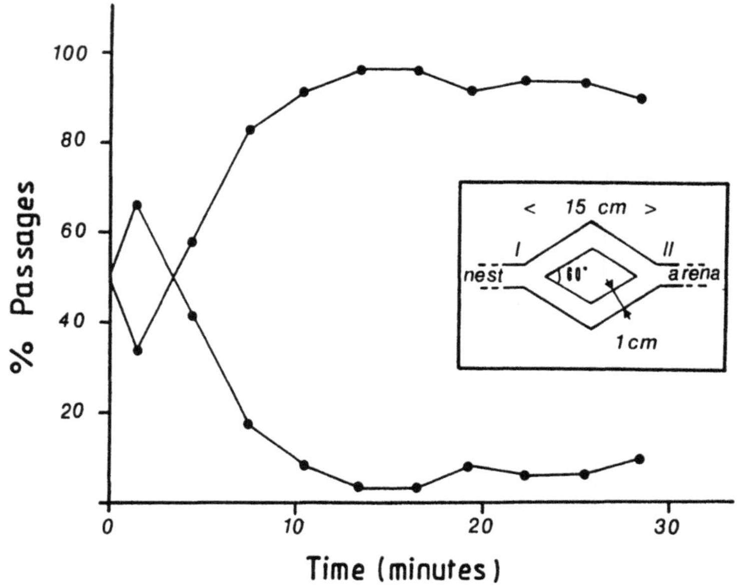

FIGURE 2. Percentage of ants per 3-minute period passing on the 2 branches of the bridge (insert). Colony of 1000 workers. (from Deneubourg et al. 1989).

terrain. Within seconds, stimulated workers leave the nest and begin trail-following. The occurrence of workers leaving the nest without any direct contact with recruiters suggests that the trail pheromone itself and/or the increased nest activity could be sufficient to induce recruitment.

Observation of explorers crossing the bridge between the nest and the new area showed trail-laying behavior both by those moving from and returning to the nest. The gaster was bent downwards and the ventral side dragged along the ground for varying lengths. The same behavior was reported by Van Vorhis Key and Baker (1986) during foraging recruitment. This strongly suggests that the same trail pheromone (Z-9-hexadecenal) is used during both food and exploratory recruitment. Z-9-hexadecenal was previously known to be a constituent of the ventral Pavan's gland secretion (Cavill et al. 1979, 1980) and its inducement of trail-following activity has been documented (Robertson et al. 1980; Van Vorhis Key and Baker, 1982a,b).

An analysis of video recordings suggests that workers crossing the bridge towards the nest lay significantly more pheromone than those moving in the opposite direction (Aron et al. 1989). They leave longer marks, with less distance between successive streaks. A theoretical study of the streak lengths' distribution showed that trail-laying approximates but is not a single-chain Markovian process.

GENERATION OF THE COLLECTIVE EXPLORATORY PATTERN

To study trail formation behind the exploratory front, the colonies were given access to a new sandy arena via a diamond-shaped bridge (Fig. 2, insert). This allowed a simple choice between one binary left/right path (in the arena, they have a much larger choice of directions). The dynamics of the ants' cumulative choice between both branches was then quantified. Initially, the flow of workers on both branches grows at a more or less equal rate. Soon, however, this symmetry is broken and one branch becomes more frequently used than the other (Fig. 2). The transition to an asymmetrical distribution can be considered the formation of the exploratory trail's latest segment at the trailing edge of the exploratory front.

Following an osmotropotactic trail-following model (Hantgartner, 1967; Pasteels et al. 1986), the workers' choice between the two branches may be a function of their relative pheromone concentrations. The more concentrated the pheromone on a branch, the more attractive for explorers, who also add to the pheromone concentration. Each passing ant modifies the following ant's probability of choosing left or right - positive feedback that, after initial fluctuation, rapidly leads to one of the two forks becoming completely preferred to the other.

This exploration on the bridge was modelled using two simplifications which do not influence the generated pattern (Deneubourg et al. 1989a). Firstly, it was assumed that the ants lay an equal quantity of pheromone when leaving or returning to the nest. Secondly, the pheromones' mean life-time may be estimated from previous data (Van Vorhis Key and Baker 1982a) as approximately 30 minutes. Its disappearance may therefore be ignored because plateau values are reached after 10-25 minutes in the experiments. This allows the quantity of pheromone laid on a branch (L and R) to be measured by the number of workers that have passed on that branch, independent of their direction.

Each ant at a choice point has a probability pl or pr of travelling along the left or the right branch, depending on the respective pheromone concentrations L and R, and adds one pheromone unit to the branch chosen. The unit for L and R is the average quantity of pheromone laid by an ant passing the choice point.

$$pl = \frac{(k + L)n}{(k + L)n + (k + R)n} \qquad (1)$$

with pl + pr = 1, and L + R = no. of previous ant crossings.

n is a coefficient of non-linearity: the higher n, the higher the probability of choosing the more concentrated branch, even if there is only a slight difference between both branches. k represents the degree of an unmarked branch's "attractiveness": the greater k, the greater the marking necessary for the choice to become significantly non-random. Figure 3 shows the results of 20 experiments on different colonies, with somewhat variable recruitment dynamics, as a function of the number of ant passages. The choice function (1) is fitted empirically by matching the experimental results with

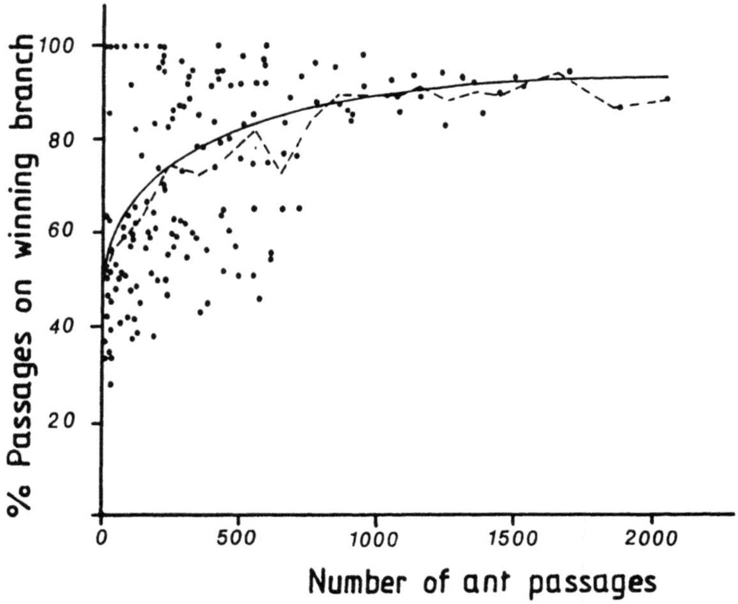

FIGURE 3. Percentage of ants passing on the collectively selected branch of the diamond-shaped bridge. 20 experiments, 9 colonies of 150-1200 workers, measured every 3 minutes for 30 minutes (dots). The horizontal axis represents the cumulative number of ant passages on the bridge (i.e. on both branches). The dashed curve represents the average values calculated every 100 ant passages. The solid curve represents the average of 200 Monte-Carlo simulations, based on eq. (1) with $n = 2$, $k = 20$. (from Deneubourg et al. 1989.)

Monte-Carlo simulations based on this model (Fig. 3, dashed curve), giving $n \approx 2$ and $k \approx 20$.

Two dimensional exploration was simulated by extending the diamond-shaped bridge model to an interconnecting network of binary choices (Fig. 1b, insert). Two further working hypotheses were made. Since trailing ants move faster than those at the front, the probability of moving at each step increases sigmoidally with the pheromone quantity at the points ahead of the ant. Secondly, the explorers leave the nest, moving forward and then returning, according to a simplified logistic-type approximation of the observed recruitment dynamics. At each point the ants move and mark according to equation (1).

The 2-D simulation illustrated in Figure 1b reproduces the two major characteristics of the ants' exploration of a new area, namely the exploratory front and the exploratory trail. This shows how interactions between individuals, whose trail-laying and following behavior are reduced to their simplest expression, can generate the complex exploratory pattern.

Territorial Marking

During collective exploration the workers' trail-laying behavior in the exploratory zone probably constitutes a form of territorial marking, allowing the workers to know whether or not they are in a previously explored zone. The following classical series of experiments tested the relationship between exploration and territorial marking. Two sheets of paper (5 x 5 cm) were placed close to the nest entrance. At least one week later, one of them was replaced by a virgin, chemically unmarked paper.

The number of explorers on the virgin paper was significantly higher than on the familiar one, ants moved onto it more frequently than onto the familiar paper and stayed longer. This corresponds to definite "coming-and-going" behavior exhibited when the ants come across the virgin paper, as seen at the exploratory front during a collective exploration.

After some hours the activity on the virgin paper and the "coming and going" behavior progressively decrease. *I. humilis* workers thus appear to distinguish new areas from those previously explored, most probably by means of chemical cues. Preliminary results show that these territorial marks are long lasting, as familiar papers do not elicit increased exploratory activity even after one month's isolation.

Exploratory and Food Recruitment Trails

The use of the same recruitment trail pheromone both during foraging and exploration suggests a strong correlation between the two processes. *I. humilis* societies were starved four days before inducing successively exploratory and alimentary recruitments in a $2m^2$ area. Once the exploratory pattern formed, a food source (1M sucrose) was introduced in the arena and the ants' movements followed for 20 min.

When placed about 20 cm from the exploratory trail, i.e. behind the exploratory front, the food was discovered by only a few foragers, and in many cases by none at all. Those that did find it generally left the food source and rejoined the exploratory trail, laying trail pheromone, but were unable to divert recruits to the food source from the much stronger exploratory trail.

When the food was placed 20 cm from the exploratory front, it was discovered more rapidly. Moreover, the discoverers rejoined the end of the exploratory trail, and directed exploratory recruits to the food source. The exploratory activity was replaced by exploitation of the food source and only a few scattered individuals continued exploration.

Although little is known about this behavior in nature, the number of ants participating in an exploration is probably larger than tested under laboratory conditions (about 600 ants/colony). In such a situation, the exploratory front would have to be large to discover and exploit any food present in the exploratory zone. Furthermore, there would have to be sufficient explorers to continue the collective exploration while also exploiting discovered food.

Short-Cuts

Unlike most ant species, which rely on a more highly developed individual capacity for orientation (review in Passera, 1984), the Argentine

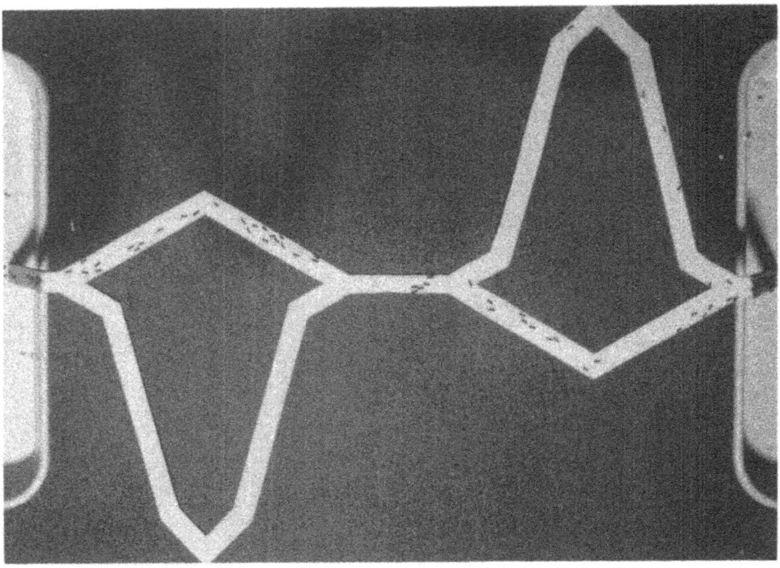

FIGURE 4. A colony of *I. humilis* selecting the short branches on both modules of the bridge (photo taken 8 minutes after the bridge was placed).

ant's navigational information is characterized by strong collective communication: workers are integrated into a network of chemical trails which regulates their external activities. Can this collective process respond efficiently to environmental cues? Beckers et al. (1988) and Deneubourg et al.(submitted) have shown that an *I. humilis* colony (and other species) can indirectly select the richer of two food sources through mass recruitment. This study will show now see how *I. humilis*' permanent trail laying behavior allows its foragers to select collectively the shortest route between nest and food source (Goss et al. submitted).

Argentine ant colonies were given access to a food source (1M sucrose) via a bridge of two identical modules, each consisting of two branches of different lengths (Fig. 4). To minimize external factors, each module's short branch was placed on opposite sides of the bridge. Furthermore, a number of experiments were performed in the dark (under red light), i.e. with no visual cues whatsoever.

The first explorers cross the bridge and discover the food source. As the ensuing recruitment proceeds, bridge traffic increases, with workers travelling equally on both branches of the modules. However, just as described above during exploratory recruitment, one branch becomes progressively more exploited than the other during any given experiment. Which branch was chosen depended on how widely the ratio (r) of their lengths varied (r is the ratio of long branch to short branch length).

When the two branches were of equal length (r=1), one or the other branch of the bridge was chosen with about equal probability (12 times the left branch out of 26 tests). In three series of experiments, in which the two branches were of different lengths, more than 80% of the foragers in more than 80% of the experiments selected the short branch (r=2 and r=1.4, 14 and 18 tests respectively) as illustrated in Figure 4, even under red light (r=2, 14 tests). This last series shows that visual cues do not influence the ants' choice of the shorter branch.

Nevertheless, if the short branch is added only after a long branch trail is well-established, the trail's strength is such that in more than 80% of the tests the colony does not switch to the shorter branch (r=2).

The ants' selection of the shorter branch may again be modelled according to choice equation (1) (see model of exploration), taking into account the difference between the times the ants took to cross the bridge via the two different branches.

The workers which chose the short branch crossed it in approximately 20 seconds. Having laid pheromone at both ends of the bridge, they influence the choice of other workers going in both directions. Those that choose the long branch take 40 seconds (with r=2) to cross it. Thus while they immediately influence the ants following them, they only influence the ants crossing in the other direction 20 seconds later, i.e. 10 seconds more than an ant that chose the short bridge. Therefore, this slight difference initially favors the short branch, and is accumulated and amplified by the autocatalytic nature of the "choose and mark" process. The short branch soon becomes more or less completely preferred to the other.

Consider: one ant arrives at each of the two choice points (j=1, 2) at each second. It chooses the short or the long branch with probability $p_{s,j}$ or $p_{l,j}$ depending on the respective quantities S_j and L_j of pheromone on the two branches. Having chosen, it adds one pheromone unit at the entry to the chosen branch. Twenty or 20r sec. later (depending on the branch chosen), it adds one pheromone unit to the exit point. The average equations are:

$$dSj/dt = P_{Sj}(t) + P_{S,j'}(t-20) \qquad (2)$$

$$j=1, j'=2 \text{ or } j=2, j'=1$$

$$dL_j/dt = P_{L,j}(t) + P_{L,j'}(t-20r) \qquad (3)$$

$$p_{s,j} = \frac{(20 + S_j)^2}{(20 + S_j)^2 + (20 + L_j)^2} \qquad (4)$$

Monte Carlo simulations of equations (2) to (4) give us two main results (Goss et al. submitted):

- the larger the difference between the short and the long branch, the greater the colony's probability of choosing the shorter branch;
- when the short branch is presented after the trail on the long branch has been established, the colony cannot switch to the short branch.

The model's results confirm that selection of the short branch is a collective decision, resulting from simple trail-laying and trail-following behavior. The model also predicts that another species whose workers only mark when returning to the nest (as in most ants species) would not select the shorter branch any more often than the long branch. In fact, the greater the

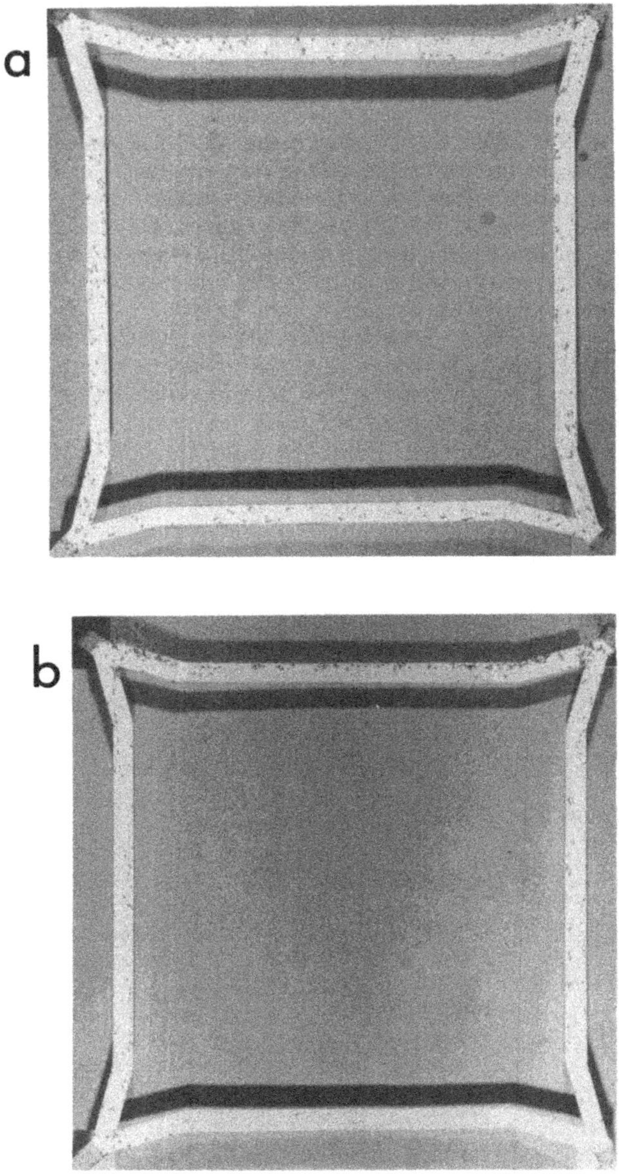

FIGURE 5. Four *I. humilis* colonies linked by four bridges arranged in a square setting up a network of communication (photos respectively taken 1hr (5a) and 10 days (5b) after the bridges were placed). After a period in which all branches were used equally, only three branches were used in the long-term.

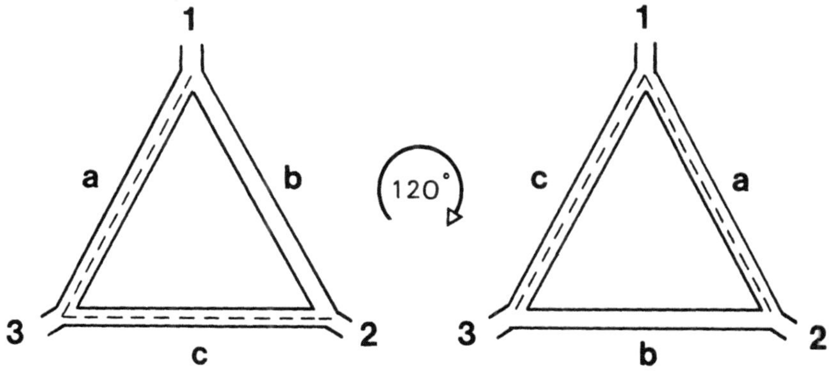

FIGURE 6. Schematic network of communication set up by three *I. humilis* colonies linked by 3 bridges arranged in an equilateral triangle (solid-lines). After rotation of 120° of the bridges, the same branches continue to be used (dashed-lines), now linking the colonies in a different way.

quantitative difference between marking by workers moving to and from the nest, the lower the probability that the colony will select the short branch.

Network

In most habitats colonized by the Argentine ant, nests are more or less evenly distributed. Each colony issued from springtime divisional migrations, nevertheless, remains connected to the others by a network of permanent trails. The workers, and more rarely some queens, move freely along these trails from one nest to the other. This behavior is also observed in the laboratory.

I. humilis colonies of about twenty to thirty thousand individuals were isolated from a common stock, each fed and watered extensively. They were connected by cardboard bridges. Two experimental devices were used: an equilateral triangle and a square with a colony at each apex. The ants' traffic on each branch of the bridge was measured every two days during the third week.

Initially, all branches of the bridge were frequented more or less symmetrically (Fig. 5a). With time, the exploration on the bridge decreases, and after two or three days an asymmetrical and permanent communication network appears (Fig. 5b). In all cases, the colonies remained inter-connected without using the most direct bilateral connections. In the triangular system, colonies 1 and 3 were linked via colony 2, and only two of the three branches used systematically. For the square, only three branches were used, colonies 1 and 4 connected via colonies 2 and 3 (Fig 5b).

When the experimenter broke one of the network's links, workers re-established a complete network by exploiting the previously less-frequented branch. Thus, a strong degree of plasticity allows the ants to maintain and connect all nests. Some preliminary experiments rotated the bridge but left the colonies as before. The ants, however, tended to maintain the traffic on the same physical branches as before, although these now linked different colonies (Fig. 6). These findings strongly suggest that chemical rather than visual cues organize branch traffic.

However, while the configuration generated by the model using equation (1) was the same for the triangular set-up, the one generated for the square without diagonals is different from that seen in Fig. 5b and consists of colonies 1-2 and 3-4 linked, with each pair isolated from the other. Thus, although chemical marking clearly plays an important role in network establishment, another unknown factor intervenes that allows linkage of all the colonies. In each experimental case, the configuration adopted was the shortest network that still allowed each colony to be linked to the others. While a number of different configurations are currently being tested, this idea is reinforced by one case where four colonies were connected with a square, complete with diagonals. The ants used both diagonals, but not the sides of the square.

DISCUSSION

Iridomyrmex humilis' exploratory behavior is peculiar in that it is a collective activity that proceeds according to a well structured characteristic pattern in which the ants keep permanent contact with each other through the medium of chemical trails.

The genesis of a massive and collective exploratory behavior corresponds to their highly opportunistic and agonistic behavior. The chemical trails allow workers to return directly to the nest, leading to rapid mobilization of a large number of individuals to freshly discovered food sources and/or new sites. Furthermore, it constitutes a massive and permanent defense against other competing ant species, while ensuring territorial expansion.

As *I. humilis* reproduces by colony fission (budding) rather than aerial dispersal of queens (Newell, 1909; Newell and Barber, 1913; Skaife, 1955), collective exploration is also the major process ensuring the species' dispersion in the environment. Upon discovery of new sources or new nest sites by explorers, Argentine ant colonies split into "sub-colonies" of hundreds or even thousands of workers, some queens and brood, establishing different colonies near different food sources, and remaining inter-connected by the equivalent of trunk trails (Skaife, 1955; Markin, 1968, 1970; Dechène, 1970; Hölldobler and Wilson, 1977).

This communication network not only allows the exchange of workers, food, brood, queens, etc., between the nests, but also could constitute a useful means of maintaining territorial control (see also Hölldobler and Wilson, 1977, Hölldobler and Lumsden, 1980, Skinner, 1980, and Rosengren, 1985). Clear trails criss-cross the territory, connect different nests and help orient and rapid mobilize foragers. Between the trails, occasional foragers maintain a minimum degree of territorial marking while searching for food. This represents an efficient way for a large colony to maintain a large territory - concentrating its exploitation on the most favorable sites and allowing a permanent and radial progression in all directions.

The probable use of the same pheromone (Z-9-hexadecenal) both in food and exploratory recruitment suggests that only one pheromone could be involved in the Argentine ants' spatial organization. Exploratory trails become, with the discovery of food, food recruitment trails. These can then be used for moving the nest closer to resources. With time, and with the evaporation/deactivation of the (Z)-9-hexadecenal, these trails no longer recruit but remain as guidelines for territorial marking, and traffic between the different nests. If confirmed, this system is both elegant and economical.

The collective patterns discussed above are clearly self-organizing, involving a large number of simple interacting individuals coordinated by trail pheromone. One might expect that a similar permanent chemical network would appear in other large trail-laying social insect colonies. *I. humilis*' exploratory behavior is also remarkably similar, on a smaller scale, to army ant-type exploratory/foraging "swarms" (Rettenmeyer, 1963; Topoff, 1972; Moffet, 1984, 1988; Franks and Fletcher, 1985), and the same basic model may be used for both (Deneubourg et al. 1989b). Highly similar exploratory patterns have been observed in the termites *Nasutitermes lujae* (Pasteels, 1965) and *Trinervitermes geminatus* (Rickli and Leuthold, 1986) as well as in the ants *I. pruinosus* and *Solenopsis* sp. (Hölldobler, pers. comm.).

The Argentine ant's life-style, in particular its high reproductive rate, adaptability and rapid colonization of new territory, has much in common with other well-known pest species such as *Solenopsis* sp., *Tapinoma melanocephalum* or *Monomorium pharaonis*.

ACKNOWLEDGMENTS

This work was supported in part by the Belgian IRSIA grant #00086, the Belgian program on inter-university attraction poles, and Les Instituts Internationaux de Physique et de Chimie.

REFERENCES

ARON, S., J.M. PASTEELS, and J.L. DENEUBOURG. 1989. Trail-laying behavior during exploratory recruitment in the Argentine ant, *Iridomyrmex humilis* (Mayr). Biol. Behav. in press.

BECKERS, R., J.L. DENEUBOURG and J.M. PASTEELS. 1988. Le recrutement de masse: un exemple de communication trop efficace. Actes Coll. Ins. Soc. (Paimpont 1987) 4: 219-226.

BENOIS, A. 1973. Incidence des facteurs écologiques sur le cycle annuel et l'activité saisonnière de la fourmi d'Argentine, *Iridomyrmex humilis* (Hym. Form.) dans la région d'Antibes. Ins Soc. 20: 267-296.

BOND, W. and P. SLINGSBY. 1984. Collapse of an ant-plant mutualism: the Argentine ant (*Iridomyrmex humilis*) and Myrmecochorous Proteaceae. Ecology 65: 1031-1037.

CAVILL, G.W.K., P.W. ROBERTSON, and N.W. DAVIES. 1979. An Argentine ant aggregation factor. Experientia 35: 989-990.

CAVILL, G.W.K., N.W. DAVIES and F.J. MC DONALD. 1980. Characterization of aggregation factors and associated compounds from the Argentine ant *Iridomyrmex humilis*. J. Chem. Ecol. 6: 371-384.

CROWELL, K.L. 1968. Rates of competitive exclusion by the Argentine ant in Bermuda. Ecology 49: 551-555.

DE BACH, P., C.A. FLESCHNER and E.J. DIETRICK. 1951. A biological check method for evaluating the effectiveness of entomophagous insects. J. Econ. Entomol. 44: 763-766.

DECHÈNE, R. 1970. Studies of some behavioral patterns of *Iridomyrmex humilis* Mayr (Formicidae, Dolichoderinae). The Wasmann J. Biology 28: 175-184.

DENEUBOURG, J.L., S. ARON, S. GOSS and J.M. PASTEELS. 1989a. The self-organizing exploratory pattern of the Argentine ant. J. Ins. Behav. in press.

DENEUBOURG, J.L., S. GOSS, N. FRANKS and J.M. PASTEELS. 1989b. The blind leading the blind: modelling chemically mediated army ant raiding patterns. J. Ins. Behav. in press.

DENEUBOURG, J.L, R. BECKERS, S. GOSS and J.M. PASTEELS. Submitted. Collective decision making through food recruitment.

DREISTADT, S.H., K.S. HAGEN. and D.L. DAHLSTEN. 1986. Predation by *Iridomyrmex humilis* (Hym: Formicidae) on eggs of *Chrysoperla carnea* (Nevichrysopidae) released for inundative control of *Illinoia liriodendri* (Hom: Aphididae) infesting *Liriodendron tulipifera*. Entomophaga 31: 397-400.

ERICKSON, M.J. 1971. The displacement of native ant species by the introduced Argentine ant *Iridomyrmex humilis* Mayr. Psyche 257-266.

FLUKER, S.S. and J.W. BEARDSLEY. 1970. Sympatric associations of three ants: *Iridomyrmex humilis*, *Pheidole megacephala* and *Anaplolepis longipes* in Hawaii. Ann. Ent. Soc. Amer. 63: 1290-1296.

FRANKS, N. and C.R. FLETCHER. 1983. Spatial patterns in army ant foraging and migration: *Eciton burchelli* on Barro Colorado Island, Panama. Behav. Ecol. Sociobiol. 12: 261-270.

GOSS, S., S. ARON, J.L. DENEUBOURG and J.M. PASTEELS. Submitted. Self-organised short-cuts in the Argentine ant.

HANEY, P.B., R.F. LUCK and D.S. MORENO. 1987. Increases in densities of the citrus red mite, *Panonychus citri* (Acarina: Tetranychidae) in association with the Argentine ant, *Iridomyrmex humilis* (Hymenoptera: Formicidae), in Southern California citrus. Entomophaga 32: 49-57.

HANTGARTNER, W. 1967. Spezifität und inaktivierung des spurpheromon von *Lasius fuliginosus* Latr. und orientierung des arbeiterinner im duftfeld. Z. Vergl. Physiol. 57: 103-136.

HÖLLDOBLER, B. and E.O. WILSON. 1977. The number of queens: an important trait in ant evolution. Naturwissenschaften 68: 8-15.

HÖLLDOBLER, B. and C.J. LUMSDEN. 1980. Territorial strategies in ants. Science 210: 732-739.

MARKIN, G.P. 1968. Nest relationship of the Argentine ant, *Iridomyrmex humilis* (Hymenoptera: Formicidae). J. Kans. Entomol. Soc. 41: 511-516.

MARKIN, G.P. 1970. The seasonal life cycle of the Argentine ant, *Iridomyrmex humilis* (Hymenoptera: Formicidae), in Southern California. Ann. Ent. Soc. Amer. 63: 1238-1242.

MOFFET, M.W. 1984. Swarm raiding in a myrmicine ant. Naturwissenschaften 71: 588-590.

MOFFET, M.W. 1988. Foraging dynamics in the group-hunting myrmicine ant, *Pheidologeton diversus*. J. Ins. Behav. 1: 309-331.

NEWELL, W. 1908. Notes on the habits of the Argentine or "New Orleans" ant, *Iridomyrmex humilis*. J. Econ. Entomol. 1: 21-34.

NEWELL, W. 1909. The life history of the Argentine ant *Iridomyrmex humilis*. J. Econ. Entomol. 2: 174-192.

NEWELL, W. and T.C. BARBER. 1913. The Argentine ant. U.S.D.A. Bureau of Entomology 122: 98.

PASTEELS, J.M. 1965. Polyéthisme chez les ouvrières de *Nasutitermes lujae* (Termitidae Isoptères). Biol. Gabon. 1: 191-205.

PASTEELS, J.M., J.L. DENEUBOURG, J.C. VERHAEGHE, J.L. BOEVE and Y. QUINET. 1986. Orientation along terrestrial trails by ants. In Payne T.L., Birch M.C. and Kennedy C.E. (eds.), Mechanisms in Insect Olfaction. Oxford Univ. Press, 131-137.

PASSERA L. 1984. L'organisation sociale des fourmis. Privat (ed.), Toulouse, 360 pp.

RETTENMEYER, C.W. 1963. Behavioral studies of army ants. Univ. Kans. Sci. Bull. 44: 281-465.

RICKLI, M. and LR.H. EUTHOLD. 1986. Spatial organisation during exploration and foraging in the Harvester termite, *Trinervitermes geminatus*. Rev. Suisse Zool. 94: 545-551.

ROBERTSON, P.L., M.L. DUDZINSKI and C.J. ORTON. 1980. Exocrine gland involvement in trailing behaviour in the Argentine ant (Formicidae: Dolochoderinae). Anim. Behav. 28: 1255-1273.

ROSENGREN, R. 1985. Internest relations in polydomous *Formica* colonies (Hymenoptera, Formicidae). Mitt. dtsch. Ges. allg. angew. Ent. 4: 288-291.

SAMWAYS, M.J., N. MAGDA and A.J. PRINS. 1982. Ants (Hymenoptera: Formicidae) foraging in citrus trees and attending honeydew-producing homoptera. Phytophylaciica 14: 155-157.

SKAIFE, S.H. 1955. The Argentine ant, *Iridomyrmex humilis* (Mayr). Trans. Roy. Soc. South Africa 34: 355-377.

SKINNER, G.J. 1980. Territory, trail-structure and activity patterns in the wood-ant, *Formica rufa* (Hymenoptera: Formicidae), in limestone woodland in North-West England. J. Anim. Ecol. 49: 384-394.

TOPOFF, H.R. 1972. The social behavior of army ants. In Eisner, T. and Wilson, E.O. (eds.), The Insects. Scient. Amer., Freeman, San Francisco, pp. 247-262.

VAN VORHIS KEY, S.E. and T.C. BAKER. 1982a. Trail-following responses of the Argentine ant, *Iridomyrmex humilis* (Mayr), to a synthetic trail pheromone component and analogs. J. Chem. Ecol. 8 (1): 3-14.

VAN VORHIS KEY, S.E. and T.C. BAKER. 1982b. Specificity of laboratory trail following by the Argentine ant, *Iridomyrmex humilis* (Mayr), to (Z)-9-hexadecenal, analogs and gaster extract. J. Chem. Ecol. 8 (7): 1057-1063.

VAN VORHIS KEY, S.E. and T.C. BAKER. 1986. Observations on the trail deposition and recruitment behaviors of the Argentine ant, *Iridomyrmex humilis* (Hymenoptera: Formicidae). Ann. Ent. Soc. Amer. 79: 283-288.

WARD, P.S. 1987. Distribution of the introduced Argentine ant (*Iridomyrmex humilis*) in natural habitats of the lower Sacramento Valley and its effects on the indigenous ant fauna. Hilgardia 55: 1-16.

41

Foraging of Pharaoh Ants *Monomorium pharaonis* (L.) (Hymenoptera: Formicidae) in the Laboratory

K.D. Haack and S.B. Vinson

INTRODUCTION

Many studies have been done on the recruitment and foraging behavior of ants; it is the subject of at least two reviews (Wilson 1971; Carroll and Janzen 1973). Sudd (1957) discusses the recruitment activities of pharaoh ant, *Monomorium pharaonis* (L.), workers in small artificial colonies containing three to nine workers, one queen and a small amount of brood. Later, Sudd (1960) discusses the recruitment and foraging behavior of a large natural population of pharaoh ants in Nigeria. These two studies, however, do not compare the recruitment and foraging behavior of pharaoh ant workers to the major food groups: carbohydrate, protein and lipid. The purpose of this study was to compare the recruitment and foraging activities of pharaoh ant workers from medium-sized standardized colonies in response to three different representative food baits.

METHODS AND MATERIALS

Experimental Ant Colonies

Standardized ant colonies used in the experiments were composed of 250 workers, two dealate queens and 25-30 mg larvae. One-half the worker force was selected from outside the nest and one-half from inside to insure the presence of both foragers and nurses (Mirenda and Vinson 1981). Callow workers and young queens were avoided. Larvae of all sizes (ages) were present in the weighed samples. Details of the methods of manipulation are discussed by Haack (1987). These colonies were maintained in the laboratory under constant illuminated conditions with the temperature maintained at 27°C (80°F). The relatively humidity was not controlled and fluctuated between 45 and 55 percent.

Experimental Nest and Foraging Arena

The experimental nest consisted of a 35 by 10 mm clear plastic petri dish (Falcon™, Dish No. 1008; Oxnard, California U.S.A.). The inner and outer walls of the nest were coated with Fluon® AD1 (Northern Products,

Inc., Woonsocket, RI U.S.A.) to restrict ant movement to the entrance only. The entrance was a two mm diameter hole located in the side of the dish at floor level. The nest was covered by its transparent lid. One experimental nest was placed in a corner of each foraging arena, with the entrance facing toward the far end.

The foraging arenas were clear plastic shoe boxes, approximately 31 by 17 by 7 cm (Tucker Housewares, Inc., Box No. 176; Arlington, TX U.S.A.). The inside walls of these boxes were coated with Fluon AD1 to prevent ant escape.

Distilled water was continually provided in a one oz glass container with a plastic snap-on lid containing a one cm dia. by three cm long cotton dental roll. This cotton wick was placed through a hole punched in the plastic lid. One watering container was located opposite the nest on the same end of the foraging arena.

Pretreatment

Standardized experimental colonies were allowed a two-day adjustment period prior to testing. During this time, mortality and behavior of colonies were monitored. Dead workers, queens, larvae and pupae were removed and replaced with living individuals. Colonies with excessive mortality, colonies showing apparent aggression between workers, or workers and queens and colonies that were reluctant to forage were replaced. All colonies, during the two-day adjustment period, were provided insect food (one *Heliothis virescens* (F.) pupa divided with a knife to expose the contents) and 0.5 ml of a 0.1 M sucrose solution to ensure uniformly well-fed colonies. After this two-day adjustment period, all unconsumed food and dry food stored in the nest was removed.

Preliminary tests showed that a greater level of starvation (hunger) was required to entice foraging pharaoh ant workers to gather and distribute oil as opposed to a carbohydrate solution or moist particulate protein. Subsequently, standardized well-fed experimental colonies were subjected to one of two different starvation periods. Colonies for carbohydrate and protein tests were starved for two days; while colonies tested for their response to oil were starved for seven days after the two-day adjustment period.

Timed Observations

Recruitment and trophallactic habits of pharaoh ant workers were recorded after bait introduction. Observations were taken every 10 minutes for the first hour, beginning with the initial forager-food bait encounter; afterwards observations were taken at two, three, six, twelve and twenty-four hours. A stereomicroscope was positioned over the top of the nest to aid in making observations of recruitment and trophallaxis within the nest. Observations of the various foraging activities were aided by recording the behavior of workers for one minute intervals within five 8 mm diameter areas marked on the underside of each nest. These areas were distributed similarly to the spots on a die and were clearly visible from above.

Food Baits

Three standard food baits were used in these studies. The carbohydrate was formulated as a 1 M sucrose solution with 1:50 (vol.:vol.) red food dye (McCormick & Co., Inc.; Baltimore, Maryland U.S.A.) to aid our observations of trophallaxis. The protein consisted of a 1:1 mixture by weight of dry egg yolk (Sigma Chemical Co. Product No. E-0625; Saint Louis, Missouri U.S.A.) and 1:25 (vol.:vol.) McCormick red food dye in water. The oil consisted of peanut oil saturated with sudan red. All baits were made fresh daily.

During experiments, one-quarter gram of the carbohydrate or protein bait was presented to the ant colony on a 25 by 25 mm coverslip placed in the arena 20 cm away from the entrance to the nest. The oil bait, however, was placed on top of and allowed to soak into a size 00 cork stopper. Only enough oil to dampen the outer layer of cork was applied. The cork stopper was placed on a coverslip 20 cm from the nest entrance. Other methods of oil presentation were found to be inadequate. The sucrose bait dried out and the oil eventually soaked deeply into the cork. Thus, after the first hour, the old oil and sucrose baits, and coverslips were removed and replaced. Subsequently, fresh oil and sucrose solution were provided 30 minutes prior to each observation period. The particulate protein bait was not changed, since drying did not appear to deter forager recruitment.

Recruitment Behavior

The number of workers feeding over time on each bait and the number of workers in the observation areas of the nest were counted to provide a relative representation of recruitment behavior. Specifically, low worker numbers inside meant high worker numbers outside or greater recruitment activity. Means and standard deviations of the means were determined for observations within each time interval for each bait type.

Trophallactic Behavior

Trophallactic food exchanges (liquid food exchanges) between workers were observed for each of five colony replicates presented with each bait type. The number of workers ants sharing a droplet of regurgitated liquid (size of the trophallactic group) and the number of trophallactic groups of each size were pooled for each observation time and bait type. The number of nurse ants (workers within the nest) engaged in trophallaxis for each observation period was recorded. The results were pooled and the percent of workers engaged in trophallaxis was determined. Only ants within the prescribed nest observation areas were recorded.

RESULTS AND DISCUSSION

Recruitment and trophallactic behavior of pharaoh ant workers in response to a sucrose solution, an oil and a solid (particulate) protein bait were observed in standardized laboratory colonies. The number of workers that remained inside the nest in response to the three food baits differed with fewer ants remaining in the nest in response to proteins (Table 1). Consequently, the heaviest feeding activity was recorded on the protein bait

TABLE 1. Recruitment Behavior: Mean number of pharaoh ant workers observed inside the nest of standardized colonies presented with one of three different food baits.[a]

	Food Bait Type		
Time	Carbohydrate	Protein	Oil
0 min	80 (15)	73 (7)	64 (21)
10 min	85 (20)	57 (10)	64 (21)
20 min	84 (20)	46 (4)	54 (10)
30 min	68 (17)	50 (9)	58 (10)
40 min	67 (13)	45 (7)	61 (7)
50 min	61 (5)	51 (3)	59 (7)
1 hr	64 (4)	50 (6)	66 (5)
2 hr	76 (9)	46 (11)	72 (13)
3 hr	79 (9)	49 (7)	67 (8)
6 hr	86 (8)	60 (10)	79 (8)
12 hr	83 (7)	59 (8)	80 (14)
24 hr	80 (16)	68 (10)	81 (22)

[a]Mean values based on five colony replicates. Numbers in parentheses are the standard deviations of the means.

(Table 2). Trophallactic activity was common within the nest in response to the sucrose solution and oil, but was less common for the protein (Table 3, 4, and 5). However, with all three food baits, trophallaxis rarely occurred outside the nest (data not shown).

Recruitment Behavior

Worker recruitment to the three food baits is summarized in Table 1. Worker recruitment to the baits was variable. The number of workers observed in the nest was an indirect measure of the relative numbers of workers foraging (Table 1). The heaviest worker recruitment of all three baits occurred during the first hour. The greatest and least number of workers remaining in the nest occurred in response to the carbohydrate and protein baits, respectively (Table 1). These results were negatively proportional with the number of workers feeding on baits (Table 2). Worker feeding activity was also used as a measure of recruitment activity (Table 2). The heaviest and lightest feeding or food gathering activity by workers occurred in response to the protein and carbohydrate baits, respectively. Although initial recruitment to the oil bait was relatively slow (Table 2), subsequent recruitment was as high as with protein. However, in order to obtain worker recruitment, colonies tested for their response to oil initially required a longer starvation period. Therefore these responses could be, in part, due to the effect of increased starvation rather than food type. However, the apparent storage of oils within the bodies of workers, in contrast to proteins (stored within the nest) and carbohydrates (rapidly passed through the colony) (Haack and Vinson, in prep.) indicate that oils are more slowly turned over within

TABLE 2. Recruitment Behavior: Mean number of pharaoh ant workers feeding in standardized colonies presented with one of three different food baits.[a]

	Food Bait Type		
Time	Carbohydrate	Protein	Oil
0 min	0 (0)	0 (0)	0 (0)
10 min	32 (11)	*	0 (0)
20 min	20 (9)	*	*
30 min	17 (3)	*	*
40 min	12 (7)	*	*
50 min	8 (4)	*	*
1 hr	3 (1)	*	*
2 hr	0 (1)	*	*
3 hr	1 (1)	*	11[c]
6 hr	0 (0)	*	2 (2)[b]
12 hr	0 (1)	*	3[c]
24 hr	1 (2)	18 (6)[b]	1[c]

[a]Data are based on five colony replicates were used unless otherwise indicated. Numbers in parentheses are the standard deviations of the means. * = 50 or more workers feeding. [b]Data from two colony replicates. [c]Data from one colony replicates.

TABLE 3. Trophallactic behavior of pharaoh ant workers in standardized colonies presented with a liquid carbohydrate bait.[a]

	Pooled No. of workers engaged in trophallactic groups						% workers engaged in trophallaxis
	Group Sizes						
Time	2	3	4	5	6	7	
0 min	5	0	0	0	0	0	8
10 min	31	21	12	16	5	2	71
20 min	61	35	11	3	0	0	69
30 min	62	16	4	2	1	0	62
40 min	44	6	1	1	0	0	37
50 min	20	3	1	1	0	0	19
1 hr	22	3	0	0	0	0	17
2 hr	3	0	0	0	0	0	1
3 hr	5	0	0	0	0	0	2
6 hr	0	0	0	0	0	0	0
12 hr	0	0	0	0	0	0	0
24 hr	4	0	0	0	0	0	2

[a]Data are based on five colony replicates.

the colony. Feeding and food gathering activities of workers slowed after 50 minutes and continued to decrease.

The differences in recruitment activities may not only be due to differences in the consistency (solid vs. liquid) and viscosity (water based vs. oil base liquid) of the foods, but the way the foods were handled. The particulate protein bait appeared to require more time for workers to remove than the liquid baits. The ants appeared to work more to chew off a chunk of protein to carry back in contrast to soaking up liquids. The solid nature of the protein bait also elicited tunneling activities by some workers. This caused even longer visits at the protein bait. Food gathering of individual workers on the oil appeared to be relatively long in duration, much longer than at the sucrose solution. Therefore we assume workers have more difficulty consuming oils than water based fluids. Differences in recruitment to baits may also in part be due to individual preference of workers. Erpenbeck (1976) stated that there appears to be some pharaoh ant foragers which are carbohydrate specialists and other foragers that specialize or prefer proteins.

Observations on the initiation of recruitment behavior by returning pharaoh ant foragers agrees with the findings of Sudd (1957). Specifically, the great majority of workers within a nest are quiet at any given time. A returning forager, which has discovered food, entices others to follow by excitedly antennating and bumping into them. The result is an excitement of the colony with some 10 percent or more of the workers leaving the nest (Haack and Granovsky, unpublished).

Trophallactic Behavior

The trophallactic behavior of pharaoh ant workers in the presence of the sucrose solution, the solid (particulate) protein, and the oil is summarized in Table 3. Trophallaxis occurred in all colonies and in response to all foods. The greatest percentage of workers engaged in trophallaxis and the largest groups of workers sharing a droplet of food occurred in colonies presented with the sucrose solution (Table 3). The high degree of trophallactic activity in the presence of the liquid carbohydrate directly effected the duration of foraging activities; satiation of workers was rapid and the desire to forage soon ceased. Pharaoh ant foragers, like fire ant foragers, collect and distribute liquid carbohydrates rapidly (Howard and Tschinkel 1980; Sorensen et al. 1981; Sorensen and Vinson 1981; Sorensen et al. 1985). This may be due to a rapid replenishment of the high energy requirements of workers.

The lowest percentage of workers engaged in trophallaxis and the smallest groups of workers sharing food occurred in colonies presented with the solid (particulate) protein bait (Table 4). Solid food was not observed to be transferred but a clear liquid, most likely water used to moisten the protein, was shared.

The trophallactic behavior of pharaoh ants differed from that of fire ants, *S. invicta*. Fire ant foragers relay food items to an intermediate caste called repletes (Mirenda and Vinson 1981; Sorensen et al. 1981), located at the nest periphery, and rarely bring a food item into the nest directly. Pharaoh ant foragers, however, rarely exchanged or relayed food outside the nest and usually brought food items directly into the nest. This finding also agrees with the observations of Sudd (1957).

TABLE 4. Trophallactic behavior of pharaoh ant workers in standardized colonies presented with a solid (particulate) protein bait.[a]

Time	Pooled number of workers engaged in trophallactic groups: group sizes 2	3	4	5	6	Percent workers engaged in trophallaxis
0 min	5	0	0	0	1	4
10 min	3	0	1	0	0	4
20 min	2	0	0	0	0	2
30 min	2	0	0	0	0	1
40 min	1	0	0	0	0	1
50 min	1	0	0	0	0	1
1 hr	0	0	0	0	0	0
2 hr	4	0	0	0	0	4
3 hr	6	2	0	0	0	8
6 hr	11	0	0	0	0	7
12 hr	4	0	0	0	0	3
24 hr	7	0	0	0	0	4

[a]See Table 3 for explanation.

TABLE 5. Trophallactic behavior of pharaoh ant workers in standardized colonies presented with an oil bait.[a]

Time	Pooled number of workers engaged in trophallactic groups: group sizes 2	3	4	5	Percent workers engaged in trophallaxis
0 min	2	0	0	0	1
10 min	12	10	8	0	31
20 min	26	16	3	2	45
30 min	38	16	2	1	43
40 min	37	11	1	2	46
50 min	29	12	1	2	37
1 hr	33	15	3	1	42
2 hr	20	4	0	0	11
3 hr	7	3	0	0	9
6 hr	3	0	0	0	1
12 hr	2	0	0	0	1
24 hr	2	0	0	0	1

[a]See Table 3 for explanation.

CONCLUSIONS

The results of these studies indicate that protein food baits, in comparison to carbohydrates and oils, may provide superior results in the control of the pharaoh ant. The carbohydrate solution led to rapid recruitment and food distribution, but also rapid satiation. Therefore, the amount of carbohydrate bait entering a colony at any one time may be restricted. Workers, in contrast, were slow to recruit to oil and may only do so when their oil reserves are depleted. Movement of oil into starved colonies started slowly but increases rapidly. Within 20 minutes after oil exposure, trophallaxis in groups of up to five workers was observed (Table 5), and this trophallaxis activity lasts longer than the response to the carbohydrate solution (Table 3). Trophallaxis is lower with protein (Table 4), but the recruitment and sustained foraging (Table 1) indicate that the protein baits may provide a more consistent means of introducing toxicants into the pharaoh ant colony.

ACKNOWLEDGEMENTS

The authors would like to thank Drs. Awinash Bhatkar, Theodore A. Granovsky, and Les Greenberg for their reviews of this manuscript. Additional thanks to Pam Rowe, On the Double, for typing the various revisions of this manuscript. These studies were performed in the Entomology Research Laboratories, Texas A&M University, College Station, Texas U.S.A. The manuscript was approved by the Director of the Texas Agricultural Experiment Station as PA-24791.

REFERENCES CITED

CARROLL, C. R., and D. H. JANZEN. 1973. Ecology of foraging by ants. Ann. Rev. Ecol. System. 4:231-258.

ERPENBECK, A. 1976. Traceruntersuchung en zur Nahrungsverteilung in Sozietäten der Pharaoameise, *Monomorium pharaonis* (L.) (Hymenoptera, Formicidae). Ph.D. Dissertation, Bonn Univ., Bonn, Germany, 70 pp.

HAACK, K. D. 1987. Aspects of the food handling behavior of the pharaoh ant, *Monomorium pharaonis* (L.) (Hymenoptera, Formicidae). M.S. Thesis, Texas A&M Univ., College Station, Texas. 125 pp.

HOWARD, D. F., and W. R. TSCHINKEL. 1980. The effect of colony size and starvation on food flow in the fire ant, *Solenopsis invicta* (Hymenoptera, Formicidae). Behav. Ecol. Sociobiol. 7:293-300.

MIRENDA, J. T., and S. B. VINSON. 1981. Division of labor and specification of castes in the red imported fire ant, *Solenopsis invicta* Buren. Animal Behav. 29:410-420.

SORENSEN, A. A., T. M. BUSCH, and S. B. VINSON. 1985. Trophallaxis by temporal subcastes in the fire ant, *Solenopsis invicta*, in response to honey. Physiol. Entomol. 10:105-112.

SORENSEN, A. A., J. T. MIRENDA, and S. B. VINSON. 1981. Food exchange and distribution by three functional worker groups of the imported fire ant, *Solenopsis invicta* Buren. Insectes Soc. 28:383-394.

SORENSEN, A. A., and S. B. VINSON. 1981. Quantitative food distribution studies within laboratory colonies of the imported fire ant, *Solenopsis invicta* Buren. Insectes Soc. 28:129-160.

SUDD, J. H. 1957. Communication and recruitment in pharaoh's ant, *Monomorium pharaonis* (L.). Brit. J. Animal Behav. 3:104-109.

SUDD, J. H. 1960. The foraging method of the pharaoh's ant, *Monomorium pharaonis* (L.). Animal Behav. 8:67-75.

WILSON, E. O. 1971. The insect societies. Belknap Press, Cambridge, Mass. 548 pp.

42

Some Findings on Neurotoxins from the Venom of the Giant Ant, *Paraponera clavata*

C. Sevcik and C.J. Hernandez

INTRODUCTION

The ant *Paraponera clavata* is widely distributed in South America. The workers of this species are approximately 2.5 cm long and the queens are even larger. Ranging from dark brown to black, they possess a well-developed sting. This species, commonly found in rainy forests, is known to natives as "cumanagota," "la veinticuatro," "yolofa" or "yolosa" in Venezuela; "mumuri" (Weber, 1937) in Guiana; "tucandeira" (Weber, 1937) in Brazil and "hormiga bala" in Costa Rica (Weber, 1937). Despite its "ferocious" reputation, *P. clavata's* aggressivity is determined by food availability (Hermann and Blum, 1966; Bequaert, 1926). Although the ants live in groups of over a hundred individuals, they are usually found as solitary individuals and sting people who step on them with bare feet or light shoes. Human envenomation also occurs when ants fall from branches brushed against while walking in the forest. The powerful sting can even pierce heavy khaki cloth.

Indians of prehispanic times knew the toxicity of this ant. The indians of the Cumanagoto tribe (North East of Venezuela) (Martínez and Vela-Diaz, 1953) reportedly used the abdomens, and sometimes the heads, of wasp-sized black ants to prepare their paralyzing venom: the curare. Since only two ant species of this size populate the area, the species in question may well have been *P. clavata*. *P. clavata's* sting is extremely painful, the Venezuelan name, "la veinticuatro" (i.e. the 24) comes from the reputation that a single sting of this ant will produce the worst 24 hours of one's life. In seven cases of a single sting by *P. clavata*, the symptoms included excruciating pain that spread from the affected limb, difficulty to move voluntary muscles and, in one case, fibrillation of skeletal muscles (C. Sevcik, unpublished). Other victims interviewed described milder symptoms, Perhaps these stings were inflicted by ants often confused with *P. clavata*, or encounters with ants which had recently used their venom. *P. clavata* uses the venom for prey capture. With only one known exception (Weber, 1937) a consensus has been reached that local damage (necrosis or blisters) does not take place.

The common occurrence of muscle disorders in *P. clavata's* victims motivated an interest in the neurological effects of the ant's venom. This demonstrates that the venom possesses dose dependent neuromuscular effects which could explain these motor disorders.

MATERIALS AND METHODS

Preparation of the Crude Extract

Ants were collected from two localities in Venezuela, the indian village of La Culebra in the Federal Amazon Territory (3° 43' 55"N, 65° 48' 42"W) and Sucre State (10° 26' 57"N, 64° 10' 52"W) in the Turimiquire mountains. The ants were placed in liquid nitrogen immediately after capture, then transferred to a deep freezer at -70°C.

To isolate the venom gland, the ants were thawed at 0°C. Under a dissection microscope, the distal part of the abdomen, which houses the venom apparatus (segments VII to X ([2])), was removed. Freed from the chitinous cover, the venom apparatus was placed in distilled water at 0°C (40 apparatuses/25 ml) and homogenized by 7 strokes of a rotary teflon pestle (6500 rpm). The homogenate was vacuum-filtered through Whatman No. 1 filter paper and divided into 0.5 ml aliquotes (crude extract), before being stored at -70°C. When required for other purification steps, the crude extract was thawed at 4°C. All purification procedures were carried out at this temperature.

Gel Filtration Procedures

The crude extract was gel filtered on Sephadex G25 (Pharmacia, Sweden). Columns (100 cm long and 2 cm in diameter) were packed with gel, equilibrated with 1M acetic acid and kept at 4°C. Crude extract (500 μl aliquotes) was carefully layered on top of the columns and eluted at one to three ml/cm /hr. Absorbance of the eluate was monitored at 280 nm, the different peaks collected, lyophilized and resuspended in distilled water. Since fractions obtained in this manner are partially pure, concentrations used in the bioassays will be expressed as the number of ant glands per unit volume (agl/ml).

Experimental Solutions

Bioassays were carried out in frog saline of the following composition (in mM): Na^+, 118; K^+, 2.5; Ca^{+2}, 1.8; Cl^-, 121 and *N*-2-hydroxyethyl-piperazine-*N'*-2-ethanesulfonic acid, 6. Tetrodotoxin was purchased from Sigma Chemical Co., St. Louis, MO.

Electrophysiological Procedures

Fractions were bioassayed on the neuromuscular preparation of frog (*Hyla crepitans*) sartorius muscle. Standard microelectrode techniques (Sevcik and Barboza, 1983) recorded membrane potential. All electrophysiological recordings took place at room temperature (22°C). The endplate potentials were digitized with an analog to digital converter (Digital Equipment Corp. (DEC), model AD11KT) at a rate of 1 kHz and stored on disk for later processing.

Miniature end plate potentials (MEPPs) were detected automatically with the algorithm described in the Appendix. The algorithm aims to simultaneously detect MEPPs and determine the time course of each miniature, a function not achieved by noise analysis techniques (Fesce *et al., 1986*). To estimate amplitude and kinetics of MEPPs the following function was used

$$V_t = V_{max}.\alpha.t.\exp(1-\alpha.t) \qquad [1]$$

where

V_{max} = Amplitude of MEPP

t = Time
α = Kinetic constant.

Equation [1] was derived from the work of Jack and Redman (1971). V_{max} was determined from MEPP recordings, as described in the Appendix, and α was obtained by fitting equation [1] with a simplex algorithm (Nelder and Mead, 1965) (50 points at 1 kHz per MEPP) to each MEPP recorded. The absolute value of the deviations of the experimental points to the line were minimized. Alpha α is the time to the MEPP peak. The rate of release was determined by counting MEPPs for 10.24 to 30.72 sec in each endplate.

Statistical Procedures

All data was analyzed using nonparametric statistical procedures. Results are presented as the median and 95% confidence interval. Results were compared with the Mann-Whitney (Wilcoxon) test. Multiple comparisons were carried out with the Kruskall and Wallis test. The null hypotheses were rejected for confidence levels equal to or smaller than 0.05 ($P \leq 0.05$). See Hollander and Wolfe (1973) for details on all statistical procedures.

RESULTS

Effect of Crude Extract on Muscle Resting Potential

The effect of crude extract on resting membrane potential is shown in Figure 1. The resting potential prior to toxin application (0.4 agl/ml) was -87 (-92, -86) mV (n = 10) and was reduced to -36 (-51,-20) mV (n = 10)($P=2.10^{-5}$) after 50 minutes of toxin application. Toxin slowly reduces membrane potential, which is slowly and partially reversible.

Elution on Sephadex G25

The elution pattern of the crude extract on Sephadex G25 is shown in Figure 2. Biological activity was detected in the peak marked *ba*, will be known from now on as PcTX.

Effect of PcTX on Frog Sartorius Membrane Potential

The effect of PcTX (0.4 agl/ml) is shown in Figure 3. The toxin depolarized membrane potential from -89 (-94, -82) mV (n = 10) to -40 (-51, -32) mV (n = 10) ($P=10^{-5}$) when added for 20 min to the saline. The membrane potential recovered only partially when the muscle was washed with normal saline for 2 hr.

Tetrodotoxin (TTX) is a potent sodium channel blocker (Narahashi et

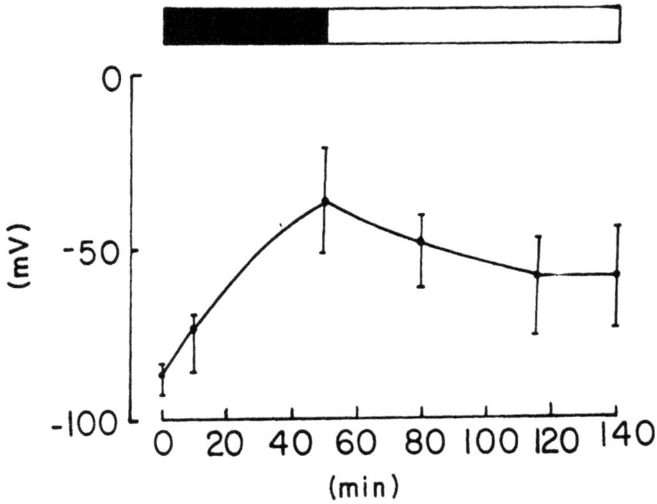

FIGURE 1. Effect of the crude extract of *P. Clavata's* venom gland on muscle membrane resting potential. Abscissa = membrane resting potential of sartorius muscles. Data presented as median (dots) and (95% confidence interval; N = 10). Leftmost point was recorded prior to toxin application. The black bar = period during which 0.4 agl/ml was applied, the open bar = application of saline without toxin.

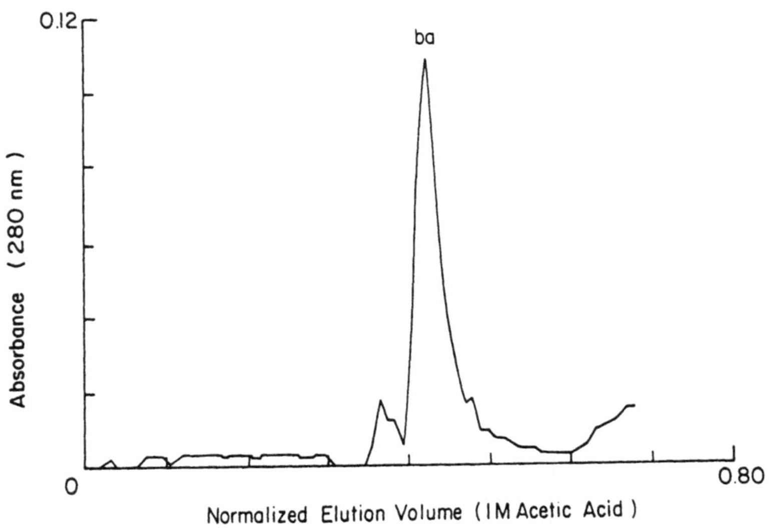

FIGURE 2. Paraponera venom gland extract elution pattern from Sephadex G25. Ordinate = absorbance of eluate at 280 nm. Abscissa = volume of 1M acetic acid eluted dividing by the total column volume. *ba* contains the biological activity. See text for details.

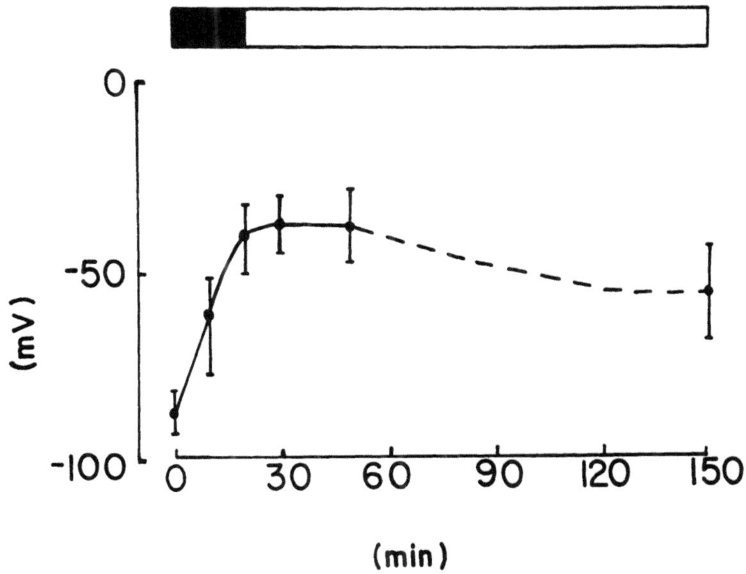

FIGURE 3. Effect of partially purified PcTX. The black bar = a period during which PcTX (0.4 agl/ml) was applied; open bar = application of saline without toxin. See Figure 1 and text for other details.

al. 1964). Figure 4 shows the results obtained in 4 muscles where the interaction of TTX and PcTX was explored. These findings demonstrate that TTX protects against PcTX only if TTX is applied first and its concentration maintained through PcTX application and through the following wash period ([2TX]). After TTX was washed out with normal saline and PcTX applied once more without TTX; this time the muscles were depolarized ($P < 10^{-5}$). The depolarizing effect of PcTX was not reversed by washing this toxin with saline containing 300 nM TTX (3TX)(P=0.788).

Surprisingly, the depolarizing effect decreases (and even disappears in some cases) when the toxin concentration is raised to 1.6 agl/ml. Figure 5 is a histogram of membrane potential in control condition, with 0.4 agl/ml ([0.4]) added to the saline and 1.6 agl/ml added to the saline ([1.6]). Although both concentrations depolarized the muscles significantly ($P<0.001$), the higher concentration depolarized the fibers to a lesser extent than the lower one ($P<5.10^{-4}$).

Effect of PcTX on Neuromuscular Transmission

However, the higher PcTX concentration (1.6 agl/ml) blocks neuromuscular transmission. The lower half of Figure 6 shows five overlapping traces showing endplate potentials (EPPs) evoked by nerve stimulation. The EPPs were recorded from a fiber on the inner surface of a frog sartorius muscle. The EPPs are reduced to <10 mV. Some twitch remained deep in the muscle where the abundance of connective tissue hinders the diffusion (Sevcik et al., 1986). Contractions within the muscle are responsible for the wavy appearance of the rightmost part of the traces.

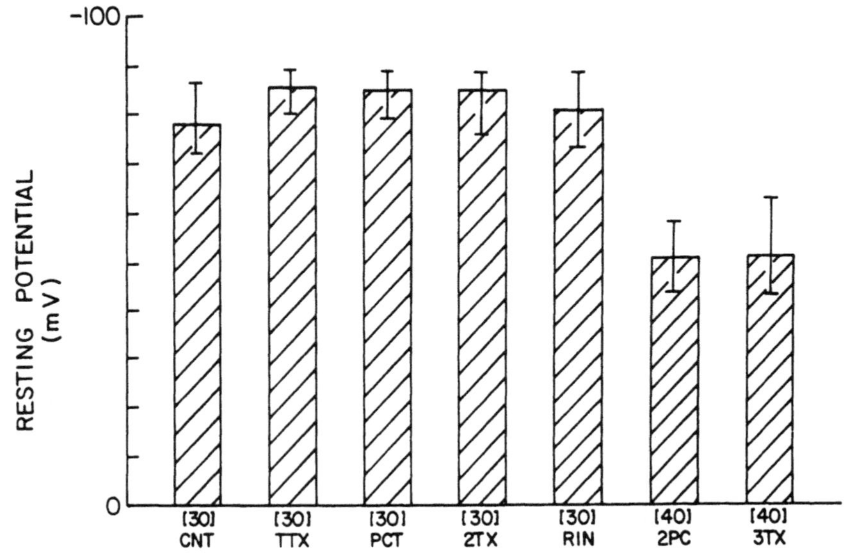

FIGURE 4. Tetrodotoxin protection against PcTX effect. The histogram presents the median (bars) and 95% confidence interval of the muscle fiber membrane potential (mV). Each group was recorded after a 20 minutes treatment, the different conditions were tested in sequence from left to right. The experimental conditions and number of fibers [ ] are indicated at the bottom of the bars: [CNT], control; [TTX], first application of 300 nM tetrodotoxin; [PCT], PcTX (0.4 agl/ml) and TTX applied together; [2TX], washout of PcTX with 300 nM TTX; [RIN], washout of TTX with normal saline; [2PC], application of PcTX alone; [3TX], second simultaneous application of PcTX and TTX. See text for other details.

TABLE 1. Amplitude and kinetic parameters of miniature endplate potentials in control condition and after adding PcTX

	V_m (μV)	α ($100.ms^{-1}$)	Time to peak (ms)
Control	70 (60, 81)	30 (26, 33)	3.3 (3.0, 3.8)
PcTX 1.6 agl/ml	65 (55, 80)	34 (29, 38)	3.0 (2.6, 3.5)

Data are presented as median and its 95% confidence interval. Data for both conditions were determined for 100 MEPPs.

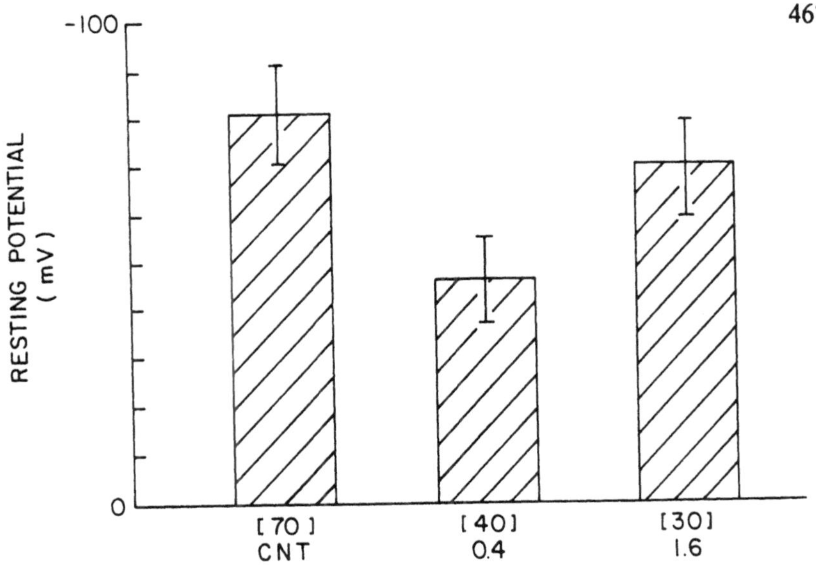

FIGURE 5. Concentration dependence of pctx effect. The Figure shows that PcTX effect is greater at a concentration of 0.4 agl/ml (middle bar) than at a concentration of 1.6 agl/ml (bar at right). Other details as in Figure 5.

The top of Figure 6 shows one of the traces digitized to study the spontaneous release of acetylcholine. The trace contains two MEPPs. The first MEPP translates as dots. The second MEPP was fitted with equation [1] and appears as a solid line.

Table 1 presents the values of the kinetic parameter α and the amplitude of MEPPs (V_m) recorded in two sets of 100 miniatures, one set in control conditions and the other after adding PcTX (1.6 agl/ml). The time to peak MEPP ($t_p = \alpha^{-1}$) is also presented in the table. The shape and amplitude of MEPPs are not modified by PcTX (P»0.05). The frequency of MEPPs was also immune to the action of PcTX.

DISCUSSION

Although not the only study on the toxicity of *P. clavata* (Schmidt et al., 1980), this is the first concerning neurotoxins from the ant's venom. Two kinds of biological activity were identified, both appearing from the same chromatographic peak from the Sephadex G25 column. One effect - only partially reversible - is to depolarize skeletal muscle membrane. The other effect, observed when the toxin was raised from 0.4 to 1.6 agl/ml, consists of blocking neurotransmitter release. The inhibiting effect of a drug or an enzyme when concentration increases (Lehninger, 1971) is called substrate inhibition. Kinetic schemes to explain this effect are complex and may be modeled with or without cooperativity. A lack of information preludes a clear choice from the possibilities. More information is necessary before it can be determined whether or not both effects of *P. clavata* venom are due to a single toxin.

Since tetrodotoxin prevents the depolarizing effect, PcTX probably

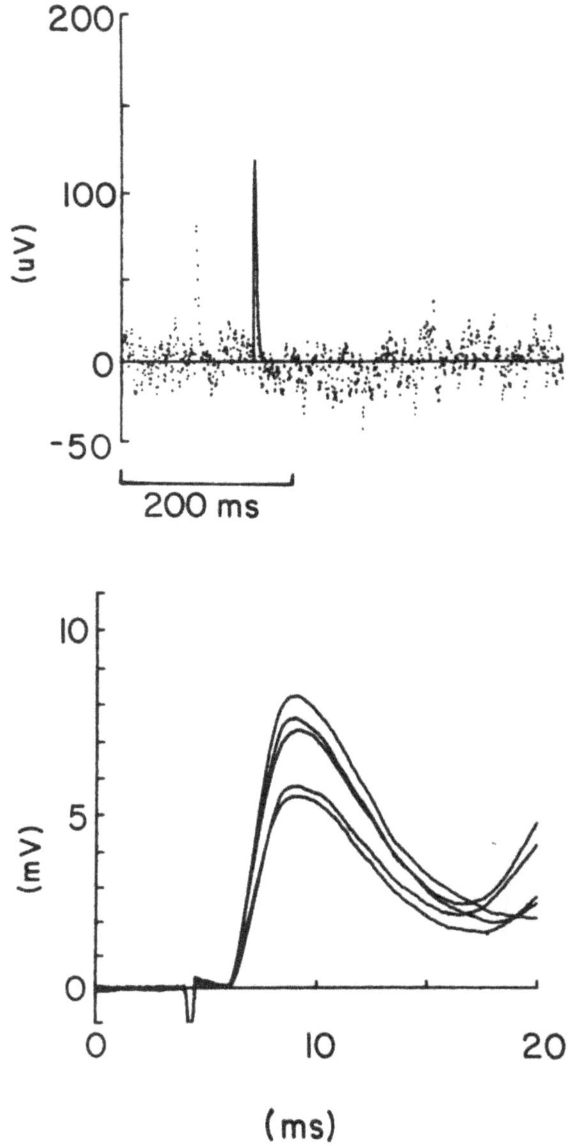

FIGURE 6. Effect of PcTX on neuromuscular transmission. Upper half: trace digitized at a high gain showing two miniature endplate potential (MEPP) recorded in 0.4 agl/ml PcTX. The second MEPP was fitted with equation [1]. Lower half: 5 overlapping evoked endplate potentials recorded in 0.4 agl/ml. Abscissa = ms in both halves. Ordinate = microvolts in the upper half and millivolts in the lower half.

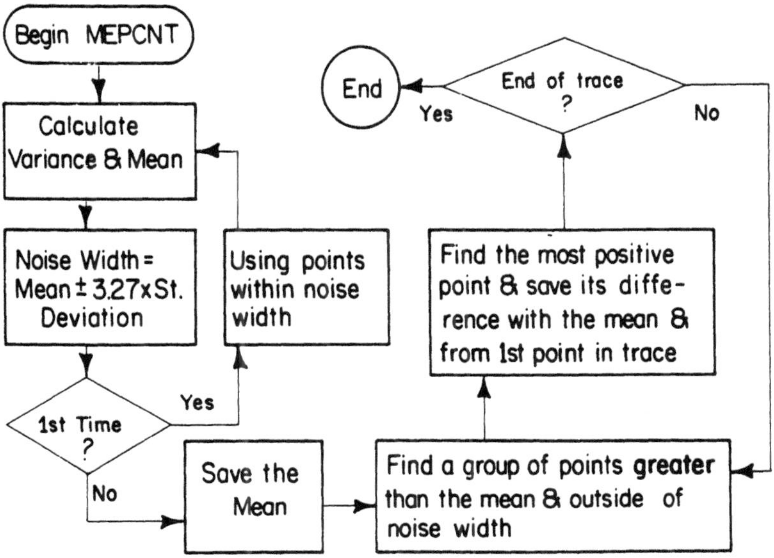

FIGURE 7. Flow chart of the algorithm used to automatically detect miniature end plate potentials. See text for other details.

depolarizes muscle membrane by increasing sodium permeability. TTX failed to reverse the depolarizing PcTX effect, probably due to the small size (25 μ in diameter) of the sartorius muscle fibers in *Hyla* (Sevcik et al., 1986). Siem-Funk and Sevcik (1983) demonstrated that a sodium channel activator, veratridine, can greatly perturb the ionic composition of the squid giant axon (diameter = 300 μ) and that this perturbation may become irreversibly large.

It is difficult to extrapolate frog muscle sensitivity to PcTX to humans. The effects of venom inoculated by a sting and the effect of toxins from gland homogenates cannot be compared using current knowledge. The results of this study indicate, however, that PcTX modifies muscle physiology, the observed depolarizing effect could provide an explanation for the spontaneous fibrillation observed in some patients suffering from *P. clavata* stings. The neuromuscular blocking effect may explain patients' inability to move muscles in the sting's vicinity. This possibly would corroborate Spanish reports, which listed the abdomens of "large black ants the size of our bees" as an ingredient in the curare used by Venezuela's Caribe indians (Martínez and Vela-Diaz 1953).

APPENDIX

Algorithm to Automatically Detect Miniature Endplate Potentials

Noise analysis can help determine amplitude and frequency of MEPPs (Fesce et al., 1986), even at high release frequencies. Noise analysis, however, does not permit the study of the time course of miniatures, which is of particular interest to us. Figure 7 is a flow chart of the algorithm which automatically detects MEPPs. The input to the algorithm is a trace digitized as indicated in Methods, consisting of *N* points. The most important assumption involved in the algorithm is that a MEPP is a *positive* voltage fluctuation having a maximum outside a region ("Noise width") of voltage fluctuations symmetric about a base line corresponding to the resting membrane potential. The fluctuations with a maximum within the region defined above are considered "noise".

Statistical criteria are used to define the base line and noise width. For this purpose the mean, m, and variance, s^2, are first estimated for all points (x_i) of each trace, according to the following expressions

$$m = \sum_{i=1}^{N} x_i / N \qquad [2]$$

and

$$s^2 = \sum_{i=1}^{N} (x_i - m)^2 / N \qquad [3]$$

$$\text{Noise width} = m \pm 3.27s \qquad [4]$$

Equation [4] implies a strictly empirical choice, which means the algorithm's accuracy in detecting MEPPs, is comparable to that of the observer's eye. However, since N is always >1000, it seems safe to assume, using the central limit theorem (Guttman and Wilks, 1965), that the points in the trace will be distributed about the mean according to the Gauss probability distribution function. If so, equation [4] defines the 99.9% confidence interval of the mean. The mean and variance of each trace were calculated twice; first using all points in the trace and again, only using points within the noise width estimated in the first pass. This double calculation reduces the contribution of MEPPs to m and s^2. The value of m is stored in a disk file for each trace.

The mean values and noise width estimated in the second pass are used to find clusters of points *greater* than the mean (i.e. positive) which lie outside the noise width. The position of the point in a cluster farthest from m (tip of the MEPP), with respect to the origin of the trace and to m, are stored on disk for each MEPP. The last point prior to the tip of each MEPP equal to m is defined as the beginning of this MEPP; the amplitude previously stored is used as V_{max}, and α calculated by fitting equation [1] with a simplex algorithm (Nelder and Mead, 1965) to the 50 points that follow the beginning of each MEPP. If the tip of a MEPP is separated by less than 50 points from the tip of the preceding MEPP, the second MEPP is discarded.

This algorithm saves considerable time, and is very reliable in detecting MEPPs and determining their kinetic properties. A high noise level and overlapping MEPPs affect the algorithm's accuracy. A signal to noise ratio ≤3 is ideal.

ACKNOWLEDGEMENTS

The authors are indebted to Dr. Edgar Martínez-Aguirre for providing some historical references and to A.J. Mijares for critically reading the manuscript. The figures were drawn by Mrs. Dhuwya Otero-Palma. This work was supported in part by Fundación José María Vargas and by Grant No. S1-1268 from CONICIT (Venezuela). The Fundación Científica Terramar S.C. provided the field facilities in the Amazon.

REFERENCES

BEQUAERT, J.C. 1926. Medical report on the Hamilton Rice Expedition to the Amazon.In conjunction with the Department of Tropical Medicine of Harvard University. 1924-25. Harvard U. Press, Cambridge, Mass.

FESCE, R., J.R. SEGAL and W.P. HURLBUT. 1986. Fluctuation Analysis of nonideal shot noise.Application to the neuromuscular junction. J. Gen. Physiol. 88: 25-27.

GUTTMAN, I. and S.S. WILKS. 1965. Introductory engineering statistics. pp. 295-297. John Wiley and Sons, Inc., New York.

HERMANN, H.R. and M.S. BLUM. 1966. The morphology and histology of the Hymenopterous poison apparatus. I. *P. clavita* (Formicidae). Ann. Ent. Soc. Amer. 59: 397-409.

HOLLANDER, M. and D.A. WOLFE. 1973. Nonparametric Statistical Methods. John Wiley and Sons, New York.

JACK, J.J.B. and S.J. REDMAN. 1971. The propagation of transient potentials in some linear cable structures. J. Physiol. 215: 283.

LEHNINGER, A.1971. Biochemistry. Worth Publishers, Inc.pp. 147-168.

MARTÍNEZ, J.M. and R. VELA-DIAZ. 1953. Contribución española a la historia del curare. Hypnos 1: 7-64.

NARAHASHI, T., J.W. MOORE and W.R. SCOTT. 1964. Tetrodotoxin blockage of sodium conductance increase in lobster giant axons. J. Gen. Physiol. 48: 965-974.

NELDER, J.A. and R. MEAD. 1965. A simplex method for function minimization. Comput. J. 7: 303.

SCHMIDT, J.O., M. BLUM and W.L. OVERAL. 1980. Comparative lethality of venoms from stinging Hymenoptera.Toxicon 18: 496-474.

SEVCIK, C. and C.A. BARBOZA. 1983. The presynaptic effect of fractions isolated from the sponge *Tedania ignis*.Toxicon 21: 191-200.

SEVCIK, C., F. ALVAREZ-VASQUEZ, J.A. SAAVEDRA and G. CORDOVEZ. 1986. Blockage of resting potassium conductance in frog muscle fibers by a toxin isolated from the sponge *Haliclona viridis*. Toxicon 24: 851-860.

SIEM-FUNK, D.J. and C. SEVCIK. 1983. The effect of temperature on veratridine action in squid giant axons.Biochem. Bioph. Acta 728: 305-310.

WEBER, N.A. 1937. The sting of an ant. Am. J. Trop. Med. 17: 765-768.

43

Prey Capture Strategy of the African Weaver Ant

A. Dejean

INTRODUCTION

Oecophylla longinoda (Latreille), the weaver ant, is an arboreal species. Its nests are built from leaves bound together by silk extracted from last instar larvae. The workers grasp the silk between their mandibles and weave their nests by shuttling to and fro (hence the name "weaver ant"). Large societies of these ants occupy the canopy of a single large tree or they may extend their territory to several trees (Way 1954). The species is characterized by extremely aggressive workers who defend their territory against all intruders, including conspecifics and other ant species. Therefore, where abundant, the weaver ant is a key element of the ant mosaic (Room 1971, Leston 1973, Hölldobler and Wilson 1978).

This aggressiveness extends to harming humans who brush against host trees. The ants bite viciously, projecting venom into the resulting wound. Thus, when they attack men who work in orchards, they are classified as pests (Weber 1949, Vanderplank 1960, Leela 1961). They also tend certain homopteran species which are vectors of plant disease (for example, the sudden death disease of the clove tree in Zanzibar; Way 1954).

The weaver ant possesses a strong predatory instinct which permits, through cooperation of major workers, the capture of a wide variety of insect prey, including the large driver ant (Hingston 1923, Way 1954, Vanderplank 1960, Gotwald 1972, Hölldobler 1983). Consequently, this species and its Asiatic homologs are sometimes put to work protecting trees of economic interest, such as the citrus (Huang and Yang 1960), coffee (Leela 1961) and mango (Way 1954).

Workers hunt mainly on two kinds of territories: the tree housing their society (the central territory) and secondary territories situated on the soil (Hingston 1923, Gotwald 1972, Dejean 1990).

This paper will study the workers' predatory behavior while on their central territory and relate this behavior to the protection of trees.

MATERIALS AND METHODS

This study was carried out in Kitwit, Zaire between 1983 and 1985 and in Yaoundé, Cameroon in 1988, on ten *O. longinoda* societies. Studies on

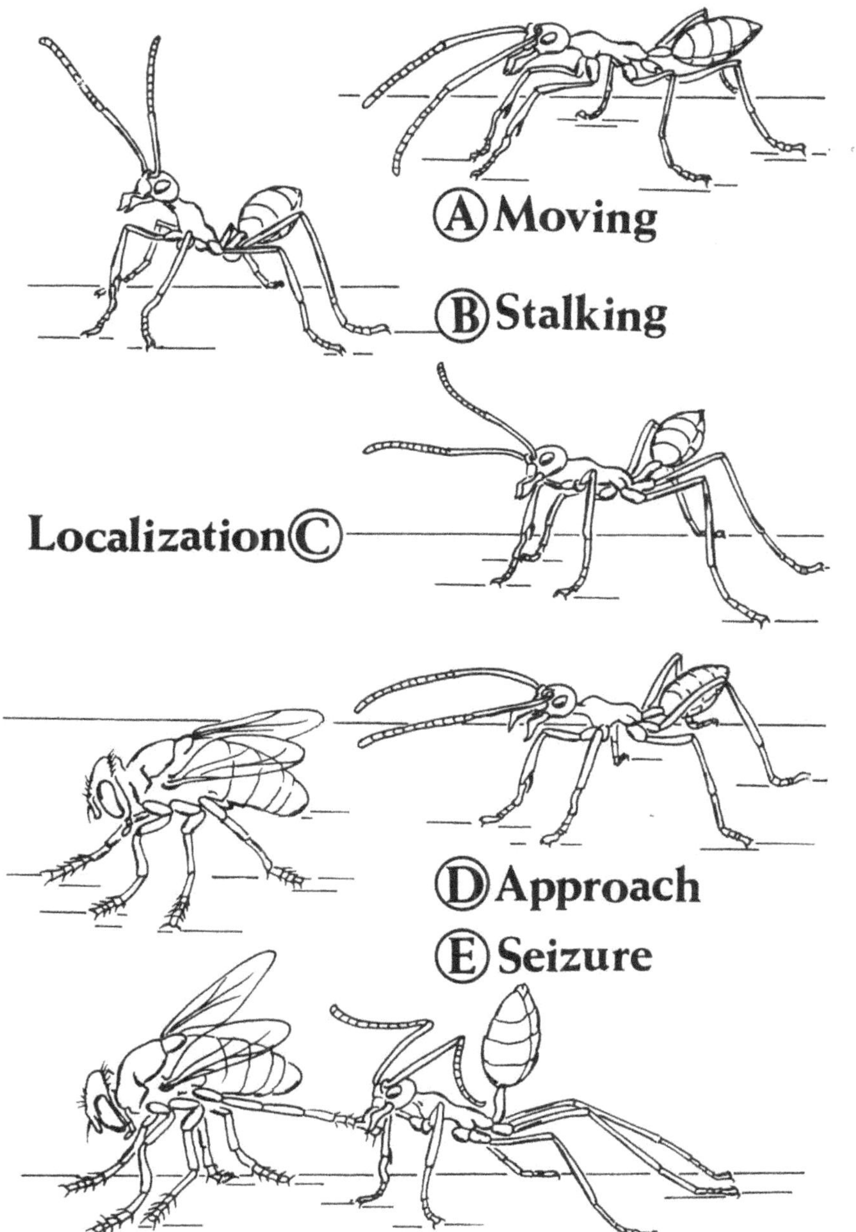

FIGURE 1. Different worker postures.

predation and territoriality were carried out with major workers (9mm < body size < 13mm), distinguished by large compound eyes which permit visual detection of prey (as in *O. smaragdina*; Hingston 1923).

The location and time periods for this study were chosen to facilitate the observations: (1) Societies nested on *Cassia alata* (Cesalpiniacea). During blossom time, buds attracted many *Drosophila* adults. Workers of *O. longinoda* hunt this small and agile prey (< 4mm) on the young inflorescences; (2) Societies nested on mango trees (*Mangilera indica*, Anacardiacea). During the time of fruiting, the Calliphoridae fly (7 < size < 10mm) proliferates; it is an agile, medium-sized prey and settles spontaneously on leaves or ripe fruit. When *O. longinoda* is absent, dozens of these flies congregate on mango leaves; (3) A society's territory includes the leaves of a *Voacanga africana* tree (Apocynacea). The large leaves of this tree permit certain types of experiments. For example, invalid grasshoppers, (15 < body size < 19mm) were thrown onto leaves occupied by workers.

RESULTS

Territorial Occupancy

In general, *O. longinoda* permanently occupy a territory (three series of observations covered a 24 hour period on three mango tree societies); however, occupancy is low at night and high during the day. The number of *O. longinoda* workers on 40 leaves from each tree was recorded between one and three a.m. (18 +- 2 workers) and between 10 to 12 a.m. (26 +- 3 workers). The degree of occupancy was significantly different for the two time periods (n = 27, comparison by paired series: t= 4.63 > 3.72; P < 0.001). The respective worker densities were $82/m^2$ and $118/m^2$. During the day, workers walk on tree branches, an uncommon practice at night. Therefore, during the day or hunting period, each worker always has a nestmate in its vicinity (118 workers/m2 corresponds to four nestmates placed in a circle with a radius of about eight cm).

During normal worker movement, the tips of the antennae are level with the substrate (33 observations) and held two and nine mm (x = 5.3 + 0.6 mm) apart.

On leaves, in most cases, the workers' posture corresponds to that of Fig 1B. (the tips of their antennae are separated by 4 to 11mm, x = 7 + 0.8 mm, 30 observations). The workers remain motionless and sometimes pivot upon themselves or make a short and slow movement. Workers are constantly replaced, but the same individual may spend up to 50 minutes on the same leaf.

PREY CAPTURING

Situation 1: The Prey is Resting When the Ant Arrives

The Prey is a Midge (39 cases). The following sequence was observed: *O. longinoda* detected and localized the prey within a short distance (Fig 1C), opened its mandibles, approached rapidly with outstretched, cranking antennae (Fig. 1D). In two cases, the prey was seized by the body and directly transported to the nest. In the other 37 cases, the prey escaped by

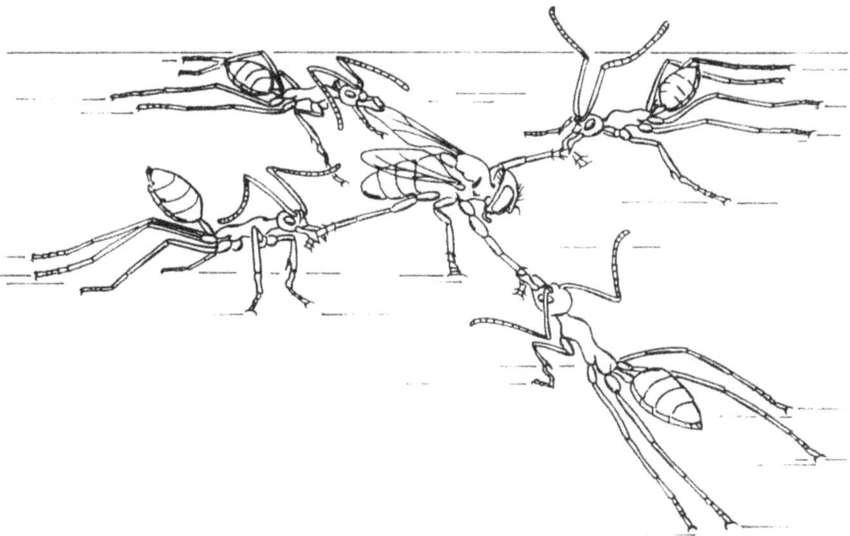

FIGURE 2. Spreadeagling of prey (tight hold on medium and posterior legs, antennae pulled back).

flying away, generally returning to the bud within 40 seconds. Following this failure, the ant made rapid movements on the bud (numerous turns and swings with mandibles open and antennae apart). Within 50 seconds, these movements began to slow, until the worker became motionless, its mandibles still open. This corresponds to the "reserve behavior" described in *Serrastruma* (Dejean 1988a).

The Prey is a Fly (410 cases). Prior to attack, the previously observed sequence was repeated. Three outcomes were observed: (A) The fly escapes before seizure (284 cases : 69.3%) and the worker adopts the reserve behavior; (B) The fly is seized (126 cases : 30.7%), the rate of seizure (by a wing or a leg) significantly higher than in midges (5.1%) X^2 Yates = 10.4 > 6.6; $P < 0.01$). This result reflects the slowness of the Calliphoridae compared to *Drosophila*. At the moment of seizure, the worker raised its gaster in 88 of 126 cases (70.6%) (Fig 1E). The fly escaped 76 of 126 times (60.3%) after initial seizure. Either a wing was ripped off (most common; 51 cases) or the ant failed to catch a leg. If seized, the leg may break off (25 cases). The "reserve behavior" occurred in only 15 of 76 cases (20%) compared to 100% when flies escape before seizure ($P << 0.001$); C) The fly was captured by the worker in 50 of 126 cases (39.7%). Within two to 25 seconds, one to three nestmates arrived and seized the prey by a leg, a wing or the abdomen and pulled back. The fly, spreadeagled for five to 18 minutes (Fig 2), lost its reflexes and was then transported to the nest by a single worker (and, rarely, two workers).

Situation 2: The Prey Lands on the Substrate Where a Worker is Settled

The Prey is a Midge (255 cases). The prey was detected at long range (more than 10 cm) and the ants followed its flight using body, head and antennae movements. Workers attacked the moment their prey landed on the bud. In 62 cases (24.3%), the midge was seized by the body and immediately transported to the nest. In the 193 other cases (75.7%), the prey escaped and the ant exhibited "reserve behavior." The rate of success was significantly higher than in Situation 1 (X^2 Yates = 6.4 > 5.4; $P < 0.02$).

The Prey is a Fly (122 cases). When a fly approaches ant territory, its path is followed by several ants which move and pivot, antennae outstretched towards their prey, their heads following the movement (Fig 3A). These workers can determine where the fly will land and in 72 cases (59%) attacked *before* the fly touched the substrate (Fig 3B). In the other cases, the workers attacked when the fly touched the substrate. Sometimes two or three workers simultaneously approached their victim.

The approach was rapid and direct (Fig 3C). Opening of the mandibles takes place during either visualization or approach.

There are three possible outcomes: (A) The fly escaped before seizure in 45 of 122 cases (36.9%), a percentage lower than for Situation 1 (X^2 Yates = 40.4 > 10.8; $P < 0.001$); (B) The worker seized the fly in 77 cases (63.1%). In 57 of the 77 cases (74%), the worker raised its gaster the moment it seized the fly by an appendage; (C) The fly escaped 27 of 77 times (35%). This rate is also significantly lower than in Situation 1 (60.3%); (X^2 Yates = 11.2 > 10.8; $P < 0.001$). The worker immobilized the fly in 50 of 77 cases (65%). In five cases, two or three workers arrived simultaneously. Generally, within two to ten seconds, one to three nestmates arrived and held the prey spreadeagle fashion. One worker then transported the prey. The success rate in seizure attempts (50 cases/122 : 41%) was significantly higher than that for midges in an equivalent situation (X^2 Yates = 11.2 > 10.8; $P < 0.001$).

The Prey is a Grasshopper (35 cases). The victim could not escape because the tibiae of its hind legs were amputated. The behavioral sequence was similar to the previous situation, except in this case four to seven nestmates spreadeagled their prey.

In five cases, one worker bent its gaster above its head and projected venom onto its prey.

Two to four workers transported the prey during the first part of the journey back to the nest and then two (69%) or even one worker (31%) completed the transport.

DISCUSSION

Because hunting occurs on permanently occupied territories, a peculiar relationship between predation and aggression exists among the *Oecophylla*. Only very small prey is captured by individual workers. Otherwise, the prey is spreadeagled even if the discovering worker can subdue it. Among other ant species which hunt individually, prey encountered near the nest entrance elicits recruitment of nestmates and is followed by group attacks, even with small prey [*Myrmecia gulosa* (Myrmeciinae) *Odontomachus troglodytes*

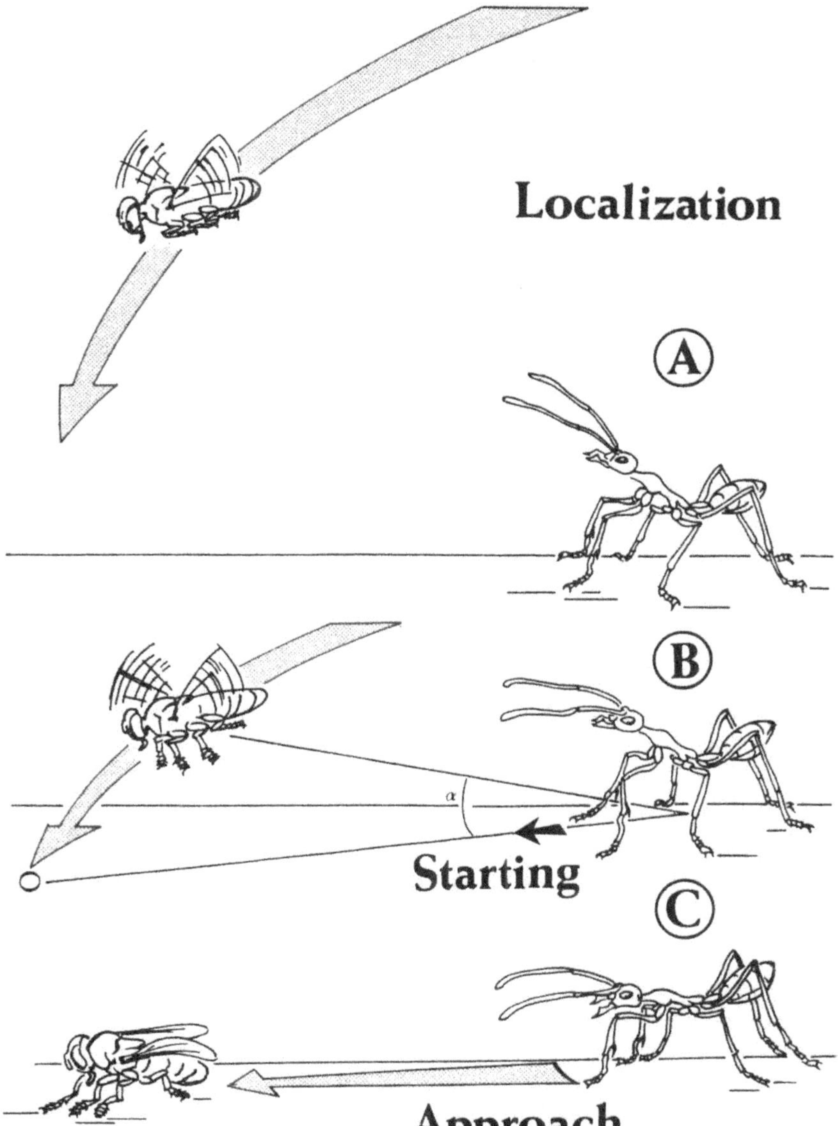

FIGURE 3. Illustration of the first part of the hunt : A. The worker in stalk detects and localizes a fly. The axis of its antennae, head and thorax is placed in the plane of the trajectory of the fly which prepares to land; B. The worker starts before the prey is in contact with the substrate; C. Approach.

(Ponerinae), *Myrmica rubra, Pheidole dentata, Dacetini* (Myrmicinae), (Robertson 1971, Cammaerts-Tricot 1973, Wilson 1976, Dejean 1982)]. Among the *Oecophylla*, aggressive behavior intensifies on a large territory. "Typical" aggression is seen during intraspecific competition for territory. Intruding workers are spreadeagled and then taken back to the nest in the same manner as prey. (Hölldobler 1979, 1983). This aggressiveness is based on the notion of "enemy specification" (Wilson 1975, Hölldobler 1979, 1983) which also applies to other ants (potential competitors or predators of the *Oecophylla*). This aggressiveness extends to all mammals who attack the tree supporting the nest. Janzen (1967, 1972) considers it a protection mechanism for the trees against herbivorous mammals. Consequently, cultivators are attacked in orchards while pruning, weeding or harvesting. This is a crucial problem in coffee plantations where the harvesting involves significant contact between the harvester's forearm and the leaves. This excites ant workers, who attack *en mass* (Fig. 4).

In Ceylon, the use of *Oecophylla* to fight against enemies of the coffee tree proved to be effective. It was nevertheless discontinued due to the ants' aggressiveness towards harvesters (Leela 1961). In Cameroon, harvesters work in pairs and without shirts, so they can remove *Oecophylla* workers from each other's backs. Cocoa tree harvesters, however, can pick the cocoa pods without disturbing *Oecophylla*.

Hingston (1923) studied *O. smaragdina*'s predatory behavior and numerous behavioral elements were given for *O. longinoda* by Way (1954), Vanderplank (1960), Gotwald (1972), Hölldobler and Wilson (1978) and Hölldobler (1979). The two species are similar. Both detect prey by sight. After the prey is caught by one worker, predation is based on cooperation between workers who spreadeagle their victim. Recruitment of nestmates by the first worker attacking its prey is accomplished visually (raising of its gaster, Hingston 1923) or through two other types of pheromone (from the sternal gland; or an odor trail from the rectal gland Hölldobler and Wilson [1978]).

Observers who believe that *Oecophylla* predatory behavior is stereotypical are mistaken because groups of workers capture prey varying in size from small (smaller than the size of workers) to large insects. Venom is rarely used. This strategy, probably intrinsic to this genus, is related to territoriality. Each hunting worker is constantly surrounded by nestmates. Although groups normally capture prey, the capacity of solitary hunters is demonstrated when they prey upon midges. This very small prey is seized by the body, whereas large prey is seized by an appendage. This has been observed in other ant species (Dejean, 1988b).

The permanent occupancy of territory by numerous workers is essential for successful implementation of group strategy by *Oecophylla* (Dejean 1990). This strategy permits the capture of different kinds of prey and the elimination of enemies. The efficiency of territoriality has been closely associated with the exploitation of permanently renewable resources (Hölldobler and Lumsden 1980; Horn 1960).

Workers of *O. loginoda* utilize two hunting strategies: the classical approach in which workers detect, localize and then attack a stationary prey, and the second approach, in which they stalk airborne prey preparing to land within their reach. Workers are more efficient while stalking, and their behavior before seizure shares common elements with that of the tiger beetle

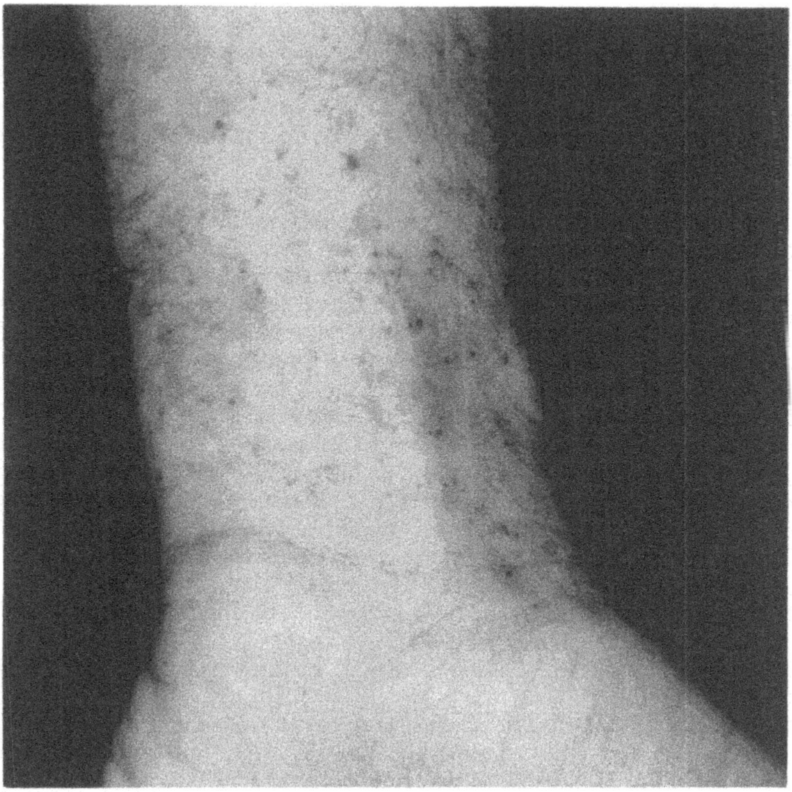

FIGURE 4. Forearm of an agronomist freshly attacked by the *Oecophylla* during a series of controls on the quality of fruits in a coffee plantation.

(Gilbert 1986). Accurate evaluation of distance between ant and prey permits a high level of success and is evident by the movement of the head when an ant follows the path of its prey. Thus, workers, as is the case with other predatory ant species, can adapt their behavior to a given situation (Dejean, 1988a and b). This includes the "reserve behavior" which permits the relocation of escaped prey. Workers remain excited after prey escapes, which is also part of "reserve behavior." The extended excitement results in quick attacks if new prey arrives and in an efficient stalking technique. In addition to the three classic predatory strategies known in ants - isolated hunting (with or without eventual recruitment of nestmates Dejean 1985), group hunting (*Megaponera* type, Longhurst et al. 1978) and column hunting (Dorylinae type, Schneirla 1957) - a fourth strategy exists known as group territorial hunting (*Oecophylla* type). This strategy is characterized by : (1) large societies, (2) hunting on territory, (3) densely occupied territory, (4) extensive visual detection of prey, (5) short range recruitment of nestmates and (6) spreadeagling of prey.

REFERENCES

CAMMAERTS-TRICOT, M.C. 1973. Pheromones agregeant les ouvrieres de *Myrmica rubra*. J. Insect. Physiol. 19: 1299-1315.

DEJEAN, A. 1982. Quelques aspects de la predation chez les fourmis de la tribu des Dacetini (Formicidae-Myrmicinae). These Doct. Etat. Toulouse 263 p.

DEJEAN, A. 1985. Etude éco-ethologique de la predation chez les fourmis du genre *Smithistruma* (Formicidae-Myrmicinae, Dacetini). III. La capture des proies chez *S. emarginata*. Insectes Soc. 32: 241-256.

DEJEAN, A. 1988a. Failure as an effective stimulus of a reserve behavior which allows the capture of alternative prey by *Serrastruma serrula* workers (Formicidae, Myrmicinae). Sociobiology 14: 325-340.

DEJEAN, A. 1988b. Behavioural plasticity of hunting workers of *Serrastruma serrula* (Hymenoptera: Formicidae, Myrmicinae) presented with different arthropods. Sociobiology 13: 191-208.

DEJEAN, A. 1990. Circadian rhythm of *Oecophylla longinoda* in relation with territoriality and predatory behavior. Physiol. Entomol. (in press).

GILBERT, C. 1986. A morphological and cinematographic analysis of tiger beetle predatory behavior (carabidae : Ciccindelinae, pp. 43-57. In P. den Boer, M. Luff, O. Messakowski and F. Weber (Ed.). Carabid beetles, their adaptations and dynamics. Fisher, Stuttgart.

GOTWALD, W.H. 1972. *Oecophylla longinoda*, an ant predator of *Anomma* driver ants. Psyche 79 : 348-356.

HINGSTON, R.W.G. 1923. The red ants. J. Bombay Nat. Hist. Soc. 29: 362-372, 681-693.

HÖLLDOBLER, B. 1979. Territories of the African weaver ant (*Oecophylla longinoda* Latreille). A field study. Z. Tierpsy chol. 51: 201-213.

HÖLLDOBLER, B. 1983. Territorial behavior in the green tree ant (*Oecophylla smaragdina*). Biotropica 15: 241-250.

HÖLLDOBLER B. AND C.J. LUMSDEN. 1980. Territorial strategies in ants. Science 210: 732-739.

HÖLLDOBLER, B. AND E.O. WILSON. 1978. The multiple recruitment systems of the weaver ant *Oecophylla longinoda* (Latreille) (Hymenoptera : Formicidae). Behav. Ecol. Sociobiol. 3: 19-60.

HORN, H.S. 1968. The adaptative significance of colonial nesting in the brewer's balckbird (*Euphaguscyanocephalus*). Ecology 49: 682-694.

HUANG, H. AND P. YANG. 1987. The ancient cultured citrus ant. Bioscience 37: 665-671.

JANZEN, D.H. 1967. Interaction of the Bull's Horn Acacia (*Acacia cornigera* L.) with an ant inhabitant (*Pseudomyrmex ferruginea*, Smith) in Eastern Mexico. Univ. Kansas Sc. Bull. 47: 315-558.

JANZEN, D.H. 1972. Protection of Barteria (Passifloracea) by *Pachysima* ants (Pseudomyrmecinae) in a Nigerian forest. Ecology. 53: 885-892.

LEELA, D.A. 1961. Notes on the biology and habits of the red tree ant *Oecophylla smaragdina* (Fabricius). Madras Agric. J. 48: 54-57.

LESTON, D. 1973. The ant mosaic-tropical tree crops and the limiting of pests and deseases. PANS 19: 311-341.

LONGHURST, C., R.A. JOHNSON AND T.G. WOOD 1978. Predation by *Megaponera foetens* (Fabr.) (Hymenoptera : Formicidae) on termites in the nigerian southern guinea savanna. Oecologia 32: 101-107.

ROBERTSON, P.L. 1971. Pheromones involved in aggressive behavior in the ant *Myrmecia gulosa*. J. Insect Physiol. 17: 691-715.

ROOM. P.M. 1971. The relative distributions of ants species in Ghana's cocoa farms. J. Anim. Ecol. 40: 735-751.

SCHNEIRLA, T.C. 1957. A comparison of species and genera in the ant subfamily Dorylinae with respect to functional pattern. Insectes Soc. 4: 259-298.

VANDERPLANK, J. 1960. The bionomics and ecology of the red tree *Oecophylla* sp, and its relationship to the coconut bug *Prendotheraptus wagi* Brown (Coreidae). J. Anim. Ecol. 29: 15-33.

WAY, M.J. 1954. Studies of the life history and ecology of the ant *Oecophylla longinoda* Latreille, Bull. Ent. Res. 45: 93-112.

WEBER, N.A. 1949. The functional significance of dimorphism in the african ant *Oecophylla*. Ecology 30: 397-400.

WILSON, E.O. 1975. Enemy specification in the alarm-recruitment system of an ant. Science. 190: 798-800.

WILSON, E.O. 1976. The organisation of colony defense in the ant *Pheidole dentata* Mayr (Hymenoptera: Formicidae). Behav. Ecol. Sociobiol. 1: 63-82.

44

The Biological Activities of Ant-derived Alkaloids

P. Escoubas and M.S. Blum

INTRODUCTION

Hymenopterans are the only animals specialized to produce alkaloids. This particularly applies to ants of the subfamily Myrmicinae. Many *Myrmicine* species produce secretions dominated by nitrogen heterocycles; the poison gland secretions of ants in the genera *Monomorium* and *Solenopsis* contain a variety of alkaloids: pyrrolidines, pyrrolines, piperidines, piperideines, pyrrolizidines and indolizidines. Most of these are dialkyl-substituted (Jones and Blum 1982). Large series of alkyl-substituted pyrazines have also been identified in ants, particularly as exocrine products of species in the subfamily *Ponerinae*. Pyrroles, pyridines and indoles have been identified in a wide range of other ant species.

The diversity of chemical structures combined with the secretions' complexities raises many questions about the roles of these compounds in formicid chemical ecology. Although frequently identified with specific defensive or communicative roles, these alkaloids should be considered multifunctional semiochemical agents. Detailed chemical ecology studies of alkaloid-producing ant species determined that the heterocycles have a much greater biological significance than previously thought. These studies have focused on their importance as evolutionary cues. In this chapter, we describe some of these compounds' diverse biological properties and emphasize their varied activities as versatile, natural products in formicid biology.

REPELLENCY

North American and European species of *Monomorium* and *Solenopsis* use their venom to repel other ant species (Urbani and Kannowski 1974; Blum et al 1980, 1989). Studies on individual alkaloids demonstrated that 2,5-dialkylpyrrolidines and 2,6-dialkylpiperidines are excellent feeding deterrents for a variety of *myrmicine* and *dolichoderine* ant species.

INSECTICIDAL ACTIVITY

Ants commonly use their venom against other insects as a chemical weapon. The high toxicity of these secretions or their alkaloidal constituents

TABLE 1. Lethal doses 50 of ant-produced alkaloids in ug/mg of termites (topical application).

COMPOUND	R_1	R_2	Species	LD_{50}/*Reticulitermes*
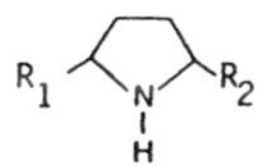	$(CH_2)_4-CH=CH_2$	$(CH_2)_7-CH=CH_2$	*Monomorium minutum*	0.23 ± 0.02
	$(CH_2)_4-CH=CH_2$	$(CH_2)_8-CH_3$	*Monomorium minutum*	2.54 ± 0.05
	$(CH_2)_5-CH_3$	$(CH_2)_8-CH_3$	*Monomorium minutum*	3.34 ± 0.48
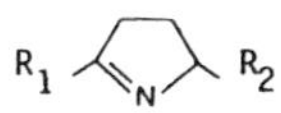	$(CH_2)_4-CH=CH_2$	$(CH_2)_7-CH=CH_2$	*Monomorium minutum*	1.07 ± 0.08
	$(CH_2)_4-CH=CH_2$	$(CH_2)_8-CH_3$	*Monomorium minutum*	0.81 ± 0.08
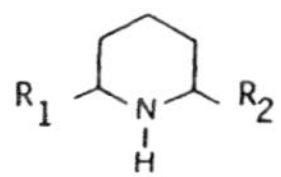	CH_3	$(CH_2)_{10}-CH_3$ trans	*Solenopsis richteri*	0.19 ± 0.04
	CH_3	$(CH_2)_{12}-CH_3$ trans	*Solenopsis invicta*	0.48 ± 0.04
	CH_3	$(CH_2)_3-CH=CH-(CH_2)_7-CH_3$	*Solenopsis richteri*	0.11 ± 0.04
	CH_3	$(CH_2)_{14}-CH_3$ trans	*Solenopsis invicta*	0.49 ± 0.15
	CH_3	$(CH_2)_5-CH\underset{Z}{=}CH-(CH_2)_7-CH_3$ trans	*Solenopsis invicta*	0.50 ± 0.05
	CH_3	$(CH_2)_8-CH_3$	*Solenopsis richteri*	0.57 ± 0.12
	CH_3	$(CH_2)_2-CH=CH-(CH_2)_6-CH_3$ trans	Synthetic	0.74 ± 0.11
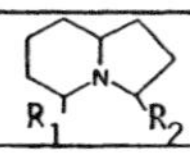	CH_3	$(CH_2)_5-CH_3$	*Monomorium pharaonis*	0.74 ± 0.05
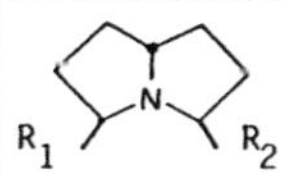	$(CH_2)_4-CH=CH_2$	$(CH_2)_4-CH=CH_2$	*Monomorium antarcticum*	0.32 ± 0.04
	CH_3	$(CH_2)_7-CH=CH_2$	*Monomorium antarcticum*	0.32 ± 0.06

has been determined in several investigations (Blum et al. 1958; Tomalski et al. unpublished data; Clément et al. 1986; Escoubas 1988). Toxicology studies of ant-derived alkaloids which used termite workers as assay organisms demonstrated that several nitrogen heterocycles work well as insecticides (Table 1).

A number of these natural alkaloids - including pyrrolidines, pyrrolines, piperidines, pyrrolizidines and indolizidines - are potent termiticides. The LD50 values range from 0.11 to 3.54 ug/mg of termite and compare favorably with several commercial insecticides (nicotine=0.5 ug/mg, methomyl=0.25 ug/mg, dieldrin=0.01 ug/mg) (Escoubas 1988).

Most saturated heterocycles were tested as mixtures of *cis* and *trans* isomers. *trans* Isomers were more toxic than *cis* isomers when evaluated separately. The presence of alkyl chains (C7-C9), in combination with terminal unsaturation, was correlated with increased insecticidal activity, at least for dialkylpyrrolidines. In limited studies, similar results were observed for both mono- and bicyclic alkaloids.

The results clearly demonstrate these compounds' potential as chemical weapons against other arthropods. In addition to being highly active against competitive ant species, the heterocycles are insecticidal against a variety of insect species, including major agricultural pests (e.g., *Spodoptera* sp.). The alkaloids' ability to penetrate insect cuticle is correlated with long hydrophobic alkyl chains, which substantially increase their lipophilicity (Escoubas 1988).

Although behavioral and neurophysiological studies strongly suggest that these alkaloids act on the central nervous system, only circumstantial conclusions can be derived from mode of action studies. Previous investigations demonstrated dialkylpiperidines' ability to selectively bind to a nicotinic-like site on the post-synaptic membrane of cholinergic synapses. However, no toxic activity was observed *in vivo* (Yeh et al. 1975; David et al. 1984). Dialkylpyrrolidines are also powerful acetylcholinesterase inhibitors *in vitro*, but *in vivo* studies eliminated this enzyme as the alkaloids' primary target (Escoubas 1988).

Trans isomers bind more efficiently than *cis* isomers and are found exclusively in venom of the pyrrolidine-producing species. These alkaloids can also function as ATPase inhibitors (Koch et al. 1977) and oxidative phosphorylation-decoupling agents (Cheng et al. 1977). However, the rapid paralysis of alkaloid-treated insects suggests that they may directly interfere with neurotransmission.

TOXICOLOGY

In addition to their potential as insecticides, ant alkaloids, especially dialkylpiperidines, are strong mammalian toxicants. Necrotic (Caro et al. 1957; Bufkin and Russel 1972), hemolytic (Adrouny et al. 1959) and allergenic activities (Read et al. 1978) are well-documented for these heterocycles. However, such activities in pyrrolines, pyrrolidines or bicyclic alkaloids remain relatively unstudied.

TABLE 2. Average antifungal efficacies of ant-produced alkaloids against thirteen fungal species (concentration of inhibitors in culture media = 800 ppm) (Cole 1974).

COMPOUND	R_1	R_2		INHIBITION
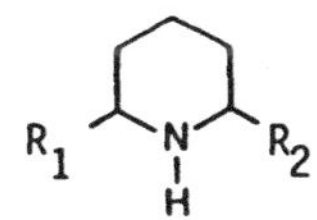	CH_3	$(CH_2)_8-CH_3$		83.9 %
	CH_3	$(CH_2)_{10}-CH_3$	trans	79.5 %
			cis	74.2 %
	CH_3	$(CH_2)_{12}-CH_3$	trans	70.9 %
			cis	67.1 %
	CH_3	$(CH_2)_3-CH=CH-(CH_2)_7-CH_3$	trans	77.9 %
			cis	77.4 %
	CH_3	$(CH_2)_{14}-CH_3$	trans	54.0 %
			cis	48.7 %
	CH_3	$(CH_2)_5-CH\underset{Z}{=}CH-(CH_2)_7-CH_3$	trans	72.2 %
			cis	66.6 %
10- Undecenoic acid (commercial)				67.4 %

TABLE 3. Wheat (Triticum aestivum) coleoptile bioassay. Readings (x3) in mm for 4 concentrations of inhibitor (12.0 = 100 % inhibition, 17.15 = control, * = significant inhibition at p = 0.01) (Cutler and Blum 1989. Unpublished data; Cutler 1984).

COMPOUND	R_1	R_2	10^{-3}M	10^{-4}M	10^{-5}M	10^{-6}M
(1) 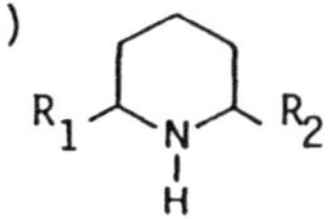	CH_3	$(CH_2)_{12}-CH_3$	12.0*	12.0*	17.2	17.1
	CH_3	$(CH_2)_{14}-CH_3$ <u>trans</u>	12.0*	12.0*	17.2	17.05
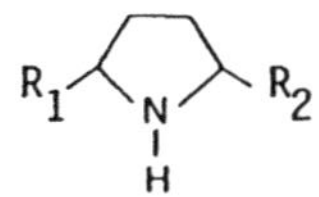	CH_2-CH_3	$(CH_2)_{11}-CH=CH_2$	12.0*	12.0*	17.35	17.0
	$(CH_2)_4-CH=CH_2$	$(CH_2)_6-CH_3$	12.0*	13.8*	17.2	17.15
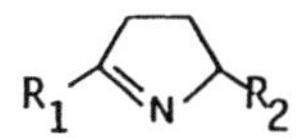	CH_2-CH_3	$(CH_2)_6-CH_3$	12.0	17.15	17.1	17.05
(2)						
Chloramphenicol			17.0	17.0	17.0	17.0
Griseofulvin®			15.0	17.2	17.2	17.2
Neomycin®			12.0*	15.0	17.2	17.2

ANTIFUNGAL AND BACTERICIDAL ACTIVITY

The antibiotic activities of ant-derived alkaloids have been determined in several investigations since the first report, Blum et al. (1958). Ant venoms often have good fungicidal and/or fungistatic activities. Several studies (Cole 1974; Escoubas 1988) demonstrated that ant dialkylpyrrolidines and dialkylpiperidines are effective antifungal agents and that activity levels of the most active compounds compare favorably to commercial fungicides (such as Griseofulvin R; Cole 1974 or Nystatin R; Escoubas 1988) (Table 2). In both studies, *trans* isomers and the compounds with a terminally unsaturated C9 alkyl chain were the most active. The results parallel those observed for insecticidal activity.

These compounds are also bactericides, effectively limiting bacterial growth (Urbani and Kannowski 1974; Jouvenaz et al. 1972). More recent studies (Cutler and Blum unpublished data) established that they inhibit the growth of Gram+ and Gram- bacteria, including *Bacillus cereus*, *B. subtilis*, *Escherichia coli*, *E. cloacae*, *Mycobacterium thermophactum* and *C. freundii*. However, the activity of mono- and bicyclic compounds was only detected at higher concentrations.

Since the alkaloids' antimicrobial activity is substantial, they could help ants limit the proliferation of microorganisms in their immediate environment (Obin and Vander Meer 1985), probably in conjunction with mechanical methods like grooming and waste disposal.

The similar results observed for both insecticidal and antifungal activities in the structure-activity studies suggests the existence of optimal structural features for biological activity. This would provide the ants with broad-spectrum antibiotics.

HERBICIDAL ACTIVITY

The *Solenopsis invicta*'s piperidine-rich venom causes chlorosis in bean leaves (Blum et al. 1958) and several ant-derived alkaloids have demonstrated herbicidal activity in the wheat coleoptile bioassay, a test designed to quickly detect natural products that inhibit or stimulate plant growth (Cutler 1986; Cutler and Blum unpublished data). Herbicidal activity often correlates with either the insecticidal or fungicidal activities of many natural products. Their activity was often superior to those observed with commercial antibiotics (Cutler 1984) (Table 3).

These results support the known biocidal properties of the alkaloids and further emphasize the broad-spectrum toxic activities of the heterocycles produced in *myrmicine* ant poison glands.

CONCLUSIONS

The nitrogen heterocycles evolved as poison gland products in *Solenopsis* and *Monomorium* ant genera have broad-spectrum biocidal activities against both prokaryotes and eukaryotes. These alkaloids may function in the colonial milieu to inhibit soil microorganisms. In addition, they serve as effective insecticides and repellents against a variety of ants. The realization that these alkaloids are versatile toxicants should spark further study and may create a new class of biocides for use against pest organisms.

REFERENCES

ADROUNY, G.A., V.J. DERBES, and R.C. JUNG. 1959. Isolation of a hemolytic component of fire ant venom. Science 130:449.

BLUM, M.S., T.H. JONES, B. HÖLLDOBLER, H.M. FALES and T. JAOUNI. 1980. Alkaloidal venom mace: offensive use by a thief ant. Naturwissenschaften, 67:144-145.

BLUM, M.S., J.R. WALKER, P.S. CALLAHAN and A.F. NOVAK. 1958. Chemical, insecticidal and antibiotic properties of fire ant venom. Science 128:306-307.

BLUM, M.S., M.D. TOMALSKI, T.H. JONES and D.M. EVERETT. 1989. Ant-derived alkaloids: activities as repellents for ants.J. Chem. Ecol. (In press).

BUFKIN, D.C. and F.E. RUSSELL. 1972. Some chemical and pharmacological properties of the venom of the imported red fire ant *Solenopsis saevissima richteri*. Toxicon 10:526.

CARO, R., V.J. DERBES and R. JUNG. 1957. Skin responses to the sting of the imported fire ant (S. saevissima). A.M.A. Arch. Derm. 75:475-488.

CHENG, E.Y., L.K. CUTKOMP and R.B. KOCH. 1977. Effect of an imported fire ant venom component on respiration and oxidative phosphorylation of mitochondria. Biochem. Pharmacol. 26:1179-1180.

CLÉMENT, J.L., M. LEMAIRE and C. LANGE. 1986. Toxicité à l'égard des termites du genre Reticulitermes des pyrrolidines et pyrrolines de la glande à poison de *Monomorium* minutum. C.R Acad. Sci. Paris 303:669-672.

COLE, L.K. 1974. Antifungal, insecticidal and potential chemotherapeutic properties of ant venom alkaloids and ant alarm pheromones. PhD Thesis. University of Georgia, 155 pp.

CUTLER, H.G. 1984. A fresh look at the wheat coleoptile bioassay. Proc. Eleventh. Ann. Meeting Plant Growth Regulator Society of America.

CUTLER, H.G. 1986. Isolating, characterizing and screening mycotoxins for herbicidal activity. In A. Putnam and C.S. Tang (Eds.) The Science of Allelopathy. John Wiley & Sons.

DAVID, J.A., P.J. CROWLEY, S.G. HALL, M. BATTERSBY and D.B. SATELLE. 1984. Actions of synthetic piperidine derivatives on an insect acetylcholine receptor/ion channel complex. J. Insect Physiol. 30:191-196.

ESCOUBAS, P. 1988. Alcaloïdes de fourmis: identification, toxicité et mode d'action. Ph.D Thesis, Université Pierre et Marie Curie. 145 pp.

JONES, T.H. and M.S. BLUM. 1982. Arthropods alkaloids: Distribution, functions and chemistry. In S.W. Pelletier (Ed.), Alkaloids, Chemical and Biological Perspectives. Wiley & Sons, 33-84.

JOUVENAZ, D.P., M.S. BLUM and J.G. MacCONNELL. 1972. Antibacterial activity of venom alkaloids from the imported fire ant, *Solenopsis invicta*. Antimicrobial Agents and Chemotherapy 24:291-293.

KOCH, R.B., D. DESAIAH, D. FOSTER, and K. AHMED. 1977. Effect of piperidines and fire ant venom on ATPase activities from brain homogenate fractions and characterization of Na+ -K+ ATPase inhibition. Biochem. Pharmacol. 26:983-985.

OBIN, M.S. and R.K. VANDER MEER. 1985. Gaster flagging by fire ants (*Solenopsis* spp.): Functional significance of venom dispersal behavior. J. Chem. Ecol. 11:1757-1768.

READ, G.W., N.K. LIND and C.S. ODA. 1978. Histamine release by fire ant (Solenopsis) venom. Toxicon, 16:361-367.

URBANI, C.B. and P.B. KANNOWSKI. 1974. Patterns in the red imported fire ant settlement of a Louisiana pasture: some demographic parameters, interspecific competition and food sharing. Envir. Ent. 3:755-760.

YEH J.Z., T. NARAHASHI and R.R. ALMON. 1975. Characterization of neuromuscular blocking action of piperidine derivatives. J. Pharmac. Exp. Ther. 194:373-383.

Applied Ecology

Ectatoma ruidum (x 6.5). Drawing by Eduardo Perez.

Overview

D.F. Williams

Ants are the most abundant eusocial insect, observed in habitats just about everywhere on earth. Generally, ants are very beneficial, helping to establish ecosystem stability, condition soil, pollinate flowering plants, and preying on destructive arthropods. Humans have learned a lot about the complexity of nature from these insects. Unfortunately, as is the case with many things, some ant species cause serious problems for humans. Although the pest species are few, the economic losses, public health and agricultural problems, and damaging effects on wildlife and the environment can be extensive. Numerous control efforts have been implemented against these pest ants, with most giving only temporary relief. In many cases, the control measures have caused undue damage to the environment and inadvertently increased the pest ant population.

This section deals with applied ecology. Although ecology has been variously defined, basically it is the study of the interrelationships of plants and animals with their environment. Therefore, applied ecology is the practical application of ecology to the well-being of society.

The nine chapters in this section deal with problems presented by major pest ants in various parts of the world, their impact on the environment, and possible methods for managing their populations.

The first two chapters discuss the fire ant *Solenopsis invicta*. Vinson and MacKay report that fire ants are an increasing problem in electrical equipment. The ants cause equipment failure by accumulating around certain electrical components where they chew and remove insulation, short circuit and ground equipment, and foul electrical contacts and circuits. However, Ervin and Tennant report that current levels of fire ant infestations in Texas parks do not appear to have affected park visitations to date. Although this implies that fire ants have caused no apparent economic loss to the state park system, it does not mean that visitors to these parks do not consider them a serious problem. It may also mean that control efforts are keeping the fire ants at low levels and they cause no problems for park visitors.

Lapointe, Garcia, and Serrano look at the effects and control of the leaf-cutting ant *Acromyrmex landoti* in improved pastures of Colombian savanna. They state that colony densities were 2x greater in improved pasture grasses than in native savanna grasses. They suggest using baits containing mirex or diflubenzuron instead of applying large amounts of aldrin and chlorpyrifos. However, they believe that the best approach for control of *Acromyrmex landoti* may be the strategic use of resistant forage grasses, such as *Brachiaria humidicola*.

Fowler describes the pest status of carpenter ants, *Camponotus* spp. and how the public is becoming increasingly concerned over this ant. Records show annual control costs of about 25 million dollars in the United States. The increased urbanization of the human population has created special pest control problems that demand different approaches for their solution.

MacKay explores the biology and economic impact of harvester ants, *Pogonomyrmex* spp. Nests can reach depths of 4 to 5 meters and range in number from 1 to 82 per hectare. Individual nests have a life expectancy of 14 to over 50 years. The ants affect rangelands by removing seeds, moving soil to the surface, and clearing vegetation around their mounds. However, because only 1% of the land area is affected, in most cases the small losses do not justify costs for control.

Buys describes the relationship between the Argentine ant, *Iridomyrmex humilis* and honeybees. The ants seldom attack strong, established colonies, because of the numerous guard bees near the hive entrance. However, exploring ants are able to enter weak colonies causing the colony to become irritated and demoralized. The bees usually leave the hive. Because the ants also inhibit bee foraging and affect the pollination of orchards, the control of ants at hive sites should become an integral part of orchard management. Although results with insecticidal baits have been disappointing, they may offer a solution to the problem.

The next three chapters focus on ant problems in various ecological and agricultural operations. Majer evaluates the role of ants in Australian land reclamation seeding operations and reports that the dispersal and survival of many Australian plant seeds is intimately linked to the corresponding ant fauna of the region. Ants were found to be the major granivores in arid Australia with far greater impact on seed resources than that of rodents. Seed taking by ants in Australian land rehabilitation programs has shown that both granivory (detrimental) and myrmecochory (beneficial) are widespread in such areas. We need to be aware of this and take measures, such as pelletization or treatment of seeds, time seeding operations or ant control measures to lessen the ant's impact in land rehabilitation.

Delabie states that pest ant control in Brazilian cocoa must take into consideration the overall consequences of those programs. Leaf-cutting ants, such as *Atta cephalotes* and *Atta sexdens sexdens*, cut young cocoa leaves and flowers and damage the pods, while the gardening ant, *Azteca paraensis bondari* directly damages the tree canopy. The little fire ant, *Wasmannia auropunctata*, one of the regions most common ant species, tend mealy bugs and/or scale insects on cocoa pods. Control measures used against some pest ants have caused a population explosion of other sometimes more destructive ants. This happened with *W. auropunctata*, now the most common stinging ant present in the area. Also, *A. cephalotes*, a typical rainforest species occurred in the central cocoa production region while *A. sexdens sexdens*, occupied the transition regions and the littoral. Today, *A. cephalotes'* numbers are declining because of successful chemical control programs, while *A. sexdens sexdens* continues to expand into many other areas.

The final chapter in this section by Samways deals with ant species in citrus and subtropical fruit orchards in southern Africa. The few pest ant species in southern Africa are highly effective mutualists with honeydew-producing Homoptera. When homopterans inhabit the trees, the ants tend them with a concomitant increase in both insect's numbers. Attempts at control of the red scale, *Aonidiella aurantii*, with natural enemies is rendered ineffective when the protective ants are present, thus biological control of red scale must be accompanied with ant control. Habitat modification or introduction of competitors are not feasible, and chemicals give only temporary or partial control. Physical barriers such as "Ant-Bar 1.2.3", an inexpensive band, are the control method of choice because they break the mutualistic relationship between the ants and homopterans.

These chapters deal with important pest ant problems in agricultural and urban environments where ecological control methods have been used. In far too many cases, declining numbers of beneficial species are resulting in an increase in pest species. The control technology of the past has been out paced by the exploding pest ant problems of the present. The future will be even more dim unless we increase and utilize our knowledge of basic biology and ecology. Thus, control methods must be based on a sound foundation of basic research dealing with both the biology and ecology of the target pest. Some of the preceding chapters indicate that we need to be more certain about what is really happening in nature because too often, the expected and actual results are far apart. Research scientists must realize that answers obtained under controlled laboratory conditions often differ greatly from the same experiment conducted under field conditions, therefore, care must be taken before we announce solutions to pest ant problems.

The challenge to science is to continue to produce answers and results, for more and more restricted environments. The urbanization of the population, a more environmentally sensitive public, and decreasing research budgets present obstacles to solving future pest insect problems. The solutions to these problems may be greatly different than those of the past. Although we are deeply committed to biotechnology and the biological control of insect pests, we must carefully consider advice given to developing countries regarding control methods banned because of perceived problems caused in "developed countries." This can be a delicate problem, since world-wide agriculture is becoming a highly competitive business. Also, in many cases, new technology has not been developed or the successful new methods are generally more expensive. However, preservation of the environment is of utmost importance, and should be a major concern around the world.

Future research involving pest insects, ants in our case, will be directed towards biotechnology and biological control. We must develop resistant plants, safer chemical controls and genetic techniques for insect control. An exciting area that has not been pursued greatly for pest ants, is the development of species-specific control systems, such as baits using pheromones or other specific attractants that will act only on the target pest ant. Many past insect control problems stem from the use of broad-spectrum chemicals and control programs that not only caused environmental damage but resulted in a population explosion of the target pest. We also must develop and refine techniques, such as augmentation of beneficial ant species, and we must devise agronomic practices that enhance beneficial ants populations, at the expense of pest species. Finally, we need to change our attitudes and learn to apply sound ecological principals to manage populations of pest ant species where problems really exist.

45

Effects of the Fire Ant, *Solenopsis invicta,* on Electrical Circuits and Equipment

S.B. Vinson and W.P. MacKay

INTRODUCTION

Since its introduction into the United States, the "imported" fire ant, *Solenopsis invicta* has caused many rural and urban problems (Vinson and Sorensen 1986). Documented problems include medical (Adams and Lofgren 1982, Clemmer and Serfling 1975, Rhoades et al. 1977), threats to domestic animals (Wilson and Eads 1949, Hunt 1976) or wildlife (Sikes and Arnold 1986, Mount 1981) and predations on crops (Adams et al. 1983, Apperson and Powell 1983). Many of the problems and economic impact due to the fire ant's expansion in the United States have been reviewed by Lofgren (1986). As Lofgren (1986) and Vinson and Sorensen (1986) observed, fire ants have been implicated in damage of electrical equipment. However, little documentation exists as to the type of damage and the types of equipment most susceptible to ant damage.

This study reports that fire ants accumulate around and cause damage to electrical equipment powered with either alternating or direct current. Damage includes direct shorting or grounding of electrical circuits and wiring, jamming of contact points in electrical equipment, and interfering with the mechanical function of electrically powered mechanical devices (when present in large numbers).

MATERIALS AND METHODS

Volunteer reports of equipment damage were accumulated with regard to equipment type, its function, location and kind of damage caused by fire ants. Whenever possible, photographs of the damage were obtained. A survey of highway signal cabinets was conducted in Bryan and College Station, Texas to determine ant damage to these cabinets and a survey of air conditioner repair companies conducted to determine whether fire ants are a problem in their equipment. Surveys were conducted from June 1985 to August 1988.

Laboratory studies were conducted with 12 electrical relays. Twelve relays were mounted on two plexi-glass frames. All relays had power to the points (120 VAC, 1.89 Amps) as occurs in traffic control cabinets. These relays were wired as either active or inactive (randomly determined) for 14 hours (night) followed by an inactive period of 10 hours (day). This regimen

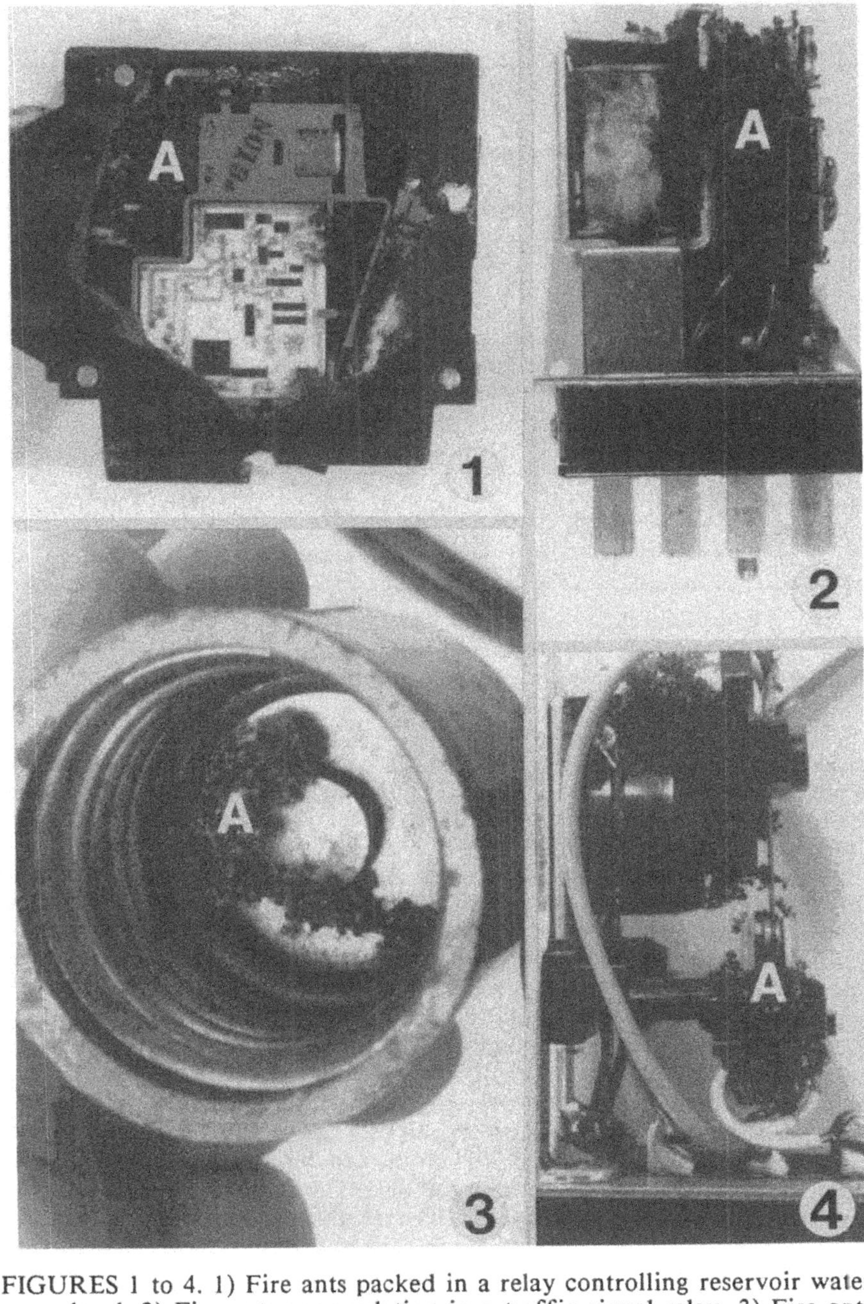

FIGURES 1 to 4. 1) Fire ants packed in a relay controlling reservoir water level. 2) Fire ants accumulating in a traffic signal relay. 3) Fire ants bridging a light socket contact. 4) Fire ants bridging a terminal block.

TABLE 1. Number of ants in relay switches in laboratory nests of the imported fire ants, after 37 days.

Condition of Relays	Number of ants in relay Avg. Number (Min. - Max)
Powered	645 (525 - 756)
Unpowered	170 (20 - 364)

TABLE 2. Estimated number of ants accumulated around commercial equipment intended for fire ant suppression

Equipment	Number of Tests[a]	Estimated Accumulated Ants[b]	
		Unpowered	Powered
Yaardvark™	10	3.4 (0-12)	403 (80-950)
Electrical Anteater™	12	2.3 (0-9)	657.5 (30-1250)

[a] Two units of each type were evaluated after two hours in a separate mound in which one unit was powered and one was not. Tests were repeated the indicated number of times. [b] Average (minimum - maximum).

simulated the use of relays in the field. The two sets of relays were placed in a plastic shoe box with Fluon®-coated sides to prevent ant escape (Vinson and Robeau 1974). The number of ants in the relays was estimated at regular intervals. Tests were repeated for a total of 37 days.

Two field tests were conducted using electrical equipment for the fire ant control. One test utilized two different commercially produced electrical units, the Yaardvark™ and the Electric Anteater™ that claim to attract and kill fire ants. While their potential in control of fire ants was not evaluated, the units' ability to attract ants was examined. Two of each type of unit was placed in a mound and left until the disturbance passed (generally 2 hours). Half of the units were then powered with 110 volts AC for 2 hours, followed by an estimation of the number of ants collected around the units. The tests were repeated 10 or more times with a different set of mounds.

A second series of field tests utilized simple light bulb sockets and electrical outlets. These were placed on the ground no closer than 1/2 meter from mounds. Some of the sockets and outlets were powered with 110 volts AC and some were not. Units were left overnight and the number of ants that accumulated in the light sockets and in the electrical outlets estimated. Tests were repeated three times with six units each.

RESULTS AND DISCUSSION

Problems with fire ants were diverse. Of 42 reports, eight involved signal systems of the highway department. This was followed by air conditioners (6 responses), telephone junction pedestals computers, electrical

FIGURE 5. A fire ant mound built in a transformer

transformers, and irrigation pipes (3 responses each), outdoor lights, electric fences, well pumps, electric meters, plastic waste liners, highway modems, and water level recorders (2 responses each). Only two single reports were received; these concerned the effects of fire ants in removing insulation and damage to airport landing lights. Of the 42 reports, 35 involved electricity. For example, ants have been found in the contacts that control a dam water level (Fig. 1), highway signal cabinet relay switches (Fig. 2) and light sockets (Fig. 3), as well as bridging a terminal block (Fig. 4) and building a nest in a transformer (Fig. 5).

A survey of highway signal cabinets in the Bryan/College Station area revealed the presence of ants in 75% of the cabinets that were not treated with insecticide. In 20% of the cabinets, the ants had caused problems that required treatment (MacKay and Vinson 1989). The most frequent problem in the cabinets concerned the accumulation of ants in relay switches, causing their failure. A survey of six air-conditioner repair services revealed that nearly one third of their summer repair calls were due to ants causing shorts in equipment, generally in the terminal block.

Since powered electrical equipment was common to most of these reports, and relays were found to be frequently shorted because of occupation by ants, this study exposed relays to laboratory colonies of fire ants. Only half

FIGURES 6 - 7. Fire ants electrocuted and accumulating around the base of the Yaardvark™ (top). Fire ants accumulating around the Electric Anteater™ (bottom). Accumulated fire ants are indicated by "A".

TABLE 3. Estimated number of ants accumulated in light bulb sockets and electrical outlets that were powered overnight within a half meter of a mound.

Equipment	Number Tests [a]	Estimated Accumulated Ants[b] Unpowered	Powered
Light Socket	9	3.6 (1.5 - 10)	74 (30-200)
Electric Outlet	9	0.3 (0-3)	0.2 (0-2)

[a] Each test consisted of one powered and one unpowered unit. [b] Average (minimum - maximum).

the relays were powered at any one time and ants were only attracted to and accumulated in the powered relays (Table 1).

The ability of the Yaardvark and Electric Anteater to accumulate ants is shown in Table 2. In both cases large numbers of ants accumulated around the base of these powered units. Ants are electrocuted as they walk across wires located at the Yaardvark's (Fig. 6). The Electric Anteater (Fig. 7) also resulted in worker death; neighboring ants waiting near the base of the unit dismembered some workers. The reason for this behavior is not clear.

Powered light bulb sockets and electrical outlets accumulated ants (Table 3), with dead ants often bridging the positive and neutral contacts (Fig. 3). In two cases the accumulation resulted in a short and loss of a fuse. The reason fewer ants accumulated in the light sockets and electrical outlets compared to the commercial Yaardvark and Electric Anteater may be a function of location or time.

This study's data suggest that ants are attracted to, and become lethargic in, electrical fields. Sometimes ants ground or short circuit the electric field resulting in their electrocution and the destruction of the component.

CONCLUSIONS

Ants accumulate in electrical equipment and thus ground and short the circuits. The ants also congregate in relays in such large numbers that mechanical action is prevented, e.g. they get between contacts so electrical connections can not be made. Their presence in relays and terminal blocks often causes problems in air conditioners, pumps, and traffic signal lights.

While electrical equipment damage from fire ants in the United States has only recently been reported (Vinson and Sorensen 1986, Lofgren 1986), the New Zealand white-footed ant, *Technomyrmex albipis*, is reported to cause relay malfunction (Little 1984). In France, Jolivet (1986) reported that *Lasius alienus* shorted a wall socket and damaged a television set and Galli and Fernandes (1988) reported a species of *Solenopsis* which damaged electrical cords.

Why ants accumulate in powered electrical equipment is unclear, but the results and observations indicate that the ants are attracted to electric fields. The accumulation of ants in powered electrical equipment demonstrates the serious threat the imported fire ant represents to exposed or accessible electrical equipment.

ACKNOWLEDGMENTS

We wish to thank James Irving and Carolyn Rose for critically reading the manuscript and Carroll Messor, Said Majdi and G. A. Brouwer for information and discussions. The research was supported in part by the Texas State Department of Highways and Public Transportation contract 2-18-86-1135. Approved as TA-24608 by the Director of the Texas Agricultural Experiment Station.

REFERENCES

ADAMS, C.T. and C.S. LOFGREN. 1982. Incidence of stings or bites of the red imported fire ant and other arthropods among patients of Ft. Stewart, Georgia U.S.A. J. Med. Entomol. 19: 366-370.

ADAMS, C.T., W.A. BANKS, C.S. LOFGREN, B.J. SMITTLE and D.P. HARLAN. 1983. Impact of the red imported fire ant, *Solenopsis invicta*, on the growth and yield of soybeans. J. Econ. Entomol. 76: 1129-1132.

APPERSON, C.S. and E.E. POWELL. 1983. Correlation of the red imported fire ant, *Solenopsis invicta* Buren, with reduced soybean yields in North Carolina. J. Econ. Entomol. 76: 259-263.

CLEMMER, D.I. and R.E. SERFLING. 1975. The imported fire ant: Dimensions of the urban problem. South. Med. J. 68: 1133-1138.

GALLI, J.C. and O.A. FERNANDES. 1988. Dano causado por formigas do genero *Solenopsis* (Hymenoptera: Formicidae) a fias electricos em Jaboticabal SP. Anais da Soc. Entomol. Bras. 17: 225-226.

HUNT, T.N. 1976. Agricultural losses due to the imported fire ant as estimated by North Carolina, U.S.A., farmers. J. Elisha Mitchell Sci. Soc. 92: 69.

JOLIVET, P. 1986. Les fourmis et la television. L' Entomol. 42: 321-323.

LITTLE, E. 1984. Ants in electric switches: a note. New Zealand Entomol. 8: 47.

LOFGREN, C.S. 1986. The economic importance and control of imported fire ants in the United States. pp. 227-256. In: "The Economic Impact and Control of Social Insects", S. B. Vinson, Ed. Praeger Press, N.Y.

MACKAY, W.P. and S.B. VINSON. 1989. Control of the red imported fire ant, *Solenopsis invicta*, in electrical equipment (Hymenoptera: Formicidae). In press.

MOUNT, R.H. 1981. The red imported fire ant, *Solenopsis invicta*, as a possible serious predator on some native southeastern vertebrates: direct observations and subjective impressions. J. Alabama Acad. Sci. 52: 71-78.

RHOADES, R.B., W.L. SCHAFER, M. NEWMAN, R. LOCKEY, R.M. DOZIER, P.F. WUBBENA, A.W. TOWER, W.H. SCHMID, G. NEDER, T. BRILL and H.J. WITHIG. 1977. Hypersensitivity to the imported fire ant in Florida. Report of 104 cases. J. Florida Med. Assoc. 64: 243-254.

SIKES, P.J. and K.A. ARNOLD. 1986. Red imported fire ant (*Solenopsis invicta*) predation on cliff swallow (*Hirundo pyrrhonota*) nestlings in east-central Texas. Southwest. Nat. 31: 105-106.

VINSON, S.B. and R. ROBEAU. 1974. Insect growth regulator effects on colonies of the Imported Fire Ant. J. Econ. Entomol. 67: 584-587.

VINSON, S.B. and A.A. SORENSEN. 1986. Imported Fire Ants: Life history and Impact. Texas Dept. Agric. Austin, TX 78711. 28 p.
WILSON, E.O. and J.H. EADS. 1949. A report on the imported fire ant areas in southern Louisiana. J. Econ. Entomol. 62: 1268-1271.

46

Red Imported Fire Ants' *(Solenopsis invicta)* Impact on Texas Outdoor Recreation

R.T. Ervin and W.T. Tennant, Jr.

INTRODUCTION

Since parks were first established, visitors to public recreational facilities have tolerated pestiferous insect, arachnid (i.e. ticks, mites, spiders, etc.), plant and other biological organisms. Most persons seeking a picnic site consider the presence of any of these pests undesirable and prefer sites free of such pests. The presence of ants not only decreases the potential enjoyment of a picnic or outing, but their bites and/or stings can result in pain, leaving a more lasting impression.

This study considers the impact of the presence of an ant pest while estimating the aggregate demand for recreational park visitation in the state of Texas. This pest is known as the red imported fire ant *Solenopsis invicta* (Buren). The *S. invicta* has become an agricultural and urban pest to much of the southern United States. Posing a risk to human health, *S. invicta* also causes damage to a wide range of objects, including machinery, crops, and livestock (Wojcik 1987).

S. invicta was first seen in Texas in 1953 (Culpepper, 1953). Since then, *S. invicta*'s range has reached approximately 198,983 sq. km, representing 29 percent of the state (Cokendolpher and Phillips 1989). The spread of *S. invicta* coincides with an increase of *S. invicta* in Texas state parks and public exposure to the pest. According to the Texas Parks and Wildlife Department, park operations change very little when a park becomes infested with *S. invicta* (Riskind, 1988). Park personnel chemically treat mounds in "use" areas. Although, due to park administration concern for protecting the environment, chemical use is minimized.

Michalson (1975) reported that Washington state recreational park visitors responding to questionnaires were willing to incur the expense and/or inconvenience to travel to parks which were not infested by Mountain Pine Beetle, *Dendroctonus ponderosae* (Hopkins). Similarly, Texas state recreational facilities infested by *S. invicta* may be losing park visitors due to visitor recognition and avoidance of this pest problem. Or, park visitors may view *S. invicta* as an unavoidable nuisance similar to flies and mosquitoes, continuing to visit parks when *S. invicta* is present. The purpose of this study is to estimate the impact of the presence of *S. invicta* on recreational park visitation in the state of Texas. An aggregate demand model representing park visitation as the dependent variable, was developed.

Predominant study methods found in current literature presenting results which explore the park visitor response to the introduction of a negative influence on their recreational enjoyment will typically use one of two approaches: the travel cost method (TCM) and the contingent valuation method (CVM). The CVM approach relies on the stated intentions of a cross-section of the affected population to pay for recreational use of resources contingent on changes in their availability (Cummings, Brookshire, and Schulze (eds.) 1986). The TCM approach to the estimation of the nonmarket value of recreation is based on observed behavior of a cross-section of users in response to direct out-of-pocket travel cost and the opportunity cost of time (see Ward and Loomis 1986). Although these methods of estimating the cross-sectional impact of introducing a pest to public recreational parks have validity, they require extensive investments of both time and money, and do not necessarily reflect actual trends. Attitudes with respect to *S. invicta* may be formed by such sensational writings as "The Ants From Hell" (Yoffe 1988). Thus, it was felt that attitudes may not realistically reflect future intentions to return to infested parks. This paper reports an estimate of *S. invicta*'s actual historical impact on park visitation in Texas.

DATA

This study used secondary data from both state and federal agencies. Texas Parks and Wildlife Department (TPWD) provided in-house summaries from each park of the annual number of visitors and revenues generated from entrance fees and concession receipts (TPWD a,b,c). The summaries were taken from park entrance registration forms. Another publication provided by TPWD described available facilities, size and location of each park in the state park system (TPWD d). Texas state-wide population and per capita income were obtained from the U. S. Bureau of Census. Population between census time periods was interpolated at an average rate between years. A study by Cokendolpher and Phillips (1989) described the movement of *S. invicta* and years in which counties became infested with the pest. It was assumed that infested counties have infested parks. Thus, county infestation dates established by Cokendolpher and Phillips (1989) were assumed to also represent park infestation dates.

Data for thirty-five parks over a 19 year period (fiscal years 1969 to 1987) were gathered. Data for fiscal year 1976 for some variables were unavailable, reducing the number of study years to eighteen over the nineteen year study period. Criteria for those parks selected were i) they represent an outdoor recreation state park facility used primarily for overnight camping, and ii) availability of complete data for the eighteen year study period.

Time is required for *S. invicta* to become introduced into an area, become established, and reproduce to a population level irritating to visitors. Therefore, the data for the thirty-five parks were separated into three categories based on whether a park: a) never had *S. invicta* during the study period (four parks); b) had *S. invicta* during the entire study period (nine parks); or c) became infested with *S. invicta* during the study period (twenty-two parks). Two analysis were then conducted. The first analysis determined whether parks infested with *S. invicta* throughout the study period had lower visitation than parks free of the pest throughout the study period.

The second analysis tested whether parks infested with *S. invicta* during the study period lost visitors due to the pest's presence. Analyzing these data was difficult because the time required for the pest to infiltrate a park and reach a population density considered pestiferous by visitors has not been determined. In an effort to account for this lack of information, the data were analyzed with the variable representing *S. invicta*'s presence lagged by successive years. It was assumed that park aesthetics, facilities and/or attractions were not changed during the study period.

MODEL SPECIFICATION AND METHODS

A regression demand model was used to estimate the impact of selected variables on the visitation levels at Texas state parks. A qualitative variable represented the presence of *S. invicta* and its statistical significance/insignificance was used to determine whether *S. invicta*'s presence altered park visits. The unit of observation was a given park in a given year. The dependent variable, park visitation (PV), was transformed from yearly park totals to a Texas per capita basis.

The aggregate visitation model was:

$$PV = f(\text{TPCI, AVGCOST, ACRES, OIL, } S.\ invicta)$$

where: PV = annual number of park visitations per capita
TPCI = Texas total per capita income, dollars per year
AVGCOST = per capita visitor expenditures developed as a ratio of gross receipts and park visitation
ACRES = park size in acres
OIL = average price per barrel of crude oil (travel cost proxy)
S. invicta = dummy variable representing the ant's presence (1) or absence (0)

It was anticipated that results would indicate a positive relationship between the dependent variable PV and independent variables TPCI and ACRES. A negative relationship was expected between the dependent variable PV and independent variables AVGCOST, and OIL. If *S. invicta* has affected park visitation, it should cause decreased visitor enjoyment and result in decreased park visitation.

To identify the most appropriate functional form necessary for this study, the data were analyzed using the semilog (the dependent variable logged), and log-log functional forms. To alleviate any problems in data transformation to the logarithmic form, the *S. invicta* dummy variable assumed the value of 2.7183 in the presence of *S. invicta*, and 1 in the absence of *S. invicta*.

The data used in this study constitute a pooled time-series and cross-section data set: park observations made over time, and observations made over groups of parks within periods. Time series data may pose problems of correlation between periods (i.e. serial correlation) while cross-sectional data may result in unequal variances (i.e. heteroskedasticity) between experimental units, or parks as in this case. The ordinary least squares (OLS) estimation of

TABLE 1. Analysis for functional form and parks infested/not infested.

Independent Variables	Functional Form		Infested/Not Infested
	Semilog	Log-Log	Log-Log
INTERCEPT	-4.6265 (-38.261)@	-11.735 (-12.877)	-16.011 (-10.422)
PER CAPITA INCOME	0.00008 (7.6250)	0.60980 (5.5300)	0.7205 (4.7356)
AVG. COST/VISIT	-1.1588 (-9.0186)	-0.34649 (-9.4088)	-0.5145 (-9.4457)
ACRES	-0.00001 (-0.1861)*	0.2420 (4.3022)	0.7067 (6.7520)
OIL	-0.00834 (-1.7807)*	-0.11227 (-1.3922)*	-0.0576 (-0.5625)*
S. invicta			0.0061 (0.0679)*
R^2	0.9289	0.8599	0.9749
NUM. OBS.	396	396	234

@Student t statistics are in parenthesis.
*not significant at the .05 level.

the model was tested for heteroskedasticity by park, and for serial correlation over time. The estimated Durbin-Watson statistic was 0.5446, indicating the presence of positive autocorrelation and the assumption of homoskedasticity was rejected using the Goldfeld-Quandt test ($P<.01$) (Pindyck and Rubinfeld, 1976, p. 104-105). To correct the data for the combined problems of autocorrelation and heteroskedasticity the data were transformed using the Prais-Winsten Two Step pooling procedures outlined by Kmenta (1986, p. 618). Therefore, differences between parks such as distance from populated areas or varying attractions will not affect analysis results.

ANALYSIS AND RESULTS

Results of testing the two nonlinear functional forms (semilog and log-log) on the data are listed in the first two columns of Table 1. Both models resulted in relatively high coefficients of multiple determination (R^2) with a difference of only .069 between the two models. The Student t-values in parenthesis beneath their respective parameters indicate that ACRES and OIL are insignificant at the 5 percent level in the semilog model, while only OIL is insignificant at the 5 percent level in the log-log model. Whether statistically significant or not, both models provide signs of the estimated coefficients consistent with *a priori* expectations. Because the log-log model

TABLE 2. Results of lagging the initial effects of S. invicta's presence using Log-Log form.

Time	Variable							
Lagged	Intercept	TPCI	AVGCOST	ACRES	OIL	S. invicta	N	R^2
Lag(0) coef.	-12.767	.74628	-.35649	.24589	-.13400	-.083394	396	.8720
(t-ratio)	(-11.419)@	(5.2728)	(-9.3822)	(4.2375)	(-1.5942)*	(-1.5088)*		
Lag(1) coef.	-12.050	.69801	-.34711	.17761	-.10947	-.042490	396	.8229
(t-ratio)	(-11.655)	(5.3590)	(-9.8097)	(3.2013)	(-1.3450)*	(-.80808)*		
Lag(2) coef.	-13.105	.77614	-.31563	.24895	-.089637	-.070136	360	.9210
(t-ratio)	(-12.537)	(6.3669)	(-9.0264)	(4.4216)	(-1.0517)*	(-1.4279)*		
Lag(3) coef.	-12.775	.72586	-.32051	.25477	-.1186	.0039266	360	.9128
(t-ratio)	(-11.643)	(5.6607)	(-9.1427)	(4.5221)	(-1.3943)*	(.070259)*		
Lag(4) coef.	-12.529	.69347	-.32041	.25412	-.11616	.028444	360	.9116
(t-ratio)	(-11.015)	(5.1574)	(-9.1852)	(4.5123)	(-1.3888)*	(.54744)*		
Lag(5) coef.	-12.395	.67493	-.31978	.25447	-.10875	.38481	360	.9123
(t-ratio)	(-10.618)	(4.8360)	(-9.1455)	(4.5096)	(-1.2867)*	(.72448)*		
Lag(6) coef.	-11.445	.56726	-.25814	.27009	-.10448	.065432	360	.8858
(t-ratio)	(-9.8479)	(3.9407)	(-7.6912)	(5.5218)	(-1.1497)*	(1.2749)*		
Lag(7) coef.	-11.799	.61858	-.25646	.26758	-.11414	.029454	360	.8864
(t-ratio)	(-10.059)	(4.2892)	(-7.6620)	(5.4497)	(-1.2654)*	(.51227)*		
Lag(8) coef.	-12.484	.69676	-.27151	.27082	-.13854	.066510	378	.9700
(t-ratio)	(-11.551)	(5.3227)	(-7.7143)	(5.3958)	(-1.7234)*	(1.3327)*		
Lag(9) coef.	-13.175	.79666	-.26396	.26751	-.15928	-.010399	378	.9779
(t-ratio)	(-10.311)	(4.9462)	(-7.5269)	(5.3481)	(-1.7321)*	(-.18660)*		
Lag(10) coef.	-12.810	.74676	-.26511	.26720	-.14935	.027988	378	.9701
(t-ratio)	(-11.897)	(5.8351)	(-7.5733)	(5.3346)	(-1.8609)*	(.59464)*		

~Indicates Years S. invicta variable is lagged. @Student t statistics are in parenthesis. *not significant at the .05 level.

held more variables as significant at the 5 percent level, the log-log functional form was accepted as the most appropriate form for the data set and used for further analysis.

The results of the test to determine whether parks infested with *S. invicta* had lower visitation due to the pest than parks free of the pest are shown in the third column of Table 1. The model provides signs of the estimated coefficients consistent with a priori expectations and Student t-values which indicate that OIL and *S. invicta* are not significant at the 5 percent level.

An analysis conducted on data representing twenty-two parks infested with *S. invicta* during the study period attempted to determine whether park visitation was affected when the presence of the pest in parks was recognized by park visitors. The data were analyzed eleven times, allowing the variable representing *S. invicta*'s presence to be lagged by successive years. Differences in number of data observations occurred between analysis of the lagged models because parks became infested at different times, causing some parks to be omitted from the data set when the lagging of the *S. invicta* variable caused them to be infested throughout the study period or not infested during the study period. The resulting parameters and Student t-values of the log-log lagged models are listed in Table 2. Each model provides signs of the estimated coefficients consistent with *a priori* expectations. The Student t-values indicate that OIL and *S. invicta* are not significant at the 5 percent level for each model. The *S. invicta* variable's lack of statistical significance is consistent with column 3 of Table 1, suggesting that either the state park system is keeping "use" areas sufficiently clear of *S. invicta* so visitors are not bothered, or visitors view *S. invicta* as a nuisance which must be tolerated if they wish to continue visiting parks.

CONCLUSIONS

The current level of *S. invicta* infestations within the Texas park system does not appear to have affected park visitations to date. This implies that *S. invicta* has caused no discernable economic impact to the state park system. Perhaps factors other than *S. invicta* are more important to park visitors if *S. invicta* are considered in the same category by visitors as flies, mosquitoes, etc. Alternatively, the results could imply that the Texas park system is successfully controlling *S. invicta* at a minimal level which satisfies their environmental concerns while also protecting visitor enjoyment. The results of this analysis apply only to those levels of infestation considered within the study. Therefore, *S. invicta* infestation levels greater than those considered cannot be estimated from the results of this study and conclusions about heavier infestations cannot be drawn.

ACKNOWLEDGMENTS

This project was supported by the Texas Department of Agriculture, and is a result of the Red Imported Fire Ant Project conducted at Texas Tech University. The authors wish to thank Drs. Bob Davis, Don Ethridge and Sujit Roy for their assistance in the preparation of this paper.

REFERENCES

COKENDOLPHER, J. C. and S. A. PHILLIPS, JR. 1989. Rate of Spread of the Red Imported Fire Ant, Solenopsis invicta (Hymenoptera: Formicidae), in Texas. Southwestern Naturalist. Forthcoming.

CULPEPPER, G. H. 1953. Status of the Imported Fire Ant in the Southern States in July 1953. Entomol. Plant Quar. USDA, Bur. Entomol. Plant Q. E-867, 8 pp.

CUMMINGS, R. G., D. S. BROOKSHIRE, and W. D. SCHULZE (eds.). 1986. Valuing Environmental Goods: An Assessment of the Contingent Valuation Method. Rowan and Allanheld Publishers, Totowa, New Jersey.

KMENTA, J. 1986. Elements of Econometrics. New York, NY: Macmillan Publishing.

MICHALSON, E. L. 1975. Economic Impact of Mountain Pine Beetle on Outdoor Recreation. So. J. Agr. Econ. 7 (Dec.): 43-50.

PINDYCK, R. S., and D. L. RUBINFELD. 1976. Econometric Models and Economic Forecasts. McGraw-Hill Book Company, New York.

RISKIND, D. 1988. Texas Parks and Wildlife Department, Austin, Texas. Personal Communication. January.

TEXAS PARKS AND WILDLIFE DEPT. (TPWDa). 1969/70-1975/76. FY 1969/70-1975/76 Parks & Wildlife Department Summary of Park Receipts - Direct Operations. Austin, Texas.

TEXAS PARKS AND WILDLIFE DEPT. (TPWDb). 1969/70-1975/76. FY 1969/70-1975/76 Texas Parks & Wildlife Department State Park Visitation. Austin, Texas.

TEXAS PARKS AND WILDLIFE DEPT. (TPWDc). 1977-1987. FY 1977-1987 Facility Utilization Summary Reports. Austin, Texas.

TEXAS PARKS AND WILDLIFE DEPT. (TPWDd). 1986. Texas State Park Information. Austin, Texas.

U. S. BUREAU OF THE CENSUS, DEPT. OF COMMERCE. 1987. Statistical Abstract of the United States 1987, December.

WARD, F. A., and J. B. LOOMIS. 1986. The Travel Cost Demand Model as an Environmental Policy Assessment Tool: A Review of Literature. Western Journal of Agricultural Economics. 11:164-178.

WOJCIK, D. P. 1987. Fire Ant Bibliography Update. Attini - An International Newsletter on Pest Ants 18:34-41.

YOFFE, E. 1988. The Ants From Hell. Texas Monthly. 16(No. 8):80-84, 142-146.

47

Control of *Acromyrmex landolti* in the Improved Pastures of Colombian Savanna

S.L. Lapointe, C.A. Garcia and M.S. Serrano

INTRODUCTION

The expansion of crops and improved pastures into eastern Colombia's well-drained, lowland savannas has been limited by edaphic and biotic factors. Most soils in this area have excellent physical characteristics but poor chemical properties, i.e., low pH (4 to 5), available phosphorous, organic matter content, and cation exchange capacity due to about 80% aluminum saturation (Cochrane et al. 1985). A major initiative of the Centro Internacional de Agricultura Tropical (CIAT) in conjunction with the Instituto Colombiano Agropecuario (ICA) has been to develop drought-resistant, high yielding grass and legume forages selected for these soil conditions. *Brachiaria decumbens* Staph cv. Basilisk is the most widely planted pasture grass on the Colombian savanna. This grass' high susceptibility to spittlebugs (Homoptera: Cercopidae) has resulted in the importation of more tolerant species such as *B. humidicola* (Rendle) Schweick. In addition, the grass *Andropogon gayanus* Kunth was released in Colombia as cultivar "Carimagua 1" (1980) and is being widely adopted by ranchers. *A. gayanus* grows well in poor soils and resists spittlebugs (Ferrufino and Lapointe 1989). These three species are the major forage grasses introduced to the Colombian savanna. *A. gayanus* has been particularly successful when associated with the legume *Stylosanthes capitata* Vog. cv. Capica, released in 1983. Since commercialization, the only major arthropod pests of *A. gayanus* have been leaf-cutting ants. *A. gayanus* is particularly susceptible to ant predation when first established, due to its small seed size. Regrowth does not occur if leaf-cutting ants harvest the germinating plants, resulting in poor, patchy establishment with subsequent weediness.

Although leaf-cutters are considered minor or unimportant forage pests in Colombia (Cherrett and Peregrine 1976), there importance will increase when agricultural activity expands into the savanna, where leaf-cutters are abundant. The development of Colombia's eastern plains depends primarily on the creation of improved pasture that will also encourage expansion of crops and mixed crop/pasture systems into the savannah ecosystem. Thus, pest control in pastures, particularly of leaf-cutters, will become more important as more intensive land use and integrated cropping systems are adopted.

At least three species of leaf-cutter ants live in the Colombian savanna.

Acromyrmex landolti (Forel) cuts only grasses and produces many small nests in pastures. *Atta cephalotes* L., confined to the gallery forests, cuts only broad leaved plants (mainly woody vegetation) and becomes a pest only if suitable crops (for example, citrus) are planted in its environment. *Atta laevigata* (Fr. Smith), which inhabits open pastures, cuts grasses and broad leaved vegetation. Both *Ac. landolti* and *A. laevigata* present significant pest problems. The small nest size and high density of *Acromyrmex* nests makes direct chemical control difficult or impossible over extensive areas. The larger *Atta* colonies can be controlled by injecting insecticide into individual nests. Aldrin, an effective formicide due to the long residual activity of its breakdown product, dieldrin (McEwen and Stephenson, 1979), has been banned in Colombia, though its use will continue until supplies are exhausted.

However, direct poisoning of *Atta* nests was always limited by the low value per unit area of extensive grazing systems. Use of toxic baits as well, if available, would probably consist of a single application over large areas. Distribution of baits would occur at planting, when deployment could be combined with sowing, thereby reducing costs associated with machinery. Unfortunately, baits for leaf-cutter control, including Mirex baits from Brazil, are unavailable in Colombia. In addition, the bait matrix used for *Atta*, orange pulp, may not work against the more parochial *Ac. landolti* (Lapointe, personal observation).

This paper presents (A) results from an *Ac landolte* population survey - conducted at CIAT's experimental station in the department of Meta, Colombia - to determine the effect of improved pasture species on *Ac. landolti* populations; (B) a study on the effect of time of soil preparation on *Ac. landolti* colony survival and activity; and (C) results that indicate host plant resistance in *B. humidicola* to *Ac. landolti*.

SURVEYS OF *AC. LANDOLTI* COLONY DENSITIES.

A survey of *Ac. landolti* colonies was conducted during the 1988 dry season on the Carimagua experiment station located in eastern Meta department. One square kilometer sites were selected randomly from a grid superimposed on a map of the area. Fifty-two sites from a total area of approximately 25,000 hectares were surveyed. At each site, three randomly located quadrats of 200 m^2 were examined for *Ac. landolti* colonies and epigeal termite nests (*Heterotermes tenuis* Hagen).

The survey results revealed a mean colony density of 567 (N=136 quadrats) and 1142 (N=6) colonies/ha in native savanna and *A. gayanus* pastures, respectively. No colonies were found on four *B. decumbens* pastures and one *B. humidicola* pasture. Fig. 1A and 1B represent the range of *Ac. landolti* nest densities found in native savanna and *A. gayanus* pastures. In general, the presence of termite mounds indicated poor drainage, confirmed by observations during the rainy season. Densities of *Ac. landolti* colonies in areas with many termite mounds were low compared to well-drained sites. Later surveys, designed to compare colony density in improved grasses, did not consider pastures with large numbers of termite mounds. No site had large numbers of both termite mounds and *Ac. landolti* nests.

To determine how improved pastures affected *Ac. landolti* populations, we counted colonies in established improved pastures infested by leaf-cutters, and surrounding savanna areas. We assumed that counts in nearby savanna

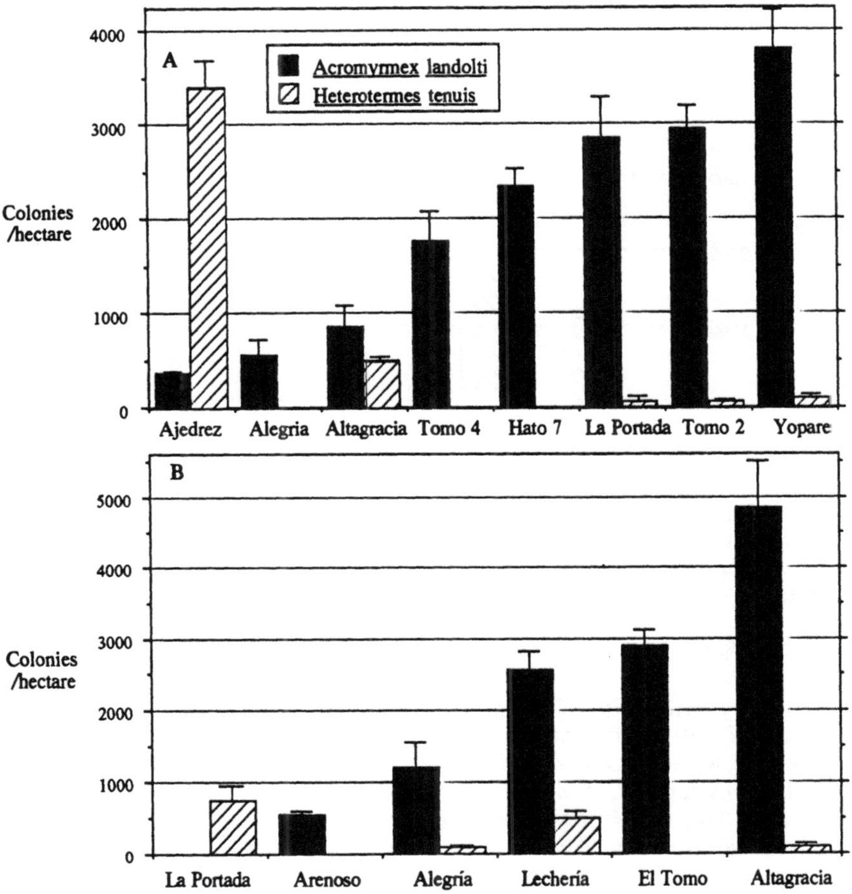

FIGURE 1. Colony density of *Acromyrmex landolti* and epigeal nests of *Heterotermes tenuis* in areas of native savanna (A), and *Andropogon gayanus* (B) at Carimagua, Meta, Colombia, 1988.

would provide an accurate estimate of background populations.

At one site (Altagracia), adjacent 50 ha plots of *A. gayanus* and *B. humidicola* were established in 1984. Although initial establishment was excellent, during 1986 *Ac. landolti* densities rose and caused severe damage to the established stand of *A. gayanus*. Counts made in January, 1988 revealed almost 5,000 nests per hectare in *A. gayanus*, whereas no *Ac. landolti* nests could be found on the adjacent *B. humidicola* plot (Fig. 2). Adjacent savanna had colony densities of ca. 750/ha. At another site (El Tomo), *B. humidicola* again had lower densities than either the savanna or *A. gayanus* (Fig. 2). Assuming current populations in contiguous savanna areas approximate baseline population levels present at establishment, these data indicate that *Ac. landolti* populations have declined in *B. humidicola* pastures but increased in *A. gayanus* pastures compared with a relatively undisturbed savanna habitat.

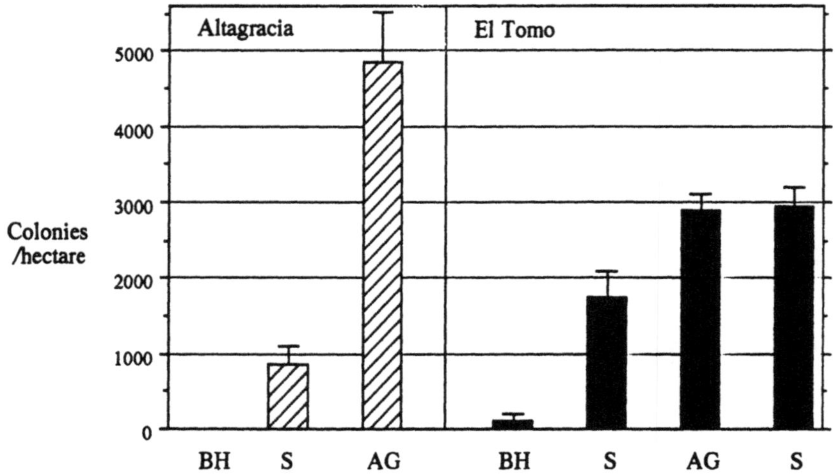

FIGURE 2. Colony density of *Acromyrmex landolti* in adjacent pastures of native savanna and three species of improved forage grasses at two sites, Altagracia and El Tomo at Carimagua, Meta, Colombia, 1988. AG = *Andropogon gayanus*, BH = *Brachiaria humidicola*, S = native savanna. Bars are standard deviation.

EFFECT OF SOIL PREPARATION AND GRASS SPECIES ON *Ac. Landolti.*

To study how timing of soil preparation affects colony density, six plots (60 x 60m) of native savanna with high populations of *Ac. landolti* and no *Atta* colonies, were seeded with *A. gayanus* in May, 1988. Three plots were prepared in December, 1987 at the beginning of the dry season (early preparation). The remaining three plots were prepared at the beginning of the rainy season in May, 1988 (late preparation). Preparation consisted of a single pass with chisels spaced 25 cm apart and 15 cm deep. At sowing, a single pass was made with an off-set disc simultaneously with fertilizer and seed application. Fertilization at planting consisted of 50, 20, 12, and 12 kg/ha of P, K, Mg, and S, respectively. Counts of *Ac. landolti* colonies were made in December, 1987, before the savanna was disturbed and at planting in May, 1988. Success of establishment was determined in July, 1988 as the percent of total row length with established seedlings.

Establishment of *A. gayanus* seedlings did not differ between plots prepared in December ($14.6 \pm 17.0\%$) and those prepared in May ($2.0 \pm 0.7\%$). Damage by *Ac. landolti* was high on all plots and establishment of seedlings was poor. However, colony density did decline on plots prepared in December compared with those prepared in May (Fig. 3). With the beginning of the rains and further preparation at planting in May, colony density decreased in both treatments (compared with undisturbed savanna) to about 15% of the initial colonies observed in December. Therefore, wherever done, plowing is a major source of colony mortality; however, in heavily infested areas, the minimal soil tillage recommended for pasture establishment does not kill enough colonies to avoid significant damage from ant predation.

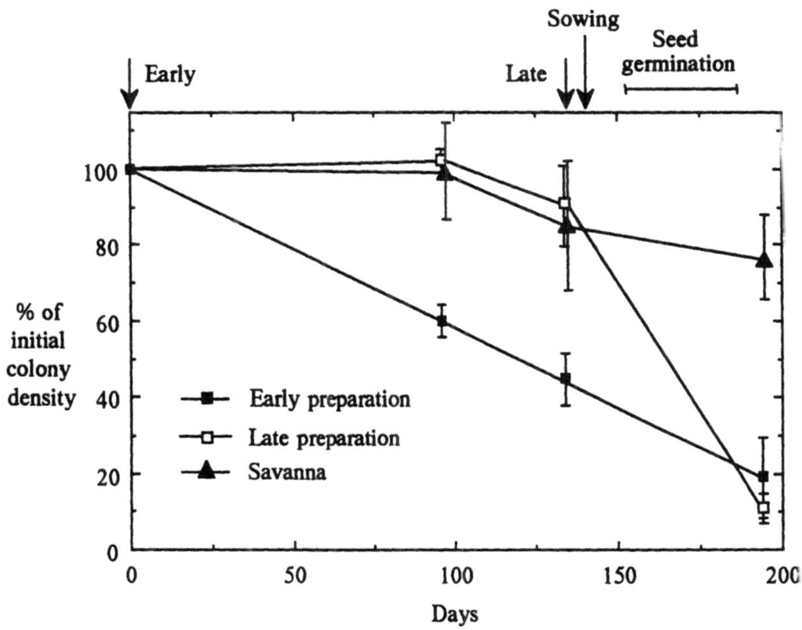

FIGURE 3. Effect of timing of land preparation on colony density of *Acromyrmex landolti* during the dry season with an adjacent area of undisturbed savanna as control. Arrows indicate timing of early and late preparation and sowing date. Day 0 = December 22, 1986. Bars are standard deviation.

Since *A. gayanus* could not be established by seed due to ant damage, the same area was used in a split plot design to test the hypothesis generated from the surveys, namely that *Ac. landolti* populations are suppressed by *B. humidicola*. Main plots (60 x 60m) consisted of early and late soil preparation, as above. Subplots (30 x 60m) were planted with either *A. gayanus* or *B. humidicola* in July, 1988. Vegetative material of both cultivars insured successful establishment.

Colony density one year after establishment in plots with early preparation was equivalent to that in the plots with late preparation (Fig. 4B). Grass species had a small but significant effect on colony density ($P=0.03$) with fewer active colonies on *B. humidicola* plots compared with plots of *A. gayanus*.

More striking, however, was the much greater colony activity in *A. gayanus* plots compared with that in *B. humidicola* plots. This was indicated by the amount of soil being excavated and deposited at nest entrances. In February 1989, the amount of soil excavated by all colonies in each of the plots was collected and measured volumetrically with a graduated cylinder. Colonies in *B. humidicola* plots were much less active compared with those in native savanna or in *A. gayanus* (Fig. 4A). We assume that the quantity of soil excavated was indicative of colony vigor. However, we do not know if excavation during the dry season is associated with a particular colony

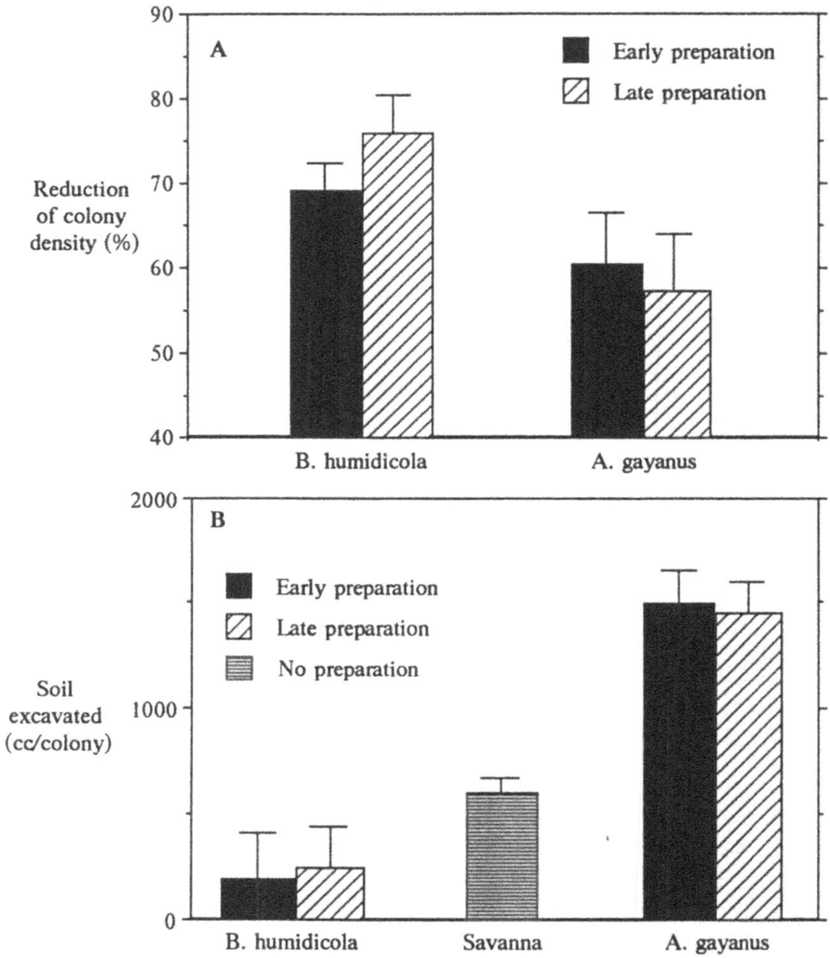

FIGURE 4. Effect of timing of soil preparation and grass species on density (A) and activity (B) of colonies of *Acromyrmex landolti* one year after preparation. Bars are standard deviation.

activity. The volume of soil excavated by *Ac. landolti* was unaffected by the timing of soil preparation but a large difference was observed in the amount of soil excavated between colonies located in plots of the two grass species (figure 4B). Mean soil volume excavated by colonies in *B. humidicola* and *A. gayanus* was 223 and 1473 cc/colony, respectively. If we assume the excavated soil had accumulated only since the last rains in December, then the amount of soil brought to the surface during a two month period by 1,000 colonies/ha in a *A. gayanus* pasture would equal 1.5 m^3/ha. The excellent physical properties and drainage of these savanna soils may be partly due to the considerable subsoil movement by leaf-cutters.

DISCUSSION

The density of *Ac. landolti* colonies in *A. gayanus* pastures was two times greater than in native savanna grasses. It was not determined if this is due to greater attractiveness of *A. gayanus* swards to founding gynes, or to greater colonization success in *A. gayanus* compared with savanna. The tendency of *Ac. landolti* to increase in *A. gayanus* limits the utility of this grass in association with grain crops such as rice. *B. humidicola*, however, resists to leaf-cutters remarkably well and may even reduce or eliminate leaf-cutter colonies in infested areas. The study of adjacent pastures of savanna, *B. humidicola* and *A. gayanus*, revealed that *Ac. landolti* colonies die out over time or migrate from *B. humidicola* swards. The greatly reduced soil volume excavated by colonies during the dry season in *B. humidicola* may indicate reduced foraging activity related to colony decline and incipient death.

We do not know what factors contribute to the resistance of *B. humidicola*. Secondary chemicals and nutrient availability may determine the palatability of woody plants to *A. cephalotes*; however, leaf toughness was not correlated with palatability (Howard 1988). Fowler and Robinson (1977) also hypothesized that chemical factors are more important than physical factors in plant selection by *Ac. landolti fracticornis*. Our observations indicate that *Ac. landolti* prefers new, tender growth. However, the characteristics responsible for palatability differences between grass species remains unknown.

Breeding plant resistance to leaf-cutters would benefit from an understanding of why *B. humidicola* is resistant. Leaf toughness may deter the ants (Cherrett 1972, Waller 1986); fresh growth of *A. gayanus* is tender compared with *B. humidicola* leaves. Differences in the two grasses' growth habits are extreme. *B. humidicola* has narrow leaves, with strongly stoloniferous and prostrate growth habit. *A. gayanus* is a tall, erect, tufted grass. Studies already underway will determine substrate preference and differentiate between the physical and chemical factors involved in resistance.

Although land preparation killed many *Ac. landolti* colonies, the timing of preparation had no effect on the number of active colonies present during the critical period of seed germination. Strategies for control of *Ac. landolti* should include the strategic use of grass germplasm based on an assessment of the area: actual ant population, conditions known to influence the ants (drainage) and knowledge of the susceptibility of various forage grasses to ant predation. Thus we need to develop economic thresholds based on colony density to help make decisions. Current efforts will identify damage thresholds for *Ac. landolti* colony densities in areas to be planted with *A. gayanus*. Simple quadrat counts of colonies in the dry season, when nests can be easily identified, would allow grazers to avoid planting susceptible species in undesirable areas.

A. laevigata at Carimagua readily accepted Mirex baits using orange pulp as a matrix (personal observation). The expected introduction of similar baits based on the chitinase inhibitor diflubenzuron is encouraging. These baits would offer a relatively safe alternative control of *Atta* colonies. In the area of tropical pastures, these baits could profitably be used in seed production and other areas where higher economic return would allow more intensive management. However, control of *Ac. landolti* colonies over extensive grazing areas remains problematic. The best approach may be strategic use of resistant forages such as *B. humidicola* based on an assessment of ant colony density prior to planting.

ACKNOWLEDGMENTS

The authors thank J. M. Cherrett for his advice and suggestions on sampling techniques, and César Cardona and Anthony Bellotti for their help with the manuscript.

LITERATURE CITED

CHERRETT, J. M. 1972. Some factors involved in the selection of vegetable substrate by *Atta cephalotes* (L.) (Hymenoptera: Formicidae) in tropical rain forest. J. Anim. Ecol. 41:647-660.

CHERRETT, J. M., and D. J. PEREGRINE. 1976. A review of the status of leaf-cutting ants and their control. Ann. Appl. Biol. 84:124-128.

COCHRANE, T. T., L. G. SÁNCHEZ, L. G. DE AZEVEDO, J. A. PORRAS, and C. L. GARVER. 1985. Land in Tropical America = La Tierra en América Tropical = A Terra na América Tropical. Centro Internacional de Agricultura Tropical (CIAT), Cali Colombia; Empresa Brasileira de Pesquisa Agropecuária, Centro de Pesquisa Agropecuária dos Cerrados (EMBRAPA-CPAC), Planaltina, D.F. Brasil, 444 p.

FERRUFINO, A,. and S. L. LAPOINTE. 1989. Host plant resistance in *Brachiaria* grasses to the spittlebug *Zulia colombiana*. Entomol. exp. appl. (in press).

FOWLER, H. G., and S. W. ROBINSON. 1977. Foraging and grass selection by the grass-cutting ant *Acromyrmex landolti fracticornis* (Forel) (Hymenoptera: Formicidae) in habitats of introduced forage grasses in Paraguay. Bull. ent. Res. 67:659-666.

HOWARD, J. J. 1988. Leafcutting ant diet selection: Relative influence of leaf chemistry and physical features. Ecology 69:250-260.

MCEWEN, F. L., and G. R. STEPHENSON. 1979. The Use and Significance of Pesticides in the Environment. John Wiley & Sons, New York. 538p.

WALLER, D. A. 1986. The foraging ecology of *Atta texana* in Texas. Ch. 12, pp 146-158 IN Fire Ants and Leaf-cutting Ants: Biology and Management. Lofgren, C. S. & R. K. Vander Meer (eds). Westview Press. 435 p.

48

Relationships Between Argentine Ants and Honeybees in South Africa

B. Buys

INTRODUCTION

Before the introduction of the Argentine ant, *Iridomyrmex humilis* (Mayr), into South Africa, other ant species were troublesome in bee colonies. Control methods for ants at that stage were based on repellent or obstructive materials rather than insecticides, e.g. Attridge (1892) recommended controlling ants by dusting naphthalene among the quilts (insulating material formerly used under the lids of hives). Legs of the hive stand could be placed in vessels containing water or oil, or rags saturated with carbolic acid could be tied around them.

IMPACT ON BEEKEEPING

Since its introduction into South Africa at the turn of the century (Lounsbury 1908), the Argentine ant has become a pest species in both urban and agricultural spheres. A decade and a half after its introduction, this ant was reported as exceedingly plentiful in the south-western Cape (Arnold 1915-1917; Mally 1918).

The Argentine ant's earliest impact on beekeeping is difficult to determine as beekeeping early this century was a relatively unorganized back-yard industry. From correspondence in the agricultural journal, it does appear that the Argentine ant had asserted itself as a beekeeping pest prior to 1925 when Pettey (Journal of the Department of Agriculture 1925) devised a control method whereby hive stands were ant-proofed with a barrier. The wooden stands were made so that the hives rested on pegs surrounded by a mercuric chloride-filled moat.

A further dimension to the pest status of this ant was its detrimental effect on pollination of fruit crops cultivated extensively in the Cape's winter rainfall region. Ants foraging in the blossoms had a repellent effect on pollinating insects, resulting in unsatisfactory pollination. In addition, the nests of pollinating insects were attacked and destroyed, resulting in fewer potential pollinators in orchards (Potgieter 1937; Durr 1952).

Argentine ants can cause a considerable amount of damage in an apiary when they enter hives, initially stealing honey from the combs, then eating the immature forms or brood of the bees and eventually even attacking the demoralized adult bees if they have not already abandoned the nest. The combs that remain after the colony has absconded are then stripped of honey by the ants and by bees from other colonies, before they are completely destroyed by the waxmoth *Galleria mellonella.* Skaife (1961), a keen observer of South African ant life, declared that beekeeping is almost impossible because hordes of ants enter the hives and irritate the bees until they desert their homes. The extent of damage to beekeeping is rarely specifically reported, although regarded as serious in a concise review by De Jong (1978) and in beekeeping handbooks (Root 1966; Eckert and Shaw 1960). Rich (1938) reports the destruction by ants of hundreds of small honeybee colonies intended for use in queen mating by commercial breeders in the southern USA. Ants wreak havoc even among full strength colonies. May (1969) lost 160 colonies over a six year period in the south-western Cape, South Africa, due to Argentine ant predations.

In the south-western Cape, members of the Western Province Beekeepers' Association and the Pollination Service Association agree that full strength production colonies are only occasionally lost to Argentine ants (less than 1%). However, weak colonies and colonies often manipulated or translocated are more susceptible to invasion. Newly established colonies are especially vulnerable. Although the ants breed year round in the south-western Cape (Durr 1956), ant colonies are more populous in the warmer season and attacks on bee colonies more prevalent. In late summer, after Sugar gum (*Eucalyptus cladocalyx*) has stopped flowering in February, the dearth of nectar would predispose populous ant colonies to raid honeybee colonies. At this time of year, manipulation of honeybee colonies, such as colony splitting, can result in heavy losses if adequate precautions against ants are not taken.

Ants seldom penetrate strong, established honeybee colonies, probably because such colonies guard the hive entrance well. Guard bees parry exploring ant foragers by turning their tails and fanning a blast of air at the intruder (Sistrank 1956). This is often accompanied by kicking behavior (Spangler and Taber III 1970).

Weak colonies, however, have fewer guard bees at the entrance so exploring ants may occasionally enter undetected. The returning forager ant communicates its successful foray to other members of its nest. New recruits, following an established odor trail, intensify their efforts to penetrate the hive, even though many are fended off by the guard bees. This persistence, after the establishment of an odor trail, gives rise to the observation that these ants co-ordinate all their invasions (Hultgren and Hultgren 1985), and that the bees are unable to keep these ants out of the hive when attacked (Root 1966). The bee colony becomes increasingly irritated and then demoralized. "Demoralized" is an accepted beekeeping term meaning that normal colony activity, such as foraging, guarding, nursing, etc., is disorganized or even stops. The colony emits a "moaning" sound, and is likely to abscond from the hive, weather permitting.

When honeybee colonies are manipulated, the seals between constituent parts of the hive, usually between the brood chamber and upper chamber or lid, break, leaving cracks. In addition, bees are often left crushed between these parts. In time the cracks are usually re-sealed by bees using plant-derived resins (propolis), but initially the cracks provide unguarded entrances for the ants. The ants take advantage of the opportunity to enter and are also attracted to the crushed bees. Such manipulated hives will likely be overrun by ants.

Argentine ants often attack translocated colonies because these hives usually are inspected *before* being moved, which results in cracks between hive parts. When a bee colony is moved, the inhabitants are disorientated and demoralized; it takes time to re-establish foraging and guarding routines. During inclement weather, colonies may actually starve before foraging is resumed. To overcome this problem, beekeepers may feed the colonies sugar syrup or dry sugar - an additional attraction to the ants. Under such conditions, ants may attack the colonies and demoralize them to such an extent that foraging may be severely inhibited.

Beekeepers must increase the number of colonies sometime during the year, either to balance out any losses, or to extend their respective enterprises. Newly established honeybee colonies suffer from most of the factors that predispose bee hives to Argentine ant attack. The bees have been manipulated and are unsettled, cracks in the hive are not adequately sealed, crushed bees may be present to attract ants, foraging and guard routines have not yet been established, and food reserves may be low or food provided by beekeepers may serve as an added attraction to the ants. If the colonies are weak and the bees have to cluster in the center of the brood nest to maintain brood nest temperature, there may be insufficient bees to adequately guard the hive entrance. Finally, the workers of the unsettled colony are not yet attached to their new home and may abandon it at the slightest provocation, such as the irritation caused by raiding ants.

IMPACT ON POLLINATION

The inhibition of foraging due to the Argentine ant is especially unfortunate when bees have been moved into orchards for pollination. In pollination, the hives are positioned on strips of land between or adjoining orchards and although ant control is practiced in the orchards themselves, it is seldom adequate where the hives are placed. This problem, which affects pollination and thus agriculture, will have to receive more attention in future. The control of ants at hive sites should become an integral part of orchard management. Furthermore, ants should preferably be controlled in advance of the beehives being placed in position for pollination, because anticides in general are also toxic to bees. In some instances, ant control among the beehives has resulted in serious bee mortality (Eckert and West 1948; R.H. Anderson, personal communication), despite adequate precautions.

The reputed repellent effect on bees when ants forage on fruit tree blossoms (Potgieter 1937; Durr 1952) has not been adequately demonstrated. From personal observations and those of other researchers (R.H. Anderson), it would appear that bees will work flowers that contain ants, but display more nervousness on their approach to and on the blossoms and may possibly spend less time there than usual. Whether this has a significant effect on the pollinating efficiency of bees is unknown. However, in view of the great economic value of the deciduous fruit crops in the area, even a slight repellent effect of ants on pollinating bees merits adequate research.

COMPETITION FOR NECTAR

Many insects visit flowers for nectar and thus compete directly with bees. Several *Eucalyptus* species are planted in the Cape as woodlots, as avenues mainly for shade and shelter, as windbreaks, and for firewood. All these plantings provide bees with valuable food sources and play an important role not only in sustaining colonies through otherwise dearth periods between pollination seasons, but also in honey production, thus providing an economic incentive for beekeeping. *Eucalyptus* species also attract the Argentine ant.

The sugar gum *Eucalyptus cladocalyx* is most prevalent in the south-western Cape area, and is of major importance to beekeeping. In two woodlots in the Stellenbosch area, Argentine ants foraged in 95% of 176 trees (Buys, unpublished data). The superabundance of ants in the flowers of *Eucalyptus* species prompted an investigation to quantify their competitive position versus bees (Buys 1987).

In this study, the foraging of Argentine ants and honeybees was observed on the Black ironbark tree *Eucalyptus sideroxylon*. The flowers of this tree species secrete nectar mainly between midnight and sunrise. Average yield per flower was 25 microliters and the concentration of sugars in the nectar rose gradually from 12% (morning) to 24% (late afternoon). Bees collected nectar from 30 minutes before sunrise until 20 minutes after sunset, whereas ants foraged continuously.

From the data, it was estimated that the ants collected about 42% of the nectar at night before the bees started foraging. What happens to the residual 58% of nectar was unclear, because in daylight the ants were unsettled by the technique. However, based on the ants' foraging rates and the time the bees needed to deplete the nectar in the flowers, a rough estimate can be made of the amounts of nectar the ants and bees would take in the daytime, namely 41% and 59% respectively. Of the total amount of nectar secreted by the flowers, ants would then collect about 66%, and bees 34%.

CONTROL OF ANTS AT THE HIVE

Unfortunately, the commercial beekeeper has no wholly satisfactory control method for the Argentine ant. Most control measures, such as those involving ant nest destruction with volatile liquids, the isolation of hives on stands and with barriers (Attridge 1923; Journal of the Department of Agriculture 1925; Davidson 1933; Taylor 1939; Lundie 1952; Beyleveld 1969a, b; Anderson et al. 1983) can only be applied by the small-scale beekeepers, who can afford the expense of frequent inspections and after treatments.

The commercial beekeeper who handles hundreds of colonies, transports them over long distances and cannot inspect the colonies for weeks on end, needs a control method that can be applied extensively and permanently. Insecticides with a long residual activity were very effective in this regard, but many of them - such as DDT and Dieldrin - have been withdrawn from the South African market due to their detrimental effect on the environment. The remaining insecticides in this category are in jeopardy for the same reason. Beekeepers are now agitating for a control method that can replace such insecticides. The current results of control with anticidal baits (Hagmann 1982) are disappointing, but it would appear that a solution to the Argentine ant problem lies in this direction.

ACKNOWLEDGMENTS

I wish to thank Dr R.H. Anderson, Plant Protection Research Institute, for valuable criticism of the manuscript. I also thank the beekeepers of the Cape for sharing their experiences of the Argentine ants with me.

REFERENCES

ANDERSON, R.H., B. BUYS, and M.F. JOHANNSMEIER. 1983. Beekeeping in South Africa. Bull. no. 394. S. Afr. Dept. Agric., Pretoria. 207 pp.

ARNOLD, G. 1915-1917. A monograph of the Formicidae of South Africa. Ann. S. Afr. Mus. 14: 1-402.

ATTRIDGE, H.L. 1892. Destructive insects in bee-hives. Agric. J. Cape Good Hope 5: 78.

ATTRIDGE, A.J. 1923. Beekeeping in South Africa. Speciality Press South Africa, Cape Town - Johannesburg. 190 pp.

BEYLEVELD, G.P. 1969a. An ant-proof hive stand. S. Afr. Bee J. 41: 2-3.

BEYLEVELD, G.P. 1969b. Praktiese byeboerdery in Suid-Afrika. Voortrekkerpers, Johannesburg. 186 pp.

BUYS, B. 1987. Competition for nectar between Argentine ants (*Iridomyrmex humilis*) and honeybees (*Apis mellifera*) on black ironbark (*Eucalyptus sideroxylon*). S. Afr. J. Zool. 22: 173-174.

DAVIDSON, G.P. 1933. Apiculture in South Africa. Central News Agency, South Africa. 80 pp.

DE JONG, D. 1978. Insects: Hymenoptera (Ants, Wasps, and Bees). pp. 138-157. In: R.A. Morse (ed.), Honey bee pests, predators, and diseases. Comstock, Ithaca and London. 430 pp.

DURR, H.J.R. 1952. The Argentine ant, *Iridomyrmex humilis* Mayr. Farming S. Afr. 27: 381-384.

DURR, H.J.R. 1956. Investigation on the seasonal history of the Argentine ant, *Iridomyrmex humilis*, under conditions of winter rainfall. Tydskr. Wet. Kuns 16: 206-209.

ECKERT, J.E., and F.R. SHAW. 1960. Beekeeping. Macmillan, New York. 536 pp.

ECKERT, J.E., and M.C. WEST. 1948. Chlordane versus bees and ants. Am. Bee J. 88: 584.

HAGMANN, L.E. 1982. Ant baits - tests with borax material show strong acceptance. Pest Control 50: 30-31.

HULTGREN, K., and R. HULTGREN. 1985. An old pest - the ant. Gleanings Bee Cult. 113: 317-318.

JOURNAL OF THE DEPARTMENT OF AGRICULTURE. 1925. Departmental Activities. Entomology. J. Dept. Agric. S. Afr. 10: 8-11.

LOUNSBURY, C.P. 1908. Report of the government entomologist for the year 1908. App. IV, pp.66-70. In: Cape of Good Hope, Report of the Department of Agriculture for the year 1908. Cape Times, Cape Town.

LUNDIE, A.E. 1952. The principle diseases and enemies of honeybees. S. Afr. Bee J. 27: 13,15.

MALLY, C.W. 1918. Natural enemies of the Argentine ant, *Iridomyrmex humilis* Mayr. S. Afr. J. Sci. 14: 245-247.

MAY, A.F. 1969. Beekeeping. HAUM, Cape Town. 266 pp.

POTGIETER, J.T. 1937. The Argentine ant. Farming S. Afr. 12: 160.

RICH, H.I. 1938. The control of ants in the apiary. Am. Bee J. 78: 120-121.
ROOT, A.I. 1966. The ABC and XYZ of bee culture. A.I. Root Company, Medina, Ohio. 726 pp.
SISTRANK, T.W. 1956. Bee defence against ants. Gleanings Bee Cult. 84: 681.
SKAIFE, S.H. 1961. The study of ants. Longmans, London. 178 pp.
SPANGLER, H.G., and S. TABER III. 1970. Defensive behavior of honey bees towards ants. Psyche, UK 77: 184-189.
TAYLOR, F. 1939. Beekeeping for the beginner. Un. S. Afr. Dept. Agric. For. Bull. no. 199. 108 pp.

49

Carpenter Ants (*Camponotus* spp.): Pest Status and Human Perception

H.G. Fowler

INTRODUCTION

Urban insects can cause two types of problems (1) Direct damage, which is quantifiable for commodity loss such as food or wood and includes time and money spent for control. (2) Psychological impact, especially the assumed threat to a city dweller's home or personal property. Since many urbanites rarely encounter animals or insects in a natural setting, they may assign pest status to non-threatening creatures or behaviors. Ethnic background will also affect reactions to the insects and their nesting activities in houses or apartments (Fowler 1984).

Carpenter ants, members of the holarctic subgenus *Camponotus* and genus *Camponotus*, damage structures by gnawing nest chambers in wood (Fowler 1986). According to recent United States and European studies, the public considers carpenter ants to be an economic problem (Fowler 1982, 1983a, 1983b, 1984) greater than subterranean termites, a major structural pest. As Akre and Hansen have discussed the biology and pest status of these ants (see chapters this volume), this paper focuses on public perceptions of the problem.

To perform an analysis of public response to the perceived threat posed by carpenter ants, I relied on the Cooperative Extension Service (CES) records in several U.S. states. These secondary sources represent unsolicited requests for information by the public and are not a planned socio-economic survey. The requests are probably a mixture of real and perceived problems (Waldron 1962).

PEST STATUS

Francoeur (1977) and Fowler (1986) have already reviewed the pest status of carpenter ants and I will touch only briefly on the salient points.

In North America, the principal pest species are *C. pennsylvanicus* (northeast) and *C. herculeanus* (northwest). In Europe, *C. herculeanus*, causes most damage. These ants generally replaced subterranean termites as the number one wood pest, in terms of money spent on control. (Fowler 1982, 1983b, 1989; Butovitsch 1976), reaching $25,000,000 in the United States each year (Fowler 1986). Other costs - such as wood services and replacement -

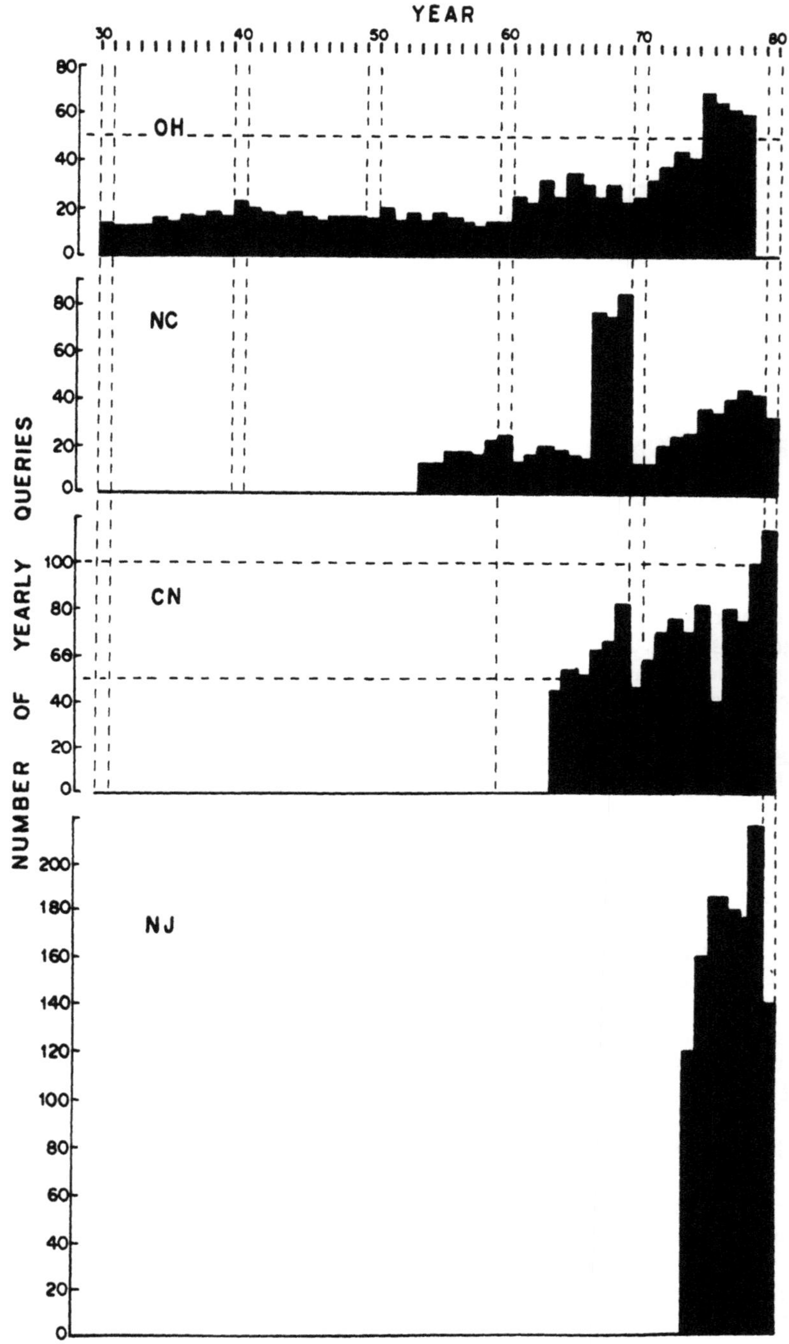

FIGURE 1. Yearly number of carpenter ant queries fielded by the Cooperative Extension Service (CES) in four Northeastern states.

probably reach an equal amount. In addition, outdoors the carpenter ant damages shade and production trees causing direct wood and esthetic losses and indirect tree replacement costs (Graham 1918; Fowler and Parrish 1982; Fowler and Roberts 1982).

If European figures match those for the United States, then each year carpenter ants cause at least $100,000,000 (U.S) of damage worldwide. Although, carpenter ants are as economically important as either fire ants or leaf-cutting ants they have been overlooked, perhaps because unlike the other groups, they are primarily a household problem.

A non-economic factor unique to the carpenter ant is its involvement with people. To properly evaluate its pest status, we must consider this interaction and its effects.

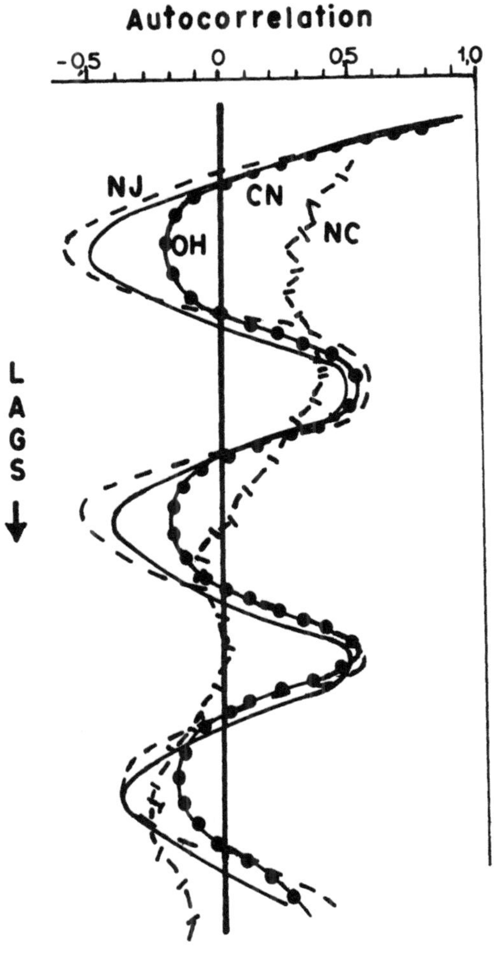

FIGURE 2. Monthly autocorrelations of carpenter ant queries fielded by the CES in various states of the United States.

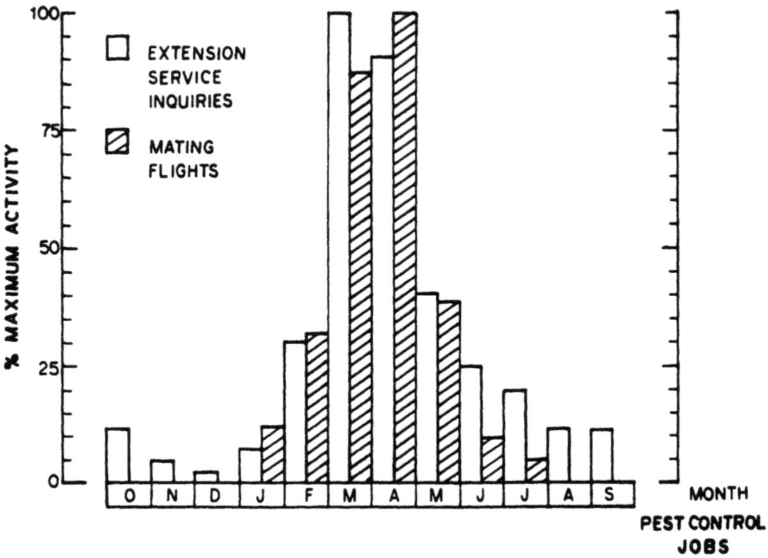

FIGURE 3. Seasonal distribution of carpenter ant queries and mating flights in New Jersey. Data from Fowler (1986); Fowler and Roberts (1982b).

HUMAN PERCEPTION

CES records clearly show an increased amount of year-round concern over carpenter ants, especially from the northern states (Fig. 1) (Fowler 1982, 1983a), whereas concern weakens in southern states like North Carolina (Fig. 2) (Fowler 1984, Fowler and Roberts 1983).

Apparently, alates arouse the most public interest (Fig. 3) (Fowler 1986), given the high correlation between queries to the CES and monthly mating flights (r=0.97) (Fig. 3). What action do people take once the CES has given them information? Most call a pest control company for control. Is this a reasonable measure?

An answer requires closer examination of the CES data. A comparison of queries about subterranean termites and carpenter ants yields one dominant pattern: a rising interest in carpenter ants (Fowler 1983a, 1983b). In cold areas with frequent snowfall, queries about the ants far outnumber those about termites (Fig. 4). Weather conditions may create the strongly seasonal patterns previously documented. Patterns for termites or ants are not strongly correlated to state surface area, population or weather variables (Fig. 5). Thus, the importance of carpenter ants relative to subterranean termites depends on the severity of the weather - carpenter ants prefer cold climes, while the frost line marks the limit of the termites' geographical distribution (Fowler 1983a, 1989).

Other trends can be interpreted by examining data from individual states. In New Jersey, carpenter ant queries continue to rise each year, while subterranean termite queries decrease (Fig. 6). These findings suggest that current control methods work more effectively against the termites than the

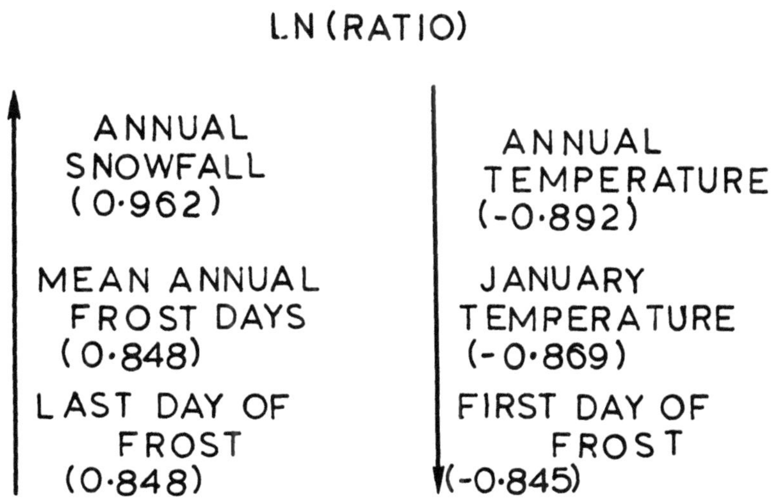

FIGURE 4. Relation between carpenter ant queries and subterranean termite queries handled by the CES in New Jersey, New York, Ohio, Connecticut, North Carolina, Rhode Island, and North Carolina.

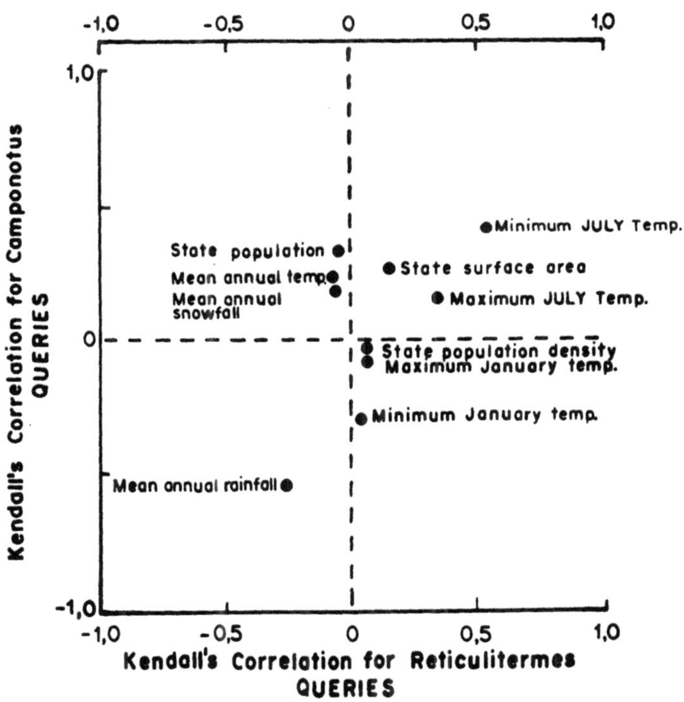

FIGURE 5. Kendall's rank correlation of the ratio of carpenter ant to subterranean termite queries in the states listed in Fig. 4.

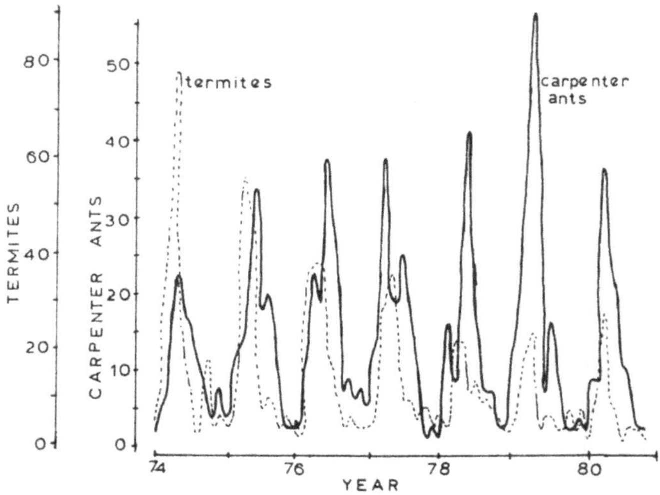

FIGURE 6. Monthly trends in carpenter ant and subterranean termite queries in New Jersey.

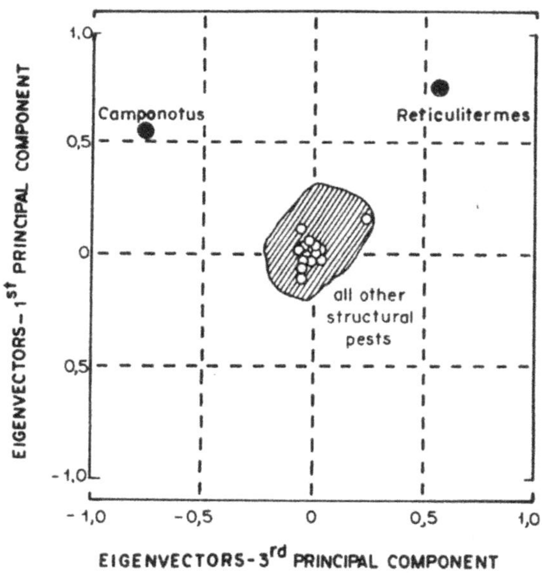

FIGURE 7. Ordination of carpenter ant and subterranean termite monthly queries with respect to all other structural pests in New Jersey. The first principal component orders annual constancy of CES queries, while the third principal component orders seasonal distribution.

ants. Indeed, structures can easily be made termite-proof because they must gain access to wood through the soil. Carpenter ants can attack any part of a structure.

Although carpenter ants and subterranean termites are less common than pests such as cockroaches, CES receives a higher number of queries about them. The public responds not simply to increased sightings, but to the possibility of suffering economic damage greater than that inflicted by other structural pests.

NEW RESEARCH DIRECTIVES

Unlike other pest ant species, carpenter ants have a social and psychological impact due to their affinity for urban settings. Entomologists, most of whom have been unable to deal with this aspect, must collect the preliminary data that will help our understanding of the situation (Fowler 1983c, 1989), and as stated by the National Research Council (1980), it is imperative that the problems posed by species such as the carpenter ant be handled by multi-disciplinary teams which include economists, sociologists, psychologists and entomologists. A new field--"Urban Entomology"-- will develop along radically different lines than standard, agriculture-based research. Due to increased urbanization and shifting economic concentrations, the demand for solutions will be great. Universities must prepare a new type of professional and create funding avenues to help tomorrow's urban pest specialist.

ACKNOWLEDGMENTS

Most of my carpenter ant research was funded by the New Jersey Pest Control Association, the New Jersey Agricultural Experiment Station, and the Graduate School of Rutgers University. My late friend and professor, Radclyffe B. Roberts, was responsible for many of my research questions, while Edwin Rajotte, Lou Vasvery, Herb Streu and Mark Parrish provided constant encouragement and criticisms. Special thanks go to the CES entomologists for permitting access to their records, and to the computing facility of Rutgers University for computing time. This manuscript was finished through assistance provided by the Special Program for the Development of Ecological Research in Argentina (SPAIDERA), and I especially thank Dr. Jorge Rabinovich for his assistance, and Abraham Willnick of the Instituto Miguel Lillo for use of facilities in Tucuman, and Silvia Claver of CRUCIT for access to facilities in Mendoza.

LITERATURE CITED

BUTOVITSCH, V. 1976. Uber Vorkommen und Schadwirklung der Rossameisen *Camponotus herculeanus* und *C. ligniperda* in Gebauden in Schweden. Mater. Organ. 11: 162-170. Ebling, W. 1975. Urban Entomology. Div. Agric. Sci., Univ. Calif., Los Angeles.

FOWLER, H.G. 1982. Public concern for carpenter ants growing. Pest Control 50: 30.

FOWLER, H.G. 1983a. Urban structural pests: carpenter ants (Hymenoptera: Formicidae) displacing subterranean termites (Isoptera: Rhinotermitidae) in public concern. Environ. Entomol. 12: 997-1002.

FOWLER, H.G. 1983b. Social insects as urban pests: an analysis of seasonality and human perception. Mater. Organ. 18: 93-105.

FOWLER, H.G. 1983c. Human effects on nest survivorship of urban synanthropic wasps. Urban Ecol. 7: 131-143.

FOWLER, H.G. 1984. Cycles of public awareness to household pests in New Jersey. J. Interdiscipl. Cycle Res. 15: 237-239.

FOWLER, H.G. 1986. Biology, economics and control of carpenter ants. pp. 272-289. In S.B. Vinson (ed.), Economic Impact and Control of Social Insects. Praeger Press, New York. 422 pp.

FOWLER, H.G. 1989. Subterranean termites and structural pests in the Northeastern United States: public perception and its analysis. In, C. Bach (ed.), Termitologia y Patrimonio Historico, Gobierno Autonomo de Andalusia, Sevilla (in press).

FOWLER, H.G. and M.D. PARRISH. 1982. Urban shade trees and carpenter ants. J. Arborcult. 8: 281-284.

FOWLER, H.G. and R.B. ROBERTS. 1982a. Carpenter ant induced wind breakage in New Jersey Shade Trees. Can. Entomol. 114: 649-650.

FOWLER, H.G. and R.B. ROBERTS. 1982b. Seasonal occurrence of founding queens and the sex ratio of *Camponotus pennsylvanicus* in New Jersey. J. N.Y. Entomol. Soc. 90: 247-251.

FOWLER, H.G. and R.B. ROBERTS. 1983. Activity cycles of carpenter ants (*Camponotus*)(Hymenoptera:Formicidae) and subterranean termites (*Reticulitermes*) (Isoptera: Rhinotermitidae):inference from synanthropic records. Ins. Soc. 30: 323-331.

FRANCOEUR, A. 1977. Synopsis taxonomique et economique des fourmis du Quebec. Ann. Soc. Entomol. Quebec 22: 205-212.

GRACE, J.K. and D.L. WOOD. 1987. Delusory cleptoparasitosis: delusions of arthropod infestation in the home. Pan Pac. Entomol. 63: 1-4.

GRAHAM, S.A. 1918. The carpenter ant as a destroyer of sound wood. 17th Ann. Report., Minn. Agric. Exp. Sta.

NATIONAL RESEARCH COUNCIL. 1980. Urban pest management. National Academy Press, Washington.

WALDRON, W.G. 1962. The role of the entomologist in delusory parasitosis (entomophobia). Bull. Entomol. Soc. Am. 8: 81-83.

50

The Biology and Economic Impact of *Pogonomyrmex* Harvester Ants

W.P. MacKay

INTRODUCTION

Ants are among the most abundant animals in most terrestrial habitats and may be the dominant insects in many ecosystems (MacKay 1981, 1989). It is not only their dominance in numbers and biomass that make these insects important in ecosystems, but also their activity. Ants can concentrate more energy at critical points than can single competitors; they use sheer numbers to construct nests in suboptimal habitats as well as defend young and retrieve food efficiently (Wilson 1985).

Seed harvesters (especially *Pogonomyrmex* spp., but also species in the genera *Pheidole, Solenopsis,* and *Messor* [= *Veromessor,* see Bolton 1982]) are especially abundant in the deserts and pasturelands of western North America (MacKay 1989). The genus *Pogonomyrmex* has existed since the Oligocene (Burnham 1978) and is distributed throughout North and South America from Canada to Patagonia (MacKay 1981). The biology and economic impact of the genus have been previously summarized by Lavigne (1969), Lavigne and Rogers (1974), Rogers (1974), Rogers et al. (1972), Whitford et al. (1976), MacKay (1981, 1985, 1989) and MacKay and MacKay (1989).

POPULATIONS

Considerable work has been done on nest densities, populations and biomass of various species in the genus *Pogonomyrmex* (MacKay 1981). Nest populations range from a few hundred, to thousands of ants. Nest density ranges from 1 to 82 nests per hectare. Nest depth can reach four or five meters. Individual nests have a life expectancy of from 14 to over 50 years (Keeler 1982, 1988; Porter and Jorgensen 1988).

PHENOLOGY OF NESTS

Egg laying in North American species begins in late April to late May (Fig. 1). Larvae appear about one week after eggs are laid, the first pupae about two weeks later. Callows (immature workers) are found in the nest five weeks after the eggs are laid, and remain pale for about three weeks. A large proportion of the first eggs become reproductives; later, eggs become workers.

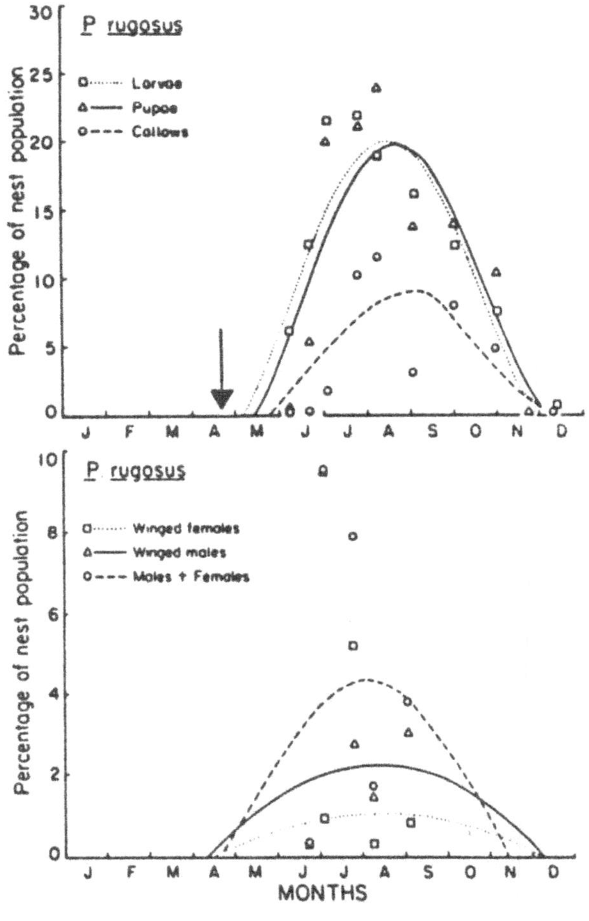

FIGURE 1a. Seasonal changes in the brood and reproductive populations of three species of *Pogonomyrmex* harvester ants. Arrows indicate when the first eggs were found in nests. (Modified from MacKay [1981].)

Larval populations peak in late July to mid August, depending on the species (Fig. 1). Pupal populations peak in August, callow populations in September. Winged reproductives remain in the nest until July or August when nuptial flights take place.

Mating flights occur either in the morning or afternoon; the timing is species-specific. Extremely aggressive during the flight, workers are present in large numbers outside the nest. The reproductives in many species meet on the highest landmark in an area. This may be a hill, a tree, a house or even a researcher. Homeowners often complain of ants flying around houses during the mating flight. Invariably, the ants gather around the tallest house in the neighborhood and nothing can be done to avoid the swarm. The female

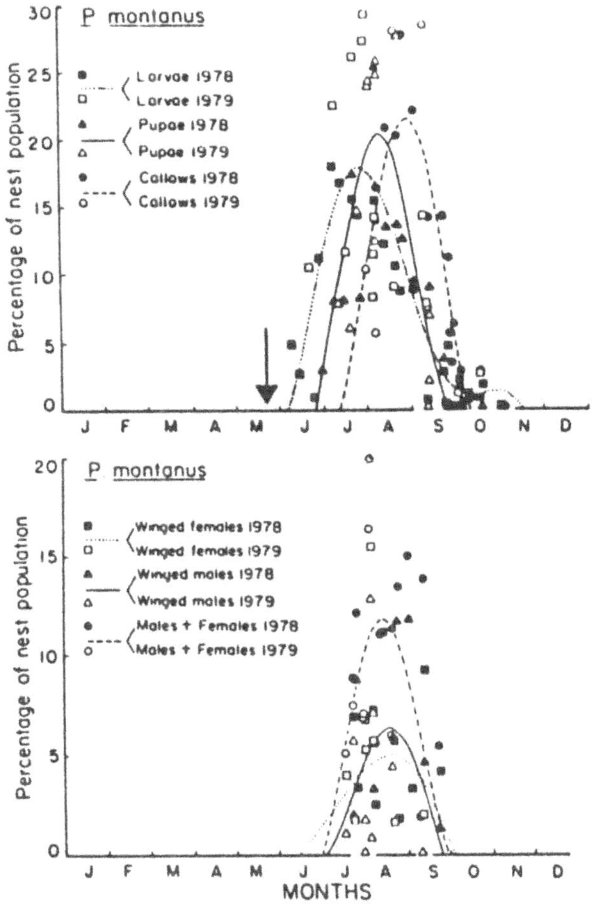

FIGURE 1b. See caption for Figure 1a.

usually mates with several males and then leaves the swarm to found a colony. The males die soon after mating.

FORAGING PATTERNS

Most species demonstrate seasonal changes in foraging patterns, with much activity in mid-summer and no activity in the winter (Fig. 2). Species which occur in more moderate environments have longer foraging seasons (MacKay 1981). Foraging increases abruptly in the spring when environmental conditions become optimum. The activity peaks in mid summer and decreases again when the temperatures drop (Fig. 2). Only about 20% of ants in the nest are foragers at any one time. Workers live for about a year

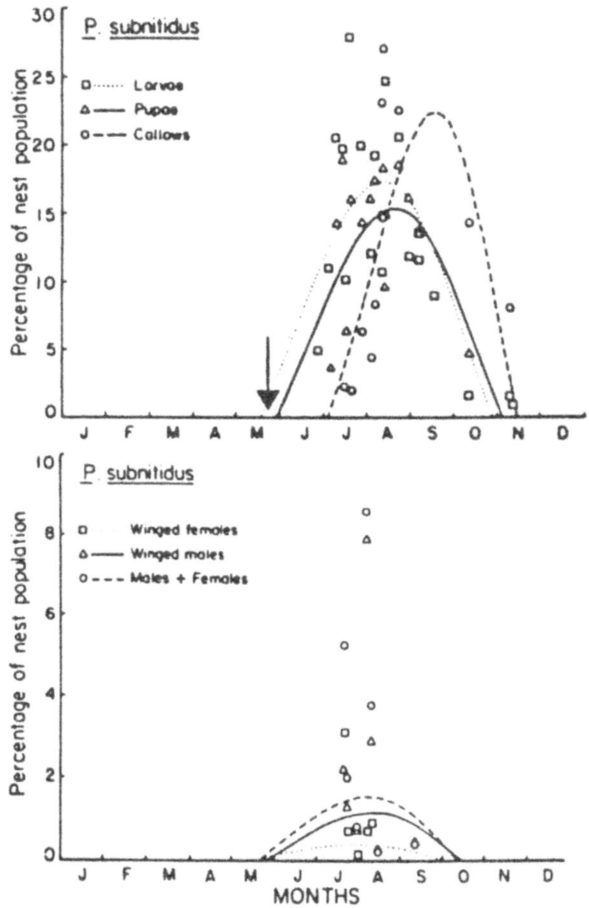

FIGURE 1c. See caption for Figure 1a.

and only become foragers at the end of their lives (MacKay 1983). They live only a few days once they become foragers (Porter and Jorgensen 1981). Each forager completes between two and ten trips each day (MacKay 1981). Harvester ants use a wide variety of food sources, e.g. vertebrate feces, insects, plant parts and seeds (Fig. 3). Most species use seeds as their primary food source (Fig. 3), especially in the most arid ecosystems. *Pogonomyrmex montanus* is exceptional as it uses dead insects and vertebrate fecal material to a large extent, but occurs in clearings of more mesic pine forests. This trend also occurs in *P. occidentalis* and *P. subnitidus* which are often found at higher elevations in more mesic environments.

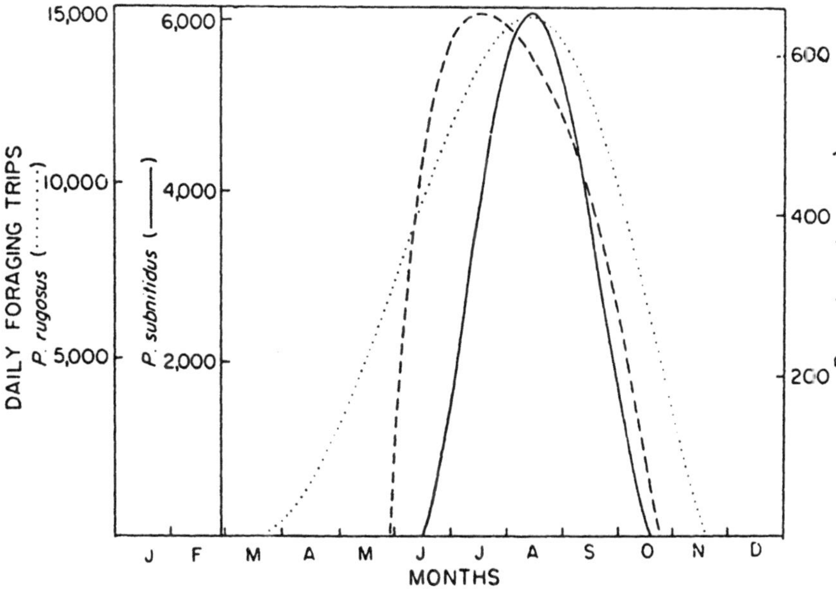

FIGURE 2. Comparison of the number of daily foraging trips in three species of *Pogonomyrmex* harvester ants. (Modified from MacKay [1985].)

SEED STORAGE

Many species of harvester ants store seeds in their nests. The mass of seeds stored changes seasonally (MacKay and MacKay 1984), with much seed storage occurring in the fall and seed consumption occurring in the spring. Seed storage may be favored in habitats with short, infrequent pulses of primary production (Marsh 1986), in which it serves to ensure survival during droughts (Marsh per. comm., Polis pers. comm.). When granaries are full, foraging ceases in some *Pogonomyrmex* (Whitford and Ettershank 1975, Bailey and Polis 1987). Many authors have assumed that seeds are eaten during the winter or are a form of insurance in an unpredictable environment. The available data do not support either of these "obvious" hypotheses (MacKay and MacKay 1984). Seed storage, in at least some *Pogonomyrmex* spp., appears to be a form of protection against predation, especially by spiders. When a nest is under siege, foragers stop above ground activity until the spiders leave.

SOIL PROCESSES

Ants are very important in soil processes because they move subterranean soil to the surface, mixing soil nutrients and lowering the bulk density of the soil. Individual nests of *Pogonomyrmex occidentalis* transport 2.8 kg of soil/colony to the surface (Rogers and Lavigne 1974). Soil turnover by ants ranges up to 842 kg/ha/yr in the Chihuahuan Desert (Whitford et al. 1986). These rates result in a deposition of from 3 mm to 2 cm of soil on the surface

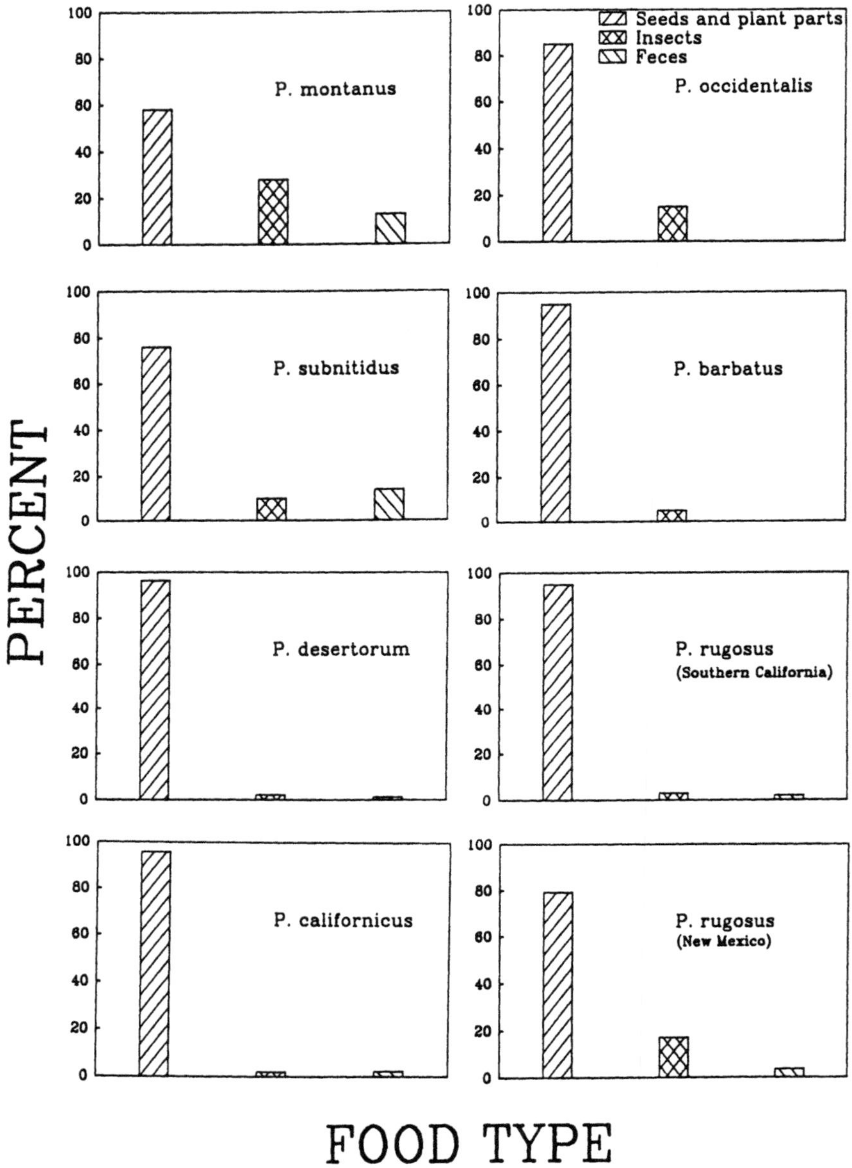

FIGURE 3. Food types of ants of the genus *Pogonomyrmex* from arid and semi-arid regions in southwestern United States. All except *P. montanus*, *P. occidentalis* and *P. subnitidus* occur in deserts. Data from Lavigne (1969), Whitford et al. (1976), Whitford (1972, 1978a) and MacKay (1981). (From MacKay [1989].)

every century (Whitford et al. 1986; MacKay 1989). This mixing of plant and animal matter with soil results in more nutrients available to plants and influences plant species composition (Whitford et al. 1986; MacKay 1989).

ECONOMIC IMPACT

Ants affect rangelands in several ways: (1) seed harvesters remove seeds, (2) ants move tremendous amounts of subsoil to the surface during nest construction, and (3) many species clear vegetation from their mounds, or the vegetation which grows on mounds is considerably different from the surrounding vegetation. They also compete with plants for soil moisture (Rissing 1988).

Pogonomyrmex also has an impact on the public. They have very toxic venoms (Pinnas et al. 1977; Schmidt et al. 1980). Mammalian toxicity is greater than that of any other insect venom, eight to ten times as toxic as honeybee venom (Schmidt and Blum 1978). It is not especially toxic to other insects, and appears to have evolved as a deterrent to vertebrate predators. They also damage highway shoulders (Crowell 1963).

Harvester ants collect between 2 to 10 % of the total annual seed production in the North American desert (Whitford 1978a). The selective removal and consumption of seeds of certain species affect the relative abundance of species (Whitford 1978b). Many species, especially *P. occidentalis, P. salinus* (=*P. owyheei* see Shattuck 1987), *P. rugosus* and *P. barbatus*, clear vegetation from areas around their mound. There are several possible explanations for this behavior. Mound humidity may become higher than the surrounding soil, due to the lack of transpiration by plants (Rogers and Lavigne 1974). Harvester ants prefer sites exposed to higher temperature (Carlson and Gentry 1973, but see MacKay et al. 1986). The removal of plants would expose the nest directly to the sun. Spiders build webs in the surrounding vegetation and trap foragers when they leave the nest (MacKay 1982). Destruction of vegetation around the nest reduces the numbers of sites that can be used by spiders. The area of soil surface denuded by harvester ants ranges from 0.02% to 8% (Sharp and Barr 1960; Kirkham and Fisser 1972; Rogers and Lavigne 1974; Sneva 1979).

Harvester ants also change the spatial distribution of desert plants by moving seeds to other locations (= myrmecochory). Survival of a plant's offspring could be increased by myrmecochory away from the parent plant where competition with the parent may reduce survivorship. Myrmecochory has also been shown to help seeds escape predation by rodents (Heithaus 1981). Several plant species have propagules (agents of reproduction) made up of diaspores (a seed with protective covering) and elaisomes (an appendage which contains oils and fats and is very attractive to ants), which are collected by ants (Drake 1981). Often the elaisome is eaten and the diaspore with seed discarded intact. Myrmecochory is found in at least 223 genera in 61 families (Nesom 1981).

Numerous studies have demonstrated the importance of ant mounds in changing an area's vegetation (Petal 1978). Soils from mounds are higher in nutrients and organic matter due to the concentration of resources by ants at a specific point (Rogers and Lavigne 1974; Davidson and Morton 1981; Beattie and Culver 1983). The highly directional dispersal of diaspores by ants to nests where nutrients are concentrated and possibly more accessible

(Davidson and Morton 1981) greatly enhances the survival of the seedlings (Beattie and Culver 1983) and increases plant production in the area surrounding the nest (Rogers and Lavigne 1974; Rissing 1986). As a consequence of seed harvesting and the change in chemical composition and physical properties of the nest's soil, flora around ant nests is quite different from flora on surrounding areas (MacKay 1989). In fact, some plant species grow exclusively on ant mounds (Davidson and Morton 1981).

CONTROL

Several methods have been used to control harvester ants (Wildermuth and Davis 1931; List 1954; Crowell 1963; Lavigne 1966; Lavigne and Fisser 1966; Borth 1986), although documentation is usually poor, especially in agricultural situations (Borth 1986). The most effective insecticide presently used is hydramethylnon (Borth 1986). Willard and Crowell (1965) recommend treatment in the spring, after foraging activity has started but before reproductives are ready to fly. This strategy kills the nest and prevents reinfestation by killing the reproductives.

CONCLUSION

Harvester ants form an interesting and important group of organisms, especially in the western United States and northern Mexico. Unfortunately, details concerning these ants' biologies are few. They play an important role in ecosystem processes such as energy flow and nutrient cycling. Any study done at an ecosystem level in these areas must include the role of harvester ants. Unfortunately harvester ants do damage pasturelands, especially by clearing vegetation from areas around their nests. Ants do not cause poor range conditions. Several authors have concluded that an abundance of harvester ant nests in a pasture is the result rather than the cause of poor range condition (ie., see Sharp and Barr 1960; Kirkham and Fisser 1972). It is evident that ants do not clear more than 1% of the land area. In addition, they increase productivity around their nests. The increase in productivity may be even greater than the mass of destroyed vegetation. In most cases the small loss in plant production and the consumption of seeds does not justify the expense of control (Sneva 1979).

ACKNOWLEDGMENTS

Emma MacKay participated in all aspects of field work. E. MacKay and C. Rose critically reviewed the manuscript. My research discussed in this paper has been supported by Sigma Xi, The Scientific Research Society of North America, The American Museum of Natural History, The Chancellor's Patent Fund of the University of California, an Irwin Newell Award of the University of California (Riverside), the Consejo Nacional para Ciencia y Tecnologia (Mexico), the Escuela Superior de Agricultura "Hermanos Escobar" (Mexico) and the National Science Foundation, and without this support the research would have been impossible. Approved for publication by the Texas Agricultural Experiment Station as number TA- 24665.

LITERATURE CITED

BAILEY, K. H. and G. A. POLIS. 1987. Optimal and central-place foraging theory applied to a desert harvester ant, *Pogonomyrmex californicus*. Oecologia 72:440-448.

BEATTIE, A. J. and D. C. CULVER. 1983. The nest chemistry of two seed-dispersing ant species. Oecologia 56:99-103.

BOLTON, B. 1982. Afrotropical species of the myrmicine ant genera *Cardiocondyla, Leptothorax, Melissotarsus, Messor*, and *Catacaulus* (Formicidae). Bulletin of the British Museum of Natural History (Entomology) 45:307-370.

BORTH, P. W. 1986. Field evaluation of several insecticides on maricopa harvester ant (Hymenoptera: Formicidae) colony activity in fallow southwestern Arizona cropland. J. Econ. Entomol. 79:1632-1636.

BURNHAM, L. 1978. Survey of social insects in the fossil record. Psyche 85:85-134.

CARLSON, D. M. and J. B. GENTRY. 1973. Effects of shading on the migratory behavior of the Florida harvester ant, *Pogonomyrmex badius*. Ecology 54:452-453.

CROWELL, H. H. 1963. Control of the western harvester ant *Pogonomyrmex occidentalis*, with poisoned baits. J. Econ. Entomol. 56:295-298.

DAVIDSON, D. W. and S. R. MORTON. 1981. Myrmecochory in some plants (F. Chenopodiaceae) of the Australian arid zone. Oecologia 50:357-366.

DRAKE, W. E. 1981. Ant-seed interaction in dry sclerophyll forest on North Stradbroke Island, Queensland. Australian Journal of Botany 29:293-309.

HEITHAUS, E. R. 1981. Seed predation by rodents on three ant-dispersed plants. Ecology 62:136-145.

KEELER, K. H. 1982. Preliminary report of colony survivorship in the western harvester ant (*Pogonomyrmex occidentalis)* in western Nebraska. Southwest. Natur. 27:245-246.

KEELER, K. H. 1988. Colony survivorship in *Pogonomyrmex occidentalis*, western harvester ant, in western Nebraska. Southwest. Natur. 33:480-482.

KIRKHAM, D. R. and H. G. FISSER. 1972. Rangeland relations and harvester ants in northcentral Wyoming. J. Range Manag. 25:55-60.

LAVIGNE, R. J. 1966. Individual mound treatments for control of the western harvester ant, *Pogonomyrmex occidentalis*, in Wyoming. J. Econ. Entomol. 59: 525-532.

LAVIGNE, R. J. 1969. Bionomics and nest structure of *Pogonomyrmex occidentalis* (Hymenoptera: Formicidae). Ann.Entomol. Soc. Amer. 62:1166-1175.

LAVIGNE, R. J. and H. G. FISSER. 1966. Controlling western harvester ants. Mountain States Regional Publication # 3, 6pp. Lavigne, R. J. and L. E. Rogers. 1974. An annotated bibliography of the harvester ants, *Pogonomyrmex occidentalis* (Cresson) and *Pogonomyrmex owyheei* Cole. Science Monogr. # 24, Univ. of Wyoming Agric. Exper. Stn. 18 pp.

LIST, G. M. 1954. Western harvester ant control tests. Tech. Bull 55, Colorado Agric. Exper. Stn. 20 pp.

MACKAY, W. P. 1981. A comparison of the nest phenologies of three species of *Pogonomyrmex* harvester ants (Hymenoptera: Formicidae). Psyche 88:25-74.

MACKAY, W.P. 1982. The effect of predation of western widow spiders (Araneae: Theridiidae) on harvester ants (Hymenoptera: Formicidae). Oecologia 53:406-411.

MACKAY, W. P. 1983. Stratification of workers in harvester ant nests (Hymenoptera: Formicidae). J. Kans. Entomol. Soc. 56:538-542.

MACKAY, W. P. 1985. A comparison of the energy budgets of three species of *Pogonomyrmex* harvester ants (Hymenoptera: Formicidae). Oecologia 66:484-494.

MACKAY, W. P. 1989. The roles of ants and termites in desert ecosystems. *In*: The Ecological Structure of Desert Communities, G. Polis ed., Univ. of Arizona Press, 213-219.

MACKAY, W. P. and E. MACKAY. 1984. Why do harvester ants store seeds in their nests? Sociobiology 9:31-47.

MACKAY, W. P. and E. MACKAY. 1989. Diurnal foraging patterns of *Pogonomyrmex* harvester ants (Hymenoptera: Formicidae). Southwestern Natur. 34: In press.

MACKAY, W. P., S. SILVA, D. C. LIGHTFOOT, M. I. PAGANI and W. G. WHITFORD. 1986. Effect of increased soil moisture and reduced soil temperature on a desert soil arthropod community. American Midland Naturalist 116:45-56.

MARSH, A. C. 1986. Microclimatic factors influencing foraging patterns and success of the thermophilic desert ant, *Ocymyrmex barbiger*. Insectes Sociaux 32:286-296.

NESOM, G. L. 1981. Ant dispersal in *Wedelia hispida* Hbk. (Heliantheae: Compositae). Southwestern Naturalist 26:5-12.

PETAL, J. 1978. The role of ants in ecosystems. Pages 293-325 *In* M. V. Brian (editor), Production ecology of ants and termites. Cambridge University Press.

PINNAS, J. L., R. C. STRUNK, T. M. WANG and H. C. THOMPSON. 1977. Harvester ant sensitivity: in vitro and in vivo studies using whole body extracts and venom. J. Allerg. Clin. Immunol. 59:10-16.

PORTER, S. D. and C. D. JORGENSEN. 1981. Foragers of the harvester ant, *Pogonomyrmex owyheei:* a disposable caste? Behav. Ecol. Sociobiol. 9:247-256.

PORTER, S. D. and C. D. JORGENSEN. 1988. Longevity of harvester ant colonies in southern Idaho. J. Range Manag. 41:104-107.

RISSING, S. 1986. Indirect effects of granivory by harvester ants: plant species composition and reproductive increase near ant nests. Oecologia 68:231-234.

RISSING, S. W. 1988. Seed-harvesting ant association with shrubs: competition for water in the Mohave Desert? Ecology 69:809-813.

ROGERS, L. E. 1974. Foraging activity of the western harvester ant in the shortgrass plains ecosystem. Environ. Entomol. 3:420-424.

ROGERS, L. E. and R. J. LAVIGNE. 1974. Environmental effects of western harvester ants on the shortgrass plains ecosystem. Environ. Entomol. 3:994-997.

ROGERS, L. E., R. LAVIGNE and J. L. MILLER. 1972. Bioenergetics of the western harvester ant in the shortgrass plains ecosystem. Environmental Entomology 1:763-768.

SCHMIDT, J. O. and M. S. BLUM. 1978. A harvester ant venom: chemistry and pharmacology. Science 200:1064-1066.

SCHMIDT, J. O., M. S. BLUM and W. L. OVERAL. 1980. Comparative lethality of venoms from stinging Hymenoptera. Toxicon 18:469-474.

SHARP, L. E. and W. F. BARR. 1960. Preliminary investigations of harvester ants on southern Idaho rangelands. J. Range Manag. 13:131-134.

SHATTUCK, S. 1987. An analysis of geographic variation in the *Pogonomyrmex occidentalis* complex (Hymenoptera: Formicidae). Psyche 94:159-179.

SNEVA, F. A. 1979. The western harvester ants: their density and hill size in relation to herbaceous productivity and big sagebrush cover. J. Range. Manag. 32:46-47.

WHITFORD, W. G. 1972. Demography and bioenergetics of herbivorous ants in a desert ecosystem as functions of vegetation, soil type and weather variables. Unpublished Manuscript.

WHITFORD, W. G. 1978a. Foraging by seed-harvesting ants. Pages 107-110. *In* M. V. Brian editor. Population ecology of ants and termites. Cambridge University Press.

WHITFORD, W. G. 1978b. Foraging in seed-harvester ants *Pogonomyrmex* spp. Ecology 59:185-189.

WHITFORD W. G. and G. ETTERSHANK. 1975. Factors affecting foraging activity in Chihuahuan Desert harvester ants. Environmental Entomologist 4:689-696.

WHITFORD, W. G., P. JOHNSON and J. RAMIREZ. 1976. Comparative ecology of the harvester ants *Pogonomyrmex barbatus* (F. Smith) and *Pogonomyrmex rugosus* (Emery). Insectes Sociaux 23:117-132.

WHITFORD, W. G., D. SCHAEFER and W. WISDOM. 1986. Soil movement by desert ants. Southwestern Naturalist 31:273-274.

WILDERMUTH, V. L. and E. G. DAVIS. 1931. The red harvester ant and how to subdue it. USDA Farmer's Bull. #1668, 21 pp.

WILLARD, J. R. and H. H. CROWELL. 1965. Biological activities of the harvester ant, *Pogonomyrmex owyheei*, in central Oregon. J. Econ. Entomol. 58:484-489.

WILSON, E. O. 1985. The sociogenesis of insect colonies. Science 228:1489-1495.

51

The Role of Ants in Australian Land Reclamation Seeding Operations

J.D. Majer

INTRODUCTION

The mining industry contributes a sizeable portion of the Australian economy. Extracted materials include coal, uranium, iron ore, bauxite and mineral sands. Most of these are extracted by open-cut mining operations. The legislative arrangements between the various mining companies and the respective State or Federal Government agencies generally require that the disturbed land be rehabilitated to an appropriate post-mining land use. In view of the fact that mining is often carried out in areas of native vegetation, restoration of the original vegetation is frequently a common goal of the rehabilitation program.

The areas of native vegetation where mining takes place are generally characterized by forest, woodland, low shrubland or open heath communities (Langkamp and Plaisted 1987). Many of the earlier rehabilitation programs attempted to revegetate the land by planting nursery-reared seedlings. However, the richness of the flora in most of the mined regions means that plant nurseries could never cope with the wide range of species which need to be replaced. Consequently, there has been a move, from the early 1970s onwards, towards seeding the mined areas with mixed native species (see reviews in Langkamp 1987).

The dispersal and survival of many Australian plant seeds is intimately lined to the corresponding ant fauna of the region. The removal of seeds (diaspores) by ants conforms to three general categories: (1) granivory - removal and subsequent predation, in which the ant removes the diaspore and consumes the embryo; (2) nest decoration - removal and incorporation of the diaspore into the nest structure; and (3) myrmecochory - dispersal of the diaspore while leaving the embryo intact. Nest decoration is not a widespread phenomenon. Fruits or seeds of myrmecochorous species possess an elaiosome, often an oil- or fat-bearing body, which is attractive to ants. After the diaspore is transported by ants, they may eat the elaiosome and discard the seed itself. This may benefit the seed by (1) dispersing it over relatively long distances (Andersen 1988); (2) isolating it from seed predators and the extreme heat of wildfires, enhancing its longevity (Berg 1975); (3) providing increased nutrients in the soil around nests (Davidson and Morton 1981a); (4) or a combination of these.

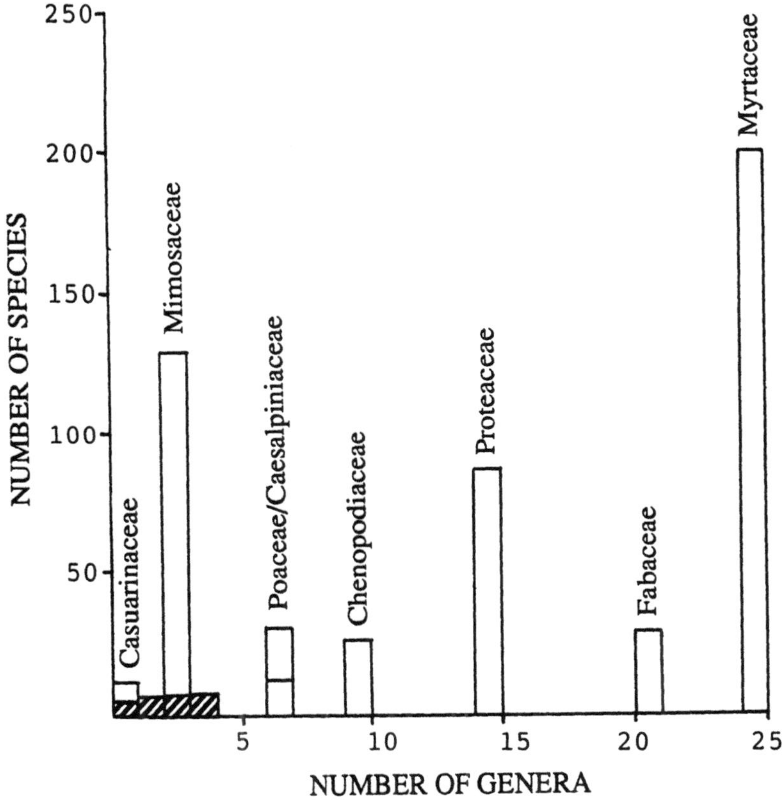

FIGURE 1. Distribution of plant families by genera and species of seed used by the Australian mining industry between May 1982 and June 1983. The hatched area represents the total of other genera and species used in the industry (from Langkamp and Plaisted 1987).

Surveys have indicated that, at least in the arid zone of Australia, the species richness of ants is greater and that of mammalian granivores less than in corresponding North American regions (Morton 1979). Subsequently, ants were found to be the major granivores in arid Australia and their impact on seed resources far outweighed that of rodents (Morton 1979). The ant fauna of mediterranean parts of Australia is also more species rich than in corresponding regions of North America (Majer 1988), indicating that trends between the arid regions of the two countries apply to other climatic zones. Myrmecochory is particularly widespread in Australia, where well in excess of 1500 plant species have diaspores with ant attracting bodies (Berg 1975). Thus, both granivory by ants and myrmecochory are particularly prevalent in most regions of Australia.

USE OF NATIVE PLANT SEED IN THE AUSTRALIAN MINING INDUSTRY

Langkamp and Plaisted (1987) surveyed the plant species used in the seeding operations of 48 mining companies throughout Australia between May 1982-June 1983. The plants, propagated from seed, were derived from 59 families and 165 genera. A breakdown of the list in terms of plant family (Fig. 1) indicated that almost all species belonged to eight major plant families. These families, ranked from highest to lowest according to the number of species used, included the Myrtaceae, Mimosaceae, Proteaceae, Fabaceae, Chenopodiaceae, Caesalpiniaceae, Poaceae and Casuarinaceae. Forty percent of the species used belonged to two genera: *Acacia* (Mimosaceae) and *Eucalyptus* (Myrtaceae).

In a preliminary grouping of Australian myrmecochorous plants, which includes a wide range of families, Berg (1975) listed certain members of the Mimosaceae, Fabaceae and Caesalpiniaceae as having myrmecochorous seeds. In addition, Davidson and Morton (1981b) recorded myrmecochory amongst members of the Chenopodiaceae; various workers have noted similar relationships between seeds of *Grevillea* spp. (Proteaceae) and ants (see references in Majer and Lamont 1985). In addition to enduring myrmecochorous relationships, many of the seeds used in the re-seeding operations, such as *Eucalyptus* spp. (Drake 1981), would also be vulnerable to granivory by ants. This would particularly apply to members of the Poaceae since experiments with grass seeds have shown that up to 100% of seeds can be removed by granivorous ants (e.g. Mott and McKeon 1977).

In conclusion, the Australian ant/seed literature indicates that many of the seeds used in the Australian mining industry may be affected by myrmecochorous relationships with ants, which generally benefit the seed, while a range of species may also be predated by granivorous ants.

ANT GENERA WHICH MIGHT AFFECT SEEDING OPERATIONS

By scanning the Australian literature it is possible to identify the genera which may be involved in granivorous or myrmecochorous relationships. Table 1 shows the genera which are likely to be involved. In view of the large number of references which were searched, reference sources have not been provided, although some genera were derived from Andersen (this volume). It should be noted that the type of seed-taking exhibited by particular ants is not rigid. For instance, *Rhytidoponera inornata*, a small ant occurring in Western Australia, generally feeds on elaiosomes, although it will consume the entire content of small seeds (Majer 1982).

Seed removal experiments were carried out in Western Australian bauxite mines (Majer 1980), bauxite and manganese mines in northern Australia (Majer 1983) and in mineral sand mines in southern Queensland (Majer 1985). Table 2 lists the genera observed taking seed within these rehabilitated areas. The most frequently observed seed-takers in the rehabilitated areas are *Iridomyrmex* spp., *Pheidole* spp., *Monomorium* spp. and *Rhytidoponera* spp., in decreasing order of importance.

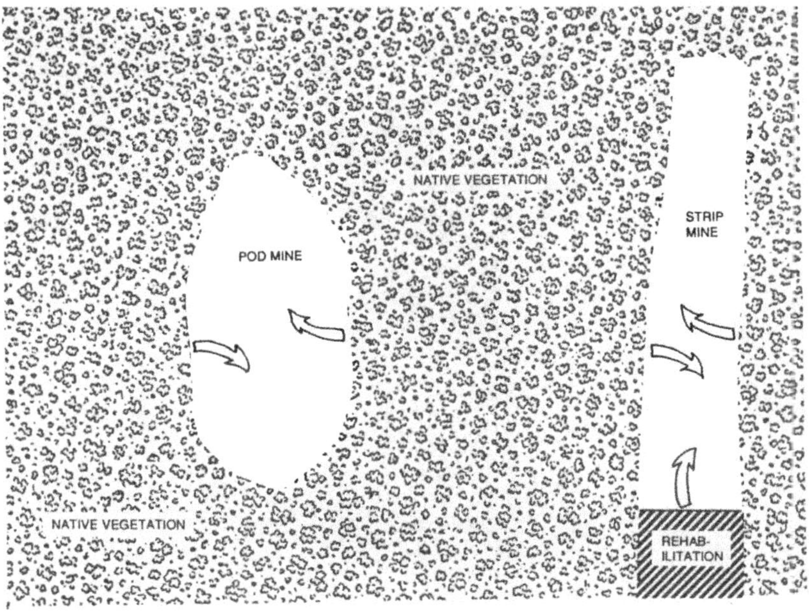

FIGURE 2. Spatial relationship between pod-type and strip-type mining showing how seed-taking ants (arrows) can exert an influence from adjacent native vegetation or previous rehabilitation.

SOURCE OF ANTS WHICH MAY AFFECT SEEDING OPERATIONS

Seeds may be taken by ants which reside within the rehabilitation area or by species present in vegetation adjacent to such areas.

Seeding operations are generally performed within weeks, and certainly within one year, of topsoil return. Consequently there is little time for species to colonize the area. A survey of the ants which colonized an unvegetated bauxite mined area in Western Australia (Majer 1981) indicated that only one ant species, an *Iridomyrmex*, had colonized the area within one year of topsoil return. If this result is typical for mined land in Australia, the influence of ants which have re-colonized the area prior to seeding is likely to be minimal (Table 1).

Mining operations within Australia generally follow the pod-type (e.g. Western Australian bauxite mines), in which irregular shaped pits are dug in areas of high ore load, or are of the linear strip-type (e.g. most Australian mineral sand mines) in which the mine progressively advances along a linear ore deposit (Fig. 2). In the former case, rehabilitation occurs next to the surrounding native vegetation, while in the latter it may abutt native vegetation on the mine's leading front and sides, and lie next to last year's rehabilitation on the trailing edge (Fig. 2).

TABLE 1. Ant genera which have been recorded in granivorous (G) or myrmecochorous (M) relationships within Australia.

Subfamily	Genus	Type of relationship
Ponerinae	*Heteroponera*	*G*
	Odontomachus	*G*
	Rhytidoponera	*M and G*
Myrmicinae	*Aphaenogaster*	*M*
	Crematogaster	*M*
	Mayriella	*G*
	Meranoplus	*M and G*
	Monomorium	*M and G*
	Pheidole	*M and G*
	Solenopsis	*M*
Formicinae	*Camponotus*	*M*
	Melophorus	*M and G*
	Notoncus	*M*
	Opisthopsis	*M?*
	Paratrechina	*M*
	Polyrhachis	*?*
	Prolasius	*G*
Dolichoderinae	*Iridomyrmex*	*M*

TABLE 2. Ant genera which have been recorded taking seed in rehabilitated areas in Australia (data from Majer 1980, 1983, 1985).

Ant genus	Western Australian bauxite mines	Northern Australian manganese or bauxite mines	Queensland mineral sand mines
Iridomyrmex	*	*	*
Melophorus	*		
Monomorium		*	*
Odontomachus		*	
Pheidole		*	
Polyrhachis		*	
Rhytidoponera	*		*
Solenopsis		*	

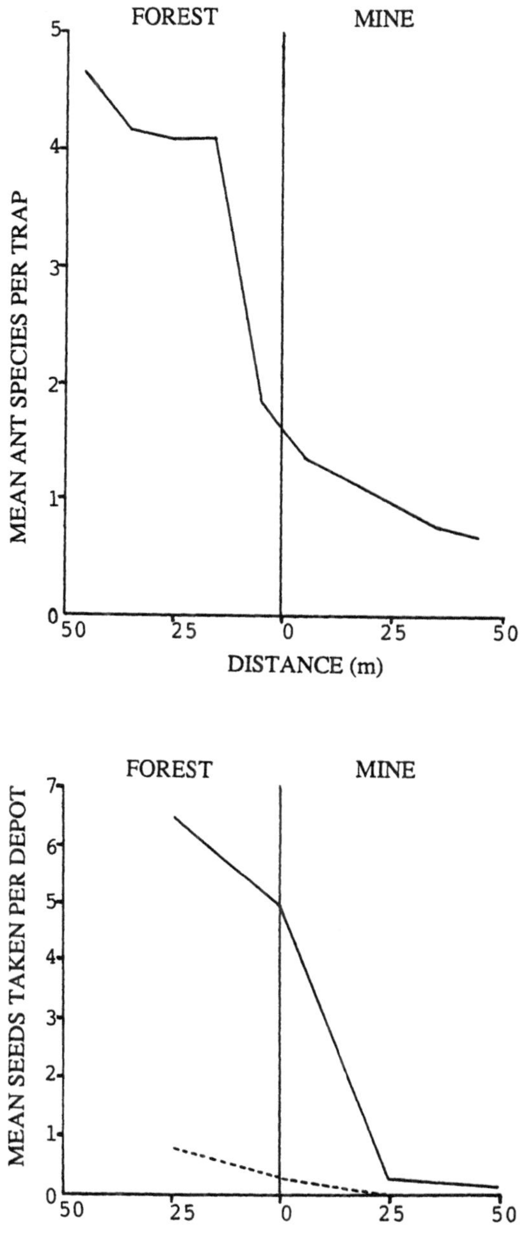

FIGURE 3. Total ant species trapped (a) and also total *Acacia extensa* (———) and *Eucalyptus marginata* (-----) seeds removed (b) along a forest/fresh topsoiled area interface of a bauxite mine at Del Park, Western Australia (from Majer 1980).

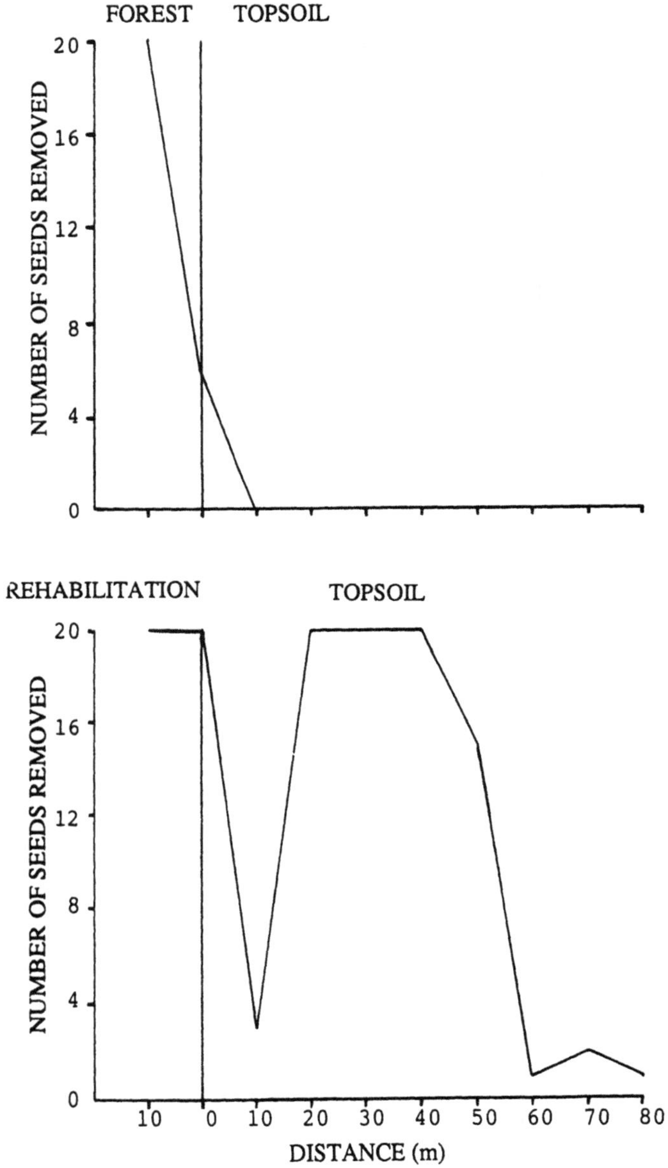

FIGURE 4. Total *Acacia concurrens* seeds removed (max. = 20) after 8 h from seed depots situated along (a) a forest/fresh topsoiled area interface, and (b) a 1.1 year-old rehabilitation/fresh topsoiled area interface of a mineral sand mine on North Stradbroke Island, Queensland (from Majer 1985).

Thus, in pod-type mines the seeds may be taken by ants from the adjacent vegetation, while in strip-type mines there could be an influence from ants in both the adjacent native vegetation and in the previous year's rehabilitation.

Transects of pitfall traps and seed sources arranged from adjoining eucalypt forest into topsoiled bauxite mined areas in Western Australia indicated that seed removal occurred 25 m into the mined area (Figure 3b) - an activity associated with forest-dwelling ants which either foraged into the mine or extended colonies into such areas (Figure 3a) (Majer 1980). However, in the Northern Territory and in Queensland, comparisons of transects run from native vegetation into topsoiled areas (Figure 4a) with similar ones from one-year old rehabilitation into topsoiled areas (Figure 4b) indicated that the impact of ants in adjacent rehabilitation was much greater than that from adjacent native vegetation (Majer 1983, 1985). The genera which foraged from the adjacent rehabilitation were *Iridomyrmex* and *Monomorium*. It appears that the high densities of seed-producing *Acacia* and grass species in rehabilitation favors the buildup of seed-taking ants and that these in turn have a dramatic effect on seeds in the adjacent rehabilitation. The distances over which this effect was noted were 80 m in the Northern Territory and 50 m in southern Queensland.

SEASONALITY OF ANTS AND TIMING OF SEEDING OPERATIONS

Ant communities throughout most of Australia exhibit strongly seasonal patterns of activity, although the nature of this seasonality depends on the climate type. Consequently, the rate of seed removal by ants differs by season (Andersen and Ashton 1985). The Western Australian bauxite mines experience a mediterranean climate with cool, wet winters and hot, dry summers. Here the ant community is extremely active from September to March but relatively inactive during the wet winter months (Figure 5). Rehabilitation seeding in this area is often carried out in May at a time when ant activity is declining. Thus, seed removal by ants may be less than if seeding was performed during the warmer months.

If adequate data on ant seasonality and seed taking by various species of ants were available, it might be possible to adjust seeding times to minimize the effects of granivory and possibly even capitalize on the beneficial effects of myrmecochory.

RESTORATION OF NORMAL ANT-SEED RELATIONSHIPS

The potentially beneficial effects of myrmecochory to seed survival and subsequent seedling growth are probably not all relevant to the initially-seeded minesite. This is because the seeds are sown into areas which have generally been fertilized and, if necessary, they have been pre-treated so that germination follows after a short period of time. However, as the rehabilitation matures it is desirable for myrmecochorous relationships to be restored so that the fate of seed produced on-site is improved. This is particularly important because some rehabilitated areas in Australia are now being subjected to accidentally-lit fires.

Seed distributed in 3-year old rehabilitation in Western Australian bauxite mines (Majer 1980) and 2.5-year old mineral sand mines in

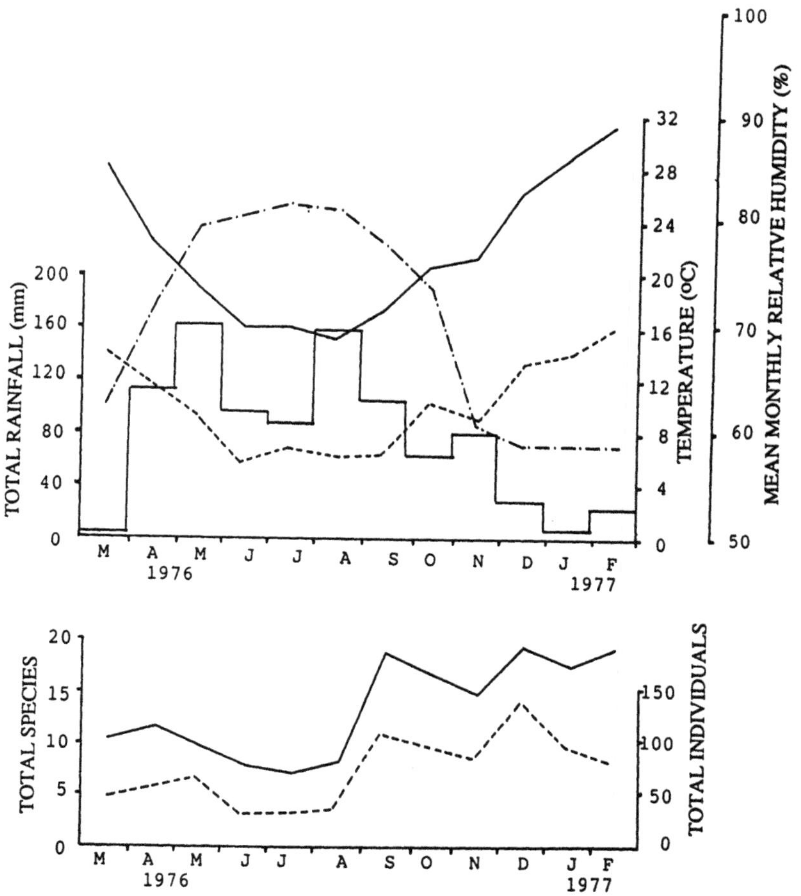

FIGURE 5. Upper graph shows the monthly rainfall (histogram), mean monthly temperatures (maximum ———, minimum -----), and relative humidities (-.-.-) at Dwellingup, Western Australia, from March 1976 to February 1977. The lower graph shows the number of species (———) and individuals (-----) of ants trapped per month at Dwellingup, Western Australia over the same period (from Majer 1982).

Queensland (Majer 1985) resulted in seed removal rates similar to those of the original native vegetation. This is encouraging as it suggests that the original broad ant-seed relationship has been partially restored. However, to support this statement it is necessary to investigate the balance between granivores, which destroy the seed, and elaiosome collectors which benefit the seed. In the Queensland study it was found that although seed was taken in approximately equal proportions by granivores and elaiosome collectors in the forest, only elaiosome collectors were involved in the mine. Thus the relative contributions of seed harvesters and elaiosome collectors in rehabilitated and native vegetated areas needs further investigation before restoration of the ant-seed relationship can be confidently claimed.

In some mines it is necessary to perform enrichment-seeding in older areas of rehabilitation without high plant diversity. Granivorous ants are a problem in some Western Australian bauxite mines which have been subjected to this treatment. The problem was overcome by treating seed with pesticide prior to their being sown (S. Ward personal communication).

CONCLUSION

This review of seed taking by ants in Australian rehabilitation programs has shown that both granivory and myrmecochory are widespread in such areas. Rehabilitation staff should be aware of this and, if necessary, take measures to minimize its impact in the initial phases of rehabilitation. Such measures could include the pelletization or pesticide-treatment of vulnerable seeds, the adjustment of seed species ratios to compensate for seed predation, selection of the appropriate timing of seeding operations, or the control of ants in hazard areas such as the interface between previous and current rehabilitation.

Some aspects of the ant-seed relationship are beneficial in the later stages of rehabilitation and, consequently, rehabilitation should be carried out in a way that encourages the long-term colonization of beneficial ant species.

ACKNOWLEDGMENTS

I would like to thank Drs Alan Andersen and Owen Nichols who commented on an earlier draft of this chapter.

LITERATURE CITED

ANDERSEN, A.N. 1988. Dispersal distance as a benefit to myrmecochory. Oecologia 75:507-511.

ANDERSEN, A.N. and D.W. ASHTON. 1985. Rates of seed removal by ants of heath and woodland sites in southeastern Australia. Australian Journal of Ecology 10:381-390.

BERG, R.Y. 1975. Myrmecochorous plants in Australia and their dispersal by ants. Australian Journal of Botany 23: 457-508.

DAVIDSON, D.W. and S.R. MORTON. 1981a. Competition for dispersal in ant-dispersed plants. Science 213: 1259-1261.

DAVIDSON, D.W. and S.R. MORTON. 1981b. Myrmecochory in some plants (F. Chenopodiaceae) of the Australian arid zone. Oecologia (Berl.) 50: 357-366.

DRAKE, W.E. 1981. Ant-seed interaction in dry sclerophyll forest on North Stradbroke Island, Queensland. Australian Journal of Botany 29: 293-309. Langkamp, P. (ed.). 1987. Germination of Australian native plant seed. AMIRA and Inkata Press, Melbourne, 236 pp.

LANGKAMP, P. and M. PLAISTED. 1987. Perspectives on the use of native plant seed in the Australian mining industry, pp. 107-114. *In* P. Langkamp (ed.), Germination of Australian native plant seed. AMIRA and Inkata Press, Melbourne, 236 pp.

MAJER, J.D. 1980. The influence of ants on broadcast and naturally spread seeds in rehabilitated bauxite mines. Reclamation Review 3: 3-9.

MAJER, J.D. 1981. The role of invertebrates in minesite rehabilitation. Forests Department of Western Australia Bulletin No. 93, 29 pp.

MAJER, J.D. 1982. Ant-plant interactions in the Darling Botanical District of Western Australia, pp. 45-61. *In* R.C. Buckley (ed.), Ant plant interactions in Australia. Dr W. Junk Publishers, The Hague, 162 pp.

MAJER, J.D. 1983. The influence of ants on seeding operations in northern Australian mined areas. Reclamation and Revegetation Research 2: 299-301.

MAJER, J.D. 1985. Recolonization by ants of rehabilitated mineral sand mines on North Stradbroke Island, Queensland, with particular reference to seed removal. Australian Journal of Ecology 10: 31-48.

MAJER, J.D. 1988. The Formicidae, pp. 219-224. *In* R.L. Specht (Ed.), Mediterranean-type ecosystems: a data source book. Kluwer Academic Publishers, Dordrecht, 248 pp.

MAJER, J.D. and B.B. LAMONT. 1985. Removal of seed of *Grevillea pteridifolia* (Proteaceae) by ants. Australian Journal of Botany 33: 611-618.

MORTON, S.R. 1979. Diversity of desert-dwelling mammals: a comparison of Australia and North America. Journal of Mammalogy 60: 253-264.

MOTT, J.J. and G.M. MCKEON. 1977. A note on the selection of seed types by harvester ants in northern Australia. Australian Journal of Ecology 2: 231-235.

52

The Ant Problems of Cocoa Farms in Brazil

J.C. Delabie

INTRODUCTION

The cocoa tree, *Theobroma cacao* L. (Sterculiaceae), is native to the Amazonian Basin. A French settler introduced the tree to Bahia from the state of Para in 1746. The first important plantation was established in 1816 near Canavieiras. As cocoa trees need shade to grow, traditional plantations used the "cabruca" system of establishment, in which the tallest, healthiest trees of the primary rainforest are selected and preserved. "Cabruca" plantations now cover about 400,000 hectares. This explains the similarities between cocoa farm and rainforest ant populations. From 1960 to the present, cocoa began to be mixed with plant species such as banana, and the following trees: *Erythrina fusca* Loureiro, *E. poeppigiana* (Walp.) Cook, *Inga spp* and *Spondias lutea* L.. This type of plantation covers approximately 200,000 hectares.

On Bahian cocoa farms, the dominant ants of the canopy are spatially distributed, in a characteristic "patchwork", or "ant-mosaic," in which each species occupies its own group of trees (Leston 1978; Winder 1978). This distribution model has been well-documented on African cocoa plantations (Majer 1972, 1976a,b; Leston 1973; Taylor 1977; Jackson 1984). In the Neotropics, the dominant ants belong to the Dolichoderinae, Ponerinae (Leston 1978) or Myrmicinae (Winder 1978) subfamilies.

Ants are an important part of the Southern Bahian cocoa agroecosystem. Leston (1978) found 130 ant species on one hectare of a secondary forest reserve, while more recent studies registered 105 soil surface species and 70 litter species on one hectare of an experimental cocoa area (Delabie, unpublished). Moreover, approximately 250 ant species have been registered for the region (Delabie, unpublished). Due to ant species diversity, interactions among themselves and other flora and fauna, the ants have enormous economic potential - both destructive and beneficial.

More than 90% of Brazil's cocoa comes from the Southern Bahian state. This paper will focus on economically-important ant species in the region, with special emphasis on current research efforts.

ANT INJURIES

Leaf-Cutting

Some fungus-growing ants (Myrmicinae, Attini) like *Atta cephalotes* (L.) and *Atta sexdens sexdens* (L.) cut young cocoa leaves - an activity particularly harmful to young plants - and flowers. Recent observations on a six year old plantation showed that continuous attacks by a mature *A. cephalotes* nest negatively affect tree development (height and trunk diameter) until the tree

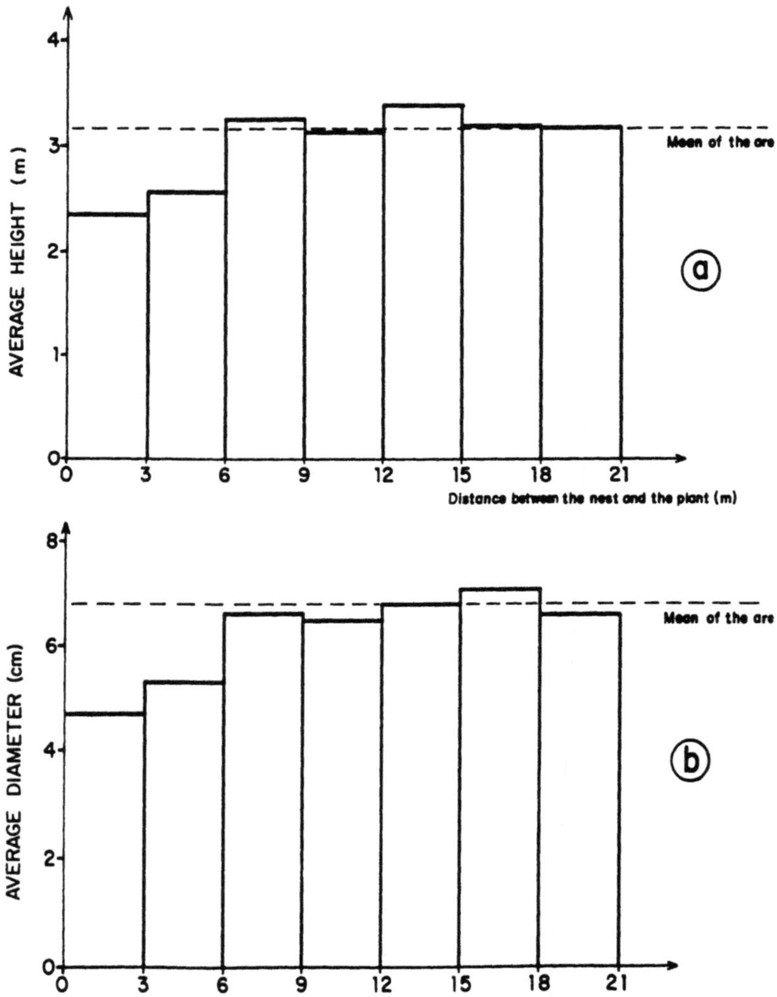

FIGURE 1. Consequences of continuous defoliation by an *Atta cephalotes* nest on the development of young cocoa trees in a 6 year old plantation.

reaches a height of six meters. Ten to 14 trees had retarded growth for two years (Fig. 1) as a consequence of the defoliation, which was inversely proportional to the tree's distance from the nest, up to about 15 meters. The results are independent of foraging trail location. In cocoa farms, nest density may reach one adult nest of *A. cephalotes* per hectare, possibly higher for *A. sexdens sexdens*, depending on vegetation type and insecticide application. *A. cephalotes* prefers dense vegetation, such as forest or well-shaded plantations, while *A. sexdens sexdens* adapts to man-made conditions, such as pasture, crop or recovering secondary forest; thus it is more frequently found on cocoa plantations. *A. cephalotes* constructs fungus-growing chambers to a depth of two meters in soils which may be flooded, while *A. sexdens sexdens* construct theirs up to five meters deep in drained areas (Abreu and Delabie, 1986).

The nests of another species--probably the most common fungus-growing ant in the region--*Acromyrmex subterraneus brunneus* Forel, are found at densities up to 150 per hectare. This often polygyne ant (Delabie, In Press) nests near the surface of the ground, at the base of trees, under rotted wood, covered by litter or beneath abandoned pieces of plants (Bondar 1923, Abreu and Delabie 1986). In addition to damaging cocoa in much the same way as *Atta* species, *Acromyrmex subterraneus brunneus* removes the skin of the pods.

Other Direct Damage to the Cocoa Tree

The gardening ant *Azteca paraensis bondari* Borgmeier (Dolichoderinae), directly damages the tree canopy by chewing or rasping buds and/or bark of the shoots above the nest, to extract gummy substances used in nest construction (Bondar 1939, Silva 1945). This action kills the branches, favors growth of associated epiphytic plants (Figure 2) and, eventually, the growth of invasor plants on the ground. Other ant species also chew the bark of buds or shoots: the "caçarema" ant, *Azteca chartifex spiriti* Forel, *Cephalotes atratus* (L.), the uncommon *Crematogaster magnifica* Santschi, (Bondar 1939), or *Solenopsis geminata* (Fabricius) (Entwistle 1972). However, except for the "caçarema" ant, the damage they cause is minimal.

The nest weight of some arboreal social insects, such as *A. chartifex spiriti* (Figure 3) or the termite *Nasutitermes spp* (Ramb.) can be damaging to cocoa (Bondar 1939, Silva and Barbosa 1948). Large and heavy, their nests may break a big branch or destabilize trees without a well-developed root system. Because of agricultural practices, a majority of cocoa pods hang on a small number of branches; these large nests can cause substantial damage.

Stinging Pests

The little fire ant, *Wasmannia auropunctata* (Roger) (Myrmicinae, Ochetomyrmicini), is among the region's most common ant species (Delabie 1988; see Ulloa-Chacon and Cherix this volume). This ant, and other, less abundant fire ants, like *Solenopsis bondari* Santschi, *S. brasiliana* Santschi and *S. geminata medusa* Mann (Myrmicinae, Solenopsidini) have the popular name of "pixixica" (Bondar 1925, 1939). All these species tend mealybugs and/or scale insects on the pods. Control of these abundant ants may be necessary to prevent harmful stings when harvesting the pods.

FIGURE 2. A nest of *Azteca paraensis bondari* with associated plants of the Gesneriaceae and Orchidaceae families.

The aggressive *Polybia versicolor* Ol. and *P. angulata* Fr. wasps must be added to this category, since they build nests near those of arboreal ants, mainly *Azteca* species (Zehntner 1917, Bondar 1939, Silva and Barbosa 1948). Aggregation of nests is common in this wasp genus (Jeanne 1978). In the presence of ants, the wasps receive a degree of protection against predators, especially birds and monkeys. The wasps probably have great utility as natural biological control agents of cocoa pests such as lepidopterous larvae. However, on cocoa plantations, field workers do not harvest pods where wasp nests are established to avoid being stung.

FIGURE 3. A nest of *Azteca chartifex spiriti*.

Homoptera-Tending

Many ants which explore and forage the canopy tend different sap-sucking insect species, such as Aleyrodoidea (genus *Aleurodicus* Douglas, *Aleurotrachelus* Quaintance and Baker and *Paraleyrodes* Quaintance), Aphidoidea (chiefly *Toxoptera aurantii* Boyer), Coccoidea (genus *Aspidiotus* Bouche, *Ceroplastodes* Ckll., *Coccus* L., *Planococcus* Ferris, *Pseudoanidia* Ckll. and *Saissetia* Depl.), and Membracoidea (genus *Aethalion* Latreille, *Bolboneta* Amyot and Serville, *Horiola* Fairmaire, *Membracis* Fabricius and *Tragopa* Germ.) (Bondar 1939, Silva 1944, 1950) on the pods, flowers or shoots. The most common tending ants belong to various subfamilies: Dolichoderinae (*A. chartifex spiriti*, *A. paraensis bondari*, *Dolichoderus attelaboides* (Fabricius), *D. decollatus* Fr. Smith and *Hypoclinea bidens* (L.)), Formicinae (*Camponotus spp* and *Paratrechina spp*), Myrmicinae (*Cephalotes atratus*, *Crematogaster spp*, *Solenopsis spp* and *W. auropunctata*) and Ponerinae (*Ectatomma tuberculatum* (Olivier) and *Odontomachus haematodus* (L.)). *E. tuberculatum* and *O. haematodus* are among the rare exceptions of

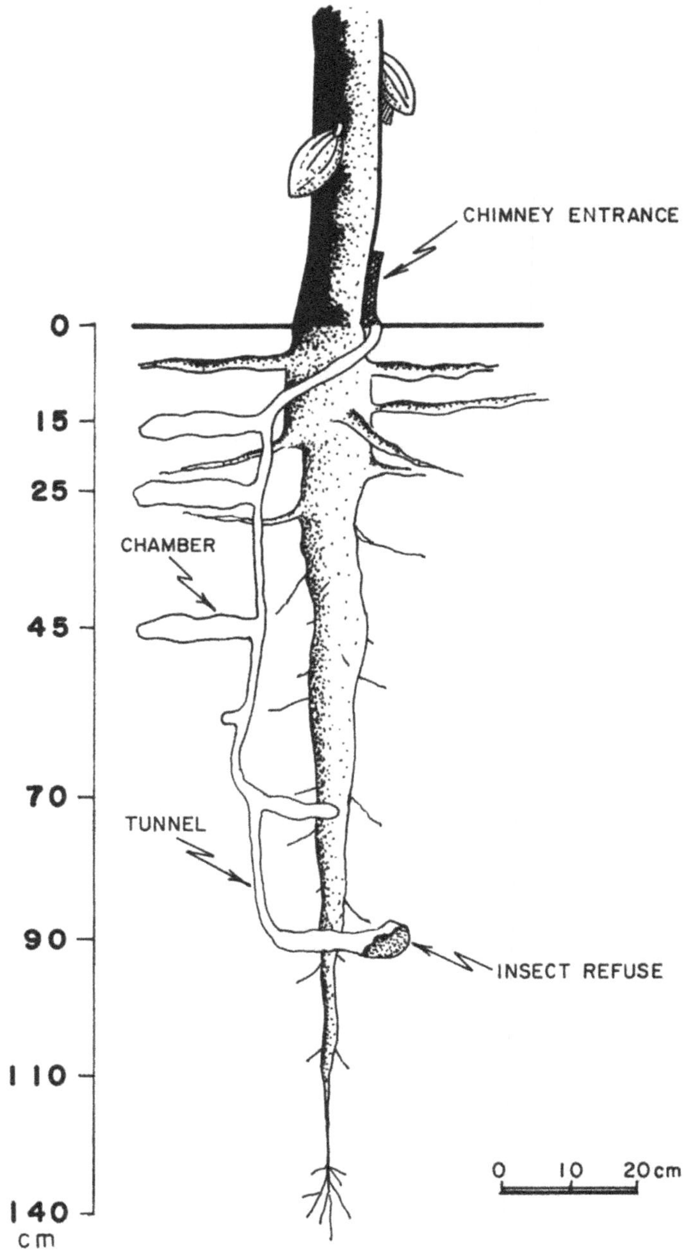

FIGURE 4. A nest of *Ectaomma tuberculatum*.

Ponerinae-tending Homoptera, since all Ponerinae were considered exclusively as predators (Bondar 1939, Weber 1946, Evans and Leston 1971, Entwistle 1972). *O. haematodus* nest in litter or rotten wood and are ground foragers. However, it occasionally nests in abandoned termite nests, shelters made of organic material or dried pods on the cocoa trees, and may tend Homoptera, chiefly treehoppers. *E. tuberculatum*, common in the cocoa agroecosystem, is a ground-nesting species. It constructs a vertical tunnel with up to five horizontal chambers near the tree tap-root (Figure 4) topped by a 10-30 cm chimney made of plant remains and applied vertically on the trunk. This ant carries excavated soil particles to the tree canopy (Bondar 1925, 1939). The colonies are polygyne and may contain several hundred workers. On some farms, 90% of the trees have a nest at their base. Some workers forage in the canopy and on the trunk, while others tend Homoptera of several families or genera, depending on the season, the tree species and the availability of flowers or fruits.

Formicinae ants (genus *Acropyga* Roger, subgenus *Rhizomyrma* Forel) rear different pseudococcid species on cocoa roots in superficial soil layers (Delabie and Mantovani, In Press), as has been reported for cocoa plantations in other countries (Wheeler 1935, Weber 1944, 1957). The association of mealybugs with these ants is obligatory. Usually, the mealybug involved is

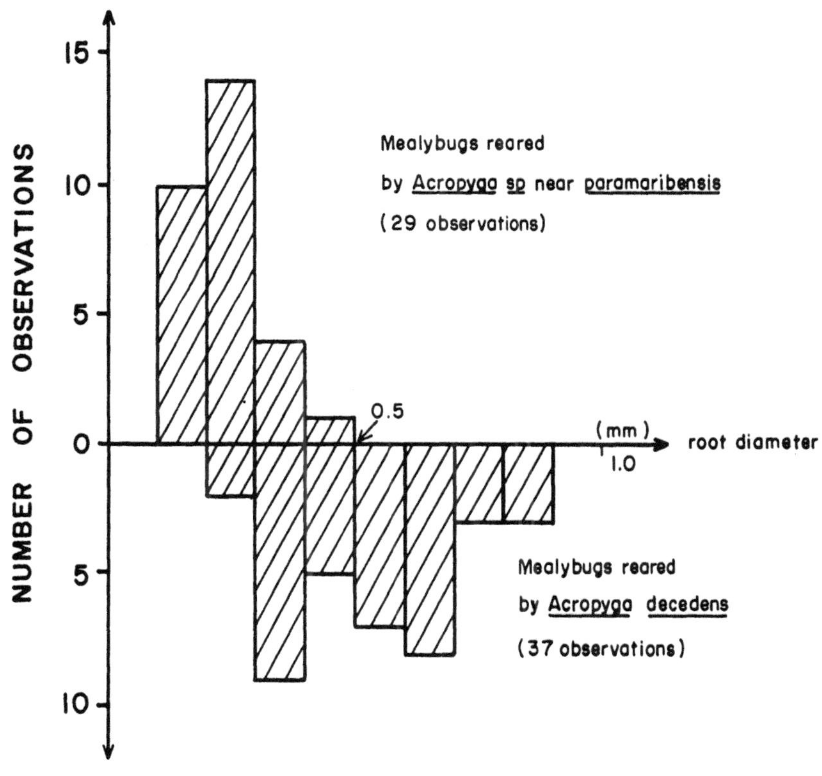

FIGURE 5. Distribution of mealybugs reared by two species of *Acropyga* in the same niche, as a function of cocoa tree root diameter.

Chavesis Balachowski (formerly referred to as *Eumyrmococcus* Silvestri), but other genera, such as *Rhizoecus* Kunckel or *Geococcus* Green may take the place of *Chavesia*. Biologically, this association is as important as that between the Attines and their fungus. Attine queens carry a mycelium fragment during the nuptial flight, while *Acropyga* female sexuals carry a fertile mealybug female (Bunzli 1935, Lima 1942, Eberhard 1978, Campos and Morais 1986). The mealybug participates in the foundation of the new ant colony, through its offspring and food flow to the ants. Colony structures have two elements: galleries which follow the plant roots where older mealybugs are established for honeydew production and "rearing-cells" where ant brood and pseudococcid young are tended by workers. It has also been noted that two *Acropyga* species coexist in the same niche by rearing mealybugs of distinct species, established on roots of different diameters (Figure 5) (Delabie and Mantovani, In Press).

There is little evidence that Homoptera tending by the ants directly affects the economics of cocoa. However, under heavy pressure by sap-sucking insects, especially coccids, the fruit buds and pods may wilt on the tree due to the high quantities of plant sap taken by the insects, e.g. *Planococcus citri* (Risso)) tended by *W. auropunctata* (Delabie 1988), tree-hoppers reared by *Dolichoderus spp* (Bondar 1939, Silva 1944) or by *Azteca spp* (Bondar 1939, Silva 1944, 1945, 1950, 1955). According to Silva (1944), cocoa tree-hoppers may destroy one-fifth of the crop.

Disease Vectors

In Africa, coccids tended by ants transmit the "Swollen-Shoot" virus disease of cocoa (Strickland 1951, 1952, Hanna et al. 1956 and Leston 1969). Although "Swollen-Shoot" and other viruses do not occur in Brazil, cocoa viruses have already been encountered in Trinidad (Posnette 1944) and other South American countries (Entwistle 1972). There is no evidence of ants participating in their propagation. *Acropyga spp* ants tending root mealybugs of cocoa are suspected vectors of root infections, as already documented for diseases coffee root diseases in Brazil (Goeldi 1892, Pickel 1927) and Surinam (Bunzli 1935).

Some African ants - in Ghana (Evans 1971, 1973) and in Nigeria (Taylor and Griffin 1981) - are directly responsible for the propagation of the "Black-Pod" of cocoa (*Phytophtora palmivora* (Butl.) Butl.), causing considerable pod losses. In Brazil, the relationship between this common disease and ants has not yet been studied, though some ant species are possible vectors of this fungus. Characteristics of ant vectors are those that use large quantities of organic material to construct their nests, such as *Azteca spp* (Silva 1955), or those which throw material originating from nest gallery excavations into the foliage, as does *E. tuberculatum*.

Flowering Inhibition

Nests with a large surface area, like those of *Azteca spp* and arboreal Termitidae, may cover important areas of the cocoa trunk and branches, preventing the development of numerous floral buds (Bondar 1939, Silva 1955, Vello and Magalhaes 1971). The two *Azteca* species build shelters made of organic material to protect their Homopteran associates (Bondar 1925,

1939, Silva 1944). These shelters are generally near the peduncule of the shoots, and may also limit the growth of new flowers.

BENEFICIAL ASPECTS OF ANTS

Biological Control

Cocoa plantation ants may control some pests by predation, repellency or by their behavior toward intruders. Efficient predators, like *Pseudomyrmex spp* (Pseudomyrmicinae) which nest in dried shoots, some Ponerinae (*Ectatomma quadridens* (Fabricius), *E. tuberculatum*, *Neoponera villosa* (Fabricius), *O. haematodus* and some Ecitoninae (especially *Labidus coecus* (Latreille) and *L. praedator* (Fr. Smith)), live on Bahia cocoa plantations. *A. chartifex spiriti*, essentially nocturnal (Benton, 1979), controls undesirable insects through its aggressive behavior or by its repellency (Silva 1955, Vello and Magalhaes 1971). Silva et al. (1982) have isolated the repellent's main component, a dialdehyde, from abdomen extracts. These ants, along with some Arachnidae species, are the main natural control agents of thrip (*Selenothrips rubrocinctus* Giard) (Zehntner 1917, Torrend 1919, Bondar 1939), as well as caterpillars of various moth species, mirid bugs (*Monalonion spp*) (Zehntner 1917, Bondar 1939), and leaf-eating beetles of the Chrysomelidae family (Ferronato 1988). For example, in *E. tuberculatum*'s lowest nest chambers (Figure 3), one meter deep, remains of arthropods caught in the cocoa environment were found--chiefly Chrysomelidae, but also Arachnidae, Gryllidae, Membracidae, Nasutermitidae and Formicidae, like *Camponotus sp* and *A. cephalotes*. The number of *A. cephalotes* workers was particularly high, however, even in an area with many *E. tuberculatum* nests, mature *A. cephalotes* colonies prospered, in contrast to Cook's opinion that the ponerines "may prevent the growth of new colonies by killing the workers as fast as they appear above ground" (in Weber 1946).

In other cocoa producing countries, ant species have been successfully utilized as biological control agents; for example, *W. auropunctata* in Cameroon (Bruneau De Mire 1969) and *Dolichoderus bituberculatus* Mayr in Indonesia (**Meer Mohr 1927, Giesberger 1983). Traditionally, some Bahian cocoa producers continue to distribute *A. chartifex spiriti* nest fragments on their plantations, knowing the pods look better on ant-infested trees. Zehntner (1917) and Torrend (1919) recommended this practice, Bondar (1925 1939) and Silva (1955) objected, mainly because of the deleterious quantities of Homoptera reared by the ants on the plants.

Pollination

For many years, until Ceratopogonidae midges were discovered to be the principal pollinators of cocoa flowers, many people believed ants or other insects with large populations in the area might have been pollinators (Harland 1925, Cope 1940, Entwistle 1972). Later, this hypothesis was rejected (Winder 1978). However, a few African ant species are responsible for a percentage of cocoa pollination (Entwistle 1972). Where *W. auropunctata* is found, its small size and high populations allow it to enter and pollinate the flowers (Billes 1941, Hernandez 1965). Moreover, studies on the productivity of cocoa trees harboring *A. chartifex spiriti* suggest that pollinating midges are attracted to odors produced by the ant (Vello 1969, Vello and Magalhaes 1971). However, further experiments are necessary.

Fertilization

The ant can serve as a soil-fertilizing agent. However, little is known about the action of the ground-nesting ants on the recycling of cocoa agroecosystem soil nutrients. Leite's (pers. comm.) preliminary work and personal observations demonstrated the important function of fungus-growing ants like *Mycocepurus goeldii* Forel and various species of *Sericomyrmex*, which occur at high densities in some plantations; other Attines are of secondary importance. These ants usually excavate their nests to a depth of 30 to 120 cm, depending on the ant species and soil type. The mineral material resulting from the excavations is abandoned near the nest entrance, where it is rapidly incorporated into the litter, principally by rain. The great number of abandoned excavations may also improve drainage during the frequent tropical rains. This is important because the soils of cocoa plantations are never mechanically mixed; without the service performed by the ants, surface organic matter would not be blended with soil minerals as fast.

The coccid-tending ants *Acropyga*, which occur in certain plantations at high densities, are restricted to the superficial rhizosphere of the cocoa tree and the excavations made by the ants represent six percent of the soil volume at this horizon (Delabie and Mantovani, In Press). This may be an important factor for oxygen, nutrient and water circulation in this layer, where 60 to 85% of the tree's radicular system grows (Cadima Z. 1970).

RECENT ALTERATIONS OF THE ANT FAUNA IN THE BAHIAN COCOA FARMS

The "pixixica" ants, initially studied by Bondar (1925, 1930) and identified by Santschi (1925), were classified in the *Solenopsis* genus. In a recent paper, Delabie (1988) showed that the commonest "pixixica" species is *W. auropunctata*. After 60 years, this ant established a strong regional presence and a reputation for frequent and nasty stings. Studies of different cocoa plantation niches revealed that this ant is always in the litter, even if it does not appear in the canopy. *W. auropunctata*, a polygyne ant, has a high intrinsic capacity for population growth and rapidly colonizes new areas (Fabres and Brown 1978, Lubin 1984). This ant becomes dominant in the canopy only if other dominants, such as *Azteca spp*, are not present. This hypothesis may explain the effect of the intensive insecticide and fungicide use in the last few decades - sprayed on the trunk and the cocoa tree canopy - and the intensive control of social insects nesting in the canopy, such as *Azteca spp*, wasps and arboreal Termitidae, initiated after Bondar's works until 1960. To illustrate this aspect, the destruction of 8,048,019 *A. paraensis bondari* nests, 28,458 termite nests and 10,752 wasp nests from January 1944, to July 1948, in 451 farms (corresponding to ca 70,000 hectares of plantations and forest) has been documented by Silva and Barbosa (1948). Even in the 700 hectares of the Ilheus Cocoa Research Center (Centro de Pesquisas do Cacau) from March to December, 1964, 68,681 nests of *A. chartifex spiriti*, 56,141 nests of *A. paraensis bondari* and 16,926 nests of Termitidae, were destroyed (Anonymous 1964).

Other current problems, with direct consequences for the ant fauna, include deforestation, increased number of cultivated areas and crop diversification. In the first half of the century, the cultivated areas were almost always situated near primary rainforest areas where the natural predator fauna found refuge. Until recently, regional crops were poorly diversified, which explains why *A. cephalotes*, a typical rainforest species, held on in the central cocoa production region while *A. sexdens sexdens*, which generally occurs in areas of low density vegetation, occupied the transition regions and the littoral (Silva et al. 1969, Abreu and Delabie 1986). Today, it seems that *A. cephalotes*' numbers are declining because of a successful chemical control program (Abreu and Silva 1973, Abreu and Delabie 1986), while *A. sexdens sexdens* continues its expansion, taking advantage of crop diversification and the opening of new cultivated areas and roads. Another consequence of reduced forest areas bordering the cocoa plantations is the relative disappearance of some ant genera while others flourish. Thus, plantations in the "cabruca" system when compared with rainforest areas have a lower diversity. The number of ants observed from the genera *Camponotus*, *Dolichoderus*, *Eciton* and *Pachycondyla*, common in the rainforest, decreases in the cocoa plantations, while the number of ants belonging to the genera *Crematogaster*, *Iridomyrmex*, *Solenopsis* and others, (Delabie et al. 1989) has increased. One of the most important consequences associated with the relative disappearance of the *Eciton* army ants in these areas is the disappearance of its associated fauna, principally birds (Willis and Oniki 1978, 1988). Some species of these very important predators could be preserved, if areas of native vegetation bordering the crop would be safeguarded.

CONCLUSION

Ant problems of the Brazilian cocoa farms as a whole must be analyzed with respect to possible consequences - including possible alteration of the agroecosystem - because the relationships between "cocoa ants" are fragile, dependent upon the ecological function of each species. There is no reason to discard the possible short-term human benefits of managing ant populations on the Brazilian cocoa farms, especially as programs have already been proposed and developed for other cocoa producer countries.

ACKNOWLEDGMENTS

The author would like to thank Joao Manuel de Abreu, Forbes P. Benton and Harold G. Fowler for critically reviewing the manuscript, Carlos Roberto F. Brandao for identifying some ants, Antonio Bispo and Gildefran A.D. de Assis for drawing the figures.

REFERENCES

ANONYMOUS. 1964. CEPEC: Mais de cem mil ninhos de pragas destruidos. Cacau Atualidades, Brazil. 1 (11-12): 10.

ABREU, J.M. and J.H.C. DELABIE. 1986. Controle das formigas cortadeiras em plantios de cacau. Rev. Theobroma, Brazil. 16 (4): 199-211.

ABREU, J.M. and P. SILVA. 1973. Controle das formigas cortadeiras *Atta cephalotes* e *Atta sexdens* na regiao cacaueira da Bahia. Rev. Theobroma, Brazil. 3 (3): 3-11.

AZEVEDO, A. de. 1925. Estado sanitario vegetal dos cacaueiros em Belmonte. Correio Agricola, Brazil. 3: 249-252, 270-274.

BENTON, F.P. 1979. Atividade diaria da formiga cacarema *Azteca charifex* Forel no cacaueiro. CEPEC-Informe Tecnico 1979, CEPLAC, p 182.

BILLES, D.J. 1941. Pollination of *Theobroma cacao* L. in Trinidad, B.W.I.. Trop. Agric., Trinidad and Tobago. 18: 151-156.

BONDAR, G. 1923. Formiga "quem-quem" *Acromyrmex subterraneus*, praga dos cacaoeiraes. Correio Agricola, Brazil. 1: 251-254.

BONDAR, G. 1925. O cacau - II. Molestias e inimigos do cacaueiro no Estado da Bahia. Salvador, Brazil, Secretaria da Agricultura, Industria, Comunicacao, Viacao e Obras Publicas. 126 p.

BONDAR, G. 1930. A formiga sauva, praga dos cacaoaes. Correio Agricola, Brazil. 8 (2): 29-32.

BONDAR, G. 1939. Insetos daninhos e parasitas do cacau na Bahia. Salvador, Brazil, I.C.B. Boletim Tecnico n 5. 112 p.

BRUNEAU DE MIRE, Ph. 1969. Une fourmi utilisèe au Cameroun dans la lutte contre les mirides du cacaoyer, *Wasmannia auropunctata* Roger. Caf'e Cacao. Thè 13 (3): 209-212.

BUNZLI, G.H. 1935. Untersuchungen uber coccidophile Ameisen aus den Kaffeefeldern von Surinam. Mitteil. Schweiz. Ent. Ges. 16 (6/7): 453-593.

CADIMA Z., A. 1970. Estudo do sistema radicular do cacaueiro em alguns tipos de solos da regiao cacaueira do Sul da Bahia. Boletim Tecnico n 5, CEPLAC, Ilheus, Bahia, Brazil. 30 p.

COPE, F.W. 1940. Agents of pollination in cacao. Ninth Annual Report on Cacao Res. - 1939, Port-of-Spain, Trinidad and Tobago. 13-19.

CAMPOS, L.A. de O. and H.C. MORAIS. 1986. Transporte de homopteros por femeas de formigas do genero *Acropyga* (Formicinae) em Vicosa, M.G. Anais do VII Encontro de Mirmecologia do Estado de Sao Paulo, F.H. Caetano ed., UNESP-FAPESP-CNPq, 52-53.

DELABIE, J.H.C. 1988. Ocorrencia de *Wasmannia auropunctata* (Roger 1863) (Hymenoptera, Formicidae, Myrmicinae) em cacauais na Bahia, Brasil. Rev. Theobroma, Brazil. 18 (1): 29-37.

DELABIE, J.H.C. 1989. Observacoes sobre a ocorrencia de poliginia em colonias de *Acromyrmex subterraneus brunneus* (Formicidae, Myrmicinae, Attini) em cacauais. Anais da Sociedade Entomologica do Brasil (In Press).

DELABIE, J.H.C., M.C. ALVES and C.I. MAURICIO. 1989. Estudo comparativo da mirmecofauna da mata atlantica e de cacauais no Sul da Bahia (Hymenoptera: Formicidae). Resumos XII Cong. Bras. Entomol Volume I. Belo Horizonte, M.G., Sociedade Entomologica do Brasil, p 74.

DELABIE, J.H.C. and J.E. MANTOVANI. 1989. Observacoes sobre a biologia de duas especies de *Acropyga* (Formicidae, Formicinae, Plagiolepidini) associadas a rizosfera do cacaueiro. Rev. Bras. Biol. (In Press).

EBERHARD, W.G. 1978. Mating swarms of a South American *Acropygia* (Hymenoptera, Formicidae). Entomol. News 89 (1/2): 14-16.

ENTWISTLE, P.F. 1972. Pests of Cocoa. London, Longman. 779 p.

EVANS, H.C. 1971. Transmission of *Phytophtora* pod rot by invertebrates. Nature 232: 346-347.

EVANS, H.C. 1973. Invertebrate vectors of *Phytophtora palmivora*, causing Black Pod disease of cocoa in Ghana. Ann. appl. Biol. 75: 331-345.

EVANS, H.C. and D. LESTON. 1971. A ponerine ant (Hym., Formicidae) associated with Homoptera on cocoa in Ghana. Bull. entomol. Res. 61: 357-362.

FABRES, G. and W.L. BROWN Jr. 1978. The recent introduction of the pest ant *Wasmannia auropunctata* into New Caledonia. J. Austral. entomol. Soc. 17: 139-142.

FERRONATO, E.M. de O. 1988. Eumolpinae associated with cacao trees (*Theobroma cacao* L.) in Southeast Bahia. Biology of Chrysomelidae. P. Jolivet, E. Petitpierre and T.H. Hsiao ed., Kluwer Academic Publishers, 553-558.

GIESBERGER, G. 1983. Biological control of the *Helopeltis* pest of cocoa in Java. Archives of Cocoa Research 2. Cocoa Research in Indonesia 1900-1950. American Cocoa Research Institute. International Office of Cocoa and Chocolate. H. Taxopeus and P.C. Wessel ed., Wageningen, The Netherlands, 90-180.

GOELDI, E.A. 1892. Relatorio sobre a molestia do cafeeiro na Provincia do Rio de Janeiro. Arch. Museu Nac. Rio de Janeiro 8: 1-121.

HANNA, A.D., E. JUDENKO and W. HEATHERINGTON. 1956. The control of *Crematogaster* ants as a means of controlling the mealybugs transmitting the Swollenshoot virus disease of cacao in the Golden Coast. Bull. entomol. Res. 47: 219-226.

HARLAND, S.C. 1925. Studies in cacao - Part I. The method of pollination. Ann. appl. Biology 12 (4): 403-409.

HERNANDEZ, B.J. 1965. Insect pollination of cacao (*Theobroma cacao* L.) in Costa Rica. Ph.D. Thesis, Madison, WI, Univ. Wisconsin, 167 p.

JACKSON, D.A. 1984. Ant distribution patterns in a Cameroonian cocoa plantation: investigation of the ant mosaic hypothesis. Oecologia 62: 318-324.

JEANNE, R.L. 1978. Intraspecific nesting associations in the Neotropical social wasp *Polybia rejecta* (Hymenoptera: Vespidae). Biotropica 10 (3): 234-235.

LESTON, D. 1969. Ants, capsids and Swollen Shoot in Ghana: interactions and the implications for pest control. Proc. III Intern. Cocoa Res. Conf., Accra, Ghana, 205-221.

LESTON, D. 1973. The ant mosaic-tropical tree crops and the limiting of pests and diseases. PANS 19 (3): 311-341.

LESTON, D. 1978. A Neotropical ant mosaic. Ann. entomol. Soc. America 71 (4): 649-653.

LIMA, A. da C. 1942. Insetos do Brasil - Homopteros. Escola Nacional de Agronomia. Serie Didatica n 4, Rio de Janeiro (Brazil), 327 p.

LUBIN, Y.D. 1984. Changes in the native fauna of the Galapagos Islands following invasion by the little red fire ant *Wasmannia auropunctata*. Biol. J. Linnean Soc. 21 (1-2): 229-242.

MAJER, J.D. 1972. Tha ant mosaic in Ghana cocoa farms. Bull. entomol. Res. 62: 151-160.
MAJER, J.D. 1976a. The maintenance of the ant mosaic in Ghana cocoa farms. J. appl. Ecol. 13: 123-144.
MAJER, J.D. 1976b. The ant mosaic in Ghana cocoa farms: further structural considerations. J. appl. Ecol. 13: 145-155.
MEER MOHR, J.-C. van der. 1927. Au sujet du role de certaines fourmis dans les plantations coloniales. Bull. Agric. Congo Belge (Belgium) 18 (1): 97-106.
PICKEL, D.B. 1927. Os parasitos do cafeeiro no Estado da Paraiba. Um novo parasito do cafeeiro, o piolho branco *Rhizoecus lendea n.sp.*. Chacaras e Quintais 36: 587-593.
POSNETTE, A.F. 1944. Virus diseases of cacao in Trinidad. Trop. Agric., Trinidad and Tobago. 21 (6): 105-106.
SANTSCHI, F. 1925. Nouveaux Formicides brèsiliens et autres. Ann. Bull. Soc. Entomol. Belgique 65: 221-247.
SILVA, A.J.R., P.M. BAKER and F.P. BENTON. 1982. Estudo da biologia da formiga cacarema (*Azteca chartifex spiriti* Forel) associada ao cacaueiro na Bahia. CEPEC - Informe Tecnico 1981, CEPLAC, Bahia, Brazil. p 87.
SILVA, P. 1944. Insects pests of cacao in the State of Bahia, Brazil. Trop. Agric., Trinidad and Tobago. 21 (1): 8-14.
SILVA, P. 1945. A formiga de enxerto. Instituto de Cacau da Bahia, Boletim de Divulgacao n 1, Livraria Duas Americas, Bahia, Brazil. 21 pp.
SILVA, P. 1950. The coccids of cacao in Bahia, Brazil. Bull. entomol. Res. 41 (1): 119-120.
SILVA, P. 1955. A formiga cacarema e o cacaueiro. Instituto Biologico da Bahia, Junta Executiva de Combate as Pragas e Doencas do Cacau, Ministerio da Agricultura, Salvador, Bahia (Brazil) 13 pp.
SILVA, P., J.M. ABREU, and SMITH F., G.E. 1969. Aspectos bioecologicos e combate as sauvas nas regioes cacaueiras da Bahia e Espirito Santo. CEPEC-Informe Tecnico 1968 e 1969, CEPLAC, 93-94.
SILVA, P. and BARBOSA, M. da R. 1948. O Instituto de Cacau e o combate a formiga de enxerto. Bahia Rural, Brazil. 16 (7): 24-26, 23.
STRICKLAND, A.H. 1951. The entomology of Swollen Shoot of cacao. I- The insect species involved, with notes on their biology. Bull. entomol. Res. 41: 725-748.
STRICKLAND, A.H. 1952. The entomology of Swollen Shoot of cacao. II- The bionomics and ecology of the species involved. Bull. entomol. Res. 42: 65-103.
TAYLOR, B. 1977. The ant mosaic on cocoa and other tree crops in Western Nigeria. Ecol. Entomol. 2: 245-255.
TAYLOR, B. and M.J. GRIFFIN. 1981. The role and relative importance of different ant species in the dissemination of Black Pod disease of cocoa. Epidemiology of *Phytophtora* on Cocoa in Nigeria, P.H. Gregory and A.C. Maddison ed., Commonwealth Mycological Institute, Kew, Surrey, United Kingdom. 114-188.
TORREND, G. 1919. As molestias dos cacaueiros em Ilheos (Bahia). Broteria, Portugal. 16: 264-278.
VELLO, F. 1969. Observacoes sobre polinizacao do cacaueiro na Bahia. Proc. III Intern. Cocoa Res. Conf., Accra, Ghana. 565-575.

VELLO, F. and W.S. MAGALHAES. 1971. Estudos sobre a participacao da formiga cacarema (*Azteca chartifex spiriti* Forel) na polinizacao do cacaueiro na Bahia. Rev. Theobroma, Brazil. 1 (4): 29-42.

WEBER, N.A. 1944. The Neotropical coccid-tending ants of the genus *Acropyga* Roger. Ann. entomol. Soc. America 37: 89-122.

WEBER, N.A. 1946. Two common ponerine ants of possible economic significance, *Ectatomma tuberculatum* (Olivier) and *E. ruidum* Roger. Proc. entomol. Soc. Washington 48 (1): 1-16.

WEBER, N.A. 1957. Costa Rican cacao insects. Communicaciones de Turrialba, Costa Rica. 58: 1-26.

WHEELER, W.M. 1935. Ants of the genus *Acropyga* Roger, with description of a new species. J. N.-Y. entomol. Soc. 43: 321-329.

WILLIS, E.O. and Y. ONIKI. 1978. Birds and army ants. A. Rev. Ecol. Syst. 9: 243-263.

WILLIS, E.O. and Y. ONIKI. 1988. Na trilha das formigas carnivoras. Ciencia Hoje 8 (47): 26-32.

WINDER, J.A. 1978. The role of non-dipterous insects in the pollination of cocoa in Brazil. Bull. entomol. Res. 68: 559-574.

ZEHNTNER, L. 1917. O "queima" dos cacaoeiros nos Municipios de Ilheos e de Itabuna. Relatorio apresentado a Associacao Commercial de Ilheos. 76 p.

53

Ant Assemblage Structure and Ecological Management in Citrus and Subtropical Fruit Orchards in Southern Africa

M.J. Samways

INTRODUCTION

Ants can be major indirect pests in citrus and subtropical fruit orchards (Flanders 1945; Ullyett 1938). Honeydew excreted by soft scales, mealybugs and aphids attracts them into the trees. This activity adversely affects the parasitoid biocontrol agents of these homopteran insects, and those of other serious pests, particularly red scale [*Aonidiella aurantii* (Maskell)]. Released from high parasitoid pressure, both types of pest increase to large numbers (Compere 1961). Only certain ants are responsible for the *A. aurantii* outbreaks - notably *Pheidole megacephala* (F.) and *Anoplolepis custodiens* (F. Smith) in southern Africa (Broekhuysen 1948; Steyn 1954a,b) - with the majority playing an insignificant role.

Since the advent of *A. aurantii* resistance to organophosphate insecticides, ant control has received accelerated attention in southern Africa (Nel et al. 1979). *A. aurantii* has become increasingly difficult to chemically control; ant control has become a vital prerequisite to biocontrol (Bedford 1968).

Southern African citrus is grown in a variety of climates from the Capensic Mediterranean conditions of the western Cape, to the winter-rainfall areas of the eastern Cape, and the summer-rainfall areas of Natal, Swaziland and the Transvaal (Fig. 1). Most ants in these areas are cosmopolitan, but some economically-significant species, such as the exotic *Iridomyrmex humilis* (Mayr) and the indigenous *Anoplolepis steingroeveri* (Forel) in the western Cape (Table 1), are currently area-specific. See Prins et al. in this volume for details of pest ants in urban and other agricultural areas of South Africa.

Despite these geographical variations, and the species differences between and within sites, the principle of control is the same: mutualism between ground-nesting ants and the tree-dwelling honeydew-producers must be broken. This review examines mutualism, as well as research which produced an economically and ecologically appropriate method of ant control. It is suggested that this method might be used extensively where ever pest ants forage in economically-important trees.

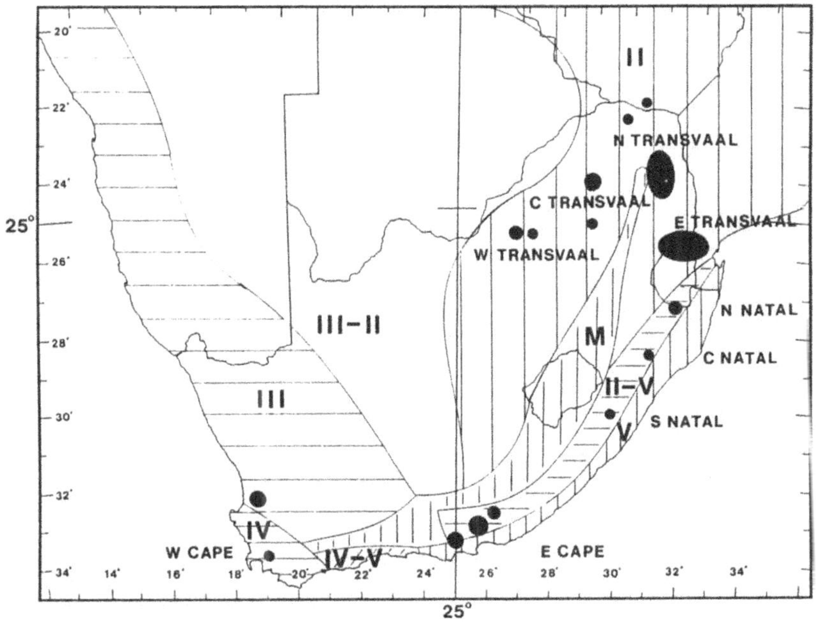

FIGURE 1. The citrus growing areas of southern Africa. The Roman numbers are zonobiomes (see Walter et al. 1975). I = Equatorial zone, II = Tropical summer-rainfall zone, III = Subtropical dry zone, IV = Transition zone with winter rainfall, V = Warm temperate climate. M = Montane areas.

INTRA- AND INTERSPECIFIC COMPETITION

Ants frequently show a high level of intraspecific (Mabelis 1979) and interspecific aggression (Hölldobler and Lumsden 1980). In both cases, the objective is to gain space and resources around a fixed and central nest. Asymetrical competition can lead to complete amensalism (Samways 1983a). Spatially, the result of this behavior is a mosaic of dominant ant species (Leston 1973; Majer 1972; Room 1975). In unusual cases, ant assemblage richness becomes reduced, with extreme dominance by one species (Fig. 2).

INTERSPECIFIC COMPETITION IN RESPONSE TO ORCHARD MANAGEMENT

Orchard pest management influences the ant assemblage pattern on the orchard floor. Samways (1981) compared three adjacent orchards: one was under parathion chemical control of *A. aurantii*, another employed only natural enemies (particularly *Aphytis africanus* Quednau) and the third was in transition from chemical to biological control (Fig. 3). Variation within the orchard was statistically zero at the chemical and biological control sites, both

TABLE 1. List of ant species foraging in citrus trees in southern Africa. 0 = not found, 1 = 1-9 trees supporting the observation, 2 = 10-99 trees, 3 = 100-999 trees, 4 = 1000-9999 trees; WC=West Cape, EC=East Cape, SN=South Natal, CN=Central Natal, NN=North Natal & Swaziland, NT=North Transvaal, WT=West Transvaal, ET=East Transvaal, CT=Central Transvaal (Data from Samways et al., 1981).

Species	Nesting in trees	Foraging in trees	Attending Coccidae	Attending Pseudococcidae	Attending Aphididae	Distribution
PONERINAE						
Odontomachus troglodytes Sant. [= O. haematoda (L.)?]	0	1	0	0	0	NT
Leptogenys nitida (Smith)	0	1	0	0	0	NT
PSEUDOMYRMECINAE						
Tetraponera natalensis (Smith)	0	2	0	0	0	EC,ET
MYRMICINAE						
Myrmicaria natalensis (Smith)	0	4	3	2	2	NN,ET,CT,NT,WT
Cardiocondyla emeryi Forel	0	1	0	0	1	ET
Pheidole akermani Arnold	0	1	1	0	0	ET
P. arnoldi Forel	0	1	0	0	0	ET
P. crassinoda Emery	0	2	1	0	0	ET,NT,WT
P. megacephala (F.)	0	4	4	3	2	EC,SN,CN,NN,ET, CT,NT,WT
P. rotundata Forel	0	2	1	1	1	ET,CT
P. sculpturata Mayr	0	4	4	3	2	EC,ET,CT,NT
P. tenuinodis Mayr	0	1	0	0	0	WC,ET,CT,NT,WT
Pheidole sp. 1	0	2	2	0	0	ET
Pheidole sp. 2	0	1	0	0	0	ET
Crematogaster ? natalensis (Forel)	0	2	0	0	0	ET

Continued.

TABLE 1. Continued.

Species	Nesting in trees	Foraging in trees	Attending Coccidae	Attending Pseudococcidae	Attending Aphididae	Distribution
C. liengmei Forel	0	1	1	1	1	EC,CN
C. ? castanea Smith	1	3	1	1	0	ET,CT,NT,WT
C. peringueyi Emery	0	1	0	0	0	CN
Monomorium delagoensis (Forel)	0	1	0	0	0	WC,EC,NN,ET,CT,NT
Tetramorium grassii Emery	0	1	0	0	1	CN
T. mossamedense Forel	0	1	0	0	0	EC,CN,ET
Xiphomyrmex weitzeackeri Emery	0	1	0	0	0	CN,CT
Cataulacus baumi Forel	0	1	0	0	0	EC,CN
DOLICHODERINAE						
Iridomyrmex humilis (Mayr)	1	3	2	2	1	WC
Technomyrmex albipes (Smith)	1	3	1	2	0	WC,EC,ET,CT,NT
FORMICINAE						
Anoplolepis custodiens (Smith)	0	4	4	4	3	WC,EC,SN,NN,ET, CT,NT,WT
A. steingroeveri (Forel)	0	3	2	2	1	WC
Plagiolepis brunni Mayr	1	1	0	0	0	NN
Acantholepis capensis? Mayr simplex Forel	0	2	0	0	1	NN,ET
A. ? capensis laevis (Sant.)	0	1	0	0	0	ET
A. ? capensis rubrovaria (Forel)	0	1	0	0	0	ET
A. ? capensis validiuscula (Emery)	0	1	0	0	0	ET
Camponotus maculatus (F.)	0	3	1	1	0	WC,EC,ET,CT,NT,WT
C. eugeniae (Forel)	0	1	0	0	1	ET,WT

Continued.

TABLE 1. Continued.

Species	Nesting in trees	Foraging in trees	Attending Coccidae	Attending Pseudococcidae	Attending Aphididae	Distribution
C. ? rufoglaucus (Jerd.) cinctellus (Gerst.)	0	1	1	1	1	ET,NT
C. ? sp. (near contraria Sant.)	0	1	0	0	1	CT
C. ? rufoglaucus flavomarginatus	0	4	3	3	2	NN,ET,CT,NT
C. fulvopilosus De Geer	0	1	1	0	1	WC,NT
C. ? pectitus Sant.	0	4	3	3	2	EC,NN,ET,CT,NT,WT
C. petersi Emery	0	1	0	0	1	CN,ET
C. mayri Forel	0	1	0	0	0	ET,CT
C. reevi Arnold	0	1	0	0	0	CN
Polyrhachis schistacea (Gerst.)	0	3	2	1	2	NN,ET,CT,NT,WT
Paratrechina sp. [near traegaordhi (Forel)]	0	1	0	0	0	CN,ET

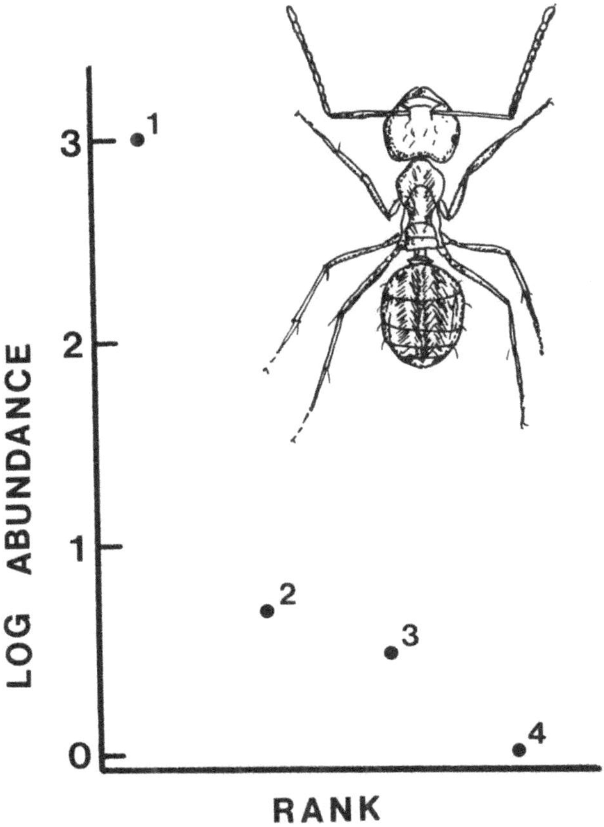

FIGURE 2. Log abundance/rank plot of an epigaeic ant assemblage in a guava orchard where there are ample mealybug (*Planococcus citri* Risso) honeydew resources supporting an extremely high population of *Anoplolepis custodiens* (F. Smith) (Species No. 1) which totally dominates the ant community, allowing only *Pheidole sculpturata* Mayr (Species No. 2), *Tetramorium mossamedense* Emery (Species No. 3) and *Solenopsis punctaticeps* May (Species NO. 4) to coexist (Redrawn from Samways et al. 1981).

of which had constant species abundances and rank-abundance sequences throughout. Within the transition site, rank-abundance sequences were constant but species abundances varied. Despite the statistical constancy of rank, within-orchard samples at the transition and chemical control sites varied in the dominant ants (*Pheidole megacephala* [F.] or *Myrmicaria natalensis* [F. Smith]) from one habitat to another. *P. Megacephala* clearly dominated the biocontrol site. This site was also the richest in number of species and individuals. Diversity, as measured by α or Q log-normal, was lowest in the chemical control site, medium in the biological control site and

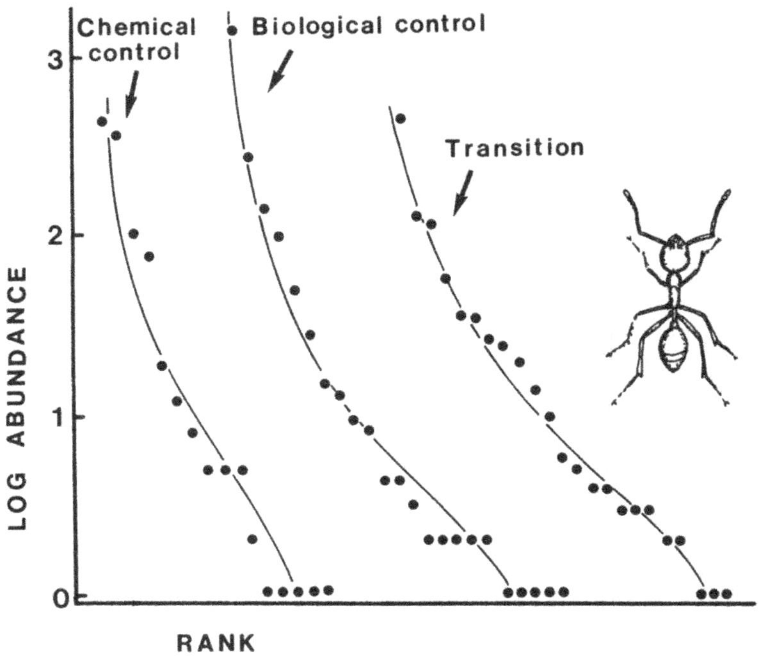

FIGURE 3. Log abundance/rank plots and fitted curves (accounting for 95% to 98% of the variance) of three epigaeic ant assemblages in three citrus orchards under different pest management practices for red scale, *Aonidiella aurantii* (Maskell). (Redrawn from Samways 1981).

highest in the transition site. These findings indicate that chemical control causes a paucity of species, lowering of overall ant abundance and creating a two- dimensional dominant-ant mosaic. During release from insecticide pressure, species richness increases dramatically and the spatial mosaic dissolves. After recovery from insecticide pressure, general ant abundance increases and a single species tends to dominate. This is a direct response to a concomitant increase in insect food resources, sparking a population surge in the potentially-dominant species.

ANT-HOMOPTERAN MUTUALISMS AND POPULATION RESILIENCE

When *A. aurantii* is under biological control, populations of other pests sometimes also increase - particularly the mealybugs *Planococcus citri* (Risso) and *Pseudococcus longispinus* (T.T.), and the soft scales *Coccus hesperidum* L. and *C. aethiopicus* De Lotto. This occurs because they are no longer suppressed by spring applications of mostly organophosphate insecticides, which are applied early in the season before parasitoids are most active. The other pests are honeydew producers, and even small numbers attract ants into the trees. Additionally, the flow of spring blossom nectar encourages an increase in foraging, especially by *P. megacephala.* The early-season,

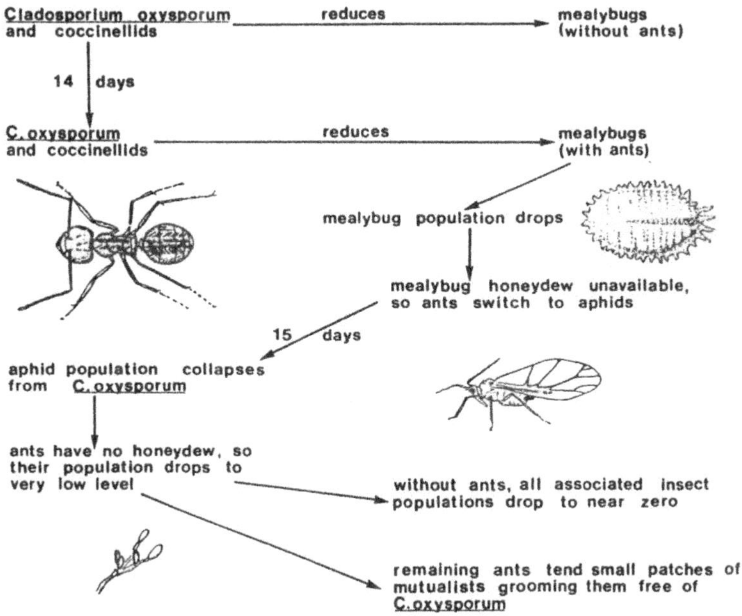

FIGURE 4. Ecological relationships associated the impact of the *Cladosporium oxysporum* (Burt. and Curt.) epizootic on the *Anoplolepis custodiens* (F. Smith)/*Planococcus citri* (Risso) mutualistic relationship.

low-level foraging and mutualism leads to an increase in both mutualists, a well-known phenomenon (Way 1963). The important point, however, is that honeydew-seeking ant populations are the most resilient. When honeydew is in short supply, either from lack of pest insects in the tree, or from absence of other insects on vegetation beneath the tree, these ants may be relatively rare. But once honeydew becomes available, the mutualistic population spiral begins, with the most responsive ant species reaching enormous numbers (Samways 1982a).

ANT-HOMOPTERAN MUTUALISMS AND FUNGAL EPIZOOTICS

On occasion, particularly in guava trees (*Psidium guajava* L.), a third party, the entomogenous fungus *Cladosporium oxysporum* (Berk. and Curt.), is also involved (Samways 1983b; Samways and Grech 1986). A mutualism between the ant *A. custodiens* and the mealybug *P. citri* raises the population levels of both. When ants are excluded from half the trees, their overall population level also drops by half. Ants deprived of their mutualists move their nests as close as possible to the nearest and largest mealybug-honeydew food source. Without their ant protectors, mealybugs are easy prey for coccinellids such as the ant-sensitive *Exochomus flavipes* (Thub.) and the ant-tolerant *Scymnus* spp. When an epizootic of *C. oxysporum* breaks out among the *P. citri*, trees without ants almost immediately lose their mealybugs

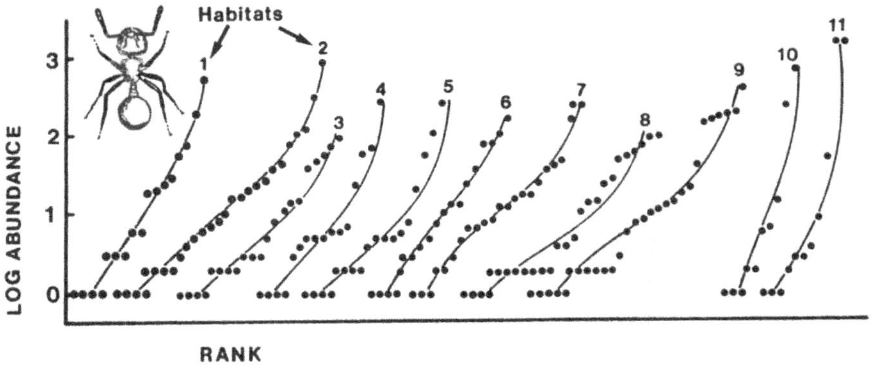

FIGURE 5. Reversed (for visual convenience) log abundance/rank plots with best-fit log-normal curves for epigaeic ant assemblages in 11 point habitats. Habitats 1, 7 and 3 = rank grassland; 2, 8 and 9 = sparse grassland; 4, 5 and 6 = old orchard; 10 and 11 = young orchard.

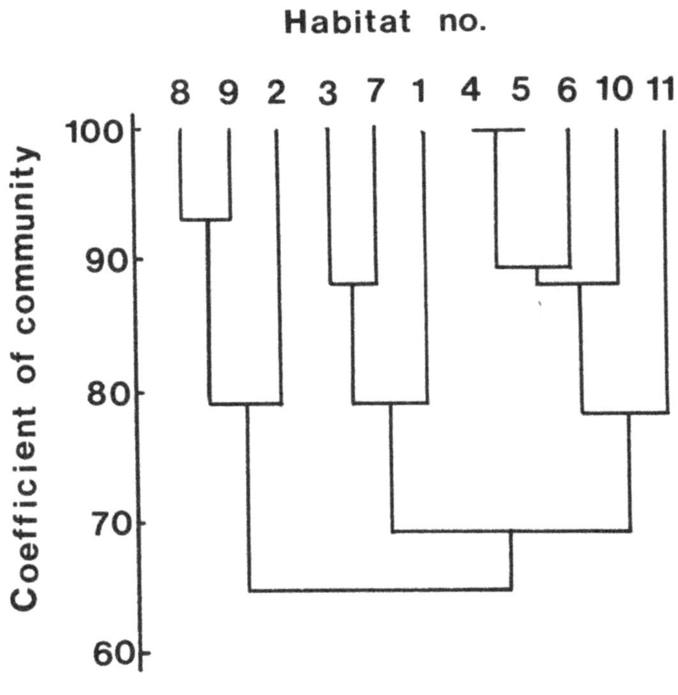

FIGURE 6. Agglomerative sorting of coefficients of community for the eleven habitats using the UPGMA method for all but single species records. Habitats 1, 7 and 3 = rank grassland; 2, 8 and 9 = sparse grassland; 4, 5 and 6 = old orchard; 10 and 11 = young orchard.

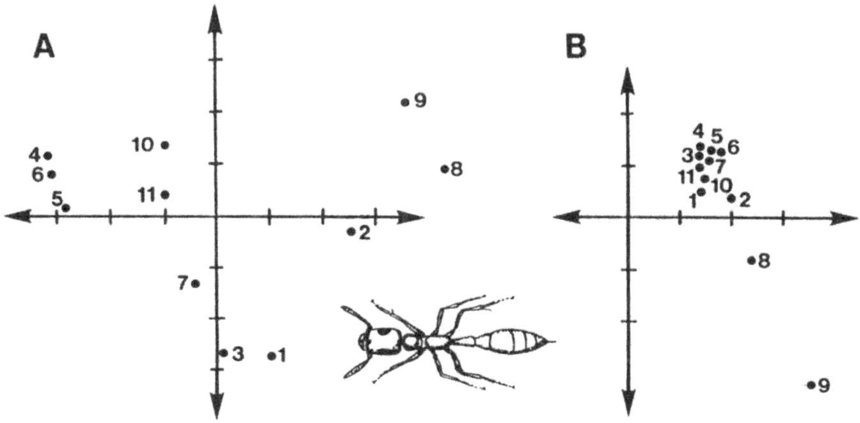

FIGURE 7. Sorting of the habitats from a 49 x 11 (species x habitats) (A) and a 6 x 11 (6 commonest species x habitats) (B) raw data matrix using Pearson's correlation coefficient and biplots. Habitats as for Fig. 6.

to the fungus. In contrast, mealybugs tended by ants avoid the epizootic for two weeks. By the 15th day, even ant-tended mealybugs become infected. Many ants then switch and tend *Aphis gossypii* Glover, though a few continue to protect isolated groups of *P. citri*. Fifteen days after the epizootic strikes the ant-attended mealybug population, it starts to devastate the aphid population. With both mealybugs and aphids at low levels, the ant population declines. However, the ant-homopteran mutualisms, although severely curtailed by the epizootic, remain intact. This frequently occurs when isolated patches of the mealybug are cleansed by the ants. The ants remove the honeydew as well as the mealybug wax covering and its attendant fungal condidia. This intensive hygiene mutualism is an important survival strategy for both ants and mealybugs (Fig. 4).

ANT RESPONSE TO HABITAT CHANGE

In South Africa, ants form a major component of the orchard floor fauna, numerous both as individuals and species. As competition is severe among these insects, it might be possible to manipulate the habitat in favor of the beneficial or neutral species.

A detailed study (Samways 1983c) of ant assemblage response to eleven habitat types revealed distinct trends. Rank-abundance plots show increasing equitability from young orchard to old orchard and rank to sparse grassland (Fig. 5). In a young citrus orchard, the dominant species account for as much as 96% of individuals, yet only 18% in sparse grassland. At these sites, the *Pheidole* genus dominates; in all but one habitat, some *Pheidole* species is dominant. In grassland, more than in orchards, ants are distributed in a two-dimensional mosaic. At the nominal level, most habitats share the majority of species. However, at the ordinal level, the rank series of ants

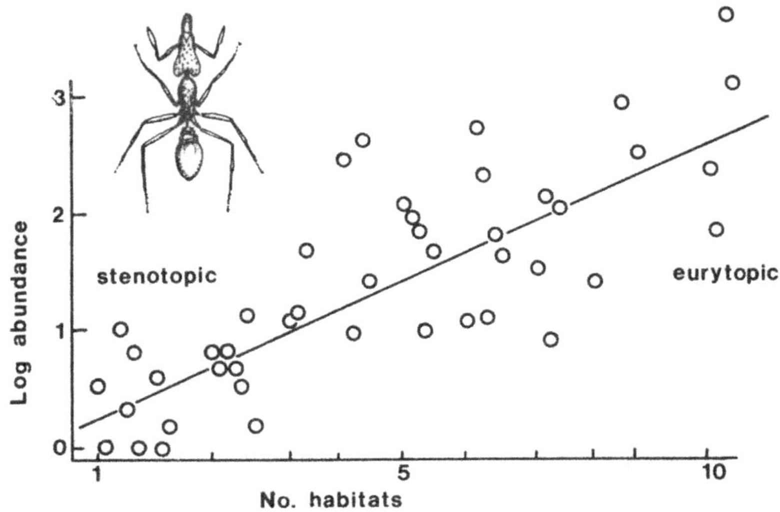

FIGURE 8. Log abundance of 49 epigaeic ant species relative to the number of habitats in which they occur ($r = 0.594$, $P<0.05$; $F = 71.2$, $P<0.0001$).

varies from one habitat to another. On the basis of ant abundance at the interval level, habitats group according to overall physiognomy and can be arranged in the following general sequence: old orchard, young orchard, rank grassland, sparse grassland (Fig. 6). This illustrates a general sensitivity to habitat type. However, this sensitivity does not apply to the most abundant ants (Fig. 7), and a significant correlation exists between the range of habitats occupied by specific ant species and their overall abundance (Fig. 8). Of course, eurytopic ants, by virtue of increased space and resources, can afford to support more members than stenotopic species.

ANT MANAGEMENT PRINCIPLES

The least abundant ants, as well as being stenotopic, are rare at all times. The most common species, particularly *P. megacephala* and *A. custodiens*, are eurytopic with high absolute population variation. These pest ants tend to drop to low population levels when resources diminish, yet rise when resources are abundant depending on their reproductive rate and the area's carrying capacity. Thus, the pest ants can be most abundant, the most competitive and demographically the most volatile, in addition to being the least selective of habitat.

This suggests that habitat modification would be an ineffective method of control, as well as logistically difficult. Also, control by interspecific competition is unfeasible; introduction of a more common and aggressive ant species would entail considerable environmental and economic risk. Chemical control (Bot et al. 1988), including ant baits (Samways 1985), is effective against small ants (*P. megacephala*), but not against the larger and more resourceful species (*A. custodiens*). Even so, control must be applied at regular intervals and use effective applications. The ants must be locally eradicated, but continuous chemical control is not economically feasible and is

FIGURE 9. Damage done to citrus trees c.v. Valencia from direct application of a polybutene-and-wax sticky ant barrier. Phytotoxicity occurs principally where the trunk is exposed to the sunshine (A), but can also occur when the tree canopy affords good shade (B). If left in position, the band can cause extensive bark damage (C), leading eventually to death of the tree (D).

environmentally unsafe. Partial control only temporarily lowers the population. Because of these pest ants' ability to breed quickly, their population soon returns to previous levels.

Paradoxically, dominant and harmful ants in the tree, may be beneficial on the ground (Samways et al. 1981). For example, *A. custodiens* and *M. natalensis* collect honeydew, but while on the ground they prey on lepidopterous larvae such as *Heliothis armigera* (Hübn.) and *Cryptophlebia leucotreta* Meyr., just prior to pupation.

In citrus orchards where ant populations are highly sensitive to variations in resources, control should aim to reduce resources, especially honeydew-producing insects. Localized high-population patches can be spot-treated with the appropriate insecticide (Bot et al. 1988) and the pest management program in the orchard organized for maximum utilization of natural enemies. This workable approach is now widely used on southern African citrus farms.

Control should focus on the mutualistic association between groundnesting ants and the tree-dwelling honeydew producers. Breaking the mutualism solves the problem. This can be achieved by a physical barrier which prevents the ants from entering the tree.

FIGURE 10. Bark phytotoxicity on a young citrus tree c.v. navel caused by a direct application of an acrylic sticky ant band.

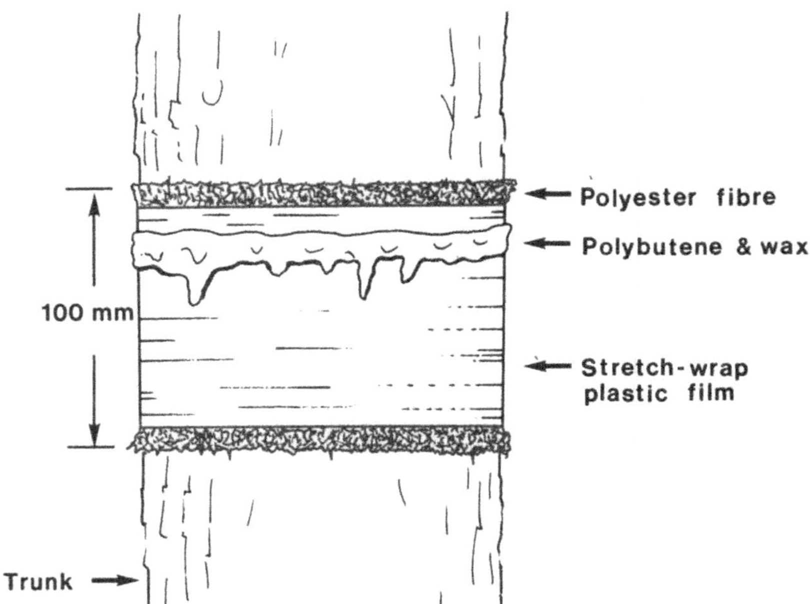

FIGURE 11. The recommended ant barrier for citrus trees ('Ant-Bar 1.2.3).

CONTROL MEASURES

The checkered history of physical ant barriers began with the direct application of a sticky polybutene and wax band. For several years this method was used, until the sunny side of the trunks, even of old trees, started to suffer phytotoxicity (Fig. 9a). Later, some shaded trunks also showed damage (Fig. 9b). In some cases, when the sticky bands were not removed, they caused severe bark necrosis (Fig. 9c), eventually leading to the tree's death (Fig. 9d). Guava trees shed their bark naturally during growth and thus were not subject to this problem (Samways 1982b).

The damage caused widespread concern within the citrus industry and alternative compounds/methods were urgently sought. At one time, researchers tried to substitute an acrylic band for the polybutene/wax band. This band was easier to apply than the highly viscous polybutene/wax compound, but dust rendered it ineffective, and it too was eventually found to be phytotoxic (Fig. 10).

An intensive search for an alternate band was undertaken (Samways and Tate 1984), this time with rigid criteria designed to ensure its acceptance by the grower community Samways and Tate 1985, 1986). It had to be cheap, easy to apply, totally effective, long-lasting, non- phytotoxic and tolerant of tree expansion. Such a band, 'Ant-Bar 1.2.3.' (available from J & S Distributors, Stallchem, Nelspruit 1200, South Africa), was subsequently recommended (Samways and Buitendag 1986).

FIGURE 12. 'Ant-Bar 1.2.3.' in position on young citrus tree.

TABLE 2. Ant control schedule using trunk barriers.

1st year winter	Identify problem areas. Skirt trees and cut weeds from under the canopy.
1st year spring	Apply bands correctly and carefully.
1st year summer	Check bands monthly, reapply sticky band if necessary. Remove weed bridges and trim branches touching the ground.
2nd year winter	Skirt trees and clear weeds from under canopy.
2nd year spring	Peel off old stretch-wrap plastic, apply a new layer, and also a new sticky layer. Some trees that have outgrown the length of polyester fiber, may need this layer replacing also.
2nd year summer	Check bands monthly, reapply sticky band if necessary. Remove weed bridges and trim branches touching the ground.

The band consists of three layers:

1. Polyester fiber layer (Bidim U24). This material, manufactured for banding in rolls of 25m with a width of 100mm, is light and easily carried. The polyester fiber is wrapped around the trunk as a single layer with a 20-30mm overlap and then cut from the rest of the roll with scissors (Fig. 11). If the tree has crevices or small undulations, the back of the polyester fiber layer can be fluffed by hand. In the rare cases where the tree has severe fluting, the hollows are filled with loose balls of the fiber.

2. Stretch-wrap plastic film layer (e.g. Gladwrap, Clingfilm). The proprietary rolls, which are 300mm wide and hold a length of 30m, are cut into three with a sharp knife to yield banding rolls with a width of 100mm. The polyester fiber layer is held in position by two or three windings of stretch-wrap plastic in such a way that 10-15mm fiber is left exposed above and below the wrapping. This is most important because the fluffy edges act as wicks taking away any moisture accumulating under the stretch-wrap plastic. The band is now ready to receive the third and sticky layer on the impermeable plastic surface (Fig. 11).

3. Outer sticky layer (Ant-Bar). This layer is the sticky barrier that actually stops the ants from moving up the tree. It is a polybutene and wax compound that remains sticky for several months. The material is manufactured in 280g cartridges and applied as a ring around the top edge of the stretch-wrap plastic with a gun applicator. This is a quick, clean and efficient method of

application. Once applied, the polybutene/wax ring begins to curtain downwards, but never runs off the plastic base. This becomes a highly effective sticky barrier (Fig. 11).

One kit, containing all the above materials, treats 500 trees with a trunk diameter of 80mm. Cost of banding is proportional to tree diameter. The control schedule for application of the band is given in Table 2. Special attention must be given to tree skirting and removal of weed bridges which might otherwise afford the ants an alternative route up the tree.

The advantage of the method is that besides its simplicity and effectiveness, it is environmentally safe since no insecticides are involved.

CONCLUSION

Few ant species in southern Africa are pests. Those species which are pests have great population variation. With abundant food resources, their numbers soar. Indeed, they are highly effective mutualists with honeydew-producing Homoptera. When such homopterans inhabit the trees, the ants attend them, and as a result both insects' population levels increase rapidly. During this mutualistic spiral, natural enemies of the key pest *Aonidiella aurantii* are rendered ineffective. This results in population outbreaks of the pest. As *A. aurantii* now resists organophosphate insecticides, control depends on the use of natural enemies. For biocontrol to be effective, ant control is vital.

As the pest ants are insensitive to habitat and highly competitive, methods of control based on habitat modification or introduction of competitors are not feasible. Similarly, chemical control only gives temporary or partial control, even if applied in baits. Physical barriers then become ecologically appropriate, because they break the mutualistic relationship between ants nesting in the ground and the homopterans in the tree.

The selected band consists of three layers: a polyester fiber closest to the tree to take up the bark's roughness, then a stretch-wrap plastic film which in turn is a basement layer for the sticky polybutene and wax barrier. This inexpensive band, Ant-Bar 1.2.3., expands with the tree, is non-phytotoxic, and repels ants for many weeks without rejuvenation. Above all, it contains no pesticides and is relatively non-contaminatory to the environment.

ACKNOWLEDGMENTS

Financial support was generally provided by the South African Co-operative Citrus Exchange, University of Natal Research Fund, and the Foundation of Research Development (Council for Scientific and Industrial Research). Mrs. Ann Best kindly typed the manuscript. Permission to reproduce Figure 3 was given by CABI International, Wallingford, Oxfordshire, U.K.

REFERENCES

BEDFORD, E.C.G. 1968. The biological control of red scale, *Aonidiella aurantii* (Mask.) on citrus in South Africa. J. ent. Soc. sth. Afr. 31: 1-15.

BOT, J., S. SWEET, M. KRAUSE and N. HOLLINGS. 1988. A guide to the use of pesticides and fungicides in the Republic of South Africa. 33rd Edition. Dept. Agric. & Water Supply, Pretoria. 319pp.

BROEKHUYSEN, G.J. 1948. The brown house ant (*Pheidole megacephala* F.). Bull. Dep. Agric. Un. S. Afr. No. 266: 1-40. Compere, H. 1961. The red scale and its insect enemies. Hilgardia 31: 173-178.

FLANDERS, S.E. 1945. Coincident infestations of *Aonidiella citrina* and *Coccus hesperidum*, a result of ant activity. J. econ. Ent. 38: 711-712.

HÖLLDOBLER, B. and C.J. LUMDENS. 1980. Territorial strategies in ants. Science, N.Y. 210: 732-739.

LESTON, D. 1973. The ant mosaic-tropical tree crops and the limiting of pests and diseases. PANS 19: 311-341.

MABELIS, A.A. 1979. Distribution of red wood ants (*Formica polyctena* Forst.) over the foraging area of their nest, and the influence of a conspecific neighbouring population. Netherlands J. Zool. 29: 221-232.

MAJER, J.D. 1972. The ant mosaic in Ghana cocoa farms. Bull. ent. Res. 62: 151-160.

NEL, J.J.C., L. DE LANGE and H. VAN ARK. 1979. Resistance of citrus red scale, *Aonidiella aurantii* (Maks.), to insecticies. J. ent. Soc. sth. Afr. 42: 275-281.

ROOM, P.M. 1975. Diversity and organisation of the ground foraging ant faunas of forest, grassland and tree crops in Papua New Guinea. Aust. J. Zool. 23: 71-89.

SAMWAYS, M.J. 1981. Comparison of ant community structure (Hymenoptera: Formicidae) in citrus orchards under chemical and biological control of red scale, *Aonidiella aurantii* (Maskell) (Hemiptera: Diaspididae). Bull. ent. Res. 71: 663-670.

SAMWAYS, M.J. 1982a. Pest management within the sphere of ecology. Bull. S. Afr. Inst. Ecol. 1: 11-23.

SAMWAYS, M.J. 1982b. Ecologically-sound and commercially acceptable control of ants in guava trees. Subtropica 3: 19-20.

SAMWAYS, M.J. 1983a. Asymmetrical competition and amensalism through soil dumping by the ant, *Myrmicaria natalensis*. Ecol. Ent. 8: 191-194.

SAMWAYS, M.J. 1983b. Inter-relationship between an entomogenous fungus and two ant-homopteran (Hymenoptera: Formicidae- Hemiptera: Pseudoccidae and Aphididae) mutualisms on guava trees. Bull. ent. Res. 73: 321-331.

SAMWAYS, M.J. 1983c. Community structure of ants (Hymenoptera: Formicidae) in a series of habitats associated with citrus. J. appl. Ecol. 20: 833-847.

SAMWAYS, M.J. 1985. Appraisal of the proprietary bait 'Amdro' for control of ants in southern African citrus. Citrus Subtrop. Fr. J. No. 621: 14-17.

SAMWAYS, M.J. and C.H. BUITENDAG. 1986. Recommendations for control of ants on citrus trees using trunk barriers. Citrus Subtrop. Fr. J. No. 624: 9-11.

SAMWAYS, M.J. and N.M. GRECH. 1986. Assessment of the fungus *Cladosporium oxysporum* (Berk. and Curt.) as a potential biocontrol biocontrol agent against certain Homoptera. Agric., Ecosyst., Environ. 15: 231-239.

SAMWAYS, M.J., M. NEL and A.J. PRINS. 1982. Ants (Hymenoptera: Formicidae) foraging in citrus trees and attending honeydew-producing Homoptera. Phytophylactica 14: 155-157.

SAMWAYS, M.J. and B.A. TATE. 1985/6. A highly efficacious and inexpensive trunk barrier to prevent ants from entering citrus trees. Citrus Subtrop. Fr. J. No. 622: 12, 13, 18.

SAMWAYS, M.J., A.J.S. WEAVING and M. NEL. 1981. Efficacy of chemical and sticky banding in preventing ants entering guava trees. Subtropica 2: 13-15.

STEYN, J.J. 1954a. The pugnacious ant *Anoplolepis custodiens* (Smith) and its relation to the control of citrus scales at Letaba. Mem. ent. Soc. S. Afr. No. 3: 1-96.

STEYN, J.J. 1954b. The effect of the cosmopolitan brown house ant (*Pheidole megacephala* F.) on citrus red scale (*Aonidiella aurantii* Mask.) at Letaba. J. ent. Soc. sth. Afr. 17: 252- 264.

ULLYETT, G.C. 1938. Ants and beneficial insects. Fmg. S. Afr. 13: 120. Way, M.J. 1963. Mutualism between ants and honeydew-producing Homoptera. A. Rev. Ent. 8: 307-344.

WALTER, H., E. HARNICKELL and D. MUELLER-DOMBOIS. 1975. Climate- diagram maps of the individual continents and the ecological climate regions of the earth. Springer-Verlag, Berlin. 36pp. + 9 maps.

Control

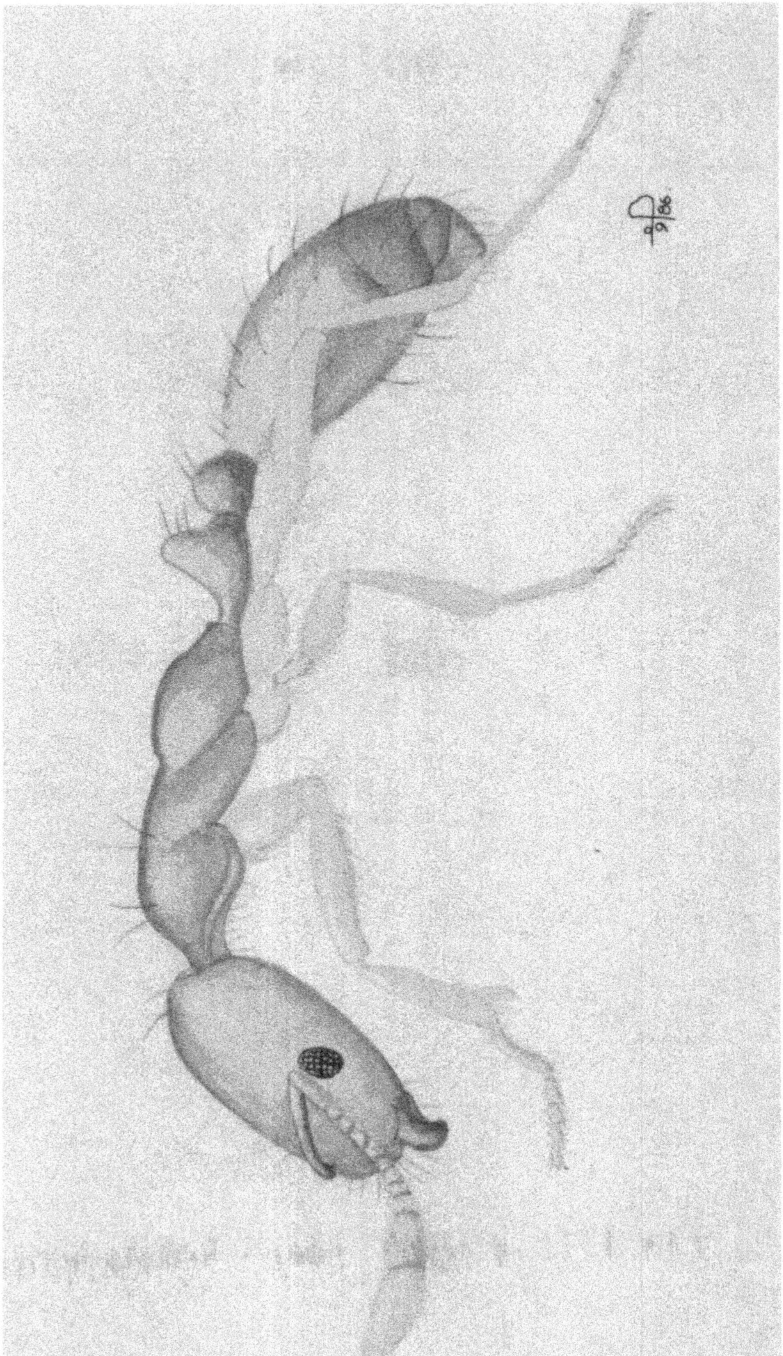

Monomorium pharaonis (x 42). Drawing by Eduardo Perez.

Overview

J.M. Cherrett

When ants cause the significant losses to man outlined in the last section, steps have to be taken to reduce, or 'control' this damage. Controlling ants is a specialized technology often very different from that required to control other insect pests. Ants, in common with some of the other social insects present particular control problems, and also particular control opportunities which are well illustrated in the 14 papers which make up this section. These differences need more detailed examination, as they are responsible for the uniqueness of pest ant control methods.

CONTROL PROBLEMS

Central Place Foraging

Ants forage over and cause damage to a much larger area than that in which they reside and breed. Thus a leaf-cutting ant nest of *Atta cephalotes* with a diameter of 10m. may utilize an approximately circular foraging area of diameter 200m. Consequently, killing foraging workers where damage occurs is not usually effective, as many more workers can be recruited. Akre and Hansen emphasize that to protect the wood in houses during construction, carpenter ant nests need to be removed from more than the immediate area.

Concealed Nests

Nests are often concealed in inaccessible places in buildings, as Eichler has shown for Pharaoh's ants. They may be built in rubbish or within the stems of plants in the case of 'hormiga loca' (Zenner-Polania), or they may penetrate deep underground as in the Texas leaf-cutting ant (Cameron). This renders surface spraying ineffective, and means that the nests may be difficult to find, and once found, they may be difficult to treat.

Reproductive Division of Labor

As foraging workers are sterile, it seems likely that killing a proportion of them will have a smaller impact on population size, than would killing a comparable proportion of normally reproducing insects. Little is known about the way that worker losses affect the egg-laying performance of the queen, the success with which the subsequent workers are reared or the caste profiles of the subsequent adults. There is some evidence that surviving workers can adjust their work patterns to compensate (Wilson, 1983). Certainly as Rust and Knight have shown, worker Argentine ants are much more easily killed than are the queens.

Colony Defenses

The colonial behavior associated with ants often involves bringing a large number of individuals together in very crowded conditions. These nests, containing adults, eggs, larvae and pupae provide a concentration of potential food large enough to attract not only invertebrate predators, but even the largest mammalian ones. Host location for parasitoids must be greatly simplified, and as man learned to his detriment when he became civilized and created medieval cities, the crowded conditions are ideal for disease transmission. The fact that ants are an immensely successful group, often becoming dominant insects in the tropics, indicates that they have successfully evolved defenses against the predators, parasitoids, parasites and pathogens which could otherwise be expected to devastate their nests. Predators are repelled by bites and stings, sometimes from specialized soldier castes which cooperate in nest defence as anyone who has dug out nests of fire ants, harvester ants, or leaf-cutting ants can testify. The nature of the communal defenses against parasitoids, parasites and pathogens are much less understood, although Jouvenaz provides a useful list of reasons why biological control of fire ants is so difficult. Kermarrec, discussing the susceptibility of leaf-cutting ants to nematode parasites, and Kermarrec and Mauleon discussing their response to fungal diseases stress that in both cases, individual workers isolated from their colonies in petri dishes frequently succumb, but that in the colony they appear unaffected. This seeming ability of ants to mount colony defenses against biological control agents, presents great practical problems, and should provide fertile ground for research.

CONTROL OPPORTUNITIES

Central Place Foraging

As foraging workers have to move out from their nest to the feeding sites where they cause damage, and as all worker ants are wingless, it is possible to deter their attacks by means of physical or chemical barriers. Mackay and Vinson give a simple example of physically sealing electrical relay switches which are attractive to fire ants, with epoxy resins, as a way of preventing access of the ants to the switches and interfering with their action. Alternatively chlorpyrifos can be scattered inside traffic light control boxes to prevent workers gaining entry to the circuitry. Similarly, Eichler recommends standing the legs of furniture in bowls of water to keep Pharaoh's ants from sterile areas in hospitals. Akre and Hansen record the caulking of holes to the exterior as a way of preventing carpenter ants from getting into the wooden structure of housing, while Zenner-Polania reports the exclusion of 'hormiga loca' from stables and houses by surrounding them with a 20cm wide concrete ditch filled with water, and their exclusion from fruit trees by banding the trees with grease. Chang and Ota used toxic chemicals to form a barrier around the orifices of plastic tubes used in drip irrigation systems in Hawaii, as several ant species enlarge these holes by biting, and so affect the rates of water flow. Finally, Rust and Knight recommend barrier spraying around houses as the best way of reducing Argentine ant problems.

Discrete Nest Site

Once the nest has been located, destroying it can potentially destroy the ant pest population of a considerable area. This 'spot treatment' of nests with toxic liquids, gases and dusts is consequently very efficient. Their precise placement should minimize environmental contamination and the poisoning of non-target species. Spot treatment of imported fire ant nests is reported by Banks, of leaf-cutting ants by Cameron, and of carpenter ants by Akre and Hansen. While spot treatment of large nests such as *Atta* is practical in the field, smaller nests are only worth treating in this way in small, high value gardens, or in urban premises where the considerable labor involved can be justified economically. Large numbers of small nests in the field must be controlled using other methods.

Wide and Active Foraging Behavior

The colony's foragers search a large area around the nest, and bring food back where it is consumed communally. If poison is added to food in the form of a toxic bait, then the poison can be scattered generally in the environment in low concentrations knowing that it will be collected by the foragers, and concentrated to lethal quantities in the nest. Alternatively, where damage is being experienced, or where ant trails can be found, bait can be placed where it is known that the workers will soon find it. In this variant of the spot treatment principle, the nest does not have to be found, and general environmental contamination is minimized. As our contributors have emphasized, baits seem ideally suited as the vehicle for controlling ants with toxic chemicals (Banks; Glancey, Reimer and Banks for fire ants: Cameron for leaf-cutting ants: Eichler; Williams for Pharaoh's ant: Campbell for harvester ants: Akre and Hansen with reservations for carpenter ants: Zenner-Polania for 'hormiga loca': Chang and Ota for the complex of ants attacking drip irrigation systems; and Rust and Knight for Argentine ants).

Mutual Feeding

The active foragers, which collect food, are only a proportion of the total colony. In some ant species non-foragers include castes which are too small, soldiers specializing in defence, males and fertile females specializing in reproduction and individuals too young to forage namely larvae, pupae and callows. All these feeding, but non-foraging stages have to be fed by the foraging workers, either directly on pieces of food brought in or indirectly through the trophallactic exchange of regurgitated liquid food. For a toxic bait to be effective, mortality must be sufficiently large to kill the reproductive queen or queens, and to disrupt colony organization so much, that dependent individuals which do survive direct poisoning, die from neglect and starvation. As Banks, and Glancey, Reimer and Banks point out, successful bait toxicants possess special properties. They must be non-repellent and virtually undetectable to the ants, so that they are consumed normally along with the rest of the food containing them and they must be slow-acting, so that at least the workers collecting the bait get it back to the nest, and are able to pass it on to other workers, sexuals and brood before the lethal effects disrupt behavior. Rust and Knight demonstrate that some toxicants kill Argentine ant queens better than others, but all researchers seem agreed that mirex was a most effective formicide.

Reproductive Division of Labor

The existence of only one, or relatively few reproductive individuals, has two important implications for control. (a) If the queens are killed, or their reproductive system destroyed, then the rest of the colony will eventually die. This opportunity has been exploited by the use of insect growth regulators, which somewhat unexpectedly, inhibit egg production in queens of imported fire ants (Banks; Glancey, Reimer and Banks), and in Pharaoh's ants (Williams). (b) As so few genetically distinct individuals (compared with most populations of pest insects) are exposed to pesticide poisoning, the likelihood of resistant individuals being present and so available for selection is less. None of the contributors has reported pesticide resistance. The only possible exception to this is the polygyne forms of *Solenopsis invicta* which Glancey, Reimer and Banks claim are now becoming more prevalent. They seem to be more difficult to control with toxic baits than the monogyne forms, and may represent a mutant life-history form selected by pesticide action.

Social Cohesion

The complex functioning of ant societies depends on communication between foraging workers, between workers and sexuals, and between workers and brood, and these communication systems are potentially vulnerable to disruption for control purposes. Chemicals, either pheromones, or genetically or environmentally determined odors, play a key role, and a good deal of research has been carried out into them. As yet, there has been little practical application. Kermarrec and Mauleon report studies in a leaf-cutting ant on the effects of psychotropic drugs on worker vigilance. The hope here is to undermine colony defenses to pathogens.

In all these ways, ant control suffers from problems, but enjoys opportunities different from those encountered in the control of most pest insects. In practice, the 14 papers in this section show the extent to which the use of toxic chemicals still dominates ant control with 11 largely devoted to this approach. However all of these mention, and most recommend, the environmentally preferable practice of baiting over general spraying. With the increasing costs of developing and launching new pesticides, in commercial terms, ant control is still too small a market to be a target for the major pesticide manufacturers. Moreover, formicides have to be tested for effectiveness against intact colonies, rather than against individuals, and this greatly increases costs. Culture methods for the smaller species capable of sustaining small, viable colonies, such as fire ants (Banks; Glancey,Reimer and Banks) and Pharaoh's ants (Williams) are sufficiently advanced to make toxicant testing a viable laboratory activity. With the large colones of leaf-cutting ants, testing must all be done in the field (Cameron), a time-consuming and expensive process which has held back bait development in these species, despite their undoubted importance. Where market forces are unwilling to fund the research and development costs of new formicides, we must all be grateful to the United States Department of Agriculture, which since 1977 has funded the testing of over 4000 chemicals against fire ants. Consequently, all important formicides since the 1950's have first been developed for fire ants, and have subsequently been used for other species, even though economically they are not the most serious of the ant pests. Unfortunately, bait toxicants effective for fire ants are not necessarily so for other species, especially for leaf-cutting ants, where bait is incorporated into the metabolically highly active fungus gardens, and Cameron has demonstrated this for abamectin and fenoxycarb.

Sensible applied biologists do not put all their eggs in one basket by relying totally on any one method of pest control, and as Zenner-Polania has pointed out, an integrated approach using a wide range of techniques is the more sensible. The role that biological control can play has yet to be established, and Jouvenaz provides an impressive list of failed hopes in the biological control of fire ants. Competitive displacement of pest species by more dominant, introduced ones, has been suggested, but is not without dangers. It seems that the 'hormiga loca' problem (Zenner-Polania) has resulted from a misplaced attempt in the early 1970's to control *Atta* spp. in Colombia by importing *Paratrechina fulva* from Brazil (Zenner de Polania and Nhora, 1984).

For the future it is still too early to forecast on which control direction we should concentrate. A broad research approach is required, including more specific formulations of baits containing novel toxicants; habitat manipulation as a way of reducing ant populations and of altering their species structure; the continued search for more effective predators, parasitoids, parasites and pathogens, and where possible their selective breeding, with or without biotechnology, for greater efficiency; and finally, more attempts at disrupting colony cohesion, and especially colony defenses against natural enemies. These controls need to be implemented within the context of physical and cultural controls which farmers should be encouraged to build in to their agricultural practices.

REFERENCES

WILSON, E.O. 1983. Caste and division of labor in leaf-cutter ants (Hymenoptera:Formicidae: Atta) 111. Ergonomic resiliency in foraging by A.cephalotes. Behav. Ecol. Sociobiol. 14: 47-54.

ZENNER de POLANIA, I. and R.B. NHORA. 1984. Habitos alimenticios y relaciones simbioticas de la 'hormiga loca'. Resumenes X1 Congreso de la Sociedad Colombiana de Entomologia p.37.

54

Chemical Control of the Imported Fire Ants

W.A. Banks

INTRODUCTION

The fire ants *Solenopsis invicta* Buren and *S. richteri* Forel, accidentally introduced into the southern United States ca. 50 and 70 years ago respectively, presently infest more than 100 million ha of land in eleven states and Puerto Rico. *S. invicta* colonies have been found in Arizona and California, and on St. Croix (US Virgin Islands), but have been eliminated. Lofgren (1986) gave a detailed history of fire ants in the U. S.

The first organized control of fire ants began eight years after Loding (1929) reported their presence in the United States. A cooperative Federal, State, and County program began in Baldwin Co., Alabama (1937) using calcium cyanide dust (Eden & Arant 1949).

By 1953 imported fire ants were present in 102 counties in 10 states (Culpepper 1953) and by 1957 infestations were so severe that large area control was required. The U. S. Congress, acting upon a resolution by the Southern Association of State Agriculture Commissioners, provided funds and authorized the U. S. Department of Agriculture to initiate a cooperative Federal-State control/eradication program. Large-scale insecticide treatments began in late 1957 and research was initiated to improve existing and develop new, control methods. (For details on the use of the residual chlorinated hydrocarbons, and the development and use of mirex bait in the control program see Bellinger et al. 1965, Adkins 1970, Alley 1973, Lofgren et al. 1975).

Mirex bait gave > 95% fire ant control over large areas (Lofgren et al. 1975); however, its persistence resulted in animal residues (Markin et al. 1972) and ultimately an EPA ban on its use in the U. S. (Johnson 1976). With mirex no longer available, research was accelerated to discover and develop new fire ant control technologies. This paper discusses the development of delayed-action toxicants and insect growth regulators for use in baits, and quick-acting insecticides for individual mound treatment.

TOXIC BAITS

We directed most of our research on fire ant control toward toxic baits because they represent the most economical and environmentally sound method for large-area control. Lofgren et al. (1975) reviewed early studies in the development of fire ant baits and Williams (1983) and Banks et al. (1985) discussed more recent studies.

Effective baits must have (1) a toxicant that exhibits several hours delayed toxicity so the chemical can be distributed to a substantial portion of the colony before ants begin to die, is effective over at least 10-fold, and preferably greater range of dilutions, is not repellent to ants, and is easily formulated with foods and carriers (Stringer et al. 1964, Banks et al. 1977, Williams 1983), (2) an attractive food material, preferably only to the target species and (3) a cheap, easy to obtain, formulate and apply carrier. Unfortunately, the limited availability of such materials hampered the development of baits.

Only fifteen of the over 4,000 chemicals we tested since 1977 met enough of the requirements of an effective fire ant bait toxicant to warrant field tests (Banks, unpublished data). Only three materials passed the additional tests of safety, availability, developmental costs and marketability and have been pursued commercially. These are hydramethylnon, the active ingredient in Amdro^R, (American Cyanamid Co., One Cyanamid Plaza, Wayne, NJ 07470); abamectin, the active ingredient in Affirm^R (MSD AgVet, Div. Merck & Co., Inc., P. O. Box 2000, Rahway, NJ 07065); and sulfluramid, one of a group of fluoroaliphatic sulfone insecticides (Vander Meer et al. 1985, Williams et al. 1987), which will probably be available in a commercial fire ant bait (Griffin Corp., P. O. Box 1847, Valdosta, GA. 31601) within the next year.

Insect growth regulators (IGRs), although essentially nontoxic to adult fire ants, meet the other requirements of effective toxicants. Several of the 85 IGRs tested in the laboratory caused the death of >85% of treated colonies (Banks 1986, Banks et al. 1978, 1988). Death occurred because the IGR prevented the replacement of workers through low level toxicity to immatures, reduction or cessation of egg production by the colony queen, and a shift in caste differentiation so that any eggs laid produced only sexual forms. Two IGRs, fenoxycarb and pyriproxyfen (S-31183, Sumitomo Chemical Co., Ltd., 15-5 chome, Kitahama, Higashi-ku, Osaka, Japan), caused a mated queen's ovaries to retrogress and prevented ovary development in fire ant virgin females (Glancey and Banks 1988, Glancey et al. 1989).

Fire ant colonies that ingest IGRs die much more slowly than those that ingest conventional bait toxicants. Treated colonies may survive for a year or more in the laboratory despite a constantly decreasing population. In the field, however, treated colonies survive for only four to six months. Although users sometimes object to the slow worker kill, two IGR baits have been developed and marketed for fire ant control. Logic^R (Maag Agrochemicals, Inc., P. O. Box 6430, Vero Beach, Florida 32961-6430), containing fenoxycarb, has gained acceptance as an effective fire ant control agent. ProDrone^R (ICI Americas, Inc., Wilmington, DE 19897), which contained 1-(8-methoxy-4,8-dimethylnonyl)-4-(1-methylethyl) benzene, was much less effective and is no longer available. We recently discovered that the IGR pyriproxyfen is very effective against the fire ants and development of a commercial bait is under consideration.

The ants must readily accept the food attractant in toxic baits. It must act as the solvent for the toxicant and absorb into the carrier without deleterious effects on either. It should remain acceptable to the ants through at least several weeks storage. Although omnivorous, fire ants can exhibit a preference for certain food materials depending on the circumstances and time of year (Glunn et al. 1981, Sorenson and Vinson 1981). Nevertheless, vegetable oil-based baits are readily harvested in the field and provide a high level of control when used with an effective toxicant (Hays and Hays 1959, Lofgren et al. 1961, 1964, Howard and Tschinkel 1981, Williams 1983, Banks et al. 1985, Lofgren 1986). The high ant acceptance level plus ready availability, favorable cost, ease of formulation and stability led us to choose soybean oil as the phagostimulant/toxicant solvent. Soybean oil baits give consistent results and recent studies (Vander Meer et al., unpublished) show that soybean oil contains volatiles attractive to fire ant foragers.

Finding suitable carriers for oil baits was always difficult. Lofgren et al. (1963) determined that a granular bait was most practical and, after evaluating a number of materials, found that ground corncobs provided the most consistent results and were the only material available in adequate quantities for the control program. These grits were such an excellent carrier that more than 90% of the over 100 million pounds of mirex bait used for fire ant control in the 1960s and 70s in the southern United States was formulated on corncob grits. Although a number of other materials (hardwood sawdust, pelleted wood fibers, styrofoam beads, etc) were evaluated during this period (Banks et al., unpublished), only Fur-Ag received extensive usage (Banks et al. 1970).

Although an excellent carrier for mirex, corncob grit formulations of hydramethylnon and fenoxycarb gave such poor fire ant control that a new carrier was sought (Banks et al. 1985, 1988). We restricted our search for new carriers to materials that would absorb more oil (corncob grits retained only 15%) because earlier studies had shown that certain toxicants were effective only in high oil content baits (Stringer et al. 1964, Banks et al. 1977). Tests with a variety of materials showed that the ants only accepted puffed cereal-type pellets of corn, rice, or wheat. These were also sufficiently absorptive and flowed through conventional application equipment. Hydramethylnon was more effective in field tests when formulated on extruded corn pellets than on corncob grits (Banks et al. 1985). Two similar materials, pregel defatted and pregel degermed corn grits, proved satisfactory as carriers when available supplies of the extruded pellets were found to be inadequate; the pregel defatted grit was slightly superior. Good handling procedures are necessary because pregel defatted corn grits breakdown rapidly in moisture, disintegrate with rough handling and may become infested by grain beetles or moths. But, with proper handling, they are good carriers for the oil baits. All currently marketed fire ant baits are formulated with pregel defatted grits (Illinois Cereal Mills, Inc., Paris, Illinois 61944).

As individual mound treatments, the baits are used at rates of three to seven tablespoons evenly spread in about a one meter circle around the mound. However, the best results are obtained when the bait is broadcast evenly over the ant-infested area (1.12-1.68 kg/ha).

Table 1 illustrates how effective broadcast applications of soybean oil-pregel defatted corn grit baits containing Amdro, Affirm, sulfluramid, fenoxycarb and pyriproxyfen are against fire ants. The four to eight (or more) weeks required to realize maximum kill with baits may be unacceptable when the ants are causing severe problems. In such situations, contact insecticides may be desirable.

TABLE 1. Effectiveness of baits against red imported fire ants (avg. from 30-50 plots). I = Population index.

Rate of Application		Avg. pretreatment		% reduction in I after 12-13
Bait[a] (kg/ha)	AI(g/ha)	Nests	I	weeks posttreatment[b]
Affirm (1.12-2.24)	0.12-.25	127	1810	97.4
Amdro (1.12-2.24)	9.8-19.7	52	1018	89.7
Logic (1.12-1.68)	11.2-16.8	54	1018	94.5
S-31183 (1.12-2.24)	8.4-22.4	28	607	87.7
GX-071 (1.12-1.68)	3.4-10.1	28	616	85.5
Control -	-	52	1036	31.1

[a]Baits - Affirm 0.011% abamectin; Amdro 0.88% hydramethylnon; Logic 1.0% fenoxycarb; S-31183 0.75% to 1.5% pyriproxyfen; GX-071 0.3 to 0.6% sulfluramid. [b]See Banks et al. 1988 for method of determining population index and reductions.

CONTACT INSECTICIDES

The following chemicals are registered and marketed in the southern United States as drenches, dusts, granules and aerosols or vapors for control of imported fire ants: acephate, bendiocarb, carbaryl, chlorpyrifos, diazinon, malathion, pyrethrins and certain pyrethroids. These chemicals will kill the ants by direct contact when applied or within one week through their residues. They are generally favored by homeowners and others who desire the quick elimination of a few colonies. Although primarily used for treatment of individual fire ant nests, these insecticides can be used as broadcast treatments for small areas.

Although effectiveness of these insecticides varies with formulation, soil type, size of mounds, rate and method of application, and time of year when treatments are made, all have been reported to give > 80% fire ant control in at least one test (Morrill 1976, 1977, Hillman 1976, 1977, Gardner 1980, 1981, Francke 1983, Williams and Lofgren 1983, Drees 1986, Lemke and Kissam 1987, Diffie et al. 1988). Because the formulations, rates and methods of application, and evaluation procedures vary specifics should be obtained from the cited publications.

Mound drenches with an insecticide aqueous emulsion or suspension are a common method of individual mound treatment. They can give excellent control if the volume of liquid is adequate to saturate the nest. We have found that mounds up to 20 cm in diameter require ca. four liters of solution and larger mounds require ca. eight liters. The best application method for drenches entails wetting the top of the mound, then soaking the soil in a ca. 30 cm circle around the mound's base. Finally the remaining solution is poured onto the top of the mound from a height to insure penetration. The height will vary depending upon the soil, but generally one to two meters is adequate.

Drenches work best when applied about mid-morning on sunny days after cool nights in the early spring or late fall. The ants, and the queen, are concentrated near the top of the mound at such times and are more likely to be contacted by the insecticide. Applications made during hot weather are rarely effective because the ants are usually deep in the mound. Drenches are usually more effective in heavier soils because the mound's tunnel structure remains intact, allowing the liquid to better penetrate.

Dusts and granules are scattered over the surface of the mound and the surrounding soil. They may be left undisturbed or watered into the soil with a sprinkling can or hose. Dusts or granules are generally less effective than drenches, although acephate dusts and diazinon granules sometimes give excellent kill of individual colonies (Hillman 1977, Drees 1986, Diffie et al. 1988, Lemke and Kissam 1987).

Aerosols of chlorpyrifos alone, or combined with pyrethrins, have been utilized for individual IFA mounds. For these treatments, a standard pushbutton aerosol cap, fitted with a 74 cm long x 8 mm diameter fiberglass probe, is attached to an aerosol container containing 1.0% of the insecticide. The spray is activated, the probe inserted into the mound at from one to four sites, depending on size, and the chemical released for ca. five seconds at each insertion site. Although the chemical is neatly packaged for easy handling, this application method does not give significantly better control than chlorpyrifos drenches (Horton et al. 1982, Drees 1986, Lemke and Kissam 1987, Diffie et al. 1988). The cost per mound is somewhat higher for the aerosol than for the drench.

Thermo-fumigation is a variation of the aerosol injection method. Insecticide is vaporized in a special heating chamber and then injected under pressure through a metal probe into the fire ant mound (Evans 1988). Although the system usually uses resmethrin as the insecticide, other pyrethroids can be substituted. Thorvilson et al. (1989) obtained very quick kill of fire ants with a mean colony size reduction of 98% four weeks after treatment. We did not obtain as good results in our tests with the system in 1986 (35% and 40% after 4 and 8 weeks respectively) (Banks et al. unpublished).

Methyl chloroform (1, 1, 1-trichloroethane) is registered in some of the fire ant infested states as a pour-on fumigant for individual mound treatment. In most efficacy tests the recommended rate of 60-90 ml of technical liquid per mound provided only 40-65% control (Williams and Lofgren 1983, Drees 1986, Lemke and Kissam 1987), although it gave >90% control in one study (Scarborough et al. 1982).

If individual mound treatment is less than 100% effective, the surviving ants often abandon the original mound and build one or more new mounds nearby. In such cases, additional treatment is necessary to eliminate the colonies.

As previously indicated, treatment of individual mounds is a very popular control method over small areas, such as backyards, playgrounds, small parks, cemeteries, and schoolyards. It usually provides rapid control at nominal cost and is relatively safe to humans, animals and the environment. For areas larger than 0.5 ha where there are large numbers of mounds, individual mound treatment is too labor-intensive and becomes impractical. In such situations, the quantity of insecticide applied, and the cost per unit area of treatment becomes much higher and the level of control lower. Thus, broadcast application of an effective bait should be used for wide area control.

REFERENCES CITED

ADKINS, H.G. 1970. The imported fire ant in the southern United States. Assoc. Am. Geogr. 60: 578-592.

ALLEY, E.G. 1973. The use of mirex in control of the imported fire ant. J. Environ. Qual. 2: 52-61.

BANKS, W.A. 1986. Insect growth regulators for control of the imported fire ant. *In* C. S. Lofgren and R. K. Vander Meer [eds.], Fire ants and leaf-cutting ants: biology & management. Westview, Boulder, CO., 434 pp.

BANKS, W.A., C.E. STRINGER, C.S. LOFGREN, N.W. PIERCE and F.J. BARTLETT. 1970. Fur-Ag: An alternate carrier for soybean oil-mirex bait for control of the imported fire ant. J. Econ. Entomol. 63: 1990-1991.

BANKS, W.A., C.S. LOFGREN, C.E. STRINGER and R. LEVY. 1977. Laboratory and field evaluation of several organochlorine and organophosphorus compounds for control of imported fire ants. ARS-S-169, 13 pp.

BANKS, W.A., C.S. LOFGREN and J.K. PLUMLEY. 1978. Red imported fire ants: effects of insect growth regulators on caste formation, and colony growth and survival. J. Econ. Entomol. 71: 75-78.

BANKS, W.A., D.F. WILLIAMS and C.S. LOFGREN. 1988. Effectiveness of fenoxycarb for control of red imported fire ants (Hymenoptera: Formicidae). J. Econ. Entomol. 81: 83-87.

BANKS, W.A., C.S. LOFGREN and D.F. WILLIAMS. 1985. Development of toxic baits for control of imported fire ants. pp. 133-143. *In* T.M. Kaneko and L.D. Spicer [eds.], Pesticide Formulations and Application Systems: Fourth Symposium. Amer. Soc. Test Mater., Special Tech. Publ. 875.

BELLINGER, F., R.E. DYER, R. KING and R.B. PLATT. 1965. A review of the problem of the imported fire ant. Bull. Ga. Acad. Sci. 23: 1-22.

CULPEPPER, G.H. 1953. The distribution of the imported fire ant in the southern states. Proc. Assoc. Agric. Workers 50: 102.

DIFFIE, S., D.C. SHEPPARD and T.D. CANERDAY. 1988. Imported fire ant control results, 1978-1987. Mimeo. Rpt. 2, Univ. Ga. Coop. Exten. Serv., Div. Entomol., Athens. 110 pp.

DREES, B. 1986. Red imported fire ant control result demonstrations, 1979-1986. Mimeo. Rpt. Texas Agric. Exten. Serv., College Station. 121 pp.

EDEN, W.G. and F.S. ARANT. 1949. Control of the imported fire ant in Alabama. J. Econ. Entomol. 42: 976-979.

EVANS, C.W. II. 1988. Method and apparatus for the destruction of imported fire ants of the genus *Solenopsis*. U. S. Patent 4,756,118.

FRANCKE, O. 1983. Efficacy tests of single-mound treatments for control of red imported fire ants, *Solenopsis invicta* Buren. Southwest. Entomol. 8: 42-45.

GARDNER, W.A. 1980. Imported fire ant control by mound drenches, 1979. Insectic. Acar. Tests 5: 209-210.

GARDNER, W.A. 1981. Red imported fire ant control by mound drenches, 1980. Insectic. Acar. Tests 6: 132.

GLANCEY, B.M. and W.A. BANKS. 1988. Effect of the insect growth regulator fenoxycarb on the ovaries of queens of the red imported fire ant (Hymenoptera: Formicidae). Ann. Entomol. Soc. Am. 81: 642-648.

GLANCEY, B.M., W.A. BANKS and M.S. OBIN. 1989. The effect of fenoxycarb on alates of the red imported fire ant. J. Entomol. Science 24: 290-297.

GLUNN, F.J., D.F. HOWARD and W.R. TSCHINKEL. 1981. Food preference in colonies of the fire ant, *Solenopsis invicta*. Insectes Soc. 28: 217-222.

HAYS, S.B. and K.L. HAYS. 1959. Food habits of *Solenopsis saevissima richteri* Forel. J. Econ. Entomol. 52: 455-457.

HILLMANN, R.C. 1976. Control of red imported fire ants, *Solenopsis invicta* Buren, on home grounds in North Carolina. J. Elisha Mitchell Soc. 92: 69-70.

HILLMANN, R.C. 1977. Red imported fire ant control with conventional insecticides, 1975, 1976. Insectic. Acar. Tests 2: 135.

HORTON, P.M., J.B. KISSAM, S.B. HAYS and G.W. QUERY. 1982. Chlorpyrifos aerosol mound injections for the control of the red imported fire ant. J. Ga. Entomol. Soc. 17: 478-484.

HOWARD, D.F. and W.R. TSCHINKEL. 1981. The flow of food in colonies of the fire ant, *Solenopsis invicta*: a multi-factorial study. Physiol. Entomol. 6: 297-306.

JOHNSON, E.L. 1976. Administrator's decision to accept plan of Mississippi Authority and order suspending hearing for the pesticide chemical mirex. Fed. Regist. 41(251): 56694-56704.

LEMKE, L.A. and J.B. KISSAM. 1987. Evaluation of various insecticides and home remedies for control of individual red imported fire ant colonies. J. Entomol. Sci. 22: 275-281.

LODING, H.P. 1929. An ant (*Solenopsis saevissima richteri* Forel). U. S. Dep. Agr. Ins. Pest Survey Bull. 9: 241.

LOFGREN, C.S. 1986. History of imported fire ants in the United States. pp. 36-47 *In* Fire Ants and Leaf-Cutting Ants: Biology and Management, C.S. Lofgren and R.K. Vander Meer [eds.]. Westview, Boulder, CO.

LOFGREN, C.S., F.J. BARTLETT and C.E. STRINGER. 1961. Imported fire ant toxic bait studies: evaluation of various food materials. J. Econ. Entomol. 54: 1096-1100.

LOFGREN, C.S., F.J. BARTLETT and C.E. STRINGER. 1963. Imported fire ant toxic bait studies: evaluation of carriers for oil baits. J. Econ. Entomol 56: 62-66.

LOFGREN, C.S., F.J. BARTLETT and C.E. STRINGER. 1964. The acceptability of some fats and oils as food to imported fire ants. J. Econ. Entomol. 57: 601-602.

LOFGREN, C.S., W.A. BANKS and B.M. GLANCEY. 1975. Biology and control of imported fire ants. Annu. Rev. Entomol. 20: 1-30.

MARKIN, G.P., J.H. FORD, J.C. HAWTHORNE, J.H. SPENCE, J. DAVIS, H.L. COLLINS and C.D. LOFTIS. 1972. The insecticide mirex and techniques for its monitoring. U. S. Anim. Plant Health Insp. Serv. [Rep.] 81-3, 19 pp.

MORRILL, W.L. 1976. Red imported fire ant control with mound drenches. J. Econ. Entomol. 69: 542-544.

MORRILL, W.L. 1977. Red imported fire ant control with diazinon and chlorpyrifos drenches. J. Ga. Entomol. Soc. 12: 96-100.

SORENSON, A.A. and S.B. VINSON. 1981. Quantitive food distribution studies within laboratory colonies of the imported fire ant, *Solenopsis invicta* Buren. Insectes Soc. 28: 129-160.

STRINGER, C.E., JR., C.S. LOFGREN and F.J. BARTLETT. 1964. Imported fire ant toxic bait studies: evaluation of toxicants. J. Econ. Entomol. 57: 941-945.

THORVILSON, H.G., S.A. PHILLIPS, JR. and A.A. SORENSON. 1989. An innovative thermo-fumigation technique for control of red imported fire ants. J. Agric. Entomol. 6: 31-36.

VANDER MEER, R.K., C.S. LOFGREN, and D.F. WILLIAMS. 1985. Fluoraliphatic sulfones: a new class of delayed-action insecticides for control of *Solenopsis invicta* (Hymenoptera: Formicidae). J. Econ. Entomol. 78: 1190-1197.

WILLIAMS, D.F. 1983. The development of toxic baits for the control of the imported fire ant. Fla. Entomol. 66: 162-172.

WILLIAMS, D.F. and C.S. LOFGREN. 1983. Imported fire ant control: evaluation of several chemicals for individual mound treatments. J. Econ. Entomol. 76: 1201-1205.

WILLIAMS, D.F., C.S. LOFGREN, and R.K. VANDER MEER. 1987. The red imported fire ant, *Solenopsis invicta*, control with fluoroaliphatic sulfone bait toxicants. J. Agric. Entomol. 4: 41-47.

55

Effects of IGR Fenoxycarb and Sumitomo S-31183 on the Queens of Two Myrmicine Ant Species

B.M. Glancey, N. Reimer and W.A. Banks

INTRODUCTION

After the U.S. Environmental Protection Agency (EPA) banned mirex and other chlorinated hydrocarbons, an intensive search began for other materials to control pest ants. Insect growth regulators (IGR) are generally non toxic to adult worker ants but provided effective control of certain problem species because they stopped queen egg production. Banks et al. (1983, 1988) showed that the IGR fenoxycarb (Maag Agro-chemicals, Vero Beach, Fl) was effective against *Solenopsis invicta* Buren (see Banks this volume). In recent years, Amdro has become the most popular toxic bait for control of fire ants. However, polygyne *S. invicta* (multiple nest queens), which is becoming more prevalent, is more difficult to control with Amdro than monogyne *S. invicta* (one nest queen) (Glancey et al. 1987). Recent field studies show that fenoxycarb effectively controls the polygyne form (Banks et al. unpublished). It kills *S. invicta* larvae and pupae, and - most importantly - inhibits queen egg production and shifts caste differentiation from worker to sexual forms (Banks et al. 1988). Histological studies of the queen's reproductive system reveal that fenoxycarb causes regression of the ovaries while completely suppressing the growth of the nurse cells, cytoplasm and follicular epithelium; developing eggs were reabsorbed (Glancey and Banks 1988).

Fenoxycarb's effectiveness against polygyne *S. invicta* colonies led us to test it and Sumitomo S-31183 (Sumitomo America Inc., New York, NY), a compound with similar IGR action, against another polygyne species, the big-headed ant, *Pheidole megacephala* F. This latter ant is a well known pest of domestic and institutional premises and may be a care-taker of mealybugs that transmit wilt in Hawaiian and Costa Rican pineapples. This mutualism enhances mealybug survival and creates dense mealybug populations. High mealybug populations promote the spread of wilt through the fields. Without *P. megacephala*, mealybug populations do not become established and wilt is not a problem. In Hawaii, *P. megacephala* also tend the green scale (*Coccus viridis* [Green]) in coffee and aphids in taro, chew on drip irrigation tubing in sugarcane (see Chang and Ota this Section) and cause quarantine and nuisance problems in Protea (a cut flower crop) where it occurs in flower heads (Rohrbach et al. 1988). The IGR methoprene effectively controlled *P.*

megacephala in hospitals (Edwards et al. 1981) and, although its effect on the *P. megacephala* queen or workers was not determined, methoprene may have impaired the queen's reproductive capacity.

Although distantly related, *S. invicta* and *P. megacephala* belong to the same subfamily (Myrmecine); therefore, treatment with fenoxycarb or Sumitomo S-31183 might produce similar pathological effects in both species. Herein we report the effects of the two IGRs on *S. invicta* and *P. megacephala* queens.

METHODS AND RESULTS

Fenoxycarb Effects

Solenopsis. The effects of fenoxycarb on *S.invicta colony queens* have been previously reported (Glancey and Banks 1988) and that research is summarized in this paper's introduction. Earlier studies also showed that after ingesting the compound queens and virgin dealates were unable to inhibit dealation in virgin queens (Obin et al. 1988). However, neither the queen nor the virgin dealates lost the ability to induce queen recognition.

In the present study, fenoxycarb also affected the maternal instinct of virgin queens. Ten days after eclosion, groups of untreated and fenoxycarb-treated virgin dealate queens histolyzed their alary muscles, produced an esophageal crop and began to produce the queen recognition pheromone. Thirty queens from each group were isolated in one ounce plastic cups with castone bottoms and given eggs produced by the queen from a laboratory colony. The untreated virgin queens tended the eggs immediately, whereas treated queens either ignored or ate the eggs. This behavior pattern remained consistent throughout a week of observation. In contrast, untreated queens exhibited typical egg-tending behavior and, by the end of the week, microlarvae were evident in the small colonies. *S. invicta* colonies treated with fenoxycarb produced virgin females that lacked ovaries (Glancey et al. 1989). The failure of the treated virgin queens to tend eggs suggests that maternal instinct (i.e., egg-tending) is mediated by a material produced by the ovaries, or the corpora allata with a release related to ovary development, and is thus absent in queens without ovaries.

Pheidole. Thirty-six *P. megacephala* colonies in hollow pineapple stumps were collected from 0.2 ha of a ratoon pineapple field on the island of Oahu, Hawaii. The ants, taken to a laboratory at a University of Hawaii at Manoa laboratory, were placed in plastic boxes 26.5 X 33 X 9 cm. To remove the ants from the stumps, they were offered a small black vial which contained moist tissue paper. The negatively phototrophic queens quickly moved into the darkened vials; these were then transferred to 15 X 29 X 9 cm plastic shoe boxes along with a number of workers. Each colony was provided with water, a modified Banks diet (Banks et al. 1981) and a modified Bishop nest (Bishop et al. 1980). Reduced to a single queen and comparable numbers of brood, workers and soldiers, the colony was allowed to acclimatize in the laboratory. Afterward, the colonies were counted; each consisted of one queen, 20-50 eggs, 1-2 ml brood, 200-400 workers and 20-40 soldiers.

Following four days of access only to water, 12 colonies were fed 0.5

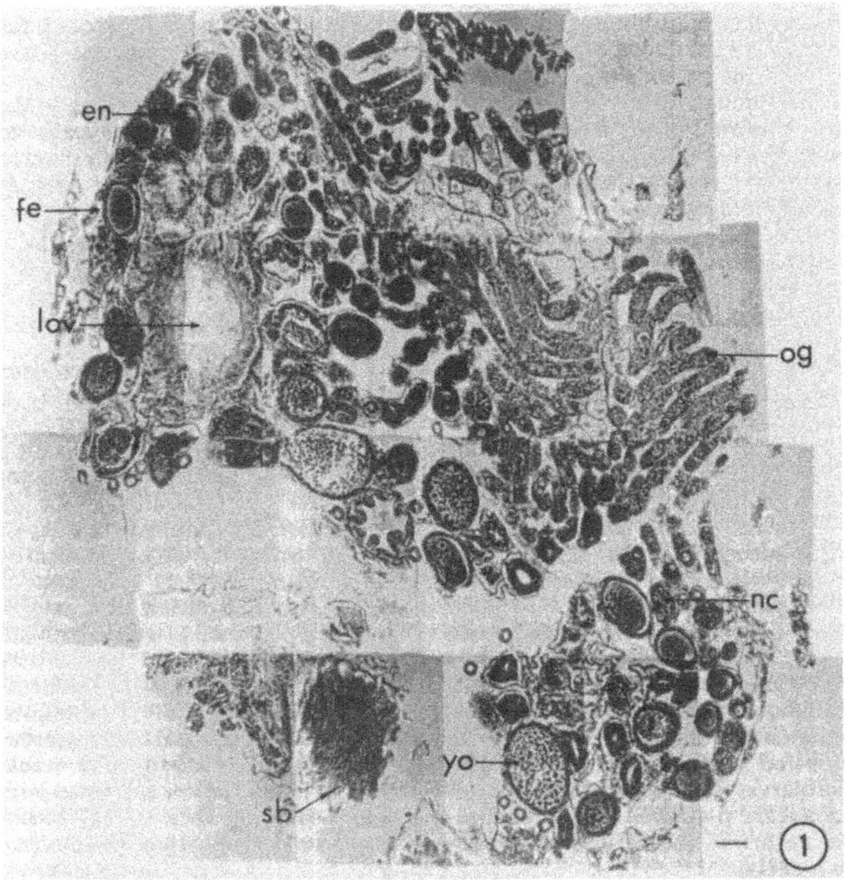

FIGURE 1. Section through the ovary of a normal *P. megacephala.* The germarium is filled with dark stained packed öogonia (og); an egg nucleus (en) is shown surrounded by the follicular epithelium (fe) with supporting nurse cells (nc), all within the zone of differentiation. A mature egg with yolkplasm (yo) is pictured along with the lateral oviduct (lov). A sperm bundle (sb) is in the spermatheca. Bar= 70u.

ml of 2.0% fenoxycarb in soybean oil and 12 colonies were fed neat soybean oil as controls. Any soybean oil not eaten was removed after 24 hours and the colonies returned to their normal diet. Every two weeks, the number of eggs, workers and soldiers and the volume of brood were measured. Ovaries of queens, removed from each of three treated and control colonies at weeks five, six, seven and eight, were fixed in Kahle's solution for 24 hours and subsequently held in 70% ethanol for embedding.

Fixed ovaries were embedded in paraffin (m.p. 57°C), sectioned at 5 um and stained with Harris's hematoxylin and eosin solution. Stained sections were examined under phase microscopy and photographed.

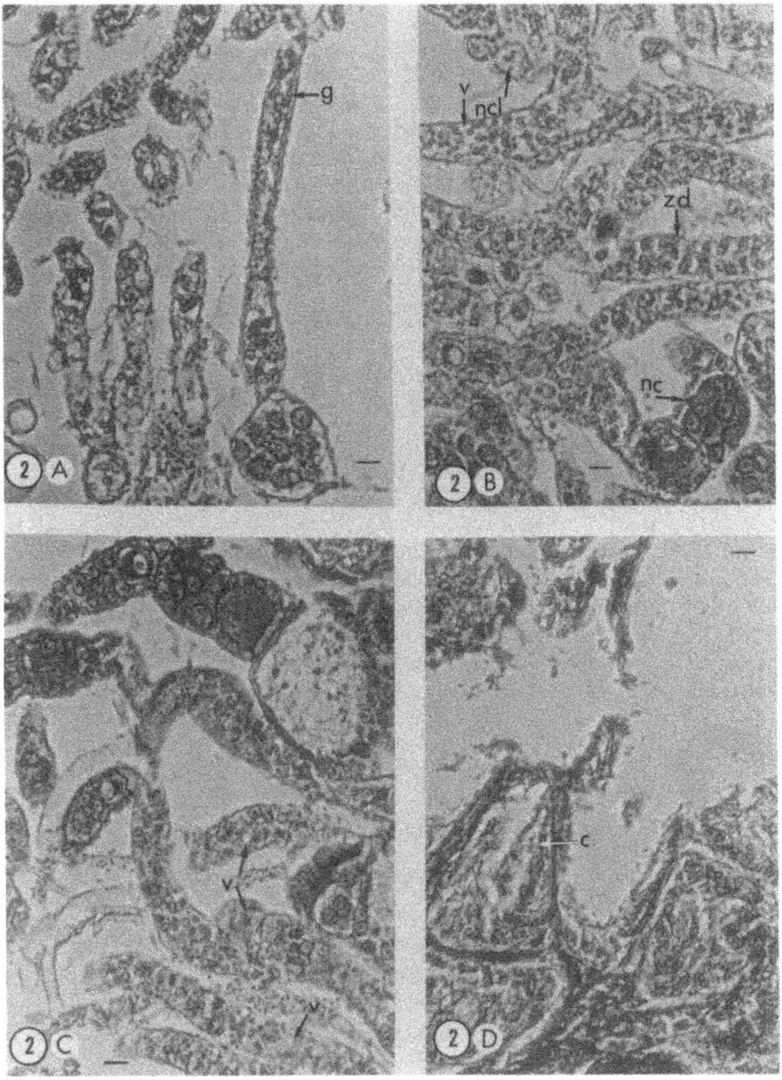

FIGURE 2. Sections through ovarioles of a *P. megacephala* queen 8 weeks after treatment with fenoxycarb. The figure shows loss of activity in the germarium (g: 2A); lack of development of nurse cells (ncl) and follicular epithelium in zone of differentiation (zd: 2B); vacuolation of lower vitellogenic zone (v: 2C) ; and remnants of the chorions of resorbed eggs (c: 2D). Bar= 40u.

Ovaries of untreated *P. megacephala* queens have fewer ovarioles but otherwise resemble the ovaries of the fire ant queens (see Glancey and Banks 1988 for a detailed discussion of the *S. invicta* ovary). Both species' ovaries nurse cells. Fenoxycarb's effects on the ovaries of Solenopsis queens are visible within two weeks after treatment. However, no effects were observed on the ovaries of *P. megacephala* queens until six weeks after treatment. This delay may have been caused by the repellent effect of fenoxycarb on *P. megacephala* workers. After six weeks, fenoxycarb affected *P. megacephala* in ways similar to *S. invicta* (Glancey and Banks 1988). Eggs laid were about one fourth normal size and few contained yolk. Resorption of many eggs left only the chorion visible. In the region of öocyte differentiation, ovarioles were highly vacuolated; nurse cells were lacking and the follicular epithelial cells lacked cytoplasm. Some ovarioles contained only a series of egg nuclei without cytoplasm and nurse cells. After eight weeks, most ovarioles had shrunk so much that the entire ovary was only one fourth its original size. Only one or two eggs with yolk could be found and they were further reduced in size. Fig. 1 shows a cross-section of a *P. megacephala* control queen's ovary. Fig. 2, A-D, shows the effect of fenoxycarb on the ovary after eight weeks.

Fenoxycarb's ultimate effect on *P. megacephala* was similar to that observed in *S. invicta* colonies (Banks et al. 1988) and is indicative of an IGR overload, which resulted in a decline in colony size and ultimately led to its death. We have previously discussed the effect of an IGR on germinal tissues: inhibition of the nutritive cells and impairment of the differentiation process (Glancey and Banks 1988). Even though *P. megacephala* and *S. invicta* are unrelated, the effects of the IGR on both species are similar. The ovarian tissues of *S. invicta* and *P. megacephala* are affected in similar ways by fenoxycarb treatment, but the effect varies as to time and degree.

Sumitomo Effects

Solenopsis. The effects of Sumitomo S-31183 on laboratory and field colonies of *S. invicta* will be reported elsewhere (Banks et al. - ms in preparation). The compound produces many of the same "symptoms" as fenoxycarb, including a disappearance of worker brood due to low level toxicity, suppression of egg production and a shift in caste differentiation so that only sexual forms are produced. Treated laboratory colonies died within 24 weeks.

We studied the effect of S-31183 on *S. invicta* queens by feeding 0.5 ml of once-refined soybean oil containing 2.0% active ingredient to each of six queenright laboratory colonies. These colonies consisted of 20-30 ml of immatures and 40,000-60,000 workers. Six similar colonies were fed neat soybean oil as controls. Ants ingested the oil solution from micropipets within 24 hours and then returned to their normal laboratory diet (Banks et al. 1981).

Five and six weeks after treatment, ovaries were removed from queens of three treated and three control colonies. The ovaries were examined, fixed in Kahle's solution, embedded in paraffin (m.p. 57° C), sectioned at 5um and stained with Harris's hematoxylin and eosin solution. Stained sections, examined under phase microscopy, were photographed.

Ovaries showed no size reduction after five weeks; however, the normal complement of yolked eggs fell from seven to three; each was approximately one half normal size. Microscopic examination of serial sections revealed some ovarioles were empty tubes, while others contained a series of egg nuclei

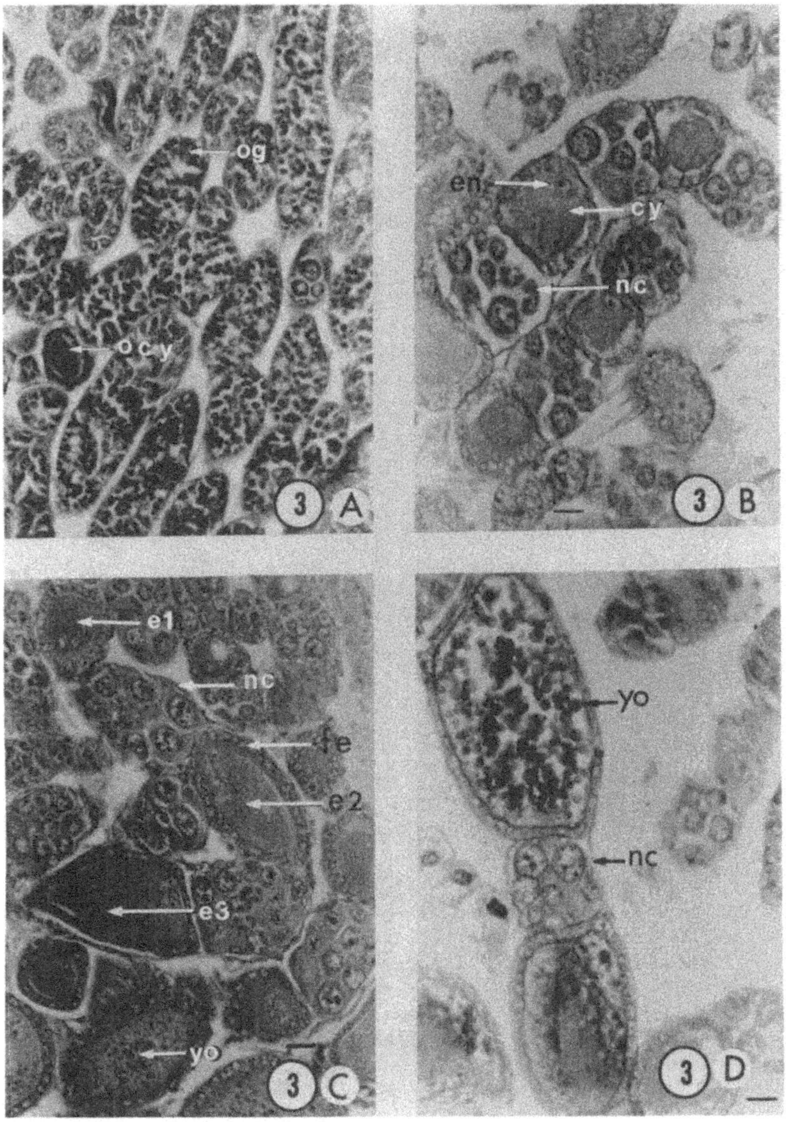

FIGURE 3. Sections through ovarioles of an untreated *S. invicta* queen (from Glancey and Banks 1988). The germarium (3A) is filled with dark stained densely packed öogonia (og) and the zone of differentiation shows differentiated nurse cells (nc) and öocytes (ocy). An egg nucleus (en) surrounded by cytoplasm (cy) is in 3B. Lower end of vitellogenic zone (3C and 3D) showing addition of yolk (yo) to eggs. Bar = 35u.

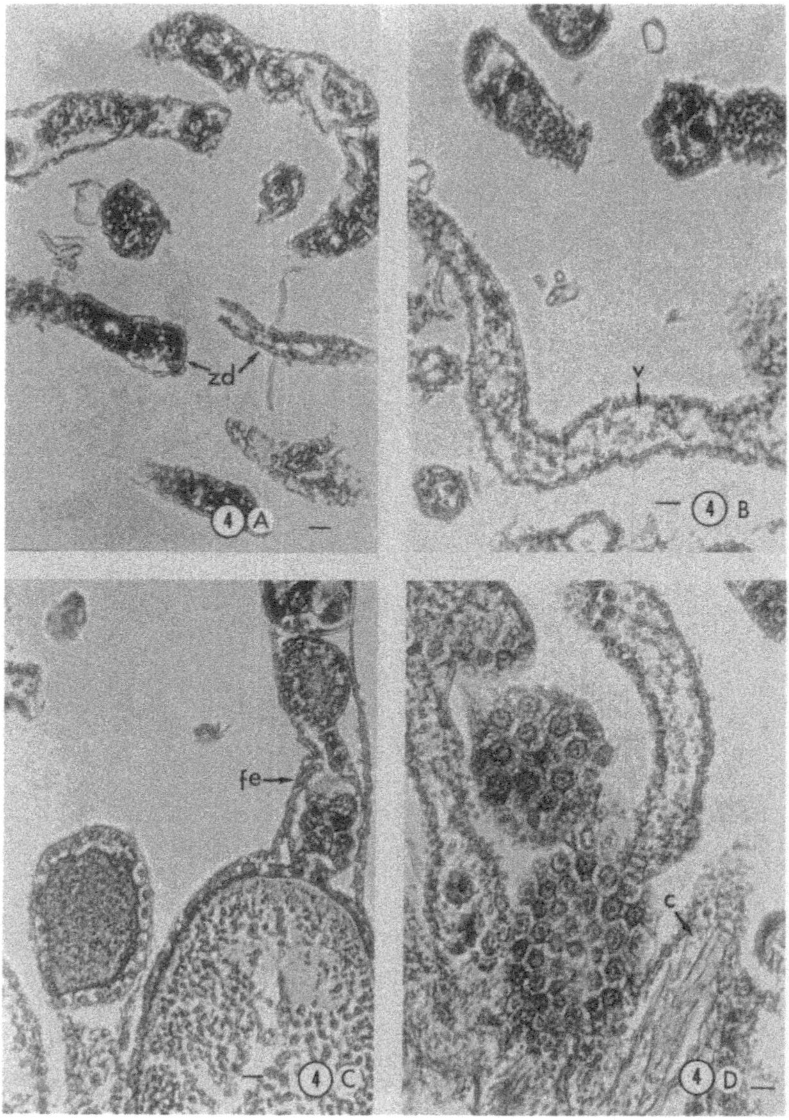

FIGURE 4. Sections through ovarioles of *S. invicta* queens 6 weeks after treatment with Sumitomo S-31183. Effects at germarial level include partial lack of zone of differentiation (4A); vacuolation of ovarioles (4B); increased thickness of follicular epithelium with failure of nurse cells to deposit cytoplasm (4C); and failure to produce large viable eggs in lower end of ovariole because of resorption of the yolk leaving the chorion (4D). Bar= 20u.

without nurse cells. The walls (tunica propria) of most ovarioles were thicker. Egg resorption left only the chorion in 5-10% of the ovarioles.

Although the ovaries in treated queens equaled the size of those in control queens six weeks after treatment, most individual ovarioles were simply tubes, somewhat enlarged at the distal end and containing one small yolked egg; other ovarioles were empty. Microscopic examination of ovarian sections revealed extensive changes. Over 90% of the ovarioles were vacuolated, with thickened walls in most cases. The differentiation zone consisted of an egg nucleus and a few nurse cells. In those instances where cytoplasm and yolkplasm were added, resorption had occurred. The few formed yolked eggs were one third normal size. Fig. 3, A-D (from Glancey and Banks 1988) shows ovarian sections from a control queen. Fig. 4, A-D, shows the effects of S-31183 six weeks after treatment.

Pheidole. Techniques for treating *P. megacephala* colonies with S-31183 were identical to those previously described for fenoxycarb. Effects of treatment on the colonies will be reported elsewhere. Ovaries of queens from treated colonies were removed at five and six weeks after treatment and handled as described for *S. invicta* queens. No *P. megacephala* queens laid eggs at five or six weeks.

Under gross examination - five weeks after treatment - ovaries of treated *P. megacephala* queens did not differ from those of control queens. Microscopic examination of serial sections showed mostly normal ovarian tissues, with a small amount of disorganization occurring at the germarial end. Gross examination six weeks after treatment revealed the same effects seen in *S. invicta* (i.e., intact ovaries), but with simple tubes instead of differentiated tissues and only one, very small yolked egg in a few ovarioles. Histological changes caused by the treatment are shown in Fig. 5, A-D. Tissue differentiation was almost absent in the ovariole. Many ovarioles were completely vacuolated; most nurse cells were undeveloped and cytoplasm was missing in those that were developed. The follicular epithelium was disrupted and the tunica propria thickened. Choria served as evidence of egg resorption.

DISCUSSION

The purpose of these studies was to determine if the IGR that controlled polygynous *S. invicta* colonies would also control the polygynous pest ant, *Pheidole megacephala.* Based on our histological studies, the tested compounds affect both species in the same manner and with the same results. However, the repellency of fenoxycarb to *P. megacephala* has proved a problem. Studies are presently being conducted in both the field and laboratory to determine if lower dosages can be successful.

We proposed in this paper that fenoxycarb affects the ants by overloading the hormonal system. S-31183 appears to be of a different modality. Fenoxycarb causes retrogression of ovarian tissues, whereas one primary effect of S-31183 is egg resorption. The effects of S-31183 on queen ovaries, particularly with respect to the thickening of the tunica propria, approximate those produced by avermectin (Glancey et al. 1982). IGRs control *S. invicta* and *P. megacephala* because they combine morphogenic effects in developing brood with activity that adversely controls egg nutrition.

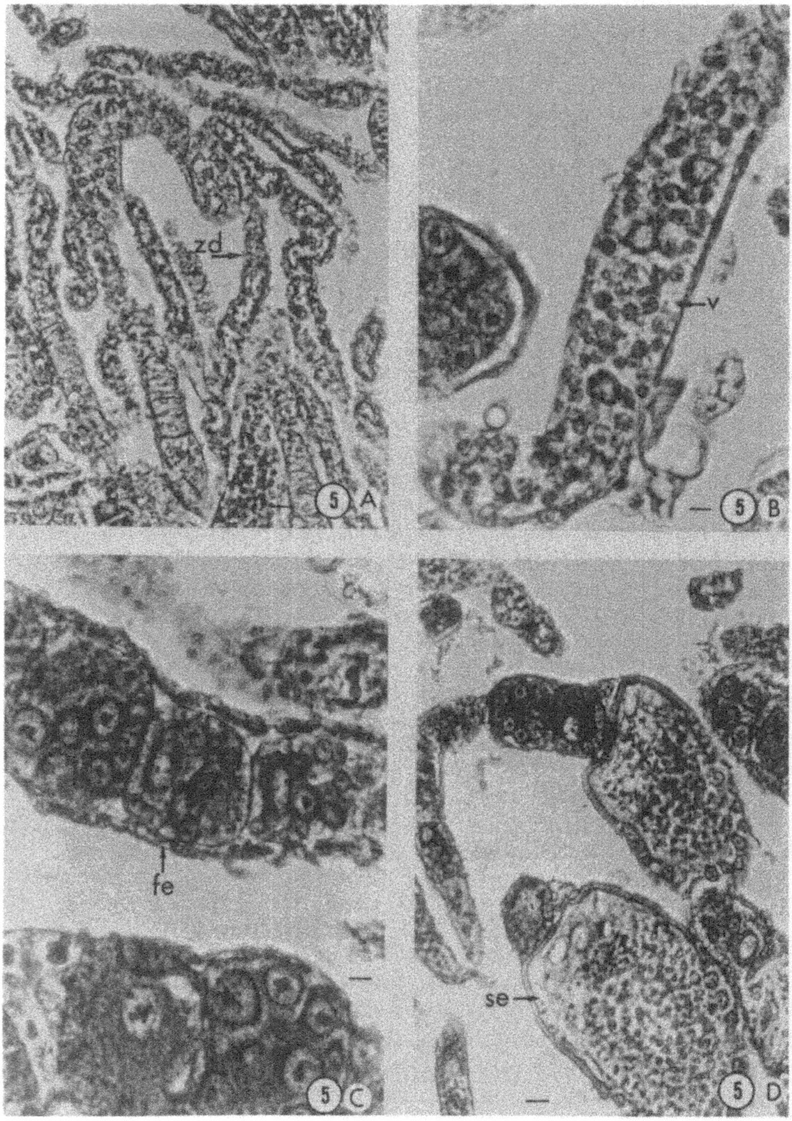

FIGURE 5. Section through ovarioles of *P. megacephala* queens 6 weeks after treatment with Sumnitomo S-31183. Effects, including small eggs (se), are the same as in *S. invicta* queen (Fig 4 A-D). Bar= 40u.

We lack the direct experimental evidence which would tell us if the action site is the corpora allata's neurosecretory cells or the ovary's cells. In any event, the queen's sterility that results from treatment with either fenoxycarb or S-31183 provides excellent control of the ants. Present evidence suggests that baits containing IGRs may be the treatment of choice to control polygyne ants.

ACKNOWLEDGMENTS

We thank T. Krueger, ARS, for his technical assistance and Dr. J. Beardsley, Chairman, Department of Entomology, University of Hawaii and The Governors Agricultural Coordinating Committee for generous support.

REFERENCES CITED

BANKS, W. A., L. R. MILES and D. P. HARLAN. 1983. The effects of insect growth regulators and their potential as control agents for imported fire ants. Fla. Entomol. 66: 172-181.

BANKS, W. A., D. F. WILLIAMS and C. S. LOFGREN. 1988. Effectiveness of fenoxycarb for control of red imported fire ants. J. Econ. Entomol. 81: 83-87.

BANKS, W. A., C. S. LOFGREN, D. P. JOUVENAZ, C. E. STRINGER, P. M. BISHOP, D. F. WILLIAMS, D. P. WOJCIK and B. M. GLANCEY. 1981. Techniques for collecting, rearing, and handling imported fire ants. USDA, AAT-S-21. 9 pp.

BISHOP, P. M., W. A. BANKS, D. F. WILLIAMS, C. E. STRINGER, J. A. MITCHELL and C. S. LOFGREN. 1980. Simple nests for culturing imported fire ants. J. Ga. Entomol. Soc. 15: 300-304.

EDWARDS, J. P., G. W. PEMBERTON and P. J. CURRAN. 1981. The use of juvenile hormone analogues for control of *Pheidole megacephala* F. and other house-infesting ants. Scientific Papers of the Institute of Organic and Physical Chemistry of Wroclaw Technical University, No. 22, Conference 7: 669-779.

GLANCEY, B. M. and W. A. BANKS. 1988. Effect of the insect growth regulator fenoxycarb on the ovaries of queens of the red imported fire ant (Hymenoptera: Formicidae). Ann. Entomol. Soc Am. 81: 642-648.

GLANCEY, B. M., C. S. LOFGREN and D. F. WILLIAMS. 1982. Avermectin B1a: Effects on the ovaries of red imported fire ant queens (Hymenoptera: Formicidae). J. Med. Entomol. 19: 743-747.

GLANCEY, B. M., W. A. BANKS and M. S. OBIN. 1989. The effect of fenoxycarb on alates of the red imported fire ant. J. Entomol. Science 24(3): 290-297.

GLANCEY, B. M., J. C. E. NICKERSON, D. WOJCIK, J. TRAGER, W. A. BANKS and C. T. ADAMS. 1987. The increasing incidence of the polygynous form of the red imported fire ant in Florida. Fla. Entomol. 70: 400-402.

OBIN, M. S., B. M. GLANCEY, W. A. BANKS and R. K. VANDER MEER. 1988. Queen pheromone production and its physiological correlates in fire ant queens (Hymenoptera: Formicidae) treated with fenoxycarb. Ann. Entomol. Soc. Am. 81: 808-815.

ROHRBACH, K. G., J. W. BEARDSLEY, T. L. GERMAN, N. J. REIMER and W. G. SANFORD. 1988. Mealybug wilt, mealybugs, and ants on pineapple. Plant Disease 72: 558-565.

56

Control of the Red Imported Fire Ant *Solenopsis invicta* in Electrical Equipment

W.P. MacKay and S.B. Vinson

INTRODUCTION

Imported fire ants, as well as numerous other ant species, are strongly attracted to electric fields (MacKay et al., unpublished). In addition, the fire ant appears to be attracted to the protection from the elements offered by structures such as traffic control cabinets. This attraction results in two basic problems: first, the ants cause short circuits and second, they interfere with the action of mechanical devices. The ants remove insulation from wires, causing shorts between lines. They also bridge contacts on terminal blocks, causing direct shorts through their bodies. The principal problem is interference when they enter relay switches. They accumulate in such large numbers that the switch cannot function or is shorted. This results in numerous equipment problems, ranging from pump motors, air conditioners to traffic signal lights. Although this is a common and widespread problem (Messor et al., 1986; Vinson and MacKay, 1989), it has not been studied to any extent. It has been reported previously in another species. The white-footed ant, *Technomyrmex albipes* enters relay switches in New Zealand and causes them to malfunction (Little, 1984). Termites may also attack plastic and lead-sheathed cables (Gay and Calaby, 1970).

For this study, a control program has been designed which uses a combination of insecticides and relay switch sealants to successfully reduce *S. invicta*'s "predations" on electrical equipment.

METHODS AND MATERIALS

The study required the treatment of 15 cabinets and 20 cabinets in the cities of Bryan and College Station, Texas, USA respectively. An additional five cabinets in Bryan and ten in College Station were used as controls. They were randomly assigned to treated or control categories. Cabinets were somewhat variable in size, with the bottoms ranging in area from 0.21 m^2 to 0.76 m^2, depending on the equipment contained in them. The majority were a standard 0.65 m^2. Each of the cabinets was mounted on a cement pad slightly larger than the base of the cabinet. Conduits entered the cabinets through holes in the bottom of the cement pad or through PVC pipe embedded in the pad.

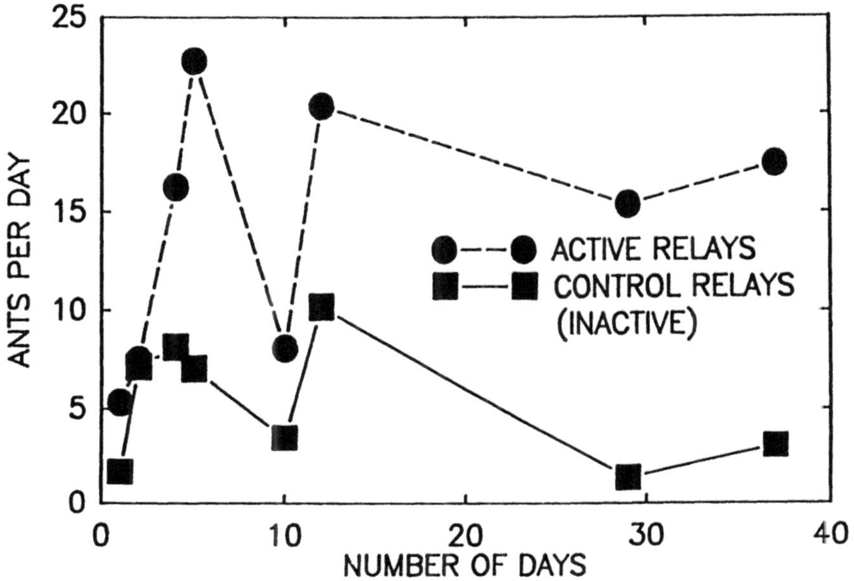

FIGURE 1. Laboratory data comparing the rate at which ants enter active and inactive traffic cabinet relays.

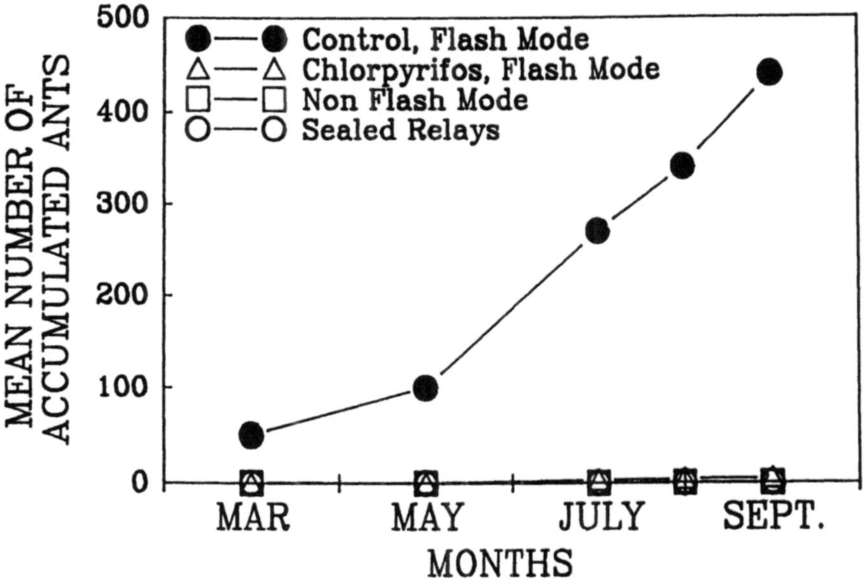

FIGURE 2. Field data comparing the accumulated numbers of imported fire ants in active and inactive relays in the city of College Station, Texas, USA.

Two treatments--the first, chlorpyrifos (0.5%) in a corncob grit, and the second a mixture of pyrethrins (0.10%), piperonyl butoxide (1.00%) and silica gel (4.00%)--were begun on 13 May 1988, after the cabinets had been cleaned. Cabinets in College Station had not been previously treated with any product except the occasional use of 1,1,1 Trichloroethane, which is not very effective. Those in Bryan had been treated with a mixture of pyrethrin (1.00%), piperonyl butoxide (10.00%) and silica gel (40.00%), whenever ants were causing problems; this treatment is also not very effective. Chlorpyrifos was used at 1/3, 2/3 and at the level recommended (915 g/m^2) by the manufacturer (five cabinets/level). Additional cabinets (five) were treated with a combination of pyrethrin, piperonyl butoxide and silica gel. The material was sprayed on the cabinet bottom, forming a uniform white cover. The cabinets of Bryan were also treated in a similar manner with three levels of chlorpyrifos on 10 May 1988 (five cabinets/level). The cabinets were monitored at monthly intervals throughout the summer season. Three other cabinets in College Station were treated with 1,1,1 trichloroethane whenever fire ant populations increased to a size which could cause problems in the circuitry. This is the most common response of signal technicians in the area (MacKay, personal observation).

The study has shown that fire ants are strongly attracted to relay switches due to their electrical field (MacKay et al., unpublished). Cabinets have from one to five relay switches, depending on the complexity of the signal lights controlled by the cabinet. Two sets of six relays were designed and then mounted on a Plexiglas frame. The relays operated (on a 1 H_z signal or one cycle/sec) at night (18:00 of one day to 08:00 of the following day) in two laboratory nests. The laboratory was dark much of this time. This schedule simulated the normal activity of the relays, which are most active at night when traffic signals are in flash mode. The circuitry designed allowed a random sequence of active and inactive relays (three controls and three active relays). which relays would be active at any specific times (three controls and three active relays). All relays had power to the points (120 VAC, 1.89 Amps) - to simulate the condition of relay switches in the field. Removed at regular intervals, ants were counted and returned to the relays to determine the rate at which ants accumulated.

The effect of several sealants on relay switches was tested. Those sealants evaluated were: roofing cement, two silicon sealers (Duncan 25 year silconized acrylic caulk and Duncan all purpose adhesive and caulk), and two types of epoxy cement (Devcon 5-minute and plastic steel). Each of the traffic cabinet relays was enclosed in a plastic box. The area around the bases of nine pins which protrude from the relay was sealed, both on the inside and outside of the switch. Sealed relays were tested in laboratory nests using relays mounted in Plexiglas and in actual field conditions.

Non-parametric statistical analysis were used as none of the distributions were normal (Kolmogorov-Smirnov one sample test) and the variances were heteroscedastic (Bartlett's test).

RESULTS AND DISCUSSION

Relay switches attract *S. invicta* workers, especially active relays (Fig. 1). Significantly more ants enter active relays in the laboratory (Wilcoxon's signed rank test, $\alpha < 0.005$), and field data reveal a similar trend (Fig. 2). Two

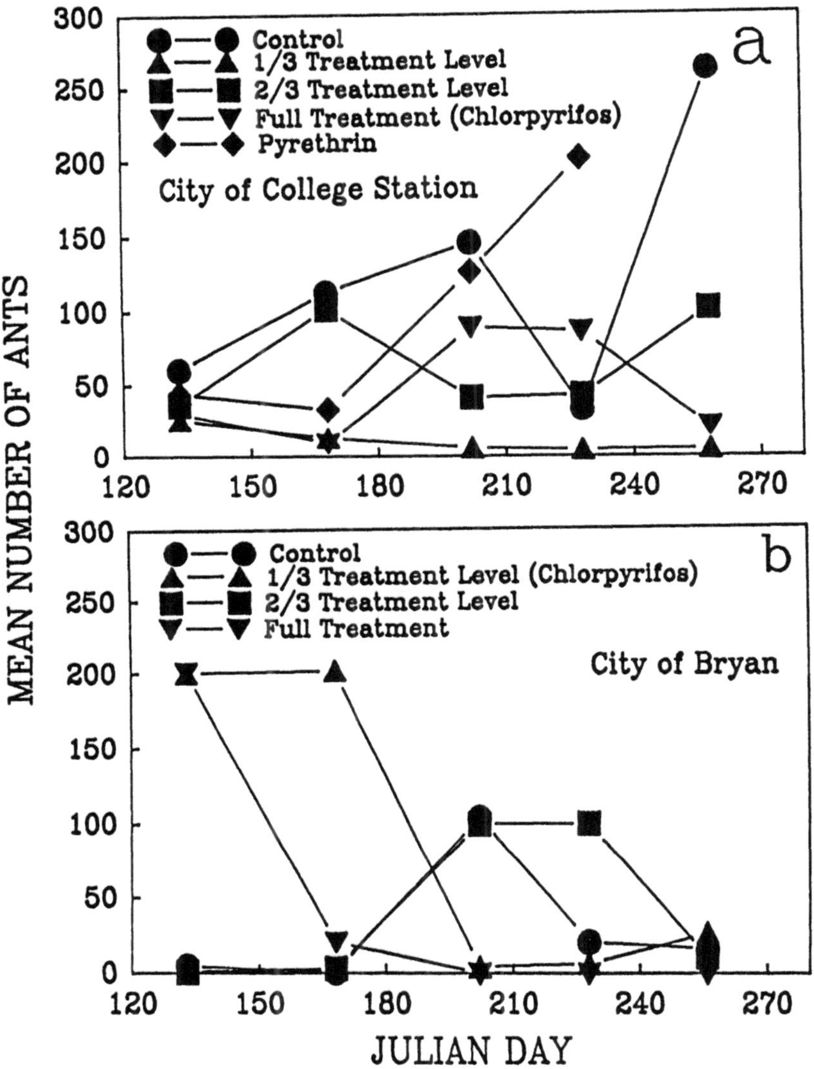

FIGURE 3. Fire ant populations in traffic signal cabinets in College Station, Texas (a) and Bryan Texas (b) during 1988. The various treatments are indicated on the figure.

cabinets were in the flash mode throughout the summer 1987 (relay switches active at all times). One of these was randomly selected to be treated with the insecticide Insect Control; the other was selected to be a control cabinet. Only these two cabinets functioned continuously in the flash mode. These two cabinets were compared with three randomly selected cabinets which were treated in the same way - one with Insect Control, the second a control, and the third equipped with sealed relays - but were not in the flash mode

TABLE 1. Wilcoxon's signed ranks tests of the effects of insecticidal treatment of traffic signal cabinets in the cities of Bryan and College Station, Texas, USA, using two-sided probabilities and a normal approximation.

COLLEGE STATION	ALPHA LEVELS		
	Ants	Level of Insecticide	Date
Ants	1.000		
Level	0.002	1.000	
Date	0.001	1.000	1.000
BRYAN	**ALPHA LEVELS**		
	Ants	Level of Insecticide	Date
Ants	1.000		
Level	0.159	1.000	
Date	0.190	1.000	1.000

throughout the summer. Each of the cabinets had three relay switches. The relays in the flash mode control cabinet accumulated large numbers of ants (Fig. 2), significantly more than the control cabinet (Wilcoxon's signed rank test, α <0.05). The treated cabinet, the cabinet with sealed relays and the cabinet not in flash mode were not statistically different from each other. Obviously, active relays accumulate ants at faster rates in both laboratory and field conditions. In addition, the treatment of a cabinet with chloropyrifos reduced the number of ants which enter the relays to levels similar to those of inactive relays. Epoxy was much more effective than any of the other sealants tested. No ants entered any of the relays, either in field experiments or lab experiments in any of the four relays sealed with epoxy. Relays sealed with other products and tested in laboratory nests accumulated hundreds or thousands of ants within a month.

S. invicta was effectively controlled in traffic signal cabinets (Fig. 3). Insecticidal treatment with chlorpyrifos significantly reduced the ant populations in College Station (Table 1, Fig. 3a), although populations in treated and control cabinets were not significantly different in Bryan (Table 1, Figure 3b). Apparently this was due to low control cabinet populations throughout the experiment. A comparison of the population levels in Bryan's treated cabinets before and after treatment demonstrated a statistically significant reduction in the populations (α = 0.022, Wilcoxon's signed ranks test). Thus, treatment of the cabinets with chlorpyrifos controlled ant populations in both cities. Orthogonal comparisons and Newman-Keuls Multiple comparisons tests showed that treatment levels did not differ significantly in controlling cabinet ant populations in either city, but ant populations were significantly lower in treated cabinets compared to control cabinets. Pyrethrins were not very effective (Fig. 3a). 1,1,1 Trichloroethane rapidly killed ants, but treated cabinets were reinfested the following day. Chlorpyrifos destroyed all of the ants in cabinets within three hours. These data are similar to those reported by MacKay (1988).

CONCLUSION

This study has demonstrated that ant problems in electrical equipment can be successfully controlled by sealing all mechanical relay switches (the most attractive electrical component) and treating the apparati with the insecticide chlorpyrifos. These procedures apply since relay switches are surrounded by a protective cover which can be sealed. In addition, electrical apparati usually have a safety cover which protects the insecticide from wetness and prevents contamination of the environment. The insecticide may be effective in some areas but not in others, depending on environmental conditions. Thus, further experimentation may be necessary before these procedures can be used in other locations and with other equipment.

ACKNOWLEDGMENTS

We would like to thank James Irving and Awinash Bhatkar for critically reading the manuscript, Jim Irving for assistance with laboratory experiments and Carroll Messor, Said Majdi and Stan Swenton for discussions of the project. John Black and Gary Lange allowed us to use the College Station cabinets and Lee Robinson, Jerry Hejl and Mark Schoenemann assisted us with the field work. T. W. Chism and George Mitchell allowed us to use the cabinets in Bryan. Bill Cargile of Rainbow Manufacturing Corporation, P. O. Box 26445, Birmingham, AL 35226 supplied the chlorpyrifos (Insect Control); Nancy Coplin of Stutton Corporation, P. O. Box 7913, Metairie LA 70010 and C. M. Carleton of State Chemical Manufacturing Co., Cleveland, Ohio 44114 supplied the pyrethrin products (JS 685 and 797A respectively). The research was supported by the Texas State Department of Highways and Public Transportation, grant number 2-18-86-1135.

LITERATURE CITED

GAY, F.J. and J.H. CALABY. 1970. Termites of the Australian Region. In: K. Krishna and F.M. Weesner (eds.). Biology of Termites Vol. 2: 393-448.

LITTLE, E. 1984. Ants in electric switches: a note. New Zealand Entomol. 8:47.

MACKAY, W.P. 1988. El control de la hormiga importada de fuego (Solenopsis invicta Buren - Hymenoptera: Formicidae) en aparatos electricos. Proc. 23rd Congreso Nacional de Entomologia (Mexico). Page 290.

MESSOR, C.J., S.B. VINSON and G.A. BROUWER. 1986. A survey of fire ant infestation in traffic signal systems. Research Report 1135-1, Texas State Department of Highways and Public Transportation, 14 pages.

VINSON, S.B. and W.P. MACKAY. 1989. The effect of the red imported fire ant on circuits and electrical equipment. In press.

57

Approaches to Biological Control of Fire Ants in the United States

D.P. Jouvenaz

INTRODUCTION

The red and black imported fire ants, *Solenopsis invicta* Buren and *Solenopsis richteri* Forel, have been extremely successful immigrants to the United States. Since the first infestation was noted at Mobile, Alabama in 1929, they have spread to ca 10^8 ha in 11 southeastern states and Puerto Rico (Lofgren, 1986). In recent years their rate of spread has slowed, for they are approaching their ecological limits in the Southeast (Francke and Cokendolpher 1986). Recently they have been accidentally moved by man across the arid southwest to Arizona (Frank 1988) and California (C.S. Lofgren, Personal communication). Should they become established in the west, their range will increase substantially. In addition, a polygynous form with denser populations, and which is sometimes more difficult to control, is spreading. However, the diminished territorial behavior of polygynous colonies may render them more vulnerable to biotic agents.

The growth of the imported fire ant infestation in the United States has not gone unopposed. As medical and agricultural pests, they have been assaulted with a succession of pesticides by Federal and State agencies at a cost (by 1976) of well over $100,000,000. Despite these efforts, the United States has more fire ants than ever, and their numbers are increasing. Pesticides, although useful for temporary local suppression, cannot provide a permanent solution to the fire ant problem in the United States. It is possible, however, that biological control may provide a permanent amelioration of the problem.

Unfortunately, fire ants are difficult subjects for biological control. Serious obstacles are posed by the longevity (Tschinkel 1987), protected environment (see "Biological Insecticides" below) and high reproductive potential (Morrell 1974) of their queens. In addition, fire ants are ecological generalists able to exploit a wide variety of habitats and environmental conditions. Their ability to exploit disturbed environments led Tschinkel (1986) to characterize them as a "weed" species. Such versatility can only add to the complexity of biological control.

APPROACHES TO BIOLOGICAL CONTROL OF FIRE ANTS

Research to date on biological control of fire ants can be divided into: 1) evaluation of non-specific natural enemies for use as biological formicides (temporary local suppression) and 2) the search for specific natural enemies for introduction into the United States (permanent reduction of the pest population).

Biological Insecticides.

Efficacious formicides are those chemical or biological agents which destroy an ant colony's reproductive capacity (kill or sterilize the queen or queens). Agents which only kill worker ants do not effectively eliminate colonies, although they may temporarily reduce their size. This requirement is a serious obstacle to the development of a biological formicide. Fire ant queens are sequestered in subterranean nests and surrounded and defended by tens of thousands of aggressive workers. They are groomed meticulously, and fed highly filtered, regurgitated liquids (latex microspheres as small as 0.88 um are filtered by workers - Glancey et al. 1981; the microflora of the queen's gut is highly restricted or absent - Jouvenaz, unpublished). As adults, they are refractory to infection by many entomopathogens. Furthermore, fire ants fumigate their nests with venom (Obin and Vander Meer 1985), which has antibacterial and antifungal properties (Jouvenaz et al. 1972; Cole 1975). Although it may be possible to overwhelm these defenses with massive doses of a biotic agent, fire ants frequently relocate their nests to escape the presence of foreign organisms, or simply in response to disturbance (Jouvenaz and Lofgren 1986).

One arthropod and many non-specific entomopathogens have been evaluated as possible biological formicides for fire ants. The straw-itch mite, *Pyemotes tritici*, was considered promising by Bruce and Le Cato (1980), and for a time was even marketed by at least two firms. However, subsequent evaluations by Jouvenaz and Lofgren (1986), Thorvilson et al. (1987), and Collins (USDA-APHIS; personal communication) failed to demonstrate a useful degree of efficacy. The mites are also hazardous to handle; they can cause dermatitis severe enough to require medical treatment or even hospitalization (Baggett et al. 1981).

The non-specific entomopathogens that have been evaluated for use as microbial formicides include the bacteria *Bacillus thuringiensis*, *B. sphaericus*, *Serratia marcescens*, *Pseudomonas aeruginosa* and *P. chloroaphis* (Broome 1974; Miller and Brown 1983; Jouvenaz, unpublished), fungi of the genera *Beauveria* and *Metarrhizium* (Broome 1974; Quattlebaum 1980; Stimac et al. 1987), microsporidia of the genera *Nosema* and *Thelohania* (Jouvenaz, unpublished), several strains of the nematodes *Neoaplectana carpocapsae* and *Heterorhabtidis heliothidis* (Quattlebaum 1980; Poole 1976; Jouvenaz, unpublished), and various baculoviruses (Broome 1974). The honey bee pathogens, *Bacillus larvae*, *B. pulvifaciens*, and honey bee paralysis virus were also tested (Jouvenaz, in press). With the possible exception of *Beauveria* spp. isolated from fire ants in Brazil (Stimac et al. 1987), none of these organisms have produced significant fire ant control under laboratory and/or field conditions.

Specific Natural Enemies.

In the United States, imported fire ants are essentially free of specific natural enemies (Jouvenaz et al. 1977). In their native South America, however, they are beset by a complex of pathogens: parasites, social parasites and symbiotic predators. Although we are aware of many of these organisms, we as yet know little of their biology, impact on individual colonies or field populations of fire ants, or their environmental safety (Jouvenaz 1986; Wojcik et al. 1987).

A large number of arthropods have been observed in fire ant colonies (see D. P. Wojcik this volume), but many appear to be commensals. Predacious symbionts such as the beetle, *Martinezia dutertrei* (Scarabaedae), and parasites such as *Orasema* spp. wasps, do destroy immature fire ants; however, their numbers are invariably very low relative to their host population, and their direct impact is therefore limited. There are sound biological reasons to suspect that some of these symbionts may be vectors of ant disease (Jouvenaz 1986).

Among arthropods, the socially parasitic ant, *Solenopsis (Labauchena) daguerri* Santschi, appears to be the most promising candidate for introduction. Queens of this workerless species integrate into fire ant colonies, where they and their offspring are cared for (apparently preferentially) by their hosts. The parasites do not contribute to the economy of the colony, and under the burden of these unproductive individuals colony vigor declines. Possibly as a result of nutritional deprivation, the oviposition rate of the host queen decreases, the proportion of major workers diminishes, and sexual forms disappear. The workers become less aggressive and allow the nest to deteriorate. Heavily infested colonies (*S. daguerri* may constitute over 70% of the total population) can readily be identified by their eroded and unrepaired nest tumuli. The process of debilitation may take two years, but ends with collapse of the colony (Silveira-Guido et al. 1973).

Solenopsis daguerri is the only natural enemy known to destroy established fire ant colonies. It is discouraging that *S. richteri* populations in Argentina can be quite high in areas where *S. daguerri* is common (Silveira-Guido et al. 1973), since this is not indicative of biocontrol. However, we do not understand the population dynamics of fire ants in their native ranges (Wojcik 1986); The parasite infestation could have been recent and later have led to collapse of the fire ant population. In a different ecosystem, *S. daguerri* might be a valuable member of a complex of natural enemies of fire ants.

Phorid flies, *Pseudacteon* spp. and an *Apodicrania* sp., are endoparasites of fire ants in South America (Williams and Whitcomb 1974). The number of ants parasitized by these flies is relatively low; however their impact on fire ant populations may be much greater than that due to simple parasitism. Feener (1981) reported that the presence of a phorid fly, *Apocephalus* sp. shifts the competitive balance between *Solenopsis texana* and *Phidole dentata* by interfering with the defensive behavior of major workers of the latter. In the absence of phorids, *P. dentata* dominated food sources. In the presence of phorids, however, *S. texana* was able to dominate food sources because *P. dentata* major workers retreated from the flies. The presence of phorids that specifically parasitize fire ants could provide an advantage to competitive native ant species.

The subject of competition between ant species brings us to the most radical strategy yet proposed for biological control of fire ants. Buren (1983) suggested that the solution to the fire ant problem in the United States is competitive displacement by a complex of deliberately introduced predator and competitor ant species. Buren called his hypothetical complex an "artificial" ant fauna. I prefer the term "selected" ant fauna.

The introduction of exotic ant species is potentially hazardous; without careful selection of candidate species based on thorough research, pest ant problems could be compounded. Replacing one dominant ant species with another is unacceptable. Rather, for ecological stability, Buren envisioned a complex of 20-30 species, the composition of which would vary geographically. Thus, the selected ant faunas of southern Florida and northern Mississippi would have some, but not all, species in common. Buren assumed a beneficial ant fauna could be selected that was at least not harmful to major elements of our native flora and fauna (some displacement of native ant fauna would be inevitable, but much displacement by imported fire ants has already occurred). All species selected would have to be beneficial, or at least innocuous, to man and agriculture.

Buren discussed the characteristics that would govern the selection of species for study and concluded that several hundred to a thousand candidates exist world-wide (it is not necessary, or perhaps even desirable, for the candidates to be sympatric with fire ants). Buren also addressed the approaches and methodologies that might be employed in evaluating these candidates. Unfortunately, the thorough evaluation of so many exotic ant species would be expensive and very time consuming. Obtaining approval to import exotic ants for release, even after extensive research, might prove difficult.

Despite obvious technical difficulties and the potential for environmental harm, Buren's proposal may have merit, especially in concert with a complex of specific natural enemies. The primary role of host-specific natural enemies may be to stress fire ants, thereby reducing their competitive fitness. Hopefully, these specific natural enemies will enable native ants to displace fire ant populations to a significant extent. Should exotic ants be required, pathogens and parasites might at least reduce the number of species needed.

CURRENT USDA BIOLOGICAL CONTROL RESEARCH

The primary goal of the USDA biological control of fire ants research program is the establishment of a complex of specific natural enemies of fire ants in the United States that will be of value in an integrated pest control program. Even a modest reduction in fire ant populations should reduce or eliminate the need for pesticide treatments in large areas. The development of biological insecticides is a secondary goal of this program.

The initial phase of USDA research was exploration - the discovery and description of the specific natural enemy biota of fire ants. The known pathogen biota now includes a virus, at least three endoparasitic fungi, six protozoans, and three nematodes; other pathogens doubtless remain to be discovered. Although the majority are probably commensals, ca 200 species of arthropods have been associated with fire ants (see Wojcik in this volume). The emphasis of research is now on the biology and impact of these organisms. A thorough understanding of their pathobiology or life cycles, host ranges (= environmental safety) and an estimate of at least their relative potential impact on fire ant populations is required for the selection of candidates for introduction. It is difficult to predict the impact of a biological control agent in a new environment; environmental safety, however, must be accurately assessed. For this reason, we are currently studying only those organisms whose host range appears to be limited to *Solenopsis* spp. For those candidates selected and approved for introduction, it will be necessary to develop techniques for mass production and quality control, dissemination, and monitoring of spread and impact. Possible interactions of the agent with control practices may also need to be assessed.

We have selected the socially parasitic ant, *S. daguerri*, the nematode *Tetradonema solenopsis* Nickle and Jouvenaz (1987), and three undescribed, endoparasitic fungi for concentrated study. We have also begun monitoring fire ant and arthropod symbiont population dynamics and epizootiology in the field through the USDA Biological Control of Weeds Laboratory, Hurlingham, Argentina.

Solenopsis daguerri, as noted above, is the only natural enemy known to destroy established fire ant colonies. As an obligate (workerless) social parasite that is at least genus specific, it is environmentally safe. However, the very specificity of *S. daguerri* for its host may present difficulties in establishing it in a new host population. In the 1960's, Uruguayan entomologists repeatedly attempted to introduce *S. daguerri* from Argentina without success. These efforts only involved moving material by air in one day from Buenos Aires to Montevideo, and then to areas of eastern Uruguay heavily infested with *S. richteri* (personal communication, M. S. Blum, University of Georgia). We do not know what procedures they employed to collect, handle, or disseminate S. daguerri; the failure may have been due to the parasite's inability to integrate into a new host population. The major mechanism of symbiont integration into fire ant colonies appears to be chemical mimicry (Vander Meer and Wojcik 1982; Vander Meer et al. 1989). A difference in the cuticular odors of parasite and new host may prevent integration. If this is the case, a technique must be developed to precondition *S. daguerri* (or a sibling species) for acceptance by North American fire ant populations.

The nematode *T. solenopsis* parasitizes *S. invicta* and was found in the states of Mato Grosso and Mato Grosso do Sul, Brazil. Adult male and female nematodes, eggs, and juveniles occur together in adult worker ants, but the life cycle and mode of transmission are unknown. Presumably juveniles emerge from the host to mature and mate in the soil, and then infect a new host. Interestingly, the distribution of nematodes in the colony is non-Poisson. Multiple (up to 35) adult females, plus many males are found in individual worker ants from colonies having infection rates of less than 5% (Jouvenaz et al. 1988). Therefore, we hypothesize that the invading inseminated parasite produces a limited number of offspring (both males and females) which become the "reproductive group." The offspring of the reproductive group then produces large numbers of offspring which infect new hosts. We cannot gain insight into the life cycle of *T. Solenopsis* from the five other known species of Tetradonematidae; the life histories of these parasites of Diptera and Coleoptera are also poorly known. *Tetradonema solenopsis* is extremely stressful to its host, and we have been unable to maintain infected colonies in the laboratory. Colonies with infection rates above 25% have been collected (most are below 5%), but infected individuals rapidly died. Our immediate goal is to culture infected colonies in soil, and transmit the infection to healthy colonies through nematode-infested soil. Success will enable us to determine host specificity and, hopefully, elucidate the life cycle. If adult queens are susceptible to infection, the nematode could destroy established fire ant colonies.

Four species of endoparasitic fungi occur as budding yeasts in the hemolymph of fire ants. At least one species appears to be an obligate parasite, for it grows very slowly *in vitro*. These fungi can be mass-produced inexpensively, have been transmitted *per os* to *S. invicta*, and we are optimistic that at least the obligate parasite will prove host-specific for *Solenopsis* spp. Our observations to date indicate that they are simply nutritional burdens to their hosts, producing no toxins, gross pathology or histopathology. They are prime candidates for genetic engineering. If a way can be found to preserve the invasive ability of these fungi (which declines during *in vitro* culture), it may be possible to alter them genetically to produce toxins of our choice, juvenile hormone or other hormones, or perhaps even semiochemicals to disrupt colony organization. We have determined that the obligate parasite is susceptable to the antibiotic hygromycin B, for which a cloned resistance gene is available. A model transformation test for hygromycin resistance will be conducted when taxonomic studies in progress are complete, and the experiment is approved by the biotechnology committee. Even without genetic modification, they may prove valuable as stressing agents in concert with other natural enemies or control practices. The endoparasitic yeasts are the only specific pathogens of fire ants which appear to have potential for both development as a microbial formicide and establishment in our ecosystem.

Biological control of fire ants is a formidable task. meeting the challenge requires time and resources for in-depth, innovative basic and applied research, particularly if permanent amelioration of the problem is to be achieved. Biotechnology, in particular, presents exciting prospects for the development of genetically engineered microbial insecticides, and for more virulent, well-adapted symbiotic microorganisms. The alternative is to limit ourselves to the recurrent expenses of temporary, local suppression, and to accept an ever-increasing fire ant problem in the United States.

REFERENCES CITED

BAGGETT, D.A., B.L. DAVIS, L.B. ELLIOT, P.V. FOURNIER, S.J. LARRO, J.A. RAWLINGS, and C.R. WEBB, Jr. 1981. Occupational dermatitis associated with grain itch mites - Texas. Centers for Disease Control Morbid. Mortal. Report 30: 590-592.

BROOME, J.R. 1974. Microbial control of the imported fire ant, *Solenopsis richteri* Forel and *Solenopsis invicta* Buren in Mississippi. Ph. D. dissertation, Mississippi State University, Mississippi State. 66 pp.

BRUCE, W.A. and G.L. LECATO. 1980. *Pyemotes tritici*: A potential new agent for control of the red imported fire ant, *Solenopsis invicta* (Acari: Pyemotidae). Internat. J. Acarol. 6: 271-274.

BUREN, W.F. 1983. Artificial faunal replacement for imported fire ant control. Fla. Entomol. 66: 93-100.

COLE, L.K. 1975. Antifungal, insecticidal, and potential chemotherapeutic properties of ant venom alkaloids and ant alarm pheromones. Ph.D. dissertation, University of Georgia, Athens, Ga. 172 pp.

FEENER, D.H. Jr. 1981. Competition between ant species: outcome controlled by parasitic flies. Science 214: 815- 817.

FRANCKE, O.F. and J.C. COLKENDOLPHER. 1986. Temperature tolerances of the red imported fire ant. Pp. 104-113. In C.S. Lofgren and R.K. Vander Meer (eds.), Fire ants and leaf-cutting ants, biology and management. Westview Press, Boulder, CO. 434 pp.

FRANK, W.A. 1988. Report of limited establishment of the red imported fire ant, *Solenopsis invicta* Buren in Arizona. Southwest. Entomol. 13: 307-308.

GLANCEY, B.M., R.K. VANDER MEER, A. GLOVER, C.S. LOFGREN and S.B. VINSON. 1981. Filtration of microparticles from liquids ingested by the imported fire ant, *Solenopsis invicta* Buren. Insectes Sociaux 28: 395-401.

JOUVENAZ, D.P., M.S. BLUM, and J.G. MacCONNELL. 1972. Antibacterial activity of venom alkaloids from the imported fire ant, *Solenopsis invicta* Buren. Antimicrob. Agents Chemother. 2: 291-293.

JOUVENAZ, D.P., G.E. ALLEN, W.A. BANKS, and D.P. WOJCIK. 1977. A survey for pathogens of fire ants, *Solenopsis* spp. in the Southeastern United States. Fla. Entomol. 60: 275- 280.

JOUVENAZ, D.P. 1986. Diseases of fire ants: problems and opportunities. pp. 327-338. In C.S. Lofgren and R.K. Vander Meer (eds.), Fire ants and leaf-cutting ants, biology and management. Westview Press, Boulder CO. 434 pp.

JOUVENAZ, D.P. and C.S. LOFGREN. 1986. An evaluation of the straw itch mite, *Pyemotes tritici* (Acari: Pyemotidae) for control of the red imported fire ant, *Solenopsis invicta* (Hymenoptera: Formicidae). Fla. Entomol. 69: 761-763.

JOUVENAZ, D.P., D.P. WOJCIK, M.A. NAVES, and C.S. LOFGREN. 1988. Observations on a parasitic nematode of fire ants, *Solenopsis* spp., from Mato Grosso. Pes. Agro. Bras. 23: 525-529.

JOUVENAZ, D.P., C.S. LOFGREN AND R.W. MILLER. In press. Steinernematid nematode drenches for control of fire ants, *Solenopsis invicta*, in Florida. Fla. Entomol.

LOFGREN, C.S. 1986. History of the imported fire ants in the United States. pp 36-47. In C.S. Lofgren and R.K. Vander Meer (eds.), Fire ants and leaf-cutting ants, biology and management. Westview Press, Boulder, CO. 434 pp.

MILLER, S.R. and L.R. BROWN. 1983. Studies on microbial fire ant pathogens. Devel. Indus. Microbiol. 24: 443-450.

MORRELL, W.L. 1974. Production and flight of alate red imported fire ants. Environ. Entomol. 3: 265-271.

NICKLE, W.R. and D.P. JOUVENAZ. 1987. *Tetradonema solenopsis* n. sp. (Nematoda: Tetradonematidae) parasitic on the red imported fire ant *Solenopsis invicta* Buren from Brazil. J. Nematol. 19: 311-313.

OBIN, M.S. and R.K. VANDER MEER. 1985. Gaster flagging by fire ants (*Solenopsis* spp.): functional significance of venom dispersal behavior. J. Chem Ecol. 11: 1757-1768.

POOLE, M.A. 1976. Survey and control of the imported fire ant, *Solenopsis richteri* Forel and *Solenopsis invicta* Buren in Mississippi. Ph.D. dissertation, Mississippi State University, Mississippi State. 92 pp. [Dissert. Abstr. Int. B, 37: 3254]

QUATTLEBAUM, E.C. 1980. Evaluation of fungal and nematode pathogens to control the red imported fire ant, *Solenopsis invicta* Buren. Ph.D. dissertation, Clemson University, Clemson, S. C. 54pp. [Dissert. Abstr. Int. B, 41: 1642].

SILVIERA-GUIDO, A., J. CARBONELL, and C. CRISCI. 1973. Insects associated with the *Solenopsis* (fire ants) complex with special reference to *Labauchena daguerri*. Proc. Tall Timbers Conf. Ecol. Anim. Control Habitat Manage. 4: 41-52.

STIMAC, J.L., S.B. ALVES, and M.T. VIEIRA CAMARGO. 1987. Sucetibilidade de *Solenopsis* spp. a diferentes especies de fungos entomopatogenicos. An. Soc. Entomol. Brasil 16: 377-387.

THORVILSON, H.G., S.A. PHILLIPS JR., A.A. SORENSON, and M.R. TROSTLE. 1987. The straw itch mite, *Pyemotes tritici*: (Acari: Pyemotidae), as a biological control agent of red imported fire ants, *Solenopsis invicta* (Hymenoptera: Formicidae). Fla. Entomol. 70: 439-444.

TSCHINKEL, W.R. 1986. The ecological nature of the fire ant: some aspects of colony function and some unanswered questions. Pp. 72-87. In C.S. Lofgren and R.K. Vander Meer (eds.), Fire ants and leaf-cutting ants, biology and management. Westview Press, Boulder, CO. 434 pp.

TSCHINKEL, W.R. 1987. Fire ant queen longevity and age: estimation by sperm depletion. Ann. Entomol Soc. Am. 80: 263-266.

VANDER MEER, R.K. and D.P. WOJCIK. 1982. Chemical mimicry in the myrmecophilous beetle, *Myrmecophodius excavaticollis*. Science 218: 806-808.

VANDER MEER, R.K., D.P. JOUVENAZ, and D.P. WOJCIK. 1989. Chemical mimicry in a parasitoid (Hymenoptera: Eucharitidae) of fire ants (Hymenoptera: Formicidae). J. Chem. Ecol. 15: 2247-2261.

WILLIAMS, R.N. AND W.H. WHITCOMB. 1974. Parasites of fire ants in South America. Proc. Tall Timbers Conf. Ecol Anim. Control Habitat Manage. 5: 49-59.

WOJCIK, D.P. 1986. Observations on the biology and ecology of fire ants in Brazil. Pp. 88-103. In C. S. Lofgren and R. K. Vander Meer (eds.), Fire ants and leaf-cutting ants, biology and management. Westview Press, Boulder, CO. 434 pp.

WOJCIK, D.P., D.P. JOUVENAZ, W.A. BANKS, and A.C. PEREIRA. 1987. Biological control agents for fire ants in Brazil. pp. 627-628. In J. Eder and H. Rembold (eds.), Chemistry and biology of social insects. Verlag J. Perperny, Munchen. 757 pp.

58

Potential Baits for Control of the Texas Leaf-cutting Ant, *Atta texana* (Hymenoptera: Formicidae)

R.S. Cameron

INTRODUCTION

Leaf-cutting ants in the genus *Atta* are among the most destructive insects in the tropical and subtropical Americas. Particularly susceptible crops include citrus, cocoa, coffee, maize, cotton, eucalyptus and pines (Cherrett 1986b). The Texas leaf-cutting ant, *Atta texana* (Buckley), has the most northerly distribution of any species in this genus and is a serious pest of first- and second-year pine plantations in western Louisiana and eastern Texas, particularly on well-drained soils (Moser 1984; Cameron and Riggs 1985).

Several chemicals and application techniques have been developed to combat leaf-cutting ants, including: (1) organochlorine insecticides (aldrin, chlordane, dieldrin, and heptachlor) applied to nests as liquids, dusts, or fogs; (2) fumigants (carbon disulfide and methyl bromide); and (3) baits containing slow-acting toxicants (aldrin or mirex), retrieved and carried by the ants into their underground nests (Mariconi 1970; Cherrett 1986a; Jaffe 1986; Vilela 1986). Currently only methyl bromide fumigation and resmethrin applied as a thermal fog (Cameron 1989) are registered by the U. S. Environmental Protection Agency to control *Atta texana*.

Baits are attractive for ant control because they are easy to apply, effectively controlling small, scattered, or multiple-queened colonies (Echols 1966) year-round. Since the EPA banned mirex bait in the U. S. A. in 1978, several baits containing hydramethylnon, abamectin or fenoxycarb have been developed to control the red and black imported fire ants, *Solenopsis invicta* Buren and *S. richteri* Forel respectively. Sulfluramids (fluoroaliphatic sulfonamides), a new class of slow acting insecticides, are also effective against fire ants (Vander Meer et al. 1985), but have yet to be cleared for commercial use.

Reported here are field acceptance, preference, and efficacy tests against *A. texana* with baits containing hydramethylnon, abamectin, fenoxycarb or sulfluramid.

MATERIALS AND METHODS

Thirteen field tests conducted in east Texas from December 1981 through January 1988 evaluated sixteen bait formulations for efficacy against *A. texana* (Table 3). Hydramethylnon was formulated on citrus pulp in a variety of extruded bait shapes and sizes and on defatted corn grits of small (ca 1.6 mm in average dimension) or large (ca 3.3 mm) particle size (Fig. 1) Abamectin and sulfluramid were also formulated on defatted corn grits. Preference and efficacy tests were conducted on a variety of sites including natural mixed hardwood/pine forests, pine plantations and open fields in Angelina, Cherokee, Nacogdoches, Polk, Shelby, and Tyler counties in east Texas.

Acceptance and Preference Tests

For purposes of this paper, acceptance is defined as bait retrieval by ants over time, without reference to other baits or standards. Preference tests record comparisons of ant retrieval rates between candidate baits and a standard substrate. Acceptance observations were made for the baits used in Tests 1-7 to determine if ants retrieved and carried baits into their nests. Bait acceptance by ants was noted at the time of application in efficacy Test 1. Numbers of bait particles removed from 1 m^2 grids in the treatment areas were recorded at 24 and 48 hr after application in Test 2. Baits employed in efficacy Tests 3-7 were evaluated for acceptance by *A. texana* using methods similar to those of Phillips et al. (1976). At least 10 bait particles were placed at six to ten positions ordered randomly along heavily travelled peripheral foraging trails or adjacent to entrance/exit holes with ant activity in the central nest area. The number of remaining bait particles was recorded at periodic intervals.

Previous studies which determined leaf-cutting ant preferences for baits or hosts used extensive counts of numbers of discs or particles retrieved, numbers of leaves cut or numbers of ants visiting baits (Lewis and Phillips 1973; Littledyke and Cherrett 1975; Rockwood 1976; Waller 1982). A more efficient method of evaluating preference was developed to screen bait formulations containing fenoxycarb or sulfluramid and to compare acceptability of the various baits used in efficacy Tests 8-13. Baits, in 3, 5, or 8 g portions placed in 100 x 15 mm plastic petri dishes, were distributed at random within each replicate along active foraging trails and/or on the central nest area near entrance/exit holes with ant activity. The dishes within each replicate were retrieved when few particles of the most attractive bait remained or at the end of the test period, usually several hours. The dishes were reweighed to determine the amounts of bait removed by ants. One to several baits and one or two standard substrates were included in each preference test. The standards were cracked corn purchased at a local feed store and/or blank defatted corn grit (No. 980 provided by Illinois Cereal Mills, Inc.). Cracked corn was used as a standard since it is highly palatable to *A. texana* (Waller 1982). Treatment means were compared statistically using the Statistical Analysis System ANOVA and Duncan's Multiple Range Test procedures (Ray 1982).

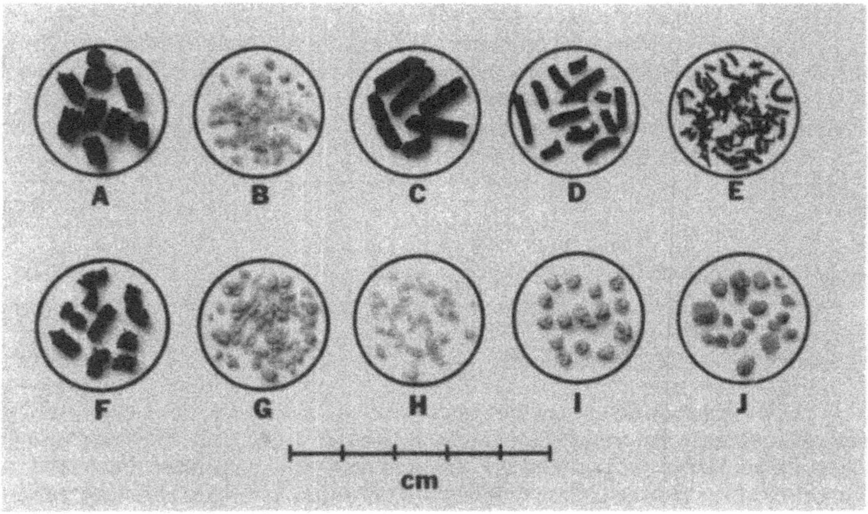

FIGURE 1. Baits evaluated in *Atta texana* acceptance and efficacy tests conducted in Texas: A) Mirex Pelleted Bait "450"; B) hydramethylnon (Amdro); C) hydramethylnon citrus pulp (AC3532-66A); D) hydramethylnon citrus pulp (AC4201-31-1,2,&3); E) hydramethylnon citrus pulp (AC4024-100A); F) hydramethylnon citrus pulp (AC4024-100B); G) hydramethylnon corn grit (AC4403-147); H) abamectin (Affirm); I) fenoxycarb oil cake bait (ACR 2913F M.2); and J) sulfluramid (GX-071 10-E230).

Efficacy Tests

For each efficacy test, a group of suitable test colonies was selected to exclude those which were large (larger than 30 x 30 m), small (less than 3 x 3 m), adjacent to each other (within ca 100m) and/or lacking an apparent distinct central nest area. Treatments were then randomly assigned to the selected ant nests.

The central nest area was defined as the above-ground portion of the nest, characterized by a concentration of entrance/exit page holes, surrounded by lose soil excavated by the ants. Scattered, peripheral entrance/exit and foraging holes were excluded from the central nest area. Application rates were based on the area (length x width) of the central nest and expressed in grams of bait per m^2 of nest area.

The areas over which baits were distributed (treated areas) varied among tests as indicated in Table 3. A Cyclone spreader spread baits uniformly over the treated areas in each test. Applications were timed to avoid wet soil conditions and rain within 24 hr following application of baits.

To evaluate the effect of treatments on ant colonies, active entrance/exit holes were counted just before and periodically following treatment (most often at 2, 4, 8, and 16 weeks). Active entrance/exit holes were defined as those with ants present, fresh deposits of soil, or fresh vegetative material in or around them. Post-treatment inspections were postponed during or immediately following prolonged cold weather or rainy periods when ants were inactive above ground. Untreated check colonies were included and monitored in each test to account for possible seasonal trends in ant activity. The percent of initial activity was calculated as the number of active holes divided by the number of initial active holes x 100 for each colony at each post-treatment check. Differences in mean percent of initial activity among treatments were tested for significance with the Kruskal-Wallis one-way analysis of variance and multiple-comparison procedures (Daniel 1978). Also, the percent of colonies totally inactive was calculated for each treatment at each post-treatment evaluation.

RESULTS AND DISCUSSION

Acceptance and Preference Tests

Few leaf-cutting ants retrieved the Amdro fire ant bait in efficacy Test 1, even though ants were actively excavating and page foraging during the application. All baits evaluated in efficacy Tests 2-7 were readily accepted by *A. texana*. In field acceptance tests, baits from most replicates were retrieved and taken into ant nests within 2-3 hr.

The preference test quickly and simply screened baits for efficacy. Baits retrieved at levels comparable to or better than cracked corn or defatted corn grits were selected for the efficacy tests (8-13). Equal amounts of fenoxycarb (RO 13-5223), Affirm and cracked corn were retrieved by ants (Table 1). These three baits were significantly more attractive than blank defatted corn grit. Significantly less Amdro fire ant bait, formulated on defatted corn grit, was retrieved than blank defatted corn grit, indicating this bait contained components repellent to ants.

Numerous preference tests were conducted to screen 13 fenoxycarb bait formulations. One formulation, moderately attractive to leaf-cutting ants, was identified and tested for efficacy. Three bait formulations containing sulfluramid were tested twice (Table 2). Since the sulfluramid formulation 10-E1540 was highly preferred, similar formulations were used in field efficacy tests.

Preferences within and between individual leaf-cutting ant colonies vary greatly. Status of the fungus garden, brood, and moisture in ant nests may influence the types of hosts preferred by foraging ants (Lewis et al. 1974; Rockwood 1976). Ants from actively foraging nests are less host selective than those from nests with low activity (Littledyke and Cherrett 1975; Rockwood 1976), and ants from substrate-starved nests show less host selectivity and retrieve more substrate than those from colonies provided adequate host material (Littledyke and Cherrett 1975). Accordingly, Waller (1982) observed that *A. texana* is less selective during the winter in Texas when host materials are scarce.

Weights of baits retrieved by ants in preference tests are influenced by particle size and density, as well as by bait ingredients. If the objective is to predict control from preference tests, measuring the weight of various

TABLE 1. *Atta texana* preference test for three bait candidates compared to defatted corn grit and cracked corn standards in Texas, January, 1986.

Treatment[a]	Mean grams removed[b]
Fenoxycarb 1% ai (RO 13-5223)	2.7a
Cracked corn	2.5a
Abamectin 0.011% ai (Affirm)	2.3a
Small defatted corn grit	1.3b
Amdro fire ant bait	0.9c

[a] 3 g/replicate, 10 replicates/treatment; exposed ca. 2.5 hr.
[b] Means followed by like letters not significantly different (P>0.05).

TABLE 2. *Atta texana* preference test for three sulfluramid (GX-071) bait formulations compared to defatted corn grit and cracked corn standards in Texas, February 1987.

Treatment[1]	Mean grams removed[2]	
	4 FEB 87	9 FEB 87
GX-071, 10-E1540	7.3a	4.6a
GX-071, 10-O230	2.0b	1.5b
GX-071, 10-S220	2.1b	2.7ab
Large defatted corn grit	-	1.7b
Cracked corn	0.6c	1.2b

[1] 8 g/replicate, 10 replicates/treatment; exposed ca 4 hr.
[2] Means followed by like letters not significantly different (P>0.05).

particles removed is required (see Table 1), especially if the baits are applied on a weight basis in the efficacy tests (g/m^2). However, if the objective is to determine which ingredients are most attractive to ants, bait particle size and density should be constant, as with the comparison of sulfluramid baits (Table 2).

Efficacy Tests

Results of efficacy tests, in which a total of 168 colonies were treated with baits, are presented in Table 3. Amdro fire ant bait did not effectively control *A. texana*, even at the extremely high rate of 45 g/m^2. A lack of bait acceptance presumably contributed to this result (see Table 1). The hydramethylnon citrus pulp bait used in Test 2 controlled up to 80% of the treated colonies at the 4.8 g/m^2 application rate, but significant treatment differences were not observed in this test due to variability among colonies and the small sample size. Ingredients in the hydramethylnon citrus pulp baits used in Test 3 varied, but bait size was uniform. Two of these baits applied at 1.9 g/m^2 significantly reduced but did not completely halt ant activity. The lack of efficacy in this test may have been due to the relatively low application rate. Higher rates were applied in subsequent efficacy tests with hydramethylnon baits. The two baits evaluated in Test 4 contained the same ingredients, but particle size varied greatly. These baits applied at 4.8 and 5.2 g/m^2, respectively, provided complete control.

Two baits containing hydramethylnon formulated with attractants on small defatted corn grit were evaluated in Tests 5-7. AC4403-147 (0.88% ai) and AC4468-143-2 (1% ai) were 100% effective within 2 and 4 wk, respectively. In Test 6, AC4468-143-1 (1% ai) significantly reduced ant activity, but only one of five colonies became inactive within 14 wk following treatment. In Test 7, all of the colonies treated with AC4468-143-2 were inactive 4 wk after treatment. AC4468-143-1 and AC4468-143-2 are separate batches of the same formulation. However, observed differences in efficacy between Test 6 and Test 7 were more likely related to a seasonal shift in ant activity rather than possible batch differences. Few ants were active during the day when the bait was applied in October, but in November, when temperatures were lower, ants retrieved the bait immediately after application. Hydramethylnon readily breaks down in UV light and may have been partially detoxified before ants retrieved the bait in the October test.

Test 8 included a large corn grit formulation with similar ingredients to those in AC4468-143 and a citrus pulp bait similar to AC4024-100B. All of the colonies treated with the corn grit bait were inactive within 8 wk following treatment, while the citrus pulp bait only controlled 36% of the treated colonies after 16 weeks. However, ant activity was significantly reduced in the colonies treated with citrus pulp bait compared to untreated colonies 8 wk after treatment.

No treatment effects were observed in Tests 9 and 10. Abamectin and fenoxycarb baits did not significantly reduce ant activity compared to check colonies, although ant activity 18 wk after treatment was minimal in the three colonies treated with fenoxycarb bait at a rate of 7.2 g/m^2. Abamectin baits have shown promise against *Atta* and *Acromyrmex* spp. in Brazil, although control was delayed compared to that of mirex baits (Vilela 1986).

Sulfluramid baits applied in March, June, and December in Tests 11-13, respectively, were highly effective against *A. texana*. Treated colonies had minimal ant activity at 2 wk following treatment and usually were totally inactive after 4 wk, even at the low rate of 0.9 g/m^2. Of the colonies treated in June, none were completely controlled at 0.3 g/m^2, but all were controlled at 1.8 g/m^2.

Seasonal and diel foraging habits may influence the effectiveness of baits for controlling leaf-cutting ants. *Atta texana* activity patterns are highly temperature dependent. During the summer, when daytime temperatures often exceed 30°C, ants are nocturnal and forage from tunnel entrances. During the winter, ants are mostly diurnal, foraging primarily from the central nest area (Waller 1982). They may undergo extended periods without foraging when temperatures remain below 10°C. During the remainder of the year ants may be active both day and night.

Attractive baits broadcast on the central nest area probably will be retrieved by *A. texana* any time the ants are actively foraging, but should be most effective during the winter when foraging is concentrated on the central

TABLE 3. Application dates, rates, and efficacy data for bait treatments applied to Atta texana colonies in east Texas, 1981-1988.

(Test no.) Treatment	Date	Applic. rate (area treated[a])	Reps	% of initial act.[b] @(wk)	% colonies inactive @(wk)[c]
Hydramethylnon - small corn grit					
(1) Amdro (0.88% ai)	Dec 81	$6.5 g/m^2$ (I)	11	33.7a (6)	9 (6)
		$45.4 g/m^2$ (I)	9	29.9a	0 (6)
		0	5	54.8a	0
Hydramethylnon - citrus pulp					
(2) AC3532-66A (1% ai)	Mar 82	$0.9 g/m^2$ (I)	5	14.5a (15)	60 (8)
		$4.8 g/m^2$	5	0.8a	80 (15)
		0	2	32.8a	0
(3) AC4201-30-1 (1% ai)	Sep 82	$1.9 g/m^2$ (I)	6	95.2ab (6)	0 (10)
AC4201-30-2 (1% ai)		$1.9 g/m^2$ (I)	6	53.9a	0 (10)
AC4201-30-3 (1% ai)		$1.9 g/m^2$ (I)	5	51.7a	0 (10)
		0	6	163.8b	0
(4) AC4024-100A (1% ai)	Mar/Apr 83	$4.8 g/m^2$ (I)	11	0.1a (10)	91 (10)
AC4024-100B (1% ai)		$5.2 g/m^2$ (I)	4	0.0a	100 (10)
		0	11	80.6b	0
Hydramethylnon - small corn grit + attractant					
(5) AC4403-147 (0.88% ai)	Feb 84	$3.6 g/m^2$ (II)	10	0.0a (2)	100 (2)
		0	10	60.5b	0
(6) AC4468-143-1 (1% ai)	Oct 84	$3.6 g/m^2$ (II)	5	3.6a (2)	20 (2)
		0	3	93.3b	0
(7) AC4468-143-2 (1% ai)	Nov 84	$3.6 g/m^2$ (II)	6	0.0a (4)	100 (4)
		0	6	61.3b	0
Hydramethylnon - large corn grit/citrus pulp					
(8) AC5560-72-2 (1% ai) corn	Jan 88	$3.7 g/m^2$ (III)	10	0.0a (8)	100 (8)
AC5560-73 (1% ai) citrus		$3.7 g/m^2$	11	36.2b	36 (16)

TABLE 3. (Continued).

(Test no.) Treatment	Date	Applic. rate (area treated[a])	Reps	% of initial act.[b] @(wk)	% colonies inactive @(wk)[c]
		0	11	90.4c	0
Abamectin - small corn grit					
(9) Affirm (0.011% ai)	Nov 85	1.1g/m² (II)	6	37.5a (6)	17 (6)
		3.6g/m²	6	32.0a	0 (27)
		0	6	55.2a	0
Fenoxycarb - oil cake bait					
(10) ACR 2913F M.2 (1% ai)	Dec 86/Jan 87	3.6g/m² (II)	12	59.4a (18)	0 (18)
		7.2g/m²	3	23.4a	0 (18)
		0	6	102.3a	0
Sulfluramid - large corn grit					
(11) GX-071 10-E230 (0.6% ai)	Mar 87	3.6g/m² (II)	10	0.0a (4)	100 (4)
		0	5	129.6b	0
(12) GX-071 10-E230 (0.6% ai)	Jun 87	0.3g/m² (II)	4	124.2ab (16)	0
		1.8g/m²	4	0.0a	100 (16)
		3.7g/m²	4	0.4a	75 (16)
		0	4	236.1b	0
(13) GX-071 10-PL2030 (0.6% ai)	Dec 87	0.9g/m² (III)	5	0.0a (8)	100 (4)
		1.8g/m²	5	0.0a	100 (8)
		3.6g/m²	5	0.0a	100 (4)
		0	5	46.8b	0

[a] Application rate: g of bait per m^2 of central nest (CN). Area treated varied among tests as follows: I - (L of CN + 12m)(W of CN + 12m), II - 4(L x W of CN), III - (L x W of CN).
[b] Mean percent of initial activity at post-treatment interval (indicated in parentheses) with maximum reduction in activity compared to check; means followed by the same letter within each test are not significantly different ($P>0.05$) according to Kruskal-Wallis one-way analysis of variance and multiple-comparison procedures (Daniel 1978).
[c] Maximum percent of colonies inactive with the time interval when first observed in parentheses.

nest and other substrates for fungal gardens are scarce. Baits should not be applied when the soil surface and vegetation are wet, when rain is expected within 24 hr, or when temperatures are expected to remain below 10°C. Hydramethylnon baits may lose their toxicity if exposed to direct sunlight for prolonged periods. Afternoon applications would be preferable during periods of hot weather. Application rates based on the central nest area may need to be increased during the summer when activity on the central nest is minimal and nest sizes may be underestimated.

In summary, baits containing hydramethylnon and sulfluramid formulated on defatted corn grit were particularly effective for controlling *A. texana* in tests conducted in east Texas. In tests conducted in December and January, hydramethylnon and sulfluramid baits distributed exclusively on the central nest area were 100% effective. This indicates that baits do not need to be distributed beyond the boundaries of the central nest to include peripheral foraging trails, at least during the winter. The application rate found to be effective in Texas for baits containing 1% ai hydramethylnon (3.6 g of bait/m^2, or 0.04 g ai/m^2), is the same as the active ingredient rate generally recommended for mirex baits (Mariconi 1970; Cherrett 1986a). Hydramethylnon is relatively safe to use and is currently registered by the U. S. Environmental Protection Agency for the control of fire ants. Sulfluramid is effective against *A. texana* at a very low rate, 0.005-0.018 g ai/m^2, but is not yet registered for ant control in the United States.

ACKNOWLEDGMENTS

I thank Herbert Pase, Carol Riggs, Alan Smith, Steve Tracy, William Upton, and Charles Ware for their help with data collection; Lee Rayburn for assistance with statistical analyses; and Ronald Billings, John Moser, Robert Nichols, and Herbert Pase for their critical reviews of the manuscript. Funds for these studies were provided by American Cyanamid Company, Maag Agrochemicals, Inc. and the Texas Forest Service. Baits were provided by American Cyanamid Company, Maag Agrochemicals, Inc., Merck Sharp & Dohme Research Laboratories, and Griffin Corporation. Defatted corn grit was provided by Illinois Cereal Mills, Inc. I thank Kirby Forest Industries, Champion International, International Paper Co., Bill Oates, Lee McNeely, Erwin Roemer, Jr., and James Brown for providing study sites.

REFERENCES

CAMERON, R.S. 1989. Control of the Texas leaf-cutting ant, *Atta texana* (Hymenoptera: Formicidae) with a thermal fog application of resmethrin, pp 236-244. *In* R.I. Alfaro and S. Glover, eds. Insects Affecting Reforestation: Biology and Damage. Proc. IUFRO Conference, XVIII International Congress of Entomol. Vancouver, B. C. July 3-9, 1988. Forestry Canada, Pacific Forestry Centre, Victoria, B.C. 256 pp.

CAMERON, S., and C. RIGGS. 1985. Distribution, impact and control of the Texas leaf-cutting ant - 1983 survey results. Texas Forest Service Publ. 139. 6 pp.

CHERRETT, J.M. 1986a. chemical control and bait formulations for leaf-cutting ants, pp. 357-368. *In* C. S. Lofgren and R. K. Vander Meer, eds. Fire ants and leaf-cutting ants biology and management. Westview Press, Boulder. 435 pp.

CHERRETT, J.M. 1986b. History of the leaf-cutting ant problem, pp.10-17. *In* C.S. Lofgren and R.K. Vander Meer, eds. Fire ants and leaf-cutting ants biology and management. Westview Press, Boulder. 435 pp.

DANIEL, W.W. 1978. Applied Nonparametric Statistics. Houghton Mifflin Co., Boston. 503 pp.

ECHOLS, H.W. 1966. Compatibility of separate nests of Texas leaf-cutting ants. J. Econ. Entomol. 59: 1299-1300.

JAFFE, K. 1986. Control of *Atta* and *Acromyrmex* spp. in pine tree plantations in the Venezuelan llanos, pp. 409-416. *In* C.S. Lofgren and R.K. Vander Meer, eds. Fire ants and leaf-cutting ants biology and management. Westview Press, Boulder. 435 pp.

LEWIS, T. and F.T. PHILLIPS. 1973. Aerial baiting to control leaf-cutting ants (Formicidae, Attini) in Trinidad. I. The bait, its production, and the effects of weathering on attractiveness and persistence of toxicants. Bull. Entomol. Res. 63: 263-273.

LEWIS, T., G.V. POLLARD, and G.C. DIBLEY. 1974. Micro-environmental factors affecting diel patterns of foraging in the leaf-cutting ant *Atta cephalotes* (L.) (Formicidae: Attini). J. Anim. Ecol. 43: 143-153.

LITTLEDYKE, M. and J.M. CHERRETT. 1975. Variability in selection of substrate by the leaf-cutting ants *Atta cephalotes* (L). and *Acromyrmex octospinosus* (Reich) (Formicidae, Attini). Bull. Entomol. Res. 65: 33-47.

MARICONI, F.A.M. 1970. As sauvas. Editora Agronomica "Ceres", Sao Paulo. 167 pp.

MOSER, J.C. 1984. Town ant, pp. 47-52. *In* T.L. Payne, R.F. Billings, R.N. Coulson, and D.L. Kulhavy, eds. History, status, and future needs for entomology research in southern forests. Proc. 10th. Anniv. East Texas For. Entomol. Sem. Texas Agric. Exp. Sta. MP-1553. 72 pp.

PHILLIPS, F.T., P. ETHERIDGE and G.C. SCOTT. 1976. Formulation and field evaluation of experimental baits for the control of leaf-cutting ants (Hymenoptera, Formicidae) in Brazil. Bull. Entomol. Res. 66: 579-585.

RAY, A.J. (ed). 1982. SAS Users' Guide: Statistics. SAS Institute, Cary, North Carolina. 584 pp.

ROCKWOOD, L.L. 1976. Plant selection and foraging patterns in two species of leaf-cutting ants (*Atta*). Ecology 57: 48-61.

VANDER MEER, R.K., C.S. LOFGREN and D.F. WILLIAMS. 1985. Fluoroaliphatic sulfones: a new class of delayed-action insecticides for control of *Solenopsis invicta* (Hymenoptera: Formicidae). J. Econ. Entomol. 78: 1190-1197.

VILELA, E.F. 1986. Status of leaf-cutting ant control in forest plantations in Brazil, pp. 399-408. *In* C.S. Lofgren and R.K. Vander Meer, eds. Fire ants and leaf-cutting ants biology and management. Westview Press, Boulder. 435 pp.

WALLER, D.A. 1982. Foraging ecology of the Texas leaf-cutting ant, *Atta texana* Buckley (Formicidae: Attini): Host choice and foraging size polymorphism. Ph.D. Thesis, University of Texas, Austin, TX. 188 pp.

59

Comparison of Susceptibility of *Acromyrmex octospinosus* Reich (Attini, Formicidae) to Two Insect Parasitic Nematodes of the Genera *heterorhabditis* and *neoaplectana* (Rhabditina, Nematoda)

A. Kermarrec, H. Mauleon and D. Marival

INTRODUCTION

More than 240 years ago, Gould (1747) observed the first nematode parasitizing an ant. Since then, numerous nematodes have been studied in relation to Hymenoptera (Bedding 1984). Generally, mermithids and simple phoretic (or even inquiline) rhabditid worms have been described.

Recently, the first Tetradonematid nematode was observed in *Solenopsis invicta* Buren in Brazil by Nickle and Jouvenaz (1987). About 25% of the worker ants of a single colony were parasitized by *Tetradonema solenopsis n.sp.* These individuals died three days later.

Very little attention has been focused on the relationships (set up over millions of years of sympatric co-existence in the soils) between ants and nematodes. Wahab (1962) provided the first review of the subject. Subsequent synecological observations on Attine ants and soil nematodes (Kermarrec 1975) showed the capability of the social insects to eliminate the helminths from the social (gardened and groomed) milieu.

Some rhaditid nematodes have shown real potential for insect pest control: they belong to two families: *Heterorhabditidae* and *Steinernematidae*. True parasitism by a steinernematid (*Neoaplectana dutkyi*, DD136 strain) was observed by Lofgren et al. (1975) on *S. invicta.* The authors concluded, simultaneously with Kermarrec (1975) testing *N. carpocapsae* (agriotos strain) against *Ac. octospinosus*, that under the conditions reported, these helminths had little effect on tested colonies. However, more studies with steinernematid nematodes have been performed since that date (Laumond et al. 1979) and the generally negative experimental results were presented and discussed at the Fire Ant and Leaf-cutting Ant Colloquium held in Gainesville, Florida (Kermarrec et al. 1986)

In this paper, we compare the capability of two related rhabditid nematodes to parasitize *Ac. octospinosus*: a steinernematid worm (*N. carpocapsae*, agriotos strain) and a heterorhabditid one (*Heterorhabditis cf. bacteriophora*, Guadeloupe strain). Both nematodes carry an entomopathogenic bacterium, *N. carpocapsae* associated with *Xenorhabdus nematophilus* and *H. cf. bacteriophora* with *X. luminescens*.

MATERIALS AND METHODS

Ants

Several stable laboratory nests of *Ac. octospinosus* provided regular supplies of ants for biological tests. They received daily fresh ornamental supplies of ants for biological tests. They received daily fresh ornamental Hibiscus leaves and flowers *ad libitum*. Sexuals were present in some of these nests. Four nests in good condition received a massive inoculation of both nematodes.

Nematodes

Neoaplectana carpocapsae (agriotos strain) and *Heterorhabditis cf. bacteriophora* (Guadeloupe strain) were produced continuously on the greater wax moth, *Galleria mellonella*, and stored at seven degrees celsius for a short time.

Parasitological Tests

Ten worker ants of various sizes were placed in a Petri dish that had blotting paper covering its bottom. The Petri dish and ants were inoculated with a dose of 20,000 infective nematode larvae. Ten replicates were performed with each nematode. The tests lasted eight days and ant mortalities were evaluated daily. Control Petri dishes received the same quantity of water (without nematodes) on the bottom. Dissected dead ants verified the occurrence of parasitism (i.e. production of a new worm generation inside the host). Mortality figures were corrected with the control ones by Abbott's formula (test - control/100 - control, per 100).

Four laboratory nests received leaf forage inoculated, during ten days, with one of the two nematodes at 20,000 infective stages/ml. No repulsive effect was observed compared with the foraging activity on clean leaves. The refuse area activity and the volume of the fungus garden were inspected each morning for six weeks.

RESULTS

1--The graphs (Figures 1 A, B, C) clearly show that *N. carpocapsae* was less efficient than *H. bacteriophora*, particularly on large workers. Mortality started earlier with the latter nematode, which was, however, less efficient on small individuals (Table 1).

TABLE 1. Lethal time for 30% of *Ac. octospinosus* workers (graphical estimates) challenged with 20,000 infective instars of *N. carpocapsae* and *H. bacteriophora*.

Lethal time 30% (Hrs)	Small workers	Large workers
H. bacteriophora	140	80
N. carpocapsae	100	240

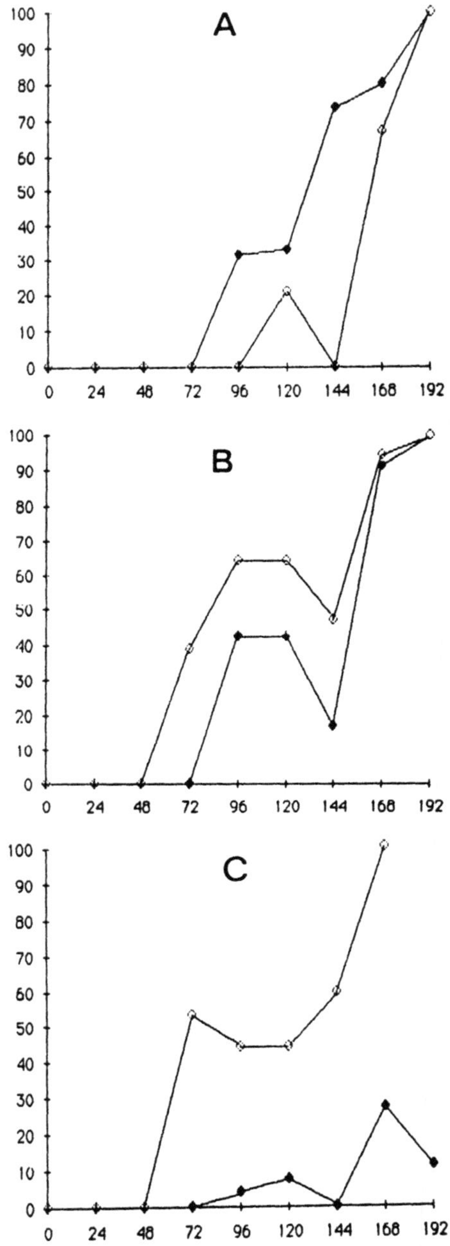

FIGURE 1. Mortality (corrected according to Abbott's formula) of workers of *Acromyrmex octospinosus* under the parasitic pressure of 20,000 infective larvae of *Heterorhabditis* (white dots) and *Neoplectana* (black dots). Small workers (A), medium workers (B) and large workers (C); time in hours.

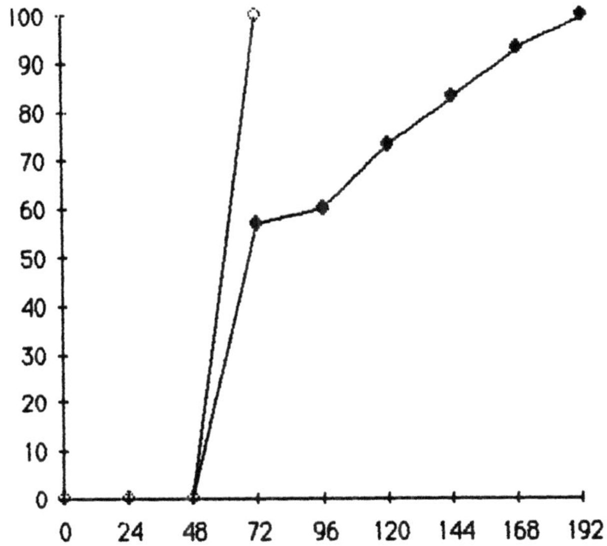

FIGURE 2. Mortality of alates (females) of *A. octospinosus* under the parasitic pressure of *Heterorhabditis* (o) and *Neoaplectana* (•). Time in hours.

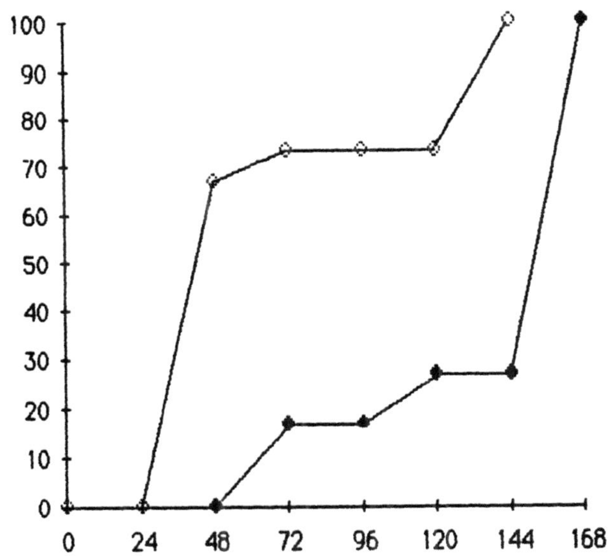

FIGURE 3. Mortality of nymphs of *A. octospinosus* large workers under the pressure of *Heterorhabditis* (o) and *Neoaplectana* (•). Time in hours.

2--Alate forms of both sexes were quickly killed by *H. bacteriophora* (LT50: 60 hours). Female mortality decreased (Figure 2) after 72 hours for *N.* carpocapsae but not the male mortality which is very similar to that observed with *H. bacteriophora.*

3--Nymphs of large workers gave the same trend (Fig. 3) as observed for the large adult workers: *H. bacteriophora* was also, in this case, more efficient than *N. carpocapsae.*

4--Massive infestations of whole nests were inefficient: the leaves were cut, introduced into the nests and integrated into the fungus garden. The dumping activity in the refuse area was not increased by the treatments. The volume of the fungus garden was somewhat affected (Fig. 4), but no abnormal mortality could be observed. One month after the last inoculation, no nematode could be detected in the fungus gardens, either from internal or external refuses.

CONCLUSION

As previously shown (Kermarrec et al. 1986), large workers are less sensitive to *N. carpocapsae*. This is clearly the converse for *H. bacteriophora*, which seems more efficient on large individuals. This experimental observation is certainly related to the only external difference in morphology between the infective instars of both nematoda genera: the presence of a dorsal sclerotized hook on the head of *H. bacteriophora* infective larvae.

Bedding and Molineux (1982) showed that infective juveniles of *Heterorhabditis spp* use a large, anterior, terminal tooth for scraping and

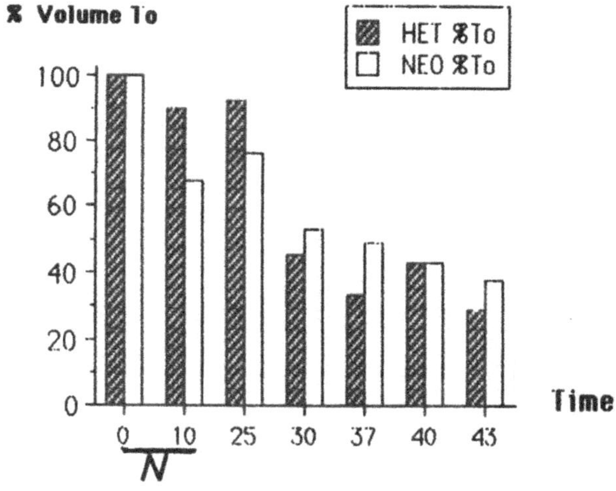

FIGURE 4. Evolution of the mean volume (in % of initial volume; mean of two nests; time in days) of the fungus gardens of *Acromyrmex octospinosus* nests treated with *H. bacteriophora* (dark columns) and *N. carpocapsae* during 10 days (phase N).

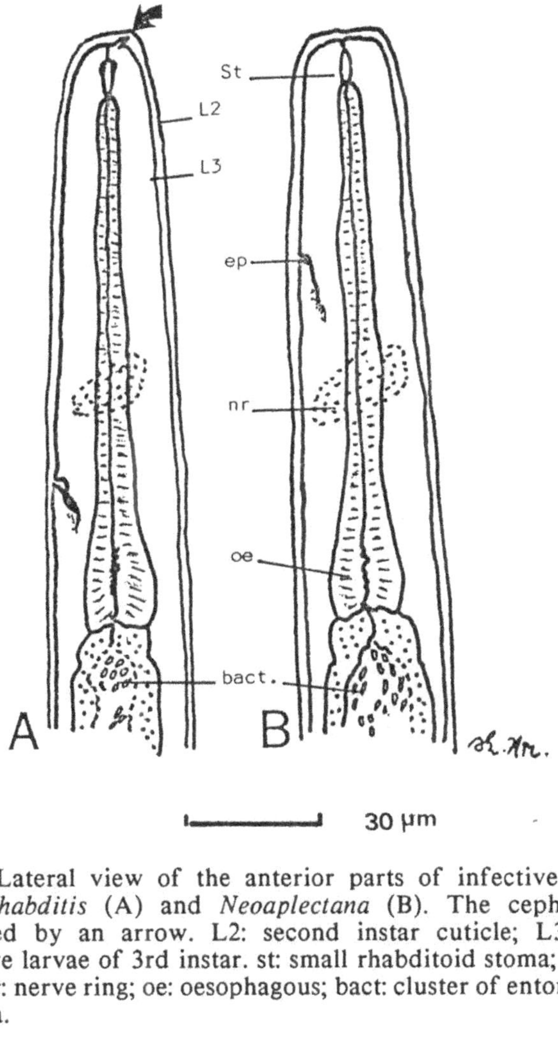

FIGURE 5. Lateral view of the anterior parts of infective juveniles of *Heterorhabditis* (A) and *Neoaplectana* (B). The cephalic tooth is indicated by an arrow. L2: second instar cuticle; L3: ensheathed infective larvae of 3rd instar. st: small rhabditoid stoma; ep: excretory pore; nr: nerve ring; oe: oesophagous; bact: cluster of entomopathogenic bacteria.

rupturing the host's cuticle (Fig. 5). The tooth has its sharp point directed ventrally, ready for use, and is inserted on an apparently rigid basal plate, both strongly sclerotized. Their conclusion is that this tool may enable entry into some insects less vulnerable to *Neoaplectana spp* which lack this hook. As the worms actively explore the crevices and folds of the insect's cuticle, they shed the enclosing L2 cuticle. Penetration of the host's body usually is preceded by the nematode forcing its head end into a cuticular fold, between joints, or other areas of sufficiently flexible cuticle where the nematode will be able to depress the surface readily. The authors also observed that, after probing movements with the head end, "the body became immobile except the head end which moved repetitively in a dorso-ventrally directed arc, so that the terminal tooth abraded or scratched the insect cuticle." These movements may last as long as half an hour.

We believe that the observed experimental differences in efficiencies of both nematodes are logically related to this anatomo-functional distinctive feature of *Heterorhabditis* infective juveniles. In this strategy, larger hosts will optimize the probability of a nematode finding a soft scratchable area on the actively explored surface of the body of its future host.

However, it remains perfectly clear, as already stated by Kermarrec et al. (1986), that the use of entomoparasitic nematodes for biological control of Attine ants does not yet seem possible. No experiment on nests (laboratory or nature) has succeeded until now and we strongly believe that grooming is active and efficient enough to minimize the consequences of any attempted artificial invasive procedure of a biocontrol agent.

RESUME: Comparaison de la susceptibilite des adultes et nymphes de *Acromyrmex octospinosus* REICH (Attini, Formicidae) a deux genres de nematodes parasites d'insectes: Heterorhabditis et Neoaplectana (Rhabditina, Nematoda). Les fourmis sont nettement plus sensibles aux larves infestantes de Heterorhabditis que de Neoaplectana. Cette observation parait liee a la presence d'une dent cephalique chez les larves infestantes du premier genre.

REFERENCES

BEDDING, R.A. 1984. Nematode parasites of Hymenoptera. In: Plant and Insect Nematodes. W.R. Nickle, ed. N.Y. U.S.A. Dekker, Inc. pp. 755-795.

BEDDING, R.A. and A.S. Molineux. 1982. Penetration of insect cuticle by infective juveniles of *Heterorhabditis spp (Heterorhabditis, Nematoda)*. Nematologica, 28, 354-359.

GOULD, W. 1747. An account of English ants. A. Millar. London.

KERMARREC, A. 1975. Etude des relations synecologiques entre les nematodes et la fourmi-manioc *Acromyrmex octospinosus* REICH. Ann. Zool. Ecol. Anim., 7: 27-44.

KERMARREC, A., G. FEBVAY and M.Y. DECHARME. 1986. Protection of leaf-cutting ants from biohazards: is there a future for microbiological control? In: Fire Ants and Leaf-cutting Ants: Biology and Management. Ed. C.S. Lofgren and R.K. Vander Meer. Westview Press. Boulder. London. Chap. 28: 339-356.

LAUMOND, C., H. MAULEON and A. KERMARREC. 1979. Donnees nouvelles sur le spectre d'hotes et le parasitisme du nematode entomophage *Neoaplectana carpocapsae*. Entomophaga, 24: 13-27.

LOFGREN, C.S., W.A. BANKS and B.M. GLANCEY. 1975. Biology and control of imported fire ants. Annu. Rev. Entomol., 20: 1-30.

NICKLE, W.R. and D.P. JOUVENAZ. 1987. *Tetradonema solenopsis n.sp. (Nematoda, Tetradonematidae)* parasitic on the red imported fire ant *Solenopsis invicta* Buren from Brazil. J. Nematol., 19, 3: 311-313.

WAHAB, A. 1962. Untersuchungen uber Nematoden in den Drusen des Kopfes der Ameisen (*Formicidae*). Z. Morphol. Okol. Tiere, 52, 33-92.

60

Psychotropic Substances Impairing the Vigilance of *Acromyrmex octospinosus* Reich (Attini, Formicidae)

A. Kermarrec and H. Mauleon

INTRODUCTION

Attine ants are sensitive to several genera of pathogenic fungi in vitro (Gosswald 1938; Schaefer 1941; Steinhaus and Marshall 1967; Weber 1972; Kermarrec and Mauléon 1975; Kermarrec et al. 1986). However, Kermarrec et al. (1986) emphasized that experimentally induced pathologies have not succeeded in a natural context. The vigilance of worker ants faced with exogenous factors is responsible for colony resistance to artificially induced or natural biohazards.

Ten years ago, Kermarrec et al.(1977) proposed that a worker's basic defensive behaviors could be impaired with psychotropic drugs added to baits which also carried pathogenic agents. These baits would otherwise be detected, discarded or even rejected from the nest.

This paper summarizes two years of research on the ability of psychotropic substances to deter worker vigilance.

General Background

Floru et al. (1969) first reviewed work in this area dealing with invertebrates; see also Witt (1956a 1956b), Mansour (1957), Sankar et al. (1960), Tamano and Kuriaki (1961) and Mercier and Desaigne (1964). All investigators administered drugs by intraperitoneal or intravascular injection, except Witt (1956) and Floru et al. (1969), both of whom used the oral route by natural or forced ingestion. Barbitals, major tranquilizers and neuroleptics - such as pentobarbital, chlorpromazine, chlordiazepoxide - were tested.

A large variety of invertebrates were tested by these authors, including *Fasciola hepatica* (liver fluke), *Zygiella x-notata* (arachnid), *Mantis religiosa*, *Tenebrio molitor* and *Bombyx mori*.

During the last two decades, presocial and social insects have also been treated with various neurotropic and psychotropic drugs, e.g. cockroaches (Schaffer and Reagan 1980), *Apis mellifica* (Mercier and Menzel 1982; Ribbans 1950 1954; Simpson 1954), *Vespa orientalis* (Floru et al. 1969), and also ants such as *Camponotus aethiops* (Rioux 1983), *Camponotus herculeanus* (Cooper and Umemoto 1971), *Camponotus vagus* (Cene-Rioux 1980), *Formica rufa* (Kostowski 1966; ibid et al. 1965 1966 1975ab; ibid and Tarchalska 1972), *Formica pratensis* (Frischknecht and Waser 1978ab 1979), and *Myrmica* sp (Weir 1957).

These authors used intraperitoneal injections or permanent aerosols to introduce drugs to the test animals; i.e. amphetamines, LSD, THC, various neuromediators and biogenic amines, apomorphine, serotonin, 5HTP, reserpine, and colchicine. Social interactions and aggressiveness were studied.

There are few references to the methodology (test organism, drugs used, and dispensing method) used in this study. Kostowski (1966) tested ingested phenobarbital and injected chlordiazepoxide and chlorpromazine on *Formica rufa*'s aggressiveness. Only chlordiazepoxide had a notable calming and ataxic effect. Floru, et al. (1969), found that ingested pentobarbital, chlorpromazine and imipramine (even in low concentrations) significantly affected hornet activity. Chlorpromazine was less effective at low dosages but had a considerable effect at higher ones.

Nothing has been tested on Attine ants and the present paper concerns the effect--under laboratory conditions--of a large spectrum of easily obtained psychotropic substances on *Acromyrmex octospinosus* Reich's vigilance. The drugs were only given orally, to avoid the trauma of injection, which would have no practical application anyway.

Preliminary Definitions

1- Classification of the tested psychotropic drugs (Table 1). According to Lambert (1979), substances are classified as psycholeptics (i.e. hypnotics, tranquilizers and neuroleptics) or psychoanaleptics (psychostimulants). No psychodisleptics were used. Benzodiazepines, common tranquilizers, are sometimes classified as psychosedatives, which refers to lowered interest and locomotor activity. The drugs should affect not only individual behavior, but also more complex social acts such as defense, grooming, tending, etc.

Some psychotropic substances induce a biphasic effect: depression of activity or interest at one dose will be rapidly followed by unexpected hyperactivity (alertness, aggressivity, locomotion). This behavior has a classical explanation; some benzodiazepines, for example, have a disinhibitory effect.

2- Vigilance. To us, vigilance is a state characterized by a high degree of behavioral efficiency, and a high level of activity and reactivity exhibited by the organism. In this way, one may refer to "levels of vigilance." It may also be postulated that the state of maximum vigilance does not necessarily correspond to optimal behavioral efficiency. Efficiency is increased by an aroused vigilence up to an optimal level, beyond which behavior will be impaired.

The level of vigilance depends on the nervous structures' state of activation, themselves submitted to various exo/endogenous activating (or inhibiting) influences. This entails alterant hyperactivation and may even enhance several pathological effects, such as agitation states, loss of coordination, deep ataxy, repeated loops of aimless, stereotyped movements.

The ants' state of alertness or watchfulness may thus be approached through the level of reactivity to external stimuli (either noxious or attractive) measurable by the response level (withdrawal, attack, vigilant walking, etc).

The boundary between individual vigilance and aggressivity is unclear: both may be increased (or inhibited) by the same stimuli.

Response to novelty (exploratory or neophobic behaviors), escape behaviors (running speed, mandibular activity, etc) and, in contrast, grooming (a fixed stereotypical behavior pattern) may help evaluate the level of individual vigilance. Levels of cooperative social behavior can be analyzed in the same way. Mutual and stigmergic cooperation between adults in certain activities aimed at the proper maintenance of the symbiotic fungus are interesting when studying Attine ants: the cleaning and digging of an agar plate to protect a piece of fungus garden is also a measurable sign of vigilance.

MATERIALS AND METHODS

Ants

Several stable, active *Acromyrmex octospinosus nests* were reared in a culture room (To= 25°C; RH= 95%) under an inverted 12:12 LD cycle using red light (wave length > 600nm). The nests were provided each morning (i.e. beginning of the ants' night) with fresh Hibiscus flowers and leaves, as well as drinking water.

The tests used large worker ants, a class more specifically concerned with defense, heavy work and, more generally, detection of any exogenous stimuli. The ants belong to the following mean class of individual fresh weight (N=40): 11.1 mg (confidence limits: 10.4 mg to 11.8 mg). The distribution of weight class frequencies (Fig 1) differs slightly from that of the ants' activity levels, evaluated by a mean work index. A mean work index (Fig. 1) was calculated for each weight class of ant using the following equation, which is based on the quantity of work done on agar plates (weight of agar cut by 10 ants per unit). Ni represents the number of ants in a given class of cutting activity (Ci) and k is an arbitrary empirical factor.

$$K \times \frac{\sum_{C1}^{Ci} (\text{Ni x modal value of class i of cut agar})}{\text{Total number of ants of the class}}$$

This means that ants of a given class perform similarly in tests of locomotion, grooming and escape.

Ants were starved for 24 hours before orally receiving the drugs. The individuals for each test were always taken from the same nest.

Psychotropic Substances

The complete list of tested drugs (Table 1) was established using Vidal's dictionary (1988 ed.) and Lambert's (1979) classification.

These substances were orally dispensed: either (exploring and grooming tests) by forced ingestion of 2ul of a water solution (or suspension) of the active ingredient, deposited carefully between the mouthparts; or (escaping tests) by an induced drinking method (Febvay et al. 1985). The ants, after a

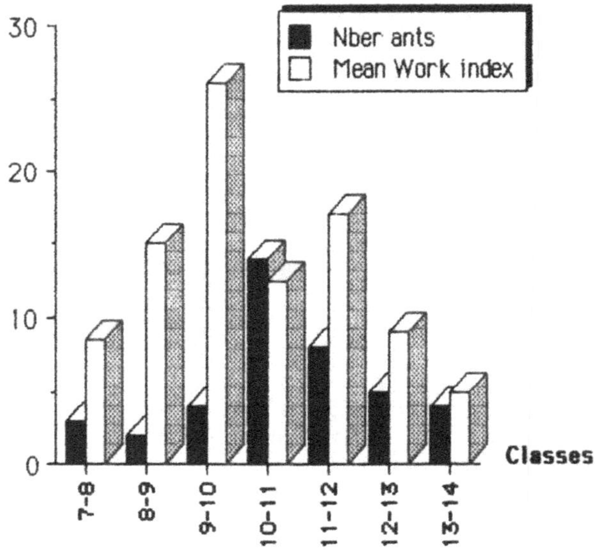

FIGURE 1. Frequency distributions of fresh weight and activity classes of *Acromyrmex octospinosus* large workers used in the behavioral tests. Black = weight classes in mg/ant; white = activity classes (mean work index- see text).

24 hour starvation period, remained for 6 hours in a Petri dish containing a piece of cotton wool soaked in a solution of: 9% glucose, 80g/l Tang (A dry chemical powder used to prepare an artificial beverage vaguely reminiscent of orange juice. Contains no orange at all. Highly appreciated, at least by leaf-cutting ants.), 2% neutral red, and the test ingredient. The estimated volume ingested was about 2ul. The neutral red control was essential as a check for possible repellency. Drug progression to the stomach was verified with neutral red added to the ingested solutions of a control batch.

Drug concentrations reached levels comparable to commercial formulations employed for medical purposes. They were measured as ug of active ingredient per ant. Several unknown additives were dispensed with the drug: in particular additives such as starch, talc, gums, cellulose and various glucanes.

Evaluation of the Behavioral Effects of Drugs

Measurement of motor activity is the accepted method for screening new stimulants or depressant compounds. Exploration, grooming and escape behaviors were used in this work.

Exploration. Exploration involves a broad category of behaviors, providing the organism with information about its external environment. The test used non-directed locomotion (i.e. generalized isotropic random walking activity).

TABLE 1. List of tested psychotropic drugs, classified according to Lambert (1979) as : Psycholeptics (Hypnotics- barbiturates or non- barbiturates; Tranquilizers- benzodiazepines; Neuroleptics- phenoriazines, thiazepines, buttrophenones) and Psycho-analeptics or Psychostimulants (tricyclic and iminpraminic derivatives). Five additional neurotropic substances are listed. French commercial names are given. * = tested behavior.

Chemical Name	Commercial Name	Class	Tested Behavior Explore	Groom	Escape
1- PSYCHOLEPTICS					
Butobarbitol	Soneryl	Barbit. hypnotic	*		
Pentobarbitol	Nembutal	" "	*	*	
Phenobarbitol	Gardenal	" "	*		*
Secobarbitol	Reposal	" "	*	*	
Thiopental	Nesdonal	" "	*		
Estazolam	Nuctalon	Non-Barb. Hypnotic	*		
Diazepam	Valium	" "	*		*
Chlordiazepoxide	Librium	" "			*
Nitrazepam	Mogadon	" "			*
Triarolam	Halcion	" "	*		
Chlorazepate	Tranxene	" "	*		
Chlorazepate + acepromazine	Noctran	" " + Phenothiazine	*		
Acepromazine	Vetranquil	Phenothz. (Aliphat.)	*	*	
Cloropromazine	Largactil	" "	*	*	*
Levopromazine	Nozinam	" "	*	*	
Haloperidol	Haldol	" (Piperaz)	*		
Propeciazine	Neuleptil	" "	*		
Thioridazine	Melleril	" "	*	*	
Clotiapine	Etumine	Thiazepine			*
Trifluperidol	Triperidol	Butyrophenone	*		
Potassium Bromide	-	CNS Depressant			*
2- PSYCHO-ANALEPTICS:					
Amineptine	Survector	tricyclic			*
Amitriptyline	Elavil	Imipraminic	*		*
Trimipramine	Surmontil	"			*
19560 RP	Experimental	Unknown			*
3- VARIOUS NEUROTROPIC SUBSTANCES:					
Gallamine	Tricuran	Phenolic ether	*		
Benzocaine	MS 222	Local Anaesthetic	*		*
Colchicine	-	Antipyretic Alkaloid			*
Serotonin (5-HT)	-	Neurotransmitter			*
Scopolamine	-	Anticholinergic			*

Inquisitive/inspective exploration (i.e. running to a source, antennating, biting) in response to various controlled stimuli (alarm pheromones, light, heat, wind, etc) was not successfully assayed, mainly because a high response variability required an excessive number of replicates.

All tests were conducted under strictly-controlled environmental conditions, e.g. red light, the ants' photoperiod, stable temperature, masked operator to avoid air disturbance and carbon dioxide excitement, attenuated substrate vibrations, etc. Under these conditions, one ant was allowed to explore a 15cm Petri dish (Fig. 2A) whose cover, sides and bottom were divided into 34 sectors. The number of explored sectors was counted over two minutes. The ant was then isolated for one hour in a smaller dish containing a piece of ant-free fungus garden. The procedure was repeated three times for each ant and the test repeated 12 times per day for five days (N=60 replicates). Subsequently, only 8 ants were tested per drug.

Grooming. It was observed (Weber 1972; Kermarrec et al. 1986) that leaf-cutting ants readily dig the agar substrate around a piece (1 ml) of fungus garden freshly introduced into a Petri dish (Fig. 2B). This stereotyped behavior was used as follows: 10 large worker ants (5 replicates) were kept in a Petri dish with only glucosated (9%) tap water for 24 hours before transfer to the agar-filled Petri dish containing the fungus garden. The cut agar was weighed after 48 hours and this number recorded per 100 mg of worker ants after mortality corrections.

As for the preceding test, drugs were dispensed perorally, by gentle forced ingestion of 2 ul of drug solution (or suspension) per ant.

Escaping. An electroacoustical automatic actograph (Turpin 1971) connected to a piezoelectrical transducer fixed on an external face of a closed box (4x2x2 cm) made of fine brass mesh (Fig. 2C). Vibrations (walking, biting) generated by 10 large workers was enhanced by the metallic substrate (Fig. 2C). The exploration behavior towards the new environment was evaluated.

A Petri dish bottom covered by a close-fitting brass grille (Fig. 2D) was also used for an actograph box. This induced obvious escape activity at the glass and metal boundary (Fig. 2D) where the ants persistently bit the metallic wires.

The final protocol consisted of 10 ants in the dish (Fig. 2D); and the vibration measurement was initiated 15 minutes later to avoid the initial worker excitement after handling. Two successive phases were recorded actographically: basic exploratory activity (B) from minute 15 to minute 60, and, from minute 60 to 75, escape activity (R) induced by a slight shifting of the metallic cover (Fig. 2E). Only the antennae could pass through the opening. This procedure combined both kinds of behavior and triggered the escape reaction (R). The tests were replicated 8 times for each ingested drug.

Due to high variability and a lack of normality of the results, all experiments were subjected to non-parametric statistical tests (Wilcoxon and U-test) according to Mann-Whitney (1907) and White (1951).

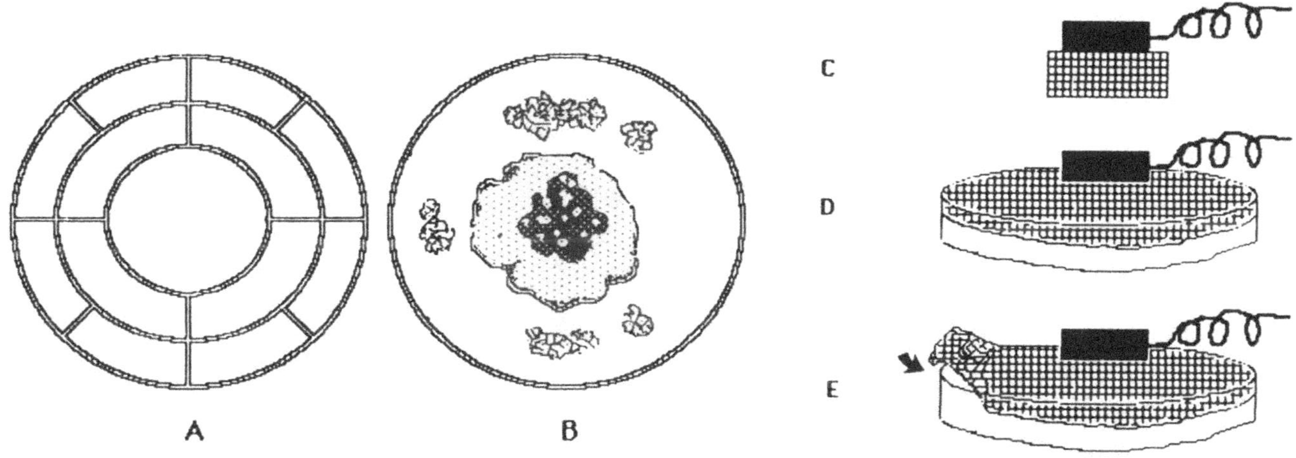

FIGURE 2. Description of the behavioral tests used. A: exploation test in an empty Petri dish with marked areas. B: grooming test in a Petri dish containing a piece of fresh fungus garden on solid agar; agar cut by the ants and to be weighed is gathered near the fungus. C: Exploratory and escape actgraphical test: a piezoelectrical transducer stuck on the brass wire cage detects the vibrations generated by the ants. D: Escape behavior is increased at the limit of the glass and brass wire. E: Test activity at a small slit opened by the operator (an improved situation D).

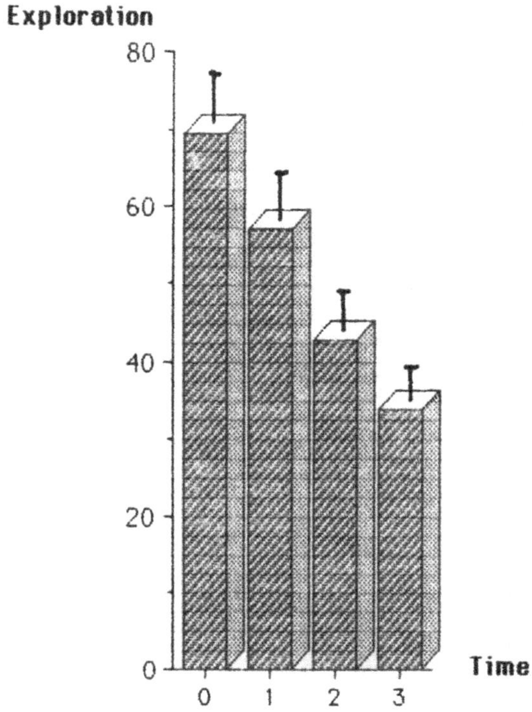

FIGURE 3. Exploratory test: number of sectors visited by one large *Acromyrmex octospinosus* worker during two minutes of each hour. Time in hours. Mean of 60 replicates; confidence intervals are drawn and all points are significantly different at $p<0.05$, Linear regression: $Y = 81.175 - 12.181X$ ($R = 1.00$).

RESULTS

Exploring tests

Control test. Pure water was dispensed perorally, on the basis of 60 replicates per hour (Fig. 3); only the third hour passed (at P: 0.09) a normality test. Individual behaviors vary greatly. The number of areas explored by an isolated ant decreased regularly hour after hour (All points were significantly different at P: 0.05).

Effect of ingested drugs. The exploring tests were conducted only during three time periods, eliminating time zero. Two opposite results may occur: deep depression of locomotion under phenobarbital (40 ug/ind), with recovery after 24 hours (Fig. 4), or hyperexcitation (Fig. 5) under estazolam (5ug/ind) and acepromazine (20ug/ind).

The effects of some 21 substances (Table 2) have been analyzed with this locomotion-exploration test and the results may be summarized as

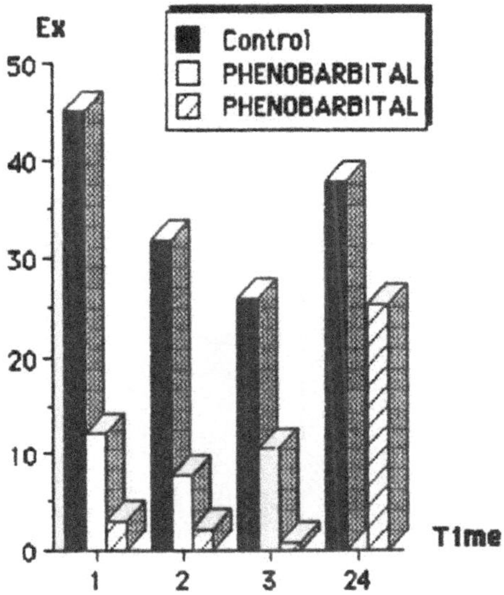

FIGURE 4. Inhibition of exploratory activity (Ex, in number of visited areas; Time in hours) of two batches of *Acromyrmex octospinosus* by phenobarbital at 40ug/ant per os and recovery of normal behavior after 24h. Mean of N=8 replicates at each hour; only one batch was compared to control at 24h.

follows. First, each test (and its control) was submitted to the summation of ranks (Wilcox on statistical analysis). Then, the differences between each rank pair (test minus control) were calculated (the more positive, the more hyperactive, and the more negative, the more depressed). Finally, an algebraic summation of the former differences (3 testing times) gave the general tendency of the results (Table 2) and their statistical significance.

Estazolam and acepromazine increased locomotion activity, while phenobarbital, thiopental, chlorazepate, haloperidol, thioridazine and benzocaine significantly depressed (at least once and at this dosage) the exploratory activity. Nevertheless, the drugs generally lowered activity, even if no significance could be found at P:0.05.

Overall, there were no significant differences from the control in 57% of the tests; 26% had one significant difference, 13% two and only 4% resulted in a significantly different response for the three behavioral test periods.

Grooming Test

Because of the importance of polyethism in evaluating the ants' activity, the interbatch homogeneity was checked: no significant difference occurred with the individual weight correlated to work.

Six drugs were tested by the oral route (Fig. 6). Pentobarbital

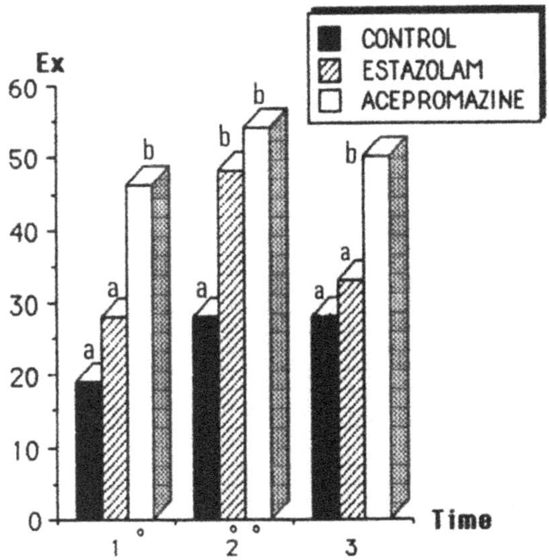

FIGURE 5. Exploration hyperactivated by Estazolam (5ug/ant) and Acepromazine (20ug/ant) for large *Acromyrmex octospinosus* workers. All differences significant at P0.07. Columns annotated with the same latter are not significantly different (P0.04) at successive test times. * = significantly different from control at each hour (P0.09).

(25ug/ant), secobarbital(50ug/ant), acepromazine (20ug/ant) and thioridazine (10 ug/ant) significantly depresses the ant's digging activity.

Levopromazine (25ug/ant) and chlorpromazine (22ug/ant) did not significantly depress the grooming activity.

Escape Test

The results of the two preliminary experiments described under "Material & Methods" are summarized in Figure 7 (A,B). The first actograph method (Fig. 2C; close wire box) gave (Fig. 7A) an initial steady exploration activity (A), followed by a five hour-long regular decrease of interest (B), ending in a final burst of escape activity preceding death by dessication.

The second actograph method (Fig. 2D) had a different starting point (phase A on Fig. 7B) characterized by a high and regular increase of escape activity. This step may be related to initial stigmergy facing a possibility of escape at the limit between two obviously different materials (glass/brass wire). The next phase (B on Fig. 7B) corresponds to 10 large worker ants progressively giving up ineffective team-work.

Finally, the combination of basic exploratory (phase A, Fig. 7A) and subsequent escape activity (phase A, Fig. 7B) was tested (Fig 2E) with control ants according to the protocol described under "Materials and Methods." The basic exploratory activity (from minute 15 to minute 60; phases B1 to B3, Fig.

TABLE 2. Effect of ingested drugs (2ul forced perorally) on the exploratory activity of *Acromyrmex octospinosus*. Dose = ug/ant; S = synthetic index (see text); No. sign. = number of significant differences with the control for the three test periods; D.eq.M. = dose equivalent for a man of 60 kg. Two hyperexcitation responses are underlined.

Chemical Name	Commercial Name	Dose	S	No. Sign.	D.eq.M.
Butobarbitol	Soneryl	100	+30	.	600
Pentobarbitol	Nembutal	50	-37	.	300
"	"	100	-23	.	600
Phenobarbitol	Gardenal	40	-141	2	240
"	"	40	-159	2	240
"	"	40	-136	3	240
Secobarbitol	Reposal	50	-23	.	300
Thiopental	Nesdonal	100	-102	2	600
Estazolam	Nuctalon	5	+75	1	30
Diazepam	Valium	40	-35	.	240
Chlorazepate	Tranxene	2	-94	.	12
Chlorazepate		300	-58	1	1800
+ acepromazine	Noctran	10	-53	.	60
Acepromazine	Vetranquil	20	+95	1	120
Cloropromazine	Largactil	50	-37	.	300
Levopromazine	Nozinam	50	-18	.	300
Haloperidol	Haldol	4	-75	1	24
Propeciazine	Neuleptil	20	-48	.	120
Thioridazine	Melleril	10	-94	1	60
Trifluperidol	Triperidol	2	-71	.	12
Amitriptyline	Elavil	100	-27	.	600
Gallamine	Tricuran	40	+23	.	240
Benzocaine	MS 222	200	-37	1	1200

8) remained constant and was consistent with the results of the preliminary experiment (Fig. 7A, phase A). The subsequent phase (R for "Reactivity") lasted 15 minutes and was dramatic, the ants trying to force the slit open at minute 60 (N=11 tests, P:0.01).

The R phase level appears correlated with that of the preceding B phase (Fig. 9): the ants' reaction to a sudden possibility of escape depends on the actual level of excitation. Only a few drugs (Table 3) affect this behavior: Diazepam completely extinguishes the exploratory and escape activities; on the other hand, phenobarbital, benzocaine and, to a lesser extent, serotonin induce hyperexcitation.

DISCUSSION

Obviously, a wide gap exists between ant and man. However, similarities do occur in the physiology of the nervous system. Floru et al. (1969) noted that psychotropic substances affect insects as in mammals.

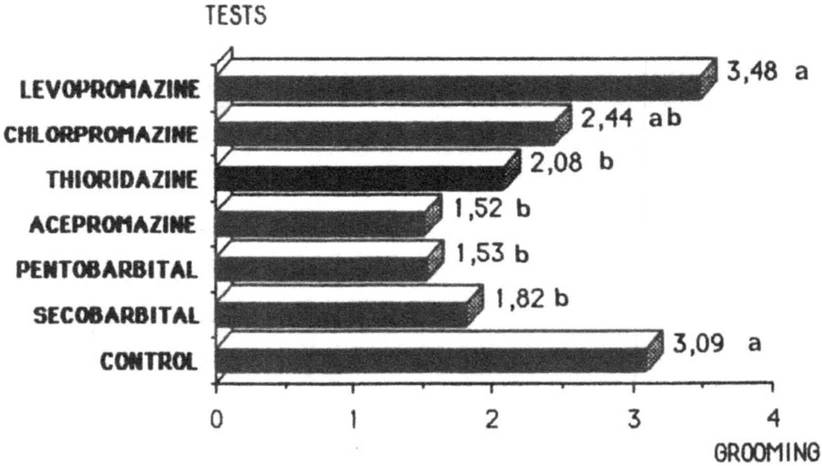

FIGURE 6. Grooming activity of large *Acromyrmex octospinosus* workers affected by some psychotropic drugs. Columns with the same letter are not significantly different at P0.05. The mean weight of cut agar is in grams.

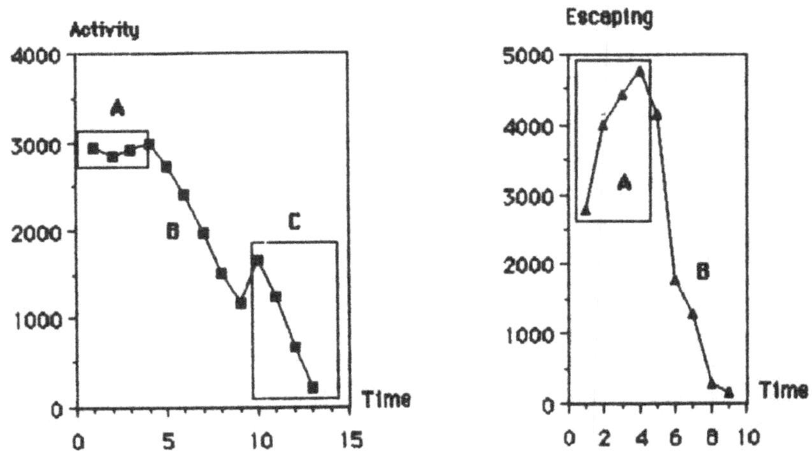

FIGURE 7. Exploratory activity of *Acromyrmex octospinosus* in a closed box (see Fig. 2). A: homogeneous metal box; B:metallic cover on Petri dishbottom. ESee text for an explanation of A, B, C. Time = hours; Activity = quantity of vibrations per unit of time.

Howse (1975), and later Burley et al. (1982) and Ibrahim et al. (1983), suggested the possible existence of cholinergic and aminergic fibers, monoamine neurotransmitters and even GABA-benzodiazepine receptors in insects. Testing insects psychopharmacologically in ways usually reserved for mice or rats may seem unrealistic when trying to elicit various "emotional behaviors." But a subject's response to novelty is a useful bioassay, since the normal neophobic response to such stimuli is withdrawal. However, cooperative social behaviors, which involve mutual cooperation of adults to help the society survive (grooming, nursing, fungus culturing) are positive parameters in psychometrical studies of attine ants.

Floru et al. (1989) emphasized the biphasic responses (hyperexcitation followed by somnolence) of chlorpromazine, chlordiazepoxide, pentobarbital

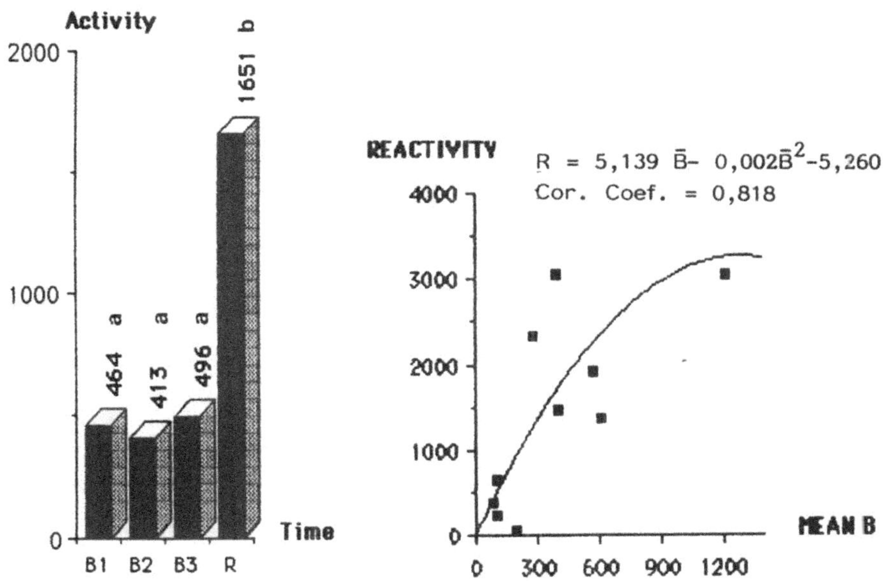

FIGURE 8 - 9. Left - Exploratory and escape activity of *Acromyrmex octospinosus* (in number of vibrations/15 min period). B1-3: basic exploratory activity; R: reaction of escape generated by a slight opening of the box. Non-parametrical test at P0.01. Columns with the same letter are not significantly different. Right - Escape activity (R, Fig. 8) depends on basic exploratory activity level (B, Fig. 8). Activities expressed as the number of vibrations per unit time.

TABLE 3. Effect of ingested drugs on the escape behavior of *Acromyrmex octospinosus* (actographical evaluation). Dose = ug A.I./ant; Mean B = mean value of the basic expoloring activity in a 45 minute period; R = reactivity or true escape activity; D,eq.M. = dose equivalent for a 60 kg man.

Chemical Name	Commercial Name	Dose	Mean B	R	D.eq.M.
Water (control)	water		458	1651	0
Phenobarbitol	Gardenal	20	1293	2693	120
Diazepam	Valium	20	9	35	120
Chlordiazepoxide	Librium	20	773	1791	120
Nitrazepam	Mogadon	20	528	1392	120
Cloropromazine	Largactil	20	360	1389	120
Haloperidol	Haldol	10	532	1581	60
Clotiapine	Etumine	20	614	1377	120
Potassium Bromide	-	20	344	1642	120
Amineptine	Survector	20	361	1025	120
Trimipramine	Surmontil	20	769	1405	120
19560 RP	Experimental	20	814	1978	120
Benzocaine	MS 222	20	1586	2170	120
Colchicine	-	20	316	1123	120
Serotonin (5-HT)	-	20	1399	1613	120
Scopolamine	-	0.25	569	1963	1.5

TABLE 4. Summary of the statisically significant effects of drugs on the behaviors of *Acromyrmex octospinosus*.

Active Ingredient	Dose ug/ant	Dispensing Method	Enhanced Activity	Depressed Activity
Pentobarbitol	25	perorally		Grooming
Phenobarbitol	40	perorally		Exploring
"	20	cotton wool	Escaping (B+R)	
Secobarbitol	50	perorally		Grooming
Thiopental	100	perorally		Exploring
Estazolam	5	perorally	Exploring	
Diazepam	20	cotton wool		Escaping (B+R)
Chlorazepate	300	perorally		Exploring
Acepromazine	20	perorally	Exploring	
"	20	perorally		Grooming
Haloperidol	4	perorally		Exploring
Thioridazine	10	perorally		Exploring
"	10	perorally		Grooming
Benzocaine	200	perorally		Exploring
"	20	Cotton Wool	Escaping (B+R)	
Serotonin	20	cotton wool	Escaping (B)	

and imipramine on hornet behavior (aggressiveness, cleaning, locomotion). They also found opposite responses to low and high dosages, the former generating unrest and hyperwakefulness. All psychotropic substances tested profoundly affected *Vespa orientalis*, even at the lowest concentrations.

We also found (Table 4) high responsiveness to lower doses of benzocaine and phenobarbital (enhanced escape at low dose as opposed to depressed exploration at higher dosage). Rioux (1983) described a similar effect, stating that apomorphine hyperexcites *Camponotus* sp at low dosages.

Moreover, some psychotropic substances affect individual behavior in ways different from social behavior. Apomorphine (op.cit.) increases individual acts (locomotion, self-cleaning, aggressivity) while lowering instances of trophallaxis, a major social altruistic behavior. With acepromazine, we showed that exploration occurred more frequently, while grooming behavior (protection of the ectosymbiotic fungus by a team of worker ants) was suppressed.

One pronounced result was the wide alteration of exploration and escape behaviors with diazepam, a major benzodiazepine. Benzodiazepines modulate GABA activity by cellular hyperpolarization, which results in a lowered CNS excitability (Iversen et al. 1977a,b). Major benzodiazepines--like chlordiazepoxide, diazepam and nitrazepam--may lower serotonin turnover (a CNS monoamine enhancing aggressiveness) and thus reduce behavioral suppression. Kostowski (1966) and Kostowski et al. (1975b) noted that after serotonin injection, *Formica rufa* showed a simultaneous increase in intraspecific aggressiveness and a decrease in intergenetic aggressiveness. For *A. octospinosis*, we demonstrated that exploration increased, but escape-oriented reactivity did not.

The literature presents many methodologies (ethotests, doses, substances, dispensing methods) representing a wide spectrum of invertebrate hosts (from flukes to ants). The results cannot be compared and often contradict each other. Nevertheless, the fact that several common psychotropic substances (such as benzodiazepines, phenothiazines or benzocaine) can impair general vigilance related behaviors sheds light on a possible new pest ant control strategy, as already proposed by Kermarrec et al. (1977). Frischknecht and Waser (1979) claim that even if drugged ants are welcomed back into the community, impairment of social behavior and of colony structure occurs when behavioral interactions or individual performances are altered by drugs. Manifestations of intraspecific aggressiveness (Floru et al. 1969) are also an interesting phenomenon because they constitute a distortion of fundamental functions designed to preserve the species.

Thus, a social insect's instinctive reflex behavior may acquire antisurvival values under the influence of drugs.

REMERCIEMENTS

À C. Juranville et B. Grenier, èléves-stagiaires IUT (1980) et ENITA (1983), pour leur collaboration technique; aux Docteurs J. Esnault et M. Manicom pour leurs conseils, interventions auprés des firmes pharmaceutiques et fournitures de psychotropes; à G. Febvay pour ses interventions toujours précieuses.

RÉSUMÉ

Les effets de barbituriques, benzodiazepines et autres drogues sont étudies chez *Acromyrmex octospinosus* sur des comportements liés a la vigilance: l'activite locomotrice exploratoire d'un milieu nouveau, l'intensité du comportement de fuite, et l'activité de groupe de protection du champignon symbiotique. Les diverses methodes et les résultats sont discutés.

Une profonde altération des comportements d'exploration et de fuite sont observés sous diazepam. *A. octospinosus* est également sensible aux substances suivantes: pentobarbital, phenobarbital, secobarbital, thiopental, estazolam, chlorazepate, acepromazine, haloperidol, thioridazine, benzocaine et serotonine.

REFERENCES CITED

ABALIS I. M., M.E. ELDEFRAWI and A.T. ELDEFRAWI. 1983. Biochemical identification of Putative GABA/Benzodiazepine Receptors in house fly thorax muscles. Pesticide Biochemistry and physiology, 20: 39-48.

BRILEY P.A., M.T. FILBIN, G.G. LUNT and J.F. DONNELLAN. 1982. Binding and uptake of glutamate and γ-aminobutryic acid in membrane fractions from locust muscle. In Neuropharmacology of Insects. Pitman, Ld. (Ciba fd. Symp. 88)pp. 176-198.

CENE-RIOX L., 1980. Action des amphétamines sur le comportement locomoteur de *Camponotus vagus*. D.E.A., Ecoéthologie, Rennes.

COOPER P. D. and L. UMEMOTO. 1971. Effects of drugs on the behavior of the Carpenter and, *Camponotus herculaenus*. Comparative and General Pharmacology, 2, 120-123.

FEBVAY G. B. HOSTACHY and A. KERMARREC. 1985. La lutte chimique contre la fourmi *Acromyrmex octospinosus* (REICH) : Test standardisé d'évaluation de l'effet retard de matières actives et nouvelle stratégie de lutte. Med. Fac. Landbouww. Rijksuniv. Gent. 50/2b, 723-734

FLORU L., J. ISHAY and S. GITTER. 1969. The influence of psychotropic substances on hornet behaviour in colonies of *Vespa orientalis* F. (Hymenoptera). Psychopharmacologia, 14, 323-341.

FRISCHKNECHT H. R. and P.G. WASER. 1978A. Actions of hallucinogens on ants *(Formica pratensis)*:I- Brain levels of L.S.D. and T.H.C. following oral administration. Gen. Pharmac. 9, 369-373.

FRISCHKNECHT H. R. and P.G. WASER. 1978b. II- Effects of amphetamine L.S.D. and Delta-9-tetrahydrocannabinol. Gen. Pharmac 9, 375,380.

FRISCHKNECHT H. R. and P.G. WASER. 1979. III- Social behavior under the influence of L.S.D. and T.H.C. Gen. Pharmac., II, 97-106.

GOSSWALD K., 1938. Uber den insektentotenden pilz *Beauveria bassiana* (Bals.) Vuill. bisher bekanntes und eigne Versuche. Arb. Biol. Reichsanstalt, Land Rorstwirts. 188, 3çç-452.

HOWSE P. E., 1975. Brain structure and behavior in insects. Ann. Rev. Entomol. 20, 359-380.

IVERSEN L. L., S.D. IVERSEN and S.H. SNYDER. (eds) 1977a. Handbook of psychopharmacology. Volume 7: Principales of behavioral Pharmacology. 453 p., Plenum Press, New York and London.

IVERSEN L. L., S.D. IVERSEN and S.H. SNYDER. (eds) 1977a. Handbook of psychopharmacology. 1979b. Volume 8: Drugs, Neurotransmitters, and behavior. 590 p., Plenum Press, New York and London.

KERMARREC A., G. MALATO and G. FEBVAY. 1977. Données nouvelles sur l'invasion de la Guadeloupe par *Acromyrmex octospinosus* REICH. (Formicidae, Attini). Nouv. agron. Antilles-Guyane, 3: 486-492.

KERMARREC A., G. FEBVAY and M. DECHARME. 1986. Protection of leaf-cutting ants from biohazards: Is there a future for microbiological control? pp 339-356. In Fire ants and leaf-cutting ants. Biology and management. C.S. Lofgren and R.K. Vander Meer (eds), Westview studies in insect biology, Westview Press, Boulder and London.

KERMARREC A. and H. MAULEON. 1975. Quelques aspects de la pathogenie d'Entomophthora coronata Cost. Kervork pour la fourmi-manioc de la Guadeloupe, *Acromyrmex octospinosus* (Formicidae: Attini). Ann. Parasitol. 50: 351-360.

KOSTOWSKI W., 1966. A note on the effects of some psychotropic drugs on the aggressive behavior in the ant *Formica rufa*. J. Pharm. Pharmac. 18: 747-749.

KOSTOWSKI W., J. BECK and J. MESZAROS. 1965. Drugs affecting the behaviour and spontaneous bio-electrical activity of the central nervous system in the ant Formica rufa. J. Pharm. Pharmacol., 17: 253-254.

KOSTOWSKI W. and B. TARCHALSKA. 1972. The effects of some drugs affecting brain 5 HT on the aggressive behavior and spontaneous electrical activity of the central nervous system of the ant *Formica rufa*. Brain Research, 38: 143-149.

KOSTOWSKI W., B. TARCHALSKA and B. WANCHOWICS. 1975a. Brain catecholamines, spontaneous bioelectrical activity and aggressive behavior in ants (*Formica rufa*). Pharmac. Bioch. Behav. 3: 337-342.

KOSTOWSKI W., B. TARCHALSKA-KRYMSKA and L. MARBOUSKA. 1975b. Aggressive behavior and brain serotonin and catecholamines in ants (*Formica rufa*). Pharmac. Biochem. Behav. 3: 717-719.

LAMBERT P.A. 1979. Pour une classification des psychotropes. Psychol. Med., 11: 1931-1939.

MANN A.B. and D.R. WHITNEY. 1947. Ann. Math. Statist. 18-50.

MANSOUR T.E. 1957. The effect of L.S.D., 5 hydroxytryptamine and related compounds on the Liver Fluke (*Fasciola hepatica*). Brit. J. Pharmacol. 12: 406-409.

MERCIER J. and S. DESAIGNE. 1964. Nouvelles observations sur l'influence exercée par divers médicaments psychotropes sur le comportement aggressif et le comportement sexuel de la Mante religieuse. C.R. Soc. Biol. (Paris) 158: 1535-1540.

MERCIER A.R. and R. MENZEL, 1982. The effect of biogenic amines on conditioned and unconditioned responses to olfactory stimuli in the honeybee *Apis mellifera*. J. Comp. Physiol. 145: 363-368.

RIBBANDS, C.R. 1950. Changes in the behavior of honeybees following their recovery from anaesthesia. J. Exp. Biol. 27: 302-310.

RIBBANDS, C.R., 1954. Nitrous anaesthesia does not encourage re-orientation of honeybees. Bee World. 35: 91-95.

RIOUX L. 1983. Influence de l'apomorphine sur les relations interindividuelles existant au sein de petits groupes de *Camponotus aethiops*. Thèse, Université de Rennes, UER Sciences de la Vie et de l'Environment, 130 pp. Ronéo.

SANKAR S., A. KOPELMAN, P. SCHILDER and E. GOLD. 1960. Effect of L.S.D. on larval growth. Nature. 187: 153-154.

SCHAEFFER E.E. 1941. A fungus of the family *Entomophthoraceae* found on sugar ant (*Componotus sp*). Bothalia. 4: 234-249.

SCHAFFER R. and P.D. REAGAN. 1981. Colchicine reversibly inhibits electrical activity in arthropod mechanoreceptors. J. Neurobiol. 12: 155-166.

SIMPSON J. 1954. Effects of some anaesthetics on honeybees: nitrous oxide, carbon dioxide, ammonium nitrate, smoker fumes. Bee World. 35: 145-155.

STEINHAUS E.A. and G.A. MARSHALL. 1967. Previously unreported accessions for diagnosis and new records. J. Invert. Pathol. 9: 436-438.

TAMANO N. and K. KURIAKI. 1961. Applicability of the silkworm for pharmacological studies on the drugs acting on the nervous system, antimitotics and insecticides. Arch. int. Pharmacodyn. 132: 49-59.

TURPIN P. 1971. Etude, réalisation et mise en oeuvre en biologie d'un actographe quantitatif. Thèse, Ecole Pratique des Hautes Etudes. Paris V.

VIDAL. 1988. (Dictionnaire), Office Pharmaceutique V., Editions VIDAL, 64ème Edition.

WEBER N.A., 1972. Gardening ants, the Attines. Mem. Am. Phil. Soc., 92, 146 pp.

WEIR J.S. 1957. Effects of anesthetics on workers of the ant *Myrmica*. J. Exp. Biol. 34: 464-468.

WHITE C. 1952. Biometrics. 8: 30.

WITT P. 1956a. Der Netzbau der Spinne als Test zur Prüfung zentral-nervös angreifender Substanzen. Arzneimittel-Forsch. 6: 628-635.

WITT P. 1956b. Die Wirkung von Substanzen auf den Netzbau der Spinne als biologischer Test. Berlin-Göttingen-Heidelberg: Springer.

61

Controlling Argentine Ants in Urban Situations

M.K. Rust and R.L. Knight

INTRODUCTION

The Argentine ant, *Iridomyrmex humilis* (Mayr), is a widespread pest of agricultural and urban situations in Mediterranean climates extending from 28 to 38°, north and south of the equator. *I. humilis* has received considerable attention as a secondary agricultural pest because it interferes with biological control programs and tends some scales and aphids that produce honeydew (Woglum and Borden 1921). It is also an important urban pest in Australia, especially Victoria, Western Australia and New South Wales (Whitehouse 1988), the United States (Ebeling 1975), as well as South Africa (Brain 1929). As far north in California as Eureka (Knight 1989), Argentine ants represent about 26% of all ants collected by pest control operators around structures.

Control recommendations for Argentine ants include treatment of nesting sites and application of residual chemical barriers and insecticidal baits (Mallis 1960; Ebeling 1975; Hadlington and Gerozisis 1985). Despite wide use of barrier treatments, there is a paucity of data concerning their effectiveness. Wright and Dupree (1984) reported that barrier sprays of 0.75 and 1% acephate provided significant reductions of ants, whereas acephate and propoxur scattered baits failed to provide control. However, baits have been an effective strategy for Argentine ant control. A community-wide campaign established in Los Angeles County (Ryan 1926) and Ventura County, Calif. (Dodds 1926) with 0.18% sodium arsenite bait in containers provided excellent control as long as the containers were serviced regularly. The use of toxic baits for Argentine ant control declined when organochlorine insecticides were developed because baits were more labor intensive (Gerhardt 1954). Baits containing hydramethylnon have shown promise in agricultural situations (Anonymous 1984) and avermectin was readily accepted by ants in laboratory tests (Baker et al. 1985). However, the major problem has been formulating toxicants into liquid sucrose baits that are highly preferred by *I. humilis* (Baker et al. 1985).

The primary objective of our research is to determine the potential efficacy of pyrethroid sprays, novel insecticides, and formulated baits against the Argentine ant.

CHEMICAL BARRIER TREATMENTS

The efficacy of barrier treatments to control Argentine ants was evaluated around single-family homes in three southern California counties: Fallbrook, San Diego County; Riverside, Riverside County; and Alta Loma, Rancho Cucamonga, San Bernardino, and Upland, San Bernardino County.

Field Applications

Two different aqueous sprays and one granular insecticide treatment were applied around each of three homes per treatment. The sprays were chlorpyrifos (Dursban 50 WP, Dow Chemical Co., Midland, MI) and cyfluthrin (Tempo 20 WP, Mobay Chemical Corp., Kansas City, MO). The granular insecticide was diazinon (Diazinon 5 G, Ford's Chemical & Service, Inc., Pasadena, TX).

Aqueous sprays were applied via three different methods to approximate actual practices used to control ants outdoors. The spray application referred to as "label" was applied as a barrier to 0.3-0.5 m up the walls or foundation and a 3-m swath of the lawn, flower beds or concrete immediately adjacent to and around the structure. For the second method, designated "PCO1," the same amount of spray was applied but in a narrower band along 0.3-0.5 m of the foundation and 0.6 m of the soil surrounding the structure. The third type of application, referred to as "PCO2," varies more than the other two methods, about 11-15 liters of spray being applied at each home. The spray was applied as a 1-m narrow band immediately surrounding the structure and the outside perimeter of the property. Nest sites and other areas likely to be frequented by ants were spot treated. Baseboards and intersections in the kitchens, bathrooms and other areas likely to be visited by ants were lightly sprayed.

Diazinon 5 G was evenly spread across a 3-m band adjacent to the house at a rate equivalent to 1.0 Kg per 93 m^2. Driveways, walkways, and large concrete surfaces were not treated. The application equipment used to apply the sprays varied somewhat because different cooperators assisted us at each site. For the PCOl method, cyfluthrin 20 WP and the label applications of chlorpyrifos 50 WP were applied with a 95-liter mechanically agitated Broyhill 1760B-2P Hypro 2 piston pump sprayer equipped with a Green Guard JD9-C spray gun at 2813 g/cm^2 (40 psi). We applied cyfluthrin label applications with a 3.8-liter stainless steel compression sprayer equipped with a fine-fan nozzle at 1407 g/cm^2 (20 psi). PCO2 type applications were made with a Cooper-Peglar CP-15 backpack sprayer equipped with a coarse-fan nozzle at 2110 g/cm^2 (30 psi).

Monitoring

Live trap captures determined the severity of ant infestations and monitored the efficacy of all treatments. The traps consisted of an inverted 5 by 5 cm plastic box lid provisioned with approximately 5 ml of honey to attract ants. Ants freely left the stations and returned with nestmates. The trap stations were placed around the perimeter of buildings next to the foundation. One station was placed at each corner and another was placed midway along each side of the home, a total of eight stations. Stations were

TABLE 1. Performance of barrier sprays and granular insecticides against Argentine ants.

Treatment	Method[a]	Home	X Initial no. ants per trap	% reduction of ants[b]	
				7 d	30 d
Cyfluthrin 20wp	PCO1[c]	1	437	100 ***	86.0*
		2	630	22.5	15.6
		3	620	60.3*	60.5*
Cyfluthrin 20WP	PCO1[d]	1	686	74.2**	54.4***
		2	393	32.3	0
		3	370	46.2	0
Cyfluthrin 20WP	PCO2	1	73	100 *	100 *
		2	30	100 *	93.3*
		3	405	100 ***	99.8***
Cyfluthrin 20WP	label	1	187	100 ***	44.9*
		2	328	99.7***	100 ***
		3	234	100 **	100 ***
Chlorpyrifos 50wp	PCO2	1	221	100 ***	100 ***
		2	515	100 **	100 **
		3	327	92.4**	100 ***
Chlorpyrifos 50wp 0.03%	label	1	398	99.0**	98.7**
		2	246	63.4	13.0
		3	429	93.7***	36.4
Diazinon 5G 1 kg/93 m^2		1	369	68.6**	38.2
		2	562	51.4*	98.6**
		3	212	38.2	0

[a]"PCO1" narrow band of spray around the home. "PCO2" narrow band around the perimeter of the home and property with some spot treatments. "Label" a 3 m wide band around the home. [b]Treatments followed by asterisks are significant at the following levels: * a = 0.05, ** a = 0.01, *** a = 0.001 (Wilcoxon's signed-ranks test). disc. = discontinued. [c]Treatment rate is based on 9.5 g AI/ 93 m^2. [d]Treatment rate is based on 19.0 g AI/ 93 m^2.

placed in the field between 1500 and 1900 hours and retrieved the following morning between 0800 and 1000 hours. The bottoms of the boxes were placed over the lid and taped together to prevent escapes. They were returned to the lab and the ants counted. In heavily infested areas, some stations were completely empty by 0800 hours and few ants remained, thereby grossly underestimating the population levels. Therefore, we substituted the average value of the five highest counts from non-empty stations for the counts at empty stations. The average counts were 708 ants for Fallbrook and 826 for homes in Riverside and San Bernardino Counties. Stations were placed out 7 and 30 days post-treatment and data analyzed using a Wilcoxon's signed-rank test (Sokal and Rohlf 1969).

RESULTS

The cyfluthrin sprays applied as the PCO2 or label methods provided significantly reduced the number of ants trapped at 7 and 30 days (Table 1). One home treated with the PCO2 style was re-treated because of continuing ant problems. The chlorpyrifos spray applied as the PCO2 method provided significant reductions at day 7, but it was necessary to re-treat all of the treatments before 30 days because of recurring ants and homeowner complaints. Two homes were re-treated with cyfluthrin and the other with chlorpyrifos; the second treatments successfully controlled the ants.

The thoroughness of barrier treatments appears important in preventing ants from entering homes. Chemical barriers of 2% chlorpyrifos painted around tree trunks prevented ant foraging as long as tree limbs did not touch the ground or other trees (Moreno et al. 1987; Philips et al. 1987). The residual deposits of chlorpyrifos applied to concrete, stucco or soil may not have lasted as long as similar deposits on wood or other cellulose debris. Wagner and Strawn (1980 a,b) found that chlorpyrifos deposits on filter paper produced 90% knockdown within 40 minutes for up to six months whereas similar deposits on concrete failed to produce 90% knockdown at one month. The rapid breakdown of chemical barriers on these surfaces might explain some of the failures.

Two of the three diazinon granule applications provided statistically significant reductions at day 7, but only provide significant reductions for 30 days. Moreno et al. (1987) found that diazinon 15 G barriers were only effective if a continuous uninterrupted ring was placed around the citrus tree. It is likely that ants found access to the monitoring stations and the structure across untreated concrete areas or from nests in wall voids or under slabs.

Baiting

The acceptability and potential efficacy of proprietary and experimental baits were evaluated in specially designed arenas against workers and queens (Knight 1989). Field tests were conducted to verify laboratory results for a barrier treatment of diazinon 5 G and a granular bait formulation of hydramethylnon.

Baits

Consumer baits included Drax Ant Kil Gel (5% boric acid, R Value, Inc., Smyrna, GA), Grants Kills Ants (0.46% arsenic trioxide, Grant Laboratories, Inc., San Leandro, CA) and Combat Ant Bait stations (0.9% hydramethylnon, American Cyanamid Co., Wayne, NJ). The laboratory-formulated baits were 0.13% mirex in peanut butter, 0.13% mirex and 0.5% sulfluramid (Griffin Chemical Co., Valdosta, GA) applied to granulated silkworm pupae (provided by American Cyanamid Co.), and 0.2% sodium arsenite in a honey bait (Mallis 1960). Other experimental baits included a 0.9% hydramethylnon in granulated silkworm pupae bait (American Cyanamid Co.) and two peanut butter-based ant baits (0.03% chlorpyrifos and 0.5% sulfluramid, S.C. Johnson

& Son, Inc., Racine, WI). The controls were untreated peanut butter and a 1:1 honey:water solution.

Laboratory Baiting Studies

Colonies of 200 workers and 1 queen were established in the laboratory (Knight 1989). They were allowed to acclimate for one week before the tests. The number of workers per colony was determined prior to each test. We did not use any colony with a mortality rate over 25% or with a dead queen.

To begin each test, 0.15 g of toxic bait or 0.15 g of the toxic bait and 0.15 g of non-toxic bait base were placed into each feeding arena. The ants fed on the baits for three days before the baits were removed and replaced with the standard laboratory diet of honey/water. Each toxicant was replicated four times with and without the alternate untreated food, with the percent mortality determined at 1, 3, and 10 d post-baiting. A two-way ANOVA was used to analyze the laboratory baiting results. The means were separated using Tukey's HSD (Tukey 1953).

Field Tests

Three apartment buildings in Grand Terrace, Cal. with severe Argentine ant infestations were used to test the efficacy of a granular formulation of 0.9% hydramethylnon and a diazinon 5G barrier treatment. The degree of ant populations and efficacy of treatments were determined by the method described above. Post-treatment ant counts were taken at 2, 30, and 60 d.

The three buildings were randomly assigned to each of the following treatments: 3.4×10^{-4} kg/m^2 hydramethylnon 0.9 G, control (no treatment), and 1.5×10^{-2} kg/m^2 diazinon 5 G. The hydramethylnon-treated silkworm pupae granules were placed in small piles along ant trails at several locations around the building's foundation. The diazinon 5 G was evenly broadcast across a 3-m barrier around the buildings. Sidewalks and driveway areas were not treated.

RESULTS

Even though hydramethylnon mirex formulated in granulated silkworm pupae provided equivalent worker mortality, mirex still killed more queens (Table 2). Queen-kill for mirex was almost always observed near the end of the observation period. Hydramethylnon exhibited toxic effects much faster than did mirex in our laboratory tests against both queens and workers. It killed 36% of workers in our colonies within 24 h, compared to 1% kill for mirex. The longer delay with mirex increased the liklihood that toxicant was passed to more workers and queens.

Sulfluramid, a relatively new toxicant, has proven successful in baits for cockroach control. We tested two bait formulations of sulfluramid, one in a peanut butter base and the other in granulated silkworm pupae; both yielded poor mortality. Chlorpyrifos was tested in a peanut butter base similar to the one used for sulfluramid. The chlorpyrifos bait produced 50% mortality and killed more queens than did any other toxicant with the exception of mirex.

TABLE 2. Acceptability and toxicity of baits against Argentine ants in laboratory arena tests.

Toxicant	% AI	Bait Base[a]	% Workers Dead at Day 10[b]	No. Dead Queens	Avg. Time to Kill Queen (days)
Hydramethylnon[c]	0.90	SP	99 A	0	-
Mirex	0.13	SP	97 A	5	10
Mirex	0.13	PB	79 AB	2	10
Chlorpyrifos	0.03	PB	50 BC	3	3
Hydramethylnon[d]	0.90	SP	49 BC	1	2
Sodium arsenite	0.20	SW	43 CD	1	4
Sulfluramid	0.50	PB	39 CDE	1	1
Arsenic trioxide	0.46	SW	30 CDE	0	-
Sulfluramid	0.50	SP	26 CDE	0	-
Peanut butter	-	PB	20 CDE	0	-
Boric acid	5.00	SW	14 DE	0	-
Honey	-	SW	7 E	0	-

[a]The following abbreviations for bait bases: PB = peanut butter, SP = granulated silkworm pupae, and SW = sugar or other sweet substance. bn=8. Percentages followed by the same letter are not significantly different (P > 0.05; Tukey's HSD). cGranulated silkworm pupae base product. [d]Silkworm pupae base bait from bait stations.

There was no indication that chlorpyrifos exhibited delayed toxicity because the workers and queens usually died within 72 hours.

The two arsenic baits and boric acid bait were not readily accepted, and killed less than 50% of the workers. This was surprising because of the reported success of arsenic compounds (Ryan 1926; Woglum and Borden 1921).

The hydramethylnon bait gave fast ant reduction, similar to our laboratory findings. Some ants were observed trailing from the untreated apartment building to the ant traps at the hydramethylnon treatment at day 30, but there were no further complaints. The diazinon 5G application produced no significant reductions in ant counts at day 2 (Table 3). At 30 and 60 days, however, a significant number of ants were killed.

DISCUSSION

We achieved the most consistent control when we applied sprays as bands around the structure and perimeter of the property, or as spot treatments to likely nest sites. Our data tends to support the use of barriers applied in a grid pattern as recommended in Australia (Whitehouse 1988: Hadlington and Gerozisis 1985). Ant trials and nests are more likely to be encountered with increased coverage. In addition, the field data corroborates Knight and Rust's (In Press) laboratory findings that cyfluthrin WP was more repellent than the chlorpyrifos WP and likely to provide a more repellent barrier in the field.

TABLE 3. Field performance of hydramethylnon bait formulated in silkworm pupae and diazinon 5G applied around apartments infested with Argentine ants.

	X no. ants per station at day[a]			
Toxicant	pre-treatment	2 days	30 days	60 days
Diazinon 5G	216	311	0*	2*
Hydramethylnon	237	1*	19*	2*
Control	427	411	590*	224*

[a] Data within columns followed by asterisk are significantly different than pre-treatment counts $P < 0.05$ (Wilcoxon's signed-ranks test).

Hydramethylnon showed promise as potential toxicants for Argentine ant baits. However, there are still major problems with formulating acceptable liquid baits.

ACKNOWLEDGMENTS

We would like to thank James Smith and Donald Reierson for their critical review and helpful suggestions, and Sue Brown for her assistance in counting traps.

LITERATURE CITED

ANONYMOUS. 1984. Toxic bait to control Argentine ant field tested. Citrograph 69(7): 188.

BAKER, T.C., S.E. VAN VORHIS KEY and L.K. GASTON. 1985. Bait-preference tests for Argentine ant (Hymenoptera: Formicidae). J. Econ. Entomol. 78(5): 1083-1088.

BRAIN, C.K. 1929. Insect pests and their control in South Africa. Die Nasionale Pers Beperk, Cape Town.

DODDS, T. 1926. How ant poison is handled in Ventura Co. control work. Calif. Citrograph 11: 446-447.

EBELING, W. 1975. Urban entomology. Univ. Calif. Press Div. Agric. Sci., Berkeley. 695 pp.

GERHARDT, P.D. 1954. Argentine ant control with sprays and bait in two California lemon groves. J. Econ. Entomol. 47(4): 591-593.

HADLINGTON, P. and J. GEROZISIS. 1985. Urban pest control in Australia. New South Wales University Press, Kensington NSW Australia. 278 pp.

KNIGHT, R.L. 1989. Urban ant pests of California and pest management strategies for the Argentine ant, *Iridomyrmex humilis* (Mayr) (Hymenoptera: Formicidae). M.S. Thesis, Univ. California, Riverside, Cal. 105 pp.

KNIGHT, R.L. and M.K. RUST. 1989. Repellency and efficacy of various insecticides against foraging workers in laboratory colonies of the Argentine ant, *Iridomyrmex humilis* (Mayr) (Hymenoptera: Formicidae). J. Econ. Entomol. (in press).

MALLIS, A. 1960. Handbook of pest control. Mac Nair-Dorland Co. 3rd Edition. New York. 1132 pp.
MORENO, D.S., P.B. HANEY, and R.F. LUCK. 1987. Chlorpyrifos and diazinon as barriers to Argentine ant (Hymenoptera: Formicidae) foraging on citrus trees. J. Econ. Entomol. 80(1): 208-214.
PHILIPS, P.A., R.S. BEKEY, and G.E. GOODALL. 1987. Argentine ant management in cherimoyas. Calif. Agric. 41: 8-9.
RYAN, H.J. 1926. Community control of Argentine ants in Los Angeles County citrus orchards. Calif. Citograph 11: 207,222,223.
SOKAL, R.R. and F.J. ROHLF. 1969. Biometry. Freeman, San Francisco. 776 pp.
TUKEY, J.W. 1953. The problem of multiple comparisons. Princeton Univ., Princeton.
WAGNER, R.E. and A.J. STRAWN. 1980a. Effectiveness of insecticides applied to concrete for Argentine ant control. Insecticide and Acaricide Test 5: 210-211.
WAGNER, R.E. and A.J. STRAWN. 1980b. Relative toxicity and residual stability of insecticides for Argentine ant control. Insecticide and Acaricide Test 5: 211.
WHITEHOUSE, S. 1988. A review of Argentine ant control in Western Australia. Agric. Protection Board WA, Baron-Hay Court, South Perth. 29 pp.
WRIGHT, C.G. and H.E. DUPREE, JR. 1984. Insect control around homes. Pest Control Technology 12: 58-60,62.
WOLGUM, R.S. and A.D. BORDEN. 1921. Control of the Argentine ant in California citrus orchards. UDSA Bulletin No. 965.

62

Health Aspects and Control of *Monomorium Pharaonis*

Wd. Eichler

INTRODUCTION

In low numbers and certain urban settings (i.e., apartments), pharaoh's ants may not be a noticeable pest. However, in other situations, pharaoh's ants are a dangerous medical threat to man. I reached this conclusion after decades of personal experience in Central European hospitals. During this time, I also learned how difficult it is to control the pharaoh's ant. This paper will discuss the health problems posed by *Monomorium pharaonis*, followed by methods of control.

HEALTH ASPECTS

The main problems caused by pharaoh's ants are: (1) skin irritation, (2) skin lesions and (3) germ transmission. Stings, which some might consider a problem, are infrequent and generally mild. I will also discuss the ants' affinity for hospitals and surgical instruments, along with the significance of social hygiene.

Skin Irritation

This results when the ants crawl under bandages, including gypsum bandages. The ants cause insufferable itching by feeding around the wounds. Such situations are often reported. For example, late one night I was called to a baby's bed in a hospital. The nurse had observed that the baby had been spitting for thirty minutes. Not long after, its bed was flooded by ants, even though a short time earlier the room harbored few ants (probably scouts). Analogous situations can occur in homes. A bedridden woman was brought home from hospital and, after only a few days, her bed was overrun by thousands of pharaoh's ants. Again, the ants were rarely observed in this room prior to the incident although they had been present in the flat for many years. This remarkable example demonstrates that the normally harmless ants became harmful to the inhabitants as soon as it was to their advantage to do so.

Skin Lesions

The ants sometimes feed directly on the undamaged skin of new-born babies and patients under heavy sedation. They occupy the incubators of premature babies, where they may cause lesions. Steinbrink (1978) observed them in Rostock (GDR) biting babies' cheeks, lips, neck folds and the backs of ears. In Riga (USSR), new-born children had been attacked by pharaoh's ants immediately after birth. The ants are more frequently found on freshly-operated persons, especially patients with head, decubitus and burn wounds. The special attraction to burn wounds may be due to the liver ointment normally applied to these wounds; liver is a well-known super-attractant of pharaoh's ants. Sweat, saliva and wound secretions are especially attractive.

Germ Transmission

Pharaoh's ants, as a result of their activities, serve as disease vectors. They collect germs from their surroundings or by consuming the bodies of dead wasps, cockroaches and *Lucilia* flies. They can carry a broad spectrum of germs. I have found the following germs on pharaoh's ants from hospitals (obtained under sterile conditions): Greening Streptococci, hemolyzing Streptococci, Enterococci, *Micrococcus pyogenes* var. *aureus*, *Micrococcus pyogenes* var. *albus*, *Proteus vulgaris*, *Escherichia coli* and *Alcaligenes faecalis*. Other authors have observed *Salmonella* spp., *Pseudomonas aeruginosa*, *Clostridium* sp., *Bordetella bronchi-septica* and mold fungi (i.e. *Mucor*). The ants have also spread cat virus disease (panlevcopenia, belonging to the parvoviruses group) and probably transmit digitalmycoses.

Agar plate tests proved that the pharaoh's ant can transport these germs to other places and deposit them on other surfaces. Thus, germs can be transferred to surgical instruments or onto the layers of "sterile" linen used in operation halls. Surgeons reported some cases in which pharaoh's ant interrupted operations.

I have constructed some hypothetical--but highly probable--infection or germ-spreading chains for pharaoh's ants in connection with hospitals (Eichler 1976). For example, contamination from waste water of a tuberculosis hospital is passed to crows who in turn contaminate a garbage depot, where flies become contaminated and enter hospital rooms. The dead flies are eaten by the ants. Many other scenarios can be envisioned.

Of extreme significance is the observation of pharaoh's ants on the oxygen supply's gauze filter in a hospital's surgical guard station. This filter prevented the direct transport of the ants into the patient's lungs. But it did not prevent the possibility of germs being blown off the ant and sucked into the patient's mouth. Some fatal cases of septic pneumonia which could not otherwise be explained had occurred at this guard station. They appear to be a direct result of the ants' presence. Other observers have seen ants gnaw through the air valve's fiberglass and Teflon-occlusion.

Affinity to Hospitals and Surgical Instruments

Pharaoh's ants are attracted to hospitals because of five factors: clean walls, excellent year-round environmental conditions, an abundance of warm spots for optimal brood development, interesting odors and sufficient food.

The ants avoid dirty walls, probably because they have difficulty moving on these surfaces; they move well on tile. Thus, in one sense the presence of pharaoh's ants indicates good sanitary conditions.

A consistent, warm environment, numerous warm spots and adequate food also allow the ants to settle in modern apartment houses. In hospitals, the ants appear to prefer pharmacies, birth wards, operation halls or the drug cupboards of station rooms for the same reasons. These areas within the hospital--a prime location already--provide ideal conditions for colony survival and growth.

The ants' affinity for surgical instruments is an extension of their preference for hospitals. They typically crawl under instrument box lids, especially if the box and instruments have just been sterilized. Once inside, they can be found in the most unexpected places; several observers have reported ants in blood transfusion tubes, inside injection syringes and in "sterile" glucose bottles "pro injectionem."

Social Hygiene Significance

Pharaoh's ants influence people's lives in small but important ways. Once I entered an administrative building overrun by ants. Their base was the restaurant kitchen in the building's basement. For weeks, the employees discussed the ants or made shallow experiments with them. This cut their work efficiency and total output significantly.

The health risk is most pronounced in hospitals. When pharaoh's ants invade a hospital, they bring that institution's standards of hygiene into question; usually, a great deal of time and money is required to rectify the situation. Control and eradication of the pharaoh's ant requires persistance and lots of patience.

CONTROL AND ERADICATION OF *MONOMORIUM PHARAONIS*

Successful control of pharaoh's ant often depends on the care taken in choosing a methodology. If suddenly confronted with a mass invasion of the ant, one could spray any contact insecticide to reduce the problem for a few days. Immediate protective results can also be gained by placing the legs of furniture in jars filled with water and detergent or by smearing vinegar on areas where ants are trailing.

However, these methods will not eradicate the colonies. Our monograph (Berndt and Eichler 1987) analyzed all methods and/or poisons determined to be valuable against pharaoh's ants. I believe that insecticide spraying for pharaoh's ant control is a form of malpractice and should be discouraged. In this section, only methods aimed at eradication will be discussed. All eradication methods use poison baits (see Williams this volume); such baits have three main components: the bait itself, the toxicant and the strategy.

The Bait

This substance should be preferred by the ants but not currently foraged by them. In bakeries, ants prefer proteins, while where proteins are abundant, they should be attracted to carbohydrates (i.e. sugar). The previously-mentioned liver, and placenta, are excellent fresh baits, but these substances decay rapidly in warm rooms and should therefore be renewed daily or even twice a day. It is important to develop and evaluate several baits with long-lasting effectiveness. Dry baits such as ground-nut butter and comparable substances are long-lasting. To discover--by trial and error--the best bait for each special infestation is the secret of successful pharaoh's ant eradication. From experience, it is evident that baits of excellent attractance in one region are sometimes refused in another region. In addition, the type of bait should be changed every two or three weeks.

The Toxicant

The toxicant in the bait must be strong enough to kill all the ants, but slowly, so that foraging workers can bring sufficient amounts to the nest and transmit it to other workers, the brood and the queen. This toxicant must be environmentally safe and ideally affect only the target species.

Sodium arsenate, from personal experience, is a reliable toxicant. The use of sodium arsenate and my strategy in employing it relies on Karl Gösswald's pioneering work (see Berndt and Eichler 1987), but this method was further developed under very different practical conditions. The strategy has a 100% success rate.

The bait formulation consists of four parts meat base to one part product, "Radox D 1," called "Ameisenfresslack" (VEB Chemierax, GDR). It contains 3.7% sodium arsenate and kills colonies by sterilizing the queen.

Chlordecone also works well, but has been banned because of carcinogenicity. New studies suggest juvenile hormone mimics Methoprene and Hydroprene could be effective, but definitive studies were lacking until recently (Eichler 1989).

Hasty conclusions about the efficacy of new bait poisons against pharaoh's ants should not be drawn. Secure testing is extraordinarily difficult and two or three trials are not adequate.

Toxicants against fire ants (*Solenopsis* spp.) seem equally effective against pharaoh's ants and vice versa (see Williams this volume). These two ants species are close relatives taxonomically.

The Strategy

Thoughtful control of pharaoh's ant requires careful positioning of the poison baits and steady observation of the results. It is worth the time to first lay out non-insecticide test baits and after two or three days replace them with the poisoned baits. This will help determine which bait (food) material currently attracts the ants. Baits should be located at loci where the ants will immediately find them. Such positions include wash basins. Investigation of ant trails may also lead to better bait placement. Furthermore, ants should be denied access to other food sources for the duration of the eradication attempt. Any possible alternative food sources should be removed.

The general rule is two poison bait boxes per 15 to 20 square meters in locations that allow easy inspection and uncomplicated renewal of the bait substance (see Berndt and Eichler 1987 for details).

After three weeks, the number of ants frequenting the baits diminishes and after the sixth week, the colony may be dead. But only after a year of such treatment and inspection can eradication be assured.

Reinfestation of the pest can be a major problem. Sometimes pharaoh's ant is permanently reintroduced; for example, from bakery goods, packages of transported food or drugs. Chocolate and cakes may also be heavily-infested - even cheese, furniture or fresh linen. If such chains of introduction are not broken, all efforts against pharaoh's ants will fail to achieve total control.

In our monograph, there is a detailed calculation of sodium arsenate's toxicity (pp. 166-167). It can serve as a model for future strategies, although we did not give suggestions for waste disposal following the experiment/procedure.

The above represents my own decades-long experiences with control and eradication of *M. pharaonis*. Local conditions may vary greatly and, again, I stress the wisdom of regional, localized solutions rather than a generalized attack, especially considering pharaoh's ant's multifarious behavior. The papers of Berndt and Eichler (1975) and of Eichler and Kleinsorge (1972) provide other details on pharaoh's ant eradication and serve as useful guides.

LITERATURE CITED

BERNDT, K.P. and EICHLER, Wd. 1987. Die Pharaoameise, *Monomorium pharaonis* (L.) (Hym., Myrmicidae). Mitt. zool. Mus. Berlin 63: 3-168 and plates I-XIV.

EICHLER, Wd. 1964. Gesichtspunkte der Gesundheitsschaedlichkeit von Insekten unter besonderer Beruecksichtigung der Pharaoameise. Wiss. Z. Humboldt-Univ. Berlin math.-nat. 13: 113-118.

EICHLER, Wd. 1973. [Die Affinitaet der Pharaoameise zu Operationssaelen und chirurgischen Instrumenten]. Trudy XIII Mezhd. ent. Kongr. Moskva 1968 3: 361-362.

EICHLER, Wd. 1976. Nachweismethoden des Auftretens von Pharaoa-meisen. Z. ges. Hyg. 22: 209-212.

EICHLER, Wd. 1978. Die hygienische Bedeutung der Pharaoameise. Memorabilia zool. 29: 23-29.

EICHLER, Wd. 1989. Zur Mittelanwendung in der Pharaoameisenbekaempfung. Prakt. Schaedlingsbekaempf. 14: 116-117.

EICHLER, Wd. and KLEINSORGE, W. 1970. Rezeptur zur Pharaoameisenbekaempfung im Arsenfrasskoederverfahren. Angew. Parasitol. 11: 47-48.

EICHLER, Wd. and KLEINSORGE, W. 1972. Pharaoameisenbekaempfung in Krankenhaeusern. Angew. Parasitol. 13: 213-225 and 14: 23-43.

EICHLER, Wd. and NEUBERT, R. 1963. Orientierende Keimueberpruefung gesundheitsschaedlicher Insekten. Angew. Parasitol. 3: 38-46.

EICHLER, Wd. and SCHEUER-KARPIN, R. 1962. Bakteriologische Befunde bei Pharaoameisen. Dtsch. Gesundheitswes. 17: 1851-1868.

STEINBRINK, H. 1978. Pharaoameisen auf einer Fruehgeburtenstation. Angew. Parasitol. 19: 28-31.

63

Effects of Fenoxycarb Baits on Laboratory Colonies of the Pharaoh's Ant, *Monomorium pharaonis*

D.F. Williams

INTRODUCTION

The pharaoh's ant, *Monomorium pharaonis* (Linnaeus), probably originated in North Africa and the Middle East, but has spread via commerce to most cosmopolitan areas of the world (Wheeler 1910). In most regions, the Pharaoh's ant is a major indoor pest (Edwards 1986). However, it nests outdoors in tropical climates and persists in temperate regions where the temperature is artificially maintained, such as refuse dumps (Kohn and Vlcek 1984). In nontropical climates, infestations usually occur in heated buildings such as large office and apartment complexes, factories, food establishments and hospitals (Edwards 1985). Especially troublesome are the problems they cause in hospitals, where they contaminate equipment and sterile packaging, penetrate I.V. solutions and tubing and feed on dressed wounds (See Eichler, this volume). Worker ants are capable of carrying Salmonella, Streptococcus, Staphylococcus, Clostridium and Pseudomonas pathogens (Beatson 1972). Infestation rates in London hospitals reach 30%; and in Prague, Czechoslovakia 86% (Edwards 1985; Rupes et al. 1983, respectively). United States medical facilities in Texas and Florida have also experienced ongoing infestations (Wilson and Booth 1981; Williams, unpublished data).

A pharaoh's ant colony consists of several nests with free worker movement and no antagonism between nests (Edwards 1981). In the cooler regions, the ants usually nest indoors, especially in inaccessible areas such as wall spaces, plumbing and duct shafts and foundations. Nests are generally polygynous and mating occurs within the nest. New colonies arise from "budding" (sociotomy) of the main colony when several workers (with or without a queen) carry eggs, larvae and pupae to a new site (Edwards 1986). Workers can live about 9 to 10 weeks while queens may live 39 weeks (Peacock and Baxter 1950), or more than 52 weeks in the lab (Edwards 1986). The oviposition rate during most of the queen's life is 25 to 35 eggs per day (Edwards 1986).

The earliest reports on pharaoh's ant control methods were published 100 years ago by Bellevoye (1889) and Riley (1889). A few years later, Lintner (1895) reported on the use of pyrethrum against *M. pharaonis*; however, not until the 1950's were inorganic and organic insecticides widely used against this pest (Rogers and Herrick 1953, Morgan and Price 1954,

Papworth 1958). Edwards (1986) gives an excellent review of control methods.

Conventional control techniques utilizing sprays and dusts are usually ineffective against pharaoh's ant. As in other social insect pests (See Banks, this section), contact insecticides primarily affect a small percentage of foraging workers. Also, the foraging habits of workers make it possible for them to largely avoid contact with the insecticide. The polygynous and polynest nature of the ant makes it more difficult to obtain total insecticide coverage. Even without foraging workers, colonies can survive for long periods on previously-stored food (Kretzschmar and Brendt 1976). Another reason new queens are less vulnerable to these control methods is that mating occurs inside the nests. Finally, because *M. pharaonis* prefer to nest in interior walls and spaces, they can infiltrate an entire building even if an insecticide covers the outer surface. Therefore, toxic baits are preferable, either alone or in combination with residual insecticide applications. Some commercially available pharaoh's ant baits are Pharorid (methoprene, Zoecon Corporation, Dallas, TX, 75234, USA), Drax (boric acid, R Value, Inc., Smyrna GA, 30081, USA), and Maxforce Pharaoh Ant Killer (hydramethylnon, American Cyanamid Company, Wayne, NJ, 07470, USA). Complete control is sometimes achieved; however, too often the control fails. Additional safe, effective chemicals and attractive bait formulations are still necessary. The basic biology and ecology of this insect must be researched for developing new control strategies. Intensive care, burn and neonatal units in hospitals, school lunchrooms, restaurants, homes and apartment complexes all come under attack by pharaoh's ants, making control vitally important.

The insect growth regulator, fenoxycarb (Maag Agrochemicals, Inc., Vero Beach, Florida 32961-6430) is presently registered as a toxic-bait (Logic) and has shown excellent control against the imported fire ant, *Solenopsis invicta* Buren (Banks et al. 1988). This paper reports our efforts to develop and evaluate fenoxycarb bait formulations against the pharaoh's ant in the laboratory and under natural conditions.

MATERIALS AND METHODS

Colony Rearing

The laboratory colonies were maintained in nesters which were plastic sweater boxes 35.6 cm long X 26.7 cm wide X 17.1 cm deep (Tucker Housewares, Arlington, TX 76011) coated with Teflon 30B (E. I. DuPont De Nemours & Co., INC., Polymer Products Department, Wilmington, Del 19898) on the inside walls (10 cm band). One or more rearing cells (depending on colony size) - a plastic petri dish (100 mm X 100 mm square X 15 mm deep) with a base of flesh-tone colored dental plaster (Dentsply/York Division, 570 West College Avenue, York, Pa 17405 USA) - were covered with red acetate and placed inside each nester. The rearing cells contained numerous queens, brood (eggs, larvae, and pupae) and workers. The colonies received the standard diet of insects (adult house flies and mosquitoes), hardboiled chicken egg yolk, and honey-agar twice weekly. They were maintained at 30 ± 2°C and 75 ± 10% RH, with a 12 hr daylight cycle. A polystyrene weighboat (100ml cap., Curtin Matheson Scientific, Inc., P.O. Box 1546, Houston, TX 77251) containing water-saturated cotton provided moisture. Water was added as needed to the cotton.

TABLE 1. Acceptability of fats and oils to Pharaoh's ants, *Monomorium pharaonis*.

Fat or oil	No. colonies	Acceptance ratio[a]
lard	12	1.388 ± 0.548a
Arrowhead Mills peanut butter oil	65	1.000 ± 0.000b
Smucker's peanut butter oil	6	0.893 ± 0.080b
Arrowhead Mills peanut butter	12	0.887 ± 0.504b
Butter	12	0.870 ± 0.156bc
Olive oil	35	0.801 ± 0.607bc
Tree of Life peanut butter oil	6	0.791 ± 0.121bc
Knudsen Family peanut butter oil	6	0.783 ± 0.159bc
Pumpkinseed oil	35	0.748 ± 0.264bc
Crisco	12	0.698 ± 0.174bc
Publix peanut butter oil	6	0.565 ± 0.125cd
Tuna oil[b]	29	0.397 ± 0.242de
Avocado oil	29	0.362 ± 0.225def
Soybean oil	6	0.353 ± 0.051def
Coconut oil	6	0.353 ± 0.083def
Almond oil	6	0.298 ± 0.255defg
Once refined soybean oil	29	0.253 ± 0.162efg
Corn oil	6	0.223 ± 0.181efg
Sunflower oil	6	0.163 ± 0.137efg
Canola oil	6	0.153 ± 0.117efg
Peanut oil	6	0.145 ± 0.167efg
Linseed oil	6	0.141 ± 0.025efg
Walnut oil	6	0.133 ± 0.121efg
Apricot kernel oil	6	0.125 ± 0.132efg
Cod liver oil	6	0.114 ± 0.106efg
Safflower oil	6	0.103 ± 0.139efg
Sesame oil	6	0.052 ± 0.058fg
Neat's foot oil	6	0.052 ± 0.017g
Mineral oil	6	0.009 ± 0.007g

[a]Acceptance ratio = (ants responding to fat or oil) / ants responding to Arrowhead Mills peanut butter oil. Means followed by different letters are significantly different at $P<0.05$ using Duncan's Multiple Range Test.
[b]tuna oil is vegetable oil removed from can of tuna.

The food acceptance tests was a modification of those described by Williams et al. 1980 and Vander Meer et al. 1988. In each test, the rearing cell contained 45 queens, 12,500 workers, and five grams of brood from a colony nester. After an hour of orientation, six foods in which one was always a standard - three at each end of the rectangular chamber - were simultaneously given to six separate colonies. To avoid spillage or contamination, each food sample was placed on a 2 cm filter paper square which was then put on a 2.54 cm aluminum foil square; the rearing cells themselves centered in the test chamber (shoe box size storage boxes 33 cm X 18.4 cm X 10.8 cm high, Sterlite Corporation, 198-T Main St., Townsend, MA 01469).

The number of ants feeding on each food was recorded at 5 minute intervals for 30 minutes. If fewer than 100 ants were recorded the 30 minutes, the results were discarded. An acceptance ratio of each candidate food to the standard was calculated by dividing the number of ants responding to the candidate food by the number responding to the standard.

Data were analyzed by the general linear model (GLM) procedures (SAS Institute 1985, 433-506) with acceptance ratio as the dependent variable. Mean values were compared with Duncan multiple range test (Duncan 1955) at the $P = 0.05$ level.

Oil removed from the top of Arrowhead Mills peanut butter was the "peanut butter" standard in fat and oil acceptance tests. All peanut butter oils used in the acceptance tests were obtained by removing the separated layer of oil on top of the peanut butter. Raw wild honey was the standard in the sugar acceptance tests.

The following primary laboratory test evaluated the effectiveness of fenoxycarb. These tests were also conducted in Sterlite storage boxes containing one rearing cell with 3 fertile queens, 0.2g of brood, and approximately 500 workers. Fenoxycarb was tested at 0.01% to 5.0% solutions in peanut butter oil. The commercial bait, Maxforce Pharaoh Ant Killer, was used at the same weight of total bait as the fenoxycarb baits. Three colonies were treated per dosage; three control colonies were given the same bait quantity as the treated colonies but without the chemical. The colonies had access to all baits for 72 hrs after which time the colonies were returned to their normal laboratory diet. Weekly observations recorded the status of the colony (normal or abnormal), queens (size, laying eggs, and acting normal), mortality in workers and other castes, morphological abnormalities, type and quantity of brood, and first appearance of winged reproductives. The last two observations are especially important when evaluating insect growth regulators. Observations continued until the colony had died or completely recovered from the treatments effects. Colonies were considered normal after the queens began normal egg-laying, all brood stages were present and worker numbers were greater than pretreatment.

RESULTS AND DISCUSSION

Out of 29 fats and oils tested, lard was better accepted than the 28 other oils and fats (Table 1). Arrowhead Mills peanut butter oil, the standard, was significantly more acceptable than 19 oils but indistinguishable from 8 others.

Table 2 indicates that of 26 substances, 11 (all but one of these were types of honey) differed little from the standard (raw wild honey), but were significantly different from 14 others. Sourwood honey and orange blossom honey showed higher acceptance ratios than the standard.

TABLE 2. Acceptability of sugars to Pharaoh's ants, *Monomorium pharaonis*.

Sugar	No. colonies	Acceptance ratio[a]
Sourwood honey	20	1.017 ± 0.296a
Orange blossom honey	20	1.014 ± 0.188a
Raw wild honey	41	1.000 ± 0.000a
Saw palmetto honey	14	0.931 ± 0.220ab
Gallberry honey	4	0.911 ± 0.125ab
Buckwheat honey	5	0.871 ± 0.436ab
Tupelo honey	5	0.844 ± 0.540abc
Sunflower honey	10	0.804 ± 0.493abcd
Alfalfa honey	5	0.763 ± 0.430abcd
Brown rice syrup	21	0.747 ± 0.285abcd
Unsulphured molasses	4	0.746 ± 0.120abcd
Avocado honey	10	0.719 ± 0.321abcd
Barley malt	26	0.685 ± 0.246bcde
Light karo syrup	5	0.559 ± 0.170cdef
Mint apple jelly	17	0.542 ± 0.252cdef
Mint jelly	5	0.532 ± 0.221def
Molasses	4	0.397 ± 0.457efg
Maple syrup	10	0.351 ± 0.224fgh
Dark karo syrup	5	0.348 ± 0.149fgh
75% honey[b]	6	0.330 ± 0.250fghi
10% turbinado sugar[c]	6	0.177 ± 0.045ghi
10% confectioner's sugar[c]	6	0.152 ± 0.041ghi
10% dextrose[ce]	6	0.070 ± 0.026hi
0.9% NaCl[ce]	6	0.069 ± 0.015hi
10% glucose[de]	6	0.069 ± 0.023hi
Coke syrup	4	0.042 ± 0.029i

[a]Acceptance ratio = (ants responding per sugar) / (ants responding to raw wild honey). Means followed by different letters are significantly different at $P<0.05$ using Duncan's Multiple Range Test; b(v:v); [c](w:v); [d](w:w); [e]hospital products.

The results of the primary laboratory evaluation tests showed that the most effective fenoxycarb concentrations in vegetable oil were 0.1%, 0.25%, 0.5%, and 1.0%. These four concentrations delayed the first appearance of sexuals longer than the other concentrations tested (Table 3). Also, 0.1%, 0.25%, and 0.5% gave the most dramatic reduction in brood during the 20 weeks; however, more than one treatment will be necessary because most of the colonies are now recovering from the affects of the IGR (Fig. 1). In general, concentrations higher than 1.0%, for example 2.5% and 5.0%, were not different than the controls and probably failed because of the chemical's repellency. Concentrations below 0.1%, for example 0.05% and 0.025% and 0.01%, caused no long-term effects, though a few colonies treated with these concentration had slight reductions in brood levels. However, these colonies quickly recovered from the effects and soon equaled the controls. After 20 weeks, the 0.25% concentration was slightly better than the 0.5% and 0.1%.

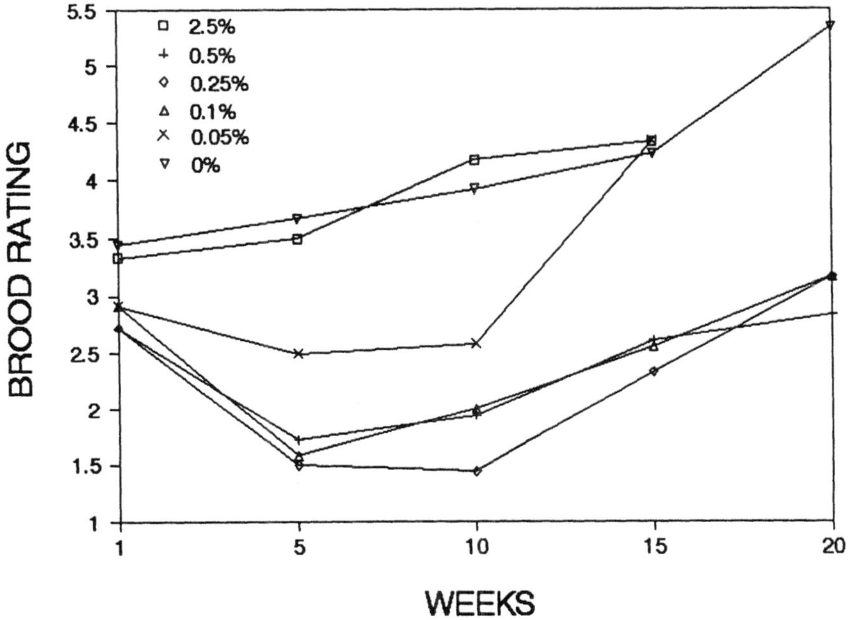

FIGURE 1. Effects of percent fenoxycarb on brood amount in Pharaoh's ant colonies. Brood rating of: 1 = 0.0115g brood, 2 = 0.0851g, 3 = 0.2752g, 4 = 1.3135g, 5 = 2.6205g, 6 ≥ 3.5g. All colonies began at brood rating = 3.

TABLE 3. First appearance of sexuals of Pharaoh's ant colonies treated with fenoxycarb.

Fenoxycarb % concentration	Colonies (N)	Appearance of sexuals[a] in weeks (X ± SD)
0.25	6	11.000 ± 4.1952a
0.10	10	10.900 ± 2.9609a
0.50	9	10.222 ± 3.6667ab
1.00	8	9.000 ± 4.5040abc
0.05	6	6.667 ± 2.8048bcd
2.50	3	6.333 ± 0.5774cd
5.00	3	5.667 ± 0.5774cd
0.00	18	4.944 ± 0.9984d
0.01	3	4.667 ± 0.5774d

[a]Means followed by different letters are significantly different at $P<0.05$ using Duncan's Multiple Range Test.

This concentration caused a more dramatic reduction in all brood levels and the appearance of intermediate castes which resembled small queens or large workers. Maxforce[R] effects were dramatic in the first two weeks, with most of the colony exhibiting high worker mortality, the loss of 1 or 2 of three queens and reduced brood levels. However, most treated colonies also returned to normal after 20 weeks. In most cases, like fenoxycarb, a single Maxforce treatment did not kill the pharaoh's ant colony.

The time required for the first appearance of winged reproductives usually indicates normal colony development. Table 3 shows that winged reproductives appeared within 4-5 weeks in normal colonies of test size. However, the effects of the 0.1, 0.25, 0.50, and 1.0% fenoxycarb significantly delayed the production of the winged reproductives in the treated colonies. The concentrations of 5.0, 2.5, 0.05, and 0.01% were not significantly different from the untreated control (normal) colonies.

CONCLUSION

This research developed laboratory procedures for testing baits for acceptance by pharaoh's ant, and for primary and secondary laboratory evaluation of bait formulation efficacy. Lard, peanut butter oil, and honey proved more acceptable to the pharaoh's ant workers than most other tested substances. Fenoxycarb at 0.5, 0.25 and 0.1 percent in Arrowhead Mills peanut butter oil caused dramatic colony brood reduction and significantly delayed production of the sexual (winged) forms in laboratory colonies. However, a single application of fenoxycarb and the other baits did not eliminate a small colony (ca 500 workers) of the pharaoh's ant; therefore, multiple applications are probably necessary to achieve complete control of these ant pests.

In the future, we plan to test the above formulations against larger laboratory colonies consisting of approximately 100+ queens, 5 grams of brood and more than 12,500 workers. Once we have determined the most effective fenoxycarb bait formulation and concentration against large laboratory colonies, the bait formulation will undergo a final evaluation. This involves field tests in infested dwellings.

ACKNOWLEDGMENTS

I am grateful to Karen Vail for excellent technical assistance. Appreciation is also given to W.A. Banks, J.A. Hogsette, J.P. Parkman, and J.T. Bridges for their advice and comments on the manuscript.

REFERENCES CITED

BANKS, W.A., D.F. WILLIAMS and C.S. LOFGREN. 1988. Effectiveness of fenoxycarb for control of red imported fire ants (Hymenoptera: Formicidae). J. Econ. Entomol. 81(1): 83-87.

BELLEVOYE, M.A. 1889. Observations on *M. pharaonis* (Latr.) Insect Life 2: 230-233.

BEATSON, S.H. 1972. Pharaoh ants as pathogen vectors in hospitals. Lancet 1:425-7.

DUNCAN, D.B. 1955. Multiple range and multiple F tests. Biometrics 11:1-42.

EDWARDS, J.P. 1981. Control of *Monomorium pharaonis* (L.) with methoprene baits: implications for the control of other pest species. The biology of social insects. Proceedings of the Ninth Congress of the International Union for the Study of Social Insects, Boulder, Colorado, August, 1982 [eds. M.D. Breed and H.E. Evans]. Boulder, Colorado, USA. Westview Press. pp 119-123.

EDWARDS, J.P. 1985. The development of Methoprene baits for the control of the pharaoh's ant, *Monomorium pharaonis*. Proc. Insect Growth Regulator Symposium.

EDWARDS, J.P. 1986. The biology, economic importance, and control of the Pharaoh's ant, *Monomorium pharaonis* [L.], pp.257-271. *In* S. Bradleigh Vinson (ed.), Economic impact and control of social insects. Praeger Publishers, New York. 421 pp.

KOHN, M. and M. VLCEK. 1986. Outdoor persistence throughout the year of *Monomorium pharaonis* (Hymenoptera: Formicidae). Entomol. Gener. 11(3/4): 213-215.

KRETZSCHMAR, K.H. and K.J. BRENDT. 1976. Zur hungeranfalligkeit von kolonien der pharaoameise im himblick auf die bedampfung. Zeit. Gest. Hyg. 22: 653-657.

LINTNER, J.A. 1895. *Monomorium pharaonis* (Linn.). 49th Ann. Report N. Y. State Museum 109-114.

MORGAN, M.T. and M.D. PRICE. 1954. Insect proofing of hospitals with the new insecticidal resins. The Hospital 767-771.

PAPWORTH, D.S. 1958. Practical experience with the control of ants in Britain. Ann. Appl. Biol. 46: 106-111.

PEACOCK, A.D. and A.T. BAXTER. 1950. Studies in Pharaoh's ant, *Monomorium pharaonis* (L.), 3. Life history. Ent. Mon. Mag. 86: 171-178.

RILEY, C.V. 1889. The little red ant. Insect Life 2: 106-108.

Rogers, T.H. and G.W. Herrick. 1953. Fieldwork against pharaoh's ants. The Sanitarian 61: 399-402.

RUPES, V., J. CHMELA, I. HRDY and J. KRECEK. 1983. Effectiveness of methoprene-impregnated baits in the control of *Monomorium pharaonis* ant populations infesting health establishment and households. Journal of Hygiene, Epidemiology, Microbiology and Immunology. 27(3): 295-303.

SAS Institute. 1985. SAS user's guide: statistics, version 5 ed. SAS Institute, Cary NC.

WHEELER, W.M. 1910. Ants: their structure, development, and behavior. Columbia Univ. Press. N.Y.

WILLIAMS, D.F., C.S. LOFGREN, W.A. BANKS, C.E. STRINGER and J.K. PLUMLEY. 1980. Laboratory studies with 9 amidinohydrazones, a promising new class of bait toxicants for control of red imported fire ants. J. Econ. Entomol. 73: 798-802.

VANDER MEER, R.K., F. ALVAREZ and C.S. LOFGREN. 1988. Isolation of the trail recruitment pheromone of *Solenopsis invicta*. J. Chem. Ecol. 14: 825-838.

WILSON, G.R. and M.J. BOOTH. 1981. Pharaoh ant control with IGR in hospitals. Pest Control 49:14-19, 74.2

64

Reducing Theft of Surface-sown Seeds by Harvester Ants

M.H. Campbell

INTRODUCTION

The major problem caused by seed-harvesting ants is that they take large quantities of pasture, lawn or crop seeds sown on or near the soil surface. For example, in temperate Australia *Pheidole* spp. took 20% to 90% of aerially sown pasture seed (Campbell and Swain 1973; Johns and Greenup 1976b); in semi-arid Australia, ants removed up to 100% of sown seed (Andersen and Ashton 1985); and in tropical, sub-tropical and Mediterranean zones, ants take large quantities of surface-sown seeds (Russell et al. 1967; McGowan 1969; Mott and McKeon 1977). The rate of pasture seed removal by *Pheidole* spp. in temperate Australia, 139/nest/hr, could exhaust the total amount of seed sown (10 kg/ha) in only 16 hours of foraging (Campbell 1966).

These losses are serious because when attempting to improve pastures, comparatively small numbers of seed are aerially sown, but these environments have large numbers of ants. For example, from 200,000 to five million pasture seeds are sown per hectare, depending on mixture, variety and rate, onto hills that may have 25 million ants per hectare. The number of ants in the above example occurred in temperate Australia and was calculated from one nest/4m^2 with 10,000 *Pheidole* sp. per nest.

The most serious losses occur when expensive seeds of pasture species ($46/ha for a 10 kg mixture) are removed by ants immediately after aerial sowing. This renders herbicide treatments, which can cost up to $80/ha, ineffective because of subsequent poor establishment. As landholders in Australia aerially distribute c. 3000 tonnes of pasture seed/annum the overall cost of seed removal by ants can be substantial.

The amount of seed removed by ants from cultivated seedbeds has not been investigated in Australia, but Anslow (1958) observed ants taking pasture seeds from cultivated seedbeds in Africa. Loss of lawn/turf seeds and vegetable and other crops sown near the soil surface is common in Australia.

The major effect of seed predation in long established pastures is population change due to selective foraging, e.g., *Veromessor pergandei* collected 50% of their seeds from species that constituted less than 10% of the vegetation, the ants ignoring less desirable seeds and climbing plants to take the preferred seeds (Tevis 1958).

SEED HARVESTING ANTS

In Australia, ants are abundant, c. 2,000 species, and highly active. Comparatively few species harvest seeds, e.g. in northern Australia only seven of 120 were true seed harvesters (Greenslade and Mott 1978) and in southern Australia only five out of 37 were seed harvesters (Briese 1982). Other species are widely-adapted, rather than specialized seed harvesters, collecting seeds when insects are scarce (Greenslade and Mott 1978) to decorate their nests, or to consume as a minor part of their diet (Briese 1982). Even though the true seed harvesters may comprise a small proportion of the species matrix, they make up a disproportionate percentage of the total population, e.g., five seed harvester species out of a total of 37 species formed 42% of the colonies in a saltbush community and 53% in an adjacent grassland (Briese 1982).

Pheidole spp. are widespread seed harvesters in Australia; they have been repeatedly observed collecting seeds in New South Wales (Campbell 1966; Briese 1982), Queensland (Russell et al. 1967), Victoria (Andersen 1982), South Australia (Greenslade and Lee 1984), Northern Territory (Mott and McKeon 1977) and Western Australia (Twigg et al. 1984). Identifications to species level are uncertain. Twigg et al. (1984) named the two principal seed-taking species in Western Australia as *Pheidole* sp. J.D.M.155 and *Pheidole* sp. J.D.M.37, while Briese (1982) in New South Wales identified *Pheidole* A, B and C based primarily on size. The largest *Pheidole* sp. A, with the minor workers 2 mm long and major workers 5 mm long. Minor workers collected seeds on trails up to 5 m, and stored them in nests with one to four entrances, cones of discarded seed litter, and had populations of c. 10,000 workers. Major workers had enlarged reddish heads and functioned as defenders. *Pheidole* spp. B and C were smaller with smaller colonies. The most widespread *Pheidole* sp. observed collecting sown pasture seeds in southern Australia (Campbell 1982) closely resembles the *Pheidole* sp. A of Briese (1982); minor workers were black and 3 mm long with a live weight of 0.5 mg, while major workers were 6 mm long. Identification of other seed harvesters is also uncertain, e.g., *Chelaner* sp. A and B (Briese 1982) and *Chelaner* sp. 1, 2, 3 (Andersen 1982).

Other seed harvesters in Australia include *Meranoplus*, *Monomorium* and *Melophorus*. Some species of *Rhytidoponera*, *Polyrhachis* and *Iridomyrmex* remove seeds to decorate their nests or only consume a small portion of the seeds taken. Greenslade and Lee (1984) contend that the species diversity of seed-harvesting ants in South Australia was determined by soil type, the vegetation's structural properties and habitat productivity factors that determine the diversity in the rest of the ant fauna; in diverse communities, seed harvesters were specialized but in areas of low diversity, broadly-adapted species predominated, notably *Pheidole*.

SEED PREFERENCE

With sown pasture seeds, Campbell (1982) concluded that Australian seed-harvesting ants preferred small to large and grass to legume seeds (Table 1). Large clover seeds, 17 times the weight of a *Pheidole* spp. worker, are difficult for the ants to carry; smaller seeds are preferred. Auld (1986) found that ants too small to drag seeds usually removed the attached eliasome *in situ* and took it to the nest, whereas larger ants took both seed and eliasome. In

TABLE 1. Preference of pasture seeds to ants in relation to species and weight (1, Campbell 1966; 2, Campbell and Swain 1973; 3, Johns and Greenup 1976a; 4, Russell et al. 1967).

Seed species	Weight (mg/seed)	Order of Preference (1=highest)
1. *Trifolium repens*	0.6	3
Dactylis glomerata	0.7	1
Phalaris aquatica	1.5	2
Trifolium subterraneum	7.5	4
2. *Dactylis glomerata*	0.7	2
Phalaris aquatica	1.5	1
Medicago sativa	2.3	3
Trifolium subterraneum	7.5	4
3. *Trifolium repens*	0.6	2
Phalaris aquatica	1.5	1
Festuca arundinacea	2.3	2
Medicago sativa	2.3	4
4. *Medicago sativa*	2.3	1
Glycine javanica	6.1	3
Sorghum almum	6.7	2
Phaseolus atropurpureus	11.9	4

Spain, Reyes Lopez (1987) found the upper load limit of *V. pergandei* was a function of its size, while in California Chew and De Veta (1980) discovered that body length and mandible gape were positively correlated with load limit. In addition, it is easier for ants to remove the palea and lemma of a grass seed than to pierce the seed coat of a legume. Within grasses, it is easier for ants to extract the caryopsis from glumes of some species, e.g. *Phalaris aquatica*, than others, thus explaining preference (Campbell 1982). Preference also depends on the available seeds, e.g. when *M. sativa* seed was distributed with smaller seeds, ants preferred the smaller seeds, but when distributed with larger seeds it was most preferred (Table 1). Risch and Carroll (1986) found that ants removed a preferred seed less frequently if it occurred among non-preferred seeds and they took non-preferred seeds more readily if occurring with preferred seeds.

Gordon (1980) suggested that since surface lipids, nutrients and caloric content are highly correlated, ants are likely to select seeds on the basis of chemical cues from monoterpenoids in the surface lipids. Certainly, ants prefer high caloric sown seeds to low caloric resident seeds because intense harvesting activity occurs within two hours of aerial sowing after which sown seeds are taken and resident seeds ignored (Campbell 1966).

FORAGING ACTIVITIES

Johns and Greenup (1976a) showed that temperature accounted for 68-78% of the variance in rates of sown pasture seed removal by *Pheidole* spp. at two sites in Australia. Most ants are paralyzed by cold (<5°C) or killed by heat (50°C; Tevis 1958); they survive by foraging during periods of moderate temperature and residing in nests during temperature extremes (Majer 1982). In Australia, *Pheidole* spp. forage in the morning and evening in summer and at midday in winter (Campbell 1968). When temperatures increase, some species switch from diurnal to nocturnal activity (Briese and Macauley 1980). Foraging by *V. pergandei* and *Pheidole* spp. ceased at 44°C and 42°C respectively (Tevis 1958; Mott and McKeon 1977).

In Western Australia, 19 species of ants foraged maximally in the warmth of summer and early autumn - a period which corresponded to the greatest supply of new seeds (Twigg et al. 1984). In semi-arid New South Wales, Briese and Macauley (1981) found all five seed harvester species most active with the production of major seed crops in spring, summer and autumn; three species were inactive in winter. Rainfall generally stops seed collection, but is not important unless the particular environment is usually wet (Ackonor 1984).

Different activities are often carried out by groups of ants of various ages. Gordon (1986) found midden work, patrolling, nest maintenance, convening and foraging were performed by four *Pogonomyrmex badius* groups for at least 35 days; midden work and patrolling were carried out by the same group. Three months later, convenors had become foragers, and nest maintenance workers patrollers. Younger callow workers were convenors and older ants midden workers and patrollers. For three California species of *Pogonomyrmex* young ants with high dry weight and fat content and low respiration rate were found at the bottom of nests; older external workers in the upper nest had low fat content and high mortality rates, an energy-saving adaptation (Mackay 1983).

In Tunisia, callow workers of *Cataglyphis bicolor* remained in the nest for two months before beginning to forage. These ants then replaced older workers as they died, each nest having a relatively constant population; daily loss of workers was 16.4% and expected half life 4.2 to 6.1 days. The duration of runs made daily by individuals increased with age and foraging experience, whereas the number of runs did not; food was found more quickly by older foragers (Schmid-Hempel and Schmid-Hempel 1984).

In the USSR, harvesting by *Formica rufa* L. was carried out by experienced foragers and observers familiar with the foraging area (Zakharov 1980). If all foragers were removed, observers resumed foraging and rapidly re-established the initial structure. Blocking foraging routes temporarily deflected ants around the obstacle and back to the original route. In order for a colony to lose all information about a foraging area, all foragers and observers had to be removed and access to foraging tracks blocked near the nest. For the *F. rufa* group in Finland, allegiance to routes was based on individual memory of spatially-organized visual cues which could be recalled after hibernation; memorized olfactory cues were used in the dark (Rosengren and Fortelius 1986).

Pheidole militicida Wheeler used trail pheromones, visual guidelines and persistent chemical cues to forage; chemical cues enabled foraging at night (Holldobler and Moglich 1980). Trail pheromones laid by workers contained a recruitment signal, and an orientation cue and were highly species-specific; they played a major part in regulating activity along foraging routes. Each *P. militicida* colony had trunk, branch and twig routes that lasted for several weeks, days or hours respectively. A trunk route was abandoned when its food supply was exhausted and a new one established. With *Camponotus sericeus* in India, as many as 16 routes radiated from nests; the routes were 2 to 153 m in warm months and 2 to 28 m in rainy months; length depended on food (Basalingappa et al. 1986). In Australia, *Pheidole* spp. foraged distances of from one (Mott and McKeon 1977) to five meters (Briese 1982).

MANAGEMENT

Time of Sowing

Because *Pheidole* spp. ants take less seed in winter than in other seasons in Australia (Johns and Greenup 1976a), sowing seed in winter could provide a method of reducing losses. However, winter sowing did not reduce losses in relatively warm environments or in unseasonably warm winters (Johns and Greenup 1976a). Campbell (1966) attributed heavy winter losses in a temperate environment to *Pheidole* spp. harvesting during the warm periods of the day; the longer the dry period after sowing in which seeds remained unswollen and ungerminated and thus attractive to ants, the greater the ant-susceptible period. This was significant even if restricted to four hours foraging/day in winter.

Pelleting

Coating seeds with lime or starch and acrylic polymers has not reduced losses due to ants (Campbell 1966, 1968, 1985a; Johns and Greenup 1976a). Ants generally take lime pelleted seeds more slowly than unpelleted. Campbell (1968) found that 100 unpelleted *Trifolium repens* and *Medicago sativa* seeds were taken in 0.3 and 1 hour respectively, compared to 3 and 12 hours for 100 pelleted seeds. Ants waited until the lime was partially removed by dew, frost or rain. The slower rate of theft of coated seeds may be due to the ant's inability to recognize the seed (Johns and Greenup 1976a) or detect surface lipids (Gordon 1980) until part of the coat is removed.

Insecticides

Research has shown that treating seeds with dieldrin (Anslow 1958; Campbell 1966), chlordane (Anon. 1964) or bendiocarb, permethrin, arprocarb, dichlorovos, (Campbell and Gilmour 1979; Campbell 1982) reduces losses due to ants. Treating seeds with dieldrin at 1.1 kg a.i./100 kg improved establishment of surface sown pasture species by reducing losses due to *Pheidole* spp. (Table 2). The magnitude of improvement was large for preferred grass or smaller legume seeds and small when rain closely followed sowings made in winter (Campbell 1966). In contrast, Russell et al. (1967) found that dieldrin, chlordane, DDT, endrin and lindane did not reduce seed losses due to ants; this lack of effect may have been due to a low rate (0.22 kg a.i./100 kg seed), a small number of seeds sown per plot (10) and tolerant ant fauna.

Insecticide treatment of seeds reduces losses by killing or repelling ants. When ants take seeds treated with dieldrin into their nest activity is markedly reduced within two hours. If sufficient seeds are taken, i.e. 10 to 80 depending on nest population, all activity ceases and dead ants are

TABLE 2. Establishment of surface sown pasture species from 10 sowings over two years (Campbell 1966).

Species sown	No. of plants to establish/m^2 -dieldrin	+dieldrin	No. of sowings where establishment improved (P<0.05) by dieldrin
Dactylis glomerata	7	41	10
Lolium perenne	3	16	7
Trifolium repens	11	25	7
Phalaris aquatica	1	4	5
Trifolium pratense	10	22	4
Trifolium subterraneum	8	13	2

TABLE 3. Effect of insecticide treatment of *Phalaris aquatica* seeds on losses due to ants (Campbell and Gilmour 1979; Campbell 1982).

Insecticide	Rate kg ai/100 kg seed	% loss of seeds in: 9 days	11 days	14 days	21 days
Permethrin	3.10	2a[+]			
	0.32		47 a		
	0.23			44 a	
	0.15/0.16		54 ab	48 a	76 a
Bendiocarb	0.32		62 b		
	0.23			37 a	
	0.20	39 b			
	0.15/0.16		69 b	53 a	94 b
Arprocarb + dichlorvos	1.6+0.3	71 c			
Chlorpyrifos	0.16				99 c
Control		100	94 c	100	99 c

[+]Values in columns not followed by a common letter differ P < 0.05

periodically discarded (Campbell 1966). Ants avoid seed treated with permethrin due, presumably, to its repellent effects and thus carry on normal activities. Bendiocarb effects are similar to dieldrin in reducing activity and killing ants (Campbell and Gilmour 1982).

Even though permethrin and bendiocarb reduce losses, ants still take substantial amounts of sown seeds over time (Table 3), perhaps due to weathering, which reduces the concentration of insecticides on the seeds. Bendiocarb has a hydrolysis half life in water of 10 days at 20°C and breaks down on contact with soil, but is not oxidized by the atmosphere or decomposed by sunlight (Reary and Whiteoak 1977). Light slowly breaks down permethrin, but moisture does not affect it (Wellcome Aust. personal communication). As seeds may remain ungerminated on the soil surface for 30 days or more after aerial sowing, an effective insecticide must maintain toxicity for at least a month.

The rates of permethrin and bendiocarb recommended for landholder use (0.13 and 0.16 kg a.i./100 kg seed respectively; Campbell 1985b) are lower than the rate of dieldrin (1.1 kg a.i./100 kg seed) used by Campbell (1966) thus reducing pollution. In addition, new insecticides are much less toxic than dieldrin.

Wettable powder formulations of permethrin and bendiocarb have no effect on seed germination or *Rhizobia* survival so they can be applied to legume seeds during the normal lime pelleting process (Campbell and Gilmour 1979). Permethrin's miscible oil formulation slightly reduces germination of some pasture species but this is due to the oil carrier (Campbell and Gilmour 1979).

Overall spraying with insecticides may kill harvesting ants and thus stop seed removal but would also kill many other ants and insects. Indiscriminate use of mirex bait killed *Pheidole* spp. as well as many non target species (Campbell 1977). Although the recent insecticides hydramethylon and nifluridide, applied as baits, are more effective in reducing the activity of *Pogonomyrmex maricopa* than carbaryl or diazanon (Borth 1986), the same criticism applies.

REFERENCES

ACKONOR, J.B. 1984. The activity pattern of ant *Cataulacus guineensis* F. Smith in a Ghanaian cocoa farm. Insect. Sci. Applic. 5: 307-24.

ANDERSEN, A.N. 1982. Seed removal by ants in the mallee of north western Victoria, pp. 31-43. *In* R.C. Buckley (ed.), Ant-plant interactions in Australia. Dr. W. Junk, The Hague. 162 pp.

ANDERSEN, A.N. and D.H. ASHTON. 1985. Rates of seed removal by ants at heath and woodland sites in Australia. Aust. J. Ecol. 10: 381-90.

ANON. 1964. Soil and pasture research on the northern tablelands, New South Wales. Agric. Res. Liaison Sect. C.S.I.R.O. Melbourne.

ANSLOW, R.C. 1958. A note on an improved technique in establishing *Setaria sphacelata* from seed. Emp. J. Exp. Agric. 26: 55-7.

AULD, T.D. 1986. Population dynamics of *Acacia suaveolens*: dispersal and dynamics of the soil seed-bank. Aust. J. Ecol. 11: 235-54.

BASALINGAPPA, S., V.B. KAREDDY, M.R. GHANDI and M. SHIVAKUMAR. 1986. Foraging paths of ants, *Camponotus sericeus*. Entomon. 11: 101-6.

BORTH, P.W. 1986. Field evaluation of several insecticides on Maricopa harvester ant (Hymenoptera: Formicidae) colony activity in fallow southwestern Arizona cropland. J.Econ. Ento. 79: 1632-6.

BRIESE, D.T. 1982. Relationship between the seed-harvesting ants and the plant community in a semi-arid environment, pp. 11-24. *In* R.C. Buckley (ed.), Ant-plant interactions in Australia. Dr. W. Junk, The Hague.

BRIESE, D.T. and B.J. MACAULEY. 1980. Temporal structure of an ant community in semi-arid Australia. Aust. J. Ecol. 5: 121-34.

BRIESE, D.T. and B.J. MACAULEY. 1981. Food collection within an ant community in semi-arid Australia with special reference to seed harvesters. Aust. J. Ecol. 6:1-19.

CAMPBELL, M.H. 1966. Theft by harvesting ants of pasture seed broadcast on unploughed land. Aust. J. Exp. Agric. Anim. Husb. 6: 334-8.

CAMPBELL, M.H. 1968. Cops and robbers. Chiasma. 6: 93-5.

CAMPBELL, M.H. 1977. No remedy for seed-stealing ants. Agric. Gaz. N.S.W. 88: 44.

CAMPBELL, M.H. 1982. Restricting losses of aerially sown seed due to seed-harvesting ants. pp. 25-30. *In* R.C. Buckley (ed.), Ant-plant interactions in Australia. Dr. W. Junk, The Hague. 162 pp.

CAMPBELL, M.H. 1985a. Effect of Terra-Sorb on seed losses due to ants. Proc. 3rd Aust. Agron. Conf. Hobart. p. 4.

CAMPBELL, M.H. 1985b. Pasture establishment using aerial techniques. N.S.W. Agriculture and Fisheries Agfact P2.2.2, Agdex 130/20.

CAMPBELL, M.H. and A.R. GILMOUR. 1979. Reducing losses of surface-sown seed due to harvesting ants. Aust. J. Exp. Agric. Anim. Husb. 19: 706-11.

CAMPBELL, M.H. and F.G. SWAIN. 1973. Factors causing losses during the establishment of surface-sown pastures. J. Range Mgmnt. 26: 355-9.

CHEW, R.M. and J. DE VITA. 1980. Foraging characteristics of a desert ant assemblage. J. Arid Environs. 3: 75-83.

GORDON, D.M. 1986. The dynamics of the daily round of the harvester ant colony (*Pogonomyrmex barbatus*). Anim. Behavior. 34: 1402-19.

GORDON, S.A. 1980. Analysis of twelve Sonoran desert seed species preferred by the desert harvester ant. Madrono. 27: 68-78.

GREENSLADE, P.J.M. and K.E. LEE. 1984. Diversity and food specificity of seed harvesting ants in relation to habitat and community structure. Proc. 3rd Australasian Conf. Grassld. Invert. Ecol. pp. 227-233.

GREENSLADE, P.J.M. and J.J. MOTT. 1978. Ants (Hymenoptera, Formicidae) of native and sown pastures in the Katherine area, N.T. Australia. 2nd Aust. Conf. Grassld. Invert. Ecol. Session 5. Palmerston North, N.Z.

HOLLDOBLER, B. and M. MOGLICH. 1980. The foraging system of *Pheidole militicida* (Hymenoptera: Formicidae). Insectes Sociaux. 27: 237-64.

JOHNS, G.G. and L.R. GREENUP. 1976a. Pasture seed theft by ants in northern New South Wales. Aust. J. Exp. Agric. Anim. Husb. 16: 249-56.

JOHNS, G.G. and L.R. GREENUP. 1976b. Predictions of likely theft by ants of oversown seed for the northern tablelands of New South Wales. Aust. J. Exp. Agric. Anim. Husb. 16: 257-64.

MACKAY, W.P. 1983. Stratification of workers in harvester ant nests (Hymenoptera, Formicidae). J. Kansas Ento. Soc. 56: 538-42.

McGOWAN, A.A. 1969. Effect of seed-harvesting ants on the persistence of Wimmera ryegrass in pastures in northeastern Victoria. Aust. J. Exp. Agric. Anim. Husb. 9: 37-40.

MAJER, J.D. 1982. Ant-plant interactions in the Darling botanical district in Western Australia. *In* R.C. Buckley (ed.), Ant-plant interactions in Australia. Dr. W. Junk, The Hague. 162 pp.

MOTT, J.J. and G.M. McKEON. 1977 A note on the selection of seed types by harvester ants in northern Australia. Aust. J. Ecol. 2: 231-5.

REARY, J.B. and J.B. WHITEOAK. 1977. Bendiocarb residue decline in soil, grass and thatch after treatment of turf with a 76 WP formulation in Mississippi in 1976. Fisons report RESID/77/43.

REYES LOPEZ, J.L. 1987. Optimal foraging in seed-harvester ants: computer-aided simulation. Ecol. 68: 1630-3.

RISCH, S.J. and C.R. CARROLL. 1986. Effects of seed predation by a tropical ant on competition among weeds. Ecol. 67: 1319-27.

ROSENGREN, R. and W. FORTELIUS. 1986. Ortstreue in foraging ants of the *Formica rufa* group-hierarchy of orienting cues and long term memory. Insectes Sociaux. 33: 306-37.

RUSSELL, M.J., J.E. COALDRAKE and A.M. SANDERS. 1967. Comparative effectiveness of insecticides, repellents and pelleting devices in the prevention of ant removal of pasture seeds. Trop. Grassld. 1: 153-66.

SCHMID-HEMPEL, P. and R. SCHMID-HEMPEL. 1984. Life duration and turnover of foragers in the ant *Cataglyphis bicolor*. Insectes Sociaux. 31: 345-60.

TEVIS, L. 1958. Interrelations between the harvester ant *Veromessor pergandei* (Mayr.) and some desert ephemerals. Ecol. 39: 695-704.

TWIGG, L., J.D. MAJER, B.A. STYNES and K.E. LEE. 1984. Influence of seed-harvesting ants in annual ryegrass and their possible effects on the epidemiology of ryegrass toxicity. Proc. 3rd Australasian Conf. Grassld. Invert. Ecol. pp. 235-243.

ZAKHAROV, A.A. 1980. Observer ants: storers of foraging area information in *Formica rufa* L. Insectes Sociaux. 27: 201-11.

65

Management of Carpenter Ants

R.D. Akre and L.D. Hansen

INTRODUCTION AND DAMAGE

For years researchers believed carpenter ants were more a nuisance than a structural pest. However, recent data indicate that carpenter ants cause vast amounts of structural and other damage (Hansen and Akre 1985; Akre and Hansen 1988).

Carpenter ants are structural pests in northern North America [western: *Camponotus modoc* Wheeler, *C. vicinus* Mayr; eastern: *C. pennsylvanicus* (DeGeer), *C. herculeanus* (L.)], but *C. abdominalis floridanus* (Buckley) is reported to be a persistent pest of honey bee colonies in Missouri (Burrill 1926) and in the South, especially in Florida (Smith 1965). Some species are also important honey bee pests, even in South America (Winston et al. 1979). Often, hives must be placed on supports immersed in oil so ants cannot kill the colony. Similarly, small cluster colonies, kept for queen production, are suspended from wires coated with sticky materials that entrap invading carpenter ants. Otherwise, most of these colonies are killed. *C. floridanus* and several other species of carpenter ants (for example *C. pennsylvanicus*) are also minor structural pests in the South (Smith 1965), but because termites rapidly move into the damaged area, carpenter ants are usually not associated with the problem.

Other nonstructural damage by carpenter ants is also important. For example, even in 1910, carpenter ants caused extensive damage in eastern U.S. telephone and telegraph poles (Snyder 1910), and this type of damage still occurs, even though most poles are chemically treated. Eight years later (Graham 1918), carpenter ants reportedly damaged white cedar harvested in Minnesota; calls to eight northern Minnesotan cedar mills in mid 1988 indicated the problem is still severe and results in the loss of much timber. A second report on timber loss showed 10% of harvested spruce and balsam fir in New Brunswick, Canada, was lost to these ants (Sanders 1964). Damage also occurs to park and ornamental trees (Fowler 1986). Indeed, 75% of sampled shade trees in New Jersey were infested; tunneling by these ants weakened the trees physically and exposed them to other pests and pathogens (Fowler and Parrish 1982). Infestation and loss of merchantable timber to carpenter ants may be a major, largely undocumented problem across the northern U. S. and southern Canada for a number of tree species, including

Douglas-fir, western red cedar, larch, jack pine, white pine, elms and maples.

In the same areas of the northern United States and Canada, carpenter ants are rapidly emerging as the single most damaging structural pest (Figs. 1,2). Hansen and Akre (1985) showed these ants were responsible for a minimum of 42,000 annual structural treatments by Washington State pest control operators. Although there are no definitive records, comparable damage may also occur in Idaho, Montana, Oregon and British Columbia. Indeed, recent information gathered by Extension personnel in Washington, Ohio, and Minnesota indicated that carpenter ants were usually the number one insect problem for which Extension received requests for information (Akre and Hansen 1988). Stinging insects and carpenter ants usually lead the structural and nuisance pest category in New York. This information, together with other less structured records from state Extension services, suggests that carpenter ants are a major problem across the entire northern U. S. from Washington to Maine, and in Southern Canada, from British Columbia to New Foundland.

Limited investigations into carpenter ant problems in northern Minnesota near Grand Rapids (Itasca County) in mid-summer 1988 indicated that infestations and damage caused by carpenter ants were comparable to that in Washington State. For example, a homeowner near Spider Lake leaned on a kitchen wall and fell completely through the wall into the bedroom. *Camponotus pennsylvanicus*, the black carpenter ant, had riddled a 3.5 m section of wall and tunneled all supporting beams. A log cabin near Squaw Lake had a column of *C. pennsylvanicus* six to ten workers wide foraging during mid-day in the surrounding vegetation. Daylight foraging, instead of the normal nocturnal forays, was probably due to the immense size and strength of the colony. Every log in the walls, ceiling, and floor was occupied by ants. The cabin was beyond repair and the ants could not be controlled. Infestations were found also in two new folding camper trailers, where ants entered between April 1st and August 1st 1988. The canvas and window supports had been destroyed in the first camper. All wooden parts in the second were destroyed and it could not be safely pulled from the site. When less dramatic cases of damage were investigated, the black carpenter ant, *C. pennsylvanicus*, was identified as the pest in almost all cases and the damage considered as common and severe as in the West (Washington). However, in several instances during the six week investigation, *C. herculeanus* were found inside structures. The ants tunneled into and nearly destroyed two large wooden sculptures in a homeowner's yard just outside Grand Rapids (Fig 3). Additionally, this species caused severe damage to a boathouse in the same area. This suggests that the species can also be a structural pest.

NEW CONSTRUCTION

Houses in western Washington, and probably in other areas of North America, are commonly built on or near heavily forested lots with minimum removal of trees. Unfortunately, nearly all northern forested areas contain carpenter ant colonies and houses are frequently invaded and damaged by satellite colonies even before their construction is complete. Damage occurs quickly, caused as it is by 5,000 to 50,000 workers. This contradicts previously published information that led homeowners to believe carpenter ant damage would occur in small increments over a number of years, and

FIGURES. (1) *C. modoc* damage to a house timber. (2) *C. modoc* galleries. Note clean, smooth passageways without debris. (3) Wooden sculpture heavily infested with *C. herculeanus*. (4) Colony of *C. modoc* in base of live Douglas-fir. (5) Sod roofed, buried, "earth" house that was heavily infested by *C. modoc*. (6) *C. modoc* trail through grass in a lawn.

only in rotten timbers. Damage was attributed to small, slowly growing colonies, within the house, that evolved from a single mated queen. We now know that large colonies or their satellites can cause extensive damage to a new house in less than three months.

Therefore, before the construction of houses or other wooden buildings in forested areas, the contractor should hire a knowledgeable professional or entomologist to inspect the site and evaluate possible problems. Trees which

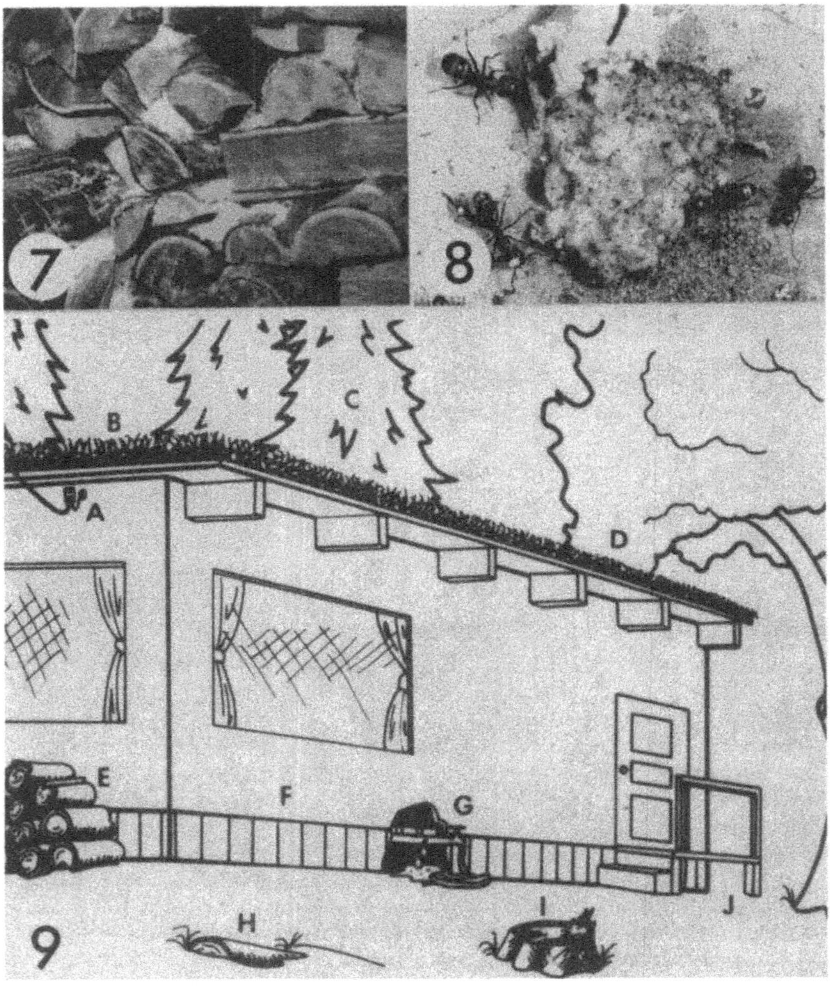

FIGURES. (7) Sawdust excavated by carpenter ants in firewood piled against a house. (8) Carpenter ants feeding on toxic bait. (9) Common carpenter ant reservoirs and entry routes. A: entry route through holes around pipes and wires; B: sod roofs are ideal nesting sites, flat roofs drain poorly and tend to rot, providing a good habitat for ants; C: forested areas harbor many colonies; D: trees, shrubs, vines, and other objects contacting the house provide excellent pathways for ants; E: log piles harbor ants; F: unventilated crawlspaces or basements provide a good environment; G: leaky pipes add moisture to unventilated areas; H: buried logs conceal nests; I: stumps are excellent nesting sites; J: wooden foundations without concrete bases attract ants.

contain large parent colonies of carpenter ants should be removed or the colonies killed (Fig. 4). This survey should include a search for satellite colonies. It is also essential that construction wood and unwanted trees be removed from the site and burned. Then, since houses in many heavily wooded sites will always be at risk from carpenter ants, a regular program of annual inspections should be instituted, and in some high risk areas, a preventative perimeter spray should be applied each spring.

One of the best times to control carpenter ants with chemicals is during house construction, before applying the drywall. Dusts should be applied to either side of sheet insulation in open beam or cathedral ceilings. A fine dust formulation of boric acid, bendiocarb, or chlorpyrifos should be placed into the wall after the electrical wiring, plumbing, and insulation are installed. The wires and top of the pipes should be liberally dusted, as should the bottom plate of the wall studs. These materials will kill carpenter ants for years unless they become wet, and, of course, wires and pipes are frequently used as ant trails to gain access to other parts of the house (Akre et al. 1984). It is also essential, at points where electrical wires and pipes enter the building, that entry holes are blocked.

Basements and crawl spaces should be properly ventilated so moisture does not collect and provide a suitable habitat for the ants. Similarly, vapor barriers are recommended for crawl spaces. It is important that all supports for the house, porch, or decking rest directly on concrete, rather than on bare soil where they will absorb moisture.

Proper ventilation must also be provided for attics. Flat roofs, which nearly always have drainage problems, are not recommended in areas of high carpenter ant infestation. Flat roofs covered with sod (earth sheltered, Fig. 5) seem particularly prone to infestation and damage. It is particularly important that the roof and house walls be kept free of tree branches and other vegetation such as vines; these can be used by carpenter ants to gain access to the house.

Proper landscaping also affects carpenter ant management. Cut trees must be removed from the lot prior to building and waste wood burned or removed from the construction site. These wastes must not be buried, a common practice in much of North America. Buried wooden debris provides an ideal nesting site. Deep piles of decorator bark (especially Douglas-fir and western redcedar) sold for landscaping purposes are one of the prime sources of new queens and small carpenter ant colonies. Occasionally even a fairly large satellite colony will nest in the interspaces of the bark; this is especially true with *C. vicinus*. The use of this bark is not recommended in high risk areas.

MAINTENANCE

In areas where structures are at risk of invasion and damage from carpenter ants, regular inspections should be instituted by a qualified professional from May to August, the peak frequency period. Frequency of inspections should be determined by risk, though no less than once a year. Inspections should include an examination of all possible nest sites on the property, especially large trees. Physical trails of carpenter ant colonies should be searched for (Fig. 6), with nocturnal inspections both inside and outside the building.

Ever since the energy crisis of 1973, people have burned an increasing quantity of wood in stoves, furnaces, and fireplaces. This wood is frequently stacked against houses and, unfortunately, frequently contains carpenter ant colonies (Fig. 7). In the Pacific Northwest, this problem is particularly acute. Loggers offer uncut log decks to homeowners. However, since firewood is nearly always old growth, dry timber (e.g. "buckskin" larch), these decks often contain ants. Thus, cities in many of the drier western areas now have infestation rates as high or higher than heavily forested areas. Therefore, firewood should be inspected for ants before it is cut or purchased and stacked in a dry shelter away from the house.

CONTROL

Physical

Repair of older or incorrectly constructed houses should be the first defense (Fig. 9). Vegetation in contact with the house should be trimmed back and all water or humidity damage repaired by correcting the ventilation, installing effective vapor barriers, or repairing broken or leaking pipes. All external holes providing access to the house, including those caused by telephone or electrical lines, or by water pipes should be caulked or otherwise closed off. Also, all wooden supports should be placed on concrete. It is usually more cost effective to dig up and eliminate the buried wooden debris that frequently constitutes the source of carpenter ant colonies, rather than trying to treat these materials chemically. This is especially true of large stumps buried in the yard.

Chemical

To achieve effective control, the satellite and parent colonies of carpenter ants must be located (Akre et al. 1984). The colonies and surrounding areas should be treated directly with an application of insecticide. If the parent colony is outdoors the insecticide can be a liquid. In wall voids only dust formulations are safe. Liquids not only cause electrical shorts, but these materials also damage the insulation and walls. The wall plates of electrical outlets and other accesses into the wall (plumbing flanges) should be removed so the dust can be applied inside the walls. In the rare event that an ant colony is located within a wall that has no convenient access by electrical outlets, small holes (three mm) can be drilled into the walls in inconspicuous locations, and the insecticide dust applied through a plastic wand. The sawdust can be saved, mixed with an epoxy resin, and applied to the wand holes to make them inconspicuous. It is not necessary to treat between all wall studs. Treatment of the wires and pipes within the wall void will usually kill all the carpenter ants as these constitute their main travel routes within the walls.

In addition to treating the in-house colony, a perimeter spray should be applied around the entire structure. This spray is applied to the foundation and extended out onto the soil (15 cm swath) near the house. If the house has a crawl space rather than a basement, this perimeter spray should also be applied to the inside of the foundation, the sill plate, and all house supports. Special care should be made to spray any pipes that enter through the crawlspace. If the newer, long-lasting synthetic pyrethroids (cyfluthrin, cypermethrin) are used for this purpose only one spray is needed during the year. Other materials such as chlorpyrifos, bendiocarb, diazinon, or propoxur can also be used, but more frequent spray applications are usually required.

Baits

The use of poisoned baits for carpenter ant control is an old concept recommended by the USDA over 30 years ago (Back 1937). However, these baits were never popular partly because they were not effective. Recent strides in bait formulations make it possible to effectively control several ant species, including Argentine and Pharaoh's ants. A protein bait, Maxforce (American Cyanamid Co., One Cyanamid Plaza, Wayne, NJ 07470), has also been tested for control of carpenter ants. The slow-acting toxicant effectively controls carpenter ants. Unfortunately, the bait is not competitive against other, natural food sources, and therefore fails in the field.

However, a protein bait attractive to both western (*C. modoc*, *C. vicinus*) and eastern species (*C. pennsylvanicus*, *C. herculeanus*) of structurally damaging carpenter ants has been developed over the last five years during our researches at Washington State University (Fig. 8). Any of at least three slow acting toxicants added to the bait kill the entire colony. The search for the best toxicant continues, but a bait should be registered for commercial and private use by 1990-1991.

LITERATURE CITED

AKRE, R.D. and L.D. HANSEN. 1988. Carpenter Ants. Proc. Second Nat. Conf. Urban Ent. College Park, MD, 23 Feb. 6 p.

AKRE, R.D., L.D. HANSEN and A.L. ANTONELLI. 1984. Carpenter ants: Their biology and control. EB 0818 Coop. Ext. Washington State University 6 p.

BACK, E.A. 1937. House ants. USDA leaflet 147. 8 p.

BURRILL, A.C. 1926. Ants that infest beehives. Amer. Bee J. 66: 29-31.

FOWLER, H.F. 1986. Biology, economics, and control of carpenter ants. Chapt. 10, pp. 272-289. In S.B. Vinson (Ed.). Economic Impact and Control of Social Insects. Praeger: NY 421 p.

FOWLER, H.G. and M.D. PARRISH. 1982. Urban shade trees and carpenter ants. J. Arboricult. 8: 281-284.

GRAHAM, S.A. 1918. The carpenter ant as a destroyer of sound wood. Minn. State Ent. Report 17: 32-40.

HANSEN, L.D. and R.D. AKRE. 1985. Biology of carpenter ants in Washington State (Hymenoptera:Formicidae:Camponotus). Melanderia 43: 1-61.

SANDERS, C.J. 1964. The biology of carpenter ants in New Brunswick. Can. Ent. 96: 894-909.

SMITH, M.A. 1965. House-infesting ants of the eastern United States: Their recognition, biology, and economic importance. ARS, USDA Tech. Bul. No. 1326. 105 p.

SNYDER, T.E. 1910. Insects injurious to forests and forest products: Damage to chestnut telephone and telegraph poles by wood-boring insects. USDA, Agric. Bur. Ent. Bul. 94, Part 1, pp. 1-12.

WINSTON, M.L., G.W. OTIS and O.R. TAYLOR, JR. 1979. Absconding behaviour of the Africanized honeybee in South America. J. Apic. Res. 18: 85-94.

66

Management of the "Hormiga Loca," *Paratrechina (Nylanderia) fulva* (Mayr), in Colombia

I. Zenner-Polania

INTRODUCTION

When population outbreaks of *Paratrechina* (*Nylanderia*) *fulva* (Mayr) (Hymenoptera: Formicidae, Formicinae) first occurred in Colombia, control measures were unknown. Initially, farmers sprayed or dusted grasslands with carbaryl, an insecticide with low residual toxicity that had to be used frequently and at high cost (ICA 1972). Aldrin was also applied, mainly at the base of coffee trees; this and other chlorinated hydrocarbons were soon banned for use in coffee, making it necessary to find efficient and environmentally safe control measures for the different types of problems caused by the ants.

A variety of methods were investigated in addition to insecticides. These methods led to the establishment of an integrated management program.

EXCLUSION

This control technique proved efficient and economical, but it was not readily accepted by farmers. Water or sticky bands were evaluated as barriers against "hormiga loca."

A concrete ditch, constructed around stables and henhouses, was permanently filled with water, inhibiting the ant's advance and providing a safe place for the animals. Every day the ditch had to be cleaned of dead leaves and other debris, which the ants could use as a "bridge." Also, care had to be taken not to bring in ants with food or other materials. A 20 cm wide ditch was found to be acceptable.

The sticky materials BIOTAC 3, BIOTAC 19 and grease were evaluated, applied in the form of bands around the tree trunk, starting at a height of at least 50 cm. Each treatment consisted of six two year old orange trees. The fluctuation of the populations of the wooly white fly, *Aleurothrixus floccosus* (Maskell) (Homoptera: Aleurodidae). The fern scale, *Pinnaspis aspidistrae* (Signoret) (Homoptera: Diaspididae) and its predators, mainly Coccinellidae (Coleoptera), Syrphidae (Diptera), Chrysopidae and Hemerobiidae (Neuroptera), were compared where "hormiga loca" were present and excluded with sticky bands. Two weeks before starting counts, the control trees were

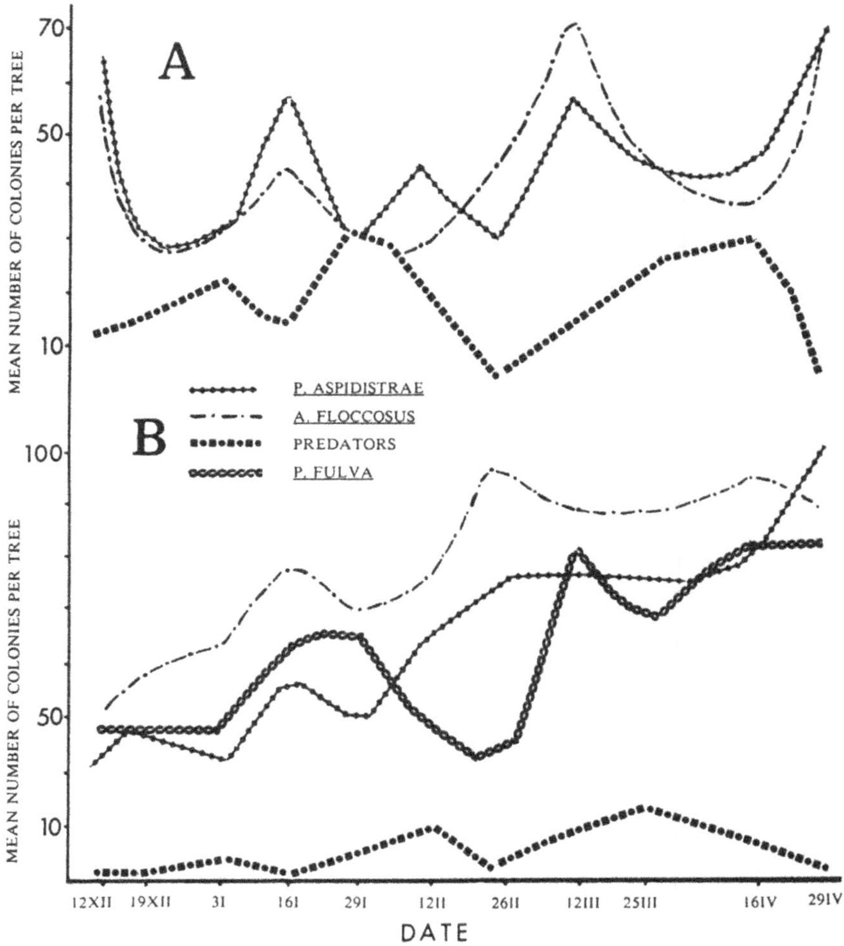

FIGURE 1. Population fluctuation of Homoptera and "hormiga loca". A = Exclusion of the ant. B = Recolonization after mechanical cleaning.

mechanically cleaned of all Homoptera and beneficial insects.

Figure 1 shows partial results for (A) the exclusion treatment with BIOTAC 19 and (B) the ant's free access. The presence of *P. fulva* influenced the Homoptera positively and negatively affected the predators. Immatures of both *A. floccosus* and *P. aspidistrae* were brought to the trees by the ant and installed; adult whiteflies were also observed flying to the trees. As pointed out in Figure 1B, their populations steadily increased, while the predators' increased slowly compared to A, where the ant was excluded. The Homoptera showed the expected fluctuation of their populations. The sticky material was efficient as a barrier and even the highest ant populations were below the economic threshold when excluded from the trees. In contrast, where the ants were present, leaves yellowed after only four months.

Grease protected the trees for only three weeks, BIOTAC 3 for six weeks and BIOTAC 19 for over three months. After these periods, although the sticky material was still effective, *P. fulva* exploited the accumulation of attached dead ants across the bands to reach the tree's canopy.

The use of sticky barriers is recommended for valuable fruit trees such as *Annona* spp.; *Psidium guayava, Citrus* spp. and *Mangifera indica.* In coffee trees, this control method is unfeasible because of the planting distance, which places branches in contact with each other, giving the ant the opportunity to walk through the canopy.

CULTURAL CONTROL

Transitory and permanent "hormiga loca" nests (see Zenner-Polania, Natural History and Biology section this volume) are found under almost all possible natural and artificial conditions, many of them provided by humans and therefore removable.

In nurseries, coffee plants, fruit trees and ornamentals, are grown in black plastic bags, which are left in the fields after being transplanted and therefore constitute a preferred nesting site. Systematic recollection and destruction of these bags had beneficial results.

Banana and plantain trees are used for shade in coffee plantations. After harvest, the fallen stem remains on the ground and slowly decays. The stem's outer layer dries, but the inner ones stay humid for months, supplying the ant with a perfect nesting site between the layers. The removal and use of these stems as cattle food or their destruction proved efficient and economically feasible . The number of permanent nests decreased by 23% on a plantation practicing good hygiene compared to another one which did not. Stem removal also facilitated other control measures like the application of toxic baits and insecticides.

The removal and destruction of all type of rubbish from around farm houses keeps the ants at low levels within buildings, where they forage mainly for sugar and oily food, becoming a severe household pest. Perfectly closed food storage containers are a must in the kitchens of houses invaded by the ant.

Cultural control alone will not keep the "hormiga loca" at subeconomic levels; it must be combined with other methods like chemical control or the use of toxic baits.

TABLE 1. Ingredients tested as possible toxic bait components for *P. fulva.*

Carrier	Attractant	Toxicant	Other Ingredient
	Bone meal		
	Fish meal	Lindane	
Sawdust	Blood meal		Water
	Molasses	Aldrin	
	Molasses + water		
Corn bran	Sugar water	Carbaryl	Propionic acid
	Pork lard		
	Formic acid		

TOXIC BAITS

Toxic baits were considered efficient and environmentally safe control measures, since they take advantage of the searching and foraging habits of the ants to distribute and accumulate the lethal dosage, while a minimum of toxicant reaches the environment. This is the only way to reach the queens and larvae, always perfectly concealed and protected against other control methods, specifically insecticide sprays.

Research to obtain an appropriate toxic agent against *P. fulva* was based on investigations conducted by Lewis et al. (1976) and Haines and Haines (1979) in the Seychelles with the ant *Anoplolepis longipes* (Jerdon). These authors obtained excellent results with baits, using as attractants boiled fish, or a mixture of dry blood, salt, sugar and yeast extract and coconut oil. Aldrin at 2.5 to 3.0% was incorporated as insecticide.

A summary of the investigations conducted by Zenner-Polania and Ruiz Bolanos (1982) to obtain an efficient toxic bait is presented here. Supported by the knowledge of the composition of the "hormiga loca's" diet - animal protein and honeydew - the carriers, attractants, poisons and other ingredients shown in Table 1 were chosen and tested. Under field conditions all possible combinations were evaluated for attractiveness and efficacy and finally those united by arrows were found to be most efficient. The statistical tests used in all experiments were Bartlett and Duncan at the 1% level.

Sawdust was discarded because of its insufficient absorption properties; blood or bone meal, although attractive to the workers was less preferred than fish meal. A sugar solution, imitating honeydew, attracted more ants than either molasses or different concentrations of molasses in water. Pork lard served both as animal protein and to impart a better consistency to the bait.

When each of the three insecticides were incorporated at a rate of 1% a.i. into the bait, the negative influence of the chlorinated hydrocarbons on the mixture's acceptability was immediately evident. Baits containing aldrin repelled workers, as did those with lindane; the effect of the latter, however, was less pronounced. The ants readily accepted baits having the toxic ingredient carbaryl.

No statistically significant difference was observed when formic acid was added. The ingredient was tested as a possible attractant, since under field conditions ants killed by humans on their arms or legs expelled this acid and typically attracted more workers, which carried away the corpses. Finally, propionic acid was added to inhibit mold.

After several field trials, it was concluded that the bait formulation indicated in Table 2, applied immediately to clean coffee plantations after the rainy season, at the rate of 12 Kg/ha, in bands 10 m apart gave good ant control for at least two months. The active ingredient of carbaryl was low enough not to kill the workers before they carried the bait to the nests, but was sufficiently high to slowly kill the larvae and the queens in the permanent nests. A week after bait distribution, the permanent nests showed a considerable decrease in activity; after two weeks, dead larvae were found and a little later the nests were completely abandoned.

The use of this control method, compared with conventional ones, is considered safer for the environment and more economic, since only 0.112 Kg a.i./ha of carbaryl is used versus the conventional rate of 1.0 to 1.5 a.i./ha (which is a 8.93 to 13.39 fold reduction in insecticide). The bait should only

TABLE 2. Toxic bait for the "hormiga loca".

INGREDIENT	QUANTITY
Corn bran (sieved through 20 mesh screen)	75 Kg
Fish meal (sieved through 30 mesh screen)	25 Kg
Sugar solution (100g sugar/1g water)	8 L
Propionic acid (100%)	0.4 L
Pork Lard	10 Kg
Carbaryl 85 PM	1% A.I.

attract ants with feeding habits similar to *P. fulva* and therefore is considered relatively harmless to beneficial insects . Also, with a DL50 of 510 to 850 gr a.i./Kg, carbaryl would not affect domestic or wild animals should they feed on it.

CHEMICAL CONTROL

The previously mentioned toxic bait has delayed activity on "hormiga loca" populations. Since the extremely high numbers of ants in coffee plantations, pasture ranges and in and around houses required a quick and substantial reduction, a series of insecticides were tested for their efficacy, knock down effect and long term residuality.

Zenner-Polania and Ruiz Bolanos (1983) carried out eight experiments around houses, grassland and coffee plantations, using a randomized block design with four replicates. The ant populations were estimated by counting the number of ants passing over a 10 x 10 cm cardboard during a period of five minutes before treatment, one day after and then every week until no control was observed. The insecticides' efficacy was calculated by using Henderson and Tilton's test.

The following materials were evaluated: pirimifos-methyl 50E, chlorpyrfos 4E, bendiocarb 80PM, diazinon 60E, carbaryl 85PM, dioxacarb 50PM, fenthion 500 gr/l E and propoxur 50PM. These products were chosen for their apparent low mammalian toxicity, residual potential and availability in Colombia.

Houses

The materials were applied to external walls up to 1 m high and to the ground surrounding the external walls (3 m. out). All garbage, stones and other materials which could have served as nesting sites were removed from the area around the house prior to treatment. All insecticides showed a good initial effect on the ant; seven days after treatment, dioxacarb (1% a.i.) failed to control the insect, while after 35 days pirimifos-methyl (1 m l/ m2), chlorpyrifos (0.5% a.i.), bendiocarb (0.15%) and fenthion (0.1% a.i.) were still effective. The highest efficacy was obtained with chlorpyrifos, which gave adequate control up to 43 days after spraying. No insecticides negatively affected the health of domestic animals like dogs, poultry and cats, which while absent during the application had access to the treated areas.

Cattle Range

Chlorpyrifos 4E, 3.51/ha offered the best control, followed by carbaryl 85PM, 2.0 Kg/ha. The other insecticides had good initial knock down properties, but lost their efficacy a week after the spray. This was probably due to their instability under the open sunny conditions of the grassland, which was not prevalent around houses, where shade trees protected the area from ultraviolet light.

To be effective, the two above mentioned insecticides need adequate soil humidity at the moment of the spray; for good coverage of the soil and base of the grass stems, the grass should not exceed a height of 30 cm and the spray volume must be around 1000 l/ha. After the required safety interval, the pasture can be maintained with very low ant populations for a normal cattle grazing cycle. Under dry conditions, however, the residual effect lasts only one week. To maintain the pasture range almost free of the ant, overly frequent applications are required, substantially increasing the production cost of beef or milk. No other satisfactory control has been found for pasture ranges.

Coffee Plantations

The plantation where the trials were carried out had good shade conditions, humid soil at the moment the insecticidal sprays were made and had been weeded recently. Only under these conditions can we hope for good efficacy of the materials.

All products showed excellent initial effects, but only chlorpyrifos (3.0 3.5 l/ha), pirimifos-methyl (3.01/ha) and diazinon (0.6 - 0.8 l/ha) gave adequate protection for three weeks. Bendiocarb, fenthion and dioxacarb were erratic in their efficacy. Taking into account product costs, the most economic insecticide to employ against the "hormiga loca" under these environmental conditions was diazinon 60E. It was recommended that the spray be used at least a week prior to any agricultural activities, i.e. harvest, weeding and fertilization.

CONCLUSION

The control strategy of *Paratrechina* (*Nylanderia*) *fulva* cannot be based on a single control method, as a combination of as many methods as possible works best.

If, for example, a coffee plantation is managed with normal agronomic practices and has a high ant population, the first step should be insecticide applications to the whole area. This tactic has to be complemented by cultural control, consisting of collection and destruction of all materials suitable for use as transitory nests by the ant. If *Musa* spp. or *Citrus* spp. are used as shade, the stems of the first plants have to be removed or cut into small pieces after harvest, and the trees of the second group have to be protected with sticky bands against the ant. If the homopteran population is already high and of economic importance, an application with an appropriate insecticide has to be made.

Finally, at the end of the rainy season, toxic bait use is recommended. Under these climatic conditions, the bait is more efficient because of the lower natural availability of Homoptera and other insects due to the previous rains. The bait has to be applied every two months in order to inhibit an economically important reinvasion of the ant.

The management strategy has to be adopted for the whole invaded area and applied at the same time. Partial control is expensive, since it has to be frequently repeated to avoid quick reinvasion by the ant from surrounding areas.

REFERENCES

INSTITUTO COLOMBIANO AGROPECUARIO. Programa de Entomologia. 1972. Bogota, Colombia. Informe Anual de Progreso. 56p.

HAINES, I. H. and J. B. HAINES. 1979. Toxic bait for the control of *Anoplolepis longipes* (Jerdon) (Hymenoptera: Formicidae). I. The basic attractant carrier, its production and weathering properties. Bull. Entomol. Res. 69: 65-75.

LEWIS, T., J. M. CHERRETT, I. HAINES, J. B. HAINES and P. L. MATHIAS. 1976. The crazy ant, *Anoplolepis longipes* (Jerd.) (Hymenoptera: Formicidae) in Seychelles and its chemical control. Bull. Entomol. Res. 66: 97-111.

ZENNER-POLANIA, I. and N. RUIZ BOLANOS. 1982. Uso de cebos contra la hormiga loca, *Nylanderia fulva* (Mayr) (Hymenoptera: Formicidae). Rev. Colombiana Entomol. 8: 24-31.

ZENNER-POLANIA, I. and N. RUIZ BOLANOS. 1983. Control quimico de la hormiga loca, *Nylanderia fulva* (Mayr). Rev. ICA, Bogota (Colombia) 18: 241-250.

67

Ant Control in Hawaiian Drip Irrigation Systems

V. Chang and A.K. Ota

INTRODUCTION

The Hawaiian sugar industry has converted about 35,000 ha of sugarcane fields from furrow to drip irrigation since 1971 (Gibson 1978). Since then, ant damage to plastic drip tubes has become a problem (Ota and Chang 1973). They chew through the tube walls and enlarge the orifices, thus flooding the damaged area and creating poor water distribution in the field. Similar damage by ants was reported in Australian drip-irrigated sugarcane fields (Chapman 1981). Sometimes damage is so extensive that drip tubes in a field must be replaced. If not corrected, this problem can result in heavy crop loss due to severe water shortage.

This paper reports the extent of ant damage and summarizes our laboratory's 18 years of research in this area.

DRIP IRRIGATION SYSTEM

The drip irrigation system uses a main line, submain lines and laterals to channel water into each field (Bui and Kinoshita 1985b). The laterals - generally 2-cm diameter low-density polyethylene tubes (drip tubes) - distribute water uniformly to plant root zones within the field. Tubes, buried about 12 cm below the soil surface between two rows of sugarcane are spaced 0.91 m apart and remain there for one crop cycle (24 months). Water is discharged through orifices (0.4 to 0.6 mm in diameter) or specially designed emitters spaced up to 61 cm apart along the length of the tube. Generally each hectare uses 3,645 m tubes with a 2.74-m line spacing.

The uniformity of a field's water distribution is measured by changes in operating pressure along selected laterals (Bui 1982). These measurements are used to estimate the orifice flow coefficient of variation (CV). CV values in a field should be less than or equal to $\leq$10% under normal operating conditions. High CV values indicate uneven water distribution. Ant damage could be a cause of irregular irrigation.

ANT DAMAGE

Four ant species in Hawaii sugarcane fields can cause tube damage: the fire ant *Solenopsis geminata* (F.), the bigheaded ant *Pheidole megacephala* (F.), the Argentine ant *Iridomyrmex humilis* (Mayr) and *Monomorium destructor* (Jerdon). All nest at least partially in soil.

S. geminata, the most destructive, can damage the tube in only two to three days. Fortunately, it is limited to the hot, dry lowlands of Oahu, Maui and Hawaii (Huddleston and Fluker 1968). It nests deep in soil, often in sunny, open areas among short grasses but also in partial shade. Before the canopy closes and sugarcane leaves overshadow a field, fire ants can be found anywhere on a field. After closing, most fire ant nests move to the field margins, usually less than 15 m from the border, or to areas between fields.

The bigheaded ant, widely distributed in shady and moist areas below the 300-m elevation, often nests near the soil surface under trash, rocks and leaf litter or in old hollow sugarcane stalks or stumps (Chang 1985). It occupies sugarcane fields throughout the crop season and high populations can cause extensive tube damage. The Argentine ant often nests in covered moist areas near the soil surface. It generally dominates areas above the 300-m elevation in Hawaii where few drip irrigated sugarcane fields are located. Damage caused by these ants is insignificant.

M. destructor has done little damage to drip irrigation tubes, since it forms small colonies and lives only in hot, dry and low areas of Oahu and Kauai.

S. geminata, *P. megacephala* and *I. humilis* leave distinctly different chew marks on the damaged area. *S. geminata*'s dull mandibles produce a rough border, while *I. humilis*'s sharp mandibles create fine scratches. *P. megacephala* makes a clean, round hole and leaves only a narrow band of mandible marks along the border.

Damage most often occurred after conversion of a field from furrow to drip irrigation because high ant populations were prevalent along the areas formerly used as ditches or flume lines. In one 32.4-ha field, we found that fire ants enlarged 26.9% of the 1,035 orifices examined, with an average increase of 260% (from the original diameter of 0.48 mm to 1.25 mm) after the tubes had been buried in the field for 19 months. Ninety-five percent of the damage occurs at the orifice, but *S. geminata* chews more frequent than *P. megacephala*.

The orifice flow rate of an undamaged dual-chamber tube averages 34.1 mL/min (at an operating pressure of 105 kPa) and a CV value of 8%. The flow rate (65 mL/min) and CV values (219%) rose after 16% of the orifices were enlarged. As a result, system operating pressure dropped and little water was emitted from undamaged orifices.

Ant behavior associated with tube orifice damage may be initiated by CO_2 released from microorganisms in the tube (Chang and Ota 1976). CO_2 stimulates digging activity by fire ants (Hangartner 1969). Ants probably enlarge the orifices to use the dark, moist inner space as a nest. Despaul (1957) reported a similar situation in which insects penetrated package wrappings and laid their eggs inside. The size of the enlargement often depends on ant population density.

Ants may chew the tube wall due to agitation caused by tube vibration during irrigation. Ants tend to stop and chew on rough spots on the tube surface; eventually they chew through the tube wall.

The amount of damage is also influenced by environmental factors. Tube orifices situated in complete darkness were enlarged twice as much as those exposed to continuous light. Damage also increased as temperature rose. Tubes held at 20°, 25°, 30° and 35°C had orifice diameter increases of 0.13, 0.81, 1.03 and 1.22 mm, respectively. Tubes generally sustain more ant damage in dry areas or when irrigation frequency is reduced because of water shortages.

CONTROL

Our attempts to minimize ant damage to the drip tubes have four areas of emphasis: (1) toxic baits, (2) toxic barriers, (3) ant-resistant drip tubes and (4) agronomic practices.

Toxic Baits

Until 1977, Mirex (Harvester Bait 300; Allied Chemical Corp., Morristown, NJ) effectively reduced any populations in drip-irrigated sugarcane fields (Ota and Chang 1979). No new ant bait has been registered for use in sugarcane fields since Mirex use was banned.

Amdro (American Cyanamid Co., Princeton, N.J.) fire ant insecticide (0.88% hydramethylnon) was registered in 1980, but only for use on non cropland. Extensive field efficacy and residue tests were conducted in Hawaiian sugarcane fields (Ota and Chang 1981) and Amdro effectively controlled *S. geminata* and *P. megacephala*. *S. geminata* population estimates based on pitfall trap data were reduced from a pretreatment count of 5.6 ants/trap day to 0.03 ant/trap day within three weeks of bait application (1.1 kg/ha). Ant counts remained low for 10 weeks. In contrast, ant counts in the untreated area increased from 2.9 ants to 5.4 ants/trap day during the same period. None of the 25 ant nests in the treated area were active five weeks after application, but 19 of 25 nests in the untreated area were still active. We found no detectable residues on live cane after seven days at an application rate of 2.24 kg bait/ha and after 14 days at 4.48 kg bait/ha. Hydramethylnon (0.15 to 0.75% in 25% soy oil bait) effectively controlled *P. megacephala* in pineapple fields (Su et al. 1980).

Fenoxycarb bait formulations (Maag Agrochemicals, Vero Beach, Florida) were repellent to ants in laboratory tests at concentrations above 0.5%. However, despite the bait's low attractiveness, 75% of the test colonies ingested enough chemical during 24 hours to kill all ants within three months. Control colonies all lived beyond three months. In a field efficacy test conducted in the spring, bait was applied twice, three weeks apart, at a rate of 1.1 kg/ha each. Ant counts in pitfall traps were the same between treated and untreated fields at 45 ants/trap day two weeks after treatment. However, ant counts increased rapidly in the untreated area to 175 ants/trap day after four months while counts in the treated area declined to 30 ants/trap day. A residue test was conducted at rates of 1.68 and 16.8 kg bait/ha; these rates represent the normal use rate and a tenfold excess. The 1.68 kg sample, collected one to 30 days following application, never contained detectable amounts of fenoxycarb. Residues were found at 16.8 kg - 10 times the normal use rate - but after an initial 10 to 12 day period, levels fell below detection limits. Registration of fenoxycarb bait formulations (Logic) for use in Hawaiian sugarcane fields is also being pursued.

Pro-Drone (Stauffer Chemical, Westport, Conn.) and Affirm (Merck & Co., Rahway, N.J.) effectively reduced ant populations (Chang and Ota 1982); however, manufacturers were not interested in registering the chemical for cropland use.

Toxic Barriers

A number of insecticides were injected into individual drip irrigation tubes buried while the system was in operation. The insecticide forms a toxic barrier around the orifices to prevent ant chewing.

Each trial consisted of six tubes (minimum length 30 m): three tubes for both the treated chemical and the control. Orifices along the entire length of the tubes dug out from the field were examined for enlargement 12 weeks after injection. The reduced percentage of enlarged orifices in the treated tube was calculated by comparison with the adjacent untreated lines. The results showed that the number of enlarged orifices in tubes treated with heptachlor at a dosage of 0.28 kg active ingredient/ha rate was reduced by 98% (Table 1). One heptachlor application protected tubes from ant damage for eight months or longer and generally lasted for the entire crop cycle (Ota and Chang 1980). Heptachlor was registered for use to control ant damage to drip tubes under "Special Local Needs" in Hawaii in 1979. However, the Environmental Protection Agency cancelled heptachlor's registration in April 1984.

Other toxicants which significantly protected the tubes were pyrethroids (permethrin, fenvalerate) and soil insecticides (isofenphos, fensulfothion, pirimiphos-ethyl) (Table 1); however, these chemicals were only effective for about three months and had to be re-applied for satisfactory control. None of these chemicals are now used for ant control in cane fields.

Ant-Resistant Drip Tubes

A standard method was developed to test different plastic tube materials for ant resistance. Discs (25 mm diameter) were punched out from tube materials and mounted atop 17 mL glass vials. A single 0.25 mm diameter "orifice" was punched out from the center of each disc by a micropunchpress. Before mounting the test discs, dead ants (25 mg) were added to each vial to stimulate ant digging and chewing response. The disc-mounted vials were then placed in laboratory *S. geminata* and *P. megacephala* colonies for at least one week. Each material was tested five times. The orifice enlargement was measured with an ocular micrometer in a dissecting microscope at weekly intervals. The average increase in orifice diameter was used as a measure of the materials' resistance to ants. The following factors affected tube resistance.

Materials. Drip tubes are commonly made from low- to medium-density polypropylene and polyethylene. Given the same material, lower density tubes are more susceptible to ant damage. Ant damage on low-density material can be reduced by increasing thickness. A 0.3-mm-thick medium-density material

TABLE 1. Control of ant damage in drip tubes with injection of toxicants.

Toxicant	Dosage (kg/ha)	No. trials	No. orifices examined	% enlarged orifices treated	% enlarged orifices control	% reduction of enlarged orifices	*t Value	*Prob.
Heptachlor	0.28	12	2522	0.30	14.35	97.91	4.04	0.001
Isofenphos	1.12	3	483	0.67	10.30	93.50	2.62	0.060
Permethrin	0.22	5	1326	1.26	8.06	84.37	2.03	0.056
Fensulfothion	1.12	3	482	1.77	10.30	82.82	3.87	0.015
Pirimiphos-ethyl	1.12	5	1138	2.24	11.40	80.35	8.32	0.001
Fenvalerate	0.22	5	1244	3.38	15.20	77.76	11.82	0.001
Methomyl	1.12	2	293	9.10	24.10	62.24	1.00	0.250
Propoxur	1.12	3	553	4.70	9.40	50.00	2.76	0.055
Chlorpyrifos	0.56	7	1731	8.81	16.65	47.09	3.05	0.112
Fonofos	1.12	5	807	9.68	16.36	40.83	2.38	0.038
Methyl parathion	1.12	2	256	14.20	22.20	36.04	1.26	0.214
Ethoprop	0.56	3	355	17.03	24.53	30.57	3.49	0.020
Oxamyl	1.12	3	509	1.60	2.30	30.43	0.27	0.407
Monocrotophos	1.12	2	293	19.60	24.10	18.67	1.00	0.250

* The t value and probability level of significance were calculated by pair comparison of the number of enlarged orifices in treated and untreated tubes from all trials for each chemical.

is equivalent in ant resistance to a 0.38 mm thick, low-density one of same material.

No materials were immune to ant damage, although poly- propylene resisted ant attack better than polyethylene. After seven days of exposure, the fire ants had enlarged orifice diameters by 36% on polypropylene material versus 536% on the same thickness of polyethylene material. Highland and Guy (1969) reported similar findings for polypropylene film. However, polypropylene deteriorates faster in sunlight than polyethylene. A tube made of 0.05-mm-thick polypropylene sandwiched between two layers of 0.076 mm of polyethylene material is commercially available. However, this combination tube is also susceptible to ant damage. Ants peeled off the first two layers and finally penetrated the tube. Among polyethylene materials, plastic resins from different manufacturers have shown variability in ant resistance. However, the resin formulation is proprietary information and was unavailable for detailed study. We found that tubes made from polyethylene films were more vulnerable to ant attack than tubes made from resin beads through an extrusion process. Tubes with cross-linking molecular structure after radiation treatment did not deter the ants.

Wall Thickness. Drip tubes made of relatively thick materials resisted ant attacks better than those of thinner materials. *S. geminata* increased orifice diameters by 87% on tubes with a wall thickness of 0.6 mm versus a 287% increase on tubes with a wall thickness of 0.35 mm. The minimum wall thickness to withstand an *S. geminata* attack for low-density polyethylene is 0.38 mm; for medium-density polyethylene, 0.3 mm. If *P. megacephala* dominates a field, 0.38 mm wall thickness will prevent ant damage.

Orifice Configuration. Orifices made by a laser were enlarged 1/3 to 1/2 more than those made by a mechanical punch. Ants passing a laser-melted glut of plastic at the orifice rim stopped and chewed more frequently than when passing a smooth orifice made by mechanical means. A slit orifice had less ant damage than a round orifice. Orifice size is determined by the hydraulic requirement of the irrigation system. However, *S. geminata* have more difficulty inserting their mandibles into small orifices (0.13 mm in diameter).

Barriers and Obstacles to the Orifices. Obstacles in the shape of a raised ring, a linear ridge or a flap overhanging the orifice reduced ant damage (Chang et al. 1980). Tubes with a pair of parallel ridges next to orifices were fabricated with different ridge heights and gap space. After 21 days, *S. geminata* increased the average orifice diameter by 94% and 519%, respectively, on tubes with and without two 1 mm parallel ridges 2 mm apart. Narrowing the space between the ridges could limit access to the orifices by large-headed ants. *S. geminata* and *P. megacephala* major workers, which inflict the most damage, have an average head width above 0.8 mm, whereas the minor workers of each species average a head width between 0.79 and 0.38 mm. A tube with a pair of parallel, 2 mm ridges (0.7 mm apart) running along its length was commercially manufactured (Guard-Wall Tubing, James Hardie Irrigation, Laguna Niguel, Calif.). Eight months after field exposure to *S. geminata* and *P. megacephala*, none of the orifices were enlarged. Tubes without ridges (Bi-Wall Tubing from the same manufacture) had 15.4% and

40.8% of the orifices enlarged by *S. geminata* and *P. megacephala*, respectively, during the same period. Although Guard-Wall Tubing resists ant damage, it is susceptible to orifice plugging when irrigation water is dirty and orifice diameter is less than 0.48 mm.

Currently, all drip tubes used by the Hawaii sugar industry have been evaluated for ant-resistance, in addition to other tests for such physical and hydraulic characteristics as durability, emitter plugging and root intrusion (Bui and Kinoshita 1985a).

Agronomic Practices

Tubes buried in soil (constant darkness) suffered significantly more damage than tubes placed on the soil surface (ambient photoperiod). Tubes placed on the surface become very hot in sunlight and are unlikely to be attacked. However, the difference in ant damage may be insignificant once the leafy canopy has closed. Hawaiian sugarcane farmers bury irrigation tubes to avoid pinching by reclining stalks. They also do not continuously irrigate their fields because of water shortages. Ants only attack the orifices when irrigation has stopped. We conducted tests to study the relationship between irrigation frequency and ant damage.

Three irrigation frequencies which simulated Hawaii plantation practice were tested: (1) one irrigation round every two days, (2) one irrigation round every four days and (3) one irrigation round every eight days. After crediting any rainfall, all treatments received the same amount of water during the test. The average durations of the off-irrigation period per irrigation cycle were 28, 56 and 112 hours for the two-, four- and eight-day treatments, respectively.

Ant damage to the tube increased with length of nonirrigation hours. The number of ant-chewed holes per 30.5 m on a susceptible tube was 5.5, 9.8 and 17.0 for the three respective irrigation cycles after a nine-month exposure in the field. Frequent irrigation reduced the damage but did not prevent it.

CONCLUSIONS

S. geminata and *P. megacephala* severely damage drip tubes, causing uneven irrigation in the field. Damage can be minimized by using ant bait to reduce the ant population in the fields, injecting an insecticide through the drip irrigation system to form a toxic barrier along the tube, or using an ant-resistant drip tube. In Hawaii sugarcane fields, drip tubes are now made with materials resistant to ant chewing and with hydraulic designs that tolerate ant damage.

LITERATURE CITED

BUI, W. 1982. A method of determining water distribution uniformity and its application in drip-irrigated fields. Hawaii. Sugar Technol. 1982 Rep., 41st Ann. Conf. pp. 57-60.

BUI, W. and C.M. KINOSHITA. 1985a. Evaluation of physical and hydraulic characteristics of drip irrigation tubing used in Hawaii. Proc. Third Int. Drip/Trickle Irrigation Congr. 1: 331-338.

BUI, W. and C.M. KINOSHITA. 1985b. Has drip irrigation in Hawaii lived up to expectations? Proc. Third Int. Drip/Trickle Irrigation Congr. 1: 84-89.

CHANG, V.C.S. 1985. Colony revival and notes on rearing and life history of the big-headed ant. Proc. Hawaii. Entomol. Soc. 25: 53-58.

CHANG, V.C.S. and A.K. OTA. 1976. Fire ant damage to polyethylene tubing used in drip irrigation systems. J. Econ. Entomol. 69: 447-450.

CHANG, V.C.S. and A.K. OTA. 1983. Laboratory studies with bait chemicals to control ants. Hawaii. Sugar Technol. 1982 Rep., 41st Ann. Conf. pp. 61-62.

CHANG, V.C.S., A.K. OTA and D. SANDERS. 1980. Parallel ridge barrier to control ant damage to orifices of drip irrigation tubes. J. Econ. Entomol. 73: 403-406.

CHAPMAN, L.S. 1981. Sub-surface irrigation for the Australian sugar industry. Australian Cane Grower, Dec. 1980-Jan. 1981. pp. 15-19.

DESPAUL, J.E. 1957. Penetration of protective coverings by insects. J. Milk & Food Technol. 20: 100-104.

GIBSON, W. 1978. Drip irrigation of sugar cane--The Hawaiian story. Int. Sugar J. 80: 362-367.

HANGARTNER, W. 1969. Carbon dioxide, a releaser for digging behavior in *Solenopsis geminata* (Hymenoptera: Formicidae). Psyche 76: 58-67.

HIGHLAND, H.A. and R.H. GUY. 1969. Insect resistance of polypropylene film overwrap on cigarette packages. Tobacco Sci. 13: 121-122.

HUDDLESTON, E.W. and S.S. FLUKER. 1968. Distribution of ant species of Hawaii. Hawaii. Entomol. Soc. 20: 45-69.

OTA, A.K. and V.C.S. CHANG. 1974. Insect-related problems in drip-irrigation systems. Hawaii. Sugar Technol. 1973 Rep., 32nd Ann. Conf. pp. 56-58.

OTA, A.K. and V.C.S. CHANG. 1979. Status of control of ant damage to drip irrigation tubing. Hawaii. Sugar Technol. 1978 Rep., 37th Ann. Conf. pp. 79-82.

OTA, A.K. and V.C.S. CHANG. 1980. Insecticides for control of ant damage to drip tubes. Hawaii. Sugar Technol. 1979 Rep., 38th Ann. Conf. pp. 87-91.

OTA, A.K. and V.C.S. CHANG. 1981. Progress in reducing ant damage to drip irrigation tubes. Hawaii. Sugar Technol. 1980 Rep., 39th Ann. Conf. pp. 111-112.

SU, T.H., J.W. BEARDSLEY and F.L. McEWEN. 1980. AC-217,300, a promising new insecticide for use in baits for control of the bigheaded ant in pineapple. J. Econ. Entomol. 73: 755-756.

Contributors

Akre, R.D.
Department of Entomology
Washington State University
Pullman, WA 99164-6432.

Andersen, A.N.
CSIRO
Division of Wildlife and Ecology
Tropical Ecosystems Res. Centre
PMB 44
Winnellie, NT 0821, Australia

Aron, S.
Unit of Behavioural Ecology
CP 160, Université Libre de Bruxelles
50, Avenue F.D. Roosevelt
1050 Bruxelles, Belgium

Aulino da Silva, O.
Instituto de Biociências
Universidade Estadual Paulista
13.500 Rio Claro, SP, Brazil

Banks, W.A.
USDA/ARS
P.O. Box 14565
Gainesville, FL 32604.

Beardsley, J.W.
Entomology Department
University of hawaii
3050 Maile way
Honolulu, Hawaii 96822

Bento, J.M.S.
Departamento de Biologia Animal
Unversidad Federal de Vicosa
36.570, Vicosa, MG, Brazil

Bernardi, J.V.E.
Departamento de Ecologia
Instituto de Biociências
UNESP
13500 Rio Claro, SP Brazil

Bhatkar, A.P.
Department of Entomology
Texas A&M University
College Station, TX 77843-2475

Billen, J.
Zoological Institute
University of Leuven
Naamsestraat 59, B-3000
Leuven, Belgium

Blum, M.S.
The University of Georgia
Department of Entomology
Athens, GA 30602.

Bueno, O.C.
Instituto de Biociências
Universidade Estadual Paulista
13.500 Rio Claro, SP, Brazil

Buys, B.
Plant Protection Research Institute
Private Bag X5017
Stellenbosch 7600, South Africa

Caetano, F.H.
Department of Biology
UNESP
13,500 Rio Claro, SP, Brazil

Cameron, R.S.
Pest Control Section
Texas Forest Service
P.O. Box 310
Lufkin, Texas 75901.

Camilo, G.R.
Texas Tech University
College of Agricultural Sciences
Department of Agronomy,
Horticulture and Entomology
Mail Stop: 2134
Lubbock, Texas 79409-2134

Campbell, M.H.
Dept. of Agriculture
Agricultrual Res. & Veterinatry
Centre, Forest Road
Orange, N.S.W., 2800, Australia

Cedeño, A.
Instituto de Zoologia Tropical
Facultad de Ciencias (U.C.V.)
Apartado 47058
Caracas 1041-A, Venezuela

Chang, V.
Experiment Station
Hawaiian Sugar Planters' Association
P.O. Box 1057
AIEA, Hawaii 96701-1057

Cherix, D.
Museum of Zoology
Palais de Rumine
P.O. Box 448
1000 Lausanne 17, Switzerland

Cherrett, J.M.
School of Biological Sciences
University College of North Wales
Bangor, Gwynedd LL57 2UW
Wales, England

Claver, S.
Instituto Argentino de Investigacion
de la Zonas Aridas (IADIZA)
Centro Regional de Investigaciones
Cientificas y Técnicas (CRICYT)
C.C. 507 (5500)
Mendoza, Argentina

Da Cruz-Landim, C.
Departamento de Biologia
Instituto de Biociências de Rio Claro
13.500 Rio Claro, SP, Brazil

Dejean, A.
Laboratoire de Zoologie
Universite de Yaounde
B.P. 812, Yaounde, Cameroon

Delabie, J.C.
Ceplac/Cepec/Entomologia
Caixa Postal 07
Itabuna Bahia 45.600, Brazil

Della-Lucia, T.M.C.
Departamento de Biologia Animal
Unversidad Federal de Vicosa
36.570, Vicosa, MG, Brazil

Deneubourg, J.L.
Unit of Behavioural Ecology
CP 160, Université Libre de
Bruxelles
50, Avenue F.D. Roosevelt
1050 Bruxelles, Belgium

Di Romagnano, L.F.T.
Departamento de Ecologia
Instituto de Biociências
UNESP
13500 Rio Claro, SP, Brazil

Dos Anjos, N.
Departamento de Biologia Animal
Unversidad Federal de Vicosa
36.570, Vicosa, MG, Brazil

Eichler, Wd.
Museum für Naturkunde der Humboldt-Universität zu Berlin
Invalidenstrasse 43
Berlin, 1040, GDR

Ervin, R.T.
Department of Agricultural Economics
Texas Tech University
Lubbock, Texas 79409

Escoubas, P.
The University of Georgia
Department of Entomology
Athens, GA 30602.

Fernandez, J.B.
Instituto de Química
Universidade Federal de Sao Carlos
SP, Brazil

Forti, L.C.
Departamento de Defesa Fitossanitária
Faculdade de Ciências Agronômicas
UNESP
18600 Botucatu, SP, Brazil

Fowler, H.G.
Departamento de Ecologia
Instituto de Biociências
UNESP
13500 Rio Claro, SP Brazil

García, C.A.
C.I.A.T.
Apartado Aéreo 6713
Cali, Colombia

Glancey, B.M.
USDA/ARS
P.O. Box 14565
Gainesville, FL 32604.

González, A.
Estación Experimental de Cítricos
San Ignacio #02
Torriente, Matanzas, Cuba

Goss, S.
Unit of Behavioural Ecology
CP 160, Université Libre de Bruxelles
50, Avenue F.D. Roosevelt
1050 Bruxelles, Belgium

Greenberg, L.
Department of Entomology
Texas A&M University
College Station, TX 77843.

Guajara, M.S.
Departamento de Biologia Animal
Unversidad Federal de Vicosa
36.570, Vicosa, MG, Brazil

Haack, K.D.
Granovsky Associates, Inc.
3206 Wilderness Road
Bryan, TX 77802

Hansen, L.D.
Visiting Research Associate, WSU, and Instructor
Spokane Falls Community College
W. 3410 Fort Wright Drive
Spokane, WA 99204

Harada, A.Y.
INPA, Departamento de Ecologia
C.P. 478
69011 Manaus, Brazil

Hebling-Beraldo, M.J.A.
Instituto de Biociências
Universidade Estadual Paulista
13.500 Rio Claro, SP, Brazil

Hernandez, C.J.
Laboratory on Cellular Neuropharmacology
Centro de Biofísica y Bioquímica
Instituto Venezolano de Investigaciones Científicas (IVIC)
Apartado 21827
Caracas 1020A, Venezuela

Howse, P.E.
Department of Biology & Chemistry
Chemical Entomology Unit, Bldg. 19
The University of Southampton
Southampton S09 5NH England

Jaffe, K.
Departamento de Biologia de Organismos
Apartado 80659
Universidad Simon Bolivar
Caracas 1080-A, Venezuela

Jahn, G.
Department of Entomology
University of Hawaii
Honolulu, Hawaii 96822

Jolivet, P.
67 Boulevard Souet
F-75012-Paris, France

Jouvenaz, D.P.
USDA/ARS
P.O. Box 14565
Gainesville, FL 32604.

Kermarrec, A.
INRA
Station de Zoologie et Lutte Biologique
BP 1232 Pointe-à-Pitre Cédex
Guadeloupe (French Antilles)

Knapp, J.J.
Department of Biology & Chemistry
Chemical Entomology Unit, Bldg. 19
The University of Southampton
Southampton S09 5NH England

Knight, R.L.
Boyle-Midway Household Products, Inc.
South Avenue & Hale Streets
Cranford, NJ 07016

Lapointe, S.L.
C.I.A.T.
Apartado Aéreo 6713
Cali, Colombia

Lattke, J.E.
Fundacion Terramar
Apartado 89.000
Caracas 1080-A, Venezuela

León, M.J.
Instituto de Zoologia Tropical
Facultad de Ciencias (U.C.V.)
Apartado 47058
Caracas 1041-A, Venezuela

Lofgren, C.S.
USDA/ARS
P.O. Box 14565
Gainesville, Florida 32604.

MacKay, W.P.
Texas A & M University
Dept. of Entomology
College Station, TX 77843-2475.

Majer, J.D.
School of Biology
Curtin University of Technology
GPO Box U1987
Perth, Western Australia, 6001

Marival, D.
INRA
Station de Zoologie et Lutte Biologique
BP 1232 Pointe-à-Pitre Cédex
Guadeloupe (French Antilles)

Martínez, J.F.
Estación Experimental de Cítricos
San Ignacio #02
Torriente, Matanzas, Cuba

Mauleon, H.
INRA
Station de Zoologie et Lutte Biologique
BP 1232 Pointe-à-Pitre Cédex
Guadeloupe (French Antilles)

McNeil, J.N.
Université Laval
Faculté des Sciences et de Génie
Département de Biologie
Cité Universitaire
Québec, Canada G1K 7P4

Mintzer, A.
1550 S. Marmora
Tucson, AZ 85713

Moreira, D.D.O.
Departamento de Biologia Animal
Unversidad Federal de Vicosa
36.570-Vicosa-MG-Brazil

Morel, L.
Department of Immunology and
Medical Microbiology
University of Florida
JHMHC Box J-266
Gainesville, Florida 32610

Morgan, E.D.
Department of Chemistry
University of Keele
Keele, Staffordshire. ST5 5BG.
England

Moser, J.C.
Southern Forest Experiment Station
USDA Forest Service
Pineville, LA 71360

Obin, M.S.
Pathology Research Division
Brigham and Women's Hospital
75 Francis Street
Boston, MA 02115

Ota, A.K.
Experiment Station
Hawaiian Sugar Planters' Association
P.O. Box 1057
AIEA, Hawaii 96701-1057

Overal, W.
Museu Paraense Emili Goeldi
Caixa Postal 399
66000 Belem, Para, Brazil

Pagnocca, F.C.
Instituto de Biociências
Universidade Estadual Paulista
13.500 Rio Claro, SP, Brazil

Passera, L.
Laboratoire d' Entomologie
Université de Paul-Sabatier
Toulouse, France

Pasteels, J.M.
Unit of Behavioural Ecology
CP 160, Université Libre de
Bruxelles
50, Avenue F.D. Roosevelt
1050 Bruxelles, Belgium

Pereira-da-Silva, V.
Instituto de Biociências
UNESP
Botucatu, SP, Brazil

Pérez Perera, L.F.
Estación Experimental de Cítricos
San Ignacio #02
Torriente, Matanzas, Cuba

Phillips, S.A. Jr.
Texas Tech University
College of Agricultural Sciences
Department of Agronomy,
Horticulture and Entomology
Mail Stop: 2134
Lubbock, Texas 79409-2134

Porter, S.D.
USDA/ARS
P.O. Box 14565
Gainesville, FL 32604

Prins, A.J.
Department of Entomology
South African Museum
P.O. Box 61
Cape Town 8000, South Africa

Prins, A.
Department of Entomology
South African Museum
P.O. Box 61
Cape Town 8000, South Africa

Reed, J.
School of Biological Sciences
University College of North Wales
Bangor, Gwynedd LL57 2UW
Wales, England

Reimer, N.
Department of Entomology
University of Hawaii
Honolulu, Hawaii 96822

Robertson, H.G.
Department of Entomology
South African Museum
P.O. Box 61
Cape Town 8000, South Africa

Rust, M.K.
Department of Entomology
University of California, Riverside
Riverside, CA 92521-0314

Salinas, P.J.
Facultad de Ciencias Forestales
Centro de Estudios de Postgrado
Universidad de Los Andes
Edo. Merida, Venezuela

Salzemann, A.
Laboratoire d'Evolution des Etres Organises
Universite P.et M. Curie
105 Boulevard Raspail
75006 Paris, France

Samways, M.J.
University of Natal
Dept. of Zoology and Entomology
P.O. Box 375
Pietermaritzburg 3200, South Africa

Serrano, M.S.
C.I.A.T.
Apartado Aéreo 6713
Cali, Colombia

Sevcik, C.
Laboratory on Cellular Neuropharmacology
Centro de Biofísica y Bioquímica
Instituto Venezolano de Investigaciones Científicas (IVIC)
Apartado 21827
Caracas 1020A, Venezuela

Tennant, W.R. Jr.
Department of Agricultural Economics
Texas Tech University
Box 4169
Lubbock, Texas 79409-2132

Therrien, P.
Ministére Energie et Resources (S.P.I.M)
1297 Boul Charest Ouest
Québec, P.Q., Canada G1N 2C9

Thompson, C.R.
Division of Plant Industry
Gainesville, FL 32605

Ulloa-Chacon, P.
Departamento de Biologia
Universidad del Valle
AA 25360 Cali, Colombia

Vander Meer, R.K.
USDA/ARS
P.O. Box 14565
Gainesville, FL 32604.

Vargo, E.L.
Brackenridge Field Laboratory and Department of Zoology
The University of Texas at Austin
Austin, TX 78712-1064

Vasconcelos, H.L.
INPA, Departamento de Ecologia
C.P. 478
69011 Manaus, Brazil

Veeresh, G.K.
Department of Entomology
University of Agricultural Sciences
G.K.V.K. Campus
Bangalore-560 065, India

Vieira, P.C.
Instituto de Química
Universidade Federal de Sao Carlos
SP, Brazil

Vilela, E.F.
Departamento de Biologia Animal
Unversidad Federal de Vicosa
36.570, Vicosa, MG, Brazil

Vinson, S.B.
Department of Entomology
Texas A&M University
College Station, TX 77843.

Waller, D.A.
Department of Biological Sciences
Old Dominion University
Norfolk, VA 23529

Weiss, B.A.
Psychology, Sociology, and
Anthropology Department
Drexel University
Philadelphia, PA 19104.

Williams, D.F.
USDA/ARS
P.O. Box 14565
Gainesville, FL 32604.

Williams, H.J.
Department of Entomology
Texas A&M University
College Station, TX 77843.

Wojcik, D.P.
USDA/ARS
P.O. Box 14565
Gainesville, FL 32604.

Zenner-Polania, I.
Sección Entomolgía ICA-Tibaitatá
AA 151123 Eldorado
Bogotá, Colombia

Taxonomic Index

Subject Index

For Product Safety Concerns and Information please contact our EU
representative GPSR@taylorandfrancis.com
Taylor & Francis Verlag GmbH, Kaufingerstraße 24, 80331 München, Germany

www.ingramcontent.com/pod-product-compliance
Ingram Content Group UK Ltd.
Pitfield, Milton Keynes, MK11 3LW, UK
UKHW022208190926
475702UK00003B/266
9780367013240